World – Political

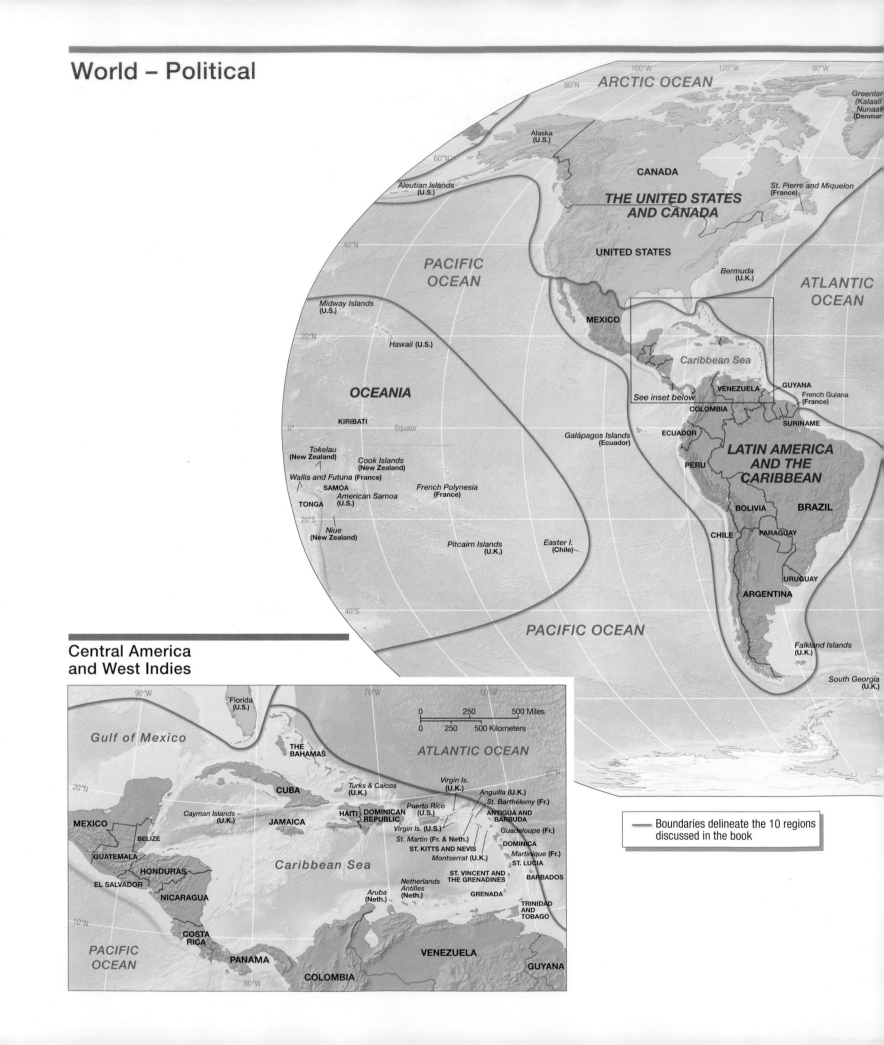

ARCTIC OCEAN

Greenland
(Kalaallit
Nunaat)
(Denmark)

Alaska
(U.S.)

CANADA

St. Pierre and Miquelon
(France)

THE UNITED STATES
AND CANADA

Aleutian Islands
(U.S.)

PACIFIC
OCEAN

UNITED STATES

ATLANTIC
OCEAN

Bermuda
(U.K.)

Midway Islands
(U.S.)

MEXICO

Caribbean Sea

Hawaii (U.S.)

VENEZUELA

GUYANA

French Guiana
(France)

See inset below

OCEANIA

COLOMBIA

SURINAME

KIRIBATI

Equator

Galápagos Islands
(Ecuador)

ECUADOR

Tokelau
(New Zealand)

Cook Islands
(New Zealand)

PERU

LATIN AMERICA
AND THE
CARIBBEAN

Wallis and Futuna (France)

SAMOA

French Polynesia
(France)

TONGA

American Samoa
(U.S.)

BOLIVIA

BRAZIL

Niue
(New Zealand)

CHILE

PARAGUAY

Pitcairn Islands
(U.K.)

Easter I.
(Chile)

URUGUAY

ARGENTINA

PACIFIC OCEAN

Falkland Islands
(U.K.)

South Georgia
(U.K.)

Central America
and West Indies

Florida
(U.S.)

Gulf of Mexico

THE
BAHAMAS

ATLANTIC OCEAN

| 0 | 250 | 500 Miles |
| 0 | 250 | 500 Kilometers |

CUBA

Turks & Caicos
(U.K.)

Virgin Is.
(U.K.)

Anguilla (U.K.)

St. Barthélemy (Fr.)

MEXICO

Cayman Islands
(U.K.)

JAMAICA

HAITI

Puerto Rico
(U.S.)

DOMINICAN
REPUBLIC

ANTIGUA AND
BARBUDA

BELIZE

Virgin Is. (U.S.)

Guadeloupe (Fr.)

St. Martin (Fr. & Neth.)

GUATEMALA

ST. KITTS AND NEVIS

DOMINICA

Montserrat (U.K.)

Martinique (Fr.)

HONDURAS

Caribbean Sea

ST. LUCIA

EL SALVADOR

NICARAGUA

Netherlands
Antilles
(Neth.)

ST. VINCENT AND
THE GRENADINES

BARBADOS

Aruba
(Neth.)

GRENADA

TRINIDAD
AND
TOBAGO

PACIFIC
OCEAN

COSTA
RICA

PANAMA

VENEZUELA

GUYANA

COLOMBIA

—— Boundaries delineate the 10 regions
discussed in the book

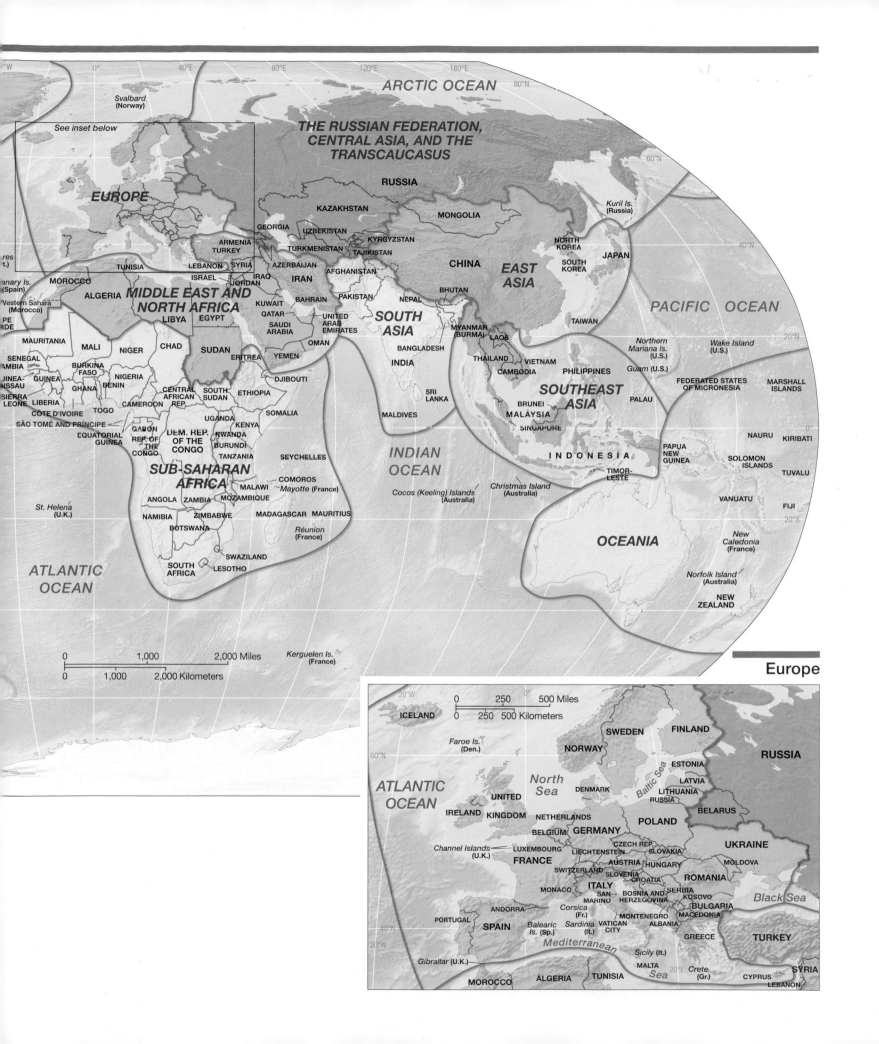

Saigon Bam
Instructor Copy

World Regions in Global Context

PEOPLES, PLACES, AND ENVIRONMENTS

FIFTH EDITION

Sallie A. Marston
University of Arizona

Paul L. Knox
Virginia Tech

Diana M. Liverman
University of Arizona; Oxford University

Vincent J. Del Casino, Jr.
University of Arizona

Paul F. Robbins
University of Wisconsin, Madison

PEARSON

Boston   Columbus   Indianapolis   New York   San Franciso   Upper Saddle River
Amsterdam   Cape Town   Dubai   London   Madrid   Milan   Paris   Montréal   Toronto
Delhi   Mexico City   São Paulo   Sydney   Hong Kong   Seoul   Singapore   Taipei   Tokyo

Geography Editor: Christian Botting
Senior Marketing Manager:
    Maureen McLaughlin
Geography Project Editor: Anton Yakovlev
Senior Market Development Manager:
    Michelle Cadden
Development Editor: Erin Mulligan
**Director of Development, Geosciences and
    Chemistry:** Jennifer Hart
Assistant Editor: Sean Hale
Editorial Assistant: Bethany Sexton
Senior Marketing Assistant: Nicola Houston
Media Producer: Tim Hainley
**Managing Editor, Geosciences and
    Chemistry:** Gina M. Cheselka
Project Manager, Science: Wendy Perez
Composition/Full Service:
    PreMediaGlobal USA Inc.

Production Editor, Full Service:
    Jenna Gray
Art Studio: International Mapping Associates
Photo Research Manager: Maya Melenchuk
Photo Researcher: Kerri Wilson,
    PreMediaGlobal
Text Permissions Manager: Alison Bruckner
Text Permissions Researchers: Melinda Durham
    and Anna Waluk, Electronic Publishing Services
Design Manager: Laura Gardner
Interior and Cover Design: Carmen
    DiBartolomeo, Red Kite Productions
Operations Supervisor: Michael Penne
Cover Image Credit: © Top Photo
    Corporation/Thinkstock

Credits and acknowledgments borrowed from other sources and reproduced, with permission, in this textbook appear on the appropriate page within the text or in the back matter.

Library of Congress Cataloging-in-Publication Data is available upon request from the Publisher.

5 6 7 8 9 10—V011—16 15

www.pearsonhighered.com

ISBN-10: 0-321-82105-X; ISBN-13: 978-0-321-82105-8 (Student Edition)
ISBN-10: 0-321-86230-9; ISBN-13: 978-0-321-86230-3 (Instructor's Review Copy)

BRIEF CONTENTS

CONTENTS

3   The Russian Federation, Central Asia, and the Transcaucasus 86

4   Middle East and North Africa 128

5 Sub-Saharan Africa 174

6 The United States and Canada 222

7 Latin America and the Caribbean 262

8 East Asia 304

9 South Asia 344

PREFACE

"The real act of discovery consists not in finding new lands but in seeing with new eyes."

Marcel Proust

We live in a world of global interconnection. This means that if we want to understand the human condition or the changing environment, we have to look not just at our local community but also at the wider world. *World Regions in Global Context* provides a framework for understanding the global connections that affect relationships within world regions, while also recognizing that the events that take place within world regions can have an impact on a global scale. Of course, no textbook can provide the answers to all the complex questions about the forces that fuel these global connections and local changes. But *World Regions in Global Context* can shed some light on the dynamic and complex relationships between people and the worlds they inhabit. This book gives students the basic geographical tools and concepts they need to understand the complexity of today's global geography and the world regions that make up that geography.

NEW TO THE FIFTH EDITION

The fifth edition of *World Regions in Global Context* has been thoroughly revised by the authors and editorial team based on reviews from teachers and scholars in the field. Every line and graphic in the book has been reviewed and edited for maximum clarity and effectiveness. The new edition includes significant changes as well as a number of new features that make the revised text more accessible and engaging.

- **Emerging Regions** are regions that cross the boundaries of the ten world regions covered in this text. In this feature, we highlight six emerging regions: the Arctic; the United States and the Caribbean; BRICS (Brazil, Russian, India, China, and South Africa); the Pacific Rim; the Greater Mekong Region; and the Alliance of Small Island States (AOSIS). The *Emerging Regions* features ask not only how the world is organized today but also how it might be organized differently in the future. The exercise of discussing these regions in a new way in this edition reflects a trend in the field of geography. We ask the reader to consider how the growing connections between world regions and countries are creating new regional formations, which ultimately impact global relations.

- Geography is strongly invested in the use of maps and other visual data. The new **Visualizing Geography** feature builds on and extends that tradition with extensive use of visualizations and maps to focus on issues such as global sea-level rise; the proliferation of oil and gas pipelines; and the migration of Muslim populations into Europe.

- The **maps, images, graphs, and tables** that make up the text's visual program have been dramatically revised and improved. Readers will notice that many maps now include topography, and others use new projections to better highlight key geographic processes. Maps in the fifth edition are easier to read, and all the information and data in the maps has been carefully reviewed and updated to support and mirror text discussion. The photo program for this edition has also been heavily revised; newer and more exciting photos will help readers gain a deeper understanding of the world regions portrayed in the book.

- We have **updated the histories, stories, and current events** in each chapter. As readers know, the world has changed a lot since the previous edition of this book. To respond to these changes, we have included stories on the European economic crisis; the Arab Spring; the recent Russian presidential election; natural disasters in Southeast Asia; and the growing connections between China and Africa, for example. New and updated information has been added to all the special feature material as well, including all the new **Geographies of Indulgence, Desire, and Addiction** features on marijuana; perfumes and fragrances; coffee; and animals, animal parts, and exotic pets.

- Each chapter begins with a new set of **Learning Outcomes** and ends with **Learning Outcomes Revisited**. These two features reinforce the main concepts, ideas, and issues that students should be able to understand and discuss after reading each chapter.

- We also include **Apply Your Knowledge** questions throughout each chapter. After key sections of the text, readers are asked a question that encourages them to synthesize the information they have just read and apply it to real-world issues and questions.

- **Chapter 1 has been extensively revised** and streamlined to focus on the key concepts that will help readers to understand the issues covered in each regional chapter.

- The book has been updated with **current data** from the latest science, statistics, and associated imagery, including data from the 2010 U.S. Census and the 2012 Population Reference Bureau World Population Data.

- The fifth edition is now supported by **MasteringGeography™**, the most widely used and effective online homework, tutorial, and assessment system for the sciences. Assignable media and activities include MapMaster™ interactive maps, *Encounter World Regional Geography* Google Earth explorations, geography videos, coaching activities on the toughest topics in geography, end-of-chapter questions, reading quizzes, Test Bank questions, and select Geoscience Animations.

OBJECTIVES AND APPROACH

World Regions in Global Context has two primary objectives. The first is to provide a body of knowledge about world regions and their distinctive political and economic practices, cultural and environmental landscapes, and sociocultural attributes. The second is to emphasize that although there is diversity among world regions, all world regions are connected.

In *World Regions in Global Context*, we are not only interested in understanding the internal dynamics of a world region; we are interested in that region's interconnections to other regions around the globe. This approach informs this book's thematic structure, which is organized to engage readers in a discussion of environmental, social, historical, economic, and territorial change as well as cultural practices and demographic shifts.

THEMATIC STRUCTURE

The conceptual framework of this book is built on a single opening chapter that describes the basics of a global geography perspective and introduces the key concepts that are deployed throughout the remaining regional chapters. The ten regional chapters that follow explore and elaborate the concepts laid out in Chapter 1. In each chapter, we balance discussions of global interconnections with local realities. To do this systematically, we divide each regional chapter into four major categories, each highlighting a set of themes that are central to understanding world regions.

Environment and Society

We begin each chapter with a discussion of the physical and environmental context of the region; this includes a discussion of climate, adaptation, and global change; geological resources, risks, and water; and ecology, land, and environmental management. Our aim is to demonstrate how environment is shaped by, and shapes, the region's inhabitants over time.

History, Economy, and Territory

This section focuses on the historical geographic context for each world region and illustrates how the economies and territories that make up each world region have evolved over time. Included are discussions of historical landscapes and legacies; economy, accumulation, and the production of inequality; and territory and politics.

Culture and Populations

This section explores the cultures and populations found in each world region, asking, for example, how certain cultural practices (e.g., world music or electronic gaming) have been globalized and changed locally. This section also emphasizes the relationships between population change and settlement patterns, while emphasizing the importance of urbanization in each region. This section is broken down into three subsections focusing on culture, religion, and language; cultural practices, social differences, and identity; and demography and urbanization.

Future Geographies

In keeping with the theme of this textbook, which emphasizes ongoing change, each chapter concludes with a brief discussion of some of the key issues facing each world region, projecting how they are likely to develop in the coming years and decades. Discussions touch on the future of the European Union; the struggle between democracy and plutocracy in Russia; peace and stability in the Middle East and North Africa; and the challenge of environmental stewardship in the United States and Canada.

FEATURES

The book includes a number of important pedagogical devices to help readers understand the complex processes that connect our world and make it different.

Key Facts

At the beginning of each chapter, we include *Key Facts* under the opening regional map. This information provides some broad regional contextual information—describing subregions and major physiogeographic features along with some important statistics that help readers both understand the region and place it in relation to other world regions.

Learning Outcomes and Learning Outcomes Revisited

On the opening pages of each regional chapter, we provide a list of *Learning Outcomes*. This list directs students to the key "take-away" points in the chapter. They are intentionally broad, drawing from a number of different discussions throughout each chapter. At the end of the chapter, we return to these learning outcomes and offer brief comments on them. The *Learning Outcomes Revisited* section helps readers grapple with some of the larger conceptual material and focuses student review.

Apply Your Knowledge

Apply Your Knowledge questions ask readers to synthesize the information in the text and respond to applied questions that link back to the chapter's broad learning outcomes. Readers will find six to eight of these questions in each chapter.

Visualizing Geography

In each chapter, we use cutting-edge cartography and data visualization techniques to introduce readers to a current geographic issue. Visual data provide a powerful way to convey information and analyze geographic processes in action, encouraging students to ask, "What types of geographic data can I use to answer the pressing questions of the day?"

Emerging Regions

This feature emphasizes global and local change and underscores the importance that these new regions have now and may have in the future. Readers are encouraged to explore *Emerging Regions* with an eye toward asking how world regional geography changes over time and how it might look different in the future.

Geographies of Indulgence, Desire, and Addiction

This feature links people in one world region to people throughout the world through a discussion of the local production and global consumption of regional commodities, helping students appreciate the links between producers and consumers around the world as well as between people and the natural world.

CONCLUSION

This book is the product of conversations among the authors, colleagues, students, and the editorial team about how best to teach a course on world regional geography. In preparing the text, we have tried to help students make sense of the world by connecting conceptual materials to the most compelling current events. We have also been careful to represent the best ideas that the discipline of geography has to offer by mixing cutting-edge and innovative theories and concepts with more classical and proven approaches and tools. Our aim has been to show how important a geographic approach is for understanding the world and its constituent places and regions.

ACKNOWLEDGMENTS

We are indebted to many people for their assistance, advice, and constructive criticism in the course of preparing this book. Among those who provided comments on various drafts and editions are the following professors:

Donald Albert, *Sam Houston State University*; Martin Balinsky, *Tallahassee Community College*; Brad Baltensperger, *Michigan Technological University*; Max Beavers, *University of Northern Colorado*; Richard Benfield, *Central Connecticut State University*; William H. Berentsen, *University of Connecticut*; Keshav Bhattarai, *Central Missouri State University*; Warren R. Bland, *California State University, Northridge*; Brian W. Blouet, *College of William and Mary*; Sarah Blue, *Northern Illinois University*; Pablo Bose, *University of Vermont*; Jean Ann Bowman, *Texas A &M University*; John Christopher Brown, *University of Kansas*; Stanley D. Brunn, *University of Kentucky*; Joe Bryan, *University of Colorado: Boulder*; Michelle Calvarese, *California State University, Fresno*; Craig Campbell, *Youngstown State University*; Xuwei Chen, *Northern Illinois University*; Jessie Clark, *University of Oregon*; David B. Cole, *University of Northern Colorado*; Jose A. da Cruz, *Ozarks Technical Community College*; Tina Delahunty, *Texas Tech University*; Cary W. de Wit, *University of Alaska, Fairbanks*; Catherine Doenges, *University of Connecticut-Stamford*; Lorraine Dowler, *Pennsylvania State University*; Dawn Drake, *Missouri Western State University*; Brian Farmer, *Amarillo College*; Caitie Finlayson, *Florida State University*; Ronald Foresta, *University of Tennessee*; Gary Gaile, *University of Colorado*; Roberto Garza, *University of Houston*; Jay Gatrell, *Indiana State University*; Mark Giordano, *Oregon State University*; Qian Guo, *San Francisco State University*; Devon A. Hansen, *University of North Dakota*; Julie E. Harris, *Harding University*; Russell Ivy, *Florida Atlantic University*; Rebecca Johns, *University of Southern Florida*; Kris Jones, *Saddleback College*; Tim Keirn, *California State University, Long Beach*; Marti Klein, *Mira*

Costa College; Lawrence M. Knopp, *University of Minnesota, Duluth*; Debbie Kreitzer, *Western Kentucky University*; Robert C. Larson, *Indiana State University*; Alan A. Lew, *Northern Arizona University*; John Liverman, *independent scholar*; Max Lu, *Kansas State University*; Donald Lyons, *University of North Texas*; Taylor Mack, *Mississippi State University*; Chris Mayda, *Eastern Michigan University*; Eugene McCann, *Simon Fraser University*; Tom L. McKnight, *University of California, Los Angeles*; M. David Meyer, *Central Michigan University*; Sherry D. Moorea-Oakes, *University of Colorado, Denver*; Barry Donald Mowell, *Broward Community College*; Darla Munroe, *The Ohio State University*; Tim Oakes, *University of Colorado*; Nancy Obermeyer, *Indiana State University*; J. Henry Owusu, *University of Northern Iowa*; Rosann Poltrone, *Arapahoe Community College*; Jeffrey E. Popke, *East Carolina University*; Henry O. Robertson, *Louisiana State University, Alexandria*; Robert Rundstrom, *University of Oklahoma*; Yda Schreuder, *University of Delaware*; Anna Secor, *University of Kentucky*; Daniel Selwa, *Coastal Carolina University*; Christa Ann Smith, *Clemson University*; Richard Smith, *Harford Community College*; Barry D. Solomon, *Michigan Technical University*; Joseph Spinelli, *Bowling Green State University*; Kristen Sziarto, *University of Wisconsin-Milwaukee*; Liem Tran, *Florida Atlantic University*; Syed (Sammy) Uddin, *William Paterson University/ St. John's University*; Samuel Wallace, *West Chester University*; Matthew Waller, *Kennesaw State University*; Gerald R. Webster, *University of Alabama*; Julie Weinert, *Southern Illinois University*; Mark Welford, *Georgia Southern University*; Keith Yearman, *College of Du Page*; and Anibal Yanez-Chavez, *California State University, San Marcos*.

Special thanks go to our editorial team at Pearson Education, Michelle Cadden, Anton Yakovlev, Christian Botting, Tim Hainley, Sean Hale and Bethany Sexton; to our "rock star" developmental editor, Erin Mulligan, and our project managers, Jenna Gray at PreMediaGlobal and Wendy Perez at Pearson; to Kerri Wilson at PreMediaGlobal for photo research; to Kevin Lear and International Mapping for their creative work with the art program; and to Alison Bruckner, Melinda Durham, and Anna Waluk for their work on permissions for text and line art. We would also like to thank our excellent research assistants, Jennifer McCormack and Christina Greene. Finally, a number of colleagues gave generously of their time and expertise in guiding our thoughts, making valuable suggestions, and providing materials: Stephen Cornell, *Udall Center for Studies in Public Policy*, and David Liverman, *Memorial University, Newfoundland*.

Sallie A. Marston
Paul L. Knox
Diana M. Liverman
Vincent J. Del Casino Jr.
Paul F. Robbins

THE TEACHING AND LEARNING PACKAGE

In addition to the text itself, the authors and publisher have been pleased to work with a number of talented people to produce an excellent instructional package of both traditional supplements and new digital resources.

FOR TEACHERS AND STUDENTS

MasteringGeography™ with Pearson eText

The **Mastering** platform is the most widely used and effective online homework, tutorial, and assessment system for the sciences. It delivers self-paced tutorials that provide individualized coaching, focus on the teacher's course objectives, and are responsive to each student's progress. The Mastering system helps teachers maximize class time with customizable, easy-to-assign, and automatically graded assessments that motivate students to learn outside of class and arrive prepared for lecture.

MasteringGeography offers:

- **Assignable activities** that include MapMaster™ Interactive Map activities, Encounter World Regional Geography Google Earth™ Explorations, Geography Video activities, Geoscience Animation activities, Map Projection activities, coaching activities on the toughest topics in geography, end-of-chapter questions and exercises, reading quizzes, and Test Bank questions.
- **A student Study Area** with MapMaster™ Interactive Maps, Geography Videos, Geoscience Animations, "In the News" RSS Feeds, Web links, glossary flashcards, chapter quizzes, an optional Pearson eText that includes versions for iPad and Android devices and more.

Pearson eText gives students access to the text whenever and wherever they can access the Internet. The eText pages look exactly like the printed text, and include powerful interactive and customization functions, including links to the multimedia.

Practicing Geography: Careers for Enhancing Society and the Environment

by the Association of American Geographers (0321811151) This book examines career opportunities for geographers and geospatial professionals in business, government, nonprofit, and educational sectors. A diverse group of academic and industry professionals share insights on career planning, networking, transitioning between employment sectors, and balancing work and home life. The book illustrates the value of geographic expertise and technologies through engaging profiles and case studies of geographers at work.

Teaching College Geography: A Practical Guide for Graduate Students and Early Career Faculty

by the Association of American Geographers (0136054471) This two-part resource provides a starting point for becoming an effective geography teacher from the very first day of class. Part One addresses "nuts-and-bolts" teaching issues. Part Two explores being an effective teacher in the field, supporting critical thinking with GIS and mapping technologies, engaging learners in large geography classes, and promoting awareness of international perspectives and geographic issues.

Aspiring Academics: A Resource Book for Graduate Students and Early Career Faculty

by the Association of American Geographers (0136048919) Drawing on several years of research, this set of essays is designed to help graduate students and early career faculty start their careers in geography and related social and environmental sciences. *Aspiring Academics* stresses the interdependence of teaching, research, and service—and the importance of achieving a healthy balance of professional and personal life—while doing faculty work. Each chapter provides accessible, forward-looking advice on topics that often cause the most stress in the first years of a college or university appointment.

Television for the Environment Earth Report Videos on DVD

(0321662989) This three-DVD set helps students visualize how human decisions and behavior have affected the environment, and how individuals are taking steps toward recovery. With topics ranging from the poor land management promoting the devastation of river systems in Central America to the struggles for electricity in China and Africa, these 13 videos from Television for the Environment's global *Earth Report* series recognize the efforts of individuals around the world to unite and protect the planet.

Television for the Environment Life World Regional Geography Videos on DVD

(013159348X) From the Television for the Environment's global *Life* series, this two-DVD set brings globalization and the developing world to the attention of any world regional geography course. These ten full-length video programs highlight matters such as the growing number of homeless children in Russia, the lives of immigrants living in the United States trying to help family still living in their native countries, and the European conflict between commercial interests and environmental concerns.

Television for the Environment Life Human Geography Videos on DVD

(0132416565) This three-DVD set is designed to enhance any human geography course. These DVDs include 14 full-length video programs from Television for the Environment's global *Life* series, covering a wide array of issues affecting people and places in the contemporary world, including the serious health risks of pregnant women in Bangladesh, the social inequalities of the "untouchables" in the Hindu caste system, and Ghana's struggle to compete in a global market.

FOR TEACHERS

Instructor Resource Manual Download (0321862449) The *Instructor Resource Manual*, originally created by one of this book's coauthors, Vincent Del Casino Jr., and now updated by Jessie Clark of the University of Oregon, follows the new organization of the main text. Strategies for Teaching Key Concepts provide teachers with a focused plan of action for every class session. Web Exercises tie in with associated Interactive Maps, and Additional Resources such as journals and Websites are provided.

TestGen/Test Bank download (0321862341) TestGen is a computerized test generator that lets teachers view and edit *Test Bank* questions, transfer questions to tests, and print the test in a variety of customized formats. Authored by Christa Ann Smith of Clemson University, this *Test Bank* includes approximately 1000 multiple-choice, true/false, and short answer/essay questions. Questions are correlated to the book's Learning Outcomes, the U.S. National Geography Standards, and Bloom's Taxonomy to help teachers to better map the assessments against both broad and specific teaching and learning objectives. The *Test Bank* is available in Microsoft Word® and also importable into Blackboard.

Instructor Resource DVD (0321862333) Everything teachers need, where they want it. The *Instructor Resource DVD (IRC)* helps make teachers more effective by saving them time and effort. All digital resources can be found in one, well-organized, easy-to-access place. The IRC DVD includes:

- All textbook images as JPEGs, PDFs, and PowerPoint™ Presentations
- Pre-authored Lecture Outline PowerPoint™ Presentations, which outline the concepts of each chapter with embedded art and can be customized to fit teachers' lecture requirements
- CRS "Clicker" Questions in PowerPoint™ format, which correlate to the book's Learning Outcomes, the U.S. National Geography Standards, and Bloom's Taxonomy
- The TestGen software, *Test Bank* questions, and answers for both MACs and PCs
- Electronic files of the *Instructor Resource Manual* and *Test Bank*

This Instructor Resource content is also available completely online via the Instructor Resources section of **www.MasteringGeography.com** and **www.pearsonhighered.com/irc**.

FOR STUDENTS

Goode's World Atlas 22nd Edition (0321652002) Goode's World Atlas has been the world's premier educational atlas since 1923, and for good reason. It features over 250 pages of maps, from definitive physical and political maps to important thematic maps that illustrate the spatial aspects of many important topics. The 22nd edition includes 160 pages of new, digitally produced reference maps as well as new thematic maps on global climate change, sea-level rise, CO_2 emissions, polar ice fluctuations, deforestation, extreme weather events, infectious diseases, water resources, and energy production.

Pearson's Encounter Series

Pearson's **Encounter** series provides rich, interactive explorations of geoscience concepts through GoogleEarth™ activities, exploring a range of topics in regional, human, and physical geography. For those who do not use MasteringGeography, all chapter explorations are available in print workbooks as well as in online quizzes at **www.mygeoscienceplace.com**, accommodating different classroom needs. Each Exploration consists of a worksheet, online quizzes, and a corresponding Google Earth™ KMZ file.

- *Encounter World Regional Geography* Workbook and Website by Jess C. Porter (0321681754)
- *Encounter Human Geography* Workbook and Website by Jess C. Porter (0321682203)
- *Encounter Physical Geography* Workbook and Website by Jess C. Porter and Stephen O'Connell (0321672526)
- *Encounter Geosystems* Workbook and Website by Charlie Thomsen (0321636996)
- *Encounter Earth* by Steve Kluge (0321581296)

Dire Predictions: Understanding Global Warming by Michael Mann, Lee R. Kump (0136044352) For any science or social science course in need of a basic understanding of IPCC reports. Periodic reports from the Intergovernmental Panel on Climate Change (IPCC) evaluate the risk of climate change brought on by humans. But the sheer volume of scientific data remains inscrutable to the general public, particularly to those who may still question the validity of climate change. In just over 200 pages, this practical text presents and expands upon the essential findings in a visually stunning and undeniably powerful way to the lay reader. Scientific findings that provide validity to the implications of climate change are presented in clear-cut graphic elements, striking images, and understandable analogies.

ABOUT THE AUTHORS

SALLIE A. MARSTON

Sallie Marston received her Ph.D. in geography from the University of Colorado, Boulder. She is currently a professor in the School of Geography and Development at the University of Arizona. Her teaching focuses on the political and cultural aspects of social life, with particular emphasis on sociospatial theory. She is the recipient of the College of Social and Behavioral Sciences Outstanding Undergraduate Teaching Award as well as the University of Arizona's Graduate College Graduate and Professional Education Teaching and Mentoring Award. Her teaching focuses on culture, politics, globalization, and research methods. She is the author of over 75 journal articles, book chapters, and books and serves on the editorial board of several scientific journals. She has co-authored, with Paul Knox, the introductory human geography textbook, *Human Geography: Places and Regions in Global Context*, also published by Pearson.

PAUL L. KNOX

Paul Knox received his Ph.D. in geography from the University of Sheffield, England. After teaching in the United Kingdom for several years, he moved to the United States to take a position as professor of urban affairs and planning at Virginia Tech. His teaching centers on urban and regional development, with an emphasis on comparative study. He has written several books on aspects of economic geography, social geography, and urbanization and serves on the editorial board of several scientific journals. In 2008, he received the Association of American Geographers Distinguished Scholarship Award. He is currently a University Distinguished Professor at Virginia Tech, where he also serves as Senior Fellow for International Advancement.

DIANA M. LIVERMAN

Diana Liverman received her Ph.D. in geography from the University of California, Los Angeles. Born in Accra, Ghana, she is the co-director of the Institute of the Environment and Regents Professor of Geography and Development at the University of Arizona. She has taught geography at Oxford University, Pennsylvania State University, and the University of Wisconsin–Madison. Her teaching and research focus on global environmental issues, environment and development, and Latin America. She has served on several national and international advisory committees dealing with environmental issues and climate change and has written about topics such as natural disasters, climate change, trade and environment, resource management, and environmental policy.

VINCENT J. DEL CASINO JR.

Vincent J. Del Casino Jr. received his Ph.D. in geography from the University of Kentucky in 2000. He is currently associate dean of the College of Social and Behavioral Sciences and professor in the School of Geography and Development at the University of Arizona. He was previously professor and chair of Geography at California State University, Long Beach. He has held a visiting research fellow post at the Australian National University and completed National Science Foundation–supported research in Thailand. His research interests include social and health geography, with a particular emphasis on HIV transmission, the care of people living with HIV and AIDS, and homelessness. His teaching focuses on social geography, geographic thought, and geographic methodology. He also teaches a number of general education courses in geography, including world regional geography, which he first began teaching as a graduate student in 1995.

PAUL F. ROBBINS

Paul Robbins received his Ph.D. in geography from Clark University in 1996. He is currently the director of the Nelson Institute for Environmental Studies at the University of Wisconsin–Madison. Previously, he taught at the University of Arizona, Ohio State University, the University of Iowa, and Eastern Connecticut State University. His teaching and research focus on the relationships between individuals (e.g., homeowners, hunters, professional foresters), environmental actors (e.g., lawns, elk, mesquite trees), and the institutions that connect them. He and his students seek to explain human environmental practices and knowledge, the influence the environment has on human behavior and organization, and the implications this holds for ecosystem health, local community, and social justice. Robbins's past projects have examined chemical use in the suburban United States, elk management in Montana, forest product collection in New England, and wolf conservation in India.

ABOUT OUR SUSTAINABILITY INITIATIVES

Pearson recognizes the environmental challenges facing this planet, as well as acknowledges our responsibility in making a difference. This book is carefully crafted to minimize environmental impact. The binding, cover, and paper come from facilities that minimize waste, energy consumption, and the use of harmful chemicals. Pearson closes the loop by recycling every out-of-date text returned to our warehouse.

Along with developing and exploring digital solutions to our market's needs, Pearson has a strong commitment to achieving carbon-neutrality. As of 2009, Pearson became the first carbon- and climate-neutral publishing company. Since then, Pearson remains strongly committed to measuring, reducing, and offsetting our carbon footprint.

The future holds great promise for reducing our impact on Earth's environment, and Pearson is proud to be leading the way. We strive to publish the best books with the most up-to-date and accurate content, and to do so in ways that minimize our impact on Earth. To learn more about our initiatives, please visit **www.pearson.com/responsibility**.

PEARSON

A Comprehensive Learning Framework

Using a consistent thematic structure to organize each regional chapter—Environment and Society; History, Economy, and Territory; Culture and Populations; and Future Geographies—this text balances the twin forces of globalization and regionalization, which together generate a world of regions that are globally interconnected and locally differentiated.

3 The Russian Federation, Central Asia, and the Transcaucasus

LEARNING OUTCOMES

- Describe how the location and vast size of the Russian Federation, Central Asia, and Transcaucasus region have influenced its climate and how climate change may influence the region's future.
- Identify the topographic areas of the region and how they are influenced by the larger global process of plate tectonics.
- List and describe the region's five major environmental zones and provide examples of how people have adapted to and modified the physical landscapes in these zones.
- Compare and contrast the regional environmental impacts of Soviet era planning with those of the post-Soviet transition.
- Describe how economic transitions in the region are playing out in the postsocialist era.
- Analyze the changing political trends in the region in terms of fragmentation and realignment.
- Summarize the cultural legacies and new developments in the region in terms of high, revolutionary, and popular forms of culture.
- Describe and explain how birth and death rates have changed relative to one another over the recent history of the region and discuss the implications of recent demographic contractions.

Thousands of Russian citizens took part in the "million person march" against the inauguration of Vladimir Putin as President of Russia in Moscow in May of 2012.

IN MARCH 2012, AN AMAZING THING HAPPENED in the Russian Federation (or Russia, as it is commonly known). Former President Vladimir Putin (2000–2008), who was serving as prime minister at the time of the election, was elected to a new, six-year term as president. Putin stepped down as president in 2008 because of a constitutional requirement that Russian presidents serve no more than two consecutive terms; however, as the prime minister working closely with President Dmitry Medvedev (Putin's own pick for president), Putin remained at the center of Russian politics. Under Putin's and then Medvedev's leadership between 2000 and 2012, Russia's economic growth rate accelerated, making it the world's ninth-largest economy in 2011. At that time, the country's overall debt was under 10% of the country's total gross domestic product (GDP). This stands in stark contrast to other European and North American economies, some of which now have debt that exceeds 100% of total GDP.

But opposition forces in Russia remain concerned that Putin, a former intelligence officer in the old Soviet Union, has too much power and control. Critics argue that Putin's rule has resulted in limited political participation in this expansive and diverse country. In April 2012, a group of political opposition leaders walked out as Putin addressed the Russian parliament to protest what they described as voting irregularities in Putin's victorious 2012 elections. Some observers also warn that Russia's recent economic successes are based, in large part, on the country's massive oil reserves. This dependence on primary commodity (raw material) production puts Russia at risk, should global oil prices slip in the future.

For many of the protesters, Putin's return to the presidency clearly raises questions about the future of political pluralism in Russia. Putin's return to the presidency also suggests that Russia might play a larger role on the world stage, where Putin has always been an important player. As Putin embarks on the first of what could be two more terms as president, Russia will likely strive to expand its regional and global significance as a political and economic power. ■

86

87

Updated chapter-opening vignettes recount contemporary regional "day in the life" human interest stories.

NEW! Learning Outcomes are listed at the beginning of each chapter to prioritize essential issues and content.

NEW! Learning Outcomes Revisited at the end of each chapter summarize and reinforce the most important concepts for students, before they move on to homework, tests, or MasteringGeography.™

Updated end-of-chapter **Thinking Geographically** questions allow students to apply and synthesize their understanding of concepts and themes.

LEARNING OUTCOMES REVISITED

■ **Describe how the location and vast size of the Russian Federation, Central Asia, and Transcaucasus region have influenced its climate and how climate change may influence the region's future.**

This region's climate is influenced by its mostly northern location, its position at the eastern edge of Europe and inland Asia, its absence of mountainous terrain (except in the far south and east), and the lack of any significant moderating influence of oceans and seas. Three climate regions dominate the landscape from the arid and semiarid regions in the south to the continental/midlatitude and polar regions to the north. As climate change opens the northern areas of the region, previously unusable resources and transportation routes in the Arctic will become increasingly critical to regional growth and change.

■ **Identify the topographic areas of the region and how they are influenced by the larger global process of plate tectonics.**

The position of the region within a single plate has resulted in the

the region appear as uniform. But there is quite a bit of variation throughout the region, a result of uplift in the southern portion of the region, for example.

■ **List and describe the region's five major environmental zones and provide examples of how people have adapted to and modified the physical landscapes in these zones.**

From north to south, the regions are the tundra, the taiga, mixed forested regions, the steppe, and open deserts. The natural resources of these regions, including oil and timber, have been exploited and harvested with limited concern for sustainability. In the drier zones, modern farming methods have also impacted the land, and irrigation methods in particular have made the region more vulnerable to droughts and other extreme weather events.

■ **Compare and contrast the regional environmental impacts of Soviet era planning with those of the post-Soviet transition.**

Soviet central planning placed strong emphasis on industrial output,

THINKING GEOGRAPHICALLY

1. How has Europe benefited from its location and its major physical features?

2. What key inventions during the period from 1400 to 1600 helped European merchants establish the basis of today's global economy? Why?

3. Which imports from the colonies helped transform Europe? Focus on natural resources and new crops.

4. How did the European Union (EU) develop? Why is the EU's Common Agricultural Policy (CAP) so important?

5. What migration patterns characterized Europe during the 19th and 20th centuries? Consider movement within Europe as well as movement to and from Europe.

APPLY YOUR KNOWLEDGE List three possible connections that might exist between the spread of HIV, drug use, and minority and ethnic politics in Southeast Asia.

NEW! Apply Your Knowledge conceptual review questions at the end of each thematic section help students check their comprehension and extend their knowledge as they read through the chapter.

Provocative Topics and Features that Engage

Edgy, conceptual features are designed to challenge conventional thinking about current world regions and provoke lively discussion.

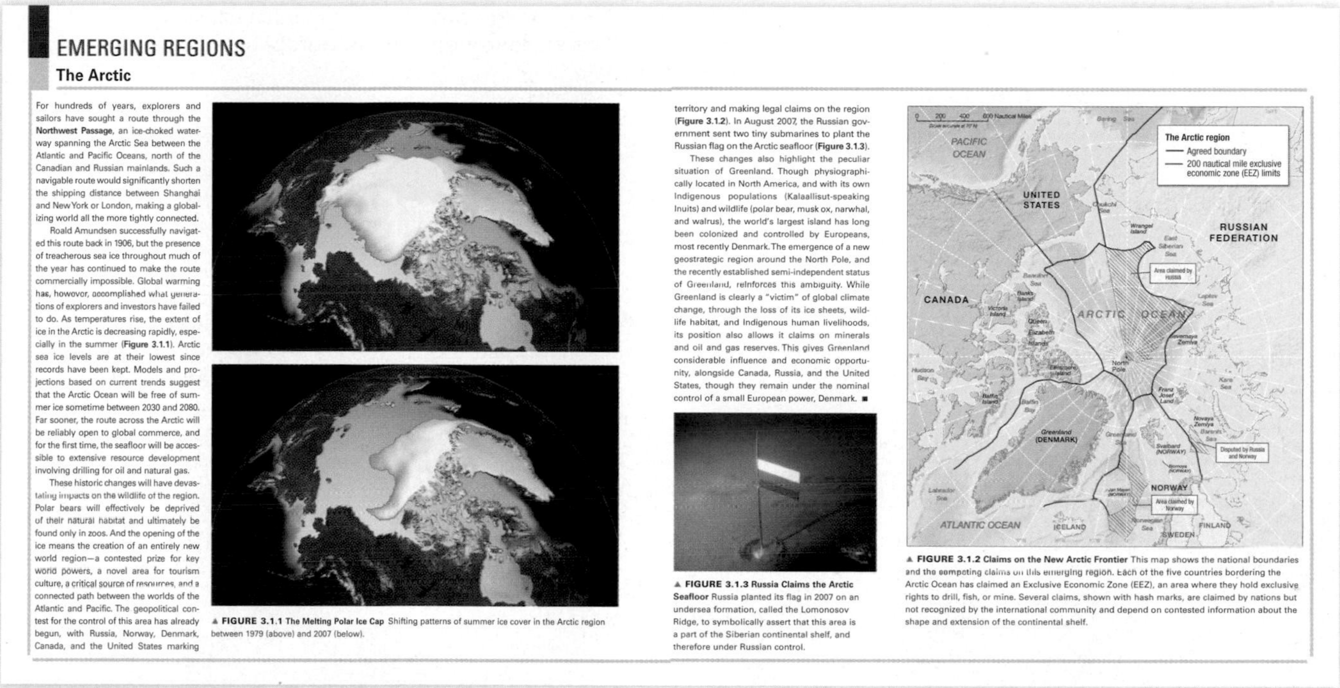

EMERGING REGIONS

The Arctic

For hundreds of years, explorers and sailors have sought a route through the **Northwest Passage**, an ice-choked waterway spanning the Arctic Sea between the Atlantic and Pacific Oceans, north of the Canadian and Russian mainlands. Such a navigable route would significantly shorten the shipping distance between Shanghai and New York or London, making a globalizing world all the more tightly connected.

Roald Amundsen successfully navigated this route back in 1906, but the presence of treacherous sea ice throughout much of the year has continued to make the route commercially impossible. Global warming has, however, accomplished what generations of explorers and investors have failed to do. As temperatures rise, the extent of ice in the Arctic is decreasing rapidly, especially in the summer (**Figure 3.1.1**). Arctic sea ice levels are at their lowest since records have been kept. Models and projections based on current trends suggest that the Arctic Ocean will be free of summer ice sometime between 2030 and 2080. Far sooner, the route across the Arctic will be reliably open to global commerce, and for the first time, the seafloor will be accessible to extensive resource development involving drilling for oil and natural gas.

These historic changes will have devastating impacts on the wildlife of the region. Polar bears will effectively be deprived of their natural habitat and ultimately be found only in zoos. And the opening of the ice means the creation of an entirely new world region—a contested prize for key world powers, a novel area for tourism culture, a critical source of resources, and a connected path between the worlds of the Atlantic and Pacific. The geopolitical contest for the control of this area has already begun, with Russia, Norway, Denmark, Canada, and the United States marking territory and making legal claims on the region (**Figure 3.1.2**). In August 2007, the Russian government sent two tiny submarines to plant the Russian flag on the Arctic seafloor (**Figure 3.1.3**).

These changes also highlight the peculiar situation of Greenland. Though physiographically located in North America, and with its own Indigenous populations (Kalaallisut-speaking Inuits) and wildlife (polar bear, musk ox, narwhal, and walrus), the world's largest island has long been colonized and controlled by Europeans, most recently Denmark. The emergence of a new geostrategic region around the North Pole, and the recently established semi-independent status of Greenland, reinforces this ambiguity. While Greenland is clearly a "victim" of global climate change, through the loss of its ice sheets, wildlife habitat, and Indigenous human livelihoods, its position also allows it claims on minerals and oil and gas reserves. This gives Greenland considerable influence and economic opportunity, alongside Canada, Russia, and the United States, though they remain under the nominal control of a small European power, Denmark. ■

▲ **FIGURE 3.1.1 The Melting Polar Ice Cap** Shifting patterns of summer ice cover in the Arctic region between 1979 (above) and 2007 (below).

▲ **FIGURE 3.1.3 Russia Claims the Arctic Seafloor** Russia planted its flag in 2007 on an undersea formation, called the Lomonosov Ridge, to symbolically assert that this area is a part of the Siberian continental shelf, and therefore under Russian control.

▲ **FIGURE 3.1.2 Claims on the New Arctic Frontier** This map shows the national boundaries and the competing claims on this emerging region. Each of the five countries bordering the Arctic Ocean has claimed an Exclusive Economic Zone (EEZ), an area where they hold exclusive rights to drill, fish, or mine. Several claims, shown with hash marks, are claimed by nations but not recognized by the international community and depend on contested information about the shape and extension of the continental shelf.

NEW! Emerging Regions features challenge conventional views of the world map and show geography to be a fluid, developing discipline with new locations and groups of countries developing greater internal coherence. Examples include the Arctic, BRICS, the Mekong River Delta, and others.

Geographies of Indulgence, Desire, and Addiction include new and updated regional case studies of commodity consumption, including wide-ranging topics like coffee, the exotic animal trade, and opium and methamphetamines.

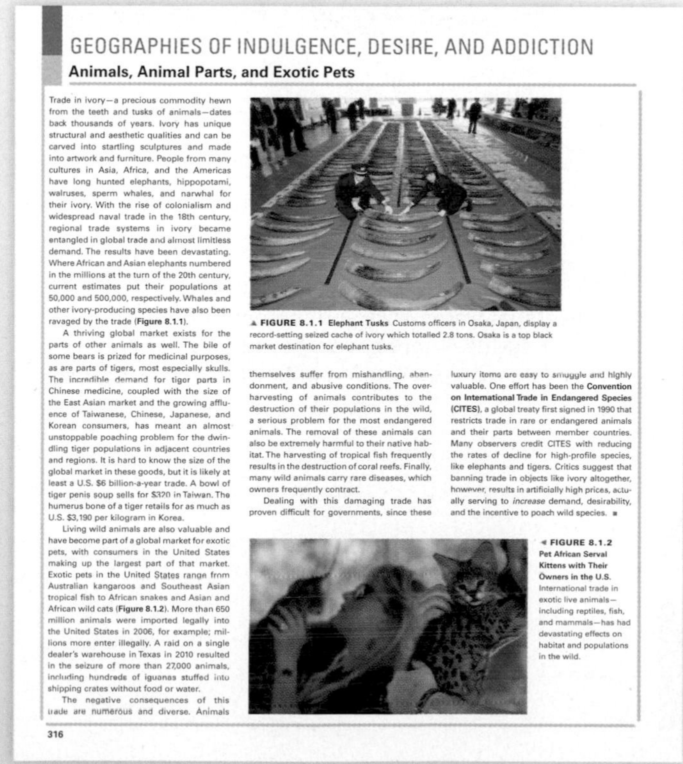

GEOGRAPHIES OF INDULGENCE, DESIRE, AND ADDICTION

Animals, Animal Parts, and Exotic Pets

Trade in ivory—a precious commodity hewn from the teeth and tusks of animals—dates back thousands of years. Ivory has unique structural and aesthetic qualities and can be carved into startling sculptures and made into artwork and furniture. People from many cultures in Asia, Africa, and the Americas have long hunted elephants, hippopotami, walruses, sperm whales, and narwhal for their ivory. With the rise of colonialism and widespread naval trade in the 18th century, regional trade systems in ivory became entangled in global trade and almost limitless demand. The results have been devastating. Where African and Asian elephants numbered in the millions at the turn of the 20th century, current estimates put their populations at 50,000 and 500,000, respectively. Whales and other ivory-producing species have also been ravaged by the trade (**Figure 8.1.1**).

A thriving global market exists for the parts of other animals as well. The bile of some bears is prized for medicinal purposes, as are parts of tigers, most especially skulls. The incredible demand for tiger parts in Chinese medicine, coupled with the size of the East Asian market and the growing affluence of Taiwanese, Chinese, Japanese, and Korean consumers, has meant an almost unstoppable poaching problem for the dwindling tiger populations in adjacent countries and regions. It is hard to know the size of the global market in these goods, but it is likely at least a U.S. $6 billion-a-year trade. A bowl of tiger penis soup sells for $320 in Taiwan. The humerus bone of a tiger retails for as much as U.S. $3,190 per kilogram in Korea.

Living wild animals are also valuable and have become part of a global market for exotic pets, with consumers in the United States making up the largest part of that market. Exotic pets in the United States range from Australian kangaroos and Southeast Asian tropical fish to African snakes and Asian and African wild cats (**Figure 8.1.2**). More than 650 million animals were imported legally into the United States in 2006, for example; millions more enter illegally. A raid on a single dealer's warehouse in Texas in 2010 resulted in the seizure of more than 27,000 animals, including hundreds of iguanas stuffed into shipping crates without food or water.

The negative consequences of this trade are numerous and diverse. Animals themselves suffer from mishandling, abandonment, and abusive conditions. The overharvesting of animals contributes to the destruction of their populations in the wild, a serious problem for the most endangered animals. The removal of these animals can also be extremely harmful to their native habitat. The harvesting of tropical fish frequently results in the destruction of coral reefs. Finally, many wild animals carry rare diseases, which owners frequently contract.

Dealing with this damaging trade has proven difficult for governments, since these luxury items are easy to smuggle and highly valuable. One effort has been the **Convention on International Trade in Endangered Species (CITES)**, a global treaty first signed in 1990 that restricts trade in rare or endangered animals and their parts between member countries. Many observers credit CITES with reducing the rates of decline for high-profile species, like elephants and tigers. Critics suggest that banning trade in objects like ivory altogether, however, results in artificially high prices, actually serving to *increase* demand, desirability, and the incentive to poach wild species. ■

▲ **FIGURE 8.1.1 Elephant Tusks** Customs officers in Osaka, Japan, display a record-setting seized cache of ivory which totalled 2.8 tons. Osaka is a top black market destination for elephant tusks.

◄ **FIGURE 8.1.2 Pet African Serval Kittens with Their Owners in the U.S.** International trade in exotic live animals—including reptiles, fish, and mammals—has had devastating effects on habitat and populations in the wild.

316

Updated **Future Geographies** features close out each chapter with a glimpse ahead to possible future conditions in the region.

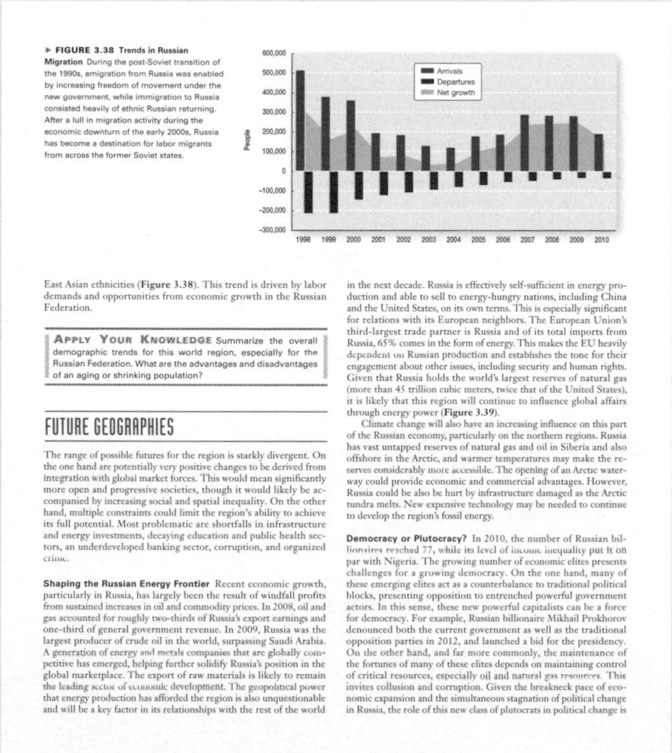

► **FIGURE 3.38 Trends in Russian Migration** During the post-Soviet transition of the 1990s, emigration from Russia was enabled by increasing freedom of movement under the new government, while immigration to Russia consisted heavily of ethnic Russian returning. After a lull in migration activity during the economic downturn of the early 2000s, Russia has become a destination for labor migrants from across the former Soviet states.

East Asian ethnicities (**Figure 3.38**). This trend is driven by labor demands and opportunities from economic growth in the Russian Federation.

Apply Your Knowledge Summarize the overall demographic trends for this world region, especially for the Russian Federation. What are the advantages and disadvantages of an aging or shrinking population?

FUTURE GEOGRAPHIES

The range of possible futures for the region is starkly divergent. On the one hand are potentially very positive changes to be derived from integration with global market forces. This would mean significantly more open and progressive societies, though it would likely be accompanied by increasing social and spatial inequality. On the other hand, multiple constraints could limit the region's ability to achieve its full potential. Most problematic are shortfalls in infrastructure and energy investments, decaying education and public health sectors, an underdeveloped banking sector, corruption, and organized crime.

Shaping the Russian Energy Frontier Recent economic growth, particularly in Russia, has largely been the result of windfall profits from sustained increases in oil and commodity prices. In 2008, oil and gas accounted for roughly two-thirds of Russia's export earnings and one-third of general government revenue. In 2009, Russia was the largest producer of crude oil in the world, surpassing Saudi Arabia. A generation of energy and metals companies that are globally competitive has emerged, helping further solidify Russia's position in the global marketplace. The export of raw materials is likely to remain the leading sector of economic development. The geopolitical power that energy production has afforded the region is also unquestionable and will be a key factor in its relationships with the rest of the world in the next decade. Russia is effectively self-sufficient in energy production and able to sell to energy-hungry nations, including China and the United States, on its own terms. This is especially significant for relations with its European neighbors. The European Union's third-largest trade partner is Russia and of its total imports from Russia, 65% comes in the form of energy. This makes the EU heavily dependent on Russian production and establishes the tone for their engagement about other issues, including security and human rights. Given that Russia holds the world's largest reserves of natural gas (more than 45 trillion cubic meters, twice that of the United States), it is likely that this region will continue to influence global affairs through energy power (**Figure 3.39**).

Climate change will also have an increasing influence on this part of the Russian economy, particularly on the northern regions. Russia has vast untapped reserves of natural gas and oil in Siberia and also offshore in the Arctic, and warmer temperatures may make the reserves considerably more accessible. The opening of an Arctic waterway could provide economic and commercial advantages. However, Russia could be also be hurt by infrastructure damaged as the Arctic tundra melts. New expensive technology may be needed to continue to develop the region's fossil energy.

Democracy or Plutocracy? In 2010, the number of Russian billionaires reached 77, while its level of income inequality put it on par with Nigeria. The growing number of economic elites presents challenges for a growing democracy. On the one hand, many of these emerging elites act as a counterbalance to traditional political blocks, presenting opposition to entrenched powerful government actors. In this sense, these new powerful capitalists can be a force for democracy. For example, Russian billionaire Mikhail Prokhorov denounced both the current government as well as the traditional opposition parties in 2012, and launched a bid for the presidency. On the other hand, and far more commonly, the maintenance of the fortunes of many of these elites depends on maintaining control of critical resources, especially oil and natural gas resources. This invites collusion and corruption. Given the breakneck pace of economic expansion and the simultaneous stagnation of political change in Russia, the role of this new class of plutocrats in political change is

Visualizing Real-World Data, Relevant Applications

Using the latest data and current examples, this text accurately captures the geography, issues, and concerns of emerging and established world regions.

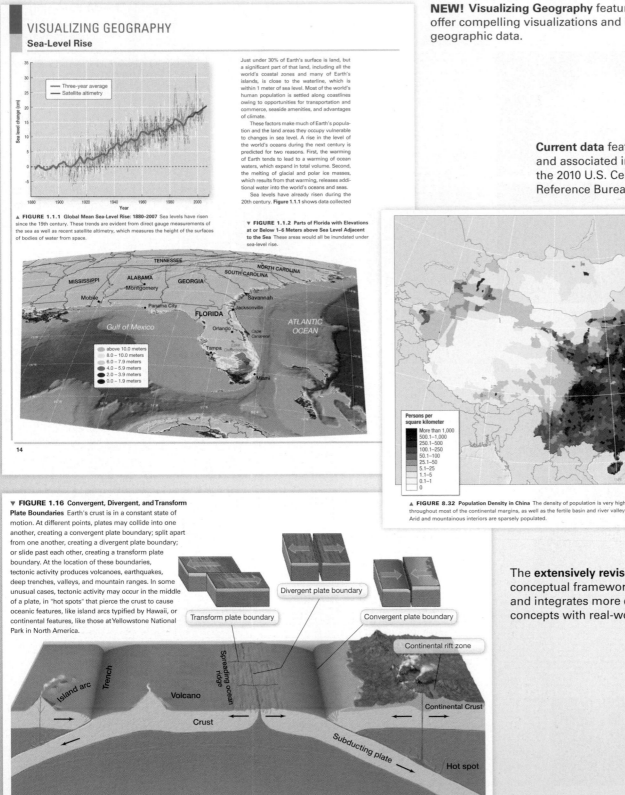

VISUALIZING GEOGRAPHY
Sea-Level Rise

▲ **FIGURE 1.1.1 Global Mean Sea-Level Rise: 1880–2007** Sea levels have risen since the 19th century. These trends are evident from direct gauge measurements of the sea as well as recent satellite altimetry, which measures the height of the surfaces of bodies of water from space.

Just under 30% of Earth's surface is land, but a significant part of that land, including all the world's coastal zones and many of Earth's islands, is close to the waterline, which is within 1 meter of sea level. Most of the world's human population is settled along coastlines owing to opportunities for transportation and commerce, seaside amenities, and advantages of climate.

These factors make much of Earth's population and the land areas they occupy vulnerable to changes in sea level. A rise in the level of the world's oceans during the next century is predicted for two reasons. First, the warming of Earth tends to lead to a warming of ocean waters, which expand in total volume. Second, the melting of glacial and polar ice masses, which results from that warming, releases additional water into the world's oceans and seas.

Sea levels have already risen during the 20th century. **Figure 1.1.1** shows data collected

▼ **FIGURE 1.1.2 Parts of Florida with Elevations at or Below 1–6 Meters above Sea Level Adjacent to the Sea** These areas would all be inundated under sea-level rise.

14

NEW! Visualizing Geography features in each chapter offer compelling visualizations and infographics of applied geographic data.

Current data feature the latest science, statistics, and associated imagery, including data from the 2010 U.S. Census and the 2012 Population Reference Bureau World Population Data.

▲ **FIGURE 8.32 Population Density in China** The density of population is very high throughout most of the continental margins, as well as the fertile basin and river valleys. Arid and mountainous interiors are sparsely populated.

▼ **FIGURE 1.16 Convergent, Divergent, and Transform Plate Boundaries** Earth's crust is in a constant state of motion. At different points, plates may collide into one another, creating a convergent plate boundary; split apart from one another, creating a divergent plate boundary; or slide past each other, creating a transform plate boundary. At the location of these boundaries, tectonic activity produces volcanoes, earthquakes, deep trenches, valleys, and mountain ranges. In some unusual cases, tectonic activity may occur in the middle of a plate, in "hot spots" that pierce the crust to cause oceanic features, like island arcs typified by Hawaii, or continental features, like those at Yellowstone National Park in North America.

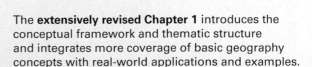

The **extensively revised Chapter 1** introduces the conceptual framework and thematic structure and integrates more coverage of basic geography concepts with real-world applications and examples.

Vibrant Visual Program Adds Color and Context

Completely overhauled cartography and photo program tells a visually compelling, current, and accessible story of global connections.

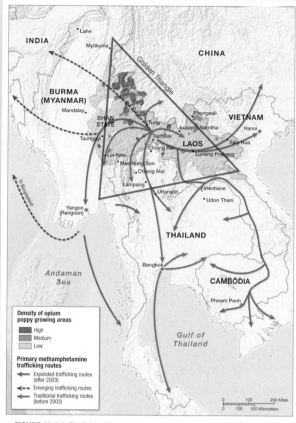

▲ **FIGURE 10.3.1 The Golden Triangle** This map highlights the density of poppy production in the region known as the Golden Triangle. The map of Thailand, which engaged in an active campaign to reduce poppy production, clearly shows the success of that program. Methamphetamine production takes place across Burma and parts of Laos, while Thailand serves as a key trans-shipment site for the drug.

The **completely revised cartography and illustration program** includes easier-to-read maps and art using the latest data, more engaging and consistent styles, textures, and colors, and a solid region-by-region comparative framework.

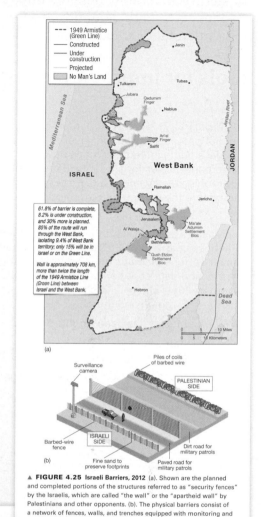

▲ **FIGURE 4.25 Israeli Barriers, 2012** (a). Shown are the planned and completed portions of the structures referred to as "security fences" by the Israelis, which are called "the wall" or the "apartheid wall" by Palestinians and other opponents. (b). The physical barriers consist of a network of fences, walls, and trenches equipped with monitoring and surveillance devices.

▲ **FIGURE 8.7 The 2011 Tsunami** The devastating wave that hit the Japanese coast in March 2011, seen here crashing over Miyako City, was set off by an offshore earthquake.

▲ **FIGURE 1.4 The Emergence of a New Country** The emergence of the Republic of South Sudan in 2011 was celebrated in many ways. In this photo a citizen of the new country waves a flag as part of the independence celebration.

Expanded visuals support the narrative with **90% new photographs.**

MasteringGeography™

www.masteringgeography.com

MasteringGeography delivers engaging, dynamic learning opportunities—focusing on course objectives and responsive to each student's progress—that are proven to help students absorb world regional course material and understand difficult geographic concepts.

Tools for improving geographic literacy and exploring Earth's dynamic landscape

MapMaster is a powerful interactive map tool that presents assignable layered thematic and place name interactive maps at world and regional scales for students to test their geographic literacy and spatial reasoning skills, and explore the modern geographer's tools.

MapMaster Layered Thematic Interactive Map Activities act as a mini-GIS tool, allowing students to layer various thematic maps to analyze spatial patterns and data at regional and global scales and answer multiple-choice and short-answer questions organized by region and theme.

NEW! MapMaster has been updated to include:

- 90 new map layers
- Zoom and annotation functionalities
- Current U.S. Census, U.N., and Population Reference Bureau Data

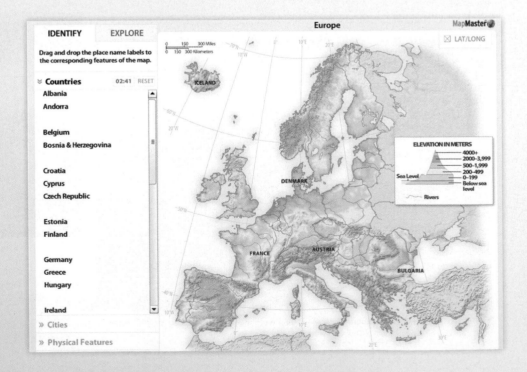

MapMaster Place Name Interactive Map Activities have students identify place names of political and physical features at regional and global scales, explore select recent country data from the CIA World Factbook, and answer associated assessment questions.

Help students develop spatial reasoning and a sense of place

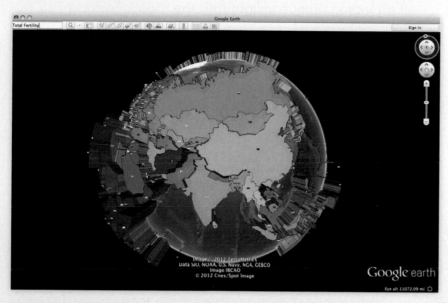

Encounter Activities provide rich, interactive explorations of geoscience concepts through Google Earth™ activities, exploring a range of topics in world regional geography. Dynamic assessment includes questions related to core physical geography concepts. All explorations include corresponding Google Earth KMZ media files, and questions include hints and specific wrong-answer feedback to help coach students towards mastery of the concepts.

Geography videos provide students a sense of place and allow them to explore a range of locations and topics related to world regional and physical geography. Covering issues of economy, development, globalization, climate and climate change, culture, etc., there are 10 multiple-choice questions for each video. These video activities allow teachers to test students' understanding and application of concepts, and offer hints and wrong-answer feedback.

Thinking Spatially and Data Analysis Activities help students master the toughest concepts and develop spatial reasoning and critical thinking skills by identifying and labeling features from maps, illustrations, graphs, and charts. Students then examine related data sets, answering multiple-choice and increasingly higher order conceptual questions, which include hints and specific wrong-answer feedback.

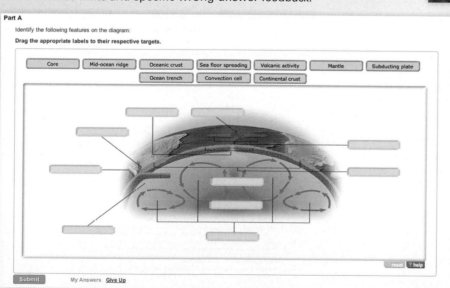

Student Study Resources in **MasteringGeography** include:

- MapMaster™ interactive maps
- Practice quizzes
- Geography videos
- Select Geoscience Animations
- "In the News" RSS feeds
- Glossary flashcards
- Optional Pearson eText and more

Callouts to MasteringGeography appear at the end of each chapter to direct students to extend their learning beyond the textbook.

MasteringGeography™ www.masteringgeography.com

With the Mastering gradebook and diagnostics, you'll be better informed about your students' progress than ever before. Mastering captures the step-by-step work of every student—including wrong answers submitted, hints requested, and time taken at every step of every problem—all providing unique insight into the most common misconceptions of your class.

Quickly monitor and display student results

The **Gradebook** records all scores for automatically graded assignments. Shades of red highlight struggling students and challenging assignments.

Diagnostics provide unique insight into class and student performance. With a single click, charts summarize the most difficult questions, vulnerable students, grade distribution, and score improvement over the duration of the course.

With a single click, **Individual Student Performance Data** provides **at-a-glance statistics** into each individual student's performance, including time spent on the questions, number of hints opened, and number of wrong and correct answers submitted.

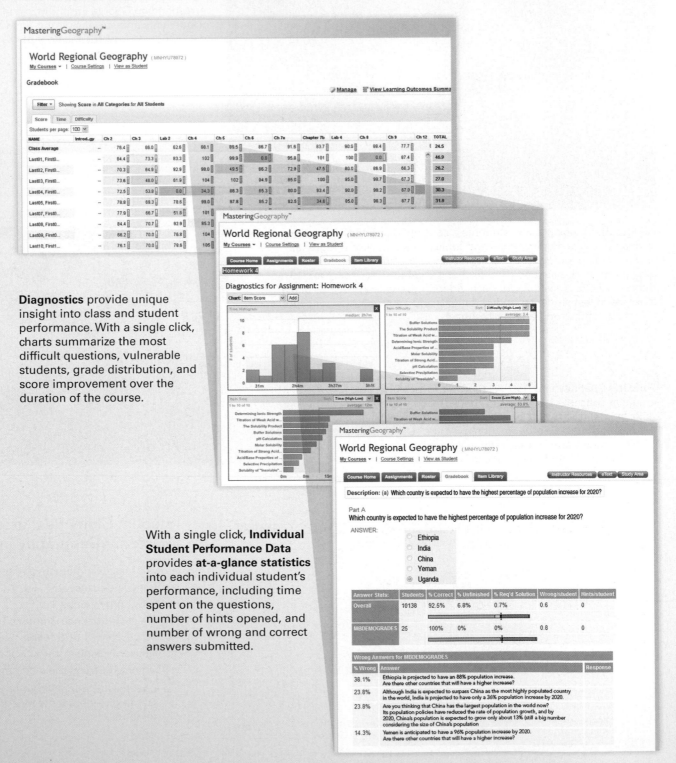

Easily measure student performance against your Learning Outcomes

Learning Outcomes

MasteringGeography provides quick and easy access to information on student performance against your learning outcomes and makes it easy to share those results.

- Quickly add your own learning outcomes, or use publisher-provided ones, to track student performance and report it to your administration.
- View class and individual student performance against specific learning outcomes.
- Effortlessly export results to a spreadsheet that you can further customize and/or share with your chair, dean, administrator, and/or accreditation board.

Easy to customize

Customize publisher-provided items or quickly add your own. MasteringGeography makes it easy to edit any questions or answers, import your own questions, and quickly add images or links to further enhance the student experience.

Upload your own video and audio files from your hard drive to share with students, as well as record video from your computer's Webcam directly into MasteringGeography—no plug-ins required. Students can download video and audio files to their local computer or launch them in Mastering to view the content.

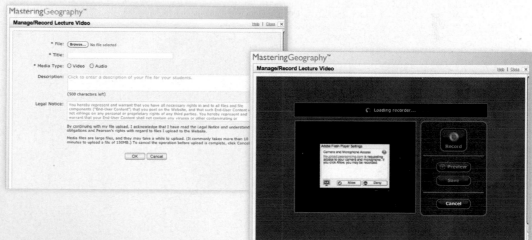

Pearson eText gives students access to *World Regions in Global Context: Peoples, Places, and Environment,* **Fifth Edition** whenever and wherever they can access the Internet. The eText pages look exactly like the printed text, and include powerful interactive and customization functions. Users can create notes, highlight text in different colors, create bookmarks, zoom, click hyperlinked words and phrases to view definitions, and view as a single page or as two pages. The Pearson eText also links students to associated media files, enabling them to view an animation as they read the text, and offers a full-text search and the ability to save and export notes. The Pearson eText also includes embedded URLs in the chapter text with active links to the Internet.

NEW! The Pearson eText app is a great companion to Pearson's eText browser-based book reader. It allows existing subscribers who view their Pearson eText titles on a Mac or PC to additionally access their titles in a bookshelf on the iPad or an Android tablet either online or via download.

1 | World Regions in Global Context

Chinese workers assemble mobile phones at the factory of Shenzhen City, Guangdong Province, in September 2010. Investment in manufacturing in the region has precipitated an increase in migration to the city. Many women are drawn into the industry and away from their rural homes.

HERE IS AN EXPERIMENT YOU REALLY SHOULDN'T TRY. Take your cell phone, throw it on the ground, stomp on it, and pick through the pieces. Amid the junk, you might find the entire world. The liquid crystal display (LCD) that formed the screen was manufactured in Mexico. The microprocessor chip you pry from the base was assembled in a factory in China, which is owned by a company in South Korea, which is funded by investment money from the United States. A programmer in India wrote the "apps" code. The pieces are made from elements gathered from copper mines in Chile, and the toxic lead that soldered together the circuit board comes from Australia. Your cell phone cannot exist without the resources and knowledge of all these different world regions. As objects such as cell phones become more common, connections develop among all the places involved in producing them. These accelerating global interconnections are central to the process of globalization. Globalization reflects a world where places and people are increasingly connected. Thanks to these connections, not only products, but ideas, languages, culture, and music flow from place to place, often making places seem more and more *similar*. And yet places remain strikingly *different* in spite of these similarities. Why?

LEARNING OUTCOMES

- Compare and contrast globalization and regionalization and explain the relationship between the two.

- Identify the main climatic, geological, and ecological forces that shape world regions and how each force is changing in the current era.

- Differentiate between forms of economic activity and explain why these forms vary around the globe.

- List contemporary economic development trends and identify the main measures used to assess social and economic advancement.

- Explain the implications of globalization for modern states and identify the actors emerging in this process.

- Specify how the global distribution of languages and religions is changing.

- Explain how globalization and regionalization affect key elements of culture.

- Explain how and why regional population growth rates rise and fall.

If you visited all the places integral in the production of your phone, you would find that the tech corridor of Bangalore, India, is populated by well-educated, relatively high-paid technicians, living in suburban homes. The LCD factory in Mexico is a highly indus-trialized site, populated by workers who have migrated from rural areas. The Chilean copper source is actually an enormous surface pit mine, three miles wide and a half-mile deep. In Australia, you might find a contested space, where Indigenous people struggle against mining operations. These places are becoming *more* different precisely because of the economic and cultural activities and environmental transformations required in manufacturing. This is regionalization—the process through which distinctive areas come into being. Regionalization reflects a world where novel cultures, ideas, and products emerge from the mix of elements into new unique regions. The conclusion you must draw from this impractical experiment is: *places are different because they are con-nected.* A major goal of this book is to explore both the connections and the differences between regions. ∎

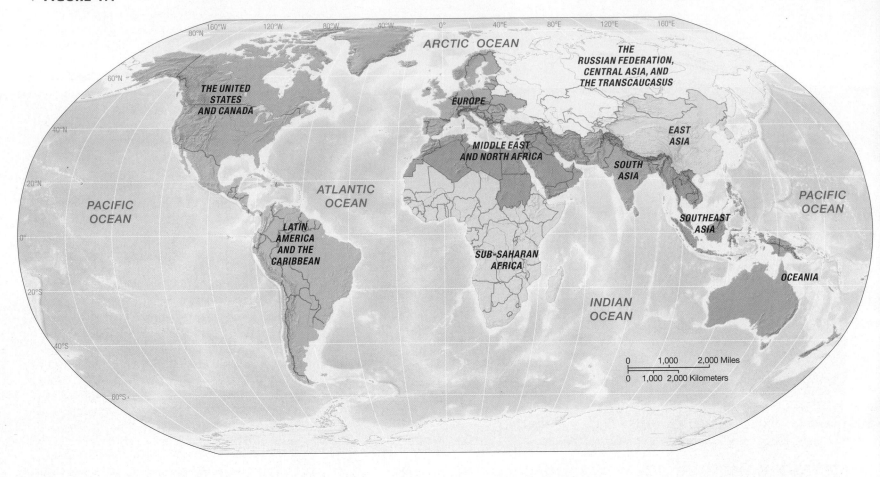

World Regions Key Facts

At the beginning of each chapter, the reader is provided with a number of "Key Facts" about the region. These facts represent the various human and physical geographic characteristics of each region or, in the case of Chapter 1, the world. Some of these key indicators will be familiar—*total area, population, urbanization*—but others may require some explanation:

- GNI PPP per Capita (per Person): Gross national income (GNI) per person in relation to purchasing power parity (PPP). This number represents the total value of goods and services a person could buy in the United States with his or her local income.

- Total Fertility Rate: The average number of children each woman will have over the course of her lifetime.

- Access to Improved Drinking Water Sources: A measure of how much of a population has water from a protected or safe source.

- Energy use: The total use of oil per capita before the oil is transformed into other products.

- Ecological Footprint per Capita: Represents the total number of hectares (a measure of land) each person consumes on a yearly basis compared to the total capacity of the planet.

- Total Area (total sq km): 510 million

- Population (2011): 6.987 billion; Population under Age 15 (%): 27; Population over Age 65 (%): 8

- Population Density (per square km) (2011): 51

- Urbanization (%) (2011): 51

- Average Life Expectancy at Birth (2011): Overall: 70; Women: 72; Men: 68

- Total Fertility Rate (2011): 2.5

- GNI PPP per Capita (current U.S. $) (2009): 10,240

- Population in Poverty (%, < $2/day) (2000–2009): 48

- Internet Users (2011): 2,095,006,005; Population with Access to the Internet (%) (2011): 30; Growth of Internet Use (%) (2000–2011): 480

- Access to Improved Drinking Water Sources (%) (2011): Urban: 96; Rural: 77

- Energy Use (kg of oil equivalent per capita) (2011): 1802.57

- Ecological Footprint (hectares per capita consumed/hectares per capita available, global scale) (2011): 2.7/1.8

THINKING GLOBALLY

Geography is the study of global interconnections involving everything from how people earn a living to how they interact with the environment. **Geography** comes from the Greek word *geographia*, which translates as "writing the world." Geographers map, travel, and measure the world to provide careful and rich accounts and descriptions of Earth's characteristics. But geography does more: it investigates the physical features of Earth and its atmosphere, the spatial organization and distribution of human activities, and the complex interrelationships between people and the natural and human-made environments in which they live. The power of geography lies in its ability to describe and explain global geographic processes, while at the same time explaining *why* certain patterns emerge on Earth.

This book has been written to help you learn about your world by exploring relationships between **regions**. Regions (like sub-Saharan Africa or South Asia) are those large areas of the world that share similar cultural, environmental, economic, or political characteristics (**Figure 1.1**). The text summarizes the important characteristics of different regions. What languages do people speak there? What do people do for a living? What religions are practiced there? Are there mountains or beaches? Does it snow? By studying the geography of world regions, you will have a far better grasp on conditions around the world and be a better-informed global citizen when you travel abroad.

But this book has a further purpose, which is to explain how these regions became the way they are. World regions are always changing. Some regions that we take for granted now would have made no sense to people in the past. Europe, for example, is a coherent region, with a mild climate and, for the most part, a unified currency (for the moment!), but such a single region would never have been recognized by ancient Celts or Romans living on the same land a thousand years ago. And earlier still half the region was buried in ice. How did Europe become the Europe we know today?

The larger object of studying world regions is to understand how things change, with some places becoming more similar through the sharing of common culture, language, or climate, while others are developing new or unique environmental or economic conditions. Many of the differences between places around the world are a product of the relationships of these places to one another. This is as true for temperature or rainfall as it is for manufacturing or food culture. In this book, a *global geography* is about understanding the variety and distinctiveness of world regions without losing sight of the interdependence between them. That *places are different because they are connected* is the single central lesson of this book.

Exploring the interconnections among world regions not only helps explain the contemporary world (**Figure 1.2**), but it also allows us to think about where the world might go from here. The world we grew up in, and all the regions we know now, will not and cannot be the ones we will inhabit during the next decades. New regions are emerging as places in the world connect in new ways. While no future is likely to be free of problems or violence, we firmly believe such a future world might be a better one, with more human equality and possibility and more thoughtful stewardship of Earth's environments.

This chapter introduces the basic tools and fundamental concepts that geographers use to study the world and describes the conceptual framework that informs the subsequent chapters.

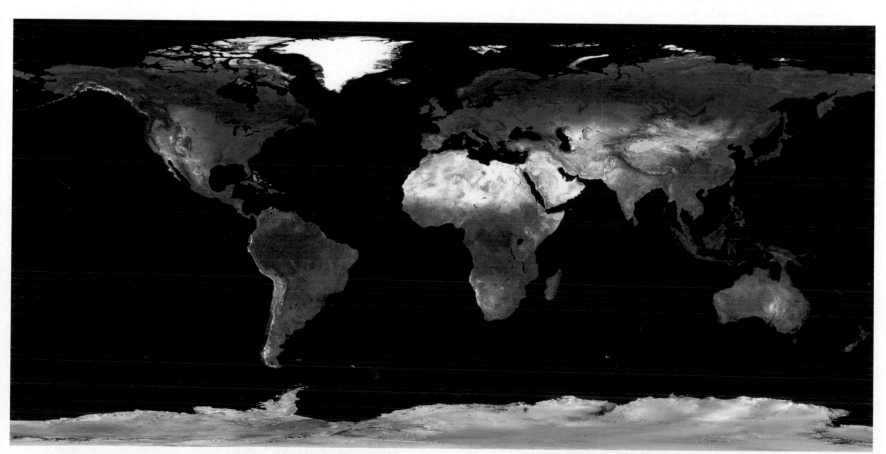

▲ **FIGURE 1.2 The World from Space** This satellite photo highlights the diversity of the world's ecosystems.

Systems, Interdependence, and Change

This book focuses on the development and organization of world regions, emphasizing how regions are related and how they change. This approach stresses the themes of *systems, interdependence, and change*:

Systems: World regions are best understood by considering how regions interact and develop as part of wider global political–economic sociocultural, and environmental **systems**. A system is a set of elements linked together so that changes in one element often result in changes in another. For example, in an economic world system, a decrease in the flow of oil from Iran may result in higher energy prices in China, changing what people pay for Chinese-made goods in the United States. Such systems characterize many of the relationships between regions, and govern how resources, products, and even ideas flow from one place to another. By thinking systemically, geographers gain a better grasp of how and why world regions come to be.

Interdependence: World regions are interdependent. That is, they affect, and are affected by, each other. In the world economic system (continuing with our example of energy prices) Iran is heavily *dependent* on China to consume its oil. At the same time, China works feverishly to try to find new sources of energy, precisely because China, in turn, is concerned about being *too dependent* on Iran. By thinking about interdependencies, geographers can explain and predict how one region may respond to disruptions and transformation of other regions and the broader system.

Change: Today's world regions are only the current configuration of an ongoing process of change. These regions emerged over time and new regions will take their place in the future. For example, if large increases in Chinese industrial production occur in the next decade, geographers will ask: What will this mean for the United States? Or for Iran? As connections within and between regions evolve over time, the characteristics of places within those regions also evolve. Similarly, the changing physical environments of the world result from interactions among complex environmental systems, as well as from human activity.

These three themes are intertwined in the processes of globalization and regionalization, the twin forces that generate a world of regions that are *both* globally interconnected and locally differentiated.

Globalization and Regionalization

The world has always been global. Since *Homo sapiens* walked out of Africa and long after the moment when McDonald's began to appear in malls in South Africa (**Figure 1.3**), the environments, economies, and societies of the globe have been tied together. In today's world, these connections have speeded up and become more widespread in a process geographers call **globalization**. Some scholars predict that the most recent wave of globalization will result in unprecedented consolidation and homogenization of the world's ecologies, economies, and societies. They stress that globalization is a process that breaks down boundaries, makes places similar, and connects them by encouraging the flow of ideas, products, and practices.

And yet parts of the world retain their uniqueness, and new regions also emerge—precisely as a result of globalization. We use the

▲ **FIGURE 1.3 A Mall in South Africa** A shopping mall is more than just a place of consumption. It is an iconic marker of a certain form of development. The concept of the mall has been globalized over the last 50 years, and malls can now be found throughout the world. Most malls provide goods and services tied to global products as well as goods unique to the local market. Malls also play valuable roles as public spaces.

term **regionalization** to describe how, when, and why new regions emerge. As we will see, it is the process of making new global connections that allows or causes regions to change. Consider the cell phone discussed at the beginning of this chapter. The regional economy of LCD screen production in Mexico is directly related to the global cell phone market found within Mexico as well as in the broader region of Latin America, the United States and Canada, Europe, as well as throughout Asia and Africa.

Often, the specific qualities of a region cause it to be linked to other regions through globalization. For example, new trade connections between East Asia and other regions led to the expansion of manufacturing in new parts of China. Manufacturing firms were drawn to East Asia to open factories for a reason: the cost of labor there was cheaper than elsewhere. In this way, globalization becomes an engine of regionalization. At the same time, regional differences can contribute to globalization. This process is evident both in economic systems and in environmental and cultural systems. In economic systems—think again about your cell phone—we have seen how globalization advances as labor from a few regions is used to develop a product eagerly consumed in every world region.

Environmental systems also change in response to the forces of globalization and regionalization, as in the introduction of species of plants and animals from one region into another region where they are not native. Rabbits and feral house cats in Australia, or Burmese pythons in the Florida Everglades, are examples of invasive, exotic species that have drastically altered the ecosystems into which they were introduced. Over time, new yet unique ecosystems will emerge based on new combinations of plants and animals.

Globalization of culture can occur in many ways. One familiar example is the spread of similar kinds of fast food around the world. But in the process, new mixes of native and foreign foods may result in totally new regional cuisines.

Globalization and regionalization operate together. It is the process of making new global connections—through trade, migration, or environmental exchange—that allows or causes regions to change. These connections have far-reaching effects: they create global and regional trade networks, migrant communities, ethnic neighborhoods in cities, and even new consumer products and ways of shopping. They may lead to the formation of new ecological communities or new agricultural systems based on imported crops and animals.

By studying world regions, we can understand why and how differences emerge, even as global processes connect the world's regions in new and important ways. The global geography of world regions gives us glimpses of the future in which today's youth and the next generation will live.

A World of Regions

World regions—such as Latin America, South Asia, and Central Asia—are geographic divisions based on both physiographic and human historical processes. (**Physiography** is another term for **physical geography**, which is the branch of geography dealing with natural features and processes.) Physical geography is concerned with climate, weather patterns, landforms, soil formation, and plant and animal ecology. **Human geography** deals with the spatial organization of human activity and with people's relationships with their environments. This book uses the perspectives of human and physical geography in exploring how globalization and regionalization shape and reshape world regions (**Figure 1.4**).

The ten regions we discuss in this book are shown in Figure 1.1 on p. 4. The regions and their characteristics are the result

of long-term historical global relationships. These relationships, and the regional connections they produce, tie some places more closely together than others. In this text, we attempt to explain how and why these regional connections emerged and how they may change in the future.

To highlight the changing nature of regions, consider that the regions described in this edition of this book are already different from those in the previous edition published only three years ago. The emergence of new nations (such as South Sudan), regional changes in politics and government (as in the Arab Spring's successful and attempted revolutions), and new regional possibilities (such as those in the Arctic Ocean as a result of global warming) demonstrate the ever-changing nature of regions.

In an effort to address the emerging and future topics that affect each region, we have developed several new *Emerging Regions* features throughout the text. An **emerging region** is an area where loosely connected locations are developing shared characteristics that differentiate them from other regions, past and present. These areas may become increasingly important to global relationships or systems. For example, the Arctic region, which has often been viewed in fragments in the past (as part of a number of different world regions, such as the United States and Canada, Europe, and Russia), has now come to be linked closely together through human migration, international trade, and shared environmental problems. An emerging region may also be non-contiguous—it might not share borders with other partners in the region. This is the case for new regions, such as BRIC—Brazil, Russia, India, and China—which have strong regional connections even though they are spread widely across the planet.

APPLY YOUR KNOWLEDGE Identify three examples of how globalization has affected your local community. Examine those examples for how they may have been modified by regionalization processes in that same community.

Organizing and Exploring the World's Regions

The world region concept is a useful tool for organizing and understanding information about the world. Accordingly, the framework for the study of world regions in this chapter provides the structure for the ten regional chapters that follow. Each chapter is organized around a set of themes common to every region, though unique in each:

- **Environment and Society:** how environments change and are changed by people
- **History, Economy, and Territory:** how history, economics, and politics evolve over time
- **Culture and Populations:** how peoples and cultures all around the world interact and change
- **Future Geographies:** how contemporary regional differences and new global forces are likely to impact important real world issues in coming years

You will find that within each of these areas of analysis, global systems connect world regions and, as a result, produce differences between them. The remainder of this chapter explores the core concepts of this thematic framework.

▲ **FIGURE 1.4 The Emergence of a New Country** The emergence of the Republic of South Sudan in 2011 was celebrated in many ways. In this photo a citizen of the new country waves a flag as part of the independence celebration.

ENVIRONMENT AND SOCIETY

The environment—understood as places and their physical and biological characteristics—is critical to the study of regions because the environment creates constraints on, and opportunities for, the activities of a society. Environmental characteristics that affect human activities include rainfall, temperature, vegetation, wildlife, and landforms. As people have settled regions, they have transformed the environment. For example, in the deserts of the United States, inhabitants have built complex irrigation systems and massive cities. In Brazil, they have logged forests on the jungle frontiers. Environments influence and help create regions, even while people transform the environments in which they live.

Throughout our history as a species, humans have adapted to the challenges of new or changing environments—in many cases environments that humans themselves had altered. Today, this process of transformation and adaptation is occurring on a global scale.

Climate, Adaptation, and Global Change

Weather and climate are ever present aspects of the environment that impact human activities. **Weather** is the current state of temperature and precipitation (for example, a rainy or a freezing day) at a particular time and place. **Climate** is the typical conditions of the weather expected at a place over a long-term average. Climate describes temperature and precipitation during different seasons (for example, a place with wet, cool winters and hot, dry summers). Weather and climate are the product of the **climate system**—all those interactions of air, water, and the Sun's energy circulating around the globe. The climate system makes places globally interconnected yet locally differentiated.

Consider the climates of Phoenix, Arizona, and Manaus, Brazil. Phoenix has a dry, desert climate with very hot summer days and mild winters. Most precipitation falls as rain during summer thunderstorms. Manaus, on the Amazon River, has hot temperatures year round and abundant precipitation except for a short, somewhat drier season. Although many factors explain the differences in climate between Arizona and Brazil, the key explanation is that the cities are *connected*. As the Sun warms the air at the Equator, that air rises and releases its moisture over the Amazon. This rising air also flows north, becoming drier and eventually sinking over the southwestern United States. There, the air dries further and forms an air mass that limits rainfall for much of the year. Connected by the climate system, Phoenix is dry *because* Manaus is wet (**Figure 1.5**).

Climate involves more than the ceaseless movement of air and moisture above Earth's surface. Climate is the result of the interaction of several Earth systems. For example, geological processes within Earth move continents and form mountains over millions of years, creating local climate variability as landmasses interact with the atmosphere and oceans. The activities of living things on Earth's surface and in the oceans also affect climate, as when plant leaves release moisture into the atmosphere.

The characteristics of regions are partly the result of the interaction of these systems, which are dynamic and ever changing. One consequence of this is that regional climates will not necessarily be the same in the future. For example, in the last few centuries, human activities have created unprecedented changes in Earth systems. Since 1950, the human impact on Earth has accelerated dramatically.

(a)

(b)

▲ **FIGURE 1.5 Climate in the Southwestern United States and the Amazon Region in Brazil** Global climatic patterns of atmospheric circulation link (a) the dryness of Phoenix, Arizona, and its historic sandstorms with (b) the wet, green spaces of the Amazonian rainforest.

Increasing rates of consumption, population growth, and technological change have caused rapid increases in pollution, changes in land use, and increased resource consumption. This new age is termed the **Anthropocene (the "human era")** in reference to the leading role humans are playing in changing the Earth system (**Figure 1.6**). As you will see, changes in regional climates are becoming an increasingly important aspect of life in the Anthropocene.

Atmospheric Circulation At a basic level, **atmospheric circulation** is the global movement of air that transports heat and moisture and explains the climates of different regions. A simple model of atmospheric circulation takes into account variations in the input

◀ **FIGURE 1.6 Dubai and the Anthropocene** Dubai has been built in an age of globalization. Oil wealth facilitated massive investment in the built environment of the city-state of Dubai. Labor and resources from many parts of the world have modified this city in a desert, creating a luxury environment of tourism, hotels, beaches, and hospitality where human comfort is maintained through air conditioning, water transfers, and imported food.

of energy from the Sun and the configuration of the major continents and mountain ranges. The spherical shape of Earth, the tilt of its axis, and its revolution around the Sun mean that sunlight does not hit all parts of Earth's surface at the same angle or for the same period or amount of time each day. At any given time, some places have the Sun directly overhead, others have less sunlight and shorter days, and still others are in darkness (**Figure 1.7**). As Earth moves around the Sun, the angle at which sunlight hits Earth varies according to the seasons. The Sun's rays hit Earth most directly and focus the greatest solar energy and heat at the Equator in March and September, at the latitude of the Tropic of Cancer (23.5°N) in the Northern Hemisphere in June and at the latitude of the Tropic of Capricorn (23.5°S) in the Southern Hemisphere in December.

The constant high input of solar radiation at the Equator produces warm temperatures throughout the year. This warmer air has a tendency to rise into the atmosphere, creating low pressure at ground level, and cooling and condensing into clouds that eventually generate heavy rainfall. In the Amazon, for example, the constant heat associated with the region's location near the Equator causes heavy precipitation. This is typical of the Equatorial climate, which is characterized by high temperatures and rainfall year-round.

The cooler air rises high into the atmosphere, moves out from the Equator toward the poles, and eventually sinks over tropical latitudes (about 30° north

and south latitude), creating a zone of high pressure in regions such as the southwestern United States and the Sahara in Africa (**Figure 1.8**, p. 10). As the air moves toward Earth's surface, it becomes warmer and drier, holding little moisture by the time it reaches ground level. In places such as Phoenix, Arizona, the high pressure locks out precipitation throughout much of the year. These regions are characterized by very low rainfall, sparse vegetation, and warm, dry conditions of desert climates.

When the sinking air reaches ground level, it spreads out, and some of the air flows back toward the Equator, where it converges with the heated air and rises again. This vertical circulation of air from the Equator to the tropics back to the Equator is called a *Hadley cell*. The **intertropical convergence zone (ITCZ)** is the region where air flows together and rises vertically as a result of intense solar heating at the Equator. The result is frequent heavy rainfall near the Equator, as shown in Figure 1.8. The ITCZ is not static, however, and moves north and south of the Equator with the seasons and the angle of the Sun. The winds associated with the ITCZ also shift as this belt of pressure moves north and south of the Equator throughout the year. This creates important global and local climatic effects. For example, *monsoon rains* result from the ITCZ drawing winds from across warm oceans. Dryness results from the

▶ **FIGURE 1.7 Seasonal Incidence of the Sun's Rays by Latitude** The Sun's rays hit Earth most directly and focus the greatest solar energy and heat at the Equator in March and September; at the latitude of the Tropic of Cancer in the Northern Hemisphere in June; and at the latitude of the Tropic of Capricorn in the Southern Hemisphere in December.

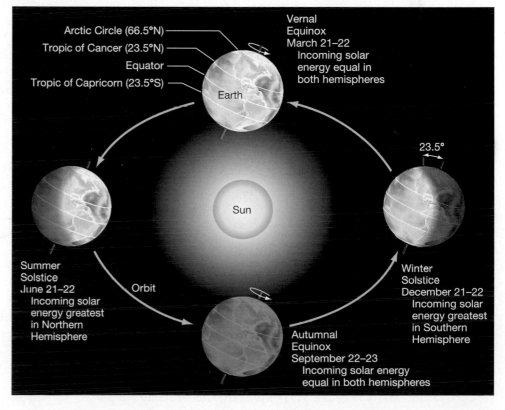

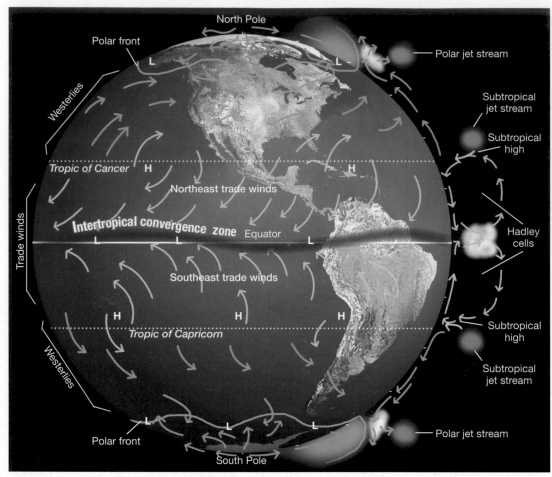

▲ **FIGURE 1.8 Atmospheric Circulation** Atmospheric circulation is based on air moving from regions of high to low pressure. The intertropical convergence zone (ITCZ) is a band of low pressure where winds converge at the Equator, creating a consistent band of thunderstorms and monsoon rains.

Atmospheric circulation produces a pattern of global climate with warmer and wetter regions nearer the Equator. East coasts in the tropics and west coasts in temperate regions also tend to be wetter, as do the complex regions where warmer air flowing toward the poles meets colder air flowing toward the Equator. Large continental land areas in the tropics drive monsoon circulation where winds blowing from warm oceans to cooler land areas bring heavy rainfall. This general pattern helps explain the climates of each of the major world regions, which, in turn, influence human activity.

Climate Zones Patterns of precipitation and temperature create a mosaic of climate regions across Earth (**Figure 1.10**). Each of these climates is distinct and has an impact on the character of a region, influencing the plants and animals that thrive there and affecting the kinds of economic activities that have emerged there. Climate even inspires regional culture. For example, in South Asia, the torrents of monsoon rainfall that characterize the region's climate are memorialized in mythology and song.

■ **Tropical** *Tropical Climates* nearest the Equator are characterized by heavy rainfall and persistent warm temperatures. These include those areas that have no dry season and are wet year-round. Also in the wet tropics are areas that have a significantly uneven seasonal rainfall pattern. In *monsoonal climates*, much of the rain may come in a few months, with long periods with little or no rain (see Chapters 9 and 10 for examples from South and Southeast Asia).

■ **Arid and Semiarid** North and south of these tropical wet areas are the high pressure *arid and semiarid climates*, which typically occur between 28 and 30 degrees north and south of the Equator. These dry zones include deep deserts and open dry grassland. Arid climates also occur in other areas and for other reasons, as in the case of the dry regions of western South America, for example, where cold water currents create a dry regional climate. Similarly, regions in the rain shadow of mountains may be arid or semiarid due to the orographic effect. Between arid zones and wet tropics come *tropical savannas*, complex transition zones with distinct wet and dry seasons that span the areas between the desert edges up to the wet forests. The African Sahel (see Chapter 5) is a belt of semiarid savanna south of the Sahara and north of the deep wet tropical regions of Central Africa.

■ **Mediterranean** Farther north and south beyond the tropical zone, climates are greatly affected by the side of continent on which they lie. *Mediterranean climates* typically occur on the western sides of continents between 30° and 45° (either north or south). These climates are warm due to their proximity to the Equator, but they have more moderate temperatures because of their proximity to the sea.

fact that the ITCZ attracts winds from dry inland areas. You can see this pattern in many parts of the world, including Thailand, Southeast Asia, where people experience roughly six months of monsoon rainfall from May through October and six months of dryness from November through April.

The rotation of Earth tends to drag air flowing back from the tropical latitudes to the Equator into a more east-to-west flow, creating a major wind belt called the *trade winds* that blow from east to west between the dry tropics and the Equator. The trade winds are also called *easterly winds* or "easterlies," as they come *from* the east. Some of the air descending over the tropics also flows poleward and is pulled by Earth's spin into a major west-to-east flow. These are called *westerly winds* or "westerlies", because they flow *from* the west.

When winds blow across warmer oceans, they tend to pick up moisture. Winds play an important role in precipitation patterns. When this moisture-laden air encounters a landmass, especially coastal mountains, it condenses into rainfall or snow as it is forced to rise over them. This phenomenon is called orographic precipitation. **Orographic precipitation** results in the formation of a dry rain shadow region on the inland, or lee, side of the mountains, where the now dry air sinks. The easterly trade winds flow across the oceans in tropical latitudes and frequently produce rain on east-facing coasts (**Figure 1.9**). Similarly, the westerly winds bring rain as they blow from oceans onto western coasts.

(a)

(b)

▲ **FIGURE 1.9 Orographic Rainfall on a Trade Wind Coast** (a) Winds blowing across the oceans pick up moisture. When they rise over coastal mountains, cooling causes the moisture to condense and fall as rainfall or snow. Orographic rainfall is common where winds, such as the trades, cross warm oceans, such as the Caribbean, and rise over mountains near coasts. Heavy orographic rainfall also occurs where the westerlies rise over coastal mountains, such as the Andes in Chile, or where high mountains emerge from generally drier regions, such as East Africa and the western United States. (b) Mount Waialeale, on the Hawaiian Island of Kauai, is one of the Earth's rainiest points.

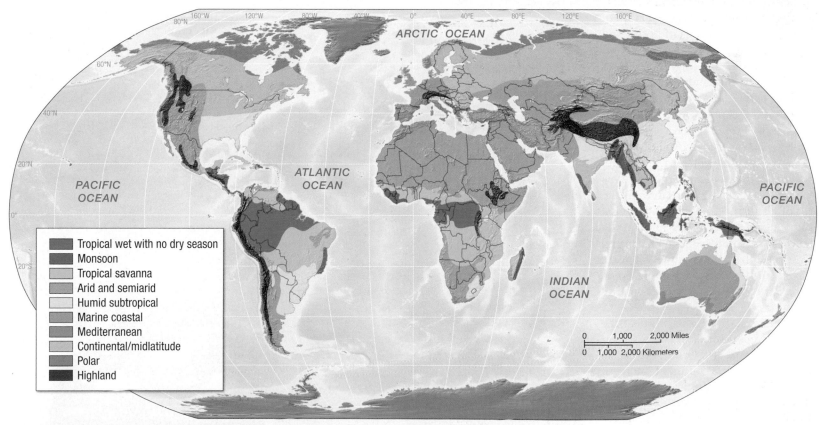

▲ **FIGURE 1.10 Major Climate Regions** The climate system of the world can be generalized into ten major types based on temperature and precipitation. This map is based on a simplification of the climate classification of German climatologist Wladimir Köppen.

- **Humid Subtropical** *Humid subtropical climates*, on the other hand, tend to fall on the southeast side of continents in a similar range of latitudes. Because of their place on the receiving side of predominant easterlies or trade winds, they are far more wet and humid throughout the year than Mediterranean climates.

- **Marine West Coast** Farther north and south are more temperate climates. *Marine coastal climates*, which occur between approximately 45° and 50° north and south, tend to be persistently wet and somewhat colder. Owing to westerly winds and oceanic currents, these are usually found on the west coasts of continents, just poleward from Mediterranean climates.

- **Continental** On the temperate interiors of the North American and Eurasian continents are *continental climates*, which can be somewhat drier and often have long cold winters.

- **Polar and Highlands** In the extreme north and south are *polar climates*, which are dominated by the cold that results from their distance from the Equator. In mountainous areas, including the Rocky Mountains, the Himalayas, and the Andes, among others, the topography is complex. In such places the climate of the mountaintops tends to be far different than the nearby valleys. The term *highland* is used to designate these unpredictable zones.

Climate Change Although the general circulation of the atmosphere remains the same over centuries, considerable evidence suggests that global and regional climates have varied over time. Landscapes show evidence of wetter or drier conditions in the past in the form of plant and animal remains associated with different climates. Human history records periods of hotter and cooler climates and their impact on harvests, health, and migrations. Most dramatically, the landscapes of many regions show the marks of ice cover, erosion, and deposition from periods when it was so cold that rivers of ice (called *glaciers*) or massive sheets of ice (*ice caps*) covered much of the world. Scientists believe that the ice ages were caused by slight changes in the tilt of Earth's axis and its orbit around the Sun and associated changes in the amount of solar radiation reaching Earth. Global and regional climates have also been affected by geological activity, especially by volcanic eruptions.

Human activity can also affect the climate by altering the composition of the atmosphere. Of greatest concern is **global warming,** an increase in world temperatures and a change in climate associated with increasing levels of carbon dioxide (CO_2), methane, and other trace gases. These so-called **greenhouse gases** result from human activities such as burning fossil fuels, cement production, and deforestation. Greenhouse gases act to trap heat within the atmosphere, resulting in the warming of the atmosphere and surface. We have already seen many indicators of this (**Figure 1.11**). Human activities that produce greenhouse gases are increasing dramatically as human population and consumption increase (**Figure 1.12**).

National differences in greenhouse gas emissions illustrate the roles that different countries play in global warming. The largest overall emitters are China, the United States, and Brazil. At more than 20 tons of greenhouse gas emissions per person, the United States leads both Brazil (12 tons) and China (4 tons). In per capita terms, the oil states of Qatar and Kuwait are responsible for more than 30 tons per person, while poor African countries such as Mali or Tanzania account for less than 1 ton per person.

Human-caused increases in carbon dioxide may result in an overall global temperature increase of about 3°C (5.5°F) over the next 50 years, as well as regional changes in the amount and distribution of precipitation. Projections vary, but worst-case scenarios are sobering (**Figure 1.13**). Evidence indicates that such changes are already occurring and include a warming of ground, ocean, and near-surface air temperatures; a decline in sea ice, snow and ice cover; a retreat of mountain glaciers; and an increase in precipitation in high latitudes.

The new regional climates that will emerge from global change are difficult to predict. It is possible to imagine whole areas of agricultural production (the wheat belt of the United States or the wine regions of California, for example) shifting northward or being eliminated altogether by drought and rising regional temperatures. Some areas are likely to become dramatically drier, while others may experience far higher levels of rainfall. Sea-level rise resulting from ice

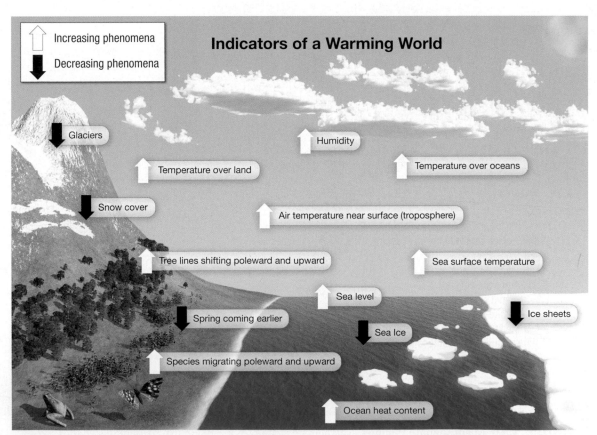

▲ **FIGURE 1.11 Major Indicators of Climate Changes** This diagram shows the major observed indicators of climate change, including higher temperatures over land and oceans, higher humidity, and higher sea levels.

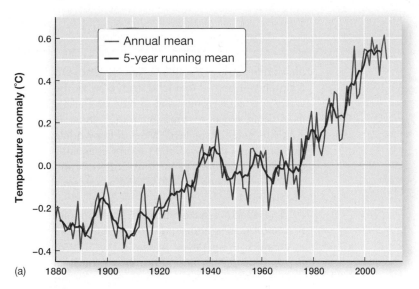

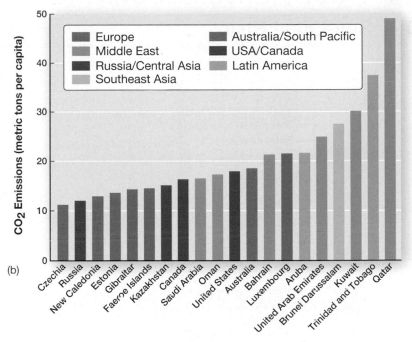

▲ **FIGURE 1.12** (a) Global Temperature Change over Time
(b) Carbon Emissions per Capita for Top Producers

melt and sea water expansion may change the outlines of continents (See "Visualizing Geography: Sea-Level Rise," p. 14). The impacts of climate change will depend on the ability of people and the wider environment in different regions to adapt to climate change and on the choices humans make about greenhouse gas emission reduction.

APPLY YOUR KNOWLEDGE The impact that humans have on the planet varies by where they live. Visit the Global Footprint Network website (**http://www.footprintnetwork.org**), select "Footprint for Nations" from the "Footprint Basics" menu, and choose three countries. Compare their ecological footprints. What makes one country's footprint different from another?

Geological Resources, Risk, and Water Management

Like the global climate system, the geologic system of Earth helps produce regions over long periods by uplifting mountains, forming water drainage and river systems, and presenting diverse resources and hazards around the world.

At a global scale, the main driver of this geologic system is **plate tectonics**. Earth's crust, which is from 50 to 100 kilometers (31 to 62 miles) thick, is composed of about a dozen large "plates" of solid rock floating on a layer of molten material. The theory of plate tectonics explains how these plates move very slowly relative to each other and interact at their boundaries, causing most of the world's volcanic,

▶ **FIGURE 1.13** Worst-case Projections of Global Temperature Increase by 2080 (in °C) Assuming ongoing levels of current emissions of greenhouse gases, scientists predict sustained warming. Temperature increases between 4° and 8°C would result in the melting of glaciers and ice caps, the inundation of island nations, and the loss of large swaths of agricultural potential.

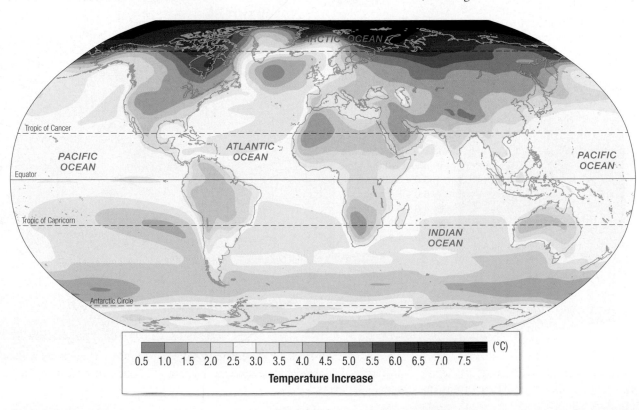

VISUALIZING GEOGRAPHY

Sea-Level Rise

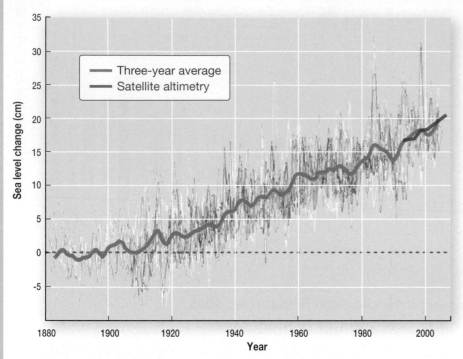

▲ FIGURE 1.1.1 Global Mean Sea-Level Rise: 1880–2007 Sea levels have risen since the 19th century. These trends are evident from direct gauge measurements of the sea as well as recent satellite altimetry, which measures the height of the surfaces of bodies of water from space.

Just under 30% of Earth's surface is land, but a significant part of that land, including all the world's coastal zones and many of Earth's islands, is close to the waterline, which is within 1 meter of sea level. Most of the world's human population is settled along coastlines owing to opportunities for transportation and commerce, seaside amenities, and advantages of climate.

These factors make much of Earth's population and the land areas they occupy vulnerable to changes in sea level. A rise in the level of the world's oceans during the next century is predicted for two reasons. First, the warming of Earth tends to lead to a warming of ocean waters, which expand in total volume. Second, the melting of glacial and polar ice masses, which results from that warming, releases additional water into the world's oceans and seas.

Sea levels have already risen during the 20th century. **Figure 1.1.1** shows data collected

▼ FIGURE 1.1.2 Parts of Florida with Elevations at or Below 1–6 Meters above Sea Level Adjacent to the Sea These areas would all be inundated under sea-level rise.

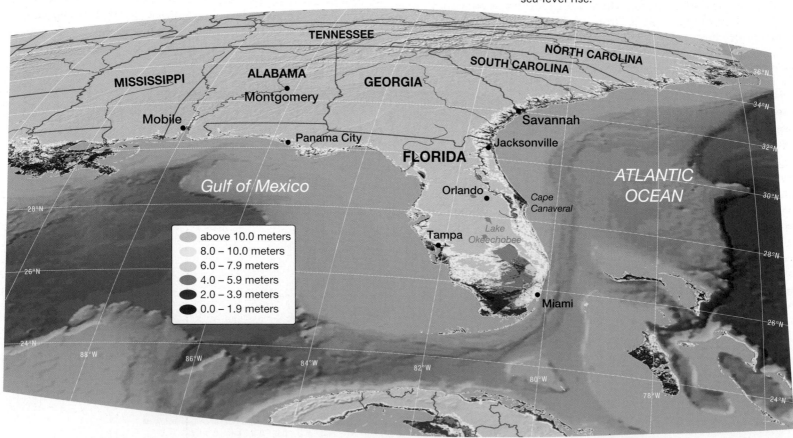

from tidal gauges—devices from around the world that measure changes in sea level over time. Tidal gauges have been continuously monitored since the late 1800s. These measurements suggest that sea level has risen almost a foot during the last century. Predictions for the future vary, with an average change of about a half-meter predicted to occur over the next century. Some estimates predict a change as high as 1 meter or more during that time period.

A rise of just 2 feet would mean that Florida would lose approximately 10% of its land area, including the homes of 1.5 million people. **Figure 1.1.2** shows a visualization of how the coastline of Florida might change should sea-level rise occur. Even a modest rise would remove the farthest tip of the state. Further transformation would make the coastline almost unrecognizable.

At a more local scale, such changes create further problems. They make many areas more vulnerable to strong storms and hurricanes, lead to the destruction of wetlands and other important ecosystems, and interfere with freshwater supplies. It is conceivable that vast seawalls might be constructed to hold back the rising oceans, modeled on the dams from places like the Netherlands (see Chapter 2). The number and total area of potentially inundated cities, however, and the degree to which these cities lie in poorer nations and regions, present a daunting challenge. **Figure 1.1.3** shows the probabilities of inundation by the ocean for the city of Miami, Florida, in the event of 1 meter of seal level rise. Conservative "low" estimates would remove most of the barrier islands, while more realistic "central" and "high" estimates suggest the sea would wash away Miami Beach and most of the middle parts of the city. Large-scale population migrations would be necessary to adapt to this new geography. Not content to wait for such changes to bankrupt them with multiplying claims, major insurance corporations around the world are carefully studying the new configurations of settlement, property, and liability, employing geographers and climatologists to help them predict where insured clients are most likely to experience damage or catastrophe. ∎

Area at risk of inundation from 1 m (3.3 ft.) rise in sea level (storm surge 17 in)

- Current sea level
- Low estimate
- Central estimate
- High estimate

▲ **FIGURE 1.1.3 Risk of Inundation in Miami, Florida, under Sea-Level Rise** Most of Miami, like many world cities, lies within a meter or two of sea level.

earthquake, and mountain-building activity (**Figure 1.14**). The continents sit on the plates and emerge from the oceans where Earth's crust is thicker or where the rocks are less dense and therefore more buoyant. The slow movement of the plates across Earth's surface causes land masses to split apart and collide over time, creating the current physical features of Earth's surface, or its **geomorphology**. For example, the collision of the Indian Plate with the Eurasian Plate uplifted the Himalayas (**Figure 1.15**).

Where plates pull apart there are **divergent plate boundaries**. A system of divergent boundaries thousands of kilometers long winds through Earth's oceans. Volcanic activity along these boundaries, also called spreading centers or mid-ocean ridges, produces the rock of Earth's ocean floors, or oceanic crust. Divergent boundaries can also form on land, where they can produce rift valleys as the plates pull apart. The Great Rift Valley in eastern Africa is a well known example. As plate motions continue, a rift valley formed at a divergent boundary may sink below sea level, forming a new ocean basin.

Where plates collide at **convergent plate boundaries**, different scenarios may unfold depending on the type of crust involved in the collision:

- A collision of continental landmasses, such as India and Asia, produces a great mountain range.

- When oceanic crust collides with another plate, it sinks downward at a deep-ocean trench. This process, called *subduction*, "recycles" oceanic crust as it sinks into the mantle.

The areas adjacent to the deep-ocean trenches where subduction occurs—known as **subduction zones**—are found along the edges of ocean basins. The interactions of plates in subduction zones are responsible for many earthquakes and volcanoes. For example, subduction zones make up the "Ring of Fire," the region of powerful earthquakes and large volcanoes that partly surrounds the Pacific Ocean.

Finally, just as plates come together and pull apart, they also slide past each other at **transform plate boundaries**. The San Andreas

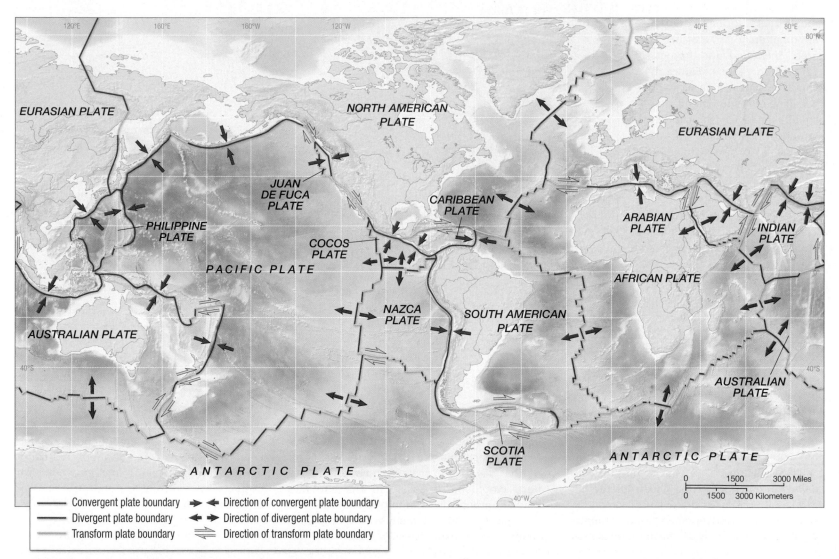

▲ **FIGURE 1.14 Major Tectonic Plates** Earth's crust is broken into a dozen or so rigid slabs or tectonic plates that are moving relative to one another. Arrows and lines represent plate movement and direction. Not all plates move at the same rate; for example, the Pacific and Nazca plates are moving more rapidly than others. The purple arrows indicate areas where plates are moving away from each other, while red arrows show *subduction zones*, where one plate sinks underneath another. The yellow lines mark *transform boundaries*, where plates are slipping past each other.

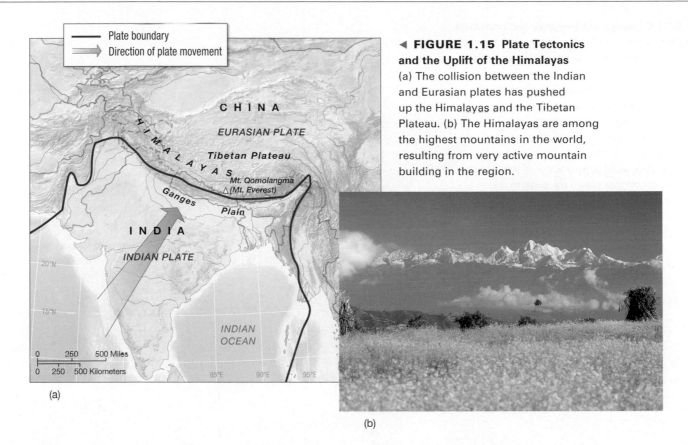

◀ **FIGURE 1.15 Plate Tectonics and the Uplift of the Himalayas** (a) The collision between the Indian and Eurasian plates has pushed up the Himalayas and the Tibetan Plateau. (b) The Himalayas are among the highest mountains in the world, resulting from very active mountain building in the region.

Fault in California is one of the most famous transform boundaries. Along the fault line, two plates slowly move past each other and sometimes quickly jerk apart, causing large-scale earthquakes. Plate tectonics is a process that is continually and actively shaping the world's regions as well as our responses to those regions (**Figure 1.16**, p. 18).

Plate Tectonics and Natural Resource Distribution The heating, compression, and folding of the crust that occur because of tectonic movements help to form oil, coal, and natural gas as well as other mineral resources. Energy-rich fossil fuels, made of hydrocarbons, remain vital to the globalized economy. However, the complex geologic processes that formed these resources also distributed them unevenly. This uneven distribution has important economic and geopolitical implications. Though both Jordan and Iraq are located near the northern boundary of the shifting Arabian Plate, Iraq is rich with oil, whereas Jordan has none (see Chapter 4, p. 128). Fossil fuels are not the only riches unevenly distributed across regions. Other mineral resources are, too. The ancient rocks of sub-Saharan Africa, for example, contain copper, gold, and diamonds.

Earthquakes and Volcanic Hazards Tectonic boundaries are also critical to the distribution of hazards, including earthquakes and volcanic activity. Areas of plate convergence give rise to massive volcanoes that have created island chains, such as the islands of Japan. In such regions, these hazards encourage the development of earthquake-resistant building designs and evacuation plans to cope with sudden volcanic eruptions or earthquakes. Two recent disasters underscore the need for these efforts. In 2004, the convergence of two plates in the Indian Ocean produced a great earthquake that triggered enormous waves of water, called tsunamis. When the tsunamis came ashore throughout South and Southeast Asia, they killed hundreds of thousands of people and left millions displaced (see Chapter 10, p. 382). Another major earthquake produced a similarly devastating

tsunami in Japan in 2011 when the Pacific Plate thrust under the North American Plate off the coast, creating an underwater earthquake through the process of subduction (**Figure 1.17**, p. 18).

River Formation and Water Management The formation of highlands and mountains through tectonic uplift creates a tilted surface over which rain falls and forms streams that erode surfaces and create river basins. These rivers carry fertile soils for agriculture and create pathways for transport and trade. To protect farming land and settlements, the banks of rivers are commonly reinforced with dams and levees. To facilitate barge traffic and transport, as well as water diversion for agriculture, they are frequently dammed.

Chang Jiang (Yangtze River) in China provides a key example. Flowing out of the massive uplifted regions of the Himalayas, where the Indian and Eurasian Plates collide, Chang Jiang originates from glacial melt. It is then fed by rainfall before emptying into the East China Sea. Owing to its enormous size and capacity, the river carries a massive amount of sediment that fosters agricultural productivity. This makes the area along its banks important for global trade in grain and goods. These merits, however, also cause problems for humans, as the river is prone to catastrophic flooding and is not navigable along its entire length. The enormous Three Gorges Dam project (see Chapter 8, p. 304) is a feat of engineering undertaken by the government to maximize the benefits of the river's power and control its flow. The resulting landscapes are entirely new ones, with cities and agriculture transformed by the dam.

Water management along great rivers such as Chang Jiang is a good example of how diverse forces (tectonic uplift and erosion, rainfall-driven flooding, and global commerce) interact and of how global circulations (of soil, water, and money) converge to create regional differences. This example also illustrates the way human activities interact with physical conditions to produce new environmental conditions. The character of world regions is formed through this kind of ongoing interaction.

▼ **FIGURE 1.16 Convergent, Divergent, and Transform Plate Boundaries** Earth's crust is in a constant state of motion. At different points, plates may collide into one another, creating a convergent plate boundary; split apart from one another, creating a divergent plate boundary; or slide past each other, creating a transform plate boundary. At the location of these boundaries, tectonic activity produces volcanoes, earthquakes, deep trenches, valleys, and mountain ranges. In some unusual cases, tectonic activity may occur in the middle of a plate, in "hot spots" that pierce the crust to cause oceanic features, like island arcs typified by Hawaii, or continental features, like those at Yellowstone National Park in North America.

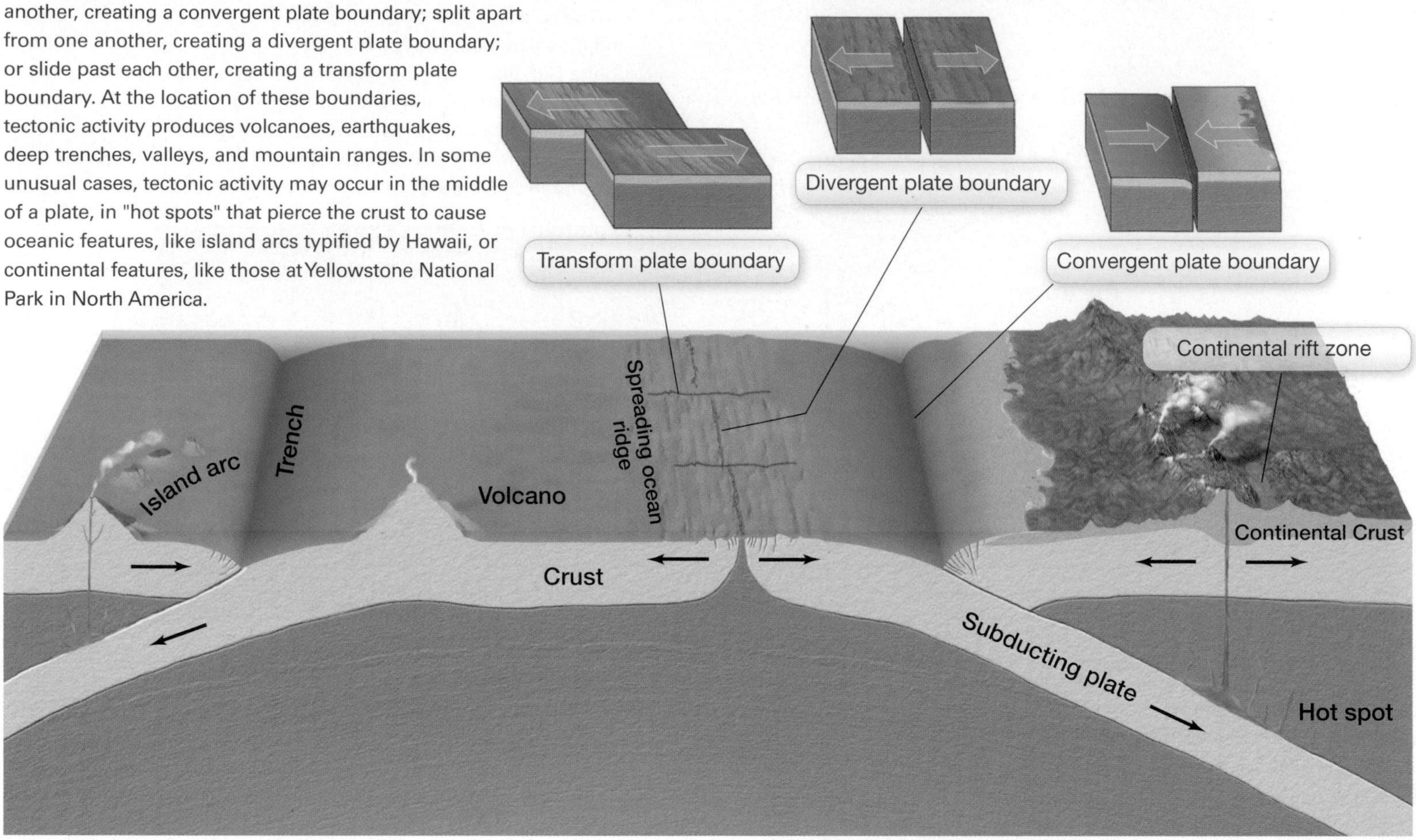

Divergent plate boundary

Transform plate boundary

Convergent plate boundary

Continental rift zone

Island arc

Trench

Volcano

Spreading ocean ridge

Crust

Continental Crust

Subducting plate

Hot spot

APPLY YOUR KNOWLEDGE Identify three major rivers that have recently been dammed for hydroelectric power production. Use the Internet to compare the different impacts these dams have had on local people and environments. How do these impacts reflect the relationships between geological forces, environmental conditions, and human activities?

Ecology, Land, and Environmental Management

The interactions of climate, species development, geomorphology, and human activity are the major influences on the global distribution of **ecosystems**—the complexes and interactions of living organisms and their environments in specific places. For example, the high levels of rainfall and warm temperatures of tropical climate zones foster the lush vegetation of tropical rainforest ecosystems. The hot, dry conditions of arid climates are associated with sparser shrubs and drought-adapted plants and animals of desert ecosystems. Colder deserts with slightly more rainfall produce the short grasses typical of steppe or prairie ecosystems. The cold, drier conditions of regions nearer the poles correspond to a tundra ecosystem of frost-resistant mosses, shrubs, and grasses. An ecosystem of scrub and dry forest is found in Mediterranean climate regions where the seasonal shift of the westerly winds brings rainfall only in winter. In marine coast climates, wet coastal rains are associated with towering coniferous forest ecosystems.

▲ **FIGURE 1.17 Japan after the 2011 Tsunami** This photo taken on March 14, 2011, in Sendai, Myagi Prefecture, Japan, shows the devastation that a tsunami can cause.

Similarly, the trade winds of Humid Subtropical climates produce rainforests on the southeast coasts of many continents.

Ecosystem and Biodiversity Decline These various ecosystems sustain the wealth of Earth's species and genetic variability. This **biodiversity**—the rich variety and differences in the types and numbers of species in different regions of the world—has been a boon for human beings for more than 10,000 years, at least since the dawn of agriculture, when people began to manipulate the plants and animals around them to produce new environments and support new forms of social organization. Extensive human manipulation of Earth through settlement and agriculture has led to a series of critical problems. For example, there is an evident decline in overall biodiversity across the globe. Although perhaps tens of millions of species exist on Earth, countless numbers disappear every day. It is impossible to estimate the exact rate of loss, but based on the number of species evaluated by the International Union for Conservation of Nature (IUN), 23% of vertebrates, 5% of invertebrates, and 70% of plants are designated as endangered or threatened.

Much of this loss is associated with a decline in the unique places and conditions where specific plants, insects, and animals have evolved and thrive. For example, over the course of the past 150 years in Latin America, forest cover declined from 1,445 million hectares (3570 million acres) in 1700 to 1,151 million hectares (2845 million acres) by 1980. Such **deforestation** is ongoing in many places in the world

TABLE 1.1 Change in Cover of Forest and Wooded Land by World Region		
	1990–2010 (in thousands of hectares)	Change (as % of 1990)
Africa	−57,770	−7.71%
Asia	+16,402	2.84%
Europe	+15,330	1.57%
North and Central America	−2,990	−0.42%
Oceania	−7,360	−3.70%
South America	−82,103	−8.67%
Total World	−135,339	−3.25%

SOURCE: Adapted using data from the United Nations Food and Agriculture Organization's 2010 Global Forest Resources Assessment (http://www.fao.org/forestry/fra/fra2010/en/).

(**Table 1.1**). Although there has been some modest return of forests in Europe after centuries of deforestation, the forests in many world regions remain in a state of critical decline.

Environmental hazards that transform ecosystems include the destruction of tropical rainforests; widespread pollution; the degradation of soil, water, and marine resources; and acid rain (**Figure 1.18**).

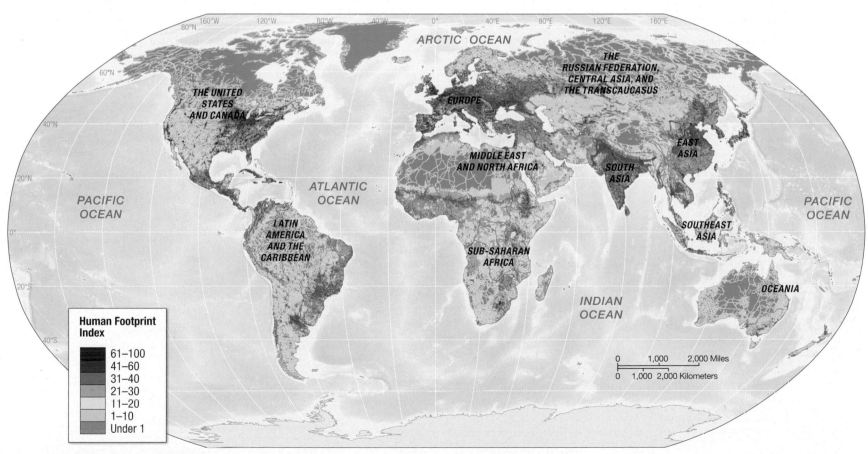

▲ **FIGURE 1.18 Scale of Human Transformation of Earth's Surface** Approximately 83% of terrestrial land has been altered to some degree by human activity, including farming, forest removal, and urbanization. The Human Footprint Index indicates that amount of human influence an area has experienced. The index ranges from 0, which denotes the least affected areas, to 100, which indicates the areas that are most influenced by humans.

Most of these threats are greatest in the world's most economically marginal regions, where the catastrophe of daily environmental pollution and degradation will continue to unfold, in slow motion, in the coming years.

Human-Influenced Ecologies Human activities are by no means always ecologically destructive. Wherever people travel or migrate, they facilitate new mixes of species and landscapes. These human actions may be intentional or accidental. For example, researchers intentionally extract genetic resources from the pool of existing biodiversity to further human health and agricultural goals. Notably, genetically engineered varieties of cotton and corn that are pest-resistant have come to replace previous varieties in many places. Roughly one-third of planted corn is biologically engineered. The possible implications of these crops for native agricultural diversity and their impacts on local ecosystems are hotly debated, but there is no doubt that biotechnology showcases the creative power of human action.

Humans may not be acting intentionally when species from one place "hitchhike" to new locations on human transport. Nonetheless, these introduced species can thrive at their destination, interacting with native species and habitats to create new ecological mixes, land covers, and ecosystems. Invasive plant and animal species can choke pastures, disrupt water systems, drive other species to extinction, and wreak expensive and unanticipated havoc on important local plant or animal resources. The total costs of losses due to these **invasive species** in the United States exceeds $138 billion annually (**Figure 1.19**). Zebra mussels from the Black Sea are now thriving in the Great Lakes of the United States, while water hyacinth from the Amazon has spread throughout the tropics of Africa and South Asia. These species can be highly problematic in their interactions with existing plants and animals, but many integrate in complex ways into their new ecosystems over time. As globalization propels new species around the world, the result is ecological regionalization: the emergence of new regional ecosystems modified by human actions.

Sustainability There is a growing concern that the recent rate of change in climate and ecology has brought us close to thresholds that, if exceeded, could lead to rapid and irreversible changes and to serious risks for much of humanity. We are reaching danger zones with regards to climate change, biodiversity loss, and pollution that present serious problems for water and land use at the regional level. For more than 40 years, there have been calls for global solutions to environmental

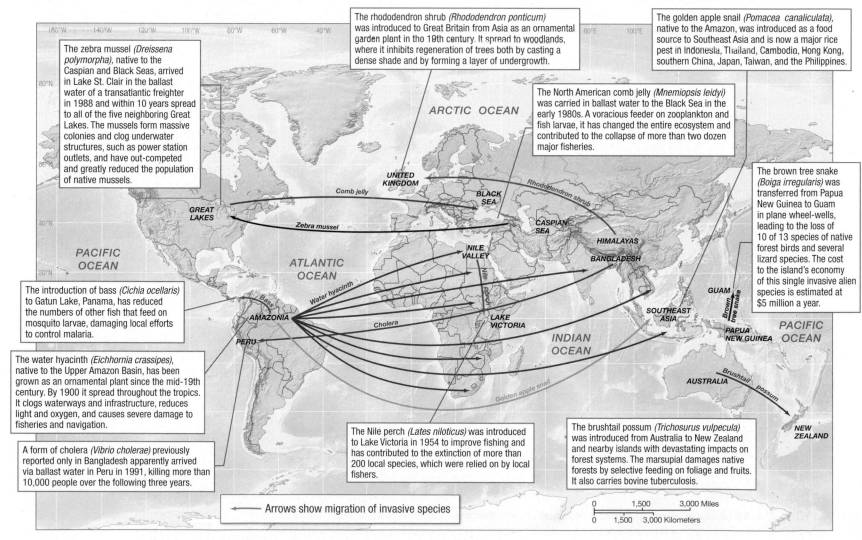

The zebra mussel (Dreissena polymorpha), native to the Caspian and Black Seas, arrived in Lake St. Clair in the ballast water of a transatlantic freighter in 1988 and within 10 years spread to all of the five neighboring Great Lakes. The mussels form massive colonies and clog underwater structures, such as power station outlets, and have out-competed and greatly reduced the population of native mussels.

The rhododendron shrub (Rhododendron ponticum) was introduced to Great Britain from Asia as an ornamental garden plant in the 19th century. It spread to woodlands, where it inhibits regeneration of trees both by casting a dense shade and by forming a layer of undergrowth.

The golden apple snail (Pomacea canaliculata), native to the Amazon, was introduced as a food source to Southeast Asia and is now a major rice pest in Indonesia, Thailand, Cambodia, Hong Kong, southern China, Japan, Taiwan, and the Philippines.

The North American comb jelly (Mnemiopsis leidyi) was carried in ballast water to the Black Sea in the early 1980s. A voracious feeder on zooplankton and fish larvae, it has changed the entire ecosystem and contributed to the collapse of more than two dozen major fisheries.

The brown tree snake (Boiga irregularis) was transferred from Papua New Guinea to Guam in plane wheel-wells, leading to the loss of 10 of 13 species of native forest birds and several lizard species. The cost to the island's economy of this single invasive alien species is estimated at $5 million a year.

The introduction of bass (Cichla ocellaris) to Gatun Lake, Panama, has reduced the numbers of other fish that feed on mosquito larvae, damaging local efforts to control malaria.

The water hyacinth (Eichhornia crassipes), native to the Upper Amazon Basin, has been grown as an ornamental plant since the mid-19th century. By 1900 it spread throughout the tropics. It clogs waterways and infrastructure, reduces light and oxygen, and causes severe damage to fisheries and navigation.

A form of cholera (Vibrio cholerae) previously reported only in Bangladesh apparently arrived via ballast water in Peru in 1991, killing more than 10,000 people over the following three years.

The Nile perch (Lates niloticus) was introduced to Lake Victoria in 1954 to improve fishing and has contributed to the extinction of more than 200 local species, which were relied on by local fishers.

The brushtail possum (Trichosurus vulpecula) was introduced from Australia to New Zealand and nearby islands with devastating impacts on forest systems. The marsupial damages native forests by selective feeding on foliage and fruits. It also carries bovine tuberculosis.

Arrows show migration of invasive species

0   1,500   3,000 Miles
0   1,500   3,000 Kilometers

▲ **FIGURE 1.19 Worldwide Invasive Species** This map shows a sampling of the many invasive species that are changing the world's ecosystems.

◄ **FIGURE 1.20 Solar Panel Array in New Mexico, United States** Alternative energy production is on the rise in many countries. Solar farms have the potential to provide clean renewable energy well into the future.

problems, including international agreements to reduce pollution and protect species. Following the principle of **sustainability**, many people advocate a more benign relationship between nature and society. Sustainability is a term with numerous definitions, but here it is intended to refer to meeting current and future human needs, while simultaneously preserving our world's precious environmental resources.

Sustainability also means that economic growth and change should occur only when environmental impacts (both costs and benefits) are benign or manageable and are fairly distributed across social classes, regions, and generations. This involves developing technologies that use resources more efficiently and managing renewable resources (those that replenish themselves, such as water, fish, and forests) to ensure replacement and continued yield (**Figure 1.20**).

Sustainability policies of major international institutions promote reforestation, energy efficiency and conservation, and birth control and poverty programs to reduce the environmental impact of rural populations. Many businesses are also demonstrating their commitment to sustainability by reporting and reducing their environmental impacts. At the same time, however, the expansion and globalization of the world economy has resulted in overall increases in resource use and inequality that contradict many of the goals of sustainability.

Regional Environments and Change Human interaction with the environment often has a profound influence on regional outcomes. The environmental problems of recent years are a result of human interactions; solutions to these problems will also be a product of human interventions and imagination. Much the same can be said for economic and political problems, to which we turn next.

APPLY YOUR KNOWLEDGE Using the definition of sustainability presented in this section, research a corporation that is attempting to produce sustainable products. What are the challenges of developing a sustainable product? What are the benefits?

HISTORY, ECONOMY, AND TERRITORY

Coffee sales are booming around the world. So why does a coffee *farmer* working in rural Guatemala face radically turbulent prices, booms and busts, and cycles of debt, while a coffee *seller* in urban Columbus, Ohio, operates in a relatively stable economic environment

of ongoing growth? There are a number of important local factors that explain this outcome, including the problem of historic political instability in Guatemala relative to the placid political environment of Ohio. Also, farming is environmentally more unstable than retailing, owing to forces such as drought. The key explanation to this economic difference, however, is that the two places are *connected* through a **commodity chain**: a network of labor and production processes that starts with production of raw materials like coffee beans and ends with sellers of coffee. In between are numerous people who buy, sell, and process brewed coffee. Their actions have very different impacts on the economies located at the two ends of the commodity chain.

For example, coffee companies that buy Costa Rican beans have an advantage over the farmers and the sellers; one buyer buys from many sellers, so the buyers can demand low prices. When production of coffee beans increases around the world, the glut results in a perilous and rapid fall in the price of beans. This situation occurred between 2000 and 2004 and resulted in mass bankruptcies for coffee producers throughout Latin America. At the buying end of the chain, these fluctuations are felt far less directly, since so many actors (buyers, sellers, roasters, processers) have a role to play between the farmer and the consumer. The economy for coffee consumers is stable, in part because the situation of growers is so unstable, and vice versa. The regional economies are different *because* they are connected.

Beginning with coffee growers in the tropics, the commodity chain that brings you your morning cup of coffee in Columbus, Ohio, shows how the globalized economy connects the most far-flung places. But the connections between places in our globalized world are deeper than that. The historical, economic, and political geography of the world is a story of interconnection and isolation, movement and settlement, growth and retreat. Places in the world have responded differently over time to changing economic, territorial, and political events. In this section, we consider the geographic complexity of the past and the present. We do this by highlighting how the processes of globalization and regionalization foster global interconnection and regional economic and political conditions in the emergence of the modern world.

Historical Legacies and Landscapes

For most of the world's history, movement, not settlement, has been the distinguishing geographic feature. Long-term movements are an essential part of the human story (**Figure 1.21**, p. 22). As people have migrated from place to place, they have developed and spread new knowledge

WORLD HISTORY AND GLOBAL GEOGRAPHY

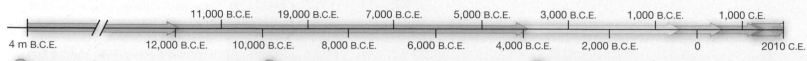

Paleolithic
4 million B.C.E.– ~12,000 B.C.E.
Old Stone Age: This was the longest period in human history and includes human evolution to the present human species. People in this period used stone tools and relied almost exclusively on hunting and gathering. Through local and global migrations, humans came to occupy all major landmasses with the exception of Antarctica.

Neolithic
~12,000 B.C.E.–4,000 B.C.E.
New Stone Age: Social and spatial organization changed as humans, developed advanced techniques of animal husbandry (breeding and herding) and early agriculture. While some people, particularly in harsher climates, turned to pastoralism, others settled into small towns and villages, using agriculture to sustain growing kinship-based populations.

Ancient
~4,000 B.C.E.–500 B.C.E.
Emerging Complex Urban Life: Urban life emerged as surplus food supplies were created in large river valleys throughout the world. The early use of metal tools in agriculture facilitated this growth, allowing communities to take advantage of the rich soils of some of the world's largest river systems. Other technological advancements, such as writing and mathematics, as well as the development of law and ethical codes, advanced social organization beyond kinship and into urban societies.

Early Modern
~1500 C.E.– ~1800 C.E.
Resource Extraction and Direct Colonization: Europeans driven by mercantilism and, later, capitalism expanded control over large territories in the Americas and Asia and coastal Africa. This resulted in massive demographic shift movements of slaves from Africa to the Americas and the rapid decline of Indigenous populations in the Caribbean. Global trade focused on the export of raw materials from newly established colonial holdings to Europe. Mainland Asia remained relatively strong during this period and largely isolated from direct European colonization, with some powers, such as China, maintaining control in the region. Most of interior Africa also remained isolated from European interference.

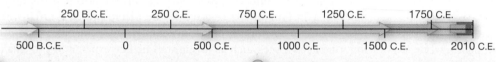

Classical
~500 B.C.E.–500 C.E.
Expanding Global Empires: As urban life took hold, competition for resources as well as the desire to control trade, people, and land pushed the expansion of largescale globalizing empires across Eurasia and Africa. This intensifed global connections through exchanges of people, cultural practices, and technologies. This period also witnessed periods of increased violence, conflict, and disease.

Postclassical
~500 C.E.–1500 C.E.
Intensification of Global Connections: Global connections increased across wider global spaces. As global trade networks expanded, commercial connections and relationships developed that included the exchange of money, the rise of merchant classes, and the spread of global religions (e.g., Christianity, Islam, Hinduism, and Buddhism), further facilitating economic and cultural connections.

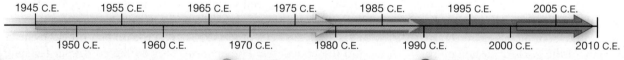

Late Modern
~1800 C.E.– ~1980 C.E.
Establishing Capitalism and the Natio-State: The rise of capitalism helped expand European hegemony as large parts of the Americas, Asia, Africa, and Oceania were brought into European systems of extraction, exploitation, and direct colonization. Other imperial powers, such as Japan, also emerged, while many parts of the Americas engaged in fights for decolonization. Cartographers carved the world into discrete political units. Many former colonial boundaries remain in place to this day, while others continue to be contested.

Cold War
1945 C.E. – 1989 C.E.
Global Conflict and Tension: After World War II, a "bipolar" world emerged, with two "superpowers" as well as a number of "nonaligned" nations. This was also a period of rapid decolonization and the emergence of new "independent" states, some of which remained subordinate to former colonizers. While the nation–state remained the basic unit of geopolitics, supranational organizations (e.g., NATO, the European Union, etc.) emerged as important global players as well.

Post-Industrial
1980 C.E. – Present
Global Realignment and Reorganization: The collapse of the Soviet Union and the Berlin Wall led to realignments in the arena of economic, cultural, and social globalization. Far from reducing global tension, the end of the Cold War resulted in new tensions, while other conflicts – based on ethnic, religious, and other social differences – escalated. An increasingly international division of labor occurred, as manufacturing continued to move from the industrialized nations of North America and Europe to other world regions.

Post-9/11
2001 C.E. – Present
Global Tension and New/Old Local Struggles: People often say that we live in a "post-9/11" world, although the United States appears at the center of global relations. With this caveat in mind, it is clear that in the post-9/11 period, new challenges face the planet, as issues of security, global climate change, and cultural conflict continue to escalate. Radical and fundamentalist movements, based in religion, ethnicity, and race have created new visions about local autonomy and placed issues, such as immigration, national identity, and borders, at the center of numerous local and global conflicts.

▲ **FIGURE 1.21 World History and Global Geography** Periods provide a broad framework to discuss geographic change and global connections over time. In this figure, certain periods overlap because the end of one period and the start of another is not a clear process in many cases and in most places. Throughout this text, we use the secular terms B.C.E., before common era, and C.E., common era.

systems—hunting and gathering, language, religion, and animal and plant domestication—and created unique sets of cultural and social practices tied to their understanding and interpretation of the world.

Global connections forged over the centuries prompted many people to explore the world beyond their shores. Beginning in the late fifteenth century these connections were intensified when Europeans undertook a series of voyages that established long-distance ocean trade routes between Europe and Asia and subjected the New World of the Americas, and its Indigenous peoples, to European conquest and exploitation. Within a few decades, places that had been separated by almost insurmountable geographical barriers came into close, and often violent, contact. The interactions that have occurred among peoples from different world regions have had a lasting impact on the landscapes and places we study today.

Colonialism, Capitalism, and the Industrial Revolution In about 1500 C.E., a rapid and decisive reconfiguration of regions occurred, with implications for the present day. It was at this point that

Europeans circumnavigated the globe, and their expanded reach into the Americas as well as parts of Africa and Asia became a significant driver of regional integration. Europeans dominated how regions were connected for several centuries thereafter. **Colonialism** is a political and economic system in which regions and societies are legally, economically, and politically dominated by an external or alien society. In this uneven connection, the colonizer works to reshape the regional economy and politics of the colonized region to suit its own needs. European colonialism took place in two major waves, and each wave had different regional and global implications. The first wave, which took place from the late 1400s through the early 1800s, was marked by European investment in port cities and trading networks (**Figure 1.22**). Led primarily by Spanish and Portuguese colonizers, with most of the extensive activity in South and Central America, this period was characterized by investment in the development of **plantation economies**, the classic institution of colonial agriculture. These economics were extensive, European-owned, operated, and financed enterprises in which single crops were produced by local

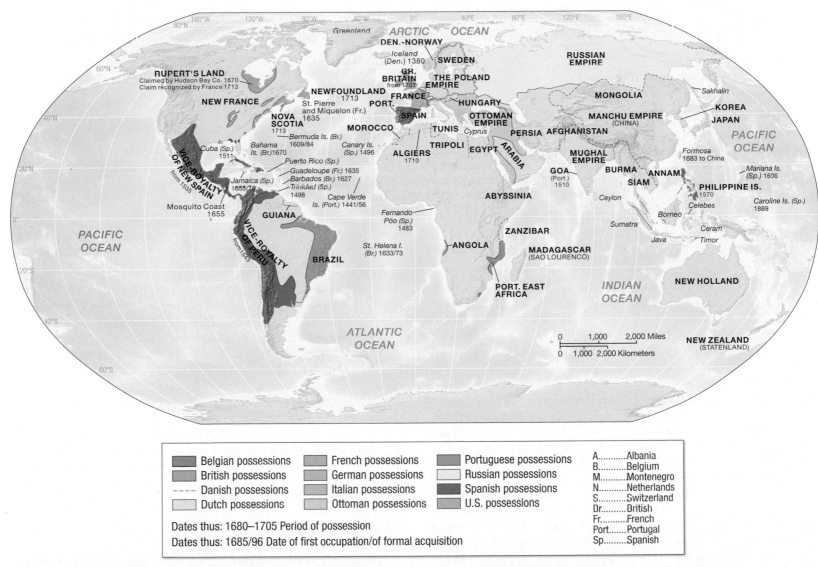

▲ **FIGURE 1.22 European Colonialism in 1714 C.E.** This map and the one shown in Figure 1.27 illustrate the transformation in the political geography of the globe that occurred during the 200 years that mark the most intense period of European global imperialism and colonialism. By the middle of the first wave of colonialism, European possessions amounted to less than 10% of the world's land area and only 2% of the world's population.

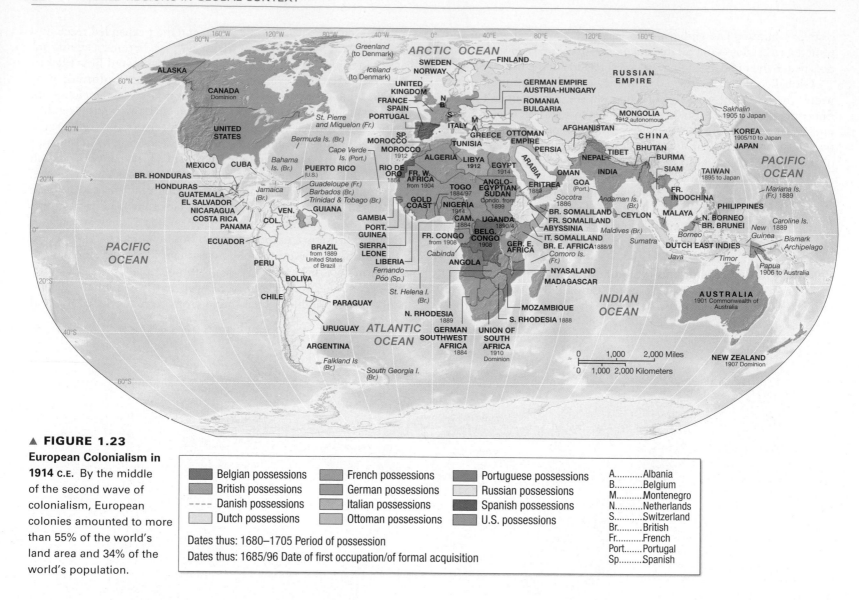

▲ **FIGURE 1.23**

European Colonialism in 1914 C.E. By the middle of the second wave of colonialism, European colonies amounted to more than 55% of the world's land area and 34% of the world's population.

Belgian possessions	French possessions	Portuguese possessions
British possessions	German possessions	Russian possessions
Danish possessions	Italian possessions	Spanish possessions
Dutch possessions	Ottoman possessions	U.S. possessions

A............Albania
B............Belgium
M...........Montenegro
N...........Netherlands
S............Switzerland
Br..........British
Fr...........French
Port.......Portugal
Sp..........Spanish

Dates thus: 1680–1705 Period of possession
Dates thus: 1685/96 Date of first occupation/of formal acquisition

or imported labor for a world market. The massive sugar plantations in the Caribbean were signature examples of colonialism. The enormous demand for labor to work these plantations led to the forced movement of millions of Africans into slavery, creating a connection between West Africa and the Americas that had dramatic implications for the contemporary world.

The second wave of colonialism began in the early 1800s and lasted through the mid-20th century; this wave incorporated the majority of the African continent as well as large parts of Asia, Australasia, and the South Pacific (**Figure 1.23**). The growth of **imperialism**, which is the extension of the power of a state through the direct or indirect control of the economic life of other territories, reorganized the world around a number of colonial powers and their colonies. This period of imperialism was characterized by heavy exploitation of agricultural and mineral resources in colonized regions and direct and total control over local affairs (**Figure 1.24**). This period also coincided with innovations and the development of institutions such as credit systems and stock companies, which stabilized and enabled the spread and evolution of **capitalism**. Capitalism is a system of social and economic organization characterized by the profit motive and individual and corporate ownership of productive goods and resources. The perceived underdevelopment of places like India and

Africa during this period left them open for exploitation and was crucial for the development of economic sophistication in Europe; Asia and Africa's economic stagnation and England's economic expansion were created out of their *connection* to one another.

Capitalist productivity and European colonial expansion helped fuel the **Industrial Revolution**, the rapid development of mechanized manufacturing that gathered momentum in the early 19th century. In the second half of the 19th century, there was a vast increase in the number of colonies and the number of people under colonial rule. Over the long run, the costs of maintaining this kind of power and influence weakened the dominant colonial powers. At the same time, as part of this imperial process, a growing European presence throughout the world restructured the political geographies of many regions, as colonial powers established defined and demarcated boundaries around their colonial holdings. Many of the world's political boundaries and regional formations today have their seeds in the period of European colonial expansion.

European powers, however, were not the only nations to establish new colonial and imperial formations. Throughout the first half of the 20th century, Japan also established itself as a regional and eventually global imperial power, expanding its control over large parts of East Asia, Southeast Asia, and the Pacific. As a result of Japan's

ARCTIC OCEAN
PACIFIC OCEAN
ATLANTIC OCEAN
INDIAN OCEAN
See inset below
$1.3 billion
$4 million
$268 million
$661 million
$1 billion
$215 million
$33 million
$538 million
Opium imports to China (annual average number of chests in thousands)
Opium exports from India ($ milllions per year)
Imports to China 1876
$32 million
$45 million
Cotton goods
Opium
Other
$35 million
MALWA
BENGAL
Canton
Shanghai
Ningpo
Foochow
Amoy
Malacca
A Senegambia
B Sierra Leone
C Windward Coast
D Gold Coast
E Bight of Benin
F Bight of Biafra
British and French slave trade (annual average number of slaves where region of origin is known)
Silver exported to Spain (annual average in $ millions)
Gold exported to Portugal (annual average in $ millions)
Amsterdam
London
Paris
Lisbon
Madrid
Profits from slave trade
British ships
French ships
Silver exports to Spain and Portugal (1531–1810)
Gold exports to Spain and Portugal (1503–1800)
Profits from use of slave labor in British West Indies (18th century)
Profits from Dutch spice trade (1650–1780)
Profits from opium trade (1800–1840)
India–China opium trade
Drain of land revenue from India (1760–1810)
Voyage of the slave trader Enterprise 1804, Liverpool to Bonney, Nigeria; 1804–1805, Bonney to Havana, Cuba, carrying 194 men, 32 men-boys, 66 boys, 42 women, 36 women-girls. Net profit on voyage, $130,000.

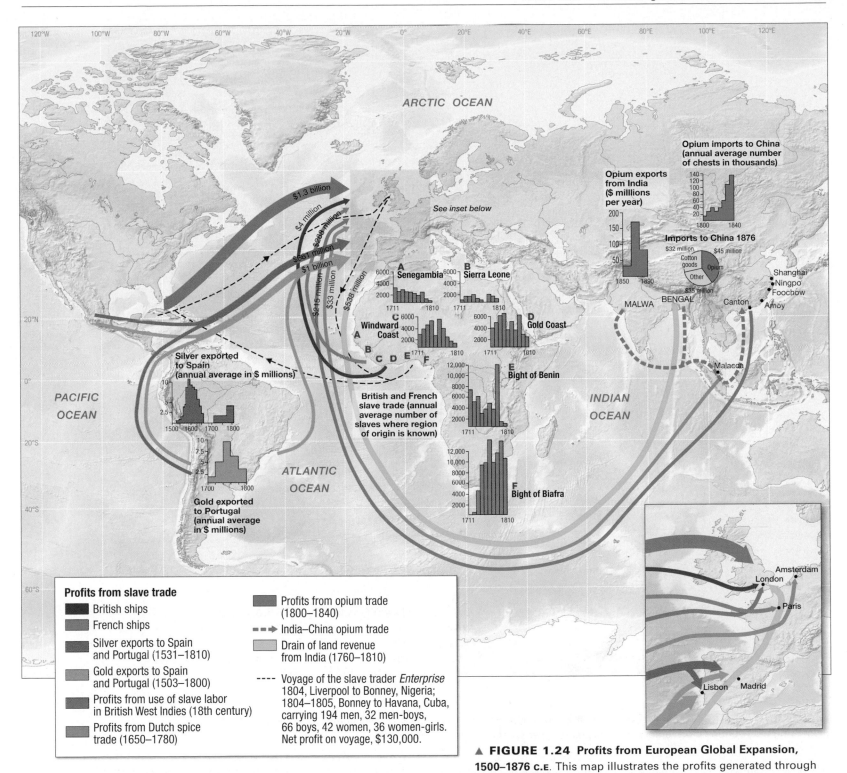

▲ **FIGURE 1.24 Profits from European Global Expansion, 1500–1876 C.E.** This map illustrates the profits generated through European plunder of global minerals, spices, and human beings over more than three centuries.

imperialism, new nations and regional formations took hold in places such as Southeast Asia.

Communism and the Cold War New political and economic philosophies emerged in the 19th and 20th centuries, establishing an alternative to the imperial expansion of Europe's capitalist states, leading to the creation of new world regions. Indeed, **communism**, a form of economic and social organization characterized by the common ownership of industry, transportation, agricultural land, and other key economic and social resources, spread in different forms to many parts of the globe. Communism influenced the establishment of a variety of economic systems in countries as diverse as China, Czechoslavakia, and Cuba.

The emergence of communism, along with the massive destruction of much of Europe during World War II (1939 C.E. to 1945 C.E.), facilitated the rapid decolonization of much of the world in the 1950s, 1960s, and 1970s and another reconfiguration of world

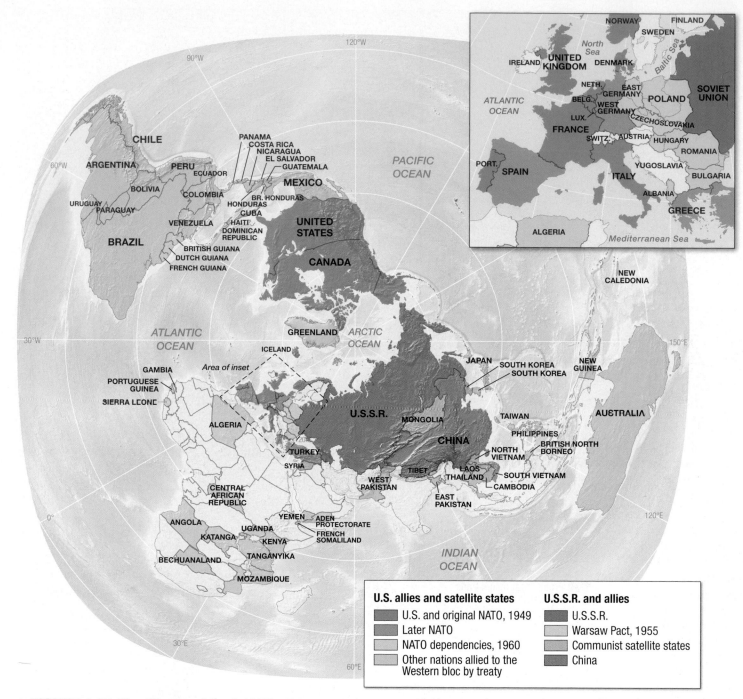

▲ **FIGURE 1.25 The Alliances of the Cold War** This map depicts the complex global geography of the Cold War, illustrating how various blocs emerged in this global system of conflict and tension.

regional organization. The challenge of communism also established what many defined as the **Cold War**—a period in which struggles between the major global power blocs, centered in large part around the United States and the Soviet Union, took place through a variety of conflicts in places such as Korea and Vietnam in Asia, Angola in sub-Saharan Africa, and Nicaragua in Central America (**Figure 1.25**).

These global contests produced new regional configurations. Western Europe, for example, became heavily economically integrated with the United States and Canada, while Eastern Europe was more tightly linked to Russia and parts of Asia. The end of the Cold War, in 1991, resulted in yet another radical reconfiguration of political and economic connections.

The world regional system we know today is not the first such system to have existed. At other moments in the past, different regions held the center of the world stage. Indeed, today's centers of economic power (Europe, for example) were marginal areas in other periods. Changes in the relative economic and political power of world regions have resulted from changes in their *connections* and *relations*. The rise of Europe over the last 500 years was not a product of something unique to European culture, resources, or politics. Instead, it was Europe's relationship with its colonies that allowed it to leverage a central place in recent history. The implications of this are important. The future regional centers of economic and political power may come to reside far from where they are today, in entirely new regions, as is clear from the rise of China.

Economy, Accumulation, and the Production of Inequality

In spite of the enormous growth of the global economy in the last several decades, there remains a huge gap between relatively wealthy *developed* countries—mainly in North America, Europe, and East Asia—and poorer *developing* countries. Developing countries are a diverse group, ranging from poverty-stricken states in sub-Saharan Africa and South Asia to rapidly growing economic powerhouses such as China, Brazil, and India. While some developing nations have made economic progress, what explains the intractable split in world regions between the "haves"—made up of countries that have accumulated wealth—and the remaining "have-nots"? The history of colonialism provides some clues. More recent economic trends provide others.

Toward the end of the Cold War period, a majority of the world's colonies had become independent nations, and many countries began to move away from communist forms of economic organization. Even prior to the demise of the Soviet Union in 1991, there were clear shifts in the global economy, as a series of oil crises in the 1980s changed how and where manufacturing took place. Since the early 1980s, manufacturing production has moved dramatically from the imperial centers of the Cold War period—North America, Europe, and Japan—toward mainland East Asia, Southeast Asia, Africa, and Latin America as corporations have sought out less expensive labor markets.

▶ **FIGURE 1.26 Four Sectors of the Global Economy** Geographically, the division of labor between people working in these sectors varies, as people in some parts of the world live in areas dominated by primary sector economies, while others live in places that have mostly service sector and information sector jobs.

This is evidenced by the specialization of different people, regions, and countries concentrated in certain kinds of economic activities.

Some regions are dominated by extractive industries or agriculture, specializing in farming, fishing, or mining raw materials. Others specialize in manufacturing. Some have little of either of these activities and provide services, like banking, while others specialize exclusively in computer software or telecommunication. Economic geographers categorize these economic activities into four types: **primary** (extractive), **secondary** (manufacturing), **tertiary** (services), and **quaternary** (information) sectors (**Figure 1.26**). When we look at these four types of activities, we see some regional shifting.

Extractive

Primary activities are any form of natural resource extraction. These include agriculture, mining, fishing, and forestry.

Manufacturing

Secondary activities are concerned with manufacturing or processing. These include steelmaking, food processing, furniture making, textile manufacturing, and garment manufacturing.

Services

Tertiary activities are those that involve the sale and exchange of goods and services. These include warehousing, retail stores, personal services such as hairdressing, commercial services such as accounting and advertising, and entertainment.

Information

Quaternary activities include handling and processing knowledge and information. Examples include data processing, information retrieval, education, and research and development (R&D).

The United States and Canada were almost exclusively a site of primary sector activities in 1865. By 1960, this world region had become a largely secondary industrial economy, only to become largely a tertiary economy, with few overall jobs in manufacturing by the year 2000.

APPLY YOUR KNOWLEDGE Do research and provide specific examples of jobs in industries from each of the four sectors of the global economy. Provide examples of how the four sector examples you have chosen are connected to each other through global exchange.

This change in the international division of labor is often attributed to the shift of some regions—the United States and Canada, Europe and parts of East Asia, South Asia, Southeast Asia, and Australia, New Zealand, and the Pacific—from manufacturing to service industries. That economic shift has reverberated throughout the global economy, as manufacturing has moved into several new key world regions as a result: Latin America and the Caribbean, sub-Saharan Africa, Southeast Asia, and parts of South Asia and East Asia. Ongoing global economic shifts complicate these trends further, as some regions have diversified their activities. For example, China has taken on a great deal of manufacturing for the global market, and is also now a significant economic player in other areas of the global economy, including banking. These shifts have altered the playing field of economic power relations both within East Asia and in the world more generally (**Figure 1.27**).

Specializations and diversifications of regional economies have an enormous influence on people's everyday lives. In the last half-century, globalization has led to a consolidation of the global economy. This consolidation concided with intensified disparities between the rich and the poor. The inequality resulting from the global economy is reflected—and reinforced—by many aspects of human well-being. It is important to understand how geographers and other social scientists measure the "gaps" between rich and poor, because many measures can be partial or misleading.

Measuring Economic Development At the global scale, levels of economic development are often measured using national economic indicators—numbers that provide a snapshot of the overall economy. **Gross domestic product (GDP)** is one such number. It is an estimate of the total value of all materials, foodstuffs, goods, and services produced by a country in a particular year. To standardize for countries' varying sizes, the statistic is normally divided by total population, which gives an indicator, *per capita* GDP, that provides a measure of relative levels of overall economic development. **Gross national income (GNI)** also includes the net value of income from abroad—flows of profits or losses from overseas investments, for example.

In making international comparisons, GDP and GNI can be problematic. They are calculated using national currencies (the

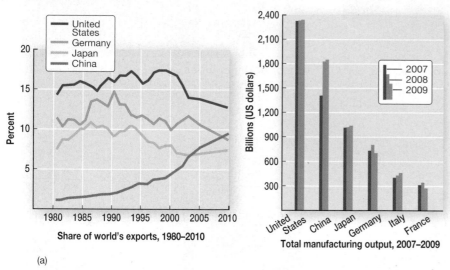

(a)

Share of world's exports, 1980–2010

Total manufacturing output, 2007–2009

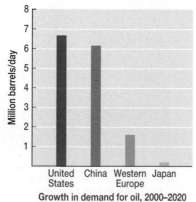

Growth in demand for oil, 2000–2020

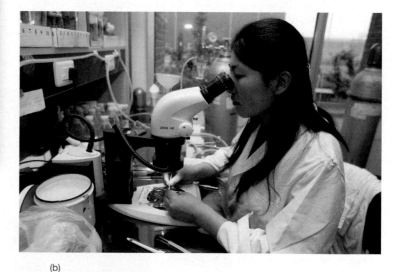

(b)

◀ **FIGURE 1.27 China's Changing Role in the World Economy** (a) Most forecasts indicate that by 2020, China's economy will exceed those of the world's core economic powers except the United States. China's reemergence as a significant player in the world economy is reflected in trends of inward investment, manufacturing output, and oil consumption. (b) A worker in a Chinese biotech firm engages in research.

value of which can be distorted by speculation and government policies), and they do not include nonmarket goods and services. A nonmarket good or service does not have an observable monetary value. Examples of a nonmarket good include wildlife or a coral reef. An example of a nonmarket service is the preparation of food for a family done by a parent. Indicators such as GDP and GNI also obscure subnational differences. They do not measure differences between women and men's access to resources or participation in the economy. They say nothing of people's health. One corrective to these more traditional economic indicators is the development of an indicator that compares national currencies based on **purchasing power parity (PPP) per capita**. In effect, PPP measures how much of a common "market basket" of goods and services each currency can purchase locally, including goods and services that are not traded internationally.

As **Figure 1.28** shows, most of the highest levels of economic development are found in northern latitudes (very roughly, north of 30°N), which has given rise to popular shorthand for the world's economic geography: the division between the **global north** and the **global south**. According to the United Nations Development Programme (UNDP), the income gap between the poorest fifth of the world's population and the wealthiest fifth increased more than

threefold between 1960 and 2000. In 55 countries, per capita incomes fell during the 1990s. In sub-Saharan Africa, economic output fell by one-third during the 1980s and stayed low during the 1990s, so that people's standard of living there is now, on average, lower than it was in the early 1960s. At the beginning of the 21st century, the fifth of the world's population living in the highest-income countries had 74% of world income while the bottom fifth had just 1%. Such enormous differences have led many people to question the fairness of geographical variations in people's levels of affluence and well-being.

Patterns of Social Well-Being Global inequality is reflected in measures other than income. For example, patterns of life expectancy, a reliable indicator of social well-being, show a dramatic difference between the global south and global north. For adults in industrial countries in the north, life expectancy is high and continues to increase. In contrast, life expectancy in the poorest countries is dramatically shorter. As explained previously, important development indicators like the GDP are not useful for explaining these discrepancies. One useful alternative metric is the United Nations Development Programme's **Human Development Index (HDI)**, which is based on measures of life expectancy, educational attainment, and

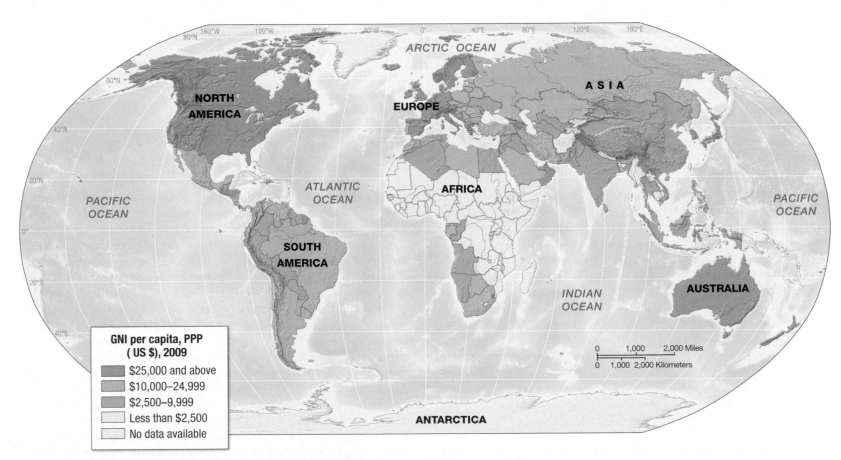

▲ **FIGURE 1.28 GNI per Capita, Purchasing Power Parity (PPP)** PPP per capita is one of the best single measures of economic development. This map, based on 2009 data, shows the tremendous gap in affluence among the countries of the world. The world's average annual per capita PPP is $10,700. The gap between the higher per capita PPPs ($62,298 in Luxembourg, $37,738 in Ireland, and $37,670 in Norway) and the lower ones ($548 in Sierra Leone and $608 in Malawi) is huge: in 2000, the average of the bottom ten nations was about 1/50th of the average of the top ten.

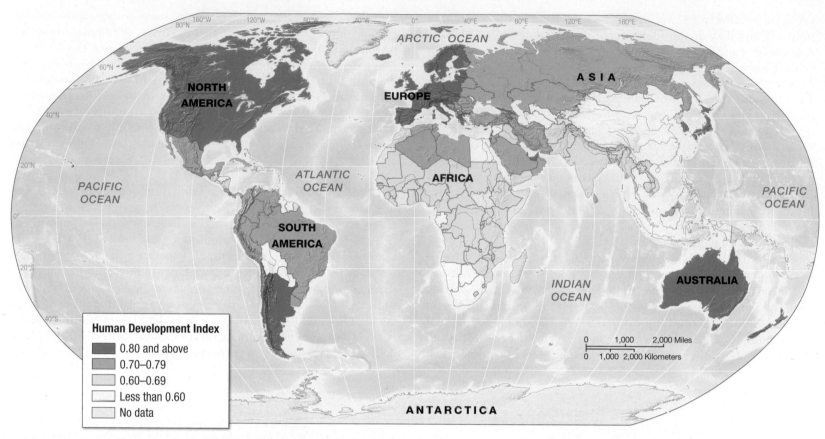

▲ **FIGURE 1.29 The Human Development Index** This index is based on measures of life expectancy, educational attainment, and personal income. A country that had the best scores among all of the countries in the world on all three measures would have a perfect index score of 1.0, whereas a country that ranked worst in the world on all three indicators would have an index score of 0. Most of the affluent countries have index scores of 0.8 or more. The worst scores—those less than 0.6—are concentrated in Africa.

personal income. **Figure 1.29** shows the international map of human development in 2011. Norway, Australia, and the Netherlands had the highest overall levels of human development (0.94, 0.93, and 0.91, respectively), whereas Niger and the Democratic Republic of Congo (0.30 and 0.28, respectively) had the lowest levels.

Another powerful metric is the United Nations's **Gender-Related Development Index (GDI)**, a composite indicator of gender equality assessing the standard of living in a country. The GDI tracks the inequalities between men and women in the following areas: long and healthy life, educational attainment and literacy, and a decent economic standard of living. While many wealthier countries score well on the GDI, others do not. So there is no direct relationship between wealth and GDI. That said, GDI scores tend to be lower for poorer countries, although recent trends point to some global improvements.

Explaining and Practicing "Development" There is a general worldwide interest in explaining the above-mentioned areas of inequality, so they might be alleviated. Regrettably, however, there is no universal agreement over the causes of inequality. As a result, there is no consensus over strategies that should be used to achieve **development**, which is an improvement in the economic well-being and standard of living of people. Despite a lack of consensus, many different development experiments have been attempted.

On one side of this discussion are economists and politicians who hold a strong belief that an unfettered market, unconstrained by financial regulations, is the strongest tool for generating wealth in poor countries and reconciling these differences in poverty and wealth. For these observers, it is the *presence* of government subsidies, rules, and interventions that cause poverty. This approach is sometimes called **neoliberalism**. For those who adhere to this belief and approach, the best policy is one in which state budgets are reduced, public ownership of industries or utilities is turned over to private parties, and laws regulating working conditions and the environment are minimized. Such an approach has been championed by international institutions in the past. Most notably, the **International Monetary Fund (IMF)**, a powerful international financial institution that provides emergency loans to countries in financial trouble, has encouraged these kinds of reforms. The IMF informs potential borrowing countries of conditions that they must accept if they are to receive emergency financial assistance. These conditions include policies that emphasize privatization, limited restrictions on imports, smaller state budgets, and liberalized trade. Many countries around the world have attempted these policies, often with poor results. Sometimes new economic development was hampered and working and environmental conditions in target countries became worse. These policies have also tended to stress *primary* economic activities in these already poorer countries. In short, the implementation of

▲ **FIGURE 1.30 Poverty and Wealth in Bangkok, Thailand** The juxtaposition of poverty and wealth in urban areas is apparent in many cities of the world where informal residences, such as the one pictured here, are built alongside high-rise apartments housing members of an affluent upper-middle class.

these policies to decrease economic inequality often has only reinforced it instead (**Figure 1.30**).

There are a number of alternative ideas about how development can or should occur that contrast with neoliberalism. These ideas come from many quarters. Feminists, concerned about the growing economic gaps between men and women, stress a focus on inequalities within regional economies and between men and women. Postcolonial scholars seek to break the dependence of underdeveloped postcolonial nations on developed ones, which are former colonial and imperial powers. These kinds of ideas and groups all call into question the basic tenets of neoliberalism such as a universal faith in markets, a focus on reducing government programs and roles, and the assumption that development is a purely economic process. Some organizations seek to empower grassroots movements and promote local knowledge, which they believe can repair the damage critics feel has been done by neoliberal development projects that have focused largely on the production of wealth (i.e., money) and not on quality of life (i.e., a social good). The alternatives to neoliberalism tend to stress the autonomy of local communities, innovation of appropriate technologies rooted in local knowledge, and new connections between people and groups around the world.

Ultimately these critiques challenge the assumption that institutions in the global north (such as the IMF) know best how to solve the problems of the global south. The "Honey Bee Network" founded in India, for example, seeks to share clever, nonmarket, local solutions to problems of poverty between poor countries. They link up people in Africa, Latin America, and South Asia to develop innovations such as a bicycle-powered hoe, a micro-windmill battery charger, and a pedal-operated washing machine. People in places like India and Costa Rica, critics of neoliberalism argue, have been *disconnected* from one another by colonialism and capitalism. Only

by creating new global connections can real, meaningful, long-term development occur, linked to both human development and environmental sustainability.

The problems of economic disparities and development underline the way regional differences can have dramatic and disturbing implications. People living in grinding poverty and those accumulating wealth are often spatially separated across the globe. But it is often the *connection* of these people that is the driving force of that inequality. Configurations inherited from the Cold War and colonialism often stubbornly persist into the 21st century, with implications for development and the quality of people's lives. The challenge of global development is one of changing the configurations between regions and the people who live in them.

APPLY YOUR KNOWLEDGE Use the Internet to search for "alternative development" schemes or programs. Choose one example and describe how that development scheme provides an alternative to more traditional forms of economic neoliberal development.

Territory and Politics

Like the world economy, the political geography of world regions has its roots in history. At any given time, territories reflect past global configurations and relations. With the rise of modern states and nations over the last 400 years, new regional configurations emerged.

States and Nations A **sovereign state** exercises power over a territory and people and is recognized by other states. The independent power of a sovereign state is codified in international law. A **nation** is a group of people sharing common elements of culture, such as language and religion. Ultimately, a national identity is built on a common sense of history, geography, and purpose such that individuals feel compelled to defend the nation and to further the objectives of the state. In addition to enabling the creation of a stable democracy, the construction of a nation also facilitates the organization of a more extensive and coherent market where buyers and sellers can communicate and all have an investment in the success of the economic enterprise.

In rare instances, a state boundary aligns with one national (or ethnic) group. The term **nation–state** is used to describe this ideal state, one consisting of a somewhat homogeneous group of people living in the same territory. Today there are very few true nation-states. Mobility and history have resulted in states that govern highly multicultural populations. Even so, **nationalism**, the feeling of belonging to a nation as well as the belief that a nation has a natural right to determine its own affairs, remains a powerful organizing force of the current state system (**Figure 1.31**, p. 32). Many observers argue that one response to globalization has been a rise in nationalism manifested in both peaceful and conflictual ways.

Over the last several centuries, the sovereign state and national identities have become the norm across the globe. Today, there are over 200 sovereign states. States are the building blocks of the modern world regions that are found in this book. States regulate the flow of people, goods, and ideas across world regions as well as across particular national boundaries. States also facilitate greater

▲ **FIGURE 1.32** **International Border in Mae Sai, Thailand, Talicheck, Burma** International borders are often lively sites of everyday exchange. People move back and forth exchanging goods and services. As concerns over violence increase in certain regions, border checkpoints, such as this one, are more highly regulated.

▲ **FIGURE 1.31** **Anti-European Union Protests in Croatia, 2011** A demonstrator holds a Croatian national flag to protest the signing of the treaty that will allow the Balkan country to become the European Union's 28th member in 2013.

or lesser regional cohesion, as state members struggle to maintain their own position within the changing context of an ever-globalizing world. Today's political and economic connections are negotiated through international laws and a state system that controls all kinds of movements. Border crossings, citizenship, trade laws, and responsibilities for the environment and resources are all managed through rules and laws within and between states (**Figure 1.32**). States, like world regions, are not static, however. They are subject to change and even failure. In 2009, there were 20 states around the world, most of them in Africa, where the central government was so weak that it has effectively no control over its territory. At the same time, new states emerge, as was the case in 2011, with the creation of South Sudan.

Political Globalization The world is highly interconnected through economics and, to a certain extent, through global cultural practices (as we will discuss in the section that follows). However, the

development of a comprehensive system of "world government," outside of states, that could address global problems, remains elusive. Policymakers, for the most part, lack an adequate framework for coping with the consequences of globalization. Although institutions and treaties such as the World Criminal Court, the World Trade Organization, and the Kyoto Agreement constitute the beginnings of a global governance framework for crime, the economy, and the environment, their effectiveness is limited by the persistence of individual states that are at odds with the goals of such organizations. In addition, trade policy has come to be governed by powerful transnational corporations. This severely hinders the ability of national governments to address many transnational issues.

Instead of a coordinated system of global governance to accompany economic globalization, **supranational organizations** have emerged. These organizations are collections of individual states with a common goal that may be economic or political and that diminish individual state sovereignty in favor of member states' interests (**Figure 1.33**). Supranational organizations include the European Union (EU), the North American Free Trade Agreement (NAFTA), and the Association of South East Asian Nations (ASEAN), among others. The political puzzle of globalization confronts these organizations constantly. For example, the Eurozone within the European Union has had a common currency since 1995, but it does not have a single unified bank, a single overarching set of financial rules, or a single budget. This created a crisis in 2010–12, when some member economies such as Greece's began to fail. Too unified to separate from Greece's economic woes, EU countries faced economic hardship as a result of their connection. Not unified enough to develop a single legal remedy or policy to address the problem, members of the European Union became mired in a set of lengthy, complex negotiations. What this example demonstrates is that the new political

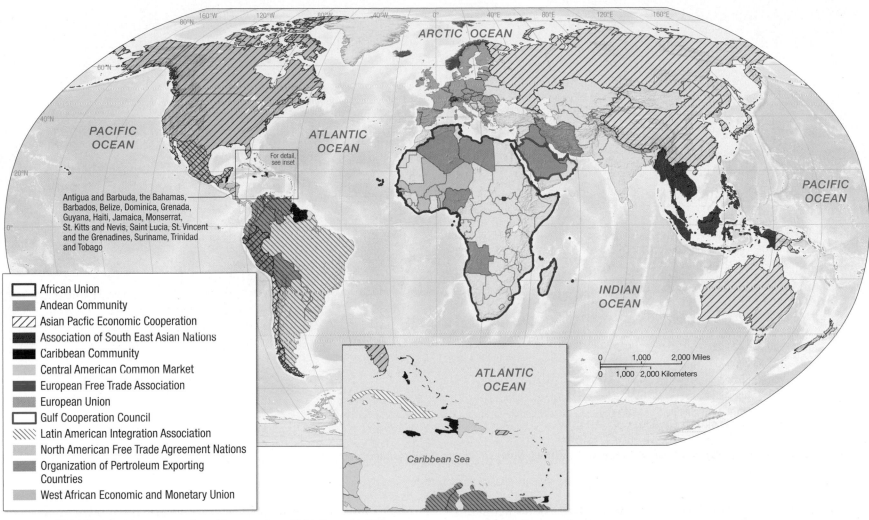

Antigua and Barbuda, the Bahamas, Barbados, Belize, Dominica, Grenada, Guyana, Haiti, Jamaica, Monserrat, St. Kitts and Nevis, Saint Lucia, St. Vincent and the Grenadines, Suriname, Trinidad and Tobago

Legend:
- African Union
- Andean Community
- Asian Pacfic Economic Cooperation
- Association of South East Asian Nations
- Caribbean Community
- Central American Common Market
- European Free Trade Association
- European Union
- Gulf Cooperation Council
- Latin American Integration Association
- North American Free Trade Agreement Nations
- Organization of Pertroleum Exporting Countries
- West African Economic and Monetary Union

▲ **FIGURE 1.33 Transnational Integration, 1945–Present** This map shows a sampling of some of the most significant supranational organizations around the globe.

connections that make up the EU created complex and uneven outcomes within the region.

Founded in 1945, the **United Nations (UN)** is another supranational organization aimed at facilitating cooperation in international law, security, economic development, human rights, and world peace. In the absence of a deliberately organized global governance system, the United Nations has become something of a *de facto* replacement. With headquarters in New York City, the United Nations is composed of five administrative organs: the Security Council, which includes as permanent members the United States, Great Britain, China, France, and Russia; the General Assembly, which includes nearly all of the world's internationally recognized sovereign states (Vatican City, Kosovo, and Taiwan are a few exceptions) (**Figure 1.34**, p. 34); the Economic and Social Council; the Secretariat; and the International Court of Justice, operating since 1946 in The Hague in the Netherlands.

At the same time that formal political organizations like the United Nations have been expanding their role, informal political organizations and movements have also increased in overall numbers and membership, and become more globally connected in the last 50 years. Known as **social movements**—large, informal groups of

individuals and organizations that focus on social or political issues—these groups are part of a worldwide array of voluntary civic and social organizations and institutions representing the interests of citizens against the power of formal states and markets. Social movements embrace a spectrum of issues, including women's rights, the environment, human rights, climate change, and others. The issues they choose are ones that affect people very directly and around which people organize to obtain formal political responses.

Social movements are sometimes represented by more formally constituted organizations that are managed through some kind of coordinated administration. These **nongovernmental organizations (NGOs)** are among the most visible actors in global politics; their reach is international, and they work across and between regions. Notable examples include Greenpeace (headquartered in Great Britain), Genocide Watch (headquartered in the United States), and the Green Belt Movement (headquartered in Kenya).

The political institutions that make up the system of global politics—including nations, supranational organizations, social movements, and NGOs—are both challenged and strengthened by globalization. As objects move more swiftly and freely through free trade and Internet communication, they cross national boundaries

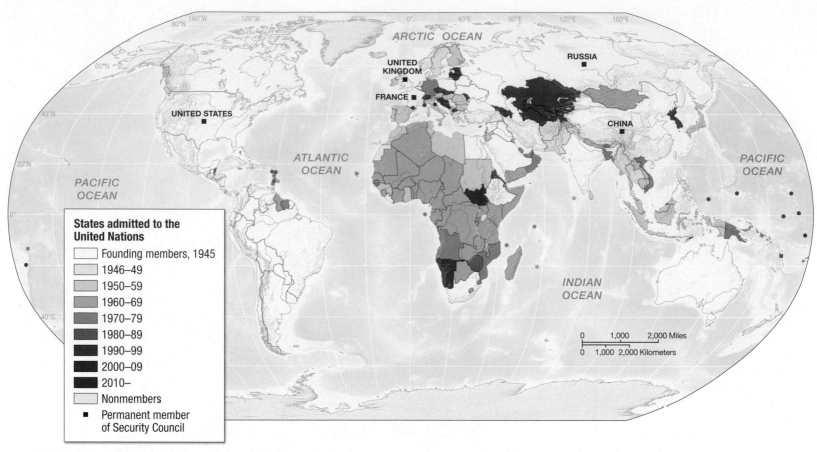

▲ **FIGURE 1.34 The United Nations Member Countries** Nearly all the sovereign states on Earth are members of the United Nations.

and elude the control of state authorities. On the other hand, the globalization of information has reinforced the power of nongovernmental organizations and elements of civic society by building new connections between states and regions. New political regions and relationships can potentially emerge out of these kinds of changes.

APPLY YOUR KNOWLEDGE Conduct research and provide a current or historical example of the relationship between nationalism and migration policies for a specific nation. Discuss the impact this relationship has had on the nation, its region, and other affected regions. (Hint: You might search for news stories that link the rise of nationalist-oriented political parties and migration policies.)

CULTURE AND POPULATIONS

Economics and politics are not the only aspects of social relations that determine the character of regions. The distinctive cultures and "styles" of the populations inhabiting world regions also play an important role.

Not many things so fully convey a sense of unique regional style and culture as music. The signature sounds of West African music and its styles and instruments are immediately identifiable by anyone who has heard them before, especially the sound of the "Talking Drum." Similarly, rock and roll, with its electric instrumentation, booming vocals, and gritty lyrics, couldn't seem more different from West African drum music. And yet the two sounds are the product of one geographic thread, which carried shuffling rhythms and a set of chord progressions and scales across the Atlantic during the slave trade in the 1500s, to emerge as blues music and subsequently as rock and roll in the 1950s. A reverse flow of ideas, energy, and styles has made the return trip in recent years, as West African music is being transformed by artists like Vieux Farka Touré, a Malian performer who is shaking up local African styles and performances with rock music. These two musical traditions are transformed by their connections.

The world is deeply affected by the circulation of cultural ideas and practices as well as by the impact of changing demographic patterns. It is often argued that instantaneous communication is making the world a uniform place. But these arguments don't take into account how culture creates global linkages—like hip-hop—that connect the world's people. The aim of this section is to identify some of the many cultural and demographic processes that bind the world, not into a homogenous global space but into a world in which difference and similarity operate hand-in-hand.

Culture, Religion, and Language

The term *culture* is often used to describe the range of activities that characterizes a particular group, such as working-class culture, corporate culture, or teenage culture. Although this understanding of culture is accurate, it is incomplete. Broadly speaking, culture is a shared set of meanings lived through the material and symbolic practices of everyday life. The material aspects of culture include such things as dress and house styles, while cultural symbols are objects that represent something else, such as a cross as a symbol of Christianity. Culture is not necessarily tied to a place; it is the material and symbolic connections among people and places that can be altered and are always changing. These changes are sometimes subtle and other times more dramatic. The shared set of meanings that constitutes culture can include values, beliefs, practices, and ideas about religion, language, family, food, gender, sexuality, ethnicity, and nationality as well as other sets of identities (**Figure 1.35**). At the same time, cultural values, beliefs, ideas, and practices are—because of globalization—increasingly subject to reevaluation and redefinition. In this section, we focus on two of the most important unifying cultural forces in the world today, language and religion.

Geographies of Religion The term **religion** is difficult to define and includes a number of different kinds of beliefs or bodies of thought, which include worship of local spirits to very structured social systems tied to regularized ritual, such as attendance at church. We use the term in this text to include belief systems and practices that recognize the existence of powers higher than humankind. Although religious affiliation is on the decline in some parts of the world, it still acts as a powerful shaper of daily life, from eating habits and dress codes to coming-of-age rituals and death ceremonies, holiday celebrations, and family practices (**Figure 1.36**). Religious beliefs and practices change as people develop different interpretations or adopt new spiritual influences.

From the Arab invasions following Muhammad's death in 632 c.e. to the Christian Crusades of the Middle Ages, to the onset of the modern period in the 15th century c.e. and into the present, religious

▲ **FIGURE 1.36 Family Practices around the Table** This Malaysian-American family celebrates Eid-al-Fitr, the feast at the end of Ramadan (the annual month of fasting that all Muslims practice), reinforcing their religious practices with their eating practices.

missionizing—propagandizing and persuasion—as well as forceful and sometimes violent conversion have been key elements in changing geographies of religion. In the past 500 years, some religions have become dramatically dislocated from their sites of origin, not only through missionizing and conversion but also by way of **diaspora**—the spatial dispersion of a previously homogeneous group—and emigration. Both diaspora and emigration involve the involuntary and voluntary movement of people who bring their religious beliefs and practices to new locales.

The processes of global political and economic change that have led to the massive movement of the world's populations over the last five centuries have resulted in the dislodging and spread of the world's many religions from the regions where they originated. Many religions have spread to so many different places that it is a challenge to present a map of the contemporary global distribution of religion. **Figure 1.37**, p. 36, identifies the contemporary distribution of the world's major religions—those that have the largest number of practitioners. There is also a historically significant and growing global population of agnostics, who remain unsure of the reality of divine beings and typically do not adhere to any particular religious faith, as well as atheists, who have no belief in divine beings. Agnostics and atheists—who may make up more than 15% of the global population—share a strong belief in the maintenance of a strict division between church and state, religion and politics.

Figure 1.38, p. 36, illustrates how the world's major religions originated and spread from two fairly small areas of the globe. The first of these areas, where Hinduism, Buddhism, and Sikhism began, is an area of lowlands in the subcontinent of India drained by the Indus and Ganges rivers. The second area, where Judaism, Christianity, and Islam originated, is in the deserts of the Middle East. During the colonial period religious missionizing and conversion flowed from Europe to the rest of the world. In the postcolonial period, however, an opposite trend has occurred. For example, the fastest-growing religion in the world today is Islam. A great number of people are

▲ **FIGURE 1.35 Japanese Anime Convention in Los Angeles, 2011** Japanese anime, a popular art form, makes its presence known in Los Angeles, California. Japanese cultural practices and images have been globalized since the late 1970s.

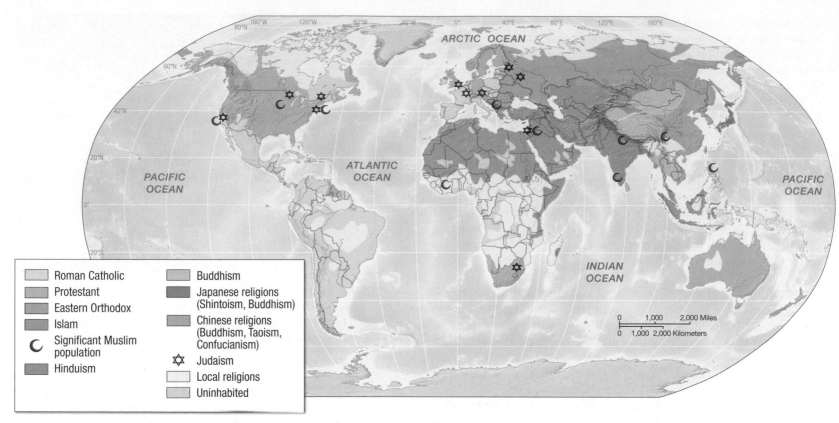

▲ **FIGURE 1.37 World Distribution of Major Religions** Most of the world's people are members of these religions. Not evident on this map are the local variations in practices, as well as the many local religions that are not easily shown on a world map.

► **FIGURE 1.38 Origin and Diffusion of Four Major Religions** The world's major religions originated in the Middle East and South Asia. Christianity began in present-day Israel and Jordan. Islam emerged from western Arabia. Buddhism originated in India, and Hinduism originated in the Indus region of Pakistan. These places are also the source areas of agriculture and urbanization.

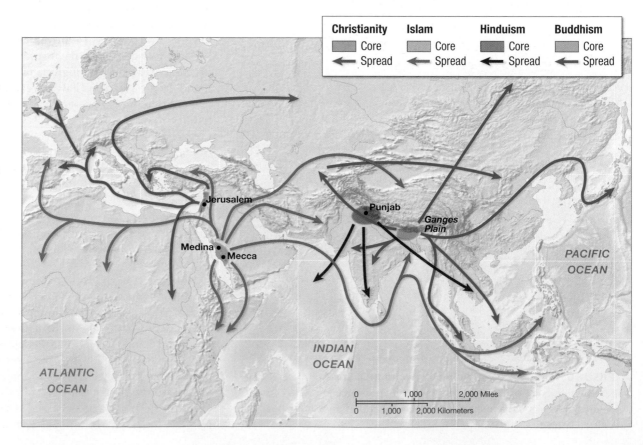

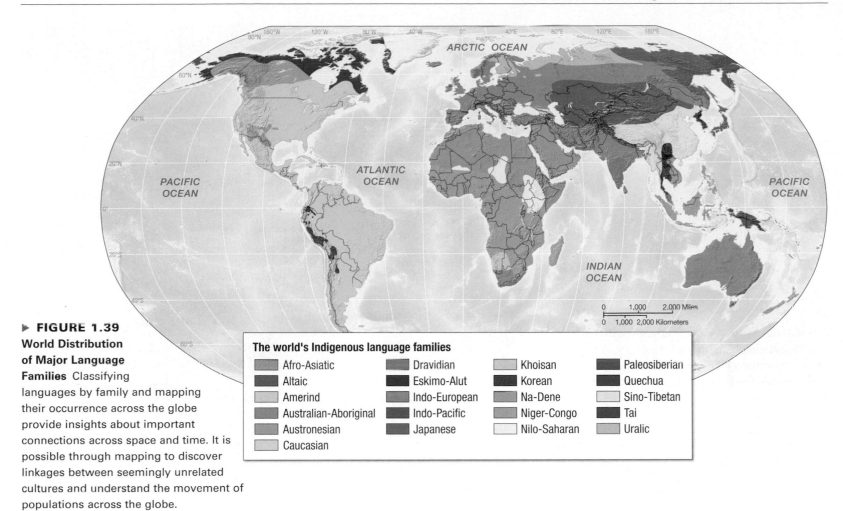

▶ **FIGURE 1.39**
World Distribution of Major Language Families Classifying languages by family and mapping their occurrence across the globe provide insights about important connections across space and time. It is possible through mapping to discover linkages between seemingly unrelated cultures and understand the movement of populations across the globe.

The world's Indigenous language families

- Afro-Asiatic
- Altaic
- Amerind
- Australian-Aboriginal
- Austronesian
- Caucasian
- Dravidian
- Eskimo-Alut
- Indo-European
- Indo-Pacific
- Japanese
- Khoisan
- Korean
- Na-Dene
- Niger-Congo
- Nilo-Saharan
- Paleosiberian
- Quechua
- Sino-Tibetan
- Tai
- Uralic

converting to Buddhism in regions such as the United States and Canada as well as Europe.

Geographies of Language The concept of **language**, like religion, is a central aspect of cultural identity. Language both reflects and influences the ways that different groups understand and interpret the world around them by providing each group with often-distinct concepts and vocabulary. Language can bind people together into systems of shared beliefs and practices, resulting in the rise of nationalism and national identity, for example. The distribution of languages reflects the changing history of human geography and the forces of globalization and regionalization on culture.

Through institutions such as schools and courts, governments often promote the use of standard languages, also known as official languages. Spoken alongside standard languages, **dialects** represent regional variations of those languages. Dialects feature place-based differences in pronunciation, grammar, and vocabulary.

At the most general level, languages are grouped into families, which are collections of languages believed to be related in their prehistorical origins. These collections are further divided and subdivided. **Figure 1.39** shows the locations of the world's Indigenous languages or language families.

Language diversity can present challenges in a globalizing world that depends on linguistic communication and interconnections.

Consider the diversity of languages and dialects in a region like South Asia, which makes communication and commerce among different language speakers difficult. In an effort to prevent this, developing states often create one national language to facilitate communication and enable the efficient conduct of state business. Unfortunately, where official languages are put into place, Indigenous languages may eventually be lost (**Figure 1.40**, p. 38).

Yet the actual unfolding of globalizing forces—such as official languages—works differently in different places and in different times. Although the overall trend appears to be toward the loss of Indigenous languages (and other forms of culture), there are movements around the world to revive dying languages by introducing them into schools and offering courses for adults. In this way, language has become an important vehicle for new cultural regionalizations in places like Belgium (among the Flemish people), Spain (within the Basque separatist movement), and Canada (among those involved in the Québecois movement).

In other cases, regional formations come to rely on a common language, or *lingua franca*, to communicate across linguistic boundaries. As an example, Swahili operates as the *lingua franca* of East Africa, allowing peoples from different backgrounds to engage each other in various forms of economic and sociocultural exchange. It is quite common, therefore, to find someone in East Africa who is bilingual, speaking their local language and the regional *lingua franca*.

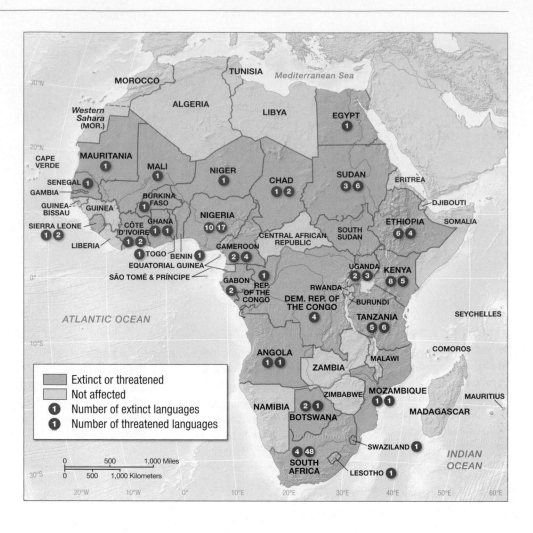

▶ **FIGURE 1.40** **African Countries with Extinct and Threatened Languages** It is not absolutely certain how many languages are currently being spoken worldwide, but the estimates range between 4,200 and 5,600. Although some languages are being created through the fusion of an Indigenous language with a colonial language, such as English or Portuguese, Indigenous languages are mostly dying out. Indigenous languages are dying out throughout the Americas and Asia as well.

APPLY YOUR KNOWLEDGE Using a newspaper search engine, find articles that show examples of the relationships among religion or language and national identity. List two or three of the key themes that emerge in these stories. (Hint: For example, what are some of the reactions to bilingual laws in countries such as Canada and New Zealand? Or, what are the concerns about the expansion of Muslim mosques in countries such as the United States or France? What themes do news stories on these topics have in common? How do the themes in the stories differ?)

Cultural Practices, Social Differences, and Identity

Airports, offices, and international hotels across the globe have become notoriously alike in recent decades; their similarities of architecture and interior design are reinforced by near-universal dress codes of the people who frequent them. The business suit, especially for men, has become the norm for office workers throughout much of the world. Meanwhile, jeans, T-shirts, and athletic shoes are the universal attire for both young people and those in lower-wage jobs. Similar automobiles can be seen on the streets of cities throughout the world (although sometimes the exact same automobiles have different model names); the same popular music is played on local radio stations; the same movies are shown in local theaters; and the same brand-name products appear in stores and restaurants. As a result, popular commentators have observed that cultures around the world are being Americanized, or "McDonaldized." According to some pundits, a single global culture is emerging that is based on material consumption, with the English language as its medium.

Globalization, Regionalization, and Culture Globalization produces a homogenization of culture through the language of consumer goods. This material culture is predicated on Airbus jets, CNN, music video channels, cell phones, the Internet, Gap clothing, Nikes, iPods, Toyotas, Disney, and formula-driven Hollywood movies and fueled by Coca-Cola, Budweiser, and McDonald's. Yet neither the widespread consumption of U.S. and U.S.-style products, nor the increasing familiarity of people around the world with global media and international brand names, adds up to the emergence of a single global culture. The processes of globalization *are* exposing the globe's inhabitants to a common set of products, symbols, myths, memories, events, cult figures, landscapes, and traditions (**Figure 1.41**). But people living in Tokyo, Japan; Tucson, Arizona; Turin, Italy; or Timbuktu, Mali, may be familiar with these shared aspects of global culture without necessarily using or responding to them in uniform ways. Often these new consumers of global products harness them to emerging, highly regional cultures. It is important to recognize that cultural flows take place in all directions, not just outward from the United States. Think, for example, of European fashions in U.S.

▲ **FIGURE 1.41 Bollywood** Bollywood is the name given to the regional site of film production in Mumbai, India. The Bollywood dance form has been picked up in popular culture across the globe, impacting not just dance but fashion as well.

stores; Chinese, Indian, Italian, Mexican, and Thai restaurants in U.S. towns and cities; and U.S. and European stores selling exotic craft goods from global south countries.

Exchanges and interactions across cultures create new cultures and also allow people to transcend some of the traditional cultural differences around the world. It may be easier on some level to identify with people who use the same products, listen to the same music, and appreciate the same sports or music stars. At the same time, however, deep cultural differences are opening up.

Several reasons account for the appearance of these new differences. One is the release of sociocultural, political, and economic pressure brought about by the end of the Cold War. The shift away from a bipolar political world allowed people to focus on other threats to their traditional culture. Another is the globalization of culture itself. The more people's lives are homogenized through their jobs and their material culture, the more many of them want to establish a distinctive cultural identity. For the poorest people in the world, a different set of processes is at work.

The contrast of poverty, environmental stress, and crowded living conditions with the luxurious lifestyles of wealthy elites creates a fertile climate for gangsterism. The same juxtaposition also provides the ideal circumstance for the spread and intensification of religious **fundamentalism** and for fundamentalist-inspired terrorism. In general, fundamentalism is belief in or strict adherence to a set of basic principles that often arises in reaction to perceived doctrinal compromises with modern social and political life. Religious fundamentalism occurs across all religions and is an important force in global geographies as it is taken up as a political cause in regions as diverse as the United States and the Middle East.

As a result, some metropolitan areas face a prospect of communities polarized between wealthier and poorer people, with outright cultural conflict suppressed through means such as electronic surveillance (**Figure 1.42**), security posts, and urban design featuring security fences and gated streets. This scenario requires a certain level of affluence to meet the costs of keeping the peace. In the unplanned metropolises of less-affluent world regions, where unprecedented numbers of migrants and refugees are thrown together, the prospect of anarchy and intercommunal violence is real.

APPLY YOUR KNOWLEDGE Research regional religious diversity for the area in which you live. What range of religions are practiced? To what degree are these religions publicly evident in practices and architecture? Are some religions more in evidence than others?

Culture and Identity Culture impacts society through its influence on social organization. Social categories such as family, gang, or generation, or some combination of these categories, figure more or less prominently in cultural groups, depending on geography. The importance of these social categories may change over time as the group interacts with internal and external forces.

▲ **FIGURE 1.42 Surveillance Cameras in New York City** In the post-9/11 era, there has been an increased investment in security measures in cities such as New York. These new surveillance systems are designed not only to capture particular illegal practices but to also deter such behaviors.

Generally, dominant forms of social organization have persisted for hundreds of years, if not longer. But both subtle and dramatic changes within these forms occur. Global media technologies, such as satellite television and the Internet, increasingly shape new potential social forms and reconfigure old ones. Many cultural groups are using their identities to assert political, economic, social, and cultural claims. These identities may be based on socioeconomic class, ethnicity, race, gender, and sexuality.

Ethnicity is a socially-created system of rules about who belongs to a particular cultural group. Ethnicity might be based on actual commonalities, such as language or religion, or on commonalties that are only perceived. For geographers, ethnicity can be used as a powerful marker of territory or space, in that people define and demarcate spaces through their practice of their ethnicity. And particular spaces, such as an ethnic neighborhood, can reinforce certain ethnic practices and identities. Imagine a neighborhood in Paris, France, that is largely occupied by people of North African descent (particularly, Algeria, Morocco, and Tunisia). You would likely find that people infuse this neighborhood with the markers of North African identity, such as food (**Figure 1.43**). In response, that space and the actions of people in it serve to promote and reinforce what it means to be North African, at least in the context of Paris, France. In other cases, ethnic identities become segregated from the wider society in ghettos or ethnic enclaves.

▲ **FIGURE 1.43 A Neighborhood for North Africans in Paris** The strong connection between France and its former colonies is represented in the modern-day migration of North and West African peoples to France. The result of such migrations has been the emergence of new neighborhoods and services dedicated to this North African population such as food shops.

Like ethnicity, **race** is a problematic and illusory classification of human beings based on skin color and other physical traits. Present-day, popular notions of race began to emerge in the 19th century during the era of colonialism. Europeans put forward "scientific" research supposedly demonstrating how people could be classified racially based on physical characteristics. Even though race is a completely social category, with no consistent or meaningful basis in biological reality, people often insist on the reality of race and racial differences. Social scientists use the concept of **racialization** to describe the practice of creating unequal castes where white skin color or some other dominant identity is considered the norm. Cultural perceptions and practices like this can create problematic and discriminatory racialized places—such as in the homelands of South Africa or in the ghettos of Europe and the Russian Federation.

Like ethnicity and race, groups form around the identity of **gender**. Gender should not be confused with sex, a biological category; gender is a category applied to people and linked to expectations about what different kinds of people can and ought to do. Gender varies across places and times. While *gender* is a category reflecting the social differences between men and women, it is not something people essentially are because of a given set of physical characteristics. People may be born as biological males (a physically given condition), but they learn how to act as men (a culturally imposed condition). Certain places are gendered through the inclusion and exclusion of certain people based on their performance of their gender. Consider, for example, the bars or barbershops in many regions frequented almost exclusively by men.

Gender interacts with other forms of identity and can intensify power differences among and between groups. For instance, women's subservience to men is deeply ingrained within South Asian cultures, and it is manifest most clearly in the cultural practices attached to family life, such as the custom of providing a dowry. The preference for male children in parts of East Asia is reflected in the widespread (but illegal) practice of selective abortion and female infanticide. Within marriages, some women are routinely neglected and maltreated across all world regions. Women in rich countries, on average, earn less money than men do in the same jobs. In many regions, women have organized for their rights and successfully secured improvements to their daily lives in terms of access to education, safe drinking water, land, and birth control.

The increasing attention being paid to human rights issues around the globe includes sexual rights as a political cause (**Figure 1.44**). **Sexuality** is a set of practices and identities related to sexual acts and desires. Sexuality is understood to be learned and expressed, or "performed," through practices that most people tend to view as "normal" or "natural." Where gay or straight identities can be performed plays a central role in who occupies those spaces, how they are occupied, and even what sexual identity a person might be performing when occupying one space or another. Issues of sexuality and space go beyond the performance of sexual identity to the emergence of new political cultures constructed to protect the rights of lesbian, gay, and transgendered people.

This survey of cultural concepts reminds us that culture is indeed regional, in the sense that similar languages or practices may prevail over large areas. At the same time, it is clear that within these regional cultures, many areas with different characteristics are also in evidence. This suggests that many of the distinctive characteristics of any region's culture are not timeless or unchanging. These characteristics may, for

▲ **FIGURE 1.44** *Kathoey* **appearing in a Transgendered Cabaret Show in Bangkok, Thailand** Sexuality remains a complex and contested issue across the globe, as does gender. In the context of Thailand, it is common to discuss not two but three genders—men, women, and *kathoey*. Bangkok is home to a robust cabaret industry of transgendered and transsexual performers.

example, be the result of migrations, innovations, or interactions. Cultures are ultimately the product of interactions between different traditions and places.

> **APPLY YOUR KNOWLEDGE** Keep a log of your activities for one full day, including where you go and what you do. Once complete, review the log and examine how your experience of particular places intersects with your own social identities (e.g., your gender, race, sexuality, ethnicity, religion, age, or class).

Demography and Urbanization

The overall distribution of people across the face of Earth is, in part, a product of the many other distributions we have reviewed already in this chapter, including environmental, historical, economic, and cultural patterns. And like these other patterns, the patterns of population distribution are highly uneven. As the world population density map demonstrates, some areas of the world are very heavily inhabited, while others are only sparsely populated (**Figure 1.45**).

Almost all the world's inhabitants live on just 10% of the world's land. Most live near the edges of land masses—near the oceans or seas or along rivers with easy access to a navigable waterway. Approximately 90% live north of the Equator, where the largest proportion of the total land area (63%) is located. Most of the world's

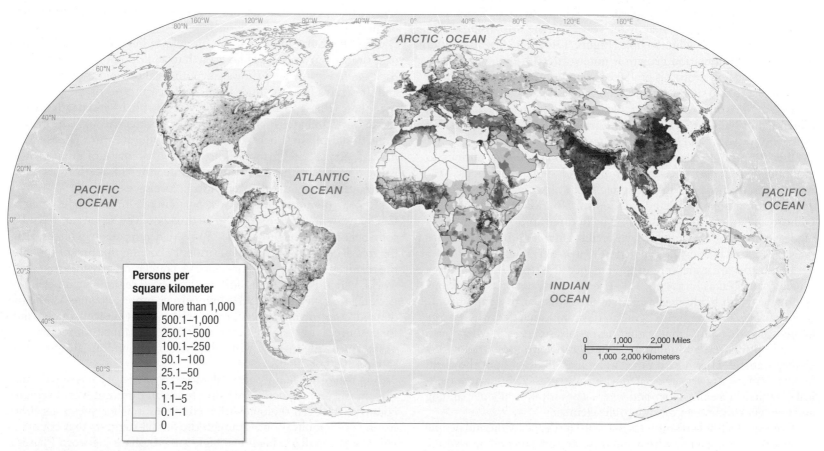

Persons per square kilometer

- More than 1,000
- 500.1–1,000
- 250.1–500
- 100.1–250
- 50.1–100
- 25.1–50
- 5.1–25
- 1.1–5
- 0.1–1
- 0

▲ **FIGURE 1.45 Population Distribution** The world's population is unevenly distributed across the globe.

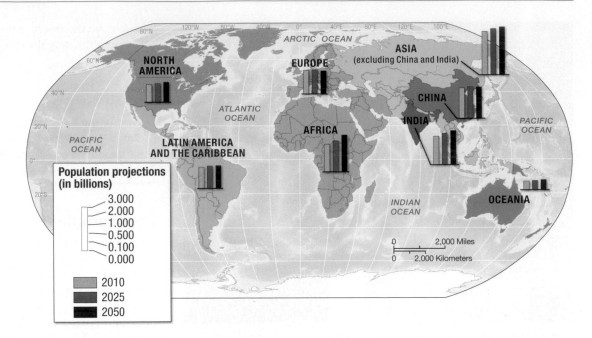

▶ **FIGURE 1.46 Population Geography of the Future** Population projections for 2025 and 2050 show very marked disparities among world regions: parts of the world will grow very little while others will continue to grow rapidly.

population lives in temperate, low-lying areas with fertile soils. Climate and Earth's topography explain a great deal of this distribution. Harsh, hot or cold places with a dearth of life and resources tend not to foster large populations.

Within this limited area, human population continues to grow. In 2011 the world's population passed the 7 billion mark. The population division of the United Nations Department of Social and Economic Affairs projects that the world's population will increase by 1.2% annually to midcentury. This means that by the year 2050, the world is projected to contain nearly 9.3 billion people (**Figure 1.46**). The distribution of this projected population growth is noteworthy. Over the next half-century, population growth is predicted to occur overwhelmingly in the regions least able to support it. Just six countries will account for half the increase in the world's population: Bangladesh, China, India, Indonesia, Nigeria, and Pakistan. Meanwhile, Europe and North America will experience very low and in some cases zero population growth.

It is increasingly clear that human population will not continue to grow indefinitely, even in countries with already-high populations. Analysts suggest that many of the economic, political, social, and technological transformations associated with industrialization and urbanization point to a demographic transition. The **demographic transition** is a model of population change, which predicts that high birth and death rates are replaced by low birth and death rates over time.

Birthrates are a measure of the number of births in a population, usually expressed as births per 1,000 people per year, or as a percentage figure. High birthrates (as high as 40 births per thousand people or more) are associated with agrarian (farming) societies, where large families are critical for labor, gathering, or subsistence, and where birth control options are rudimentary. **Death rates** are measured with a similar statistic. High death rates (as high as 40 deaths per thousand people or more) are also associated with agrarian societies, where medical care may be rudimentary and where work is physically demanding and involves continued exposure to the elements.

Conversely, low birthrates (as low as 10 births per thousand people or fewer) are associated with contemporary urban societies, where children are less crucial for labor, birth control is widely available, and women are significant participants in the paid workforce. Low

death rates are also common in contemporary urban settings, largely due to improved health care.

Where birthrates and death rates are roughly the same, no matter how high or low these might be, little or no growth in a population occurs. Most societies throughout history have not actually experienced high and sustained rates of population growth because their high birthrates were typically countered by comparable death rates. Only when death rates become lower than birthrates does growth occur.

The demographic transition model predicts patterns of change over time: from a period of no growth to a period of high growth and back again. **Figure 1.47** shows a hypothetical demographic transition and illustrates how the high birth- and death rates of the preindustrial phase (Phase 1) are replaced by the low birth- and death rates of the industrial phase (Phase 4) only after passing through the critical transitional phase (Phase 2) and more moderate rates (Phase 3) of natural increase and growth. The transitional phase of rapid growth is the direct result of early and steep declines in death rates while birthrates remain at high, preindustrial levels.

Recent demographic trends, particularly in countries such as Germany, where population growth is actually negative, have led some scholars to suggest a Phase 5, where deaths outstrip births, leading to a decrease in overall population. Following this line of thought, the relatively recent period of explosive population growth around the world, which is now being followed in some regions by a period of decline, is a reflection of radical modern economic change. While the demands of growing populations are daunting (in terms of demands for resources, for example), the demographic transition model suggests worldwide growth will "play itself out," ending altogether as soon as 2050 or 2060.

On the other hand, recent demographic history has shown that this transition may occur very differently in different world regions, countries, or even regions within countries, raising questions about its universal application. Although the model suggests that countries will pass through a transitional high-growth phase (between Phases 2 and 3), many observe that some regions become stalled there, in a "demographic trap."

▶ **FIGURE 1.47 Demographic Transition** This model of economic development is based on the idea of successive stages of growth and change. Each stage is seen as leading to the next, though different regions or countries may require different periods of time to make the transition from one stage to the next.

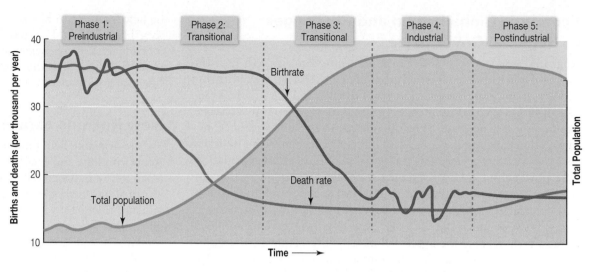

It should be noted that the demographic transition model is based solely on the experience of countries in the global north and is thought by many experts to be less useful in explaining the demographic trends in the global south, whose entire development experience is quite different. Most importantly, both population geographers and policymakers now also recognize that a close relationship exists between women's status and fertility. Women who have access to education and employment tend to have fewer children because they have less of a need for the economic security and social recognition that children are thought to provide. Success at *damping* population growth in less-developed countries appears to be tied to enhancing the possibility for a good quality of life and to empowering people, especially women, to make informed choices.

As should be evident now, none of these regional trends occurs in a vacuum, and what happens demographically in one region may have a large influence on what happens in another. For example, accelerated trade relations between Mexico and the United States in recent years had a significant impact on Mexican society. Increased demand for Mexican products led to a large increase in female factory workers, which resulted in later marriages, and separation between spouses, which unquestionably slowed birthrates in some parts of the Mexican population. The demographic patterns of the world are subject to the complex interplay of globalization and regionalization.

FUTURE GEOGRAPHIES

The world is in transition, and the distribution of people, money, resources, opportunities, and crises is necessarily changing as a result. Each of the chapters in this book contains a "Future Geographies" section that speculates about the way unfolding events and current trends are likely to shape the future of different regions. In addition, "Emerging Regions" sections introduce new regions as they emerge from amid these trends. Though the future is impossible to predict with any precision, we do have access to extensive data on human and physical resources, climate change, economic growth trends, technological innovation, and other key components that contribute to global geography. Extrapolating from what is known, it is possible to make predictions about what might emerge in the future. Each global trend portends very different implications for the future of specific regions.

Unprecedented Resource Demands and Emerging Resource Regions

The expansion of the global economy and the globalization of industry will boost the overall demand for raw materials of various kinds, and this will spur the development of some previously underexploited but resource-rich regions in Africa, Europe, and Asia. Raw materials, however, will only meet a fraction of future resource needs. The main issue, by far, will be energy resources. World energy consumption has been increasing steadily over the recent past (**Figure 1.48**). As the global south becomes industrialized and its population increases further, the demand for energy will expand rapidly. The International Energy Agency estimates that developing-country energy consumption will more than double by 2025, increasing total world energy demand by almost 50%.

However, concerns about global warming may cause countries to choose less carbon-intensive energy sources in the future, and the development of new materials may reduce the growth of demand for energy and resources. In addition, technological breakthroughs may improve energy efficiency or make renewable energy sources (such as wind, tidal, and solar power) commercially viable. Even so, the rising demand of emerging economic powers means the emergence of new resource regions, especially in Africa, each with its own geopolitical relationship to rising powers like China and India.

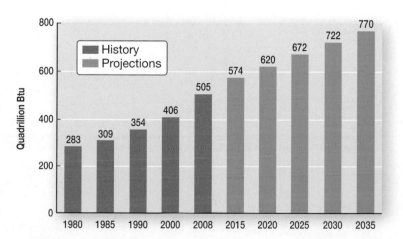

▲ **FIGURE 1.48 Trends in Energy Consumption** Global commercial energy consumption is expected to continue to rise steadily over the next two decades, as peripheral countries continue to develop economically and require a larger share of the world's energy resources.

Economic Globalization and Challenges to Regional Governance

Regional and supranational organizations such as the EU, the North American Free Trade Agreement (NAFTA), Organization of Petroleum Exporting Countries (OPEC), and the World Trade Organization (WTO) are increasingly molding the world into a seamless trading area unhindered by the rules that regulate national economies. The increasing importance of these organizations is an indicator that the world is experiencing a global geopolitical transformation. The powers and roles of modern states are changing, as they are forced to interact with these organizations. As a result, future politics will be directed to the international arena, rather than the national one. This is so much the case that even city governments and local interest groups are in the process of making international connections and conducting their activities beyond as well as within the boundaries of their own states. At the same time, however, there are likely to be new fractures in historic coalitions. The case of the EU is indicative. The euro tied EU states closely together for two decades. But as individual EU states began to face severe fiscal crises, it is increasingly clear that the region must either innovate a more unified, interstate, financial governance regime or risk disintegration.

New Regions of Insecurity and Conflict

We are arguably living in the most peaceful period in human history. Areas with long and even very recent histories of conflict and violence (such as the Balkans) have experienced a shift toward relative peace, and civilians everywhere are less likely to be the victims of military violence than in any other time. Nevertheless, in some regions, weak governments, lagging economies, religious extremism, and increasing numbers of young people will align to create a "perfect storm" for internal conflict, with far-reaching repercussions for security elsewhere (**Figure 1.49**).

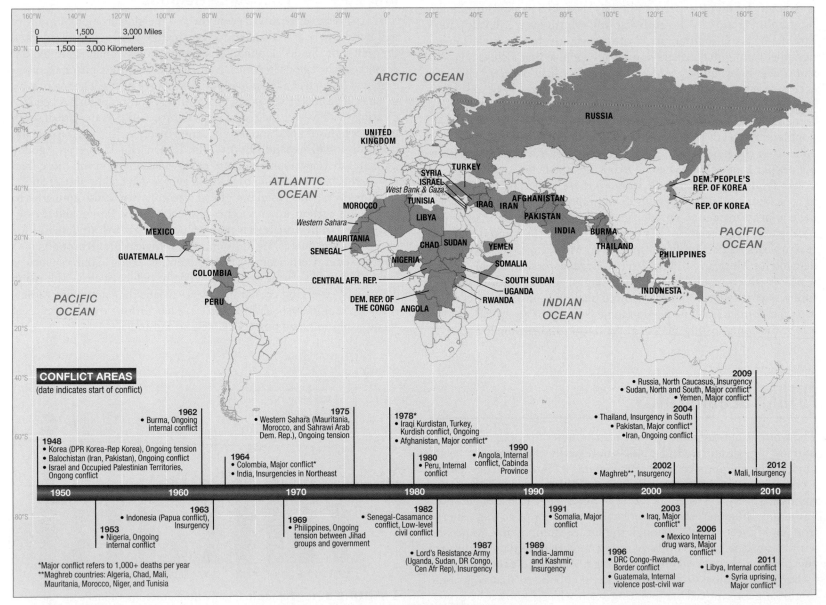

▲ **FIGURE 1.49 Ongoing Conflict across the Globe** A number of ongoing conflicts are both internal to particular countries and engage multiple players. Even if they only involve one country, these conflicts are always global, as international organizations and other countries are very often directly or indirectly involved.

The geographies of violence and stability are shifting. Information technology, allowing for instant connectivity, communication, and learning, has enabled terrorist threats to become increasingly decentralized, evolving into an eclectic array of groups, cells, and individuals that do not need a stationary headquarters to plan and carry out operations. Training materials, targeting guidance, weapons know-how, and fundraising have become virtual (existing online).

Transnational crime syndicates also reflect the globalizing influences on regional security. These organizations operate across national boundaries, distributing harmful materials, weapons, and drugs; exploiting local communities; disrupting fragile ecosystems; and controlling significant economic resources. The bulk of crime syndicate revenue comes from drug trafficking, but other significant sources include environmental products—everything from protected plants and animals to hazardous waste and banned chemicals. Trafficking in humans—for labor, sex work, and even organ harvesting for transplant purposes—is another aspect of transnational crime. The U.S. State Department estimates that at least 600,000 to 800,000 people are sold internationally each year.

Future Environmental Threats and Global Sustainability

The world currently faces a daunting list of environmental threats: the destruction of tropical rainforests and the consequent loss of biodiversity; widespread, health-threatening pollution; the degradation of soil, water, and marine resources essential to food production; and the effects of greenhouse gas emissions on global climate. Most of these threats are greatest in the world's poorest regions, where daily environmental pollution and degradation will continue to unfold in the coming years. Environmental problems are inseparable from processes of demographic change, economic development, and human welfare. In addition, it is becoming clear that environmental problems are increasingly enmeshed in matters of national security and regional conflict. The prospect of civil unrest and mass migrations resulting from the pressures of rapidly growing populations, deforestation, soil erosion, water depletion, air pollution, disease epidemics, and intractable poverty is real.

At the same time, the interconnectivity of global economies and politics offers avenues for environmental innovation. New international environmental agreements might foster the adoption of alternative sources of energy. Nations have begun to renegotiate their geopolitical commitments to one another with increased exchange of technology and ideas. Some countries are even considering large-scale engineering options to combat climate change, such as blocking incoming sunlight with chemicals or water vapor or trying to store carbon on a massive scale (**Figure 1.50**).

APPLY YOUR KNOWLEDGE Apply the concept of climate change to the question of future global sustainability. Research and explore one geoengineering option and list its pros and cons.

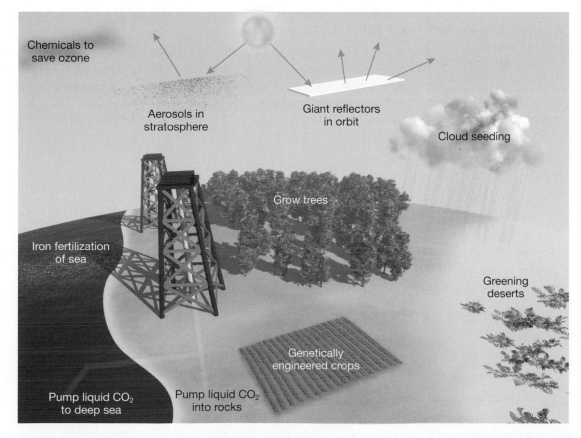

Chemicals to save ozone

Aerosols in stratosphere

Giant reflectors in orbit

Cloud seeding

Grow trees

Iron fertilization of sea

Greening deserts

Genetically engineered crops

Pump liquid CO$_2$ to deep sea

Pump liquid CO$_2$ into rocks

◄ **FIGURE 1.50 Geoengineering** A number of proposals have been made to counteract global warming. These include interventions that would block sunlight from reaching Earth's surface or remove greenhouse gases, such as CO$_2$, from the atmosphere. By injecting tiny aerosol particles into the highest levels of the atmosphere, sunlight could be reflected back into space. Seeding the oceans with iron might increase their absorption of of carbon dioxide.

LEARNING OUTCOMES *REVISITED*

■ **Compare and contrast globalization and regionalization and explain the relationship between the two.**

While globalization suggests that the world is becoming more interconnected, regionalization demonstrates that global relations are organized differently in different regions. Global connections that have increased the flow of people, ideas, animals, and plants over time influence regions differently depending on where they touch down.

■ **Identify the main climatic, geological, and ecological forces that shape world regions and how each force is changing in the current era.**

The world continues to change, even as humans seek to modify the environment around them. Climatic, geological, and ecological change shape how people live and manage their own long-term sustainability.

■ **Differentiate between forms of economic activity and explain why these forms vary around the globe.**

Over the last several hundred years, economic activities have diversified away from primary (extractive) and secondary (manufacturing) to include large-scale tertiary (service) and quaternary (information) economies. The distribution of these four sectors is tied to the global international division of labor, which is driven by the processes of capitalist development. In this context, some regions are more heavily invested in primary activities, while other regions are more heavily invested in service economies, for example.

■ **List contemporary economic development trends and identify the main measures used to assess social and economic advancement.**

Contemporary economic development trends indicate that the inequalities within and across world regions have become greater over the last several hundred years. Measures of human development demonstrate that these inequalities are driven by access to health care, education, and income.

■ **Explain the implications of globalization for modern states and identify the actors emerging in this process.**

Globalization has created new issues for modern states as they have had to manage the increasing flow of not only people but also products, ideas, plants, animals, and even diseases. Actors that have emerged to grapple with the increasing political and economic interconnectivies of the world include supranational organizations, social movements, and NGOS.

■ **Specify how the global distribution of languages and religions is changing.**

The maps of language and religion are, like all global and regional processes, dynamic and changing. In some contexts, long-standing languages in regions, such as local ethnic languages in sub-Saharan Africa or Native American languages in the United States and Canada, are nearing extinction, while other languages—ones that are tied to the global economy, such as English or Chinese—are spreading. Religious geographies continue to change as well, as religions such as Christianity and Islam have used trade, empire, and missions to spread their religious beliefs, and areas that were once homes to certain religions have seen their decline. A classic example of this latter trend is Buddhism, which now exists much more strongly in places outside its historic homeland in India.

■ **Explain how globalization and regionalization affect key elements of culture.**

Globalization has taken place not only on the political and economic scales but on the cultural and social scales as well. For some, this has produced the McDonaldization of the world's culture, with common features appearing globally, from jeans and T-shirts to music and movies. The globalization of culture flows in many directions, as popular cultural practices in East or Southeast Asia, for example, find their way into the cultural imagination of people living in Latin America, the United States and Canada, or Europe.

■ **Explain how and why regional population growth rates rise and fall.**

Population rates rise and fall as economic, political, social, and cultural conditions change in particular places. As life expectancies are extended, for example, and access to leisure time increases, birthrates tend to decline. Cultural practices, tied to the importance of family size, for example, also have an impact on population growth rates over time.

KEY TERMS

Anthropocene (the "human era") (p. 8)

atmospheric circulation (p. 8)

biodiversity (p. 19)

birthrates (p. 42)

capitalism (p. 24)

climate (p. 8)

climate system (p. 8)

colonialism (p. 23)

commodity chain (p. 21)

convergent plate boundaries (p. 16)

culture (p. 39)

death rates (p. 42)

deforestation (p. 19)

demographic transition (p. 42)

development (p. 30)

dialects (p. 37)

diaspora (p. 35)

divergent plate boundaries (p. 16)

ecosystem (p. 18)

emerging region (p. 7)

ethnicity (p. 40)

fundamentalism (p. 39)

gender (p. 40)

Gender-Related Development Index (**GDI**) (p. 30)

geography (p. 5)

geomorphology (p. 16)

globalization (p. 6)

global north (p. 29)

global south (p. 29)

global warming (p. 12)

greenhouse gases (p. 12)

gross domestic product (**GDP**) (p. 28)

gross national income (**GNI**) (p. 28)

Human Development Index (**HDI**) (p. 29)

imperialism (p. 24)

Industrial Revolution (p. 24)

International Monetary Fund (**IMF**) (p. 30)

intertropical convergence zone (**ITCZ**) (p. 9)

invasive species (p. 20)

language (p. 37)

lingua franca (p. 37)

nation (p. 31)

nationalism (p. 31)

nation–state (p. 31)

neoliberalism (p. 30)

Neolithic (p. 22)

nongovernmental organizations (**NGOs**) (p. 33)

orographic precipitation (p. 10)

physical geography (p. 7)

physiography (p. 7)

plantation economies (p. 23)

plate tectonics (p. 13)

primary sector (p. 27)

purchasing power parity (**PPP**) **per capita** (p. 29)

quaternary sector (p. 27)

race (p. 40)

racialization (p. 40)

region (p. 5)

regionalization (p. 6)

religion (p. 35)

secondary sector (p. 27)

sexuality (p. 40)

social movements (p. 33)

sovereign state (p. 31)

subduction zone (p. 16)

supranational organizations (p. 32)

sustainability (p. 21)

system (p. 6)

tertiary sector (p. 27)

transform plate boundaries (p. 16)

United Nations (**UN**) (p. 33)

weather (p. 8)

THINKING GEOGRAPHICALLY

1. What is geography, and what can studying it provide beyond a description of the world?

2. What is global climate change, and what is the evidence that it may be happening? What might its impacts include?

3. How are the world regions we recognize today different from those prior to 1500 C.E.?

4. How do economic activities differ from region to region? What factors accounts for these differences?

5. What worldwide demographic trends may emerge over the next 25 years?

6. Globalization has a complex impact on cultural identity. Describe how groups seek to protect themselves from it while others embrace it.

7. What are supranational organizations and how do they function with respect to formal states?

MasteringGeography™

Looking for additional review and test prep materials? Visit the Study Area in MasteringGeography™ to enhance your geographic literacy, spatial reasoning skills, and understanding of this chapter's content by accessing a variety of resources, including **MapMaster** interactive maps, videos, RSS feeds, flashcards, web links, self-study quizzes, and an eText version of *World Regions in Global Context*.

Workers inspect construction on the Gotthard Base Tunnel through the Alps; the tunnel will shorten travel times and reduce carbon emissions for rail traffic in Europe.

ALMOST 2 KILOMETERS (1 MILE) UNDER THE ALPS IN SWITZERLAND, more than 2,500 workers are completing the excavation of 23 metric million tons of rock. Their work is expected to be completed in 2017, when a new 57-km (35-mile) high-speed rail tunnel, the Gotthard Base Tunnel, will open for traffic. The Gotthard Base Tunnel is actually a pair of tunnels, each 10 meters (32 feet) in diameter, one for northbound trains and the other for southbound trains.

An engineering feat of the highest order, the total cost of the project will be in excess of U.S. $12 billion. Once the technical infrastructure inside the tunnels has been completed, the project will link the high-speed rail network of northwestern Europe with that of Italy, with the potential to connect to planned high-speed networks in southeastern Europe. It will relieve the acute congestion and traffic gridlock of trans-Alpine road routes and slash rail journey times through the Alps by over an hour. The new tunnel will also accommodate longer freight trains, allowing annual freight-transportation capacity to more than double. One benefit of the tunnel will be that high-speed rail will increase

LEARNING OUTCOMES

- Understand how Europe's landscapes developed around regions of distinctive geology, relief, landforms, soils, and vegetation.

- Explain the rise of Europe as a major world region and understand how Europeans established the basis of a worldwide economy.

- Describe the reintegration of Eastern Europe.

- Summarize the importance of the European Union and its policies.

- Identify the principal core regions within Europe and describe the pattern of urban development.

- Assess the importance of language, religion, and ethnicity in shaping ongoing change within Europe.

the volume of trade and travel without increasing carbon emissions. But of broader significance still is the impact of increased trade and travel in integrating the economies of northern and southern Europe.

The Alps have always been the most significant internal barrier in Europe. Hannibal, a Carthaginian military commander, famously required elephants to help move his armies across the Alps to invade Roman Italy in 217 B.C.E. For centuries after that, this great mountain range effectively separated (and protected) a succession of prosperous Mediterranean empires from less-developed territories to the north. After the Industrial Revolution of the 19th century, the situation was reversed, and the logistical problems of crossing the Alps effectively turned much of southern Europe into an economic periphery. The Gotthard Base Tunnel will link major economic centers on both sides of the Alps and be pivotal in the high-speed train network that has become an important element of Europe's modern infrastructure (see p. 76). ■

▼ FIGURE 2.1

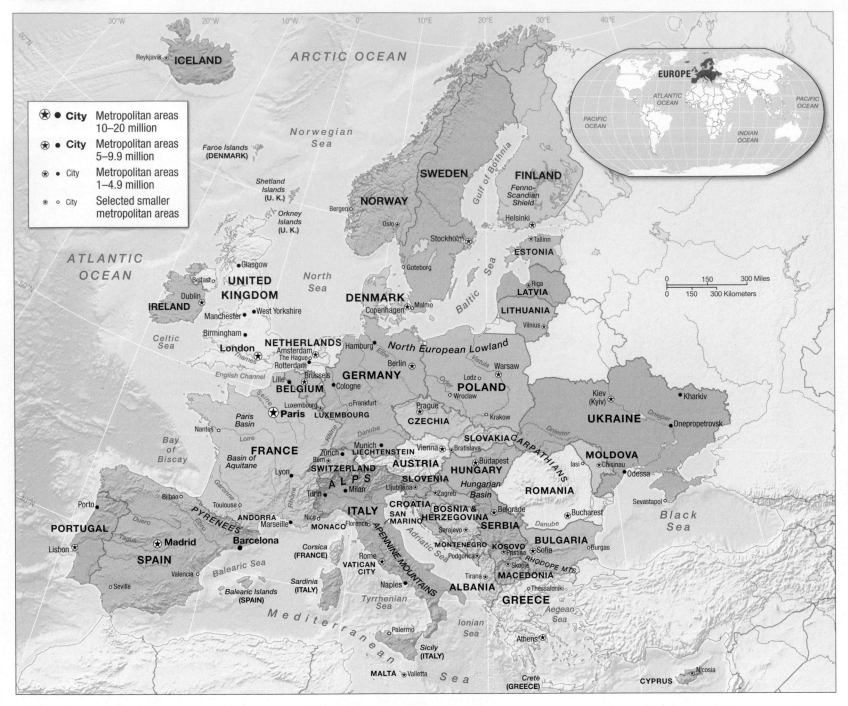

Europe Key Facts

- Major Subregions: Eastern Europe, northern Europe, southern Europe, Western Europe

- Major Physiogeographic Features: Alps, Pyrenees, and Carpathian mountains; longest river in Europe is Danube at 2.872 km; largest island in the region is Great Britain at 220,000 square km. Temperate climate due to influence of Gulf Stream.

- Major Religions: Roman Catholicism, Orthodox Christianity, Protestantism, Islam. Atheism and secularism are also strongly represented in Europe.

- Major Language Families: Most languages in Europe stem from the Indo-European family, which includes Slavic, Germanic, and Romance branches. Other families include Finno-Urgic and Turkic.

- Total Area (total sq km): 10 million

- Population (2011): 588 million; 82 million in Germany, 64 million in France, 62 million in the United Kingdom, and 60 million in Italy

- Population Density (per sq km): 107

- Urbanization (%) (2011): 64

- Average Life Expectancy at Birth (2011): Overall: 75; Women: 79; Men: 72

- Total Fertility Rate (2011): 1.4

- GNI PPP per Capita (current U.S. $) (2009): 20,897; highest = Luxembourg, 77,227; lowest = Moldova, 3,108

- Population in Poverty (%, < $2/day) (2000–2009): 3

- Internet Users (2011): 399,814,875; Population with Access to the Internet (%) (2011): 66; Growth of Internet Use (%) (2000–2011): 306

- Access to Improved Drinking Water Sources (%) (2011): Urban: 91; Rural: 90

- Energy Use (kg of oil equivalent per capita) (2009): 3,482

- Ecological Footprint (hectares per capita consumed)/(hectares per capita available, global scale) (2011): 4.9/1.8

ENVIRONMENT AND SOCIETY

Two aspects of Europe's physical geography have been fundamental to its evolution as a world region and have influenced the evolution of regional geographies within Europe itself. First, as a world region (**Figure 2.1**), Europe is situated between the Americas, Africa, and the Middle East. Second, as a satellite photograph of Europe reveals, the region consists mainly of a collection of peninsulas and islands at the western extremity of the great Eurasian landmass (**Figure 2.2**).

The largest of the European peninsulas is the Scandinavian Peninsula, the prominent western mountains of which separate Atlantic-oriented Norway from continental-oriented Sweden. Equally striking are the Iberian Peninsula, a square mass that projects into the Atlantic, and the boot-shaped Italian Peninsula. In the southeast is the broad triangle of the Balkan Peninsula, which projects into the Mediterranean, terminating in the intricate coastlines of the Greek peninsulas and islands. In the northwest are Europe's two largest islands, Britain and Ireland.

The overall effect is that tongues of shallow seas penetrate deep into the European landmass. This was especially important in the pre-Modern period, when the only means of transporting goods were sailing vessels and wagons. The Mediterranean and North seas, in particular, provided relatively sheltered sea lanes, fostering seafaring traditions in the people all around their coasts. The penetration of the seas deep into the European landmass provided numerous short land routes across the major peninsulas, making it easier for trade and communications to take place in the days of sail and wagon. As we shall see, Europeans' relationship to the surrounding seas has been a crucial factor in the evolution of European—and, indeed, world—geography.

Europe's navigable rivers also shaped the human geography of the region. Although small by comparison with major rivers in other regions of the world, some of the principal rivers of Europe—the Danube, the Dneiper, the Elbe, the Rhine, the Seine, and the Thames—played key roles. In addition, the low-lying **watersheds** (drainage areas) between the major rivers of Europe's plains allowed canal building to take place relatively easily, thereby increasing the mobility of river traffic.

Whereas the western, northern, and southern limits of Europe as a whole are clearly defined by surrounding seas—the Atlantic Ocean to the west, the Arctic Ocean to the north, and the Mediterranean Sea to the south—the eastern edge of Europe merges into the vastness of Asia and is less easily defined. Geographers sometimes use the mountain ranges of the Urals to mark the boundary between Europe and Asia, but the most significant factors separating Europe from Asia are human and relate to race, language, and a common set of ethical values that stem from Roman Catholic, Protestant, and Orthodox forms of Christianity. As a result, the eastern boundary of Europe is often demarcated through political and administrative boundaries, rather than physical features.

▲ **FIGURE 2.2 Europe from Space** This composite of satellite images clearly illustrates one of the key features of Europe: the many arms of the seas that penetrate deep into the western extremity of the great Eurasian landmass.

Climate, Adaptation, and Global Change

The seas that surround Europe strongly influence the region's climate (**Figure 2.3**). In winter, seas cool more slowly than the land and, in summer, they warm up more slowly than the land. As a result, the seas provide a warming effect in winter and a cooling effect in summer. Europe's arrangement of islands and peninsulas contributes to an overall climate that does not have great seasonal extremes of heat and cold. The moderating effect is intensified by the North Atlantic Drift, which carries great quantities of warm water from the tropical Gulf Stream as far as the United Kingdom.

Given its latitude (Paris, at almost 49°N, is the same latitude as Winnipeg and Newfoundland in Canada), most of Europe is remarkably warm. It is continually crossed by moist, warm air masses that drift in from the Atlantic. The effects of these warm, wet, westerly winds are most pronounced in northwestern Europe, where squalls and showers accompany the passage of successive eastward-moving weather systems. Weather in northwestern Europe tends to be unpredictable, partly because of the swirling movement of air masses as they pass over the Atlantic and partly because of the complex effects of the widely varying temperatures of interpenetrating bodies of land and water. Farther east, in continental Europe, seasonal weather tends to be more settled, with more pronounced extremes of summer heat and winter cold. In these interior regions, local variations in weather are influenced a great deal by the direction in which a particular slope or land surface faces and its elevation above sea level.

The Mediterranean Basin has a different and quite distinctive climate. Winters are cool, with an Atlantic airstream that brings overcast skies and intermittent rain—though snow is unusual. Low-pressure systems along the northern Mediterranean draw in rain-bearing weather fronts from the Atlantic. When low pressure over the northern Mediterranean coincides with high pressure over continental Europe, southerly airflows spill over mountain ranges and down valleys, bringing cold blasts of air. These events have local names: the mistral, for example, which blows down the Rhône valley in southern France, and the bora, which blows over the eastern Alps toward the Adriatic region of Italy. In spring, the temperature rises rapidly, and rainfall is more abundant. Then summer bursts forth suddenly as dry, hot, Saharan air brings three months of hot, sunny weather. There is

▶ **FIGURE 2.3 Climates of Europe**
Northern Europe is dominated by midlatitude climates. Westerly winds and storm tracks bring Marine coastal wetter conditions to most of northwest Europe for most of the year. Eastern Europe, farther from the moderating effects of the oceans, has colder and drier conditions associated with the Continental midlatitude climate type. Southern Europe, along the shores of the Mediterranean, has drier conditions and winter rains associated with a Mediterranean climate type. Regions farther from the sea in Spain and north of the Black Sea have dry conditions for most of the year and are classified as semiarid climates. Parts of Italy and the Balkans are wetter than the Mediterranean climate but warmer than climates to the north and are classified as humid subtropical climates.

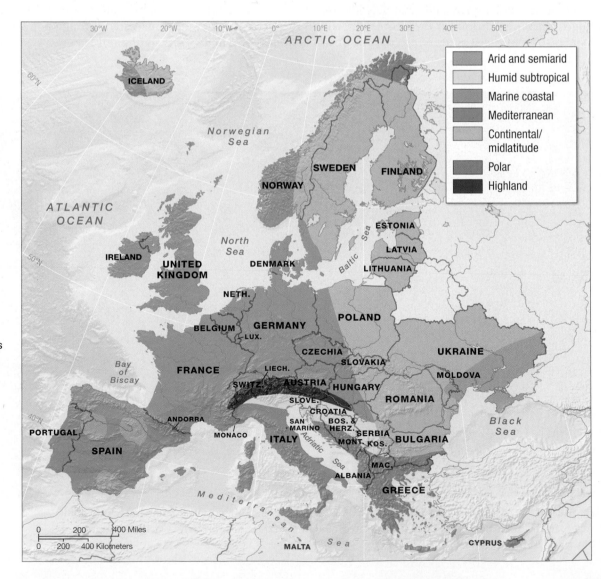

no rain save an occasional storm; the soil cracks and splits and is easily washed away in the occasional downpours. In October, the temperature drops, and deluges of rain show that Atlantic air prevails once more.

In such conditions, delicate plants cannot survive. The Mediterranean climate precludes all plant species that cannot tolerate the range of conditions—cold as well as heat and drought as well as wet. The result is a distinctive natural landscape of dry terrain dotted with cypress trees, holm oaks, cork oaks, parasol pines, and eucalyptus trees. These same conditions make agriculture a challenge. The crops that prosper best include olives, figs, almonds, vines, oranges, lemons, wheat, and barley. Sheep and goats graze on dry pastureland and stubble fields. Irrigation is often necessary, and in some localities it sustains high yields of fruit, vegetables, and rice.

Europe has warmed significantly more than the global average: the average temperature for the European land area for 2001–2010 was 1.2°C above the 1850–1899 average; most of those years were among the warmest since 1850. High-temperature extremes like hot days, tropical nights, and heat waves have become more frequent, while cold spells and frost days have become less frequent.

Highly urbanized and industrialized, Europe is a major contributor to the carbon dioxide emissions that drive global climate change. Under the **United Nations Kyoto Protocol**, a legally binding global agreement to reduce greenhouse gas emissions, industrialized nations are obliged to reduce the amount of greenhouse gases being released into the atmosphere. The European Union (27 countries that account for most of the region; see Figure 2.16, p. 69) has attempted to reduce emissions of greenhouse gases by establishing an **Emissions Trading Scheme (ETS)**, that allows energy-intensive facilities (e.g., power generation plants and iron and steel, glass, and cement factories) to buy and sell permits that allow them to emit carbon dioxide (CO_2) into the atmosphere. Companies that exceed their individual limit are able to buy unused permits from firms that have successfully reduced their emissions. Those that exceed their limit and are unable to buy spare permits are fined. The system has had mixed success in reducing the total amount of CO_2 released into the atmosphere but can probably be counted as successful in limiting the rate of growth in CO_2 emissions.

APPLY YOUR KNOWLEDGE Find a European city at the same latitude as a city that is familiar to you. Use the Internet to find climate data (monthly temperatures and rainfall statistics, for example). Describe and provide a reason for the differences between the two sets of data. (Hint: Good sources are http://www.worldclimate.com/ and http://www.bestplaces.net/climate/.)

Geological Resources, Risk, and Water Management

The physical environments of Europe are complex and varied. It is impossible to travel far without encountering significant changes in physical landscapes. There is, however, a broad pattern to this variability, and it is based on four principal **physiographic regions** that are characterized by broad coherence of geology, relief, landforms, soils, and vegetation. These regions are the Northwestern Uplands, the Alpine System, the Central Plateaus, and the North European Lowlands (**Figure 2.4**).

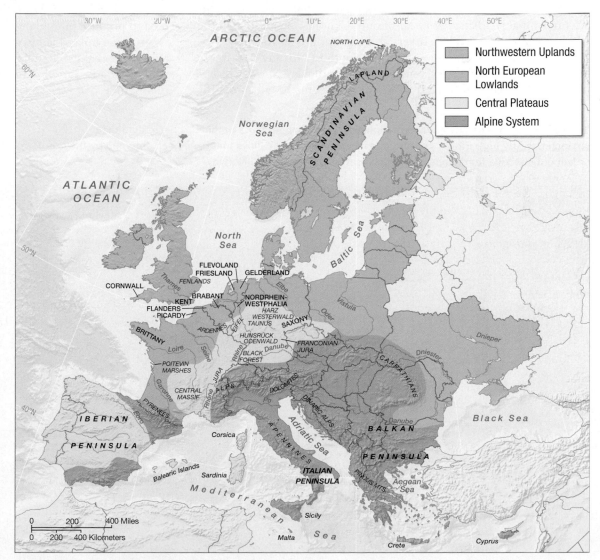

▲ **FIGURE 2.4 Europe's Physiographic Regions** Each of the four principal physiographic regions of Europe has a broad coherence in terms of geology, relief, landforms, soils, and vegetation.

Northwestern Uplands The Northwestern Uplands are composed of the most ancient rocks in Europe, the product of the Caledonian mountain-building episode about 400 million years ago. Included in this region are the mountains of Norway and Scotland and the **uplands** (high, hilly land) of Iceland, Ireland, Wales, Cornwall (in England), and Brittany (in France). The original Caledonian mountain system was eroded and uplifted several times, and following the most recent uplift, the Northwestern Uplands have been worn down again, molded by ice sheets and glaciers. Many valleys were deepened and straightened by ice, leaving spectacular glaciated landscapes. There are **cirques** (deep, bowl-shaped basins on mountainsides, shaped by ice action), glaciated valleys, and **fjords** (some as deep as 1,200 meters—about 3,900 feet) in Norway; countless lakes; lines of **moraines** that mark the ice sheet's final recession; extensive deposits of sand and gravel from ancient glacial deltas; and peat bogs that lie on the granite shield that forms the physiographic foundation of the region. Since the last glaciation, sea levels have risen, forming fjords and chains of offshore islands wherever valleys have been flooded by the rising sea (**Figure 2.5**).

This formidable environment is rendered even more forbidding in much of the region as a result of climatic conditions. In the far north, the summer sun shines for 57 days without setting, but the winter nights are interminable. Oslo sees a *total* of only 17 hours of sunshine in the whole month of December. Rivers and lakes in the far north are frozen from mid-October, and even farther south in Norway they freeze at the end of November and remain frozen until May. Snow, which is permanent in parts of Iceland and Lapland, begins to fall across the rest of the Northwestern Uplands toward the middle of September and covers much of the landscape from October through early April.

Much of the Northwestern Uplands is covered by forests. In the southern parts of the region, conifers (mostly evergreen trees such as pine, spruce, and fir) are mixed with birches and other deciduous trees; further north, conifers become entirely dominant, and in the far north, toward the North Cape in Lapland, the forest gives way to desolate treeless stretches of the **tundra** with its characteristic gray lichens and dwarf willows and birches. Not surprisingly, forestry is a major industry in much of the north. The forests supply timber for domestic building and fuel and produce the woods that are in greatest demand on world markets: pine and spruce for timber and the paper industry, birch for plywood and cabinetmaking, and aspen for matches.

The farmers in these mountain subregions depend on **pastoralism**, a system of farming and way of life based on keeping herds of grazing animals, supplementing with a little produce grown on the valley floors. In the less mountainous parts of Scandinavia, as in much of Baltic Europe, with their short growing season and cold, acid soils, agriculture supports only a low density of settlement. Landscapes reflect a mixed farming system of oats, rye, potatoes, and flax, with hay for cattle. Oats, the largest single crop, often has to be harvested while it is still green. More than half of the milk from dairy cattle is used to produce butter and cheese.

In these upland landscapes, farms and hamlets are casually situated on any habitable site, their buildings often widely scattered. The countryside is dotted with trim wooden houses roofed with slate, tiles, shingles, or even sods of turf. Often, the buildings of a particular

▼ **FIGURE 2.5 The Northwestern Uplands** The northern parts of the European region are characterized by high mountains and deeply eroded glacial valleys that have been drowned by the sea, creating distinctive fjord landscapes such as this one in western Norway.

district are distinguished by some special stylistic feature. In some places, the old weatherboard houses that were used as shelters in bad weather still exist. They date back to the Middle Ages, as does the custom of storing reserve stocks of food or hay in a special isolated building, the *stabbur*, decorated with beautifully carved woodwork. The predominant color of the buildings is gray, the natural color that wood acquires with age. Nearer to towns, the houses are painted brighter colors: yellow, dark red, and mid-blue.

The more temperate subregions of the Northwestern Uplands (in Ireland and the United Kingdom) are dominated by dairy farming on meadowland, sheep farming on exposed uplands, and arable farming (mainly wheat, oats, potatoes, and barley) on drier lowland areas. In these areas, dispersed settlement, in the form of hamlets and scattered farms, is characteristic, and stone is more often the traditional building material.

Alpine Europe The Alpine System occupies a vast area of Europe, stretching eastward for nearly 1,290 kilometers (about 800 miles) across the southern part of Europe from the Pyrenees, which mark the border between Spain and France, through the Alps and the Dolomites and on to the Carpathians, the Dinaric Alps, and some ranges in the Balkan Peninsula. The Apennines of Italy and the Pindus Mountains of Greece are also part of the Alpine System. The Alpine System is the product of the most recent of Europe's mountain-building episodes, which occurred about 50 million years ago. Its relative youth explains the sharpness of the mountains and the boldness of their peaks. The Alpine landscape is characterized by jagged mountains with high, pyramidal peaks and deeply glaciated valleys (**Figure 2.6**). The highest peak, Mont Blanc, reaches 4,810 meters (15,781 feet). Most of the rest of the mountains are between 2,500 and 3,600 meters (about 8,200 and 11,800 feet) in height. Seven of the peaks in the western Alps exceed 4,000 meters (13,123 feet) in height. The Alps pose a formidable barrier between northwestern Europe and Italy and the Adriatic, with only a few passes—including the Brenner Pass, the Simplon Pass, the Saint Gotthard Pass, and the Great Saint Bernard Pass—allowing transalpine routes. Modern transportation has required expensive tunnels, like the Gotthard Base Tunnel described at the beginning of this chapter, to be driven through the mountain range.

The dominant direction of the Alps and their parallel valleys is roughly southwest to northeast. The major Alpine valleys thus have one sunny, fully exposed slope that is suitable for vine growing and a shaded side rich with orchards, woods, and meadows. The mountains and valleys of the Alps proper are surrounded by glacial outwash deposits that provide rich farmland. The limestone of the Alpine region is widely quarried for cement, whereas mineral deposits—lead, copper, and iron—and small deposits of coal and salt have long been locally important throughout the region. In addition, the Alps are a valuable source of hydroelectric power, providing about 65 billion watt-hours in Switzerland (60% of the country's electricity consumption), about 72 billion watt-hours in France (15% of the country's electricity consumption), and about 45 billion watt-hours in Austria (85% of the country's electricity consumption).

The traditional staple of the economy, however, has been agriculture, and farming has given the Alpine region its distinctive human landscape: a patchwork of fields, orchards, vineyards, deciduous woodlands, pine groves, and meadows on the lower slopes of the valleys, with broad Alpine pasture above. In these pastures, which are dotted with wooden haylofts and summer chalets, dairy cattle wander far and wide. Farmers attach bells around the necks of their animals to be able to locate them, and the consequent effect is a resonant pastoral "soundscape" of clanking cowbells. Farms and hamlets tend to cling to lower elevations, the chalet-style architecture drawing on timber or rough-cast stone construction, with overhanging eaves, tiers of windows, and painted ornamentation.

The landscapes of the Alpine fringes are lusher. Lavender and fruit have been introduced to enrich and give variety to the mixed farming system there, which features vine and wheat growing, along with dairy cattle. The higher slopes, which receive more rainfall, provide excellent pastures that have made the region famous for its rich cheeses, such as Gruyère.

▲ **FIGURE 2.6 The Alpine System** Jagged peaks and glaciated valleys are typical of the Alpine ranges. Shown here are the Dolomite ranges near Sankta Magdalena in Italy.

The principal industry of the Alpine region is tourism. Attractive rural landscapes, together with magnificent mountain scenery, beautiful lakes, and first-class winter sports facilities, have attracted tourists to this region since the 1800s. Lakeside resorts such as Lucerne and Lugano, Switzerland; mountain resorts such as Chamonix, France, and Innsbruck, Austria; and winter sports resorts such as Val d'Isere, France, and Davos, St. Moritz, and Zermatt, Switzerland, are all well established, with an affluent clientele from across Europe. With the growth of the global tourist industry, Alpine resorts have attracted a new clientele from North America and Japan.

Central Plateaus Between the Alpine System and the Northwestern Uplands are the landscapes of the Central Plateaus and the North European Lowlands. The Central Plateaus are formed from 250- to 300-million-year-old rocks that have been eroded down to broad tracts of uplands. Beneath the forest-clad slopes and fertile valleys of these plateaus lie many of Europe's major coalfields. For the most part, the plateaus reach between 500 and 800 meters (1,640 and 2,625 feet) in height, though they rise to more than 1,800 meters (5,905 feet) in the Central Massif of France (**Figure 2.7**). *Massif* refers to a compact chain of mountains that is separate from other ranges. The Central Plateau landscape is characterized by rolling hills, steep slopes and dipping vales, and deeply carved river valleys.

In central Spain, the landscape is dominated by plateaus and high plains that go on for hundreds of miles, with long narrow mountain ranges—the *cordillera*—stretched out like long cords along the edges. The flat landscape of the region is a result of immense horizontal sheets of sedimentary rock that cover an extensive block, or massif, of ancient rock, with a general appearance of tables ending in ledges—hence the term *meseta* (mesa means table in Spanish), which is generally used to designate the center of Spain. These dry, open lands are covered in grain crops and flocks and herds of sheep and cattle that in summer are driven up to the cooler mountains. Olive trees dominate the shallow valley slopes, and in irrigated valley bottoms vines and market gardens flourish.

Farther east—in the Massif Central in France and the Eifel, Westerwald, Taunus, Hunsruck, Odenwald, and Franconian Jura in Germany—where the climate is wetter, the landscape is dominated by gently rounded, well-wooded hills, with villages surrounded by neat fields and orchards in the vales. These hills rarely rise above 700 meters (2,300 feet), and the landscape includes many attractive and fertile subregions of low hills planted with vines and shallow valleys with orchards and meticulously maintained farms. The hills are covered with beech and oak forests, interspersed with growths of fir. The Black Forest in southwestern Germany is much higher, its bare granite summits reaching 1,493 meters (about 4,900 feet) in the south. It is scored by steep, narrow valleys with terraces of glacial outwash. The northeastern reaches of the Black Forest form an immense, silent, solid mass of fir forests. To the southwest, near the Rhine River, are small fields and meadows with prosperous farms at the edge of white fir forests and market towns nestling in tributary valleys. The Rhine has cut deep, scenic gorges through the higher plateau lands, but for much of its course across the central plateaus it is majestic and calm, with gentler slopes covered with vines and the bottomlands of the valley a busy corridor of prosperous towns surrounded by industrial crops and market gardens.

North European Lowlands The North European Lowlands sweep in a broad crescent from southern France, through Belgium, the Netherlands, and southeastern England and into northern Germany, Denmark, and the southern tip of Sweden. Continuing eastward, they broaden into the immense European plain that extends through Poland, Czechia, Slovakia, and Hungary, all the way into Russia. Coal is found in quantity under the lowlands of England, France, Germany, and Poland and in smaller deposits in Belgium and the Netherlands. Oil and natural gas deposits are found beneath the North Sea and under the lowlands of southern England, the Netherlands, and northern Germany. Nearly all of this area lies below 200 meters (656 feet) in elevation, and the topography everywhere is flat or gently undulating. As a result, the region has been particularly attractive to farming and settlement. The fertility of the soil varies, however, so that settlement patterns are uneven, and agriculture is finely tuned to the limits and opportunities of local soils, landscape, and climate.

The western parts of the North European Lowlands are densely settled and intensively farmed, with the moist Atlantic climate supporting lush agricultural landscapes (**Figure 2.8**). Surrounding the highly urbanized regions of the North European Lowlands are the fruit orchards and hop-growing fields of Kent, in southeastern England; the bulb fields of the Netherlands; the dikes and rectangular fields of the reclaimed marshland of North Holland, Flevoland, and Friesland along the Dutch coastal plain; the pastures, woodlands, and forests of the upland plateaus of the Ardennes, the Eifel, the Westerwald, and the Harz; and the meadows and cultivated fields separated by hedgerows and patches of woodland that characterize most of the remaining countryside.

Farther east, the Lowlands are characterized by a drier continental climate and lowland river

▲ **FIGURE 2.7 The Central Plateaus** This photograph shows the Vallee du Falgoux in the Massif Central region of southern France.

▲ **FIGURE 2.8 The North European Lowlands** Shown here is the gently rolling coutryside of Hampshire, England.

basins with a rolling cover of sandy river deposits and **loess** (a fine-grained, extremely fertile soil). The hills are covered with oak and beech forests, but much of the region consists of broad loess plateaus where very irregular rainfall averages about 16 inches—approximately 40 centimeters—a year. There are no woods, and irrigation is often necessary to sustain the typical two-year rotation of corn and wheat. In some parts, several meters of loess and rich, sandy soil rest on the rocky substratum. Stone and trees are so scarce that houses are built of *pisé*, a kind of rammed-earth brick. The scarcity of trees forces storks to build nests atop chimneys and telegraph poles. The rich soils produce high yields of wheat and corn, together with hops, sugar beets, and forage crops for livestock.

The area also features residual regions of **steppe**, semiarid, tree-less, grassland plains with landscapes that are infinitely monotonous. This land, which is too dry or too marshy to have invited cultivation, was once the domain of wild horses, cattle, and pigs, but today huge flocks of sheep find pasture there. The climate, though, is harsh, with seasonal extremes of burning hot and freezing cold, with the winter easterlies blowing down from midcontinent Russia. Population densities are low, and there are few villages.

The mosaic of regions and landscapes within Europe is both rich and detailed. Within these broad physiographic divisions, marked variations exist. Physical differences are encountered over quite short distances, and numerous specialized farming regions exist, where agricultural conditions have influenced local ways of life to produce distinctive landscapes. In detail, these landscapes are a product of centuries of human adaptation to climate, soils, altitude, and **aspect** (exposure) and to changing economic and political circumstances. Farming practices, field patterns, settlement types, traditional building styles, and ways of life have all become attuned to the opportunities and constraints of regional physical environments, resulting in distinctive regional landscapes.

Ecology, Land, and Environmental Management

Temperate forests originally covered about 95% of Europe, with a natural ecosystem dominated by oak, together with elm, beech, and linden (lime). By the end of the medieval period, Europe's forest cover had been reduced to about 20% of what it was at its peak, and today it is around 5%. Between 1000 C.E. and 1300 C.E., a period of warmer climate, together with advances in agricultural knowledge and practices, led to a significant transformation of the European landscape. The population more than doubled, from around 36 million to more than 80 million, and a vast amount of land was cultivated for the first time. By about 1200, most of the best soils of Western Europe had been cleared of forest, and new settlements were increasingly forced into the more marginal areas of heavy clays or thin sandy soils on higher ground and **heathlands**. (Heath is open land with coarse soil and poor drainage.) Many parts of Europe undertook large-scale drainage projects to reclaim marshlands.

The Romans had already demonstrated the effectiveness of drainage schemes by reclaiming parts of Italy and north-western Europe. Under Roman colonization, land was often subdivided into a checkerboard pattern of rectilinear fields. This highly ordered system was known as *centuriation*, and the pattern can still be seen in some districts today—in parts of the Po Valley, for example. Elsewhere, across large tracts of the Mediterranean, the soil was suitable for cultivation only on a large scale, and its poor quality necessitated that vast areas were left as un-tilled pasture lands. In these areas, successive conquerors, from the Greeks, Phoenicians, and Carthaginians to the Ottoman Turks and Christian Crusaders, carved out huge estates, known as **latifundia,** on which they set peasants to work. Land that did not belong to these big estates was often subdivided by independent peasant farmers into very small, intensively cultivated lots, or **minifundia,** most of which were barely able to support a family.

Another important legacy of the Romans is the doctrine of public trust, which asserts public rights in navigable waters, fisheries, and tidelands. It is reflected today in the European Union's approach to water management, which features the most progressive overall system of water management, organized by river basin instead of according to administrative or political boundaries. However, since the 1980s the rising dominance of neoliberal policies has led to the **privatization** (the transfer of ownership of businesses from government agencies to privately owned entities) of water services and certain aspects of water management, especially in France and the United Kingdom.

In the 12th and 13th centuries, extensive drainage and resettlement schemes were developed in Italy's Po Valley, in the Poitevin marshes of France, and in the Fenlands of eastern England. In Eastern Europe, forest clearances were organized by agents acting for various princes and bishops, who controlled extensive tracts of land. The agents would also arrange financing for settlers and develop villages and towns, often to standardized designs.

This great medieval colonization came to a halt nearly every-where in about 1300. One factor was the so-called **Little Ice Age**, a period of cooler climate that significantly reduced the growing season—perhaps by as much as five weeks. Another factor was the catastrophic loss of population during the period of the Black Death

(1347–1351) and the periodic recurrences of the plague that continued for the rest of the 14th century. The Little Ice Age lasted until the early 16th century, by which time many villages, and much of the more marginal land, had been abandoned.

The resurgence of European economies from the 16th century onward coincided with overseas exploration and trade, while at the same time, domestic landscapes were significantly affected by repopulation, by reforms to land tenure systems, and by advances in science and technology that changed agricultural practices, allowing for a more intensive use of the land. In the Netherlands, a steadily growing population and the consequent requirement for more agricultural land led to increased efforts to reclaim land from the sea and to drain coastal marshlands. Hundreds of small coastal barrier islands were slowly joined into larger units, and sea defense walls were constructed to protect low-lying land. The land was drained by windmill-powered water pumps and the excess water was carried off into a web of drainage ditches and canals.

The resulting **polder** landscape provided excellent, flat, fertile, and stone-free soil. Between 1550 and 1650, 165,000 hectares (407,715 acres) of polderland were established in the Netherlands, and the sophisticated techniques developed by the Dutch began to be applied elsewhere in Europe—including eastern England and the Rhône estuary in southern France. Although most of these schemes resulted in improved farmland, the environmental consequences were often serious. In addition to the vulnerability of the polderlands to inundation by the sea, large-scale drainage schemes devastated the wetland habitat of many species, and some ill-conceived schemes simply ended in widespread flooding.

These environmental problems were but a prelude to the environmental changes and ecological disasters that accompanied the industrialization of Europe, beginning in the 18th century. Mining—especially coal mining—created derelict landscapes of spoil heaps; urbanization encroached on rural landscapes and generated unprecedented amounts and concentrations of human, domestic, and industrial waste; and manufacturing, unregulated at first, resulted in extremely unhealthy levels of air pollution. Much of Europe's forest cover was cleared, and remaining forests and woodlands suffered from the **acid rain** (rain that contains dilute sulfuric and nitric acids derived from burning fossil fuels). Many streams and rivers also became polluted, and the landscape everywhere was scarred with quarries, pits, cuttings, dumps, and waste heaps.

HISTORY, ECONOMY, AND TERRITORY

Perhaps more than any other world region, Europe has a geography that is the product of its world-spanning economic and demographic systems. Beginning in the 16th century, Europeans became, as University of Wisconsin economic historian Robert Reynolds put it, the "leaders, drivers, persuaders, shapers, crushers and builders"[1] of the

world's economies and societies. It was as a result of these processes that the core areas of Europe emerged with a strong comparative advantage in the modern world system. For several centuries, other world regions came to play a subordinate economic role to Europe. Nevertheless, the preindustrial trajectories of other parts of the world had often eclipsed that of Europe and sometimes influenced events in Europe itself. Today, Europe is totally integrated into the world economy, with ethnically diverse populations that reflect the region's past economic and political history.

The foundations of Europe's human geography were laid by the Greek and Roman empires. Beginning in about 750 B.C.E., the ancient Greeks developed a series of fortified city–states (called *poleis*) along the Mediterranean coast, and by 550 B.C.E. there were about 250 of these trading colonies. **Figure 2.9** shows the location of the largest of these, some of which subsequently grew into thriving cities (for example, Athens and Corinth), whereas others remain as isolated ruins or archaeological sites (such as Delphi and Olympia). The Roman Republic was established in 509 B.C.E. and took almost 300 years to establish control over the Italian Peninsula. By 14 C.E., however, the Romans had conquered much of Europe, together with parts of North Africa and Asia Minor. Most of today's major European cities had their origin as Roman settlements. In quite a few of these cities, it is possible to find traces of the original Roman street layouts. In some, it is possible to glimpse remnants of defensive city walls, paved streets, aqueducts, viaducts, arenas, sewage systems, baths, and

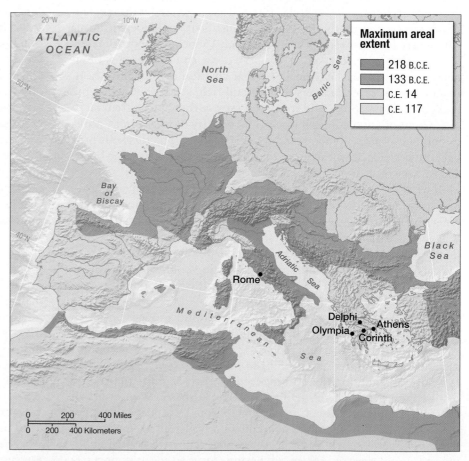

Maximum areal extent	
	218 B.C.E.
	133 B.C.E.
	C.E. 14
	C.E. 117

▲ **FIGURE 2.9 Greek Colonies and the Extent of the Roman Empire** This map shows the distribution of Greek *poleis* (city–states) and Carthaginian colonies and the spread of the Roman Empire from 218 B.C.E. to C.E. 117.

[1]R. Reynolds, *Europe Emerges: Transition Toward an Industrial World-Wide Society* (Madison: University of Wisconsin Press, 1961), p. vii.

public buildings. In the modern countryside, the legacy of the Roman Empire is represented by arrow-straight roads, built by Roman engineers and maintained and improved by successive generations.

The decline of the Roman Empire, beginning in the 4th century C.E., was accompanied by a long period of rural reorganization and consolidation under feudal systems, a period often characterized as uneventful and stagnant. **Feudal systems** were forms of economic organization where wealth was appropriated though a hierarchy of social ranks by means of institutionalized political or religious coercion. In fact, the roots of European regional differentiation can be traced to this long feudal era of slow change. Feudal systems were almost wholly agricultural, with 80% or 90% of the workforce engaged in farming and most of the rest occupied in basic craft work. Most production was for people's immediate needs; very little of a community's output ever found its way to wider markets.

By 1000 C.E., the countryside of most of Europe had been consolidated into a regional patchwork of feudal agricultural subsystems, each of which was more or less self-sufficient. For a long time, European towns were small, their existence tied mainly to the castles, palaces, churches, and cathedrals of the upper ranks of the feudal hierarchy. These economic landscapes—inflexible, slow motion, and introverted—nevertheless contained the essential preconditions for the rise of Europe as the dynamic hub of the world economy.

Historical Legacies and Landscapes

A key factor in the rise of Europe as a major world region was the emergence of merchant capitalism in the 15th century. **Merchant capitalism** refers to the earliest phase in the development of capitalism as an economic and social system, when the key entrepreneurs were merchant traders and merchant bankers. The immensely complex trading system that soon came to span Europe was based on long-standing trading patterns that had been developed from the 12th century by the merchants of Venice, Pisa, Genoa, Florence,

Bruges, Antwerp, and the Hanseatic League (a federation of city–states around the North Sea and Baltic coasts that included Bremen, Hamburg, Lübeck (**Figure 2.10**), Rostock, and Danzig).

Trade and the Age of Discovery In the 15th and 16th centuries, a series of innovations in business and technology contributed to the consolidation of Europe's new merchant capitalist economy. These included several key innovations in the organization of business and finance: banking, loan systems, credit transfers, company partnerships, shares in stock, speculation in commodity futures, commercial insurance, and courier/news services. Meanwhile, technological innovations began to further strengthen Europe's economic advantages. Some of these were adaptations and improvements of Oriental discoveries—the windmill, spinning wheels, paper manufacture, gunpowder, and the compass, for example. In Europe there was a unique passion for mechanizing the manufacturing process. Key engineering breakthroughs included the more efficient use of energy in water mills and blast furnaces, the design of reliable clocks and firearms, and the introduction of new methods of processing metals and manufacturing glass.

The geographical knowledge acquired during this Age of Discovery was crucial to the expansion of European political and economic power in the 16th century. In societies that were becoming more and more commercially oriented and profit conscious, geographical knowledge became a valuable commodity in itself. Gathering information about overseas regions was a first step to controlling and influencing them; this in turn was a step to wealth and power. At the same time, every region began to open up to the influence of other regions because of the economic and political competition unleashed by geographical discovery. Not only was the New World affected by European colonists, missionaries, and adventurers, but the countries of the Old World also found themselves pitched into competition with one another for overseas resources.

These changes had a profound effect on the geography of Europe. Before the mid-15th century, Europe was organized around two

◀ **FIGURE 2.10 Lübeck, Germany** The Hanseatic League was a federation of city–states founded in the 13th century by north German towns and affiliated German merchant groups abroad to defend their mutual trading interests. The League, which remained an influential economic and political force until the 15th century, laid the foundations for the subsequent growth of merchant trade throughout Europe.

subregional maritime economies—one based on the Mediterranean and the other on the Baltic. The overseas expansions pioneered first by the Portuguese and then by the Spanish, Dutch, English, and French reoriented Europe's geography toward the Atlantic. The river basins of the Rhine, the Seine, and the Thames rapidly became focused on a thriving network of **entrepôt** seaports (intermediary centers of trade and transshipment) that transformed Europe. These three river basins, backed by the increasingly powerful states in which they were embedded—the Netherlands, France, and Britain, respectively—then became engaged in a struggle for economic and political hegemony. Although the Rhine was the principal natural routeway into the heart of Europe, the convoluted politics of the Netherlands allowed Britain and France to become the dominant powers by the late 1600s.

Colonialism As we noted in Chapter 1, the first wave of European colonialism began in the late 1400s. A combination of factors led to European colonial exploitation. The large number of impoverished aristocrats—second and subsequent sons—produced by western European inheritance laws and by expensive local wars and religious Crusades was one important factor. Discouraged from commercial careers by sheer snobbery and encouraged by a culture that romanticized the fighting man, these poor but ambitious men provided a plentiful supply of adventurers. These nobles were willing to die for glory and often even more willing to exercise greed and cruelty in the name of god and country, encouraged by the evangelical zeal of the Catholic Church and the political competitiveness of European monarchies. Fundamentally, however, colonialism was fuelled by the logic of capitalism. Europe's growth could only be sustained as long as productivity could be improved, and, after a point, increased productivity required food and energy resources that could only be obtained by the conquest—peaceful or otherwise—of new territories.

And this first phase of colonial exploitation could not have been achieved without the development of a remarkable combination of European innovations in shipbuilding, navigation, and naval ordnance that literally had the most far-reaching consequences for this important phase of world history. Over the course of the 15th century, full-rigged ship designs were developed, enabling faster voyages in larger and more maneuverable vessels that were less dependent on favorable winds. Meanwhile, the quadrant (1450) and the astrolabe (1480) were invented, and a systematic knowledge of Atlantic winds had been acquired. By the mid-1500s, England, Holland, and Sweden had perfected the technique of casting iron guns, making it possible to replace bronze cannon with larger numbers of more effective guns at lower expense. Together, these advances made it possible for the merchants of Europe to establish the basis of a worldwide economy in the space of less than 100 years (**Figure 2.11**).

Europeans soon destroyed most of the Muslim shipping trade in the Indian Ocean and captured a large share of intra-Asian trade. It was the gold and silver from the Americas, however, that provided the first major economic transformation, allowing Europe to live above its means. In effect, the plundered bullion was converted into demand for goods of all kinds—textiles, wine, food, furniture, weapons, and ships—thus stimulating production throughout Europe and creating the basis for a "Golden Age" of prosperity for most of the 16th century. Meanwhile, overseas expansion made available a variety of new and unusual products—cocoa, beans, maize, potatoes, tomatoes, sugar cane, tobacco, and vanilla from the Americas, tea from the Orient—which opened up large new markets to enterprising merchants.

266   PUNCH, OR THE LONDON CHARIVARI.   [December 10, 1892.

THE RHODES COLOSSUS
STRIDING FROM CAPE TOWN TO CAIRO.

▲ **FIGURE 2.11 British Empire** This cartoon from *Punch* magazine in 1892 depicts British businessman and colonialist Cecil Rhodes, who had announced plans for a telegraph line connecting the British colonies from Cape Town, in Cape Colony (now part of modern South Africa), to Cairo, in British-ruled Egypt.

As European traders came to monopolize trade routes, they were able to identify foreign articles and ship them home to Europe, where skilled workmen learned to imitate them. In time, overseas expansion stimulated further improvements in technology and business techniques, thus adding to the self-propelling growth of European capitalism. Further developments were achieved in nautical mapmaking, naval artillery, shipbuilding, and the use of sail. Ultimately the whole experience of overseas colonization provided a great practical school of entrepreneurship and investment. Most important of all, perhaps, was the way that the profits from overseas colonies and trading posts flowed back into domestic agriculture, mining, and manufacturing. This contributed to an accumulation of capital that was undoubtedly one of the main preconditions for the emergence of industrial capitalism in Europe in the 18th century.

APPLY YOUR KNOWLEDGE Identify and research three cities that have been important entrepôt seaports. Use the Internet to compile profiles of their present-day economies.

Industrialization In the wake of colonialism, Europe's regional geographies were comprehensively recast once more by the new technology systems that marked the onset of the Industrial Revolution (in the late 1700s). **Technology systems** are clusters of interrelated energy, transportation, and production technologies that dominate economic activity for several decades at a time—until a new cluster of improved technologies evolves. What is especially remarkable about technology systems is that they have come along at about 50-year intervals throughout modern history. Since the beginning of the Industrial Revolution, we can identify four of them:

- **1790–1840:** early mechanization based on water power and steam engines; development of cotton textiles and ironworking; development of river transport systems, canals, and turnpike roads
- **1840–1890:** exploitation of coal-powered steam engines; steel products; railroads; world shipping; and machine tools
- **1890–1950:** exploitation of the internal combustion engine; oil and plastics; electrical and heavy engineering; aircraft; radio and telecommunications
- **1950–1990:** exploitation of nuclear power, aerospace, electronics, and petrochemicals; development of limited-access highways and global air routes

A fifth technology system, still incomplete, began to take shape in the 1980s with a series of innovations that are now being commercially exploited:

- **1990 onward:** exploitation of solar energy, robotics, microelectronics, biotechnology, advanced materials (fine chemicals and thermoplastics, for example), and information technology (digital telecommunications and geographic information systems, for example)

The production technologies of the first industrial technology systems, based on more efficient energy sources, helped raise levels of productivity and created new and better products that stimulated demand, increased profits, and generated a pool of capital for further investment. Transportation technologies enabled successive geographic expansion that completely reorganized the geography of Europe. As the application of new technologies altered the margins of profitability in different kinds of enterprise, the fortunes of places and regions shifted.

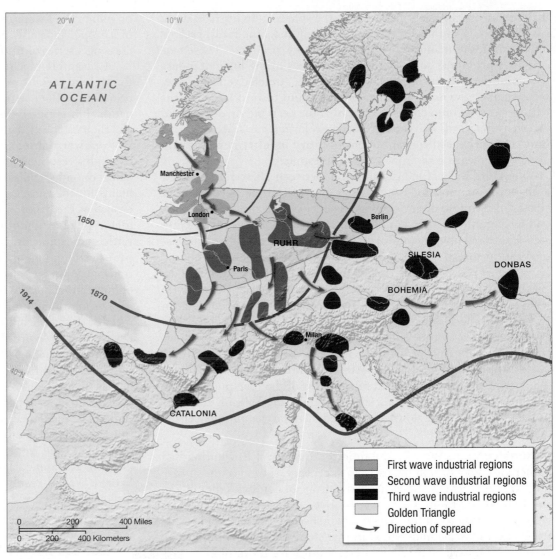

▲ **FIGURE 2.12 The Spread of Industrialization in Europe** The first wave of European industrialization began with the emergence of small industrial regions in several parts of Britain. As new industrial and transportation technologies emerged, industrialization spread to other regions with the right attributes: access to raw materials and energy sources, good communications, and large labor markets. A third wave of industrialization, after 1870, saw industrial development spread to southern and Eastern Europe and southern Scandinavia.

There was, in fact, not a sudden, single Industrial Revolution but three distinctive transitional waves of industrialization, each with a different degree of impact on different regions and countries (**Figure 2.12**). The first, between about 1790 and 1850, was based on the cluster of early industrial technologies (steam engines, cotton textiles, and ironworking) and was highly localized. It was limited to a few regions in Britain where industrial entrepreneurs and workforces had first exploited key innovations and the availability of key resources (coal, iron ore, and water).

The second wave, between about 1850 and 1870, involved the diffusion of industrialization to most of the rest of Britain and to parts of northwest Europe, particularly the coalfield areas of northern France, Belgium, and Germany. New opportunities were created as railroads and steamships made more places accessible, bringing their resources and their markets into the sphere of industrialization. New materials and new technologies (steel, machine tools) created opportunities to manufacture and market new products. These new activities prompted some significant changes in the logic of industrial location. Railway networks, for example, attracted industry away from smaller towns on the canal systems and toward larger towns with good rail connections. Steamships for carrying on coastal and international trade attracted industry to larger ports. At the same time, steel produced concentrations of heavy industry in places with nearby supplies of coal, iron ore, and limestone.

The third wave of industrialization, between 1870 and 1914, saw a further reorganization of the geography of Europe as yet another cluster of technologies (including electricity, electrical engineering, and telecommunications) brought different resource needs and created additional investment opportunities. During this period, industrialization spread for the first time to remoter parts of the United Kingdom, France, and Germany and to most of the Netherlands, southern Scandinavia, northern Italy, eastern Austria, Bohemia (in what was then Czechoslovakia), Silesia (in Poland), Catalonia (in Spain), and the Donbas region of Ukraine, then into Russia. The overall result was to create the foundations of a core-periphery structure within Europe, with the heart of the core centered on the **Golden Triangle** stretching between London, Paris, and Berlin. The peripheral territories of Europe—most of the Iberian peninsula, northern Scandinavia, Ireland, southern Italy, the Balkans, and east-central Europe—were slowly penetrated by industrialization over the next 50 years.

Imperialism and War By the time the Industrial Revolution had gathered momentum in the early 19th century, several of the most powerful and heavily industrialized European countries (notably the United Kingdom, Germany, France, and the Netherlands) were competing for influence on a global scale. Europe's industrialization must be understood in this context. The ascent of Europe's industrial core regions could not have taken place without the foodstuffs, raw materials, and markets provided by the rest of the world. In order to ensure the avaiabiity of the produce, materials, and markets

on which they were increasingly dependent, the industrial nations of Europe vigorously pursued a second phase of overseas expansion, creating a series of **trading empires**.

A scramble began for territorial and commercial domination through **imperialism**—the deliberate exercise of military power and economic influence by powerful states to advance and secure their national interests. European countries engaged in expansion to protect their established interests and to limit the opportunities of others. Each power wanted to secure as much of the world as possible—through a combination of military oversight, administrative control, and economic regulations—to ensure stable and profitable environments for their traders and investors. A second phase of colonialism ensued as the major economic and military powers embarked on the inland penetration of mid-continental grassland zones in order to settle and exploit them for grain or stock production. The detailed pattern and timing of this exploitation was heavily influenced by new transportation technologies—especially railways—and the invention of innovations such as barbed wire and refrigeration.

During the first half of the 20th century, the economic development of the whole of Europe was disrupted twice by major wars. The devastation of World War I was immense. The overall loss of life, including the victims of influenza epidemics and border conflicts that followed the war, amounted to between 50 and 60 million. About half as many again were permanently disabled. For some countries, this meant a loss of between 10% and 15% of the male workforce.

Just as European economies had adjusted to these devastations, the **Great Depression**—a severe decline in the world economy that lasted from 1929 until the mid 1930s—created a further phase of economic damage and reorganization throughout Europe. World War II resulted in yet another round of destruction and dislocation (**Figure 2.13**). The total loss of life in Europe this time was 42 million, two-thirds of whom were civilian casualties. The German persecution of Jews and others—the **Holocaust**, Nazi Germany's systematic genocide of various ethnic, religious, and national minority groups—resulted in approximately 2 million Jews being put to death in extermintion camps such as Auchwitz and Treblinka, with up to 2 million more, along with gypsies and others, being exterminated

▲ **FIGURE 2.13 London, England** London sustained heavy bomb damage during World War II.

elsewhere. The German occupation of continental Europe also involved ruthless economic exploitation. By the end of the war, France was depressed to below 50% of its prewar standard of living and had lost 8% of its industrial assets. The United Kingdom lost 18% of its industrial assets (including overseas holdings), and the Soviet Union lost 25% in the war. Germany lost 13% of its assets and ended the war with a level of income per capita that was less than 25% of the prewar figure. In addition to the millions killed and disabled during World War II, approximately 46 million people were displaced between 1938 and 1948 through flight, evacuation, resettlement, or forced labor. Some of these movements were temporary, but most were not.

After the war, the Cold War rift between Eastern Europe (countries dominated by the Soviet Union) and Western Europe (the rest) resulted in a further handicap to the European economy and, indeed, to its economic geography. Ironically, this rift helped speed economic recovery in Western Europe. The United States, whose leaders believed that poverty and economic chaos in Western Europe would foster communism, embarked on a massive program of economic aid under the **Marshall Plan**, a program of loans and other economic assistance provided by the U.S. government between 1947 and 1952 to bolster European allies whose weakness, it was believed, made them susceptible to communism. This pump-priming action, together with the backlog of demand in almost every sphere of production, provided the basis for a remarkable recovery. Meanwhile, Eastern Europe began an interlude of **state socialism**, a form of economy based on principles of collective ownership and administration of the means of production and distribution of goods dominated and directed by state bureaucracies.

Eastern Europe's Legacy of State Socialism After World War II, the leaders of the Soviet Union felt compelled to establish a **buffer zone** between their homeland and the major Western powers in Europe. The Soviet Union rapidly established its dominance throughout Eastern Europe; Estonia, Moldova, Latvia, and Lithuania were absorbed into the Soviet Union itself, and Soviet-style regimes were installed in Albania, Bulgaria, Czechoslovakia, East Germany, Hungary, Poland, Romania, and Yugoslavia. The result was what Winston Churchill called an **Iron Curtain** along the western frontier of Soviet-dominated territory, a militarized frontier zone across which Soviet and East European authorities allowed the absolute minimum movement of people, goods, and information (**Figure 2.14**). In addition, Soviet intervention resulted in the complete nationalization of the means of production, the collectivization of agriculture, and the imposition of rigid social and economic controls within the eastern European **satellite states** (countries that are under heavy political, economic, and military influence of another state—in this case, the Soviet Union).

The economies of the Soviet Union and its satellites were *not* based on true socialist or communist principles in which the working class had democratic control over the processes of production, distribution, and development. Rather, these economies evolved as something of a hybrid, in which state power was used by a **bureaucratic class** (nonelected government officials) to create **command economies** in the pursuit of modernization and economic development. In a

▲ **FIGURE 2.14 Berlin Wall** The Berlin Wall was a physical reminder of the separation between eastern and western Europe created by the Iron Curtain after World War II.

command economy, every aspect of economic production and distribution is controlled centrally by government agencies.

The **Communist Council for Mutual Economic Assistance** (CMEA, better known as **COMECON**) was established to reorganize eastern European economies in the Soviet mold—with individual members each pursuing independent, centralized plans designed to produce economic self-sufficiency. This quickly proved unsuccessful, however, and in 1958 COMECON was reorganized. The goal of economic self-sufficiency was abandoned, mutual trade among the *Soviet bloc*—the Soviet Union plus its eastern European satellite states—was fostered, and some trade with Western Europe was permitted. Meanwhile, Yugoslavia had been expelled from the Soviet bloc in 1948 (because of ideological differences over the interpretation of socialism), and Albania had withdrawn from the Soviet bloc in pursuit of a more authoritarian form of communism inspired by the Chinese revolution of 1949 (see Chapter 8).

The experience of the East European countries under state socialism varied considerably, but, in general, rates of industrial growth were high. The industrialized landscapes of Eastern Europe came to be dominated by the localization of manufacturing activity, by regional specialization, and by contrasts in levels of economic development. The geography of industrial development under state socialism, as in democratic capitalism, was heavily influenced by the uneven distribution of natural resources and by the economic logic of initial advantage, specialization, and the cost advantages clustering together for firms in similar or related industries.

The most distinctive landscapes of state socialism were those of urban residential areas, where mass-produced, system-built apartment blocks allowed impressive progress in eliminating urban slums and providing the physical framework for an **egalitarian society**—one based on belief in equal social, political, and economic rights and privileges. The price paid for this progress however,

▲ **FIGURE 2.15 Socialist Housing** The socialist countries of Eastern Europe eradicated a great deal of substandard housing in the three decades following World War II, rehousing the population in mass-produced, system-built apartment blocks. Although this new housing provided adequate shelter and basic utilities at very low rents, space standards were extremely low, and housing projects were uniformly drab. This example is from the suburban district of Drumul Taberei in Bucharest, Romania.

was uniformly modest dwellings and strikingly sterile cityscapes (**Figure 2.15**).

The Reintegration of Eastern Europe Eventually, the economic and social constraints imposed by excessive state control and the dissent that resulted from the lack of democracy under state socialism combined to bring the experiment to a sudden halt. By the time the Soviet bloc collapsed in 1989 (see Chapter 3), Poland and Hungary had already accomplished a modest degree of democratic and economic reform. By 1992, East Germany (the German Democratic Republic) had been reunited with West Germany (the German Federal Republic); Estonia, Latvia, and Lithuania had become independent states once more; and the whole of Eastern Europe had begun to be reintegrated with the rest of Europe. In 1991, COMECON was abolished, and one by one, eastern European countries began a complex series of reforms. These included the abolition of controls on prices and wages, the removal of restrictions on trade and investment, the creation of a financial infrastructure to handle private investment, the creation of government fiscal systems to balance taxation and spending, and the privatization of state-owned industries and enterprises. After more than 40 years of state socialism, such reforms were difficult and painful. Indeed, the reforms are still by no means complete in any of the countries, and economic and social dislocation is a continuing fact of life.

Nevertheless, the reintegration of Eastern Europe added a potentially dynamic market of 130 million consumers to the European economy. Within a capitalist framework, Eastern Europe has the comparative advantage of relatively cheap land and labor. This has attracted foreign investment, particularly from transnational corporations and from German firms and investors, many of whom

have historic ties with parts of Eastern Europe. In some ways, the transition toward market economies has been remarkably swift. It did not take long for Western-style consumerism to appear on the streets and in many of the stores in larger eastern European cities. On the flip side, it also did not take long for inflation, unemployment, and homelessness to appear. Overall, Eastern Europe is increasingly reintegrated with the rest of Europe, but for the most part as a set of economically peripheral regions, with agriculture still geared to local markets and former COMECON trading opportunities and industry still geared more to heavy industry and standardized products than to competitive consumer products. The service sector in general and knowledge-based industries in particular are still only weakly developed.

In detail, the pace and degree of reintegration vary considerably across Eastern Europe. Ethnic conflict in Bosnia and Herzegovina, Croatia, and the former republic of Yugoslavia has severely retarded reform and reintegration, whereas Albania, Bulgaria, Macedonia, Moldova, and Romania suffer from the combined disadvantages of having relatively poor resource bases, weakly developed communications and transportation infrastructures, and political regimes with little ability or inclination to press for economic and social reform. In Ukraine, which has a much better infrastructure, a significant industrial base, and the capacity for extensive trade in grain exports and advanced technology, reintegration has been retarded by a combination of geographical isolation from Western Europe, continuing economic and political ties with Russia, and a surviving political elite that has little interest in reform.

The Baltic states of Estonia, Latvia, and Lithuania have been more successful in reintegrating with the rest of Europe. Their small size and relatively high levels of education have made them attractive as production subcontracting centers for western European high-technology industries. They are reviving old ties with neighboring Nordic countries; in this regard Estonia is particularly well placed because its language belongs to the Finnish family, and it was part of the Kingdom of Sweden when it was annexed by the Russians in 1710. The best-integrated states of Eastern Europe are Czechia, Hungary, Poland, and Slovenia. All have a relatively strong industrial base, and Hungary has a productive agricultural sector as well.

APPLY YOUR KNOWLEDGE Research what happened to Berlin during the Cold War. What has happened to the Berlin Wall and the adjacent "no man's land" since the reunification of Germany and the reintegration of Eastern Europe? What have been the problems and opportunities that Berlin has faced as a result of reintegration?

Economy, Accumulation, and the Production of Inequality

Contemporary Europe is a cornerstone of the world economy with a complex, multilayered, and multifaceted regional geography. In overall terms, Europe, with about 12% of the world's population, accounts for almost 35% of the world's exports, almost 43% of the world's imports, and 33% of the world's aggregate GNP. The bulk of Europe's exports are machinery, motor vehicles, aircraft, plastics, pharmaceuticals, iron and steel, nonferrous metals, wood pulp and paper products, textiles/fashion, meat, dairy products, wine, and fish. Europe has long been an important exporter of specialized luxury products such as perfumes (see "Geographies of Indulgence, Desire, and Addiction: Perfumes and Fragrances," p. 66). Europe's inhabitants, on average, consume about twice the quantity of goods and commercial services they did in 1975. Purchasing power has risen everywhere to the extent that food and clothing now account for only about 30% of household expenditures, leaving more resources for leisure and consumer durables. Levels of material consumption in much of Europe approach those of households in the United States.

Although Europe is a relatively affluent world region, there are, in fact, persistent and significant economic inequalities at every geographic scale. Regional income disparities within many European countries are increasing. In northwestern Europe, this is generally a result of the declining fortunes of "Rust Belt" regions and the relative prosperity of regions with high-tech industry and advanced business services. In southern and eastern Europe, it is a result of differences between regions dominated by rural economies (generally poorer) and those dominated by metropolitan areas (generally more prosperous). The resulting disparities are significant, with annual per capita GDP (in PPP) in 2010 ranging from U.S. $3,092 in Moldova and U.S. $6,698 in Ukraine to U.S. $51,959 in Norway and U.S. $81,466 in Luxembourg. Poverty and homelessness exist in every European country, though poverty as measured on a global scale (U.S. $1 or U.S. $2 a day per person) is virtually unknown within Europe.

The development of European **welfare states** (institutions with the aim of distributing income and resources to the poorer members of society) has helped maintain purchasing power during periods of recession and ensured at least a tolerable level of living for most groups at all times. Levels of personal taxation are high, but all citizens receive a wide array of services and benefits in return. The most striking of these services are high-quality medical care, public transport systems, **social housing** (where the dwellings are owned by a government or nonprofit agency), schools, and universities. The most significant benefits are pensions and unemployment benefits.

▶ **FIGURE 2.16 The Expansion of the European Union** The advantages of membership in the European Union led to a dramatic growth in its size, transforming it into a major economic and political force in world affairs. Economic union among very different economies has, however, created significant problems for the member countries of the eurozone, which share the same currency, the euro.

The European Union: Coping with Uneven Development Contemporary Europe is a dynamic region that embodies a great deal of change. Formerly prosperous industrial regions have suffered economic decline, but some places and regions have reinvented themselves to take advantage of new paths to economic development. Meanwhile, the former Soviet satellite states have been reintegrated into the European world region, and much of Europe has joined in the **European Union (EU)**, a supranational organization founded to recapture prosperity and power through economic and political integration. The EU has its origins in the political and economic climate that followed World War II. The initial idea behind the EU was to ensure European autonomy from the United States and to recapture the prosperity Europe had forfeited as a result of the war. Part of the rationale for its creation was also to bring Germany and France together in a close association, which would prevent any repetition of the geopolitical problems in Western Europe that had led to two world wars.

In the 1950s, several institutions were created to promote economic efficiency through integration. These were subsequently amalgamated to form the European Community (EC), which was in turn expanded in scope to form the European Union. EU membership has expanded from the six original members of the EC—Belgium, France, Italy, Luxembourg, the Netherlands, and West Germany—to 27 countries (**Figure 2.16**). Member states combined

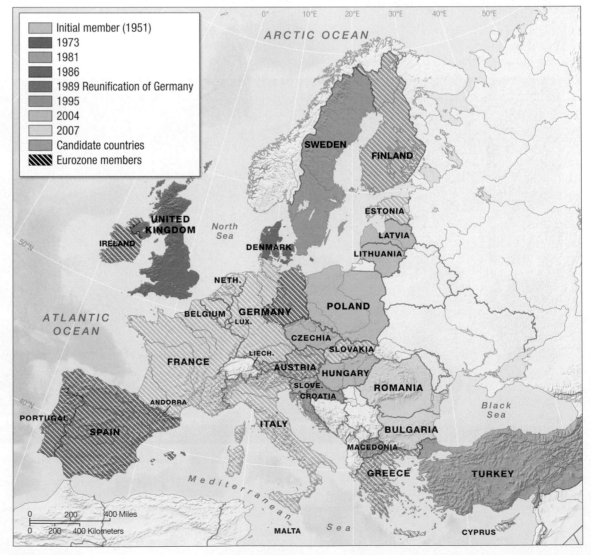

GEOGRAPHIES OF INDULGENCE, DESIRE, AND ADDICTION
Perfumes and Fragrances

From the earliest times, perfumes and fragrances have been prized commodities. The ancient Greeks and Romans were keen users of scents and perfumes. The ancient Greeks believed that fragrances had powers to heal and protect from evil, while the Romans used perfumes for seduction and used herbs as aphrodisiacs. With the fall of the Roman Empire in the 4th century C.E., there was a decline in the use of perfumery, but Europeans were introduced to sophisticated perfumery from the East during the 12th and 13th century Crusades. Subsequently, Venice became the center of the perfume trade, and perfume became a luxury good popular with the wealthiest classes as a means of masking the odors of unwashed bodies, both their own and others'. Bathing was unpopular at the time because it was thought to open the pores up and allow diseases in.

Catherine de Medici, wife of King Henri II, is credited with having introduced a fashion for perfumes into France in the mid-16th century. By this time, knowledge of Arab methods of extracting fragrances from flowers had spread into Europe. These methods involved either maceration (soaking flowers in a liquid that will absorb fragrances) or distillation. The resulting concentrates are known as "essential oils," and it is from these that perfumes are blended and made. Both processes require enormous quantities of aromatic flowers. The mild Mediterranean climate of southern France is particularly suited to horticulture, and it became the center of production of key ingredients of perfumery such as jasmine, lavender, myrtle, roses, and mimosa. Production of essential oils was centered in and around the small town of Grasse, in the Alpes-Maritimes department, northwest of Nice. Grasse also became home to the highly skilled and specialized professions involved in designing appealing perfumes from combinations of essential oils.

In the late 1800s, the Paris-based Houbigant company introduced *Fougère Royal*, a new type of fragrance based on a combination of oakmoss, geranium, bergamot (oil from the peel of bitter oranges), and coumarin (a chemical compound that occurs naturally in many plants, including sweet clover, strawberries, and cherries, but is difficult and expensive to extract). The significance of Houbigant's new fragrance was that the coumarin had been synthetically produced. The advent of synthetics meant that perfume became a mass-market indulgence, a product that was marketed as an agent of desire. By the early 1900s, synthetic ingredients were being used on a regular basis. In 1921, Chanel No. 5 was introduced, the first fragrance that was dominated by the use of synthetic ingredients. At about the same time, gas chromatography and mass spectrometry enabled producers to analyze the secret formulas of traditional fragrances.

▲ **FIGURE 2.1.1 Designer Perfumes** A wide variety of European designer fragrances and perfumes are available worldwide.

Perfume manufacture became a branch of the globalized petrochemical industry. Fragrances were added to all sorts of products, from detergents to cigarettes, and fragrances became part of brand identity. Perfumes themselves became powerful brands, extended to other product lines such as shampoos, lotions, and soaps. Perfumes also became an important component of the fashion industry, with designer brands marketing their own fragrances (**Figure 2.1.1**). Recently, this has extended to sensory branding, whereby retailers and hotel chains deploy their own distinctive ambient scenting through sophisticated dispersion technologies in stores, lobbies, and bedrooms.

Although the geography of production and consumption of perfumes has become worldwide, the traditional association of place and product means that French perfumes have retained a comparative advantage in the marketplace, accounting for about 30% of the world market. Similarly, Paris remains an important corporate center for the nexus of high-end fashion, cosmetics, and perfumery. The production of the very finest perfumes has also retained a strong geographic identity, reflecting the importance of agglomeration economies and the clustering of specialized labor, research institutes, corporate research and development (R&D) laboratories, and equipment manufacturers. Grasse is still the heart of the French perfume industry and the surrounding Provence-Alpes-Côte d'Azur region is still the world's leading area for the production of lavender and other aromatic flowers (**Figure 2.1.2**). The perfume industry in Grasse involves some 60 different companies and employs almost 3,500 people. In the Provence-Alpes-Côte d'Azur region as a whole there are more than 400 fragrance, cosmetics, and flavor-related companies; 120 essential oil distilleries; and 2,250 growers of aromatic plants, supported by 14 public sector and 8 private sector research laboratories. ■

▲ **FIGURE 2.1.2 Fields of Lavender** Lavender, which is a key component in many high-end perfumes manufactured in Europe, grows in field near the perfumery manufacturing area of Grasse.

have a population of nearly 503 million, and a gross domestic product (GDP) of U.S. \$16.2 trillion in 2010 (compared to the 2010 U.S. GDP of U.S. \$14.5 trillion). The EU has developed into a sophisticated and powerful institution with a pervasive influence on patterns of economic and social well-being within its member states. It also has a significant impact on certain aspects of economic development within some nonmember countries.

The origin of the organization was a compromise worked out between the strongest two of the original six members. West Germany wanted a larger but protected market for its industrial goods, whereas France wanted to continue to protect its highly inefficient but large and politically important agricultural sector from overseas competition. The result was the creation of a tariff-free market within the EU, the creation of a unified external tariff, and a **Common Agricultural Policy (CAP)** to bolster the agricultural sector.

Some of the most striking changes in the regional geography of the EU have been related to the operation of the CAP. Although agriculture accounts for less than 3% of the EU workforce, it was the CAP that dominated the EU budget from the beginning. For a long time, it accounted for more than 70% of the EU's total expenditures. The CAP has had a significant impact on rural economies, rural landscapes, and rural standards of living, and it has even influenced urban living through its effects on food prices.

The basis of the CAP is a system of EU support of wholesale prices for agricultural produce. This support has the dual effects of stabilizing the price of agricultural products and of subsidizing farmers' incomes. The overall result has been a realignment of agricultural production patterns, with a general withdrawal from mixed farming. Ireland, the United Kingdom, and Denmark, for example, have increased their specialization in the production of wheat, barley, poultry, and milk, whereas France and Germany have increased their specialization in the production of barley, maize, and sugar beets.

The reorganization of Europe's agricultural landscapes under the CAP brought some unwanted side effects, however, including environmental problems that have occurred as a result of the speed and scale of farm modernization, combined with farmers' desire to take advantage of generous levels of guaranteed prices for crops. Moorlands, woodlands, wetlands, and hedgerows have come under threat, and some traditional mixed-farming landscapes have been replaced by the prairie-style settings of specialized agribusiness.

Meanwhile, the rest of the world economy had changed significantly, intensifying the challenge to Europe. By the early 1980s, the U.S. and Japanese economies were becoming increasingly interdependent and prosperous on the basis of globalized producer services and new, high-tech industries. London's once preeminent financial services were losing ground to those of New York and Tokyo. Even the former West Germany, Europe's healthiest economy at the time, was facing the prospect of being left behind as a producer of obsolescent capital goods and consumer goods. In response, the European Community relaunched itself, beginning in the mid-1980s with the ratification of the **Single European Act of 1985 (SEA)**, which affirmed the ultimate aim of economic and political harmonization within a single supranational government. This relaunching represents an impressive achievement, particularly because it had to be undertaken at a time when there were major distractions: having to manage a changing relationship with the United States through trade renegotiations, having to cope with the reunification of Germany and the breakup of the former Soviet sphere of influence in Eastern

Europe and, not least, having to cope with a widespread internal resurgence of nationalism. The Single European Act was followed in 1999 by the creation of a common European currency, the euro. To join the currency, member states had to qualify by meeting strict conditions in terms of national budget deficits, inflation, and interest. Of EU members at the time, the United Kingdom, Sweden, and Denmark declined to join the currency, leaving a "eurozone" that was smaller than the EU as a whole (See Figure 2.16 on p. 65).

Overall, EU membership has brought regional stresses as well as the prospect of overall economic gain. Existing member countries have found that the removal of internal barriers to labor, capital, and trade has worked to the clear disadvantage of peripheral regions and in particular to the disadvantage of those farthest from the Golden Triangle (between London, Paris, and Berlin), which is increasingly the European center of gravity in terms of both production and consumption. This regional imbalance was recognized by the Single European Act, which included "economic and social cohesion" as a major policy. The SEA doubled its grant funding for regional development assistance and established a Cohesion Fund to help Greece, Ireland, Portugal, and Spain achieve levels of economic development comparable to those of the rest of the EU. In its 2014 to 2020 budget cycle, the EU has allocated €376 billion (U.S. \$487 billion)—about one-third of the total EU budget—to projects and policies designed to improve economic and social cohesion within and among its member countries.

APPLY YOUR KNOWLEDGE Provide an example of European Union policies that are directed toward (a) the environment, and (b) regional policy. What can you find out about the effectiveness of these policies? (Hint: A good starting point is the European Union's own website: http://europa.eu/pol/index_en.htm.)

Growth, Deindustrialization, and Reinvestment Europe provides a classic example of long-term shifts in technology systems leading to regional economic change. Innovations associated with new technology systems generate new industries that are not yet tied down by enormous investments in factories or tied to existing industrial agglomerations. Combined with innovations in transport and communications, this creates windows of opportunity that can result in new industrial districts and in some towns and cities that eventually grow into dominant metropolitan areas through new rounds of investment. Within Europe, the regions that have prospered most through the onset of the most recent technology system are the Thames Valley to the west of London, the Île-de-France region around Paris, the Ruhr valley in northwestern Germany, and the metropolitan regions of Lyon–Grenoble (France), Amsterdam–Rotterdam (Netherlands), Milan and Turin (Italy), and Frankfurt, Munich, and Stuttgart (Germany).

Just as high-tech industries and regions have grown in parts of Europe, the profitability of traditional industries in established regions has declined. **Disinvestment** has taken place in the less-profitable industries and regions, with production moving not so much to the growth regions of Europe but to emerging manufacturing regions in other world regions. Disinvestment involves selling off assets such as factories and equipment. Widespread disinvestment leads to **deindustrialization** in formerly prosperous industrial regions. Deindustrialization involves a relative decline (and in extreme

▲ **FIGURE 2.17 Deindustrialization** The loss of manufacturing jobs in older industrial regions has resulted in widespread dereliction. This abandoned cotton mill is in the Ancoats district of Manchester, England.

cases, an absolute decline) in industrial employment in core regions as firms scale back their activities in response to lower levels of profitability. This is what happened to the industrial regions of northern England, (**Figure 2.17**) South Wales, and central Scotland in the early part of the 20th century; it is what happened to the industrial region of Alsace-Lorraine, in France, and to many other traditional manufacturing towns and regions within Western Europe in the 1960s and 1970s; and it is what has been happening to the industrial centers of Eastern Europe since the 1990s.

Regional Development The economic and political integration of Europe has intensified core-periphery differences within Europe. The removal of internal barriers to flows of labor, capital, and trade has worked to the clear disadvantage of geographically peripheral regions within the EU, whereas core metropolitan regions have benefited. The character and relative prosperity of Europe's chief core region, the Golden Triangle, stem from four advantages: 1) It is well situated for shipping and trade. Its geographic situation provides access to southern and central Europe by way of the Rhine and Rhône river systems and access to the sea lanes of the Baltic Sea, the North Sea, and, by way of the English Channel, the Atlantic Ocean. 2) Within it are the capital cities of the major former imperial powers of Europe. 3) It includes the industrial heartlands of central England, northeastern France, and the Ruhr district of Germany. 4) Its concentrated population provides both a skilled labor force and an affluent consumer market.

The advantages have been reinforced by the integrative policies of the European Union, whose administrative headquarters are situated squarely in the heart of the Golden Triangle, in Brussels, Belgium. They have also been reinforced by the emergence of Berlin, Paris, and, especially, London as world cities, in which a disproportionate share

of the world's economic, political, and cultural business is conducted (**Figure 2.18**). As these world cities have come to play an increasingly central role in the world economy, they have become home to a vast web of sophisticated financial, legal, marketing, and communications services. These services, in turn, have added to the wealth and cosmopolitanism of the region.

Agriculture within the Golden Triangle tends to be highly intensive and geared toward supplying the highly urbanized population with fresh dairy produce, vegetables, and flowers. Industry that remains within the Golden Triangle is rather technical, drawing on the highly skilled and well-educated workforce. Most heavy industry and large-scale, routine manufacturing has relocated from the region in favor of cheaper land and labor found elsewhere in Europe or beyond. In the Golden Triangle, a tightly knit network of towns and cities is linked by an elaborate infrastructure of canals, railways, and highways. The region has reached saturation levels of urbanization, as exemplified in Randstad Holland, the densely settled region of the western Netherlands that includes Rotterdam, The Hague, Haarlem, Amsterdam, and Utrecht. In addition to London, Paris, and Berlin, major cities of the Golden Triangle include Amsterdam, Antwerp, Birmingham, Brussels, Cologne, Dortmund, Düsseldorf, Frankfurt, Hamburg, Hannover, Lille, Portsmouth, and Rotterdam. Nevertheless, the Golden Triangle still contains fragments of attractive rural landscapes, together with some unspoiled villages and small towns. These have survived partly because of market forces: they are very attractive to affluent commuters. Equally important to their survival, though, has been the relatively strong role of environmental, land use, and conservation planning in European countries.

The affluence and dynamism of the region have attracted large numbers of immigrants. London, in particular, has become a city with a global mix of populations and subcultures. Almost one-third of

▲ **FIGURE 2.18 Global Business** Office workers on lunch break in Broadgate, an office complex in the banking district of the City of London.

London's current residents—2.2 million people—were born outside England, and this total takes no account of the contribution of the city's second- and third-generation immigrants, many of whom have inherited the traditions of their parents and grandparents. Altogether, the people of London speak more than 300 languages, and the city has at least 50 nonindigenous communities with populations of 10,000 or more. Elsewhere in the Golden Triangle, immigrant groups have also added a new ethnic dimension to city populations. Most have settled in distinctive enclaves, where they have retained powerful attachments to their cultural roots. In the United Kingdom, the Netherlands, and Germany, immigrant populations have tended to concentrate in older inner-city neighborhoods, whereas in France the pattern is one of concentration in suburban public housing projects. Many immigrant groups now face high unemployment and low wages as the postwar boom has leveled off.

The **Southern Crescent**, which stretches south from the Golden Triangle, is a secondary, emergent, core region that straddles the Alps, running from Frankfurt in Germany through Stuttgart, Zürich, and Munich, and finally to Milan, Turin, and northern Italy. The prosperity of this Southern Crescent is in part a result of a general decentralization of industry from northwestern Europe and in part a result of the integrative effects of the EU. Some of the capital freed up by the deindustrialization of traditional manufacturing regions in northwestern Europe has found its way to more southerly regions, where land is less expensive and labor is both less expensive and less unionized. Frankfurt and Zürich are global-scale business and financial centers in their own right, whereas Milan (**Figure 2.19**) is a center of both finance and design, and Munich, Stuttgart, and Turin are important centers of industry and commerce.

This Southern Crescent stretches across a great variety of landscapes, from the plateaus of central Germany, across the Alps, and into northern Italy and the Apennines. Overall, these landscapes are much less urbanized than those of the Golden Triangle. However, the *rate* of urbanization is much higher. Urban growth is taking place in smaller towns and cities that are part of new-style industrial subregions. They represent a very different form of industrialization based on loose spatial agglomerations of small firms that are part of one or more leading industries. Small firms using computerized control systems and an extensive subcontracting network have the advantage of being flexible in what they produce and when and how they produce it. Consequently, the new industrial districts with which they are associated are often referred to as **flexible production regions.** Within each of these regions, small firms tend to share a specific local industrial culture that is characterized by technological dynamism and well-developed social and economic networks.

Northern Italy is home to a number of flexible production regions. Here, regional networks of innovative, flexible, and high-quality manufacturers make products that include textiles, knitwear, jewelry, shoes, ceramics, machinery, machine tools, and furniture. Other examples of flexible production regions based on a similar mixture of design- and labor-intensive industries include the Baden–Württemberg region around Stuttgart in Germany (textiles, machine tools, auto parts, and clothing) and the Rhône-Alpes region around Lyon in France.

The Golden Triangle and the Southern Crescent are linked by the trans-European high-speed rail system (**Figure 2.20**). The first high-speed rail lines in Europe, built in the 1980s and 1990s, were designed to improve travel times within major national corridors of urban development. Since then, several countries have built more extensive high-speed networks, and there are now several cross-border high-speed rail links. Railway operators run frequent international services, and tracks are continuously being built and upgraded to international standards (a minimum of 200 kilometers per hour [125 miles per hour]). The Gotthard Base Tunnel described at the beginning of this chapter is a small part of this overall strategy. Since 2005, high-speed passenger traffic in Europe has been increasing by 10% each year. In 2011, the system carried more than 135 billion passenger-kilometers.

Developing a trans-European high-speed rail network is a stated goal of the EU, and most cross-border railway lines receive EU funding.

With relatively short distances between major cities, Europe is ideally suited for rail travel and less suited, because of population densities and traffic congestion around airports, to air traffic. High-speed trains are also considered more energy efficient than other modes of transit per passenger-kilometer. In terms of possible passenger capacity, high-speed trains also reduce the amount of land used per passenger when compared to cars on roads. In addition, train stations are normally smaller than airports and can be located within major cities and spaced closer together, allowing for more convenient travel. Allowing for check-in times and accessibility to terminals, travel between many major European cities is already quicker by rail than by air. Improved locomotive technologies and specially engineered tracks and rolling stock make it possible to offer passenger rail services on some routes at speeds of 275 to 350 kilometers per hour (180 to 250 miles per hour) (**Figure 2.21**). In addition, new tilt-technology railway cars, which are designed

▲ **FIGURE 2.19 Milan** Milan has long been a major regional center and today has developed into a prosperous city of global importance in finance, fashion, and industrial design.

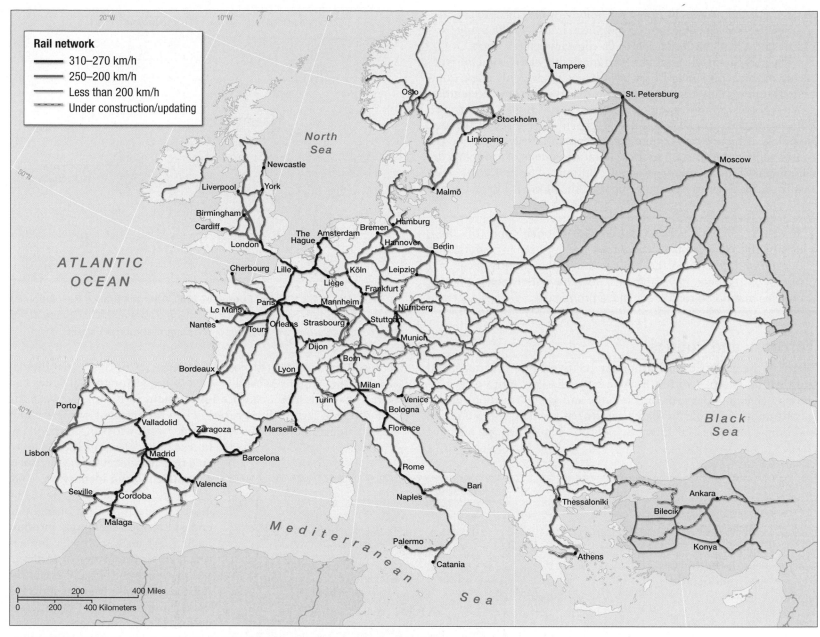

▲ **FIGURE 2.20 Europe's High-Speed Rail Network** The expansion of the high-speed rail network marks a major achievement of the European Union and is an important factor in the integration of European economies.

◄ **FIGURE 2.21 High-Speed Train** A third-generation ICE (Inter City Express) train, designed to run at speeds up to 320 km/h (200 mph), stands at the station at Frankfurt International Airport.

to negotiate tight curves by tilting the train body into turns to counter-act the effects of centrifugal force, have been introduced in many parts of Europe to raise maximum speeds on conventional rail tracks.

These high-speed rail routes will inevitably cause some restructuring of the geography of Europe. They will have only a few time-tabled stops because the time penalties that result from deceleration and acceleration undermine the advantages of high-speed travel. Places that do not have scheduled stops will be less accessible and, then, less attractive for economic development. Places linked to the routes will be well situated to grow in future rounds of economic development. This is why, for example, billions of Swiss francs (more than U.S. $12 billion) have been spent on the Gotthard Base Tunnel.

APPLY YOUR KNOWLEDGE Choose one of the geographically peripheral countries of northern or southern Europe and determine the problems that the country faces in terms of regional development. How have European Union policies attempted to address these problems?

Territory and Politics

Many of the countries of Europe are relatively new creations, and political boundaries within Europe have changed quite often, reflecting changing patterns of economic and geopolitical power and a continuous struggle to match territorial boundaries to cultural and ethnic identities. In 1500, the political map of Europe included

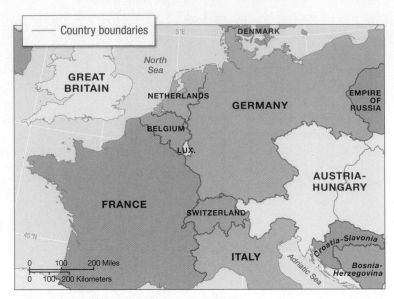

▲ **FIGURE 2.23 Central Europe in 1900** After the French Revolution at the end of the 18th century and the Napoleonic Wars at the beginning of the 19th century, Europe was reordered in a pattern of modern states.

scores of microstates (**Figure 2.22**), legacies of the feudal hierarchies of the Middle Ages. The idea of modern national states can be traced to the subsequent **Enlightenment** in Europe between the late 17th and late 18th centuries, when the ferment of ideas about human rights and democracy, together with widening horizons of literacy and communication, created new perspectives on allegiance, communality, and identity. In 1648, the Treaty of Westphalia, signed by most European powers, brought an end to Europe's seemingly interminable religious wars by making national states the principal actors in international politics and establishing the principle that no state has the right to interfere in the internal politics of any other state.

Gradually, these perspectives began to undermine the dominance of the great European continental empires controlled by family dynasties—the Bourbons, the Hapsburgs, the Hohenzollerns, the House of Savoy, and so on. After the French Revolution (1789–1793) and the kaleidoscopic changes of the Napoleonic Wars (1800–1815), Europe was reordered, in 1815, in a pattern of modern states (**Figure 2.23**). Denmark, France, Portugal, Spain, and the United Kingdom had long existed as separate, independent states. The 19th century saw the unification of Italy (1861–1870) and of Germany (1871) and the creation of Belgium, Bulgaria, Greece, Luxembourg, the

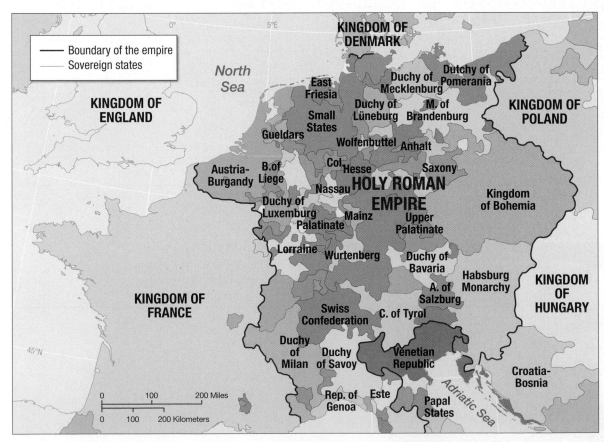

▲ **FIGURE 2.22 Central Europe in 1500** In the late Middle Ages, the political map of Europe included scores of microstates, legacies of the feudal system.

Netherlands, Romania, Serbia, and Switzerland as independent national states. Early in the 20th century they were joined by Czechoslovakia, Estonia, Finland, Latvia, Lithuania, Norway, and Sweden. Austria was created in its present form in the aftermath of World War I, as part of the carving up of the German and Austro-Hungarian empires. In 1921, long-standing religious cleavages in Ireland resulted in the creation of the Irish Free State (now Ireland), with the six Protestant counties of Ulster remaining in the United Kingdom.

Regionalism and Boundary Disputes The European concept of the nation–state has immensely influenced the modern world. As we saw in Chapter 1, the idea of a nation–state is based on the concept of a homogeneous group of people governed by their own state. In a true nation–state, no significant group exists that is not part of the nation. In practice, most European states were established around the

concept of a nation–state but with territorial boundaries that did, in fact, encompass substantial ethnic minorities (**Figure 2.24**). The result has been that the geography of Europe has been characterized by regionalism and boundary disputes throughout the 20th century and into the 21st.

Recall the example of Basque regionalism in Spain and France in Chapter 1 (see p. 37). Other examples of regionalism include regional independence movements in Catalonia (within Spain), Scotland (within the United Kingdom), and the Turkish Cypriots' determination to secede from Cyprus. Examples of **irredentism**— the assertion by a government that a minority living outside its borders belongs to it culturally and historically—include Ireland's claim on Northern Ireland (renounced in 1999), the claims of Nazi Germany on Austria and the German parts of Czechoslovakia and Poland, and the claims of Croatia and Serbia and Montenegro on

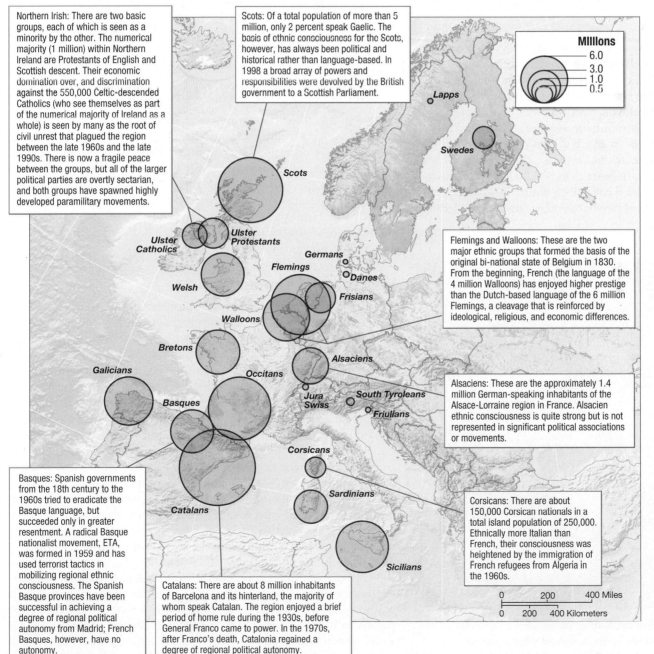

Northern Irish: There are two basic groups, each of which is seen as a minority by the other. The numerical majority (1 million) within Northern Ireland are Protestants of English and Scottish descent. Their economic domination over, and discrimination against the 550,000 Celtic-descended Catholics (who see themselves as part of the numerical majority of Ireland as a whole) is seen by many as the root of civil unrest that plagued the region between the late 1960s and the late 1990s. There is now a fragile peace between the groups, but all of the larger political parties are overtly sectarian, and both groups have spawned highly developed paramilitary movements.

Scots: Of a total population of more than 5 million, only 2 percent speak Gaelic. The basis of ethnic consciousness for the Scots, however, has always been political and historical rather than language-based. In 1998 a broad array of powers and responsibilities were devolved by the British government to a Scottish Parliament.

Flemings and Walloons: These are the two major ethnic groups that formed the basis of the original bi-national state of Belgium in 1830. From the beginning, French (the language of the 4 million Walloons) has enjoyed higher prestige than the Dutch-based language of the 6 million Flemings, a cleavage that is reinforced by ideological, religious, and economic differences.

Alsaciens: These are the approximately 1.4 million German-speaking inhabitants of the Alsace-Lorraine region in France. Alsacien ethnic consciousness is quite strong but is not represented in significant political associations or movements.

Basques: Spanish governments from the 18th century to the 1960s tried to eradicate the Basque language, but succeeded only in greater resentment. A radical Basque nationalist movement, ETA, was formed in 1959 and has used terrorist tactics in mobilizing regional ethnic consciousness. The Spanish Basque provinces have been successful in achieving a degree of regional political autonomy from Madrid; French Basques, however, have no autonomy.

Catalans: There are about 8 million inhabitants of Barcelona and its hinterland, the majority of whom speak Catalan. The region enjoyed a brief period of home rule during the 1930s, before General Franco came to power. In the 1970s, after Franco's death, Catalonia regained a degree of regional political autonomy.

Corsicans: There are about 150,000 Corsican nationals in a total island population of 250,000. Ethnically more Italian than French, their consciousness was heightened by the immigration of French refugees from Algeria in the 1960s.

Millions
6.0
3.0
1.0
0.5

Lapps
Swedes
Scots
Ulster Catholics
Ulster Protestants
Germans
Flemings
Danes
Welsh
Frisians
Walloons
Bretons
Alsaciens
Galicians
Occitans
Jura Swiss
South Tyroleans
Basques
Friulians
Corsicans
Catalans
Sardinians
Sicilians

0   200   400 Miles
0   200   400 Kilometers

◄ **FIGURE 2.24 Minority Ethnic Subgroups in Western Europe** Regional and ethnic consciousness now represents a strong political factor in many European countries. Within Spain, for example, the dominant population—some 27 million—is Castilian, but there are between 6 and 8 million Catalans, almost 2 million Basques, and about 3 million Galicians.

various parts of Bosnia-Herzegovina. Some cases of regionalism have led to violence, social disorder, or even civil war, as in Cyprus. For the most part, however, regional ethnic separatism has been pursued within the framework of civil society, and the result has been that several regional minorities have achieved a degree of political autonomy. For example, the United Kingdom created regional parliaments for Scotland and Wales.

Ethnic Conflict in the Balkans In contrast to regionalism, most cases of irredentism have contributed at some point in history to war or conflict. Nowhere has this been more evident than in the troubled region of the Balkans. When 19th-century empires were dismantled after World War I, an entirely new political geography was created in the Balkans. Those political boundaries survived until the 1990s, when the breakup of Yugoslavia marked the end of attempts to unite Serbs, Croats, and Slovenes within a single territory. The repeated fragmentation and reorganization of ethnic groups into separate states within the region has given rise to the term **balkanization,** which refers to any situation in which a larger territory is broken up into smaller units, especially where territorial jealousies give rise to a degree of hostility. In the Balkans themselves, the geopolitical reorganizations of the 1990s have left significant **enclaves,** culturally distinct territories that are surrounded by the territory of a different cultural group, and **exclaves,** portions of a country or of a cultural group's territory that lie outside its contiguous land area. These enclaves remain the focus of continued or potential hostility. In Romania, for example, there are more than 1.6 million Hungarians, whereas in Bulgaria there are more than 800,000 Turks. Serbian nationalism, however, has provided the principal catalyst for violence and conflict in the region. In the 1990s, Serbian nationalism led to attempts at **ethnic cleansing**.

▲ **FIGURE 2.25 Europe's Newest Nation State** Ethnic Kosovar Albanians celebrate the creation of Europe's newest nation state—Kosovo—in the center of the capital Pristina on February 17, 2008.

The most extreme example of ethnic cleansing was in the Kosovo region. In 1998, Yugoslavia's Serbian leader, Slobodan Milošević, initiated a brutal, premeditated, and systematic campaign of ethnic cleansing that was aimed at removing Kosovar Albanians from what had become their homeland. Serbian forces expelled Kosovar Albanians at gunpoint from villages and larger towns, looted and burned their homes, organized the systematic rape of young women, and used Kosovar Albanians as human shields to escort Serbian military convoys. By systematically destroying schools, places of worship, and hospitals, Serbian forces sought to destroy social identity and the fabric of Kosovar Albanian society.

International outrage at these human rights violations finally led to the declaration of war against Yugoslavia by NATO in March 1999. Within a few weeks, Slobodan Milošević and other Serbian leaders were indicted in the International Court of Justice for their roles in human rights violations. Since 2000, many Kosovar Albanian refugees have returned to their homeland in Kosovo to attempt to rebuild their lives, and in 2008 Kosovo declared its independence (**Figure 2.25**)—recognized by 22 of the EU's 27 member states, but strongly opposed by Serbia.

CULTURE AND POPULATIONS

Although its culture is distinctive at the global scale, Europe also bears the legacy of interactions with other world regions through colonialism, emigration, and immigration. Europe's cultures and populations are also characterized by some sharp internal regional variations. In the broadest terms, there is a significant north–south cultural divide, with a lingering legacy of an east–west political and cultural divide. However, contemporary processes of political and cultural change are beginning to modify some of the traditional patterns associated with European geography.

Religion and Language

The foundations of contemporary European culture were established by the ancient Greeks, who in turn had been influenced by the ideas and cultural and economic practices of ancient Egypt, Mesopotamia (centered along the Tigris and Euphrates rivers, in modern-day Iraq), and Phoenicia (centered along the coastal regions of modern-day Lebanon, Israel, Syria, and the Palestinian territories). Between 600 B.C.E. and 200 B.C.E., the Greeks built an intellectual tradition of rational inquiry into the causes of everything, along with a belief that individuals are free, self-understanding, and valuable in themselves. Greek understanding of the world was framed around mythological creatures, a pantheon of gods—from Aphrodite to Zeus—and a variety of specialized spirits. The Romans took over this intellectual tradition and added Roman law and a tradition of disciplined participation in the state as a central tenet of citizenship. Their culture was dominated by ritual and cult, with a pantheon of gods that related closely to the practical needs of daily life. From the Near East came the Hebrew tradition which, in conjunction with Greek thought, produced Judaism and Christianity. In these religions the individual spirit is seen as having its own responsibility and destiny within the creation. At the heart of European philosophy, then, are the curiosity, open-mindedness, and rationality of the Greeks; the civic responsibility and political individualism of both Greeks and Romans; and the sense of the significance of the free individual spirit that is found in the Judeo-Christian tradition.

Southern Europe has always been more traditional in its religious affiliations—not just in terms of the dominance of Roman Catholicism over Protestantism, but of the prevalence of conservative and mystical forms of Catholicism. The Roman Catholic Church, still one of the most widespread within Europe, emerged in the 4th century under the bishop of Rome and spread quickly through the weakening Roman Empire. Missionaries helped spread not only the gospel but also the use of the Latin alphabet throughout most of Europe. The Eastern Orthodox Church, under the auspices of the Byzantine Empire centered in Constantinople (present-day Istanbul), dominated the eastern margins of Europe and much of the Balkans, whereas Islamic influence spread into parts of the Balkans and, for a while, southern Spain.

With the religious upheavals of the 16th and 17th centuries, Protestant Christianity came to dominate much of northern Europe. More recently, immigrants from the Middle East, Africa, and South Asia have reintroduced Islam to Europe, adding an important dimension to contemporary politics as well as culture. Since the mid-20th century there has been a marked decrease in religiousness and church attendance in much of Europe. A recent survey by the European Union found that, on average, only 52% of the citizens of member states "believe in God." An additional 27% believe there is some sort of spirit or life force, while 18% do not believe there is any sort of spirit, God, or life force at all (3% declined to answer).[2]

The western part of southern Europe shares the Romance family of languages, the development of which was fostered by the spread of the Roman Empire. A second major group of languages—Germanic languages—occupy northwestern Europe, extending as far south as the Alps (**Figure 2.26**). English is one of the Germanic family of languages, an amalgam of Anglo-Saxon and Norman French, with Scandinavian and Celtic traces. A third major language group consists of Slavic languages, which dominate Eastern Europe.

These broad geographic divisions of religion, language, and family life are reflected in other cultural traits: folk art, traditional costume, music, folklore, and cuisine. For example, there is a Scandinavian cultural subregion with a collection of related languages (except Finnish), a uniformity of Protestant denominations, and a strong cultural affinity in art and music that reaches back to the Viking age and even to pre-Christian myths. A second distinctive subregion is the sphere of Romance languages in the south and west. A third is the British Isles, bound by language, history, art forms, and folk music, but with a religious divide between the Protestant Anglo-Saxon and Catholic Celtic spheres. A fourth clear cultural subregion is the Germanic sphere of central Europe, again with mixed religious patterns—Lutheran Protestantism in the

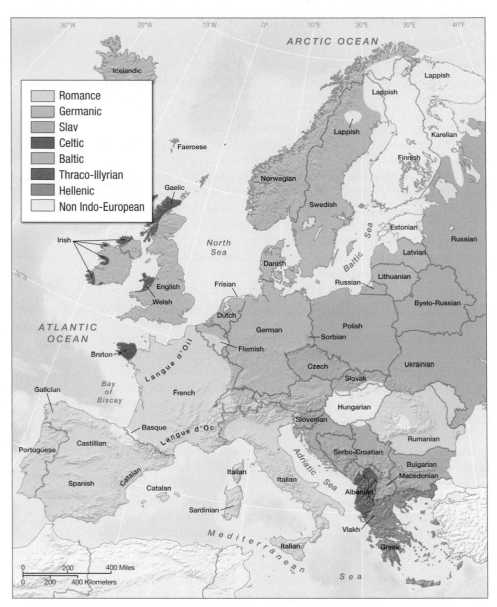

▲ **FIGURE 2.26 Major Languages in Europe** Although three main language groups—Romance, Germanic, and Slav—dominate Europe, differences among specific languages are significant. These differences have contributed to the cultural diversity of Europe but have also contributed to ethnic and geopolitical tensions.

north, Roman Catholicism in the south—but with a common bond of language, folklore, art, and music. The Slavic subregion of eastern and southeastern Europe forms another broad cultural subregion, though there is considerable diversity beyond the related languages and some common physical traits.

APPLY YOUR KNOWLEDGE Select three European countries and conduct an Internet search to investigate the rates of religious adherence of their populations. How might you account for the differences in religious beliefs among the three countries? (Hint: A good source of data is the European Commission's survey of social values: http://ec.europa.eu/public_opinion/archives/ebs/ebs_225_report_en.pdf.)

[2]European Commission (2005). "Social Values, Science and Technology," *Special Eurobarometer 225*, 2005, http://ec.europa.eu/public_opinion/archives/ebs/ebs_225_report_en.pdf.

Cultural Practices, Social Differences, and Identity

One of the single most important developments in European cultural sensibility has been the emergence of the idea of **Modernity**. Modernity is a way of thinking that was triggered by the changing world geography of the Age of Discovery; it emphasizes innovation over tradition, rationality over mysticism, and utopianism over fatalism. As Europeans tried to make sense of their own ideas and values in the context of those they encountered in the East, in Africa, in Islamic regions, and among Native Americans during the 16th and 17th centuries, many certainties of traditional thinking were cracked open. In the 18th century, this ferment of ideas culminated in the Enlightenment movement, which was based on the conviction that all of nature, as well as human beings and their societies, could be understood as a rational system. Politically, the Enlightenment reinforced the idea of human rights and democratic forms of government and society. Expanded into the fields of economics, social philosophy, art, and music, the Enlightenment gave rise to the cultural sensibility of Modernity.

The subsequent dislocations and new experiences introduced by industrialization resulted in still new ways of seeing and new ways of representing things. The places where all this was played out with the greatest intensity were the major cities of Europe. In London, painters, poets, and critics set out to reform art. In Vienna, thinkers met in cafés to discuss new ideas about art, design, psychiatry and politics. In Milan, Futurists propagated the idea that the past was a corrupting influence on society, celebrating speed, technology and youth as the keys to the triumph of humanity over nature. Paris became the seat of revolutionary ideas, an unrivaled artistic and cultural scene that included, at various times, the artists Jean-Baptiste Corot, Pierre-Auguste Renoir, Henri de Toulouse Lautrec, Paul Cézanne, Vincent Van Gogh, Georges Braque, Pablo Picasso, and Henri Matisse; and writers Victor Hugo, Honoré de Balzac, and Emile Zola.

By the late 20th century, after the decline of heavy industry and repeated episodes of economic recession; after two terrible world wars; after interludes of fascist dictatorships in Germany, Italy, Greece, Portugal, and Spain; after a protracted period of being on the front line of a Cold War that divided European geography in two; and after intermittent episodes of regional and ethnic conflict, it was not surprising that the culture of Europe had become a culture of doubt and criticism, heavily influenced by a search for radical rethinking. In this search, Europeans not only established a new cultural sensibility for themselves, but they also generated some powerful new ideas and philosophies that influenced other cultures around the world. Dismay with the side effects of capitalism and, later, horror at the results of fascism and Nazism gave a strong impetus to left-wing critiques that were powerful enough to reshape entire national and regional cultures and, with them, some dimensions of regional geographies. **Fascism**, of which Nazism was one variety, involves a centralized, autocratic government and values nation and race over the individual.

The most profound influence of all was Karl Marx (**Figure 2.27**), whose penetrating critique of industrial capitalism (written in London and drawing heavily on descriptions of conditions in Manchester, England, supplied by his colleague Friedrich Engels) inspired both a socialist political economy in Russia and a fascist countermovement in Germany. After World War II, western European left-wing critique portrayed both fascism and Soviet-style socialism as essentially

▲ **FIGURE 2.27 Karl Marx (1818–1883)** Marx was a German political scientist and economist whose concepts and theories about economic change and social justice have had a profound influence on culture and politics in Europe and beyond.

imperialist, whereas American-style capitalism was critiqued as being intrinsically exploitative in privileging the individual and property over the community and the public good.

As a result of this interchange, contemporary Europe has a distinctive set of social values, and the "European Dream" is quite distinctive from the "American Dream":

> The European Dream emphasizes community relationships over individual autonomy, cultural diversity over assimilation, quality of life over the accumulation of wealth, sustainable development over unlimited material growth, deep play over unrelenting toil, universal human rights and the rights of nature over property rights, and global cooperation over the unilateral exercise of power.[3]

A "European" Identity? Globalization has heightened people's awareness of cultural heritage and ethnic identities. As we saw in Chapter 1, the more universal the diffusion of material culture and lifestyles, the more valuable regional and ethnic identities tend to become. Globalization has also brought large numbers of immigrants to some European countries, and their presence has further heightened awareness of cultural identities.

In the more affluent countries of northwestern Europe, immigration has emerged as one of the most controversial issues since the end of the Cold War. Although the economic benefits of immigration far outweigh any additional demands that may be made on a country's health or welfare system, fears that unrestrained immigration might lead to cultural fragmentation and political tension have provoked some governments to propose new legislation to restrict immigration. The same fears have been responsible for a resurgence of popular **xenophobia**—a hate, or fear, of foreigners—in some countries. In Germany, for example, right-wing nationalistic groups

[3]J. Rifkin, *The European Dream: How Europe's Vision of the Future Is Quietly Eclipsing the American Dream* (New York: Tarcher, 2004), p. **3**.

have attacked hostels housing immigrant families and citizenship laws have prevented second-generation *Gastarbeiter* ("guest worker") families from obtaining German citizenship. In France, claims that immigrants from North Africa are a threat to the traditional French way of life have led to some success for the *National Front Party*. Asylum seekers, drawn to northwestern Europe from all parts of the globe in such large numbers that they have had to be accommodated in processing centers, have also provoked negative reactions.

All this raises the question, How "European" *are* the populations of Europe? European history and ethnicity have resulted in a collection of national prides, prejudices, and stereotypes that are strongly resistant to the forces of cultural globalization. Germans continue to be seen by most other Europeans as a little overserious, preoccupied by work, and inclined to arrogance. Scots carry the popular image of a dour, unimaginative, ginger-haired people who love bagpipe and accordion music, dress in kilts and sporrans, live on whisky and porridge, and spend as little as possible. The English are seen as a nation of lager-swilling hooligans, well-meaning middle classes, and out-of-touch aristocrats. Norwegians and Danes continue to resent the Swedes' "neutrality" during World War II, and so on. In reality, such stereotypes are, of course, exaggerations that stem from the behaviors of a relative minority, and opinion surveys show that these stereotypes, prejudices, and identities are steadily being countered by a growing sense of European identity, especially among younger and better-educated persons. Much of this can be attributed to the growing influence of the European Union.

Women in European Society Another powerful postwar movement deeply critical of the dominant structures of capitalist society was feminism, built on the ideas of Simone de Beauvoir expressed in her 1949 book, *The Second Sex*. Compared with peoples in most other world regions, Europeans have been more willing to address the deep inequalities between men and women that are rooted in both traditional societies and industrial capitalism. Still, patriarchal society and the culture of machismo remain strong in Mediterranean Europe—especially in rural areas—and working-class communities throughout Europe are still characterized by significant gender inequalities.

In northwestern Europe—and especially in Scandinavia—gender equality has improved most, as a result of both the progressive social values of the "baby boom" generation (people born in the years following World War II) and legislation that has translated these values into law. By the mid-1980s, younger men in much of northwestern Europe had acquired a new, progressive collective identity associated with ideals of gender equality—especially as they relate to men's domestic roles. More recently, however, a "men-behaving-badly" syndrome—known as "laddism" in the United Kingdom—has emerged in reaction.

In global context, Europe stands out as a region where women's representation in senior positions in industry and government is relatively high. Women in Europe generally have a significantly longer life expectancy than men and have comparable levels of adult literacy. Nevertheless, women in the European labor force tend to earn, on average, only 45% to 65% of what men earn. These statistics demonstrate a distinctly regional pattern. Broadly speaking, the gender gaps in education, employment, health, and legal standing are wider in southern and Eastern Europe and narrower in Scandinavia and northwestern Europe. In part, this reflects regional differences in social customs and ways of life; in part, it reflects regional differences in overall levels of affluence.

Demography and Urbanization

A distinctive characteristic of Europe as a whole is the size and relative density of its population. With less than 7% of Earth's land surface, Europe contains about 13% of its population at an overall density of nearly 100 persons per square kilometer (260 per square mile). Within Europe, the highest national densities match those of Asian countries such as Japan, the Republic of Korea, and Sri Lanka. On the other hand, population density in Finland, Norway, and Sweden stands at only about 15 persons per square kilometer (39 persons per square mile), the same as in Kansas and Oklahoma (**Figure 2.28**). This reflects a fundamental feature of the human geography of Europe: the existence of a densely populated core and a sparsely populated periphery. Recall that earlier in the chapter we noted the economic roots of this core–periphery contrast.

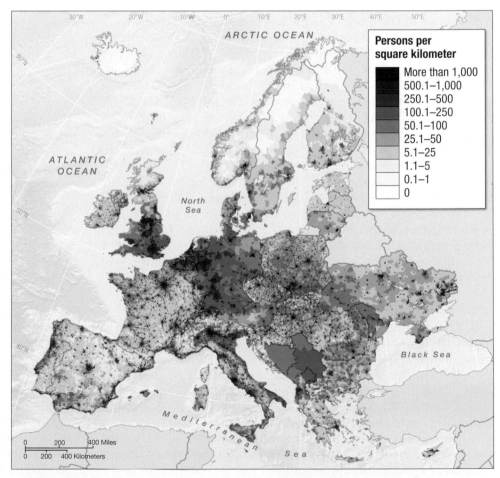

▲ **FIGURE 2.28 Population Density in Europe** The distribution of population in Europe reflects the region's economic history. The highest densities are in the Golden Triangle, the newer industrial regions of northern Italy, and the richer agricultural regions of the North European Lowlands.

Whereas the population of the world as a whole is increasing fast, the population of Europe is roughly stable. Europe's population boom coincided roughly with the Industrial Revolution of the late 18th to late 19th centuries. Today, Europe's population is growing slowly in some regions, while declining slightly in others. The main reason for Europe's slow population growth is a general decline in birthrates (though certain subgroups, especially immigrant groups, are an exception to this trend).

It seems that conditions of family life in Europe, including readily available contraception, have led to a widespread fall in birthrates. The average size of families has dropped well below the rate needed for replacement of the population (about 2.1 children per family), to about 1.5 per family in 2008. The "baby boom" after World War II was followed by a "baby bust." Meanwhile, life expectancy has increased because of improved health care, medical knowledge, and healthier lifestyles. The effect is not sufficient to outweigh falling birthrates, but it has meant a dramatic increase in the proportion of people over the age of 65, from 9% in 1950 to nearly 16% in 2010. The ratio of retirees to workers in Europe is expected to double by 2050, from four workers per retiree to two workers per retiree. By then, one-third of Europe's population will be over 60, compared to 13% who will be under 16. Germany's population (**Figure 2.29**) reflects these trends and shows the impact of two world wars.

The European Diaspora The upheavals associated with the transition to industrial societies, together with the opportunities presented by colonialism and imperialism and the dislocations of two world wars, have dispersed Europe's population around the globe. Beginning with the colonization of the Americas, vast numbers of people have left Europe to settle overseas. The full flood of emigration began in the early 19th century, with people from northwestern and central Europe heading for North America and southern Europeans heading for destinations throughout the Americas. In addition, large numbers of British left for Australia and New Zealand and eastern and southern Africa. French and Italian emigrants traveled to North Africa, Ethiopia, and Eritrea, and the Dutch went to southern Africa and Indonesia. The final surge of emigration occurred just after World War II, when various relief agencies helped homeless and displaced persons move to Australia and New Zealand, North America, and South Africa, and large numbers of Jews settled in Israel.

Migration within Europe Industrialization and geopolitical conflict have also resulted in a great deal of population movement within Europe. Three major waves of industrial development drew migrants from less-prosperous rural areas to a succession of industrial growth areas around coalfields. As industrial capitalism evolved, the diversified economies of cities offered the most opportunities and the highest wages, thus prompting a further redistribution of population. In Britain, this involved a drift of population from manufacturing towns southward to London and the southeast. In France, migration to Paris from towns all around France resulted in a polarization between Paris and the rest of the country. Some countries, developing an industrial base after the "coalfield" stage, experienced a more straightforward shift of population, directly from peripheral rural areas to prosperous metropolitan

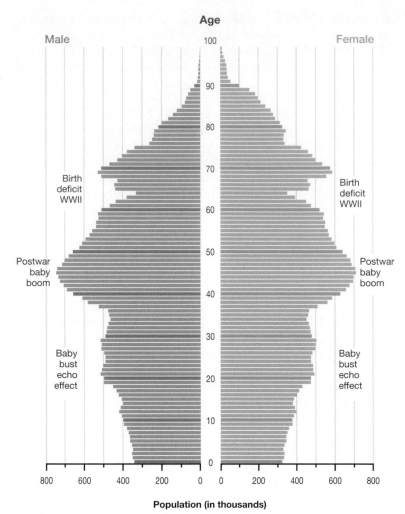

▲ **FIGURE 2.29 Population of Germany, by Age and Sex, 2010**
Germany's population profile is that of a wealthy country that has passed through the postwar baby boom and currently possesses a low birthrate. It is also the profile of a country whose population has experienced the ravages of two world wars.

regions. In this way, Barcelona, Copenhagen, Madrid, Milan, Oslo, Stockholm, and Turin all emerged as regionally dominant metropolitan areas.

Wars and political crises have also led to significant redistributions of population within Europe. World War I forced about 7.7 million people to move. Another major transfer of population took place in the early 1920s, when more than 1 million Greeks were transferred from Turkey and a half million Turks were transferred from Greece in the aftermath of an unsuccessful Greek attempt to gain control over the eastern coast of the Aegean Sea. Soon afterward, more people were on the move, this time in the cause of ethnic and ideological purity, as the policies of Nazi Germany and fascist Italy began to emerge. Jews, in particular, were squeezed out of Germany. With World War II, further forced migrations occurred that involved approximately 46 million people.

These migrations, together with mass exterminations undertaken by Nazi Germany, left large parts of west and central Europe with significantly fewer ethnic minorities than before the war. In Poland, for example, minorities constituted 32% of the population before the war but only 3% after the war. Similar changes occurred in Czechoslovakia—from 33% to 15%—and in Romania—from 28% to 12%. Southeast Europe did not experience such large-scale transfers, and, as a result, many ethnic minorities remained intermixed or surrounded and isolated, as in the former republic of Yugoslavia. The geopolitical division of Europe after the war also resulted in significant transfers of population; West Germany, for example, had absorbed nearly 11 million refugees from Eastern Europe by 1961.

Recent Migration Streams More recently, the main currents of migration within Europe have also been a consequence of patterns of economic development. Rural–urban migration continues to empty the countryside of Mediterranean Europe as metropolitan regions become increasingly prosperous. Meanwhile, cities have experienced a decentralization of population as factories, offices, and housing developments have moved out of congested central areas. Another stream of migration has involved better-off retired persons, who have congregated in spas, coastal resorts, and picturesque rural regions.

▲ **FIGURE 2.30 Immigrant District** Southall in West London, also known as Little India by some, is an area almost completely populated by people from South Asia.

The most striking of all recent streams of migration within Europe, however, have been those of migrant workers. These population movements were initially the result of Western Europe's postwar economic boom in the 1960s and early 1970s, which created labor shortages in Western Europe's industrial centers. The demand for labor represented welcome opportunities to many of the unemployed and poorly paid workers of Mediterranean Europe and of former European colonies. By the mid-1970s these migration streams had become an early component of the globalization of the world economy. By 1975, between 12 and 14 million immigrants had arrived in northwestern Europe. Most came from Mediterranean countries—Spain, Portugal, southern Italy, Greece, Yugoslavia, Turkey, Morocco, Algeria, and Tunisia. In Britain and France, the majority of immigrants came from former colonies in Africa, the Caribbean, and Asia. In the Netherlands, most came from former colonies in Indonesia. Most of these immigrants have stayed on, adding a striking new ethnic dimension to Europe's cities and regions (**Figure 2.30**).

Meanwhile, it is estimated that more than 18 million people moved within Europe during the 1980s and 1990s as refugees from war and persecution or in flight from economic collapse in Russia and Eastern Europe. Civil war and dislocation in the Balkans displaced more than 4 million people in the early 1990s. Wars in Iraq and Afghanistan have also resulted in significant numbers of immigrants to Europe, especially to Denmark and Sweden.

Until the global recession that began in 2008, labor migration continued to expand throughout Europe. Economic growth and political stability in countries that recently joined the EU—Cyprus, Hungary, Czechia, Slovakia, and Slovenia—have made them destination countries in their own right. Today, most European states are now net immigration countries (the exceptions are Lithuania and Bulgaria), and within Europe as a whole there are more than 56 million migrants (compared to about 42 million in North America).

Muslims in Europe Europe is home to approximately 15 to 20 million Muslims—between 4% and 5% of the total population of the region. The greatest concentrations of Muslims are in the Balkan countries, where Islam has been important for centuries, a legacy of the Turkish Ottoman Empire. Although the empire was dissolved at the end of World War I, Islamic culture has remained in place. In Albania, about 65% of the population is Muslim; in Bosnia-Herzegovina, the figure is 55%; in Macedonia, 29%; and in Montenegro, about 17%. Kosovo's population of 2 million is 93% Muslim (see "Visualizing Geography: Europe's Muslims," p. 80). The majority of Europe's Muslims, however, are located in the industrial cities of Western Europe, and they are a relatively recent addition to the population of the region.

Government policies in most European countries favor multiculturalism, an idea that, in general terms, accepts all cultures as having equal value. The growth of Muslim communities and their resistance to cultural assimilation, however, has challenged the European ideal of strict separation of religion and public life. Many of these immigrant groups, now facing high unemployment and low wages as the postwar boom has leveled off, have clustered together in distinctive urban neighborhoods. In Muslim neighborhoods with high concentrations of unemployment, heavy-handed policing and racial discrimination can easily trigger civil disorder. Islamist terrorist attacks against west European targets have meanwhile heightened fears and tensions between Muslim communities and host populations.

VISUALIZING GEOGRAPHY

Europe's Muslims

The mass immigration of Muslims to Europe was a consequence of Europe's post–World War II economic boom, which created labor shortages in Western Europe's industrial centers in the 1960s and 1970s and spawned programs that encouraged migrant workers. The French Muslim population of between 5 and 6 million (about 10% of the total population) is the largest in Western Europe. About 70% have their heritage in France's former North African colonies of Algeria, Morocco, and Tunisia. By contrast, Germany's Muslim population of about 3.7 million is dominated by people of Turkish origin. In the United Kingdom, significant numbers of Muslims arrived in the 1960s as people from the former colonies took up offers of work. Some of the first were East African Asians, whereas many came from south Asia. Permanent communities formed, and at least 50% of the current population of 1.6 million was born in the United Kingdom, with one-third of the Muslim population currently age 15 or less. Approximately 1 million Muslims live in the Netherlands, the majority with ties to the former Dutch colonies of Suriname and Indonesia. In the past 20 years, Greece, Italy, and Spain have received large numbers of illegal Muslim migrants. Greece has experienced migration of mainly Albanian Muslims, but also Muslims from Pakistan, Bangladesh, and Iraq. In Spain, the Muslim migrants have mostly been from Morocco and sub-Saharan Africa. In Italy, large numbers of Muslim migrants have also arrived illegally

from North Africa and Albania. In northern Europe, meanwhile, recent streams of Muslim immigration have been dominated by legal immigrants, mostly from Iraq, Somalia, Eritrea, and Afghanistan, who have arrived through refugee and asylum programs (**Figure 2.2.1**).

Since the early 1990s, migration from predominantly Muslim countries into Europe can be broadly characterized as follows:

(1) In the north of Europe, Muslim migration has been dominated by—largely— legal entry through refugee/asylum

applications and employment opportunities motivated by war and civil unrest at Europe's borders, and by associated economic push and pull factors.

(2) In the south of Europe Muslim migration has been dominated by largely illegal entry (including trafficking in human beings) as a reflection of the geographical proximity of countries with Muslim populations to southern Europe, and motivated by the same factors as migration to the north of Europe. ∎

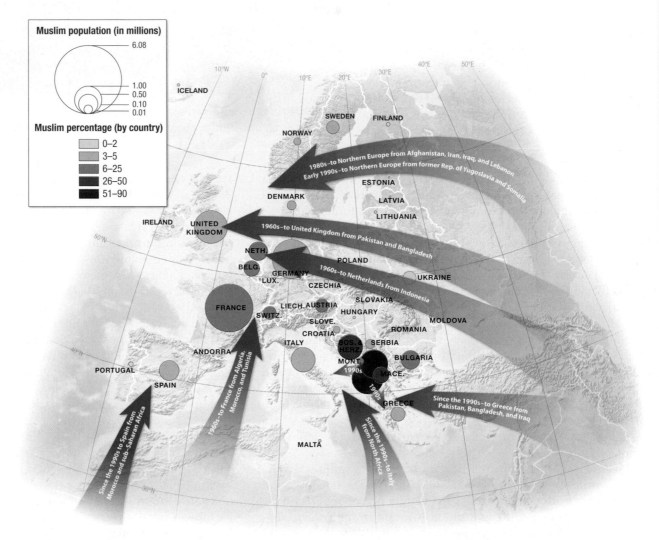

▲ **FIGURE 2.2.1 Europe's Muslims** The immigration of Muslims to Europe from several different world regions has had a significant impact on demography, culture, and politics.

▲ **FIGURE 2.31 The Politics of Immigration** This is a campaign poster (translated as "Stop mass immigration") from the right-wing Swiss People's Party (SVP), the top vote-getting party in the 2011 parliamentary elections in Switzerland. The poster promotes the SVP's anti-immigration policies.

Europe's Towns and Cities Europe as a whole is highly urbanized, with a dense network of hundreds of cities and thousands of small towns. Because of the pivotal role of Europe in shaping world trade and politics since the 1500s, a number of its cities have been of special significance in organizing and influencing spatial organization well beyond their regional or national boundaries. These are known as **world cities**. In the first stages of the growth of the modern world economy, these cities played key roles in the organization of trade and the execution of colonial, imperial, and geopolitical strategies. The world cities of the 17th century were London, Amsterdam, Antwerp, Genoa, Lisbon, and Venice. In the 18th century, Paris, Rome, and Vienna also became world cities. In the 19th century, Berlin and Manchester joined the ranks of world cities while Amsterdam, Antwerp, Genoa, Lisbon, and Venice became less influential. Today, London (**Figure 2.32**), with its financial markets, specialized office space, specialized business services, expert professionals, and high-order cultural amenities, is Europe's preeminent world city. Other contemporary European world cities include Berlin, Paris (**Figure 2.33**, p. 86), Frankfurt, and Milan, control centers for the flows of information, cultural products, and finance that collectively sustain the economic and cultural globalization of the world.

In the past decade, immigration has become a major electoral issue in Austria, Belgium, Denmark, France, the Netherlands, and Switzerland (**Figure 2.31**), and Muslim communities across Europe have faced increased resentment and hostility from host populations. Tensions remain high, and the cultural issues associated with Europe's Muslim population are likely to continue to be an important dimension of European politics, especially with the possibility of Turkey joining the European Union, which would add about 83 million Muslims to Europe's population.

At the other end of the urban spectrum are small towns with fewer than 50,000 inhabitants. Many of these are traditional market towns whose origins were in the trading networks of preindustrial Europe, but which were bypassed—literally—by the canals and railways of the industrial era. Others, in contrast, are specialized mining and manufacturing towns that only emerged during the industrial era but have never grown. Overall, more than one-fifth of Europe's population lives in small towns. Beyond the Golden Triangle, the figure is often closer to one-third and in some regions, such as Scandinavia and southern Italy, at least half of the population live

◄ **FIGURE 2.32 London** London's original river port trade gave rise to a commercial core that developed into a major financial hub in the district known as the City. The skyline of this financial precinct is dominated by the distinctive "gherkin" building.

▲ **FIGURE 2.33 La Défense** One of the largest and most successful growth centers within metropolitan Paris, La Défense was designed by planners to attract office development from the central core of the city.

in small towns. Small towns play an important role in the European urban system. They do not have the economic advantages of creativity and productivity associated with metropolitan settings but serve important functions as local and regional service centers, as places that can absorb metropolitan overspill, as specialized fishing, mining, and agricultural processing centers, as centers for tourism, and—in traditional market towns—as centers and symbols of regional culture and identity (**Figure 2.34**).

FUTURE GEOGRAPHIES

Some dimensions of Europe's geography are more certain than others. In some ways, the future is already here, embedded in the region's institutional structures and in the dynamics of its populations. We know, for example, a good deal about the demographic trends of the next quarter century, given present populations, birth and death rates, and so on. Given the long and deep history of regional development in Europe, we also know a good deal about the distribution of environmental resources and constraints, about the characteristics of local and regional economies, and about the legal and political frameworks within which geographic change will probably take place. On the other hand, we can only guess at some aspects of the future. One of the most predictable

aspects of Europe's future geographies concerns its aging populations; in contrast, one of the most speculative is the politics and economics of European integration.

Coping with Aging Populations

As we discussed earlier in this chapter, a widespread fall in birthrates is leading toward a graying population in Europe. Population growth has hit record lows, especially in southern European countries. Italy has a fertility rate of 1.2 children per woman, which puts it among the lowest in the world. Europe already has 19 of the world's 20 oldest countries in terms of average population age (**Figure 2.35**). Given current trends, about a quarter of Europe's population will be age 65 or above by 2025; and by 2050, about 20% of the population will be age 80 or above. This will place an immense burden on Europe's working-age populations and their capacity to fund health-care and pension systems at a time when economic growth rates are likely to be modest at best. The **old-age dependency ratio**—the number of people age 65 and older compared with the number of working-age people (ages 15 to 64)—will more than double by 2050, from one in every four to fewer than one in every two.

The aging of Europe's populations is unprecedented in human history and poses new challenges for government budgets and new uncertainties in national politics. The fiscal challenges that will confront governments will likely limit their capacities to address other problems, including problems of environmental quality and climate change as well as problems of uneven regional economic development. These challenges and

▲ **FIGURE 2.34 Procida, Italy** Europe has thousands of small towns that were largely by-passed by industrialization. Many, like this example of Procida, near Naples, have become attractive for tourism or retirement.

of Europe are becoming increasingly integrated. We have also seen how the European Union has grown ever larger, with a huge budget dedicated to economic integration and political cohesion. The future stability of the EU, however, depends on whether it undertakes major structural economic and social reforms to deal with the problem of its aging workforce. As noted, this will require more immigration and much better integration of immigrant workers, most of whom are likely to be coming from North Africa and the Middle East. Even if more migrant workers are not allowed in, Western Europe will have to integrate its growing Muslim population, a potential flashpoint for social and cultural conflict that could undermine the European project.

There are also concerns that the enlargement of the EU has put its stability at risk. Many of the millions of citizens in new member countries have very modest incomes yet now have an expectation of the same kinds of economic standards and social rights that those in France, Germany, and Britain take for granted. New EU citizens also find themselves bound by thousands of "harmonizing" policy directives, such as rules on food preparation and hygiene in restaurants, which are impossibly expensive to implement without help. It is likely to take several of the new EU

▲ **FIGURE 2.35 Aging Europe** These men in Todi, Italy, seem to be enjoying their retirement.

uncertainties might be mitigated if European countries were to allow significant flows of immigration. Such "replacement migration" is unlikely to be encouraged, however, while Europe's economies remain sluggish. The European Union has already stepped up the control of illegal immigration across its borders, and some countries have taken steps to impose strict limits on immigration. This leads, in turn, to questions of the politics and economics of European integration.

The EU: Holding It Together

This chapter began with a discussion of the Gotthard Base Tunnel, an example of one of the many ways in which the regions and economies

member countries at least a decade, probably two decades or more, to catch up to average EU living standards. Meanwhile, the expansion of the EU will remain a costly business; EU subsidies to new member countries currently account for more than 40% of the total budget. On top of this, the global fiscal crisis of 2008–2012 has highlighted an even greater danger to the European project; the indebtedness of countries like Greece, Portugal, Italy, and Spain, whose uncompetitive and vastly different economies have been tied to the same currency as stronger economies in the "eurozone" (**Figure 2.36**). Dealing with the stress on monetary policies is likely to influence future European geographies as much as any other single factor.

◀ **FIGURE 2.36 Stress and Dissent** Mounting national debt required Greece to impose severe cutbacks in wages and pensions in return for continued loans and guarantees from the European Union, prompting widespread public protests in 2012.

| LEARNING OUTCOMES *REVISITED* |

■ **Understand how Europe's landscapes developed around regions of distinctive geology, relief, landforms, soils, and vegetation.**

Europe's four key regions are the Northwestern Uplands, the Alpine System, the Central Plateaus, and the North European Lowlands. The characteristics of these regions are a product of centuries of human adaptation to climate, soils, altitude, and aspect and to changing economic and political circumstances. Farming practices, field patterns, settlement types, local architecture, and ways of life have all become attuned to the opportunities and constraints of regional physical environments, producing distinctive regional landscapes.

■ **Explain the rise of Europe as a major world region and understand how Europeans established the basis of a worldwide economy.**

The rise of Europe as a major world region had its origins in the emergence of a system of merchant capitalism in the 15th century, when advances in business practices, technology, and navigation made it possible for merchants to establish a worldwide economy in the space of less than 100 years. Since then, Europe's regional geographies have been comprehensively recast several times: by the new production and transportation technologies that marked the onset of the Industrial Revolution, by two world wars, and by the Cold War rift between Eastern and Western Europe.

■ **Describe the reintegration of Eastern Europe.**

The reintegration of Eastern Europe has added a potentially dynamic market of 150 million consumers to the European economy. Overall, Eastern Europe functions as a group of economically peripheral regions, with agriculture still geared to local markets and former COMECON trading opportunities and industry still geared more to heavy industry and standardized products than to competitive consumer products.

■ **Summarize the importance of the European Union and its policies.**

The European Union (EU) emerged after World War II as a major factor in reestablishing Europe's role in the world. In overall terms, Europe, with about 12% of the world's population, accounts for almost 35% of the world's exports, more than 40% of the world's imports, and about one-third of the world's aggregate GDP. The EU itself is now a sophisticated and powerful institution with a pervasive influence on patterns of economic and social well-being within its member states. It has a population of more than 500 million, with a combined GDP slightly larger than that of the United States.

■ **Identify the principal core regions within Europe and describe the pattern of urban development.**

The principal core region within Europe is the Golden Triangle, which stretches between London, Paris, and Berlin and includes the industrial heartlands of central England, northeastern France, and the Ruhr district of Germany. A secondary, emergent core is developing along a north–south crescent that straddles the Alps, stretching from Frankfurt, just to the south of the Golden Triangle, through Stuttgart, Zürich, and Munich, to Milan and Turin.

■ **Assess the importance of language, religion, and ethnicity in shaping ongoing change within Europe.**

Many of the countries of Europe are relatively new creations and political boundaries within Europe have changed quite often, reflecting changing patterns of economic and geopolitical power and a continuous struggle to match territorial boundaries to cultural, religious, and ethnic identities.

KEY TERMS

acid rain (p. 58)

aspect (p. 57)

balkanization (p. 74)

buffer zone (p. 63)

bureaucratic class (p. 63)

cirque (p. 54)

command economy (p. 63)

Common Agricultural Policy (CAP) (p. 68)

Communist Council for Mutual Economic Assistance (COMECON) (p. 63)

deindustrialization (p. 68)

disinvestment (p. 68)

egalitarian society (p. 63)

Emissions Trading Scheme (ETS) (p. 53)

enclave (p. 74)

Enlightenment (p. 72)

entrepôt (p. 60)

ethnic cleansing (p. 74)

European Union (EU) (p. 65)

exclave (p. 74)

fascism (p. 76)

feudal system (p. 59)

fjord (p. 54)

flexible production region (p. 70)

Golden Triangle (p. 62)

Great Depression (p. 62)

heathland (p. 57)

Holocaust (p. 62)

imperialism (p. 62)

Iron Curtain (p. 63)

irredentism (p. 73)

latifundia (p. 57)

Little Ice Age (p. 57)

loess (p. 57)

Marshall Plan (p. 63)

merchant capitalism (p. 59)

minifundia (p. 57)

Modernity (p. 76)

moraine (p. 54)

old-age dependency ratio (p. 82)

pastoralism (p. 54)

physiographic region (p. 53)

polder (p. 58)

privatization (p. 57)

satellite state (p. 63)

Single European Act of 1985 (SEA) (p. 68)

social housing (p. 65)

Southern Crescent (p. 70)

state socialism (p. 63)

steppe (p. 57)

technology systems (p. 61)

trading empire (p. 62)

tundra (p. 54)

United Nations Kyoto Protocol (p. 53)

uplands (p. 54)

watershed (p. 51)

welfare state (p. 65)

world cities (p. 81)

xenophobia (p. 76)

THINKING GEOGRAPHICALLY

1. How has Europe benefited from its location and its major physical features?

2. What key inventions during the period from 1400 to 1600 helped European merchants establish the basis of today's global economy? Why?

3. Which imports from the colonies helped transform Europe? Focus on natural resources and new crops.

4. How did the European Union (EU) develop? Why is the EU's Common Agricultural Policy (CAP) so important?

5. What migration patterns characterized Europe during the 19th and 20th centuries? Consider movement within Europe as well as movement to and from Europe.

MasteringGeography™

Looking for additional review and test prep materials? Visit the Study Area in MasteringGeography™ to enhance your geographic literacy, spatial reasoning skills, and understanding of this chapter's content by accessing a variety of resources, including **MapMaster** interactive maps, videos, RSS feeds, flashcards, web links, self-study quizzes, and an eText version of *World Regions in Global Context*.

3 | The Russian Federation, Central Asia, and the Transcaucasus

Thousands of Russian citizens took part in the "million person march" against the inauguration of Vladimir Putin as President of Russia in Moscow in May of 2012.

IN MARCH 2012, AN AMAZING THING HAPPENED IN the Russian Federation (or Russia, as it is commonly known). Former President Vladimir Putin (2000–2008), who was serving as prime minister at the time of the election, was elected to a new, six-year term as president. Putin stepped down as president in 2008 because of a constitutional requirement that Russian presidents serve no more than two consecutive terms; however, as the prime minister working closely with President Dmitry Medvedev (Putin's own pick for president), Putin remained at the center of Russian politics. Under Putin's and then Medvedev's leadership between 2000 and 2012, Russia's economic growth rate accelerated, making it the world's ninth-largest economy in 2011. At that time, the country's overall debt was under 10% of the country's total gross domestic product (GDP). This stands in stark contrast to other European and North American economies, some of which now have debt that exceeds 100% of total GDP.

But opposition forces in Russia remain concerned that Putin, a former intelligence officer in the old Soviet Union, has too much power and control. Critics argue that Putin's rule

LEARNING OUTCOMES

- ▪ Describe how the location and vast size of the Russian Federation, Central Asia, and Transcaucasus region have influenced its climate and how climate change may influence the region's future.

- ▪ Identify the topographic areas of the region and how they are influenced by the larger global process of plate tectonics.

- ▪ List and describe the region's five major environmental zones and provide examples of how people have adapted to and modified the physical landscapes in these zones.

- ▪ Compare and contrast the regional environmental impacts of Soviet era planning with those of the post-Soviet transition.

- ▪ Describe how economic transitions in the region are playing out in the postsocialist era.

- ▪ Analyze the changing political trends in the region in terms of fragmentation and realignment.

- ▪ Summarize the cultural legacies and new developments in the region in terms of high, revolutionary, and popular forms of culture.

- ▪ Describe and explain how birth and death rates have changed relative to one another over the recent history of the region and discuss the implications of recent demographic contractions.

has resulted in limited political participation in this expansive and diverse country. In April 2012, a group of political opposition leaders walked out as Putin addressed the Russian parliament to protest what they described as voting irregularities in Putin's victorious 2012 elections. Some observers also warn that Russia's recent economic successes are based, in large part, on the country's massive oil reserves. This dependence on primary commodity (raw material) production puts Russia at risk, should global oil prices slip in the future.

For many of the protesters, Putin's return to the presidency clearly raises questions about the future of political pluralism in Russia. Putin's return to the presidency also suggests that Russia might play a larger role on the world stage, where Putin has always been an important player. As Putin embarks on the first of what could be two more terms as president, Russia will likely strive to expand its regional and global significance as a political and economic power. ▪

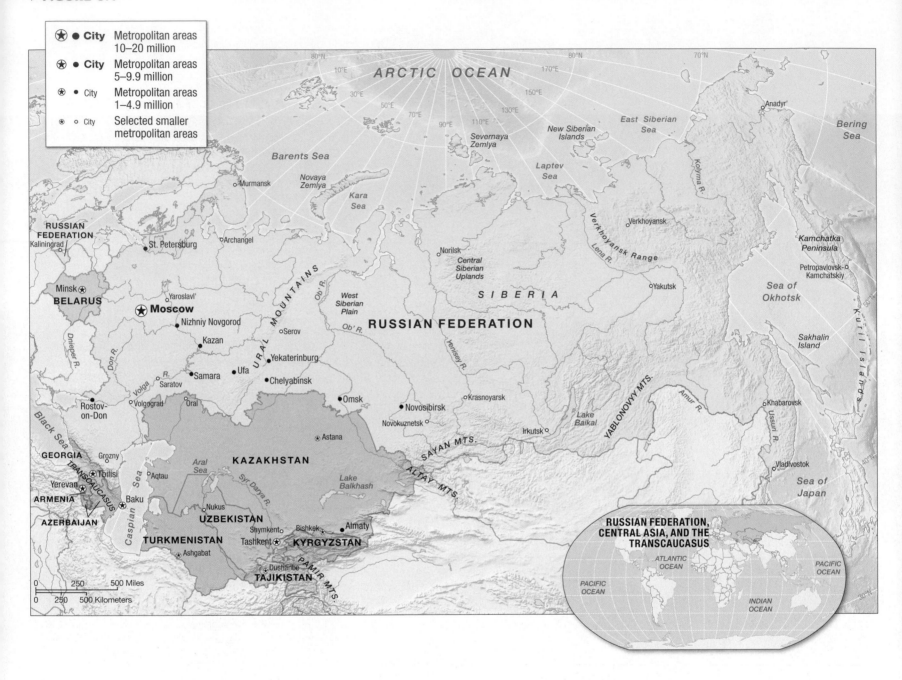

The Russian Federation, Central Asia, and the Transcaucasus Key Facts

- Major Subregions: The Russian Federation, Central Asia, and the Transcaucasus

- Major Physiogeographic Features: The Russian Plain, the Central Siberian Uplands, the West Siberian Plain, the Ural Mountains, the desert plateaus of Central Asia. The tundra is an arctic wilderness that accounts for 13% of Russia. It is a continental climate, with long and intense winters.

- Major Religions: Orthodox Christianity, Protestantism, Islam, Roman Catholicism

- Major Language Families: Indo-European (Slavic, Iranian, Armenian), Altaic, Caucasian, Uralic, Palea-Siberian

- Total Area (total sq km): 21 million

- Population (2011): 232 million; Population under Age 15 (%): 24; Population over Age 65 (%): 8

- Population Density (per sq km) (2011): 48

- Urbanization (%) (2011): 51

- Average Life Expectancy at Birth (2011): Overall: 70; Women: 74; Men: 65

- Total Fertility Rate (2011): 2.3

- GNI PPP per Capita (current U.S. $) dollars (2009): 7,456

- Population in Poverty (%, < $2/day) (2000–2009): 26

- Internet Users (2011): 88,393,133; Population with Access to the Internet (%) (2011): 32.3; Growth of Internet Use (%) (2000–2011): 2310

- Access to Improved Drinking Water Sources (%) (2011): Urban: 97; Rural: 85

- Energy Use (kg of oil equivalent per capita) (2009): 2101

- Ecological Footprint (hectares per capita consumed/hectares per capita available, global scale) (2011): 2.6/1.8

ENVIRONMENT AND SOCIETY

One way to describe the Russian Federation, Central Asia, and Transcaucasus region in terms of its physical geography is that *it is vast* (**Figure 3.1**). Stretching more than 10,000 kilometers (6,200 miles) east–west and more than 2,500 kilometers (1,550 miles) north–south, it takes a full week to cross by train from Vladivostok in the east to St. Petersburg in the west. As a result, the environments of the region are enormously diverse and include a range of climatic and physiographic settings. Despite its daunting size, the region's physical position as a low-lying set of plains between East Asia and Europe and the Middle East has made it a historic crossroads and the site of interactions between Europe and Asia. At the same time, the region has restricted access to the world's seas, as a satellite photograph (**Figure 3.2**) emphasizes. It is also bounded on the north by the icy seas of the Arctic and on the south by a mountain wall that stretches from the Elburz Mountains of northern Iran to the Altay and Sayan Mountains, which separate Siberia from Mongolia, and the Yablonovyy and Khingan ranges, which separate southeastern Siberia from northern China. People have adapted to and modified the physical geographic landscapes of this region as they have expanded their communities across the region's expanse, creating new environments, economies, and societies.

Climate, Adaptation, and Global Change

This region's mostly northern location and its position at the eastern edge of the marine climates of Europe and the deserts of inland Asia have a strong influence on its overall physical geography. The absence of mountainous terrain, except in the far south and east, and the lack of any significant moderating influence of oceans and seas mean that the prevailing climatic pattern is relatively simple (**Figure 3.3**, p. 90).

Three major climate regions dominate the landscape from the arid and semiarid regions in the southern portion of the region to the continental/midlatitude and polar regions to the north.

Severe Climates In terms of its northern position, nearly half the territory of the Russian Federation is north of 60°N. Moscow is approximately the same latitude as Juneau, Alaska, and Tbilisi, Georgia—one of the southernmost cities of the region—is approximately the same latitude as Chicago (42°N). A severe continental climate dominates, with long, cold winters and relatively short, warm summers. The cold winters become colder eastward, as one moves away from the weak marine influence that carries over from the westerly weather systems, which cross Europe from the Atlantic. Pronounced high-pressure systems develop over Siberia in winter, bringing clear skies and calm air. Average January temperatures in Verkhoyansk, a mining center in the middle of this high-pressure area, hover around –50°C (–58°F). Long and intense winters mean that the subsoil is permanently frozen—a condition known as **permafrost**—in more than two-thirds of the Russian Federation. In the extreme northeast, winter conditions can last for ten months of the year. Consequently, most ports are icebound during the long winter. Murmansk, the far north, is an important exception. Murmansk benefits from its location near the tail end of the warm Gulf Stream and is open year-round. Some ports, such as Vladivostok, on the Sea of Japan, are kept open by icebreakers.

Summer comes quickly over most of Belarus and the Russian Federation, as spring is typically a brief interlude of dirty snow and abundant mud. Because many rural roads remain unpaved, they are typically impassable until the summer heat bakes the mud. As the landmass warms, low-pressure systems develop, drawing moist air from Atlantic Europe that brings moderate summer rains. Across much of Siberia, though, summer rainfall is quite low and summers also become hotter southward. Drought is a frequent problem in the southwestern and southern parts of the Russian Federation. In Central Asia, **aridity** (a dry condition where an absence of moisture and rainfall negatively affects plant growth, especially trees) is a severe problem, with desert and semidesert covering much of Kazakhstan, Uzbekistan, and Turkmenistan. The climate in this subregion is harsh: total annual precipitation in the deserts is less than 18 centimeters (7 inches), shade temperatures can reach 50°C (122°F), and ground surfaces can heat up to 80°C (176°F). The scorching heat is aggravated by strong drying winds that blow on more than half the days of summer, often causing dust storms. In late summer and fall, the increasing temperature range between the hot days and the longer, cooler nights becomes so extreme that rocks exfoliate, or "peel," leaving the debris to be blown away by the wind.

In the Transcaucasus, climatic patterns are mainly a result of the presence of massive mountain ranges to the north

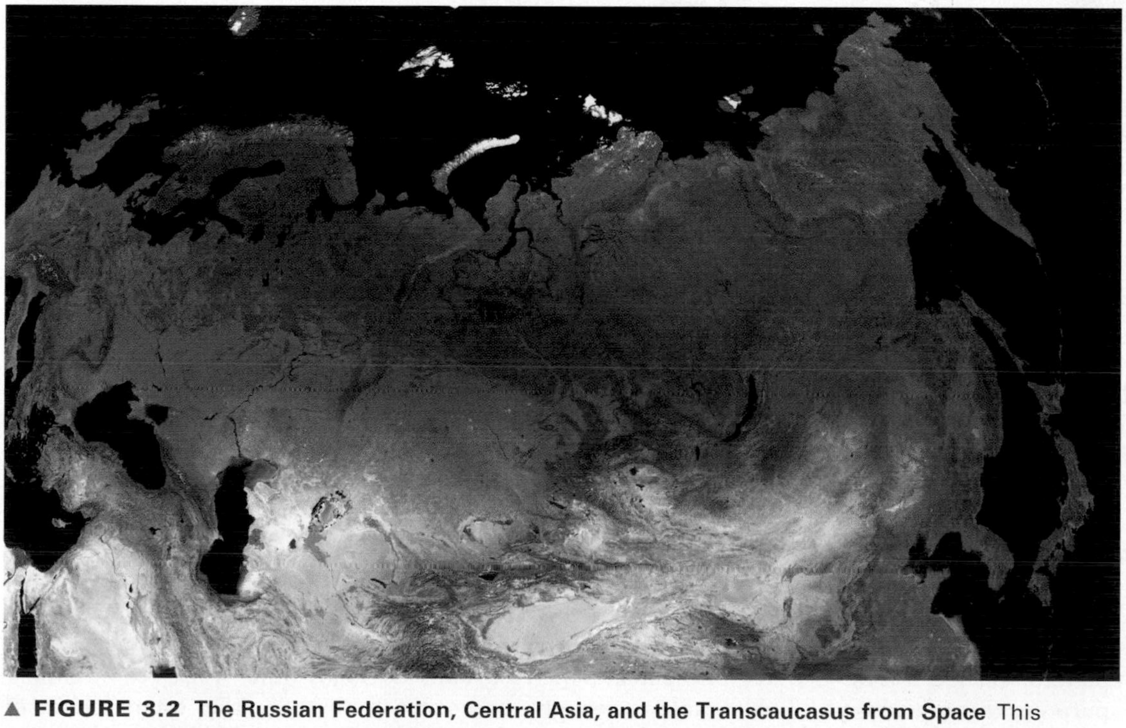

▲ **FIGURE 3.2 The Russian Federation, Central Asia, and the Transcaucasus from Space** This composite of satellite images illustrates one of the key features of the Russian Federation, Central Asia, and the Transcaucasus: its sheer size.

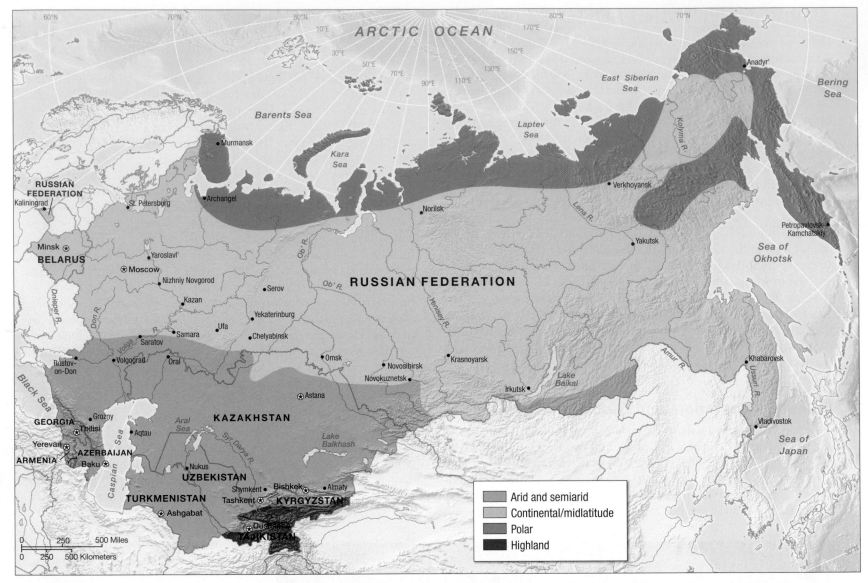

▲ **FIGURE 3.3 Climates of Russia** The Russian climate features very cold winters and summer droughts are common. Around the Arctic Circle, a very cold and dry polar climate exists. Further south, climates are dry but warmer, typical of semiarid climate types. The higher mountain regions have colder highland climates.

and south and substantial bodies of water to both east and west. A lot of precipitation falls on the windward side of the mountains, although the Transcaucasus is also influenced by the warm, dry air masses that originate over the deserts of Central Asia. The most distinctive feature of climatic patterns in the Transcaucasus, however, is the subtropical niche of western Georgia, on the shores of the Black Sea—a unique and striking feature in a region of otherwise severe climatic regimes.

Adapting to a Warming Continent and Melting Sea Ice The Russian Federation, Central Asia, and Transcaucasus region is already experiencing the effects of global climate change. The implications for northern latitudes are mixed. Milder winters mean that, as growing seasons become longer and precipitation patterns change, it is becoming possible to use lands for agricultural purposes that previously have been inhospitable for too much of the year. Farmers are also able to raise new crops and new varieties of crops in some regions. On the other hand, massive investments will be required to stabilize

roads, bridges, and the infrastructure of public utilities as the permafrost melts. Global climate change also has implications for the oil and gas industries that are vital to the region's economies. The warming climate features milder and shorter heating seasons, leading to reduced energy demand within the region (and worldwide). At the same time, increased water availability—particularly along the Siberian rivers that can be used for hydroelectric power—should result in increased domestic power production, which may also reduce the demand for oil and gas.

In the Transcaucasus and the southern parts of European Russia, agriculture is becoming more reliant on irrigation, pesticides, and herbicides (the spread of pests and disease typically accelerates as global climates increase) and, like many other world regions, more vulnerable to droughts and other extreme weather events. Some of the most affected regions are areas where socioeconomic and sociopolitical relations are unsettled. For example, the politically turbulent North Caucasus region is becoming drier and hotter and is likely to become less prosperous as an agricultural region. And in the eastern part of the region, longstanding cross-border tensions will probably intensify as water availability becomes an increasingly serious challenge in Central Asia, Mongolia, and northeastern China, prompting large-scale in-migration to the Russian Federation. Finally, warming trends across the northernmost parts of the region have begun to result in sea ice melt in the Arctic Ocean. This dramatic transformation of an area previously known only to the hardiest explorers has turned this part of the world into an important player in global commerce and energy production (see "Emerging Regions: The Arctic," on p. 92).

> **APPLY YOUR KNOWLEDGE** List two serious effects of global climate change in the Russian Federation, Central Asia, and Transcaucasus region. How are people adapting to those changes? How are the changes impacting not only the environment but the wider economy of the region?

Geological Resources, Risk, and Water Management

Remarkably, the entire Russian Federation, Central Asia, and Transcaucasus region sits atop a single tectonic plate. It occupies the eastern portion of the Eurasian Plate (see Chapter 1), which stretches unbroken between the Pacific and Atlantic Plates. The main physical geography feature that results from this location is the geologically stable monotony of the region's plains, which stretch over thousands and thousands of kilometers. At the edges of this expanse, geological upheaval at the plate boundary has formed the southernmost physiographic barriers to the region.

Mountains and Plains This large contiguous region is broken by Earth forces into several subregions. In the west, the Russian Plain, an extension of the Central European Plateau, runs from Belarus in the west to the Ural Mountains in the east, and from the Kola Peninsula in the north to the Black Sea in the south (**Figure 3.4**). East of

▲ **FIGURE 3.4 Physiographic Regions of the Russian Federation, Central Asia, and the Transcaucasus** The framework for the physical geography of the region consists of two stable shields of the ancient Ural Mountains, with a wall of young mountains that runs along the southern and eastern margins of the shields.

EMERGING REGIONS
The Arctic

For hundreds of years, explorers and sailors have sought a route through the **Northwest Passage**, an ice-choked waterway spanning the Arctic Sea between the Atlantic and Pacific Oceans, north of the Canadian and Russian mainlands. Such a navigable route would significantly shorten the shipping distance between Shanghai and New York or London, making a globalizing world all the more tightly connected.

Roald Amundsen successfully navigated this route back in 1906, but the presence of treacherous sea ice throughout much of the year has continued to make the route commercially impossible. Global warming has, however, accomplished what generations of explorers and investors have failed to do. As temperatures rise, the extent of ice in the Arctic is decreasing rapidly, especially in the summer (**Figure 3.1.1**). Arctic sea ice levels are at their lowest since records have been kept. Models and projections based on current trends suggest that the Arctic Ocean will be free of summer ice sometime between 2030 and 2080. Far sooner, the route across the Arctic will be reliably open to global commerce, and for the first time, the seafloor will be accessible to extensive resource development involving drilling for oil and natural gas.

These historic changes will have devastating impacts on the wildlife of the region. Polar bears will effectively be deprived of their natural habitat and ultimately be found only in zoos. And the opening of the ice means the creation of an entirely new world region—a contested prize for key world powers, a novel area for tourism culture, a critical source of resources, and a connected path between the worlds of the Atlantic and Pacific. The geopolitical contest for the control of this area has already begun, with Russia, Norway, Denmark, Canada, and the United States marking

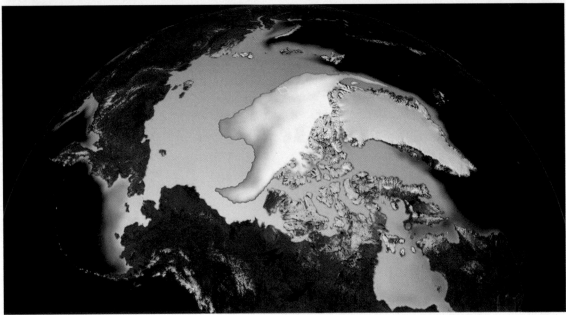

▲ **FIGURE 3.1.1 The Melting Polar Ice Cap** Shifting patterns of summer ice cover in the Arctic region between 1979 (above) and 2007 (below).

the Urals lies the West Siberian Plain, the Central Siberian Plateau, and the desert plateaus of Central Asia. The Urals themselves represent their own distinctive region, as does the mountain rim that runs along the southern and eastern margins of the region.

The Russian Plain generally has a gently rolling topography, although the major rivers that drain across these plains—the Dnieper, the Don, and the Volga—have carved a more varied topography

along their length. The West Siberian Plain is even flatter and contains extensive wetlands and tens of thousands of small lakes. Poorly drained by the slow-moving Ob' and Irtysh Rivers, the West Siberian Plain (**Figure 3.5**, p. 94) is mostly inhospitable for settlement and agriculture, though it contains significant oil and natural gas reserves. The West Siberian Plain is distinctive for its absolute flatness: across the whole broad expanse—more than 1,800 kilometers (1,116 miles)

territory and making legal claims on the region (**Figure 3.1.2**). In August 2007, the Russian government sent two tiny submarines to plant the Russian flag on the Arctic seafloor (**Figure 3.1.3**).

These changes also highlight the peculiar situation of Greenland. Though physiographically located in North America, and with its own Indigenous populations (Kalaallisut-speaking Inuits) and wildlife (polar bear, musk ox, narwhal, and walrus), the world's largest island has long been colonized and controlled by Europeans, most recently Denmark. The emergence of a new geostrategic region around the North Pole, and the recently established semi-independent status of Greenland, reinforces this ambiguity. While Greenland is clearly a "victim" of global climate change, through the loss of its ice sheets, wildlife habitat, and Indigenous human livelihoods, its position also allows it claims on minerals and oil and gas reserves. This gives Greenland considerable influence and economic opportunity, alongside Canada, Russia, and the United States, though they remain under the nominal control of a small European power, Denmark. ■

▲ **FIGURE 3.1.3 Russia Claims the Arctic Seafloor** Russia planted its flag in 2007 on an undersea formation, called the Lomonosov Ridge, to symbolically assert that this area is a part of the Siberian continental shelf, and therefore under Russian control.

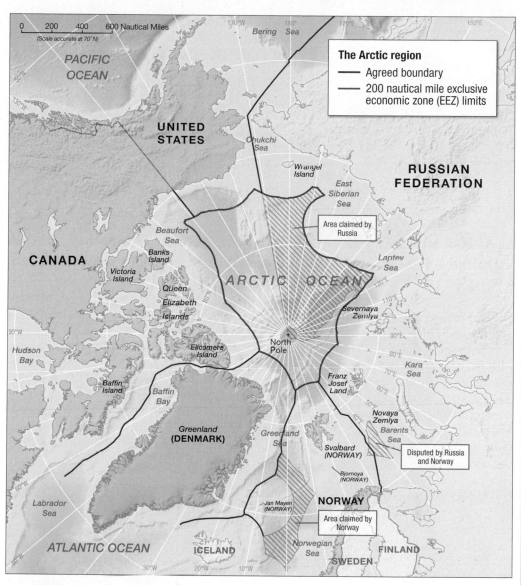

▲ **FIGURE 3.1.2 Claims on the New Arctic Frontier** This map shows the national boundaries and the competing claims on this emerging region. Each of the five countries bordering the Arctic Ocean has claimed an Exclusive Economic Zone (EEZ), an area where they hold exclusive rights to drill, fish, or mine. Several claims, shown with hash marks, are claimed by nations but not recognized by the international community and depend on contested information about the shape and extension of the continental shelf.

in each direction—relief is no more than 400 meters (1,312 feet). The monotony of the landscape is captured in this quote from Russian writer A. Bitov, describing a train journey:

Once I was traveling through the Western Siberian Lowlands. I woke up and glanced out of the window—sparse woods, a swamp, level terrain. A cow standing knee-deep in the swamp and chewing, levelly moving her jaw. I fell asleep, woke up—sparse woods, a swamp, a cow chewing, knee deep. I woke up the second day—a swamp, a cow. . . ."[1]

[1]A. Bitov, A *Captive of the Caucasus* (Cambridge: Cambridge University Press, 1993, p. 50). Quoted in A. Novikov, "Between Space and Race: Rediscovering Russian Cultural Geography," in *Geography and Transition in the Post-Soviet Republics*, ed. M. J. Bradshaw (Chichester: John Wiley & Sons, 1997), p. 45.

▲ **FIGURE 3.5 The West Siberian Plain** This photograph of marshland near Primorye, western Siberia, shows the difficult, boggy conditions that prevail in much of western Siberia in summer.

The Yenisey River marks the eastern boundary of this flat expanse and the beginning of the Central Siberian Plateau (**Figure 3.6**), an uplifted region dissected by rivers into a hilly upland topography with occasional deep river gorges. The Urals are a once-great mountain range of ancient rocks that has been worn down over the ages. They stretch for more than 3,000 kilometers (1,864 miles) from the northern frontier of Kazakhstan to the Arctic coast of the Russian Federation. The rocks of the Urals are heavily mineralized and contain significant quantities of chromite, copper, gold, graphite, iron ore, nickel, titanium, tungsten, and vanadium. As a result, the Urals are home to a number of significant industrial cities, which have developed around mining and refining these resources. Magnitogorsk (in Russian, roughly: "magnet-mountain city") for example, is a Russian frontier industrial settlement located at the foot of an enormous mountain of nearly pure iron on the eastern side of the Urals.

The mountain wall that runs along the southern and eastern margins of the region is the product of geological instability. Younger, sedimentary rocks have been pushed up in successive episodes of mountain-building along the edges of the Eurasian Plate, forming a series of mountain ranges of varying height, composition, and complexity. The highest ranges are those of the Caucasus (where Mt. Elbrus reaches 5,642 meters, or 18,510 feet) and the Pamirs and Tien Shan ranges along the borders with Iran, Afghanistan, and China. Here, many peaks reach 5,000 to 6,000 meters (or 16,404 to 19,685 feet). In the eastern part of the region, the ranges of the Kamchatka Peninsula contain numerous active volcanoes.

Central Asia connects to the rest of the region without interruption and sits atop a huge **geosyncline**, a geological depression of sedimentary rocks. The Central Asian geosyncline is of special importance as a source of energy resources. It is home to oil reserves equivalent to between 17 and 49 billion barrels—about 2.7% of the

▲ **FIGURE 3.6 The Central Siberian Plateau** The rock shield of the Central Siberian Plateau has been uplifted by geological movements. As a result, the land has been dissected by rivers into a hilly upland with occasional deep river gorges.

world's proven reserves—plus about 7% of the world's proven reserves of natural gas. Some estimates of the potential oil reserves run even higher—between 60 and 250 billion barrels.

Rivers and Seas Because the Russian Federation, Central Asia, and Transcaucasus region as a whole is located so far north, its rivers are of limited use for navigation and hydroelectric power generation. Many rivers are frozen for much of the year, and the mouths of many remain frozen through the spring, causing backed-up meltwater to flood extensive areas of wetlands. Climate change exacerbates the flooding risk in these rivers, making flow management even more difficult. These flows are made even more severe by extreme weather events, such as heavy downpours or springtime ice-clogged floods. Nevertheless, the territory sustains several historically important transport routes, which have allowed for conquest, colonization, and trade. Some of the longest rivers on Earth drain the huge Siberian landmass. Beginning in the southern mountains of Central Asia, the Lena, Kolyma, Ob', and Yenisey flow north to the Arctic Ocean, whereas the Amur flows north to the Pacific. In the western part of the Russian Federation, the rivers flow south from the Moscow and Nizhniy Novgorod area. The Dnieper flows to the Black Sea, the Don to the Sea of Azov (which in turn connects to the Black Sea), and the Volga to the Caspian Sea. The Volga, with huge reservoirs built during the Soviet era to regulate water flow and conserve spring floodwaters for the dry summer months, has become particularly important as a navigable waterway and source of hydropower.

The Black Sea is an inland sea (**Figure 3.7**), connected to the Aegean Sea and the Mediterranean by way of the Bosporus, a narrow strait, the Sea of Marmara, and then another narrow strait, the Dardanelles (see Chapter 4). The many rivers that empty into the Black Sea give its surface waters a low salinity. It is almost tideless, and below about 80 fathoms (478.3 feet; 145.8 meters), it is stagnant and lifeless. The region's other inland sea, the Caspian Sea, is the largest inland sea in the world, at 371,000 square kilometers (143,205 square miles—roughly the size of Germany). There are several lakes of significant size in the region as well, including Lake Balkhash (17,400 square kilometers; 6,715 square miles) and Lake Baykal (30,500 square kilometers; 11,775 square miles), which, with a depth of 1,615 meters (5,300 feet), is the deepest lake in the world.

APPLY YOUR KNOWLEDGE Explain how each of the key defining physiographic features of the Russian Federation, Central Asia, and Transcaucasus region has emerged in relation to global plate tectonics.

▼ **FIGURE 3.7 The Black Sea** The Black Sea is a critical component of the ecology and economy of the region. It serves as a central passageway for trade and transit between this region and the rest of the world.

▲ **FIGURE 3.8 Environmental Zones of Russia and Central Asia** The major environments of the region correspond to latitude, and span the continent, from west to east, in long bands.

Ecology, Land, and Environmental Management

The natural ecologies of the Russian Federation, Central Asia, and the Transcaucasus consist of five long, latitudinal zones that run roughly from west to east (**Figure 3.8**). These zones are very closely related to climatic patterns and are still easily recognizable to the modern traveler, despite centuries (or, in places, millennia) of human interference and modification. The region's environmental zones are controlled by temperatures that trend upward to the south and precipitation, which tends to trend downward. If one were to walk in a straight line, moving from north to south through the center of the region, dominant patterns would be evident.

The Tundra The **tundra**, which fringes the entire Arctic Ocean coastline wond part of the Pacific, is an arctic wilderness where the climate precludes any agriculture or forestry. The tundra zone covers

2.16 million square kilometers (833,760 square miles), representing almost 13% of the Russian Federation. For nine months or more, the landscape is locked up by ice and covered by snow. During the brief summer, much of the melted snow and ice is trapped in ponds, lakes, boggy depressions, and swamps by permafrost that extends, in places, to a recorded depth of 1,450 meters (4,757 feet). Water seeps slowly to streams, which drain into the slow-moving rivers that cross the tundra and drain in turn into the Arctic (**Figure 3.9**). During winter, the sun remains low in the sky and the tundra landscape has a uniformity that derives from snow cover and the somber effect of long nights and weak daylight. Herds of reindeer or occasional polar bear or fox are the only signs of life. In summer, vegetation bursts into bloom and animals, birds, and insects appear. The summer days are long—in June the sun circles the horizon, and there is no night at all. Mosses and tiny flowering plants provide color to the landscape, contrasting with the black, peaty soils and the luminous bright skies. Swans, geese, ducks, and snipe arrive on lakes and wetlands for their breeding season, as do seabirds and seals along the coast.

▲ **FIGURE 3.9 Tundra Landscape** The tundra landscape is bleak, with sparse vegetation and a surface strewn with rocks.

From the forests to the south, wolves enter the tundra in search of prey. Everywhere there are swarms of gnats and mosquitoes.

Though the tundra is a daunting environment, human life is evident. Indigenous people have herded reindeer along these coasts for centuries, managing their herds' migrations to coincide with changing seasons. The steady warming of the poles in recent years has been devastating for these cultures, however, since these herds must cross areas of seasonal ice to access forage. With delays in the freezing of the sea ice, people and animals remain stranded in overgrazed pastures, subject to starvation (**Figure 3.10**).

The Taiga To the south of the tundra, annual temperatures rise and rainfall increases, allowing the growth of larger and sturdier plant life. These conditions give rise to the most extensive zone of all—a belt of coniferous forest known as the **taiga**. The term *taiga* originally referred to virgin forest, though it is now used to describe the entire zone of coniferous forest that stretches from the Gulf of Finland to the Kamchatka Peninsula—more than 4.4 million square kilometers (1.3 million square miles) of

▲ **FIGURE 3.10 Indigenous Herders of the Russian Arctic** An ethnic Dolgan herder lassos a reindeer during a roundup in Taymar in Northern Siberia, Russia. Herding local animals is an integral part of the life and economy of Indigenous peoples in the Russian Arctic. These livelihoods are being challenged by global climate change, however.

▲ **FIGURE 3.11 Taiga Landscape** The taiga is a zone of coniferous forest that stretches from the Gulf of Finland to the Kamchatka Peninsula.

In recent decades, after decades of relentless exploitation, large swathes of Siberia, once dense and practically impassable, have been cleared. Loggers are now moving farther and farther north to cut down century-old pines. Recently, this sector has become more fully globalized, which has intensified concern over forest resource management. **Privatization** of the timber industry, which is the selling of government extraction rights to private individuals and firms, has attracted U.S., Korean, and Japanese transnational corporations. The corporations invest in "slash-for-cash" logging operations (**Figure 3.12**) in a loosely regulated and increasingly corrupt business environment. The future of central Siberia has become an issue of international concern because the region accounts for a significant fraction of the world's temperate forests, which absorb huge amounts of carbon dioxide gas in the process of photosynthesis, thereby removing a main contributor to global warming from the atmosphere.

territory (**Figure 3.11**). This topography, though broken by some deep river gorges and hills, is uniformly covered in characteristic forests made up of larches: hardy, flat-rooted trees that can establish themselves above the permafrost. Slow-growing and long-lived, larches are able to counter the upward encroachment of moss and peat by putting out fresh roots above the base of the trunk. The larches can grow to 18 meters (59 feet) or more, allowing an undergrowth of dwarf willow, juniper, dog rose, and whortleberry.

The Indigenous inhabitants of the taiga were hunters and gatherers, not farmers. Where the forest is cleared, some cultivation of hardy crops, such as potatoes, beets, and cabbage, is possible, but the poor, swampy soils and short growing season make agriculture challenging. The small populations of towns and villages are engaged in mining, administration, construction, and local services. The central Siberian taiga is one of the richest timber regions in the world. Overall, close to 90% of the territory is covered with forest, and more than a quarter of the Russian Federation's lumber production comes from the region. The timber is methodically exploited and exported, and the taiga is also commercially important for its fur-bearing animals, whose luxuriant pelts are well adapted to the bitter cold (see "Geographies of Indulgence, Desire, And Addiction: Furs," p. 100).

Mixed Forest Continuing south, annual precipitation begins to decline, leading to a more open region of mixed forest and plains. In this zone, stands of mixed woodland arise, mostly in valley bottoms. This mixed forest was cleared and cultivated early in Russian history, providing both an agricultural heartland for the emerging Russian empire and a corridor along which Russian traders and colonists pushed eastward in the 16th and 17th centuries—eventually settling the Pacific coast.

The Steppe To the south, the mixed woodlands give way to the **steppe**—the large area of flat grassland or prairie—as rainfall decreases, falling below 400 millimeters (15 inches) annually. The steppe belt stretches about 4,000 kilometers (2,486 miles) from the Carpathians to the Altay Mountains (see Figure 3.8, p. 96), covering a total area of more than 4.25 million square kilometers (1.25 million square miles). The topography of the region is strikingly flat, punctuated here and there by streams and river valleys. The natural vegetation of tall and luxuriant grasses has matted roots that trap the limited moisture available in this arid region. Trees and shrubs are restricted to valleys, where oak, ash, elm, and maple have established themselves; pinewoods and a low scrub of blackthorn, laburnum, dwarf cherry, and Siberian pea-trees grow in drier locations.

For centuries, the steppe region, which is today divided into the western steppe (roughly made up of the northern region of modern-day Kazakhstan) and the eastern steppe (an area that includes parts of southern Russia and Georgia) was the realm of nomadic people. In the late 1700s, when the Turkish Empire's hold on the steppes was broken, large numbers of colonists from the wider region began to enter the western steppe. Wheat growing rapidly expanded wherever transportation was good enough to get the grain to the expanding world market, but large flocks of sheep dominated most of the colonized steppe until the railway arrived. To the east, the flat steppe of northern Kazakhstan remained largely untouched until the 19th century, save for Kazakh nomads and their herds and a few Russian forts and trading posts. During the 19th century, settlers came from the west in increasing numbers—a million or more by 1900—displacing the Kazakh nomads, thousands of whom died in famines or in unsuccessful uprisings against the Russians.

With the arrival of modern transportation and harvesting technologies in the late 19th century, the steppe region became an enormously productive wheat belt for Russia, as it remains to this day. Since the middle of the 20th century, small farms have been consolidated into large holdings and rivers have been dammed to

◀ **FIGURE 3.12 Logging in Siberia**
More than 1.5 million cubic meters of oak, cedar, and ash are illegally logged in the far eastern Primorye region each year. Chinese traders pay U.S. $100 per cubic meter and resell the wood at prices of U.S. $400 to U.S. $500 per cubic meter.

provide hydroelectric power and irrigation for extensive farming of wheat, corn, and cotton (**Figure 3.13**). The eastern steppe, transformed by modern machinery, fertilizers, and pesticides, initially produced as much wheat annually, on average, as Canada or France.

This extensive wheat farming quickly led to dust-bowl conditions. In response, **dry-farming** techniques that allow the cultivation of crops without irrigation in regions of limited moisture (50 centimeters, or 20 inches, of rain per year) have been introduced. Such techniques include keeping the land free from weeds and leaving stubble in the fields after the harvest to trap snow. Together with irrigation schemes, dry-farming techniques now allow for the cultivation of not only wheat, but also millet and sunflowers, together with silage corn and fodder crops to support livestock in the steppe.

Semidesert and Desert Further south, where the rainfall declines even more and summer temperatures soar to extremes, are zones of semidesert and desert. Largely a feature of Central Asia—making up most of Kazakhstan, Turkmenistan, and Uzbekistan—they continue south of Siberia into Chinese and Mongolian territory. The semidesert is characterized by boulder-strewn wastes and *salt pans* (areas where salt has been deposited as water evaporates from short-lived lakes and ponds created by runoff from surrounding hills) and patches of rough vegetation used by **nomadic pastoralists**. These groups herd animals by moving from place to place and carefully and deliberately following rainfall and plant growth to maintain their flocks. The desert is characterized by bare rock and extensive sand dunes, though there are occasional oases and fertile river valleys.

The two principal deserts here are the Ust-Urt and Bek-pak-Dala. Farther south is a zone of sandy deserts: the Kara-Kum (Black Sands) and the Kyzyl-Kum (Red

Sands). The landscape is dominated by pockets of desert punctuated by patches of sand hills and by isolated remnants of worn-down mountain ranges. Several rivers—including the Amu Dar'ya, the Syr Dar'ya, and the Zeravshan—drain from the Tien Shan and Pamir mountain ranges across these deserts toward the Aral Sea. Dotted along their valleys are irrigated oases, where impenetrable thickets of hardy trees and thorny bushes thrive in the salty soils of the valley floors.

These deserts are also managed intensively, however. The last century has seen the advent of cash crops for sale on global markets, including wheat and cotton. This has resulted in the wholesale transformation of these arid areas through the engineering of irrigation canals and transportation networks. Desert farming settlements,

▲ **FIGURE 3.13 Intensive Cultivation of the Russian Steppe** Beginning in the 1950s, the eastern steppes were transformed into an extensive wheat farming region through massive investments of modern machinery, fertilizers, and pesticides.

GEOGRAPHIES OF INDULGENCE, DESIRE, AND ADDICTION

Furs

From earliest times, fur has been a prized commodity. In cold regions, fur coats, hats, and boots are valued for their warmth and as a portable form of wealth. In Russia and Europe, fur has long had royal and aristocratic connotations and, as a result, became a status symbol for all who could afford it. Fur is still seen by many status-conscious consumers as a fashionable luxury good, a clear marker of material wealth. But it has also become a controversial luxury item. In Europe and North America in particular, the market for furs is significantly affected by concerns over extinction and the realization that fur trapping and fur farming can involve unnecessary cruelty to animals.

European merchants began trading fur in the Middle Ages, and fur was the commodity that drew Russian trappers and traders to Siberia in the 16th century. In 1581, under sponsorship of the rich merchant family of Stroganov, a military expedition opened a route to Siberia for *promyshlenniks* (fur hunters) searching for sable and sea otter. The pelts of these animals were exchanged for Chinese and Indian goods, and the tax revenues from the trade were the mainstay of the Russian imperial treasury for the next 300 years. It became government policy to encourage the fur trade and to support the promyshlenniks in their ruthless displacement of Indigenous people. By the reign of Peter the Great (1682–1725), the promyshlenniks had reached the Sea of Okhotsk, and fur hunting was coming to a saturation point. In response, Peter the Great sponsored maritime expeditions to the Kamchatka Peninsula and to the offshore islands of the northeast. After Vitus Bering's expedition to the Northern Pacific in 1741–1742 established that the islands had abundant populations of sea otter, foxes, seals, and walruses, a "fur rush" drew promyshlenniks all the way across the Bering Sea to Alaska.

Meanwhile, trade in fur pelts (beavers, muskrats, minks, and martens) was fueling Europeans' initial interest in North America. Beaver, trapped by Native Americans, was a main source of barter at trading posts that later grew into Chicago, Detroit, Montréal, New Orleans, Québec, St. Louis, St. Paul, and Spokane. The Hudson's Bay Company, founded in England in the mid-17th century to trade skins for guns, knives, and kettles, gained almost total control of the North American fur trade. Today, fur farming (raising animals in captivity under controlled conditions), rather than trapping, is the principal source of furs for the world market (**Figure 3.2.1**). Fur farming was started in Canada in 1887 on Prince Edward Island. Through controlled breeding, animals with unique characteristics of size, color, or

▲ **FIGURE 3.2.1 Russian Fur Farm** Most of Russia's fur farms are located in remote settings, where there are few regulatory controls or inspections.

▲ **FIGURE 3.2.2 Antifur Protest** A PETA (People for the Ethical Treatment of Animals) member sits inside a cage during a protest in Moscow.

texture can pass on those characteristics to their offspring. The silver fox, developed from the red fox, was the first fur so produced.

Fur production is spread across the globe. In 2009, Europe farmed 65% of the world's mink and 56% of the world's fox fur. The highest levels of production are in Denmark, the Netherlands, Finland, and Sweden. Canada and Denmark together now produce more than half of the world's farmed mink. China is a growing fur producer; in 2010, it produced 12 million pelts, up one-third from the year before. Hong Kong and Greece lead the list of value-added sites, where fur dyeing, sewing, and finishing occurs for furs destined for global markets. Global production remains high and growing.

The industry's growth has recently been uneven, however, with a leveling off of sales in the last decade. The reason for this leveling-off in sales is partly a result of the recent economic downturn, but also a result of a shift in attitudes toward furs, led by animal rights activists. Antifur demonstrations targeting designers, led by entertainment icons such as Charlize Theron, Ricky Gervais, Pamela Anderson, and Angelina Jolie, have captured widespread attention (**Figure 3.2.2**). A number of well-publicized cases of maltreatment on fur farms have further reinforced the case of Western antifur activists. Some Russian fur farmers, faced with a combination of falling consumer demand because of economic recession, higher taxes, and widespread corruption, let their animals go hungry. Western visitors to Siberian fur farms found starving animals in tiny cages with no bedding, no protection against the elements, and no veterinary care. Russian fur farming declined greatly between 2000 and 2010, and Russian farmed mink represented less than 3% of the world's total in 2010. The industry now only employs approximately 200,000 people in Russia.

The antifur sentiment is not shared by everyone, of course. Most leading designers dropped furs from their fashion lines in the mid-1990s, but furs made a fashion comeback in the 2000s, led by designer labels such as Dolce & Gabbana and Louis Vuitton and promoted by celebrities such as Jennifer Lopez, Halle Berry, Maggie Gyllenhaal, Kanye West, Victoria Beckham, and even some who once took part in antifur protests, such as Madonna, Cindy Crawford, Naomi Campbell, and Kate Moss.

And although Russian fur farming has declined, regional demand has not. Russia remains the largest market for furs in the world, with China running a close second. Global demand for fur is ongoing: 2010 global retail sales of fur show a 5.4% increase from 2009, representing a U.S. $14 billion U.S. global business. This shift underlines the way a global cultural phenomenon (the antifur movement) can have a sharp regional impact (on Russian fur farmers), and also may result in other unanticipated changes in the globalized economy (such as the meteoric rise in Chinese fur farming). ∎

towns, and small cities are now supported by agricultural goods and services, as this region has become tied more closely to global commodity markets for grains and fiber. This has not come without startling environmental problems, as we will see later in this chapter.

> **APPLY YOUR KNOWLEDGE** Compare and contrast the climates and vegetation of the five zones of the physical geography of the region, starting in the north and moving to the south. How have the inhabitants of each zone adjusted to their climates?

HISTORY, ECONOMY, AND TERRITORY

The Russian Federation, Central Asia, and Transcaucasus region's agriculture and settlement have been greatly restricted by severe climatic conditions, highly acidic soils, poor drainage, and mountainous terrain. Even in the zone of the region's richer soils, low and irregular rainfall rendered agriculture and settlement marginal until large-scale irrigation schemes were introduced in the 20th century. Only in the mixed forest and the wooded steppe west of the Urals were conditions suitable for the emergence of a more prosperous and densely settled population. It was this area, in fact—from Smolensk in the west to Nizhniy Novgorod in the east, and from Tula in the south to Vologda and Velikiy Ustyug in the north—that was the Russian "homeland" that developed around the principality of Muscovy from late medieval times.

Historical Legacies and Landscapes Although relatively closed off from the people and economies of the rest of the world's regions for a good part of the 20th century as a result of Soviet policies, the character of this world region, like that of every other world region, derives in large part from world-spanning processes that intersect with its unique internal attributes and processes. Those global processes

brought this region into contact with East Asia and a wider Europe several thousand years ago, as trade networks emerged that connected empires in Europe with empires in East and South Asia.

Beginning almost 2,500 years ago, the towns of Central Asia became key nodes in the vast trading network known as the **Silk Road**—the collective name given to a network of overland trade routes that connected China with Mediterranean Europe and facilitated the exchange of silk, spices, and porcelain from the East and gold, precious stones, and Venetian glass from the West (**Figure 3.14**). The Silk Road had existed since Roman times and remained important until Portuguese navigators found their way around Africa and established the seaborne trade routes that exist to this day.

Along the Silk Road stood the ancient cities of Samarkand, Bukhara, and Khiva, places of glory and wealth that astonished Western travelers such as Marco Polo in the 13th century. These cities were east–west meeting places for philosophies, knowledge, and religions. They were known as centers for mathematics, music, art, astronomy, and Islamic architecture. In the 14th century, an empire ruled by local Central Asian peoples subsequently built up a vast empire stretching from northern India to Syria, with its capital in Samarkand. The decline of this empire in the 16th century saw the rise of nomadic people, who established smaller *khanates* (kingdoms ruled by **khans** (rulers or leaders in the region)) in the region. They prospered as traders on the caravan routes until the late 19th century, when these Central Asian khanates fell to Russian troops.

The extension of political control by the people of the Russian homeland is key to the present-day geography of the entire region. In the mid-15th century, Muscovy was a principality of approximately 5,790 square kilometers (2,235 square miles) centered on the city of Moscow. Over a 400-year period, the Muscovite state expanded at a rate of about 135 square kilometers (52 square miles) per day so that by 1914, on the eve of the Russian Revolution, the empire occupied more than 22 million square kilometers (roughly 8.5 million square miles), or one-seventh of

▲ **FIGURE 3.14 Landscapes of the Silk Road** The Shah-i Zinda complex sits at the center of the city of Samarkand, the third largest city in Uzbekistan today. As a key point along the Silk Road, this city served as an important site for the spread of goods between East Asia and Europe for well over a thousand years. It also became a vital center for scholarly study and Islamic practice in Central Asia.

the land surface of Earth (**Figure 3.15**). At first, Muscovy formed part of the Mongol–Tatar Empire, whose armies were known as the Golden Horde, and Russian princes were obliged to pay homage to the Khan, the leader of the Golden Horde. In 1552, under Ivan the Terrible, the Muscovites defeated the Tatars at the battle of Kazan—a victory that prompted the commissioning of St. Basil's Cathedral in Moscow.

Desiring more forest resources, especially furs, Muscovy expanded into Siberia. Gradually, more and more territory was colonized. By the mid-17th century, the eastern and central parts of Ukraine had been wrested from Poland. The steppe regions, though, remained very much a frontier region of the Russian Empire because of the constant threat of attack by nomads. Early in the 18th century, Peter the Great (1682–1725) founded St. Petersburg and developed it as the planned capital of Russia. Beyond the wealth and grandeur of a few cities, however, Russia was very much a rural, peasant economy. In the latter part of the 18th century, under Catherine the Great (1762–1796), Russia secured the territory that would eventually become southern

Latvia, Lithuania, Belarus, and western Ukraine. In the wake of the defeat of the Crimean Tatars in the late 18th century, the steppes were opened to colonization by Russians and by ethnic and religious minorities from Eastern and Central Europe. During this period, Russia ousted the Ottoman Turks from the Crimean Peninsula and gained the warm-water port city of Odessa on the Black Sea.

The Russian Empire The factors behind Russia's imperial expansion were the drive for more territorial resources (especially a warm-water port) and additional subjects. Russian **tsars** (rulers of the Russian Empire) annexed the vast stretches of adjacent land on the Eurasian continent instead of establishing new territories overseas as the European imperial powers had. The final phases of expansion occurred in the late 18th and 19th centuries. Finland was acquired from Sweden in 1809. In the Transcaucasus, Georgians and Armenians were "rescued" from the Turks and Persians and brought under the "protection" of the Russian Empire. In Central Asia, the Muslim khanates fell one by one under Russian

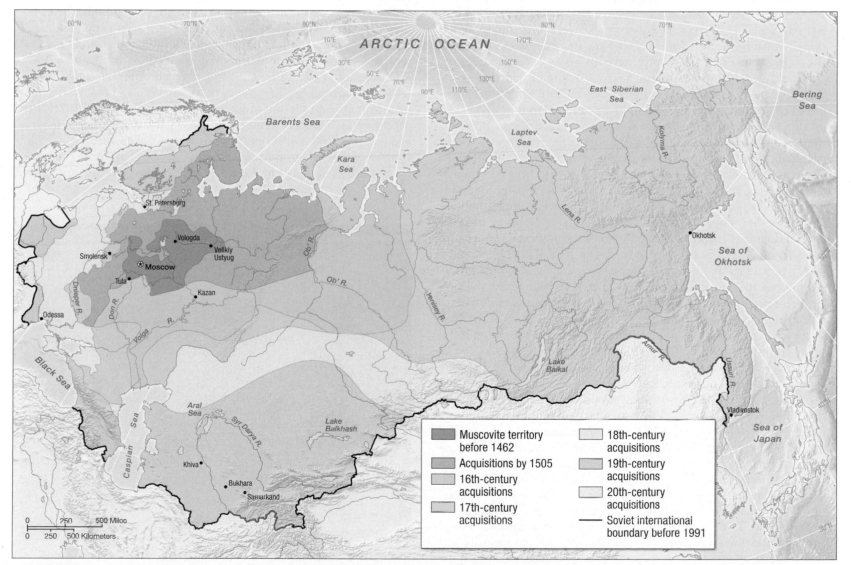

▲ **FIGURE 3.15 Territorial Growth of the Muscovite/Russian State** The Muscovite empire was vast and was acquired over the same period (15th century to the late 20th century) that corresponds to the globalization of the world economy. When the Bolsheviks came to power at the beginning of the 20th century, some of the territory was lost. Eventually, however, the Bolsheviks were able to control most of the territories formerly held by the tsars, and it was on this that they also built the Soviet state.

control. Meanwhile, the weakening of the Manchu dynasty in China prompted the Russian annexation of Chinese territory, where colonization and settlement were aided by the construction of the Trans-Siberian Railroad in the final years of the 19th century.

By 1904, when Japanese victory in Manchuria brought a halt to Russian territorial expansion, the Russian Empire contained about 130 million persons, only 56 million of whom were Russian. Of the rest, which included more than 170 distinct ethnic groups, some 23 million were Ukrainian, 6 million were Belorussian, more than 4 million were Kazakh or Kyrgyz, nearly 4 million were Jews, and nearly 3 million were Uzbek. Russia's strategies to bind together the 100-plus "nationalities" (non-Russian ethnic people) into a unified Russian state were oftentimes punitive and unsuccessful. Non-Russian nations were simply expected to conform to Russian cultural norms. Those who did not were persecuted. The result was opposition and, sometimes, rebellion and stubborn refusal to bow to Russian cultural dominance.

Revolution and the Rise of the Soviet Union Since the time of Peter the Great, Russia had been seeking to modernize. By 1861, when Tsar Alexander II decreed the abolition of **serfdom**—the practice whereby members of the lowest class were attached to a lord and his land—Russia had built up an internal core with a large bureaucracy, a substantial intellectual class, and a sizable group of skilled workers. The abolition of serfdom was designed to accelerate the industrialization of the economy by compelling the peasantry to raise crops and sell them on a commercial basis; the idea was that the profits from exporting grain would be used to import foreign technology and machinery. In many ways, the strategy seems to have been successful: between 1860 and 1900 grain exports increased fivefold, whereas manufacturing activity expanded rapidly. In 1906, further measures broke up collective agricultural production in favor of private land ownership, helping establish large, consolidated farms in place of some of the many small-scale peasant holdings, as wealthy elites were able to purchase large tracts of agricultural land. This forced many poorer people to flood to the cities and created acute problems as housing became overcrowded and living conditions deteriorated.

These problems, to which the Russian tsars remained indifferent, fueled deep discontent among the population. At the turn of the 20th century, Russia was in the grip of a severe economic recession. Inflation, with high prices for food and other basic commodities, led to famine and widespread hardship. With little political voice, peasants rioted across the countryside and, in 1905, a revolutionary outbreak of strikes and mass demonstrations took place. A network of grassroots councils of workers—called **soviets**—emerged spontaneously not only to coordinate strikes but also to help maintain public order. The unrest was eventually subdued by brute force, and the soviets were abolished by the tsars. World War I intensified the discontent of the population, as casualties mounted and the government's handling of both the armed forces and the domestic economy led to a successful revolution against the tsars in 1917 and the establishment of the Soviet Union. The revolution is often referred to as the October or Bolshevik Revolution, named after the **Bolsheviks**, the majority faction of the Russian Social Democratic Party, which was renamed the Communist Party after seizing power in 1917. The revolution was greeted with hostility by surrounding countries, and the civil war was compounded by the invasion of Russia from all sides. The conflict ended in a victory for the revolutionary government of the new

Soviet Union, but the period set a tone of hostility that influenced later global and regional geopolitical relationships between the Soviet Union and the United States as well as the Soviet Union and those states that were not ethnically Russian.

The Soviet Union's **state socialism**, in which the government (or state) controlled industries and services, was based on a new kind of social contract between the state and the people. In exchange for people's compliance with the system, their housing, education, and health care were to be provided by state agencies at little or no cost (**Figure 3.16**). By the early 1920s, **Vladimir Ilyich Lenin**, whose real name was Vladimir Ilyich Ulyanov, the revolutionary leader and head of state of the new regime in Russia, was also able to focus attention on the more idealistic aspects of state socialism. Lenin was an **internationalist**, believing in equal rights for all nations and wanting to break down national barriers and end ethnic rivalries. Lenin's solution was recognition of the many regional nationalities through the establishment of the **Union of Soviet Socialist Republics (U.S.S.R.)** in 1922. Lenin believed that this federal system would provide different nationalities with political independence.

Lenin was optimistic that once inequalities were diminished, and the many nationalities became united as one Soviet people, the federated state would no longer be needed: local nationalisms would be replaced by communism. Lenin's vision was short-lived, and following his death in 1924, the federal ideal faded. After eliminating several rivals, **Joseph Stalin** came to power in 1928 and enforced a new nationality policy, the aim of which was to construct a unified Soviet people whose interests transcended nationality. Although the federal administrative framework remained in place, nations within the U.S.S.R. increasingly lost their independence and by the 1930s were punished for displays of nationalism. **Figure 3.17** illustrates the administrative

▲ **FIGURE 3.16 The Ideals of the Soviet Union** This painting, titled *The Harvest*, was produced in 1925. It is an example of socialist realist art. This form of art depicted the worker as a hero in the new Soviet society through the literal representation of the everyday labor that supported the people collectively.

▲ **FIGURE 3.17 Soviet State Expansionism, 1940s and 1950s** World War II gave the Soviet state pretext to move westward in search of additional territories. Insisting that these countries would never again be used as a base for aggression against the U.S.S.R., Stalin retained control over Poland, East Germany, Czechoslovakia, Hungary, Romania, Bulgaria, Albania, Yugoslavia, and eastern Austria. In 1945, Stalin promised democratic elections in these territories. After 1946, however, Soviet control over Eastern and Central Europe became complete as noncommunist parties were dissolved and Stalinist governments were installed in all controlled territories.

units and nationalities that were part of the U.S.S.R. during Stalin's tenure as premier (1928–1953) and also shows how, during and immediately after World War II, Stalin expanded the power of the Soviet state westward to include Albania, Bulgaria, Czechoslovakia, the German Democratic Republic, Hungary, Poland, Romania, and Yugoslavia.

Under Stalin's leadership, a **command economy**, an economy in which all goods, services, prices, and supplies are regulated by the government, was established. Engineers, managers, and *apparatchiks* (state bureaucrats) drawn from the membership of the Communist Party ran this new economy. At this juncture, the Soviet Union chose to withdraw from the capitalist world economy as far as possible. The nation relied on its vast territories to produce the raw materials needed for rapid industrialization. Stalin's industrialization drive was founded on severe exploitation of the rural population. Peasants were relocated onto state or collective farms, where their labor was expected to produce bigger yields. The state then purchased harvests at relatively low prices so that, in effect, the collectivized peasant paid for industrialization by "gifts" of labor.

Severe exploitation required severe repression. Stalin employed police terror to compel the peasantry to comply. Dissidents, along with "enemies of the state" uncovered by purges of the army, the bureaucracy, and the Communist Party, provided convict labor for infrastructure projects to support the industrialization drive. Altogether, some 10 million people were sentenced to serve in the workforce, imprisoned, or shot.

The Soviet economy *did* modernize, however. Between 1928 and 1940 the rate of industrial growth increased steadily, reaching levels of more than 10% per year in the late 1930s—growth rates that had never before been achieved. When the Germans attacked the Soviet Union in 1941, they took on an economy that in absolute terms (though not per capita) had industrial output figures comparable with their own.

Growth, the Cold War, Stagnation, and the Breakup of the Soviet Empire The entire Soviet Union and the countries that became known as the U.S.S.R.'s "bloc" of satellite states in Eastern Europe continued to give high priority to industrialization after World War II. Between 1950 and 1955, output in the Soviet Union again grew at nearly 10% per year. Soviet economic planners sought to follow three broad criteria in shaping the economic geography of state socialism. First was the idea of technical optimization. Without free markets to provide competitive cost-minimization strategies, Soviet planners attempted to organize industry in ways that ensured both internal and external efficiencies. Perhaps the most striking result of this was the development of **territorial production complexes**, regional groupings of production facilities based on local resources that were suited to clusters of interdependent industries. Key territorial production complexes were petrochemical complexes, for example, or iron-and-steel complexes (**Figure 3.18**). Second, the planners fostered industrialization in economically less-developed subregions, such as Central Asia and the Transcaucasus.

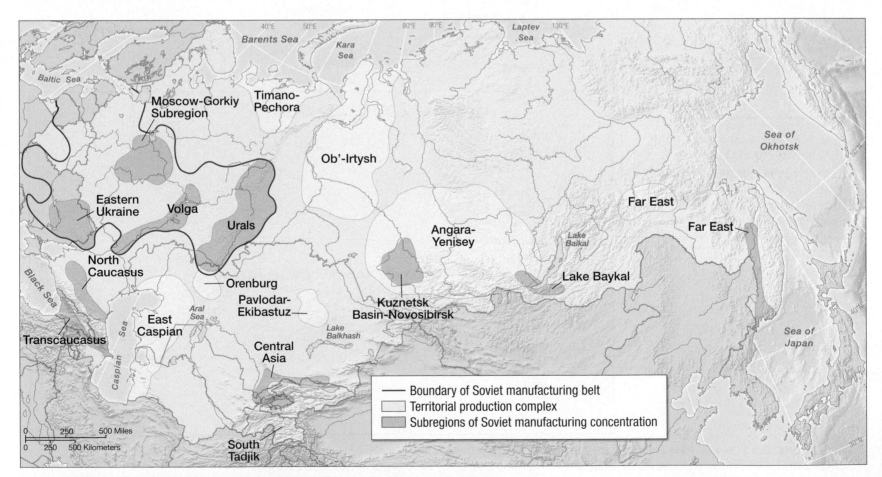

▲ **FIGURE 3.18 Industrial Regions of the Soviet Union** Soviet planners prioritized industrialization and sought to take advantage of agglomeration economies by establishing huge regional concentrations of heavy industry.

The Soviet planners' third concern was secrecy and security from external military attack. This criterion led to some military–industrial development in Siberia and to the creation of so-called secret cities. In secret cities, even the inhabitants' contact with relatives and friends was strictly controlled because of the presence of military research and production facilities.

By the 1960s, the Soviet Union had clearly demonstrated its technological capabilities with its manned space program and some of the world's most sophisticated military hardware. These successes were paralleled by the Soviet Union's broader geographical and political influence throughout the world. Armed with nuclear weapons and an economic model that challenged the dominance of capitalism, the Soviet Union posed a very real threat to U.S. power. The resulting tensions between the global power of the Soviet Union and its socialist economic model and the United States with its capitalist economic model resulted in the Cold War. A **cold war** is not a direct conflict of arms between two states but a war of ideology and proxy conflicts in which other countries are used to fight. The Cold War between the Soviet Union and the United States lasted between 1950 and 1989 and provided the principal framework for world affairs during this period. The conflict resulted in tensions and a succession of geopolitical crises in many regions of the world, including Cuba, much of the Middle East, South Asia, East Asia, Southeast Asia, and parts of South America (Chile), Central America (Nicaragua and Panama), and Africa (Angola, Libya, and Egypt). In this way, the Cold War simultaneously isolated the Soviet Union and its satellite countries from Western nations as a region while increasing their global reach and connections throughout the less-developed world (**Figure 3.19**).

Yet throughout most of the Soviet Union itself, millions of peasants worked with primitive and obsolete equipment as they toiled to meet centrally planned production targets. Most nonmilitary industrial productivity was also constrained by technological backwardness and cumbersome bureaucratic management systems. A second economy—an informal or shadow economy—of private production, distribution, and sales emerged. It was largely tolerated by the government, mainly because without it, the formal command economy would not have been able to function as well as it did. By the 1970s, the Soviet economic system was steadily being enveloped by an era of stagnation.

By the 1980s, the Soviet system was in crisis. In part, the crisis resulted from a failure to deliver consumer goods to a population that had become increasingly well-informed about the consumer societies of their foreign enemies. Persistent regional inequalities also contributed to a loss of confidence in the Soviet system. The cynical manipulation of power for personal gain by ruling elites and the drain on national resources from the arms race with the United States also undoubtedly played a role in undermining the Soviet model. The critical economic failure, however, was state socialism's inherent inflexibility and its consequent inability to take advantage of the new computerized information technologies that were emerging elsewhere.

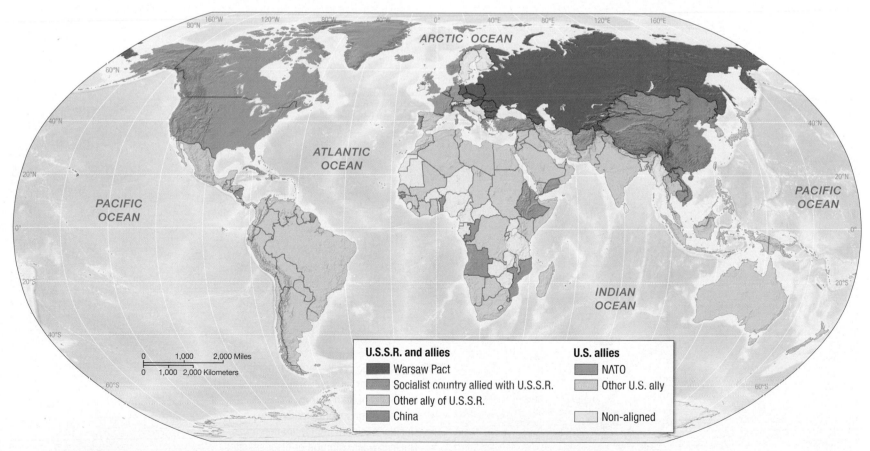

▲ **FIGURE 3.19 The Global Geopolitics of the Cold War** During the last half of the 20th century, both the Soviet Union and the U.S. and its NATO allies sought to influence counties around the world. Some nations tilted towards one side or the other, while many attempted to remain neutral and non-aligned.

Surprising even the most astute observers, the Soviet system unraveled rapidly between 1989 and 1991, leaving 15 independent countries as successors to the former U.S.S.R. The former states of Yugoslavia and Czechoslovakia were broken into smaller entities; East Germany was absorbed into Germany; Hungary, Poland, and the Baltic states (Estonia, Latvia, and Lithuania) were drawn rapidly into the European Union's sphere of influence; and Moldova and Ukraine began to show signs of a Western orientation. Belarus, the Russian Federation, and the states of Central Asia and the Transcaucasus continue to experience somewhat chaotic transitions, at different speeds, toward market economies. In the process, all local and regional economies were disrupted, leaving many people to survive by supplementing their income with informal activities, such as street trading and domestic service.

Environmental Legacies of Socialism and Transition Every country in the former Soviet region has a legacy of serious environmental problems that stem from mismanagement of natural resources and failure to control pollution during the Soviet era. Soviet central planning placed strong emphasis on industrial output, with very little regard for environmental protection. Early Soviet ideology suggested that is was possible to harness the energy of nature through the collective will and effort of the people. The tendency during the Soviet era was to squander natural resources and to "take on" and "conquer" nature through ambitious civil engineering projects. Soviet authorities saw problems of pollution and environmental degradation as an inevitable cost of modernization and industrialization, and the people most affected—the general public—had no political power or means to voice environmental concerns. Today, serious environmental degradation, which is a legacy of Soviet planning, affects all parts of the region (**Figure 3.20**).

Radioactive contamination is seen by many to epitomize the consequences of Soviet attitudes toward the environment. Both the large-scale civilian nuclear energy program and the military nuclear capability of the Soviet Union resulted in an alarming incidence of radioactive pollution. The 1986 disaster at Chernobyl, in which a nuclear reactor exploded in a power plant in the former Soviet republic of Ukraine, became emblematic of the Soviet nuclear legacy. Stories of radiation

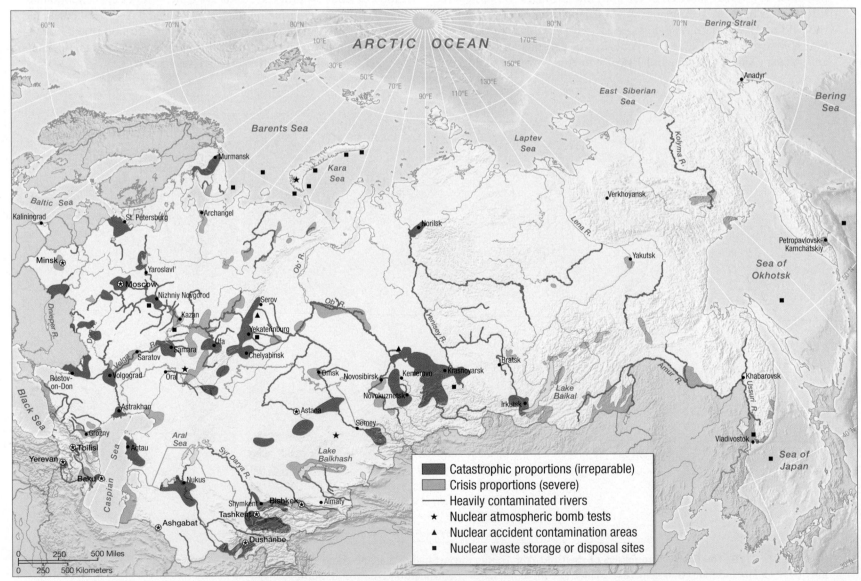

▲ **FIGURE 3.20 Environmental Degradation** Serious environmental degradation afflicts all parts of the Russian Federation, Central Asia, and the Transcaucasus.

▲ **FIGURE 3.21 Environmental Recovery after Chernobyl** These wild buffalo were rare in the region but are now reclaiming the native territory since they were reintroduced to the area following the accident. Though still radioactive in areas, the region surrounding Chernobyl has experienced ecological recovery, largely owing to the absence of human activity.

sickness and eventual ghastly deaths of the facility workers, firefighters, medical personnel, and other volunteers filled international newspapers for weeks following the meltdown. In subsequent years, radiation levels at Chernobyl have remained high, and an area surrounding

the failed reactor, 31 kilometers (19 miles) in every direction, has been established as a **"zone of alienation,"** where only a handful of residents and scientific teams continue to reside.

Nevertheless, the Chernobyl region has become a fascinating test of how the environment responds to such a disaster and of the recovery of the environment in the almost-total absence of human beings (**Figure 3.21**). In the years since the accident, forests have regrown rapidly and many wild animals, including wild horses and wolves, have recolonized the area. Though heavily irradiated, the Chernobyl area has become a unique and emblematic region of the Anthropocene, a place transformed by people and yet still changing and evolving in unexpected and managed ways.

Radioactivity is only one of several major environmental problems that have left an enduring legacy to the Soviet Union's successor states. Soviet modernization programs brought large-scale irrigation schemes to the desert and semi-desert regions of Central Asia, notably the Kara-Kum Canal, a 770-kilometer (478-mile) irrigation canal that diverts water from the Amu Dar'ya and irrigates 1.5 million hectares (4.7 million acres) of arable land and 5 million hectares (12.4 million acres) of pasture as it trails. Cotton was, and still is, the dominant crop in these irrigated lands (**Figure 3.22**). In Turkmenistan, for example, more than half the arable land is devoted to cotton, and Uzbekistan is the world's fifth-largest producer of raw cotton and third-largest exporter of cotton.

The worst consequence of the Soviet-initiated program of irrigated cotton cultivation has been the effects of excessive withdrawals of water from the main rivers that drain into the Aral Sea. The Kara-Kum Canal alone took away almost one-quarter of the Aral

▶ **FIGURE 3.22 Irrigation and the Shrinking Aral Sea** The massive increase in irrigation in central Asia, especially for cotton production, has diverted water from flowing to the Aral Sea. This vast body of water, once one of the four largest lakes in the world, is now a fraction of its original size and is heavily salinized and polluted.

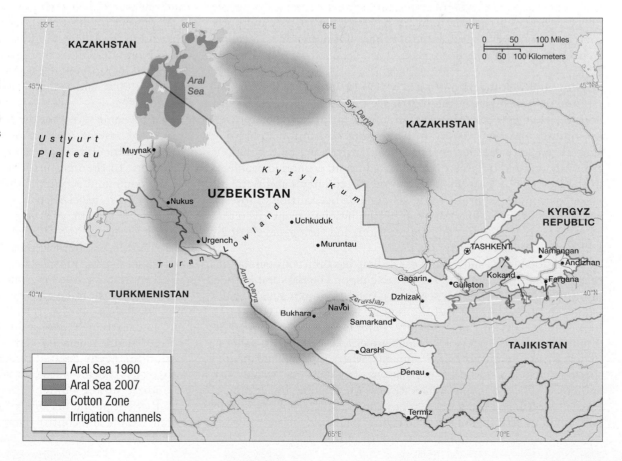

▲ **FIGURE 3.23 The Emerging Desert around the Aral Sea** As the shoreline of the Aral Sea retreats, it leaves major fishing ports and vessels stranded on dry land and exposes the seabed, which is thick with toxic salts and polluted sands. This new desert region is harsh and hazardous.

Sea's annual supply of water. The Aral Sea was the fourth-largest lake on the planet in 1960. By 2008, it had shrunk to 10% of its original size. The sea has shriveled into three major residual lakes, two of which are so salty that fish have disappeared. The level of the Aral Sea has already dropped by more than 10 meters (33 feet), and the former seabed is littered with stranded ships (**Figure 3.23**). The seabed generates a constant series of dust storms that are thought to cause unusually high levels of respiratory ailments among the people of the region. The fishing industry has been devastated, and ports such as Muynaq are stranded more than 40 kilometers (25 miles) from the retreating lakeshore.

Environmental Impacts of Russia's New Economy Not all the environmental problems in the region are the result of Soviet legacy, however. The new emerging capitalist economy of Russia has proven ecologically destructive as well. Ubiquitous corruption and strong centralized state decision making mean environmental regulations are easily ignored or circumvented. In addition, economic problems in the Russian Federation have limited the country's ability to address environmental problems which include overcutting of forests, widespread overuse of pesticides, heavy pollution of many rivers and lakes, extensive acid rain and soil erosion, and serious levels of air pollution in industrial towns and cities.

Most notably, oil and natural gas drilling has been considerably accelerated since the demise of the Soviet Union. The risks and impacts of the drilling are highest in sensitive coastal environments. Sakhalin Island, in the eastern part of the region, has become a central hub for oil and gas drilling most recently. Impacts here range from pollution of the island's waterways, which are crucial habitats for salmon, to degradation of offshore environments, where rare and endangered species—including dwindling populations of western gray whale—live and breed. These examples illustrate how the ongoing integration of Russia's globalizing economy is as environmentally problematic as its long-standing tradition of state-led central planning.

APPLY YOUR KNOWLEDGE Describe the environmental changes that were brought on by the practice of socialism in the U.S.S.R. List three examples of how socialist planning and the subsequent pursuit of capitalism have impacted the environment of the region.

Economy, Accumulation, and the Production of Inequality

Restoring capitalism in countries where it was suppressed for more than 70 years and reestablishing global connections to other parts of the global capitalist economy have proven to be problematic. Although still a nuclear power with a large standing army and a vast territory containing a rich array of natural resources, the Russian Federation is, relatively speaking, still economically weak and internally disorganized. The Soviet system left industry in the current Russian Federation with obsolete technology and low-grade product lines. Similarly, the inherited infrastructure was poorly developed, shoddy, and often downright dangerous. The Soviet authorities had deliberately suppressed investment in the development of computers and new information networks because computers, like photocopiers and fax machines, were viewed as a threat to central control. As a result, the Russian Federation's economy faced a massive task of modernization before approaching its full potential.

Building a Market Economy The economy of the Russian Federation shrank by 62% between 1991 and 1999. Global capital had begun to flow into the Russian Federation during this period, but it was targeted mainly at the fuel and energy sector, natural resources, and raw materials (which now account for about half of the Russian Federation's total exports), rather than manufacturing industry. By the end of the 1990s, the Russian Federation's economy was in crisis. About one-half of the government's budget revenue was being absorbed by repaying debts to creditor nations; the rate of inflation reached 100%; and economic output plunged to about half of what it was in 1989.

At the same time that the daunting task of establishing the institutions of business and democracy loomed, the Russian Federation found it difficult to create some of the essential pillars of a market economy. In the institutional vacuum that followed the breakup of the command economy, there were no accepted codes of business behavior, no civil code, no effective bank system, no effective accounting system, and no procedures for declaring bankruptcy. Security agencies were disorganized, bureaucratic lines of command were blurred, and border controls between the new post-Soviet states were nonexistent. The absence of these key economic elements provided enormous scope for crime and corruption and fostered regional ethnic **mafiyas** (organized crime groups) in nations such as Chechnya, Azerbaijan, and Georgia. State assets were often sold off quickly and at an extremely undervalued price, so that some well-connected entrepreneurs managed to amass huge amounts of assets and wealth within the newly created liberal market environment. These new **oligarchs**—business leaders who wielded significant political and economic power—quickly became extremely unpopular among the Russian public because of their undue political influence, extreme wealth, and control over media outlets. While many oligarchs

▲ **FIGURE 3.24 Upscale Housing in the New Russia** A new wealthy elite has emerged in Russia over the last 20 years. Like wealthy elites around the world, some people now have access to the trappings of that wealth, including large homes located in idealized landscapes, such as the ones pictured here on the Volga River in Tver, Russia.

the world and is now considered the sixth-largest economy, with the 2012 output expected to exceed U.S. $2 trillion. The biggest reason for the economic achievements is current oil prices, with Urals oil passing U.S. $125 barrel, and growing domestic consumption. High oil prices equal large export and fiscal revenues and a "bullish" stock market. Unemployment has returned to its precrisis level of 6.5%, and inflation in 2012 had fallen to 3.8%, its lowest in two decades. This does not mean that everyone in the new Russian economy benefits equally. Income disparity remains large; the earnings of richest Russians are 16 times higher than poorest.

The picture that emerges of the new Russian economy is of a fast-moving one, linked to global trade especially through an explosive energy sector. This changing economic environment produces complex political effects, including the rise of powerful elites and the persistence of high levels of inequality, and raises questions about the sustainability of growth and the possibility of reconciling the new economy with emerging democratic institutions.

suffered massive losses in the global economic crisis of 2008 and 2009, they still held considerable wealth (**Figure 3.24**). Because of their unpopularity, though, in Russia many have taken up residence outside Russia—particularly in upscale areas of London (dubbed "Moscow on the Thames" or "Londongrad").

The rise of these new elites has produced an uneasy tension within the emerging political system of the Russian Federation. While some of these oligarchs have been supported over the last decade by Vladimir Putin—the former Soviet intelligence officer who served as the President of Russia from 2000 to 2008 and was reelected to that position in 2012 (see the chapter-opening vignette, pp. 86–87)—many oligarchs are being asked to abide by new government banking and resource control policies, and a few have been prosecuted for tax evasion and other financial crimes.

Despite the dominance of these elites and their ongoing conflicts with the Russian government, recent years have seen significant economic growth. Overall, the Russian economy grew between 5% and 10% each year between 2000 and 2008, largely on the strength of its commodities exports. Growing sectors of the economy also include pharmaceuticals, furniture, cosmetics, clothing, electrical appliances, and automobiles. During the same period, real incomes more than doubled while the proportion of population living below the poverty line decreased from 30% to 14%. The average wage increased from 2,200 rubles (U.S. $90) to 12,500 rubles (U.S. $500), and the average pension from 823 rubles (U.S. $33) to 3,500 rubles (U.S. $140). Russia's superrich prospered as never before. By early 2008, Moscow had 74 billionaires—more than New York and twice as many as London.

The 2008 global financial crisis hit Russia's economy very hard but, in the period since the crisis, growth has slowly resumed in the country. By the end of 2011, Russia's economy was the ninth biggest in

Regional Development and Inequalities Although patterns of regional development are beginning to change, at present, the Central Region—an area that extends about 400 kilometers (248 miles) around Moscow—remains the core of the Russian Federation, Central Asia, and Transcaucasus region. Today, this area is highly urbanized, with about 85% of the population living in towns and cities. Moscow, with roughly 11 million people, is the largest city, although other significant urban centers also exist (**Figure 3.25**). The Central

▲ **FIGURE 3.25 Moscow and the New Russian Economy** There has been massive investment in the modern infrastructure of Moscow over the last 20 years, particularly in the development of integrated business, entertainment, and living spaces such as the Moscow International Business Center (also called Moscow City), shown here.

Region accounts for about 80% of the Federation's textile manufacturers. Cotton textiles are the most significant of these. Engineering, automobile and truck manufacture, machine tools, chemicals, electrical equipment, and food processing are also important. Overall, the Central Region accounts for about 20% of the Russian Federation's industrial production. Despite the high degree of urbanization and industrialization, much of the region has a rural flavor. About 25% of the Central Region remains forested, and there are numerous lakes and marshy areas. The traditional staple crop of the area was rye, but when the railways made it possible to import cheaper grain, farmers began to grow industrial crops, such as flax, potatoes and sugar beets, and they also began to dairy farm. Around rural settlements in the Central Region, there are orchards of apples, cherries, pears, and plums.

Market forces have introduced a much greater disparity between the economic well-being of regional winners and losers. The unevenness of patterns of regional economic development has been intensified but, after two decades of transition, many of the regional winners are the same as under state socialism. This is partly because of the natural advantages of certain regions and partly because of the initial advantage of economic development inherited from Soviet-era regional planning. However, two different kinds of regions have experienced significantly *decreasing* levels of prosperity. The first consists of regions of armed territorial conflict (such as North Ossetia in Georgia, Ingushetia, and Chechnya in the North Caucasus; Nagorno-Karabakh; and Tajikistan). The second consists of resource-poor regions—mainly in the European north, in Siberia, and in the eastern parts of the region, where conditions are harsh in both rural and urban settings (**Figure 3.26**).

APPLY YOUR KNOWLEDGE Consider the dramatic transition from one form of economy to another in the region. List two ways that the change provides opportunities or advantages for middle-class individuals or families and two ways it may present hardships.

Territory and Politics

Within the Russian Federation, there are approximately 27 million non-Russians. This number encompasses 92 different ethnonational groups. (**Ethnonationalism** is nationalism based on ethnic identity.) However, it must be noted that 25 of these groups include minority people of the north, who together number fewer than 200,000. Most of the larger ethnonational groups enjoy a fair degree of autonomy within the Russian Federation, although conflicts along border areas throughout the region are numerous (**Figure 3.27**). One of the most troubled regions is the North Caucasus, a complex mosaic of mountain people with strong territorial and ethnic identities. Soon after the breakup of the Soviet Union, Ingushetia broke away from the Chechen-Ingush Republic, and Chechnya promptly declared independence from the Russian Federation.

Chechen Independence Of the many movements that surfaced with the breakup of the Soviet Union, the Chechen independence

▲ **FIGURE 3.26 Marginal Economic Regions and the Periphery**
Many of the outlying and rural areas of the region depend heavily on primary resource extraction, especially forestry, drilling, and mining. Working conditions in these industries can be harsh, especially where health and safety regulations are limited. A mining disaster in 2010 killed 90 Russian workers.

movement has been the most bloody. In this region of the North Caucasus, clans, not territory, had been the traditional form of political organization. From the time that imperial Russia began its territorial expansion into the northern Caucasus in the late 1700s, the Sunni Muslim Chechens put up strong resistance, periodically waging holy wars against Christian Russia. When revolution came in 1917, Chechens did not look on the Soviet Union as a liberating force. Following a brief, failed attempt on the part of the people of the North Caucasus and Transcaucasus to resist Soviet domination, the Soviet began to divide and conquer by creating administrative regions encompassing mixtures of clans and ethnic groups.

In the late 1930s, tens of thousands of Chechens were murdered by Stalin as part of his purges against suspected anti-Soviet elements. Then, in 1944, after invading German forces had been forced to retreat from the North Caucasus, Stalin accused the Chechens of having collaborated with the Nazis and ordered the entire Chechen population—then numbering about 700,000—to be exiled to Kazakhstan and Siberia. Brutal treatment during this mass deportation led to the death of more than 200,000 Chechens.

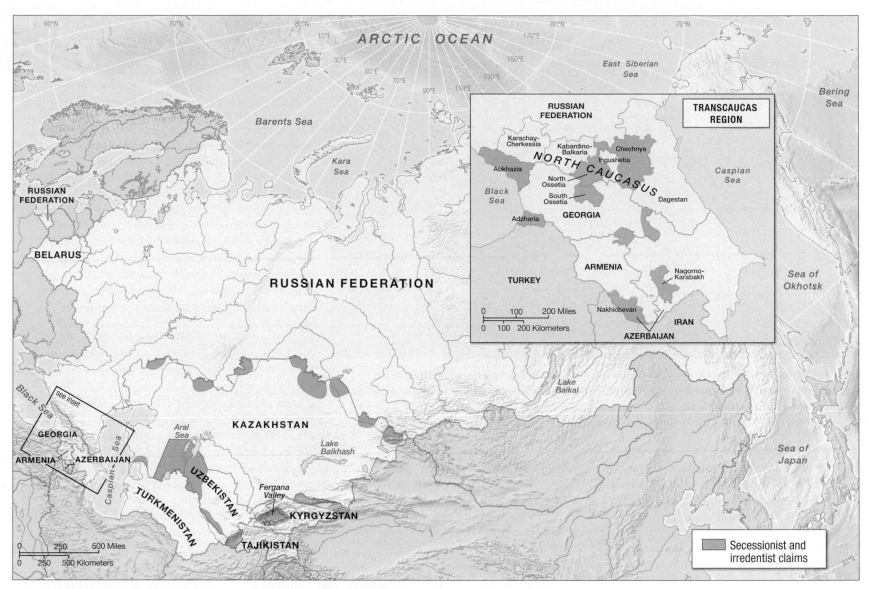

▲ **FIGURE 3.27 Conflict and Tension in the Post-Soviet States** The politics of multiculturalism is especially important in the Transcaucasus and Central Asia, where long-standing ethnic tensions, suppressed by the Soviet regime, have found renewed energy and expression.

In 1957, Nikita Khruschev—then leader of the Soviet Union— embarked on a program of de-Stalinization that included the rehabilitation of Chechens. But when Chechens returned, they found that newcomers had taken over many of their homes and possessions. Over the next 30 years, many of these newcomers withdrew and the Chechen population consolidated and grew to almost 1 million. In 1985, when Mikhail Gorbachev initiated his policy of **glasnost** (an official policy change that stressed open government and increased access to information as well as more honest discussion about the country's social issues and concerns), Chechens finally saw the possibility for self-determination, and with the breakup of the Soviet empire in 1989, Chechens wasted no time in unilaterally declaring their complete independence.

The Russian Federation at first ignored Chechnya's declaration of independence but ultimately could not tolerate the loss of

the region because the Russian Federation's major oil-refining centers and significant natural gas reserves reside within Chechnya. In December 1994, Russian troops invaded. The ensuing conflict brought terrible suffering to the Chechen population and resulted in mass migrations. Chechen resistance continued, with increased popular support because of the invasion. In 1996 the Russian Federation settled for peace, leaving Chechnya with *de facto* independence. For three years there were protracted negotiations over the nature of the peace settlement. Then, in the summer of 1999, after Chechen rebels had taken the fight to the neighboring republic of Dagestan and to the Russian heartland with a series of terrorist bombings of apartment blocks, the Russians renewed their military effort. Bitter and intense fighting ensued, during which the Russian army suffered more than 400 deaths, and nearly 1,500 wounded. Hundreds of thousands of Chechens were made homeless and several thousand

▲ **FIGURE 3.28 Grozny, Chechnya** Two women walk past a destroyed building in the Zavodskoi District of Grozny, Chechnya, in March 2011. Even though the military campaign has formally ended, the war landscape remains as a reminder the tension that exists in Chechnya. People remain displaced as a result of the conflict.

were killed or went missing. Russian troops took the Chechen capital, Grozny, in February 2000 (**Figure 3.28**).

Between early 2000 and April 2009, Russian Federation troops maintained control of Grozny, imposing restrictions such as curfews, roadblocks, periodic searches and summary detention. The special regime, which included restricted access for journalists, encouraged massive rights violations in the region. Human rights organizations documented patterns of abduction, detention, disappearances, collective punishment, extrajudicial executions, and the systematic use of torture by authorities. Chechen rebel guerillas, meanwhile, used terrorist attacks against the regime, including attacks on children. In 2009, Russia formally ended its military campaign in Chechnya, although terrorists who seek full separation from the Russian Federation have been active in the capital Grozny as late as August 2011.

Ethnonationalism in the Transcaucasus and Central Asia In the Transcaucasus, one trouble spot of ethnonationalism is the region of Nagorno-Karabakh, in Azerbaijan. At one time, Nagorno-Karabakh's population was about 90% Armenian. By the mid-1980s, the population of Nagorno-Karabakh was still more than 75% Armenian, and the breakup of the Soviet Union brought the opportunity for them to formally petition for secession from Azerbaijan. When the petition was refused, pent-up anger was unleashed in Azerbaijan against Armenians and in Armenia against Azerbaijanis. Riots and forced migrations quickly led to civil war. Armenian military forces secured Nagorno-Karabakh and established a militarized corridor as a lifeline from Azerbaijan to Armenia. Russian Federation armed forces were invited to serve a peacekeeping role, and the United States, France, and Russia have been involved in mediating

a peaceful settlement. The political situation in Azerbaijan is of broad international interest because of the area's rich natural resources and the 1,760-kilometer (1,094-mile) pipeline from Baku, Azerbaijan, to Ceyhan, Turkey (**Figure 3.29**). The pipeline opened in 2005 to carry oil from U.S. and European companies' oil fields in the Caspian Sea to Western markets via the Mediterranean (see "Visualizing Geography: Pipelines and Landscape Change in Central Asia," pp. 116–117).

Another flashpoint in the Transcaucasus is South Ossetia, which declared its independence from Georgia in 1991. It did so mainly because of strong ethnic ties to North Ossetia-Alania, a republic of the Russian Federation. In 2008, Georgia launched a massive artillery attack on the separatists of South Ossetia, prompting a brief but full-scale war that drew Russian troops into South Ossetia, where they remained at observation and security posts for several months. Since then, South Ossetia has remained independent from Georgia, but its politics remain unstable. The recent presidential election of April 2012 helped Leonid Tibilov rise to power. But he has an uphill battle; the country is marred by a legacy of mistrust and alleged corruption, as the previous government was suspected of misusing Russian aid to the country.

The breakup of the Soviet Union also led to other ethnonational movements in Central Asia. Tajikistan has been beset by conflict between ethnic Tajik and Uzbek insurgents and activists, who define themselves as patriots but assert the primacy of their own individual ethnic groups. A brief civil war ended in 1993, when the

▲ **FIGURE 3.29 Resource Development in Azerbaijan** The port city of Baku is the capital of Azerbaijan and its largest city. Located on the shores of Caspian Sea, the city has long been a cultural crossroads with a rich architectural tradition. It now sits at the center of a network for for the production and transport of Caspian oil.

ruling government accepted intervention by the Russian Federation. Between 1993 and 2005, Russian troops, in cooperation with the Tajik government, defended Tajikistan's border to counter frequent military incursions by rebels based in Afghanistan. Though these Russian troops withdrew in 2005, in 2010, the Russian government expressed an interest in resuming its military presence. Also in Central Asia, one of the most complex areas of ethnonational movements and irredentism is the Fergana Valley, where Tajik, Kyrgyz, and Uzbek nationalists living in areas outside but adjacent to their homeland states have called for the redrawing of international borders.

Geopolitical Shifts In an attempt to counter some of the economic disruption caused by the political disintegration of the Soviet Union, several successor states agreed to form a loose association, known as the **Commonwealth of Independent States (CIS)**. The CIS was designed to provide a forum for discussing the management of economic and political problems, including defense issues, transport and communications, regional trade agreements, and environmental protection. The founding members were the Russian Federation, Belarus, and Ukraine; soon these three were joined by the Central Asian states and some Transcaucasus states.

The reorientation of the Baltic and eastern European states toward Europe and the expansion of the European Union and NATO membership for some of these states has not only undercut the economic prospects of the CIS (which never really blossomed) but also weakened the geopolitical security of the Russian Federation (see Chapter 2, p. 64). The post-Soviet states are now well-integrated into the capitalist world system, and all of them have had to find markets for their uncompetitive products while at the same time engaging in domestic economic reform. Under Vladimir's Putin's government, in particular, the Russian Federation became further integrated into the global economy, demonstrating the capacity of the former Soviet Union to adjust to capitalist development practices.

In response to the changes brought about by the breakup of the Soviet Union, the Russian Federation asserted its claims in the 1990s and early 2000s in a special sphere of influence it has called the **Near Abroad**. The Near Abroad consists of the former components of the Soviet Union, particularly those countries that contain a large number of ethnic Russians. Under the leadership of Vladimir Putin in the early part of the 21st century, in particular, the Russian Federation pressed its authority in this subregion. These actions suggest that it remains an important sphere of influence for the Russian Federation. The question of how much energy and resources to focus on the Near Abroad continues to be an issue, as Russia looks to expand its economic and geopolitical role as a global player. This is perhaps best seen in the increased attention that Russia has given its relationships with China, Brazil, and India in recent years (**Figure 3.30**).

Ethnonational movements illustrate the regional dimensions of the problems involved in the transition from state socialism to market economies. If democracy is to flourish, the new states must guarantee territorial integrity, physical security, and effective

▲ **FIGURE 3.30 Putin in India** In this photo, President Vladimir Putin, consults with the Prime Minister of India, Dr. Manmohan Singh. India and Russia share a number of converging interests, including promoting trade partnerships between their globalizing economies and maintaining political stability in the wider Central Asian region.

governance. In some regions, secessionist and irredentist tensions are clearly undermining these preconditions for democracy. A second and more widespread problem concerns the vitality of civil society. **Civil society** involves the presence of a network of voluntary organizations, business organizations, pressure groups, and cultural traditions that operate independent of the state and its political institutions. A vibrant civil society is an essential precondition for **pluralist democracy**—a society in which members of diverse groups continue to participate in their traditional cultures and special interests. The Soviet state did not tolerate a civil society. Since the breakup of the Soviet Union, civil society has begun to flourish only in parts of the former Soviet empire that have reoriented themselves toward Europe. In the countries covered in this chapter, civil society is emerging only slowly, and in some regions—especially in Central Asia—democratic reform has been so limited that the emergence of civil society has been hard to detect.

APPLY YOUR KNOWLEDGE Identify three key political conflicts in the region. List two factors contributing to each of these conflicts. What role has the Russian Federation played in the regional conflict?

CULTURE AND POPULATIONS

Within the compass of this vast world region, there exists a great deal of cultural and political diversity. This section first outlines the traditional religious and linguistic geographies of the Russian

VISUALIZING GEOGRAPHY
Pipelines and Landscape Change in Central Asia

One immediate consequence of the breakup of the Soviet Union in 1989 was that the collective natural resource base was fragmented among the new states. The states of Central Asia and the Transcaucasus were particularly affected because their smaller territories and less varied physiography left each of them with a relatively narrow resource base—though the oil and natural gas reserves of Central Asia are a

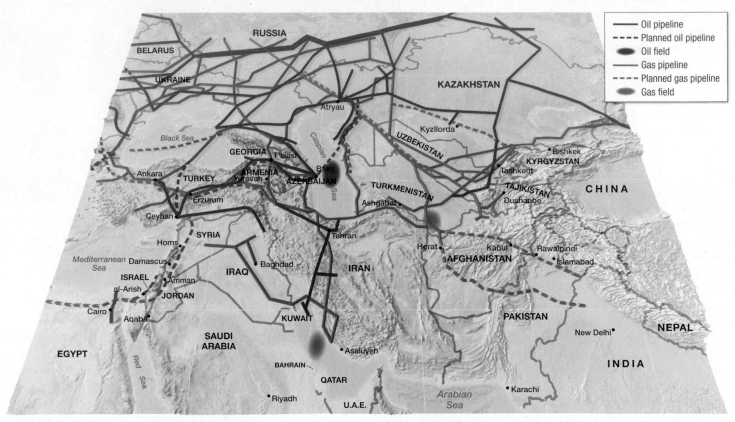

▲ **Figure 3.3.1 Resource Development Infrastructures and Pipelines across Asia** A proliferation of pipelines for oil and natural gas in Central Asia crisscross numerous states, contributing to complex politics and economics. For example, the new pipeline exporting oil across Kazakhstan to China, completed in 2009, is operated by a Chinese company, but transports oil controlled by Russian companies.

Federation, Central Asia, and the Transcaucasus, which have been shaped by the influence of past movements of people into the region from Europe, the Middle East, South Asia, and East Asia. For example, European influences have been particularly important in Belarus and the western parts of the Russian Federation, while the Arab conquests of the 8th century introduced Islam to the Transcaucasus and Central Asia. Since the region's exposure to global economic, political, and cultural systems in the post-Soviet period, traditional patterns of ethnicity and culture have become increasingly important as the basis for political identity.

Religion and Language

The patterns of religious adherence in the region reflect past global connections. Russians of all religions have enjoyed freedom of worship since the collapse of the Soviet Union; large numbers of religious buildings that were abandoned or converted during the rule of the officially atheist regime have been returned to active religious use. The dominant religion in Russia, professed by about 75% of citizens who describe themselves as religious believers, is Eastern Orthodox Christianity. The emergence of Eastern Orthodox Christianity dates from the 11th century and the "Great Schism" between Rome and Constantinople, which led to the separation of the Roman Catholic Church and the Eastern Byzantine Churches, now the Eastern Orthodox. The schism was based on political factors and cultural differences between Latins and Greeks as well as doctrinal issues. Eastern Orthodoxy adheres to ancient traditions and practices rooted in Greek, Slavic, and Middle Eastern traditions. In Russia today, about 10% of the population self-identify as Muslim. They are concentrated among the ethnic minority nationalities located to the north of the Caucasus, between the Black Sea and the Caspian Sea and in the larger cities of European Russia. Because of higher birthrates, Muslims

major asset. Kazakhstan has the bulk of the oil reserves; the natural gas fields are mainly to the south in Turkmenistan and Uzbekistan. Proven oil reserves in the Central Asian geosyncline, around and beneath the Caspian Sea, amount to between 15 and 31 billion barrels, but estimates of the potential reserves run between 60 and 140 billion barrels. Only the oil fields of the Persian Gulf states and Siberia are larger. This represents a tremendous economic asset for the Central Asian states. Exploiting these assets is beset with difficulties, however. Most of the states involved are effectively landlocked, which means that expensive pipelines have to be constructed before the oil fields and gas fields can be fully developed. But pipeline construction and routing are both risky and contentious because of political tensions and instability in the region.

The relationship between where the oil is located and where it needs to go is producing a landscape of pipelines stretching across Central Asia and into the Transcaucasus as well as the Middle East and East Asia (**Figure 3.3.1**). The intricate pipeline system— of both existing and proposed lines— illustrates how important global and regional stability is for the ongoing economic development of Central Asia, not just in Central Asia but other parts of the world. Pipelines also produce interesting new landscapes, as these above-the-ground features distinguish the region's investment in primary commodity oil and gas extraction (**Figure 3.3.2**). ■

◀ **Figure 3.3.2 Pipeline** A worker completes the last pipe links at the construction site near Almaty.

account for the fastest-growing religious grouping in Russia. About 20% of the people in the Russian Federation profess no religious affiliation—a similar figure to Germany, Italy, Spain, and the United Kingdom.

In Central Asia, the Sunni branch of Islam is the dominant religious group. The dominance of Islam in this subregion dates from the Battle of Talas in 751 C.E., when the Arab Caliphate (the chief Muslim civil and religious ruler, regarded as the successor of Muhammad, based in Baghdad) defeated the Chinese Tang Dynasty, thereby gaining control of the region. The Arab Conquest also brought Islam to the Transcaucasus in the 8th century, displacing Zoroastrianism and various pagan cults. In Azerbaijan, about 95% of the population is nominally Islamic. In Georgia, on the other hand, Eastern Orthodoxy was introduced after King David IV defeated Turkish rulers at the Battle of Didgori in 1121 C.E. Today, Eastern Orthodoxy is the dominant religion in Georgia, claimed by about 80% of the population.

Linguistic patterns and ethnicity are more complex. Dominant today throughout Belarus and the Russian Federation are Slavic people, among whom Russians represent one particular ethnic group. The Slavs are fundamentally defined by linguistic commonalities rather than territorial, racial, or other attributes. The Slavonic group of languages forms one of the major components of the great Indo-European language family, whose speakers range from north India (Hindi and Urdu) through Iran (Farsi) and parts of Middle Asia (Tajik) to virtually the whole of Europe. Written language came late to the Slavs, and when it did it was the deliberate effort of two missionaries. The missionaries— Constantine (later, as a monk, called Cyril) and Methodius— were sent by the 9th-century Byzantine emperor Michael III to the Slavic nation of Greater Moravia (which occupied much of present-day Hungary, Germany, Slovakia, and Czechia) to spread the Scriptures. The new alphabet that Constantine/Cyril devised to accommodate Slavonic speech sounds became known

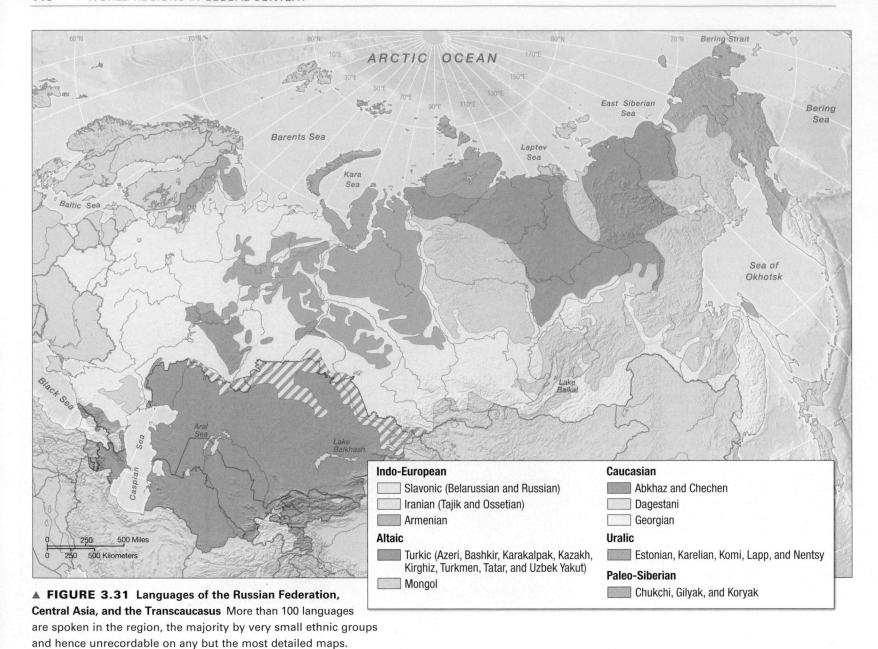

▲ **FIGURE 3.31 Languages of the Russian Federation, Central Asia, and the Transcaucasus** More than 100 languages are spoken in the region, the majority by very small ethnic groups and hence unrecordable on any but the most detailed maps.

Legend:

Indo-European
- Slavonic (Belarussian and Russian)
- Iranian (Tajik and Ossetian)
- Armenian

Altaic
- Turkic (Azeri, Bashkir, Karakalpak, Kazakh, Kirghiz, Turkmen, Tatar, and Uzbek Yakut)
- Mongol

Caucasian
- Abkhaz and Chechen
- Dagestani
- Georgian

Uralic
- Estonian, Karelian, Komi, Lapp, and Nentsy

Paleo-Siberian
- Chukchi, Gilyak, and Koryak

as the Cyrillic alphabet. Slavonic-speaking people correspond to the most densely settled parts of the region, extending along the steppe to the east (**Figure 3.31**).

A second major language group is that of Turkic languages, which belong to the Altaic family of languages. The Turkic languages are spoken by the peoples of Central Asia and parts of the Transcaucasus. They were spread into Russia through the Tatar invasion and period of rule (in about 1240–1480 C.E.). Much of northern and eastern Siberia is occupied by people who speak other branches of the Altaic language group. At the eastern edge of the region, the languages in use are part of the Paleo-Siberian language family, including Gilyak and Koryak. Finally, there are several smaller areas of Caucasian languages: Abkhaz and Chechen on the northern slopes of the Caucasus, and Georgian and Dagestani languages in the Transcaucasus.

Cultural Practices, Social Differences, and Identity

Just as the region's economies are in transition from state socialism to capitalism, cultural practices in the region have been adapting and contributing to globalization. In some ways, this is simply a resumption of the emerging cultural practices of the pre-Soviet era, when Western, Byzantine, and Oriental influences all contributed to the cultural makeup of the region, and the folkways of Russia and the Transcaucasus found their way into the "high culture" of the West through art, literature, and symphonic music.

High Culture The influence of European culture in Russia during the 17th and 18th centuries, strongly encouraged by Peter the Great and Catherine the Great, brought Russian high culture closer to the

▲ **FIGURE 3.32 Russian Ballet** Ballet was developed in Russia in the 1700s under the patronage of Empress Anna when she established the Imperial Theatre School for ballet training in 1738. Today, Russian ballet troupes, such as the Classical Grand Ballet of the St. Petersburg Ballet Theater shown here, travel the world sharing this important art form.

traditions of Western Europe. But by the end of the 19th century, uniquely Russian artistic styles had nonetheless developed. Some of these developed in conjunction with liberal forces of social reform, and some were inspired by Russian rural folk culture. The late 19th and early 20th centuries were a golden age for Russian high culture. In the performing arts there was the work of composers such as Borodin, Tchaikovsky, Rimsky-Korsakov, Rachmaninoff, Prokofiev, Stravinsky, and Shostakovich, together with the ballet impresario Sergei Diaghilev and the dancers Vaslav Nijinsky and Anna Pavlova (**Figure 3.32**). Equally influential in both Russia and the West were the novels of Dostoevsky and Tolstoy, the stories and plays of Chekhov, and the poetry of Pushkin.

Popular Culture The 1917 Revolution halted these emergent cultural dynamics. At the same time, the revolution inspired radical cultural and political movements in much of the rest of the world, eventually taking root in different forms in China, North Korea, and Cuba as well as being imposed on the Soviet satellite states of Eastern Europe. In Russia itself, the trajectories of both high culture and mass culture were repressed by revolutionary ideology. But to an emergent artistic avant-garde movement, the new Bolshevik regime seemed to promise just the sort of radicalism that they had been working toward for years. They produced political posters with a distinctive genre of graphic design and developed a unique form of modernist architecture—Constructivism—that drew on the new regime's emphasis on the importance of industrial power, rationality, and technology. The Bolsheviks saw mass culture as a way of educating and transforming the population and employed film, fiction, radio, television, and poster art to help create the "new Soviet person." After Stalin's death in 1953, the ideological controls

on culture were somewhat relaxed, and subcultures emerged that were based largely on covert jokes, underground grassroots media such as cassette tapes, and **samizdat publications** (dissident or banned literature produced through systems of clandestine printing and distribution). Western rock music, in particular, became popular in the 1960s largely through illegal copies of albums that circulated from hand to hand.

Popular culture in Russia and post-Soviet states includes a range of deeply traditional practices, which together represent the hybrid history of the region. Notably, the food traditions of Russia are a product of its intricate connections both to more western parts of Europe as well as Central and East Asia. Consider Russian *pirozhki*. Pirozhki are small pies, stuffed with savory meats that come in diverse forms. For example, they are either baked, in the Slavic method, or fried, in a way introduced by the Tatars in the 1500s. Examples such as this one illustrate the ways in which the traditional popular culture, though distinctive of the region, also reflects a complex mix of cultures that have been connected through political history (**Figure 3.33**).

Since the breakup of the Soviet Union, popular culture has gone through an enormous upheaval, thanks to the collapse of censorship, the pressures of market forces, and a flood of cultural imports from Europe, the United States, and elsewhere. The advent of new

▲ **FIGURE 3.33 Pirozhki Vendor in Moscow** Vendors selling *pirozkhi* in Moscow mix food traditions adapted from the wider region into the city's modern-day street food culture.

television programming and different ways of thinking about issues like class, identity, and sexuality opened up post-Soviet society, especially in western Russia and the larger cities. While Russians at first imported a great deal of popular culture from the West, it did not take long before a hybridized, home-grown popular culture began to emerge. No longer do American television shows and Brazilian and Mexican soap operas dominate. People now are more inclined to watch domestic programming, where their own lives are portrayed. Similarly, thousands of rock groups of all kinds—hard, soft, punk, art, folk, fusion, retro, and heavy metal—now flourish in Russia and the Transcaucasus.

One distinctive aspect of post-Soviet Russia entertainment and mass culture is its sensationally violent and abnormally graphic sexual nature. Free from censorship for the first time in Russia's history, the popular culture industry disseminates works that feature excessive and appalling details of social and moral decay. Cultural theorists have interpreted this as the popular culture industry's response to the scale of Russia's national collapse: a distraction from despair over economic woes and everyday threats.

APPLY YOUR KNOWLEDGE Search the Internet to find one current thriving example of high culture and one of more popular culture in the region. What are some distinct characteristics evident in the online descriptions of these cultural forms? How do those characteristics reflect the complex dynamics and interaction of regionalism and globalization and the region's unique history?

Social Stratification and Conspicuous Consumption Clearly, many aspects of life have changed significantly since the breakup of the Soviet Union. Restaurants, for example, were not highly developed under communism, but the post-Soviet period has seen an explosion of restaurants, cafés, and fast-food places in cities. The majority of people do not frequent restaurants often, mainly for economic reasons, but for the new business classes, dining out is part of a new pattern of conspicuous and competitive consumption. This, in turn, reflects the emergence of new class factions with distinctive identities. Although they had special privileges, most officials in the Soviet system did not accrue wealth. As discussed previously in the chapter, privatization has allowed some of them to become oligarchs by building large fortunes and taking advantage of insider status to acquire a share of direct ownership of state resources and industries. A new entrepreneurial class has also emerged, some of whose members have become significantly wealthy. More slowly, an affluent middle class is emerging in the cities, formed of an educated elite newly employed in business ventures and midlevel management. Most of the rest of the population, meanwhile, remains relatively impoverished.

Gender and Inequality When the Soviet system collapsed in 1989, the usual social and economic safety nets were dismantled, and social and economic upheaval was accompanied by an intensification of inequality. In the early 1990s, industrial production in Russia

fell by nearly half; hyperinflation devalued people's savings, leaving many destitute. Then in the late 1990s, there was a deep crisis of national finance as a result of Russia's weak trading record. The consequences for many parts of the Russian population included rising mortality rates, especially among older men. The higher mortality rates were attributed in part to the stresses surrounding economic dislocation, in part to increasing poverty, and in part to the decline in the provision of health care in post-Soviet Russia. There was also an increase in the rate of industrial accidents and alcohol-related illnesses and a rapid escalation of infant mortality rates. The modest levels of material welfare to which families had become accustomed under state socialism were increasingly difficult to sustain.

Amid this upheaval, the role of women changed significantly. Women, together with all other groups discriminated against under the tsars, were "freed" by the 1917 Bolshevik Revolution, which declared them equal and granted them all social and political rights. In theory, women were to be trained for and encouraged to take up what was previously male-only labor, such as operating agricultural machinery, working in construction, and laying and maintaining roads and rail beds. Nurseries and day-care centers were established to liberate women from child rearing. Women's increased participation in medicine, engineering, the sciences, and other fields was supposedly encouraged. In practice, however, the Soviet political system was male-dominated, its legislative organs developing into a rigid structure based on proportional representation.

After the collapse of the Soviet system, many women began in earnest to take up new opportunities presented by the transition to a market economy, venturing into small trade and opening their own businesses. Some women also took advantage of opportunities in newly established firms, quickly climbing through the ranks to become managers. The post-Soviet transition has opened up some opportunities for women, but closed down many others (**Figure 3.34**). Legal restrictions have been placed on the kinds of jobs women are allowed to do in the new Russia. As of 2012, there were 460 jobs that were legally off-limits to women, including firefighting and driving metro subway trains. Another area in which the women of the region have clearly regressed is political representation. Under state socialism, quotas ensured that one-third of the seats in parliament went to women. In 2011, 13.6% of the seats in the Russian Federation's lower house of Parliament were held by women; that same year, Valentina Matviyenko, former governor of St. Petersburg, became the first woman speaker of Russia's upper house of Parliament.

Today, the number of women holding full-time jobs (about 46%) in Russia is on par with developed countries. Nevertheless, where women's wages had been, on average, about 70% of men's during the 1980s, they had dropped to 52% by 1999, recovering only to 64% by 2010. The transition to new market economies has cost women many of the benefits they enjoyed under state socialism—such as child care, health care, equal pay, and political representation. Between 1985 and 2005, the number of working women in the Russian Federation fell by 24%. Many women who might have landed clerical or professional jobs under state socialism now find themselves forced into unskilled work to make ends meet while caring for children and keeping their family together. Often, this means working in

▲ **FIGURE 3.34 Women in Today's Russia** Members of Pussy Riot, a Russian radical feminist punk rock group, protest the policies of Vladimir Putin in Moscow in January 2012. In March 2012, three members of the band were arrested for their participation in a protest against the government held in a Russian Orthodox Church. In August of that year they were sentenced to two years in prison. Band members wear masks to draw attention to women's roles in Russia, which they view as more passive and anonymous than those of men.

the unprotected (and sometimes illegal) realm of the informal economy. A great deal of media attention has been given to the fact that tens of thousands of women have been forced into prostitution, often after being trafficked abroad on the pretense that they would work as maids or waitresses.

Demography and Urbanization

A distinctive characteristic of this world region is the relatively low density of its population (**Figure 3.35a**, p. 122). With a total population in 2011 of some 232 million and almost 14% of Earth's land surface, the Russian Federation, Central Asia, and the Transcaucasus contains about 4% of Earth's population at an overall density of only 48 persons per square kilometer (124 people per square mile). The highest national densities—106 per square kilometer in Azerbaijan (274 per square mile), 105 per square kilometer in (262 per square mile) Armenia, and 64 per square kilometer (160 per square mile) in Uzbekistan—approximately the population densities of Ohio, Pennsylvania, and Georgia. Within the Russian Federation, population density is relatively higher in the Central Region, falling off to less than one person per square kilometer in the far north, Siberia, and the far eastern parts of the region, about the same as in the far north of Canada.

Levels of urbanization reflect this same broad pattern. Most large cities are in the European part of the Russian Federation and in the Urals. These include Moscow, Nizhniy Novgorod, St. Petersburg, Volgograd, and Yekaterinburg. Most of the other cities of any significant size are found in southern Siberia, on or near the Trans-Siberian Railway, a network of railways connecting Moscow with the Russian Far East and the Sea of Japan. Overall, both Belarus and the Russian Federation are quite highly urbanized, with 72% and 76% of their total populations living in cities, according to their respective census counts in the mid-1990s. The populations of the Transcaucasus are moderately urbanized (56% to 69% living in cities), whereas those of Central Asia are more rural (only 30% to 50% living in cities).

Population Growth and Decline This world region has a relatively slow-growing population. Throughout the 20th century, there was a general decline in both birth- and death rates (**Figure 3.35b**, p. 122). Since the 1990s, the population began to register a decline as a result of more deaths than births. Viewed in greater detail, it is clear that this trend masks some important regional differences. In Belarus and the Russian Federation, population growth has for a long time been relatively modest, and it is in these countries that declines have been most pronounced. In contrast, in Central Asia and the Transcaucasus, birthrates have historically been relatively high, and rates of natural increase remain at a level comparable with those in South Asia and Southeast Asia.

Both World War I and World War II resulted in huge population losses that are still reflected in the age–sex profile of the Russian

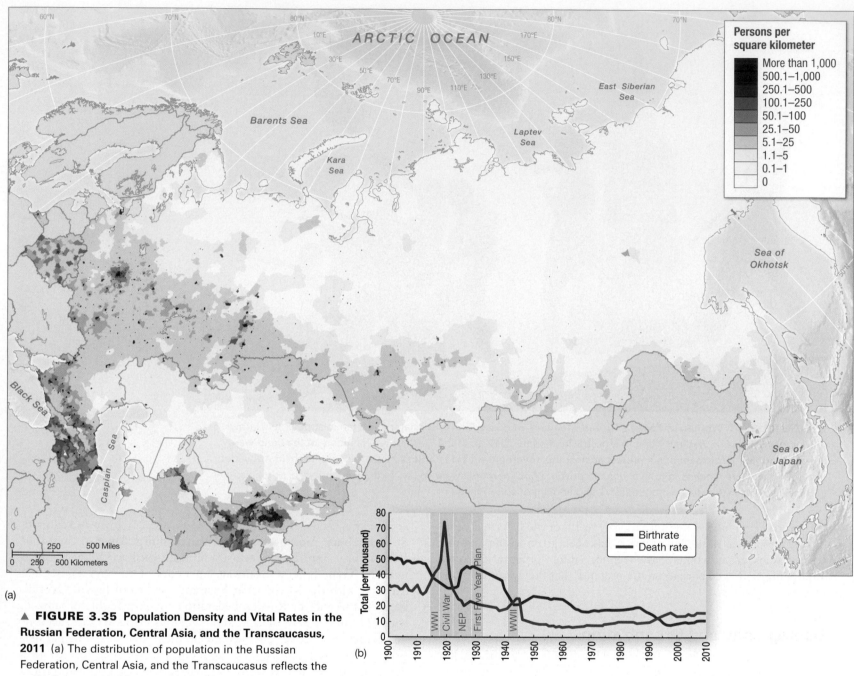

(a)

(b)

▲ **FIGURE 3.35** **Population Density and Vital Rates in the Russian Federation, Central Asia, and the Transcaucasus, 2011** (a) The distribution of population in the Russian Federation, Central Asia, and the Transcaucasus reflects the region's economic history, with the highest densities in the industrial regions of the western parts of the Russian Federation and the richer agricultural regions of the Transcaucasus. (b) This graph shows the dramatic drop in the birthrate in Russia that characterized the 1960s and 1990s. It also shows the slight rise in birthrates and the steadiness of death rates between 2000 and 2010.

Federation (**Figure 3.36**). It was not until the 1960s, however, that rates of natural population increase in the Soviet empire began to decrease significantly on a long-term basis. At the beginning of the 1960s, birthrates fell sharply. This happened as a result of a combination of things, including the legalization of abortion; a greater propensity to divorce; deferral of marriage among the rapidly expanding urban population; and a growing preference to trade parenthood for higher levels of material consumption.

In the period following the collapse of the Soviet Union, death rates began to rise. The reasons for this increase are several. Deteriorating health-care systems and the worsening health of mothers have contributed to an escalation of infant mortality rates. Meanwhile, public health standards have generally deteriorated, environmental degradation has intensified, and the rates of industrial accidents and alcohol-related illnesses have increased. In this recent period, life expectancy has declined dramatically, reaching a low of 67 in the mid-1980s (though it has since rebounded somewhat). By 2011, the average life expectancy for men in the entire region was 65 years and 74 years for women, representing a deeply troubling disparity between the sexes. Cardiovascular disease, the leading cause of death, has continued to take a heavy toll on the region, especially among men.

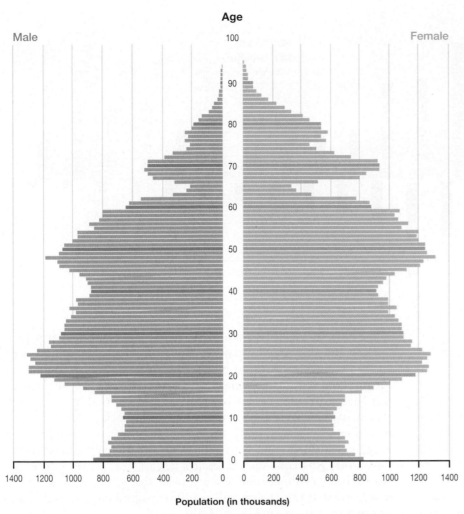

Age

Male   100   Female

<small>90</small>

<small>80</small>

<small>70</small>

<small>60</small>

<small>50</small>

<small>40</small>

<small>30</small>

<small>20</small>

<small>10</small>

<small>0</small>

1400 1200 1000 800 600 400 200 0   0 200 400 600 800 1000 1200 1400

Population (in thousands)

▲ **FIGURE 3.36 Age–Sex Pyramid for the Russian Federation, 2012** This profile of the Russian Federation's population shows the effects of World War II (the relative lack of men and women in their early 60s and the reduced number of men age 70 and older) and the reduced birthrates of the 1960s and 1990s. It is also shows the declining number of younger people in the country.

The Russian Diaspora and Migration Streams The spread of the Russian Empire from its hearth in Muscovy took Russian colonists and traders to the Baltic, Finland, Ukraine, most of Siberia, the eastern parts of the region, and parts of Central Asia and the Transcaucasus. In the late 19th and early 20th centuries, many Russians joined the stream of emigrants headed toward North America. Concentrations of Russian immigrants developed in Chicago, New York, and San Francisco. They were joined by others who had fled the civil war and Bolshevik Revolution of 1917. Over a quarter of these immigrants settled in New York City, the majority in Brooklyn, where distinctive Russian communities, such as the Brighton Beach neighborhood, are vital nodes in the Russian global diaspora (**Figure 3.37**).

During the rise of the Soviet empire, many Russians were directed and encouraged to settle in the Baltic, Ukraine, Siberia, the east of Russia, Central Asia, and the Transcaucasus. This was partly to further the Stalinist ideal of a transcendent Soviet people and partly to provide workers needed to run the mines, farms, and factories required by Soviet economic, strategic, and regional planners.

By the time of the breakup of the U.S.S.R., 80% or more of the population of Siberia and the eastern parts of the region were Russian, and the Russian diaspora had become very pronounced in most of the Soviet Union's successor states beyond the borders of the Russian Federation.

In 1989, some 25 million Russians suddenly found themselves to be ethnic minorities in newly independent countries. The largest group was in Ukraine, where more than 11.3 million Russians made up 22% of the population. In Kazakhstan, Russians represented nearly 38% of the population. Overall, the sudden collapse of the Soviet Union created havoc in the lives of many families, who suddenly found themselves living "abroad" even though they had not relocated. During the 1990s, a good number of them decided to migrate back to the Russian Federation. In the Transcaucasus, where the proportion of Russians was generally lower than elsewhere, strongly nationalistic governments of the successor states quickly enacted policies that encouraged Russians to leave—reducing the number of Russian-language schools, for example. In Central Asia, nationalistic policies were enacted with similar effect. Kyrgyzstan, Turkmenistan, and Uzbekistan dropped the use of the Cyrillic alphabet, deliberately creating institutional barriers for Russian speakers. Civil war in Tajikistan led to the departure of 80% of that country's Russian-speaking population within just 3 years of Tajikistan independence from the Soviet Union. About 17% of the Russian population of Kyrgyzstan departed in that same period, mainly because of the withdrawal of the Russian Federation's defense industry enterprises and military installations.

In the last decade, more people have migrated into Russia than out. Migration has come to include a far wider range of nationalities and ethnicities, including not only Slavic speakers, but members of Central and

▲ **FIGURE 3.37 Russian Communities in Brighton Beach, Brooklyn, New York** Brighton Beach is called "Little Odessa" by locals; several generations of Ukrainians from the city of Odessa have settled there, and the area has numerous Russian markets and restaurants.

▶ **FIGURE 3.38 Trends in Russian Migration** During the post-Soviet transition of the 1990s, emigration from Russia was enabled by increasing freedom of movement under the new government, while immigration to Russia consisted heavily of ethnic Russian returning. After a lull in migration activity during the economic downturn of the early 2000s, Russia has become a destination for labor migrants from across the former Soviet states.

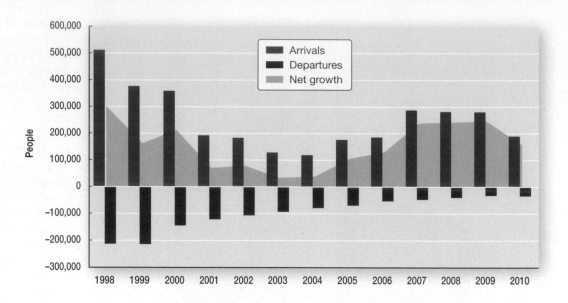

East Asian ethnicities (**Figure 3.38**). This trend is driven by labor demands and opportunities from economic growth in the Russian Federation.

APPLY YOUR KNOWLEDGE Summarize the overall demographic trends for this world region, especially for the Russian Federation. What are the advantages and disadvantages of an aging or shrinking population?

FUTURE GEOGRAPHIES

The range of possible futures for the region is starkly divergent. On the one hand are potentially very positive changes to be derived from integration with global market forces. This would mean significantly more open and progressive societies, though it would likely be accompanied by increasing social and spatial inequality. On the other hand, multiple constraints could limit the region's ability to achieve its full potential. Most problematic are shortfalls in infrastructure and energy investments, decaying education and public health sectors, an underdeveloped banking sector, corruption, and organized crime.

Shaping the Russian Energy Frontier Recent economic growth, particularly in Russia, has largely been the result of windfall profits from sustained increases in oil and commodity prices. In 2008, oil and gas accounted for roughly two-thirds of Russia's export earnings and one-third of general government revenue. In 2009, Russia was the largest producer of crude oil in the world, surpassing Saudi Arabia. A generation of energy and metals companies that are globally competitive has emerged, helping further solidify Russia's position in the global marketplace. The export of raw materials is likely to remain the leading sector of economic development. The geopolitical power that energy production has afforded the region is also unquestionable and will be a key factor in its relationships with the rest of the world

in the next decade. Russia is effectively self-sufficient in energy production and able to sell to energy-hungry nations, including China and the United States, on its own terms. This is especially significant for relations with its European neighbors. The European Union's third-largest trade partner is Russia and of its total imports from Russia, 65% comes in the form of energy. This makes the EU heavily dependent on Russian production and establishes the tone for their engagement about other issues, including security and human rights. Given that Russia holds the world's largest reserves of natural gas (more than 45 trillion cubic meters, twice that of the United States), it is likely that this region will continue to influence global affairs through energy power (**Figure 3.39**).

Climate change will also have an increasing influence on this part of the Russian economy, particularly on the northern regions. Russia has vast untapped reserves of natural gas and oil in Siberia and also offshore in the Arctic, and warmer temperatures may make the reserves considerably more accessible. The opening of an Arctic waterway could provide economic and commercial advantages. However, Russia could be also be hurt by infrastructure damaged as the Arctic tundra melts. New expensive technology may be needed to continue to develop the region's fossil energy.

Democracy or Plutocracy? In 2010, the number of Russian billionaires reached 77, while its level of income inequality put it on par with Nigeria. The growing number of economic elites presents challenges for a growing democracy. On the one hand, many of these emerging elites act as a counterbalance to traditional political blocks, presenting opposition to entrenched powerful government actors. In this sense, these new powerful capitalists can be a force for democracy. For example, Russian billionaire Mikhail Prokhorov denounced both the current government as well as the traditional opposition parties in 2012, and launched a bid for the presidency. On the other hand, and far more commonly, the maintenance of the fortunes of many of these elites depends on maintaining control of critical resources, especially oil and natural gas resources. This invites collusion and corruption. Given the breakneck pace of economic expansion and the simultaneous stagnation of political change in Russia, the role of this new class of plutocrats in political change is

unquestionably critical for deciding the region's future, but equally unpredictable.

Demographic Implosion? Demographic changes will also create challenges in the region. The populations of Russia and Belarus are aging dramatically. Between now and 2025, their populations are expected to decline by as much as 12%. By 2025, between one-fifth and one-quarter of the populations in nine eastern European and former Soviet Union countries will be 65 and older. The chances of stemming a steep decline in population age over this period are slim: the population of women in their 20s—their prime childbearing years—will be declining rapidly, falling to around 55% of today's total by 2025. However, in 2009, Russia reported its first population increase in 15 years; while much of the growth was due to a falling death rate and increasing immigration, births did rise 2.8%.

Ultimately, the labor force may be insufficient for the size of the economy unless immigration is significantly increased; the burden of dependency of the elderly on the young will intensify significantly. Within these overall trends, though, the Muslim minority population is projected to grow—from 14% in 2005 to 19% in 2030, and 23% in 2050, in Russia's case. This could provoke a nationalist backlash and even contribute to the emergence of a nationalistic, authoritarian government.

◀ **FIGURE 3.39 Russian Natural Gas Development** Pictured here is a natural gas extraction pipeline located in Tyumen, Russia, which is about 2,200 kilometers (1367 miles) from Moscow. This pipeline, which is owned by the gas monopoly, Gazprom, is tied into a sophisticated network of lines that bring natural gas to Russia as well as Kazakhstan and other parts of Central Asia and the Middle East.

LEARNING OUTCOMES REVISITED

■ **Describe how the location and vast size of the Russian Federation, Central Asia, and Transcaucasus region have influenced its climate and how climate change may influence the region's future.**

This region's climate is influenced by its mostly northern location, its position at the eastern edge of Europe and inland Asia, its absence of mountainous terrain (except in the far south and east), and the lack of any significant moderating influence of oceans and seas. Three climate regions dominate the landscape from the arid and semiarid regions in the south to the continental/midlatitude and polar regions to the north. As climate change opens the northern areas of the region, previously unusable resources and transportation routes in the Arctic will become increasingly critical to regional growth and change.

■ **Identify the topographic areas of the region and how they are influenced by the larger global process of plate tectonics.**

The position of the region within a single plate has resulted in the vast unbroken steppes of Russia and Central Asia, but also the mountainous frontiers of the region in the Transcaucasus. This makes the region appear as uniform. But there is quite a bit of variation throughout the region, a result of uplift in the southern portion of the region, for example.

■ **List and describe the region's five major environmental zones and provide examples of how people have adapted to and modified the physical landscapes in these zones.**

From north to south, the regions are the tundra, the taiga, mixed forested regions, the steppe, and open deserts. The natural resources of these regions, including oil and timber, have been exploited and harvested with limited concern for sustainability. In the drier zones, modern farming methods have also impacted the land, and irrigation methods in particular have made the region more vulnerable to droughts and other extreme weather events.

■ **Compare and contrast the regional environmental impacts of Soviet era planning with those of the post-Soviet transition.**

Soviet central planning placed strong emphasis on industrial output, with very little regard for environmental protection. The Chernobyl disaster was emblematic of the consequences of this attitude, whereas deforestation and the desertification of the Aral Sea and its surrounding region also represent large-scale environmental disasters. The emergence of a new economy has not, however, abated all ecological problems, as the rise of oil and natural gas exploitation and industrial forestry in a free-market environment also pose grave environmental impacts.

■ **Describe how economic transitions in the region are playing out in the postsocialist era.**

After more than seven decades under the Soviet system, the Russian Federation, Belarus, and the Soviet Union's other successor states in Central Asia and the Transcaucasus are now experiencing transitions to new forms of economic organization and new ways of life. An economic boom has ensued throughout the region propelled by a deep and extensive natural resource base. Growing economic inequality and the rise of economic elites pose challenges to democracy, however.

■ **Analyze the changing political trends in the region in terms of fragmentation and realignment.**

Some of the old ties among the countries of the region have been weakened or reorganized, and patterns of regional interdependence have been disrupted and destabilized. All of the new, post-Soviet states have joined the capitalist world system, and all of them have to find markets for uncompetitive products while at the same time engaging in domestic economic reform. The Russian Federation, as the principal successor state to the Soviet Union, remains a world power with a large standing army and a formidable sophisticated arsenal, including nuclear weapons; a large, talented, and discontented population; a huge wealth of natural resources; and a pivotal strategic location in the center of the Eurasian landmass.

■ **Summarize the cultural legacies and new developments in the region in terms of high, revolutionary, and popular forms of culture.**

Since the breakup of the Soviet Union, popular culture has gone through an enormous upheaval, thanks to the collapse of censorship, the pressures of market forces, and a flood of cultural imports. New television programming and different ways of thinking about issues like class, identity, and sexuality opened up post-Soviet society, especially in western Russia and the larger cities. But it did not take long before a hybridized, homegrown popular culture began to emerge.

■ **Describe and explain how birth and death rates have changed relative to one another over the recent history of the region and discuss the implications of recent demographic contractions.**

Although many parts of Central Asia have growing populations, in the core of the region, which includes the Russian Federation, death rates have risen higher than birthrates. This presents problems for maintaining the labor force, which may depend much more heavily on in-migration into the region in the future. This trend toward increasing immigration into the region reverses the historical trajectory of population mobility, which has historically been dominated by out-migration.

KEY TERMS

apparatchik (p. 106)

aridity (p. 89)

Bolshevik (p. 104)

civil society (p. 115)

cold war (p. 107)

command economy (p. 106)

Commonwealth of Independent States (CIS) (p. 115)

dry farming (p. 99)

ethnonationalism (p. 112)

geosyncline (p. 94)

glasnost (p. 113)

internationalist (p. 104)

Joseph Stalin (p. 104)

khan (p. 102)

mafiya (p. 110)

Near Abroad (p. 115)

nomadic pastoralist (p. 99)

Northwest Passage (p. 92)

oligarch (p. 110)

permafrost (p. 89)

pluralist democracy (p. 115)

privatization (p. 98)

samizdat publications (p. 119)

serfdom (p. 104)

Silk Road (p. 102)

soviets (p. 104)

state socialism (p. 104)

steppe (p. 98)

taiga (p. 97)

territorial production complexes (p. 106)

tsar (p. 103)

tundra (p. 96)

Union of Soviet Socialist Republics (U.S.S.R.) (p. 104)

Vladimir Ilyich Lenin (p. 104)

zone of alienation (p. 109)

THINKING GEOGRAPHICALLY

1. Describe two examples of how recent trade linkages with global commodity markets have created or exacerbated local and regional environmental problems in Russia and Central Asia.

2. What role did the fur trade play in the expansion of Russia?

3. How did the establishment of the Soviet bloc aid development of the Soviet Union following World War II? Discuss technical optimization, industrialization, and military security.

4. What factors led to the breakup of the Soviet empire?

5. Discuss the environmental degradation of the Aral Sea and compare and contrast it to the problems associated with the region's nuclear landscapes.

6. How have national identities been asserted in the decade since the Central Asian republics became independent countries? What cultural factors serve to unify or separate the states in this region?

MasteringGeography™

Looking for additional review and test prep materials? Visit the Study Area in MasteringGeography™ to enhance your geographic literacy, spatial reasoning skills, and understanding of this chapter's content by accessing a variety of resources, including **MapMaster** interactive maps, videos, RSS feeds, flashcards, Web links, self-study quizzes, and an eText version of *World Regions in Global Context*.

The availability of opulent leisure activities, such as skiing in Dubai, United Arab Emirates, gives a clear sense of just how far oil wealth has taken this country. Construction projects such as this one have been built by a large immigrant labor force, which makes up almost 70% of the population of Dubai today. Immigrant workers come from India, Pakistan, the Philippines, and Sri Lanka, as well as Egypt, Yemen, Sudan, and Iran.

SNOW FALLS IN THE MIDDLE EAST AND NORTH AFRICA. Most people from outside the region might not imagine this to be the case, but it is true. For example, there is snow in the mountains of Morocco, Iran, and Turkey in the winter. Snow in the mountains in winter makes sense, but one place you probably would not expect to find snow is in the Persian Gulf city of Dubai, where average monthly temperatures fluctuate from a low of 23°C (75°F) in January to a high of over 38°C (100°F) in July and August. But snow it does. The wealth of Dubai, stemming from the city's massive oil and natural gas reserves, allowed investors to create snow on the slopes of an indoor ski resort year-round. That's right, people in Dubai can go to the mall, strap on skis, and even have a hot chocolate at the resort's chalet. This icon of Dubai's wealth attracts thousands of visitors, including some of the world's most famous celebrities, who now count Dubai as a stop for fine dining, fashion, and relaxation. As a result, Dubai's airport is the 13th busiest in the world, with over 47 million passengers in 2010—an astonishing 15.4% jump from the previous year.

LEARNING OUTCOMES

- Recognize the diversity of physical geographic environments in the Middle East and North Africa as well as the importance of key resources, such as oil and water.

- Account for the diffusion of cultural and environmental practices, such as religion and agriculture, throughout the region.

- Explain the natural and historical factors that contribute to the diverse economies of the Middle East and North Africa, how they vary from country to country, and how they are tied to global and regional processes.

- Understand the geopolitical conflicts and tensions that bring instability to the Middle East and North Africa and the relationship between these conflicts and the political history of the region.

- Appreciate the differences in practices of kinship, gender, and religion in the region and how these factors affect the lives of women.

- Describe and account for the patterns of migration in the Middle East and North Africa, the push and pull factors that are involved in migration patterns, and the effects these patterns have on the region's demography.

Dubai, which is part of a larger federated state called the United Arab Emirates (U.A.E.), has a GDP per capita of just under U.S. $40,000 a year, making it one of the wealthiest countries in the world. And the leaders of the city have used that wealth to construct a massive urban space of skyscrapers and high rises. There is a dark side to this development, however. The U.A.E. rivals the United States in energy consumption and has one of the most significant ecological footprints on the planet. According to the Global Footprint Network, the U.A.E. per capita footprint was 9.5 hectares (23.2 acres) in 2010. That means that each person in the U.A.E. consumes an average of 9.5 hectares worth of the world's bio-resources in a given year. The planet can only support 1.8 hectares (4.2 acres) per person. The snowfall in Dubai comes at a price, and that price is global in scope. ∎

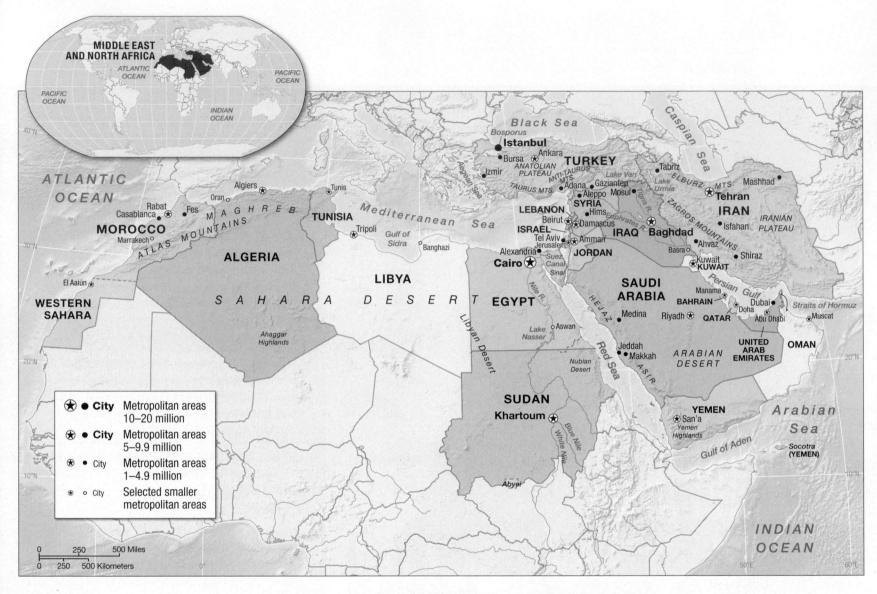

Middle East and North Africa Key Facts

- Major Subregions: Arabian Peninsula, Persian Gulf, Maghreb

- Major Physiogeographic Features: Sahara and Nubian Deserts; Atlas, Taurus, Anti-Taurus, and Yemen Highlands; Iranian and Anatolian Plateaus; Caspian, Arabian, Red, and Black Seas; Tigris and Euphrates Rivers. Climate varies throughout the region, the single unifying element being its aridity.

- Major Religions: Islam, Christianity, Judaism

- Major Language Families: Afro-Asiatic, Semitic (Arabic and Hebrew), Indo-European, and Altaic

- Total Area (total sq km*): 6.7 million

- Population (2011): 503 million; Population under Age 15 (%): 30; Population over Age 65 (%): 3

- Population Density (per square km) (2011): 270

- Urbanization (%) (2011): 77

- Average Life Expectancy at Birth (2011): Overall: 73; Women: 75; Men: 71

- Total Fertility Rate (2011): 2.9

- GNI PPP per Capita (current U.S. $) (2009): 14,098

- Population in Poverty (%, < $2/day) (2000–2009): 22

- Internet Users (2011): 159,564,042; Population with Access to the Internet (%) (2011): 37.9; Growth of Internet Use (%) (2000–2011): 2,548

- Access to Improved Drinking Water Sources (%) (2011): Urban: 95; Rural: 86

- Energy Use (kg of oil equivalent per capita) (2011): 3,715

- Ecological Footprint (hectares per capita consumed/hectares per capita available, global scale) (2011): 3.1/1.8

*In 2011, South Sudan gained its independence. Because of its strong connection to sub-Saharan Africa, we cover material on this country in Chapter 5. Since the change is so new, not all the statistics in the Key Facts in Chapters 4 and 5 have been adjusted. The following numbers do reflect the change to coverage in Chapter 4: Total area (total sq km); Population (2011); Population Density (per sq km) (2011); and Urbanization (%) (2011).

ENVIRONMENT AND SOCIETY

The Middle East and North Africa is an environmentally diverse world region (**Figure 4.1**), despite popular images in Western media and film which portray it as one giant desertscape. This region's diversity is driven by its location relative to wider global climate patterns, which help create intense dry regions (such as the arid desert landscapes of the Arabian Peninsula centered on Saudi Arabia) as well as forested regions (such as the pine forests of the Atlas Mountains in Morocco). Pine forests, vast deserts, and grass plains all can be found in relatively close proximity to each other in this region. Global plate tectonics have had a significant impact on this region; uplift has created highland environments and sources of water for some of the world's most historically significant river systems, such as the Tigris and Euphrates Rivers, which run from Turkey through Iraq. This diverse landscape of mountains, plains, and deserts is evidenced when viewed from space (**Figure 4.2**). Human adaptation to the environmental conditions of the Middle East and North Africa led to some of the earliest agricultural societies in the world, and modern management of these environments has changed the region's physical landscape in striking ways.

Climate, Adaptation, and Global Change

As in any region, global climatic systems interact with temperature, humidity, and rainfall as well as topography and large water bodies to produce climatic variability (**Figure 4.3**, p. 132) in the Middle East and North Africa. Temperatures vary by season and location, and in the desert areas temperatures exhibit extreme variation between night and day. Humidity and rainfall are also variable across the region. Although there is variability in the climate, the characteristic that unifies the region's climate is **aridity**; in many areas, the climate lacks sufficient moisture to support trees or woody plants. As in other arid lands throughout the world, summers in the lowland areas of the region are extremely hot and dry, with daily high temperatures often climbing to 38°C (100°F). The highland areas, such as the Atlas Mountains and the Iranian and Anatolian Plateaus, and coastal areas of the Atlantic Ocean and the Mediterranean, Caspian, Arabian, Red, and Black Seas experience more moderate daily summer temperatures and a predictable influx of visitors escaping the heat elsewhere. Winter temperatures are more moderate in the lowlands and colder in the highlands. The moderating effect of large bodies of water produces milder year-round climates with wet winters in coastal zones.

Arid Lands and Mountain Landscapes Except in the coastal mountain areas, precipitation in this world region is generally low but highly variable. Nearly three-quarters of the region experiences average annual rainfall of less than 250 millimeters (10 inches). Scarce rainfall means the soils in the region tend to be thin and deficient in nutrients and most agricultural land must be irrigated. There are exceptions. Agriculturalists along the coastal plains and lowlands of Turkey and the floodplains of the Nile are able to take advantage of fertile soils along rivers. The Central Highlands of Yemen experience abundant rainfall in the summer thanks to the Indian Ocean monsoon system, which brings seasonal rains from the Indian Ocean to the southern part of the Arabian Peninsula.

Most of the rain that does fall in the region is affected by mountain ranges—such as the Atlas Mountains in Morocco or the Judaean hill country of the Levant (the eastern Mediterranean, including Syria, Lebanon, and Israel and the Palestinian territories). The moisture that does fall in the region does so in the mountains. And in these mountain spaces, rainfall can be quite plentiful. In the Zagros Mountains in Iran, annual rainfall can range between 600 and 2,000 millimeters (23 and 80 inches). Snowcapped peaks are found in Turkey, Iran, and Lebanon, where spring and summer snowmelt provides water for lowland human, animal, and plant populations

◄ **FIGURE 4.2 Middle East and North Africa from Space** This composite of satellite photos highlights the extreme aridity of the Middle East and North Africa. Deserts dominate the southern part of the region. Along the coasts and in the mountains, high plateaus, and steppes, greater moisture availability means more plants (and humans) can survive and thrive. Few large rivers and lakes exist in the region, but the ones that do are crucial to human, plant, and animal life.

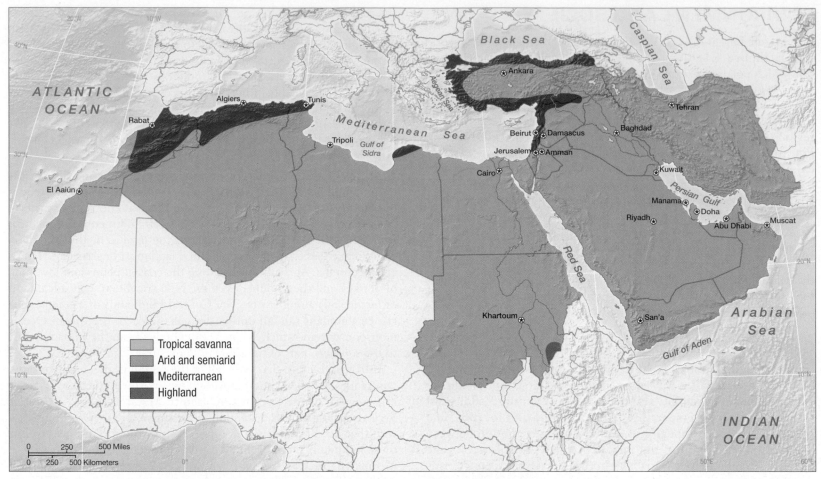

▲ **FIGURE 4.3 Climate of the Middle East and North Africa** The Middle East and North Africa is broad, stretching from the Mediterranean climates of northern Turkey to the wet, tropical climates of southern Sudan, but it is clearly dominated at midsection by a continuous swath of dry lands.

(**Figure 4.4**). In some mountainous areas where the peaks are especially high, such as the central Anatolian Plain in Turkey or the Syrian Plateau, a **rain shadow** effect occurs. The mountains cause most of the moisture contained in the air masses passing over them to condense and fall as rain before it can reach the parched interior deserts of the region. This occurs in southeastern Saudi Arabia, western Oman, and Dasht-e-Kavir (the Great Salt Desert) of Iran, where there is a complete absence of vegetation. In contrast, most of the coastal areas of the region experience between 375 and 1,000 millimeters (15 and 40 inches) of rain a year. Although most rain falls in the winter and early spring, some areas, such as the Black Sea slope of the Pontic Mountains in Turkey, experience summer rains adequate for **dry farming** techniques, which allow the cultivation of crops without irrigation in regions of limited moisture (50 centimeters, or 20 inches, of rain per year).

Adaptation to Aridity People have adapted to the aridity and high temperatures characteristic of the region through architecture, patterns of daily and seasonal activity, and dress. Regional architecture features high ceilings, thick walls, deep-set windows, and arched roofs that enable warm air to rise away from human activity. The practice of locating living quarters around a shady courtyard enables residents to move many activities to cooler outdoor spaces that are also very private. The clothing worn by Middle Eastern and North African people is also an adaptation to the hot, dry climate. Head coverings and long, flowing robes made from fabrics of light color lower body

▲ **FIGURE 4.4 Snowfall in the Atlas Mountains** Athough very close to the Equator and just north of the Sahara, the Atlas Mountains receive orographic precipitation, which results from rising moist air that is condensed as it rises over the mountains. Because of the high altitudes, it is cool enough for that moisture to turn to snow. Resorts have developed over time to cater to those interested in these cool, snow-laden places. Pictured here is the Kasbah du Toubkai Hotel in Morocco.

temperatures by reflecting sunlight. They also inhibit perspiration and thus diminish moisture loss.

Some populations, such as the Berber of North Africa, migrate to mountainous areas in the summer and warmer lowlands in the winter to avoid temperature extremes. Plants and animals also have adaptive strategies to deal with the intense heat and aridity. For example, native plant species are able to store water for long periods of time or survive on very small amounts of water by keeping their leaves and stems very small or developing an extensive root system. Animals adapt by water, salt and temperature regulation, as well as having light coloration.

Climate Change A number of countries in the region, recognizing that the arid climate makes the region vulnerable to the effects of climate change, including increasing surface temperature, decreasing precipitation, and sea-level rise and drought, have become participants in the U.N. Framework Convention on Climate Change (UNFCCC) and the Kyoto Protocol, which commits signature countries to reduce greenhouse gas emissions. The effects of global climate change in the region are most severe in coastal communities, which are seeing an inundation of saltwater in low-lying areas. A generally poor record on environmental protection at the global and regional scales has left the state of the environment in the Middle East and North Africa seriously challenged. Structural problems, such as rapid urbanization and a burgeoning population of very poor people, make solutions extremely difficult to implement. On the positive side, the growing global **carbon market** is highly attractive to countries in this region. The carbon market allows developing countries that have ratified the Kyoto Protocol agreement to receive payments as incentives for investments in climate-friendly projects that reduce greenhouse gas emissions. In this way, countries can take action to reduce pollution, increase energy efficiency, and participate in global efforts to halt climate change. Climate-friendly, internationally supported projects have been launched in Egypt, Tunisia, Jordan, Algeria, Morocco, Iran, and Saudi Arabia.

APPLY YOUR KNOWLEDGE Briefly explain the diversity of this region's climate. Provide two examples of ways global climate patterns affect the region.

Geological Resources, Risk, and Water Management

The Middle East and North Africa contain a wide variety of physical landforms (**Figure 4.5**). The region is interspersed with seas,

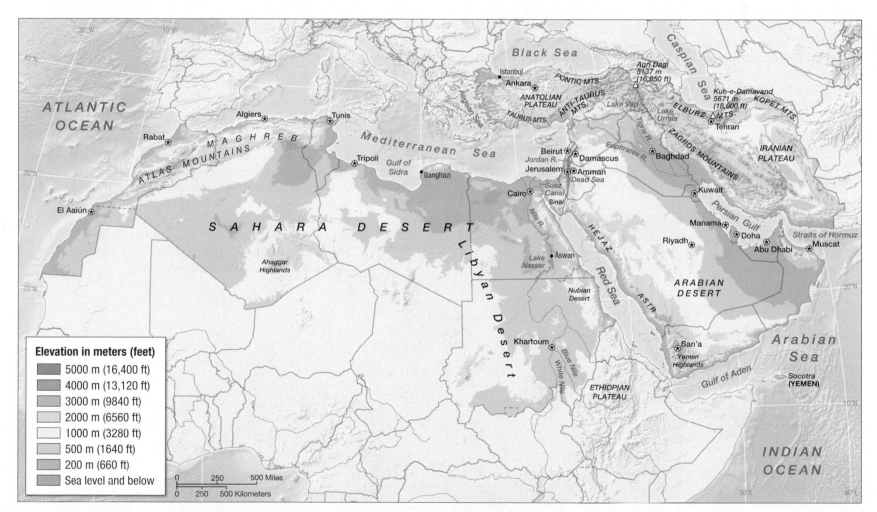

▲ **FIGURE 4.5 Physical Geographic Features of the Middle East and North Africa** Perhaps the most consistent feature of the physiographic map of the region is the way that land and water features seem to alternate in a somewhat regular pattern. Also significant is the scattering of plateaus and mountain ranges that punctuate the vast lowland areas of desert and coastal plains.

such as the Mediterranean, and is bordered by the Atlantic Ocean on the Moroccan coast and the Indian Ocean to the south and east. There are substantial mountain ranges, and two major river systems drain through the region. Even though the region possesses a wide variety of landscapes, from seaside beaches to towering mountains, the landscapes most heavily occupied by humans are the highland plateaus and the coastal lowlands, as well as the floodplains of the major rivers, where rain and surface water are most dependable.

Mountain and Coastal Environments The Arabian Peninsula, a tilted plateau that rises at its western flank on the Red Sea and slopes gradually to the Persian Gulf on its eastern flank, is part of the Arabian tectonic plate. The Arabian Peninsula was once part of the African Plate tucked against the coastal areas of present-day Egypt, Sudan, and Eritrea. The separation of the Arabian Plate as it moved eastward millions of years ago created the Red Sea, which initially formed as a rift valley and later filled with water. Both the African and the Arabian plates rub against the Eurasian Plate along the Mediterranean Sea and at the mountains that separate the Arabian Peninsula from the Anatolian and Iranian Plateaus. As **Figure 4.6** shows, the region is located at the conjunction of three continental landmasses. Plate contact means that the subregion is prone to severe earthquakes, like the one that shook eastern Turkey in October 2011, killing close to 700 people and injuring thousands more.

Three mountain ranges dominate the region. Impressive in terms of their beauty and ruggedness, these ranges generate rainfall and are the source of rivers and runoff for this arid region. The first set of ranges, which contains the most extensive and highest mountains, is the result of contact between the African, Arabian, and Eurasian Plates. This set includes the Taurus and Anti-Taurus Mountains in Turkey and the Elburz, Zagros, and Kopet Mountains in Iran. The second set includes the Atlas Mountains of northwest Africa, which stretch along the southern edge of the Mediterranean from Morocco to Tunisia. The third set of mountains border the Red Sea and are known as the Asir Mountains in Saudi Arabia and the Yemen Mountains in Yemen (see Figure 4.5).

Water Environments Water is a precious and somewhat limited commodity in the Middle East and North Africa. This is because there are only two major freshwater river systems in the region—the Nile River and the integrated Tigris and Euphrates Rivers. The Nile, which has its source in the mountains of Ethiopia and East Africa, is the world's longest river. From the Ethiopian Plateau, the Blue Nile flows northward across the Sahara, where it joins the White Nile at Khartoum, Sudan, to form the Nile River. The river proceeds northward, finally emptying into the Mediterranean north of Cairo. The Nile flows through some of the driest conditions on the planet, where no additional moisture is added and high evaporation occurs. The Tigris River originates in the Anatolian Plateau of Turkey and flows through Iraq. It is joined by the Euphrates River, which flows through Syria, into lower Iraq, where it eventually empties into the Persian Gulf (**Figure 4.7**). These two river systems have been the lifeblood for millions of the region's inhabitants over thousands of years. Because of their importance to human activity in the region, it is not surprising that they have also been and remain a source of much conflict in the region—largely having to do with access.

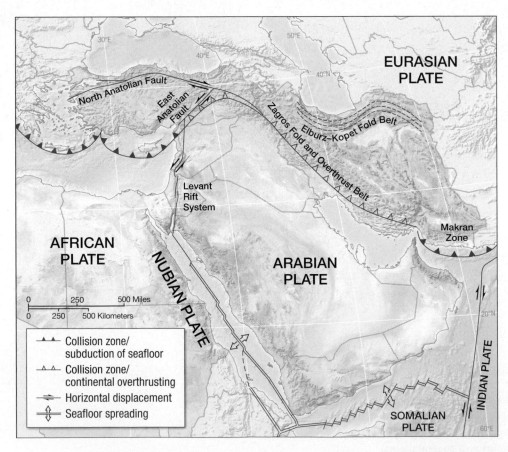

◀ **FIGURE 4.6 Generalized Tectonics of the Middle East** This generalized map shows the concentration of tectonic activity in eastern Turkey. The African and Arabian plates are made of ancient rock and are stable; little tectonic activity occurs there. Along the Eurasian Plate, folded and faulted mountains extend from western to eastern Anatolia and then south across Iran and eastward again into the Himalayas. These mountains are the result of active plates colliding.

◄ **FIGURE 4.7 The Euphrates River** The Euphrates River is a critical source of freshwater for Turkey, Iraq, and Syria. Shown here is the river as it flows through Dura Europos just north of the border between Iraq and Syria. Ruins in this area date back several thousand years and demonstrate the importance of rivers in the history of this region.

There are other sources of water in the region, but many of these are highly undependable. In the Sahara, runoff from the Atlas Mountains collects underground in porous rock layers deep below the desert surface. In some fertile places in the desert, known as **oases**, land erosion and a high water table have enabled underground water to percolate to the surface. Because the soils of oases are usually quite fertile, they can support animal and plant life and even agriculture, such as date and other fruit production. Oases also play an economic role because they serve as stopping points for caravans carrying commercial goods across the vast deserts. There are also natural springs, perennial streams, and wells in the region. Wells are drilled largely for irrigation, though some water holes are used to create artificial oases. One of the most ingenious methods for mining water in the region involves a system of low-gradient tunnels that collects groundwater and brings it to the surface through gravity flow. This **gravity system** of water mining—where holes are drilled into the ground at the foothills of a mountain or hill to tap local groundwater—is known as *qanat* in Iran, *flaj* on the Arabian Peninsula, and *foggara* in North Africa (**Figure 4.8**). These systems rely on runoff from rain and snowmelt that percolates below ground level. Under ideal circumstances, where the gravity system is carefully practiced, it provides a highly dependable source of water enabling year-round irrigation.

Desalinization—the process that converts saltwater to drinking water by removing salt—is another way of obtaining freshwater in the region. This process is controversial because it is expensive and has the potential to disturb rich ocean habitats, but many countries in the region have invested in desalinization plants, such

as Qatar, which gets about 97% of its water from desalinization. Saudi Arabia, which is expected to spend over U.S. $50 billion on desalinization technology over the next ten years, currently gets about 50% of its water through desalinization. Israel has also produced some of the most effective desalinization plants in the world.

All the states in the region are experiencing serious problems related to water quality and accessibility. In many states, excessive

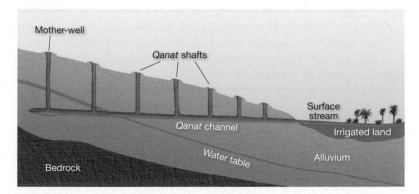

▲ **FIGURE 4.8 Simplified Diagram of *Qanat* Irrigation** Through a series of low-gradient tunnels, the *qanat* collects groundwater, bringing it to the surface by way of gravity. The diagram shows the shafts drilled into the loose soil and gravel at the foot of the mountain that reach the water table below. The water is drawn up through these shafts and directed out for use through pipes.

▶ **FIGURE 4.9 Sahara Desert**
Oasis Geological evidence suggests that between 50,000 to 100,000 years ago, the Sahara possessed a system of shallow lakes that sustained extensive areas of vegetation. Though most of these lakes had disappeared by the time the Romans arrived in the region, a few survive in the form of oases. Pictured here is the Mandara Oasis and Um El Ma Lake in Libya's portion of the Sahara.

extraction of water from oasis wells has been occurring for such a long period that oases are dying (**Figure 4.9**). The states in the Persian Gulf, in particular, regularly face severe shortages of freshwater. Some water-control schemes, such as those employed in Iraq, are destroying the natural habitat of wildlife. Along the Nile in Egypt, irrigation is increasing the amount of salt in the soil and decreasing soil fertility. Coastlines along the Persian Gulf are experiencing erosion and degradation of marine habitats caused by oil spills and the discharge of ships and industry. The problem of water security is further exacerbated by the fact that the countries of the Persian Gulf region have some of the highest per capita water use rates in the world. In fact, the per capita consumption of water in the Persian Gulf region is nearly double that of Europe.

APPLY YOUR KNOWLEDGE Identify two examples of how people have applied technological solutions to adapt to the region's water conditions. What sorts of limitations and costs do the adaptations involve?

Oil and Natural Gas Oil and natural gas are both **fossil fuels**—deposits of hydrocarbon that have developed over millions of years from the remains of plants and animals, which have been converted to potential energy sources under extreme pressure below Earth's surface. These two fossil fuels, along with coal (which is not commonly found in the Middle East and North Africa), are arguably the most important products in the world today (see "Geographies of Indulgence, Desire, and Addiction: Petroleum," p. 138). These products make the Middle East and North Africa one of the most economically significant and highly scrutinized world regions of the 20th and early 21st centuries. Almost 60% of the proven oil reserves

and just over 40% of the proven natural gas reserves in the world are located in the Middle East and North Africa. Of those totals, 17.4% of the world's proven oil reserves are located in Saudi Arabia, while 9.9% and 9.4% are found in Iran and Iraq and 6.7% and 6.4% are known to exist in Kuwait and the United Arab Emirates. In terms of natural gas, 17.3% of the proven reserve is located in Iran, while an additional 13.2% can be found in Qatar. These basic facts make this region a very important player in the global economy today. Much attention is paid to the political and economic stability of the region by the global powers that consume great amounts of these resources, such as the European Union, the United States, and Russia as well as China, India, Brazil, and Japan.

Global demand for both oil and natural gas has increased dramatically over the past 50 years, with increases predicted to continue through the 21st century. Some scientists are now estimating that we may have reached the world's "peak oil" point, meaning the world has likely depleted at least half of its finite reserves of crude oil. Some scientists have gone even further, suggesting that the world's overall oil reserve is exaggerated by as much as one-third. If this is true, natural gas demand might surpass oil supply before 2020, although many scientists now believe that natural gas production may have reached its peak as well. The more modest oil estimates suggest the supply of crude oil will be exhausted sometime toward the middle of the 21st century, with natural gas supply ending 20 to 30 years later. The peak-oil scientists, as well as many energy-policy experts, believe that the use of oil and natural gas as major sources of energy will end up being a very brief affair in global historical terms, lasting little more than a century (see **Figure 4.1.3** in "Geographies of Indulgence, Desire, and Addiction: Petroleum," p. 138). The rate at which these resources are depleted may increase, as China and India as well as other new global industrial countries increase their consumption of oil and natural gas.

Ecology, Land, and Environmental Management

After thousands of years of human occupation, the landscape of the Middle East and North Africa has been dramatically transformed, in some cases so much so that entire species of plants and animals were eliminated. Population increases in recent decades have also placed greater pressures on the existing resource base of the region. The region's arid spaces demand unique adaptations on the part of all living inhabitants, including plants, animals, and humans.

Flora and Fauna It is not surprising that most of the plant and animal life in the region today is found where most of the people are: in places where the climate offers sufficient moisture. The map of flora and fauna did not always look as it does today, however. At one time, dense and extensive forests existed throughout Turkey, Syria, Lebanon, and Iran. After several thousand years of woodcutting and overgrazing, the forests of Turkey and Syria are nearly entirely denuded; only a few small remnant areas remain. Large areas of Lebanon have also been deforested, except for the Horsh Ehden Forest Nature Reserve in the northern mountainous part of the country. This reserve is still home to several species of rare orchids and other flowering plants. Sadly, Lebanon's famous cedars continue to grow only in a few high-mountain areas (**Figure 4.10**). Iran does retain a substantial expanse of its deciduous forests, particularly in the region of the Elburz Mountains. The forests of the Atlas Mountains of Morocco and Algeria are also still being harvested for commercial purposes.

Animal life has been greatly affected by the millennia of human occupation, while changes to the diversity of animals have been tested most recently by the intensification of human settlement. Several thousand years ago, a wide variety of large mammals inhabited the region's forests, including leopards, cheetahs, oryx, striped hyenas, and caracals; crocodiles also thrived in the Nile; and lions roamed the highlands of Persia (present-day Iran). Nearly all these species are now extinct or near extinction. In their place, on can find domesticated camels, donkeys, and buffalo, which are the most ubiquitous mammals today. The highland areas of Turkey and Iran still contain a fairly wide variety of mammals similar to those found in parts of Europe, including bears, deer, jackals, lynx, wild boars, and wolves. Birds are also plentiful in the region, with different ecosystems supporting a wide variety of species.

Environmental Challenges Different parts of the region are experiencing pressing environmental problems, which are particularly severe because national governments lack the resources to either prevent or mitigate them. For example, every state in the region (except Turkey) is experiencing **desertification**—the process by which arid and semiarid lands become degraded and less productive, leading to more desertlike conditions. Tourism, a source of income for many of the states of the region, is worsening some of these environmental problems. For example, increased coastal development in Lebanon is causing the loss of precious wetlands, and heavy traffic is damaging the coral reefs in the Red Sea.

Even though the region as a whole has little surplus capital to invest in environmental protection or preservation, some important efforts have been made. Reforestation and **afforestation** programs that convert previously nonforested land to forest by planting seeds or trees are under way in several states of the Middle East and North Africa. Oman and the United Arab Emirates have begun to take a deliberate stand against desertification, and greenbelts are being planted. Israel has introduced active breeding programs to encourage the regeneration of endangered animal species as well.

◀ **FIGURE 4.10 Cedars of Lebanon** For millennia, this coniferous tree has been significant to the region for trade, medicine, religion, and habitation. The once-abundant cedar forests—the national symbol of Lebanon and shown on its flag—have been almost completely eliminated by overexploitation. Reforestation programs are underway, however, in both Lebanon and Turkey. This photo shows the last remaining cedar forest in Lebanon.

GEOGRAPHIES OF INDULGENCE, DESIRE, AND ADDICTION

Petroleum

Most commonly thought of as oil, **petroleum** is a liquid compound that can be developed into many unique and useful everyday products. Not only does it burn effectively when converted into fuel, it can be separated into a number of different components—gas and liquid—and converted into energy sources, lubricants, and waxes. Petroleum can be used in asphalt and medicine. And the discovery of petroleum jelly in 1859 led to an entire industry of petroleum byproducts.

The invention of the internal combustion engine and World War I helped establish petroleum as a foundational product of the industrial age (**Figure 4.1.1**). From that moment, petroleum took on a truly global nature, as it became one of the most sought-after commodities for businesses and governments. Indeed, the expansion of global military power was fueled, quite literally, by the use of the combustion engine in ships, tanks, and cars in the 20th century. It is not surprising, then, that governments and corporations have fought vigorously around the world to produce and distribute the world's petroleum supplies.

In the early 20th century, petroleum deposits in the Middle East and North Africa were identified as some of the most significant in the world (**Figure 4.1.2**). In the process of consolidating control over global petroleum production, five U.S.-based and two U.K.-based companies began to drill in Iran in about 1907, in Iraq in the 1920s, and in Bahrain and Saudi Arabia in the 1930s. In the United States, Standard Oil, which by mid–20th-century controlled 95% of the U.S. petroleum industry, was the most prominent global oil company. In Britain, Shell Oil dominated.

By 1960, enraged by the unilateral cuts in oil prices made by the seven big oil companies, the major oil-exporting countries in the Middle East and North Africa formed the **Organization of Petroleum Exporting Countries (OPEC)**. OPEC, as its name suggests, is a specialist economic organization

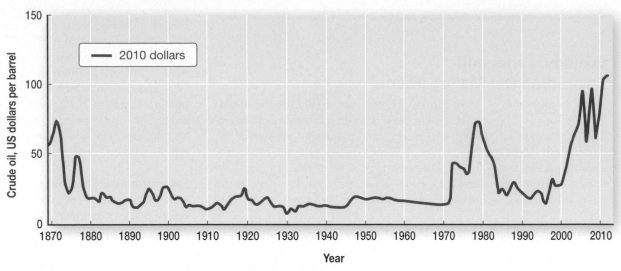

▲ **FIGURE 4.1.1 Price of Crude Oil 1870s–2012** As this graph shows, the barrel price of crude oil was relatively stable until 2003. Production, consumption, and political factors all contributed to its recent escalation and decline.

▲ **FIGURE 4.1.2 Oil Landscape of Kuwait** As seen from the air, the photograph shows the impact of oil production activities on the landscape. The wet-looking patches throughout the image are oil, not water.

whose central purpose is to coordinate the crude-oil policies of its member states. When OPEC decided to flex its muscle in the 1970s and cut back oil production, effectively raising the price of oil, it produced the world's first "oil crisis." Another "oil crisis" in 1978 precipitated further concern about the global petroleum industry by non–oil-producing countries, which had grown dependent on oil for almost all facets of their economic development. In response, many countries increased fuel efficiency standards for cars and similarly increased efficiencies in industrial production systems. The conflicts and economic problems, however, that emerged due to the global oil supply in the 1970s had made it very clear that oil was a global political issue. As the recent political crises in the Middle East and North Africa also show, petroleum politics continue to hold the attention in the global political arena.

Despite attempts to limit the world's dependence on petroleum products, global consumption continues to rise in parts of the world. The data show that crude oil consumption, for example, is increasing in the newly emerging economies of Brazil, India, China, and Russia, while consumption rates have fallen for most of the last five years in the United States (**Figure 4.1.3**). China is now the second-largest consumer of crude oil in the world, having moved from consuming approximately 4.7 million barrels of crude oil a day in 2000 to almost 9.4 million barrels a day in 2010. Between 2009 and 2010, Brazil also increased its consumption by almost 6%, while India increased its consumption over the same period by 3.6%. Even so, in 2010, the combined consumption of oil for Brazil, India, China, and Russia did not reach the total consumed by the United States. Given the fact that almost 60% of the world's proven crude oil reserves are in the Middle East and North Africa, it is clear that this region will remain an important global energy producer in the 21st century.

The challenge of the 21st century, however, is not necessarily how to increase production, but rather how to reduce consumption in order to mitigate the impacts that the burning of fossil fuels are having on the global environment. Climate change is related to the increases in carbon dioxide emitted when burning these fuels. And the impact of oil pollution on Earth's oceans and rivers is also substantial. The amount of petroleum products winding up in the ocean is estimated at 25% of world oil production. As petroleum products break down, they are ending up in animal and human food chains; this causes concern among biologists, who worry that the world's fisheries may be irreversibly damaged by petroleum and other industrial wastes in our water. ■

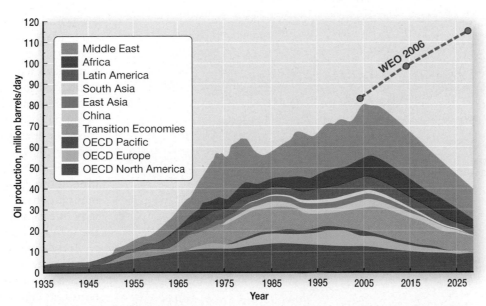

▲ **FIGURE 4.1.3 Predictions of Future Oil Production** This figure illustrates Energy Watch Group's forecast for global oil supply through 2030. The prediction is vastly different from the International Energy Agency's (IEA) "World Energy Outlook (WEO) 2006" forecasts, which are plotted on the graph in red. This divergence reflects the differences in methodology and also contradictory attitudes toward future ambiguities. Note that the IEA is opting to project a more optimistic scenario.

HISTORY, ECONOMY, AND TERRITORY

The Middle East and North Africa's first known human settlers domesticated plants such as wheat, between 10,000 and 12,000 years ago. This led to the emergence of some of the earliest large-scale societies featuring cities, networks of villages and towns, and well-organized religious and political systems. The region has a venerable history of participating in global trade, and its inhabitants developed writing, mathematical, and cultural systems that spread throughout the world. The Ottoman Empire—centered among the Ottoman society of modern-day Turkey—united parts of the region for almost 700 years between 1200 C.E. to 1920 C.E. and helped spread religious and trade systems throughout the Middle East and North Africa and into other world regions such as Europe. As the region has influenced the world, the world has also influenced the region. European colonization in the 20th century, global political conflicts, and the thirst for oil in the global economy have had an enormous impact on cultural change and the politics and economy of the Middle East and North Africa.

Historical Legacies and Landscapes

A remarkable number of culturally rich and intellectually sophisticated empires have emerged and flourished in the region over the last 6,000 years. The geographical center of these empires was the **Fertile Crescent**, a region arching across the northern part of the Syrian Desert and extending from the Nile Valley to the Mesopotamian Basin in the depression between the Tigris and Euphrates Rivers (**Figure 4.11**). Ideas and technologies generated in the Fertile Crescent diffused outward to similar nearby environments and beyond the Middle East and North Africa to reshape landscapes in Europe as well as other parts of Asia. Mesopotamia, Asia Minor (most of present-day Turkey), the Nile Valley, and the Iranian Plateau as well as Greece and later Rome all controlled parts of this region over the centuries (**Figure 4.12**). The interaction between the empires located in each of these places led to an exchange of ideas, goods, people, and belief and value systems that helped tie the region together. A significant amount of social stimulation for new ideas and practices existed and enabled technological innovations and cultural revolutions to take place, especially with respect to religion and culture, but also in terms of trade and the growth of cities.

Plant and Animal Domestication The regional empires came about because the people in this area were some of the first in the world to domesticate plants and animals. The domestication of plants and animals about 12,000 years ago in the region marked a transition from hunting and gathering to agricultural and pastoral (herding-based) societies. The transition to agriculture was made possible by technological innovations such as fire, grindstones, and improved tools to prepare and store food. In about 10,000 to 7,000 B.C.E. early neolithic—New Stone Age—farmers and herders began dry farming by relying exclusively on rainfall. Pastoralists—groups of herders—moved and traded with their herds across the region's vast grasslands as well. Agricultural needs led to the invention of numerous tools and metallurgy, which in turn allowed people to begin cultivating larger crops in the river spaces of the region in about 4,000 B.C.E. Eventually, cities were built as the region's inhabitants adapted to environmental constraints and engaged in political competition to control the region's limited resources.

Irrigation (or **wet farming**) was the key to the success of the many large-scale agricultural systems that emerged in the region. Archaeologists and other scholars argue that when farmers were able to minimize their dependence on rainfall, they were able to control larger areas and exert dominance over groups who did not have this technology. By using food as a weapon of control, some groups were able to thrive by incorporating weaker communities

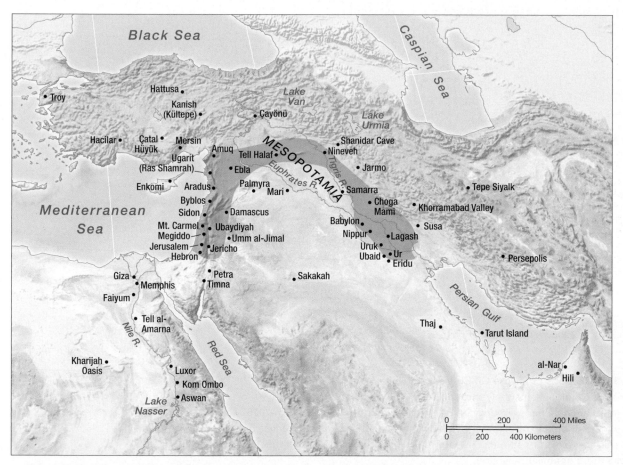

▲ **FIGURE 4.11 The Fertile Crescent** This region is one of the first places in the world where plants and animals were domesticated by humans. Historically, the Fertile Crescent has been an important center for the spread of agricultural technologies as well as writing, mathematics, and religious systems. Many urban-based societies, as highlighted here, developed in this region as a result.

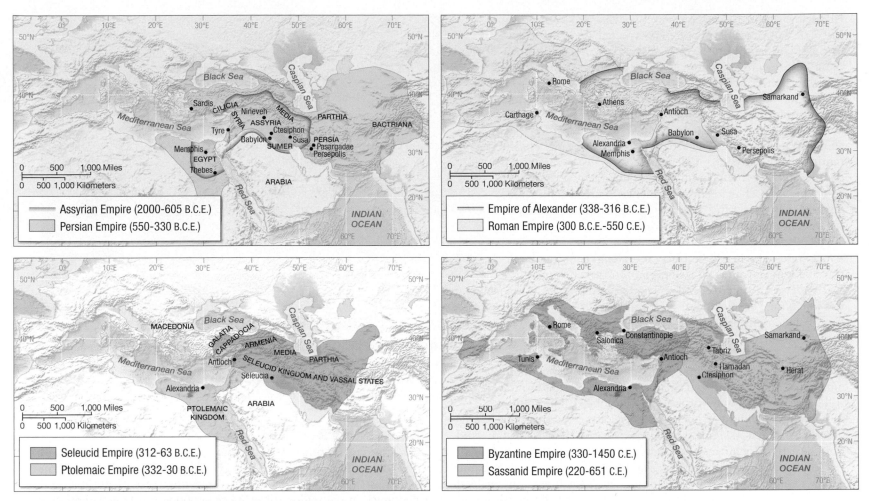

▲ **FIGURE 4.12 Empires of the Middle East from 2,000 B.C.E. to the Rise of Islam** The four maps show the extent of the most powerful regional empires (The time frames are approximate.)

into stronger ones. An example is Babylon, which was able to turn itself into a 2,000-year empire (see Figure 4.11). Babylon's dominance was achieved by systematically increasing control over regional agricultural production, building its military strength (including a walled and fortified city center), establishing a long-distance trade network (through extensive port facilities), and organizing extensive religious and symbolic political control.

Birthplace of World Religions While religion is covered more extensively later in the chapter in the section "Religion and Language," it is important to note here the historical significance of religion to the region. Three of the world's major religions—Judaism, Christianity, and Islam—all developed among **Semitic**-speaking people, those who spoke Arabic, Aramaic, and Hebrew, of the Middle East. These three **world religions**—belief systems that have adherents worldwide—are thus closely related. Judaism originated about 3,500 years ago, Christianity about 2,000 years ago, and Islam about 1,300 years ago. Although Judaism is the oldest monotheistic religion (a religion that believes in one God) in the world, it is numerically small because it does not seek new converts.

Christianity developed in Jerusalem among the Jewish followers of Jesus. The religion spread through his disciples and others who made it their mission to convert non-Christians to their religion. Christianity's successful missionizing was made possible by

the extensive road systems developed by regional empires. Imperial sponsorship of Christianity by the Roman emperor Constantine, who moved the Roman capital to Byzantium (modern-day Istanbul), further strengthened Christianity and helped it spread through Europe as well as modern-day Russia. As Christianity spread, many different versions of the religion were developed, from Roman Catholicism to Eastern Orthodoxy to Protestantism.

Similarly, Islam spread through the power of missionaries and across trade networks in the region. The religion's two major sects, **Sunni** and **Shi'a**, developed from different interpretations of Islam in the 7th century C.E. They were disseminated over the region in different ways, with Sunni Islam becoming the majority practice across the wider region and Shi'a Islam the minority sect, located predominantly in modern-day Iran and Iraq. Sunni Islam spread more widely when the Turkish Ottomans made it their main goal to convert non-Muslims to their religion; traders also expanded the religion to many parts of Asia, including Southeast Asia (see Chapter 10).

While these three world religions share origins and even core beliefs, there have been tensions between them. This includes a period of intense conflict between Christians and Muslims throughout the Middle Ages. Much more recently, particularly beginning in the 20th century, there has also been increasing tension between Jews in Israel and some Muslim communities in the region, which is discussed at length later in this chapter in the section "Territory and Politics."

Historic Urbanization Considerable evidence suggests that the oldest cities on Earth were constructed along the valleys of the Tigris and Euphrates Rivers during the 4th millennium B.C.E. in a region known as Mesopotamia. Walled towns began to appear as early as 4500 B.C.E., and early cities probably contained between 7,000 and 25,000 inhabitants. The major producers in these towns and early cities were fishers and farmers, who supported a nonproducing class of priests, administrators, traders, and artisans. These early cities consistently included three main elements: city walls; a commercial district; and suburbs, including houses, fields, groves, pastures, and cattle folds. The historical evidence of urban commercial districts throughout the region suggests that trade was an essential part of urban life in early Mesopotamia.

The first larger-scale urban centers in the region, such as Babylon, were located at crucial points along natural and well-traveled human routes. These organizational anchors of the region shaped distinctive land-use patterns and served as centers for cultural developments. The notion of urban life spread from Babylon throughout the Fertile Crescent and into areas such as the Nile Valley. The Egyptians drew on Mesopotamian knowledge systems—including mathematics, writing, and agricultural technologies—to build large cities centered on monumental tombs, temples, and palaces, the engineering of which still baffles architectural historians (**Figure 4.13**). The culture of Egypt also involved elaborate rituals and sophisticated body adornments that required significant quantities of gold, cedar, ebony, and turquoise. The Egyptians obtained these products through trade with settlements in the northern Red Sea, the upper Nile of sub-Saharan Africa, and the eastern Mediterranean. These trade relations enabled the transfer of ideas and technologies that enriched all the cultures of the region, including Greece and later Rome. As an example, Greek realist art, which included images of very detailed human bodies, was likely influenced by Egyptian knowledge of anatomy, which resulted from the practice of mummification. As the Greek and later Roman empires advanced into the Middle East and North Africa, urban design structures once again changed, producing new urban landscapes that included grain estates, large theatres, and other Greco-Roman architectural forms, including images of Greek and Roman gods.

Islamic rule has had the most visible historical impact on urban patterns in the region, as Islamic architecture and urban form were placed on top of earlier urban forms. At its greatest extent, Islamic rule reached westward as far as France; eastward beyond Turkey and the Iranian Plateau into Afghanistan, Pakistan, and India; and southward into North and West Africa. Throughout this vast area, local variations of the Islamic city still can be found today. From the 7th to the 15th centuries C.E., Islamic trade networks were so far-reaching that they linked Mediterranean Europe to parts of the Transcaucasus, Pakistan, and China (see Figure 1.24 in Chapter 1, p. 25). Such extensive trade networks are evidence that parts of the world were highly integrated—politically, economically, and culturally—long before contemporary globalization occurred.

▲ **FIGURE 4.13 Pyramids of Giza in Egypt** The ancient Egyptian belief in the continuity and stability of the cosmos was supported by a range of cultural activities from the preservation of wood, cloth, people, and animals to the construction of large-scale monuments. The pyramids are one example of this monumental architecture.

APPLY YOUR KNOWLEDGE Briefly explain the relationship between plant and animal domestication and the emergence of urban centers in the Middle East and North Africa. What key technological changes led to social developments in the region?

The Ottoman Empire The Ottomans were Turkish Muslims based on the Anatolian Plateau in Turkey. They replaced the Christian Greeks as the political power of the region after 1100 C.E. and ruled much of the region for more than 600 years. At its height, the Ottoman Empire (named after the founder of the Ottoman dynasty, Osman) extended from the Danube River in southeastern Europe (including present-day Hungary, Albania, Bosnia, and Kosovo) to North Africa and to the Arab lands of the eastern end of the Mediterranean. Within the region, only the Persian Empire (on the Iranian Plateau), the central Arabian Peninsula, and Morocco were able to resist direct Ottoman control.

By the mid-19th century, Ottoman rule was under siege from Europe, as Egypt was occupied by Britain, and Algeria and Tunisia by France. European occupation exposed the region to continental ideas about democracy, and **nationalist movements** erupted; groups of people who shared common elements of culture—such as language, religion, or history—wished to determine their own political affairs. The nationalist movements were particularly problematic for the Ottoman Empire, which had ruled by means of an elaborate imperial legal and administrative structure that tended to allow for cultural differences. As a result, by the start of World War I, the Ottoman Empire had been reduced in size, controlling only modern-day Turkey to the north and the coastal regions of the Arabian Peninsula to the south. After World War I ended, the Ottoman Empire was collapsed further into modern-day Turkey, as the other parts of the empire were carved up by Russia and other European powers (**Figure 4.14**).

The Mandate System As part of the spoils of World War I, the Arab provinces of the Ottoman Empire were divided up and became **mandates**—areas administered by a European power. The

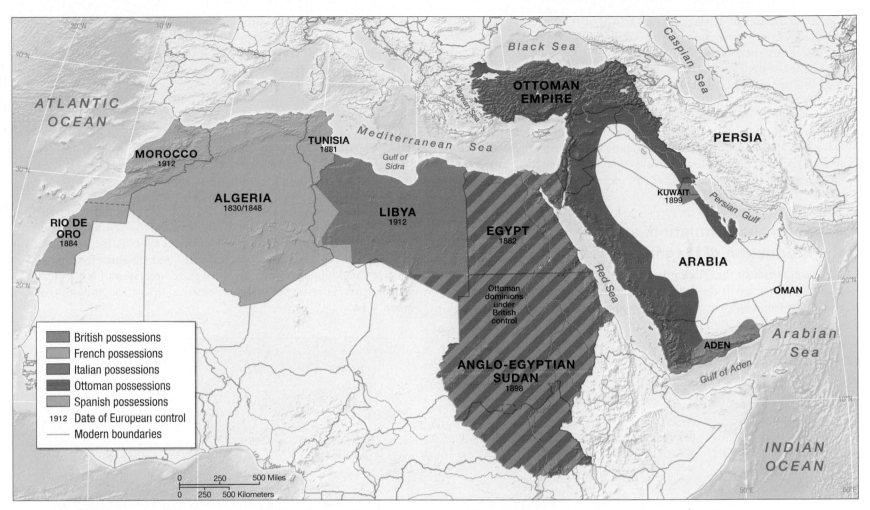

▲ **FIGURE 4.14 Europe in the Middle East and North Africa in 1914** The colonial presence of Europe in the Middle East and North Africa region was short-lived—only about 50 years—but significant. This map shows the European possession of the region before the Ottoman Empire was split up in 1920. Algeria shows two dates, with the first (1830) representing the French occupation of northern Algeria and the second (1848) representing the takeover of the entire country.

European powers were supposed to prepare the mandated regions for self-government and future independence. Syria and Lebanon were mandates of the French; the British took Iraq, Palestine, and Jordan (called Transjordan at that time). Mandate holders were required to submit to internationally sanctioned guidelines, which required that constitutional governments be established as the first step in preparing the new states for independence.

The political order that was imposed following World War I was seriously challenged throughout the region. People in Egypt, Iraq, Syria, and Palestine all revolted violently against the European presence. The negative impact of the mandate system helped foment increasing regional dissatisfaction with outside dominance. But it was not until the end of World War II, when Europe was severely damaged by war, that all the states in the region gained their independence. Winning independence from colonizers who had lost interest in their colony is not the same as gaining the allegiance of a diverse collection of new citizens. Many challenges emerged for these new states, which had to create national identities from a collection of diverse ethnic, religious, and cultural populations. This is discussed further in this chapter in the section "Territory and Politics."

APPLY YOUR KNOWLEDGE What are some of the similarities and differences between a mandate and a colony? Explain how European rule may have led to the emergence of nationalist movements in the region, and identify the long-term consequences of the mandate system on the Middle East and North Africa.

Naming the Middle East The term "Middle East" emerged over time as Western Europeans began to have increased contact with the peoples and places situated to their east. In the 13th century, Europeans began to use the term "orient" (meaning east) to describe the regions to their east. Over time, this concept of "the East" was modified, and Europeans began to use labels such as the "Near East" (to refer to the modern-day Balkan Peninsula in Eastern Europe) and the "Far East" (to refer to modern-day South Asia, East Asia, and Southeast Asia). The term "Middle East" was first formally used in 1902 by a U.S. naval officer. In the 1930s, the British adopted similar language, designating a "Middle East Command," which extended across the central Mediterranean all the way into India. Over time, the term's meaning has been narrowed to refer mainly to the countries located in the area of Turkey through Iran. The term itself was adopted by many western governments and is still used today by international organizations, such as the United Nations.

Critics of the "Middle East" label argue is a loaded term, because it is the name given to the region by Europeans and later by the United States. Many of those critics prefer the label "Southwest Asia" for this region; they feel it carries none of the colonial or postcolonial implications of the term "Middle East." The term "Middle East," however, does carry important geographic meaning because of its use in international politics and because it demonstrates the central and strategic role the region plays at the interface of three continental spaces: Africa, Asia, and Europe.

Economy, Accumulation, and the Production of Inequality

The integration of the Middle East and North Africa into the global capitalist economy in the 19th and 20th centuries brought increased wealth for some but increased poverty or reduced living standards for many others, even in the wealthy oil-producing states. The changes that occurred in the region in the 20th century, in particular, provide insight into how the region became integrated into the modern-day global economy. It also demonstrates how the Middle East and North Africa region was divided economically into a number of different subregions with somewhat different economic trajectories (**Figure 4.15**).

Independence and Economic Challenges In the 1930s, entrepreneurial states in the Middle East and North Africa (such as Turkey) adopted policies of **import substitution**, whereby domestic producers were protected because of high tariffs on competitive imports from outside the country. After World War II, nationalist movements in the region pushed many countries toward aggressive state-led development approaches. Countries such as Iran, Turkey, Egypt, Syria, Iraq, Tunisia, and Algeria adopted **nationalization** of economic development, which involved the conversion of key industries from private operation to governmental operation and control (**Figure 4.16**).

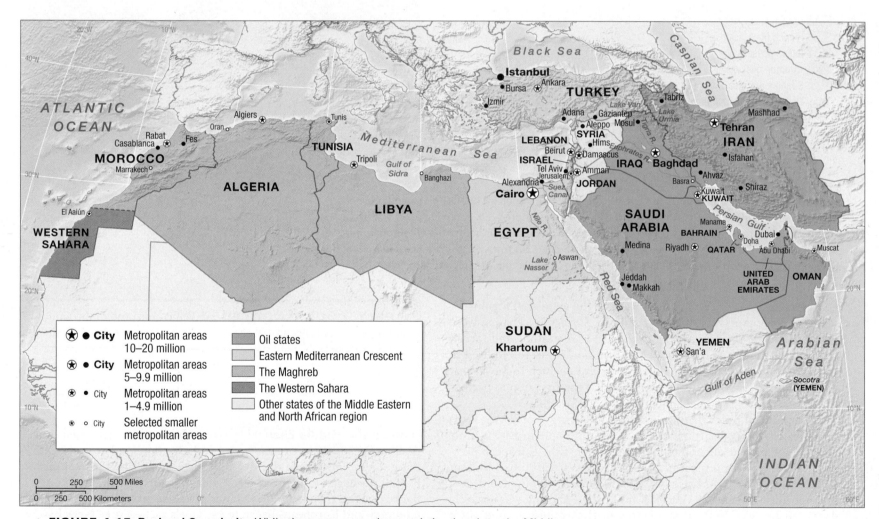

▲ **FIGURE 4.15 Regional Complexity** While there are many characteristics that draw the Middle East and North Africa together, there are also some stark economic and cultural differences. This map depicts some subregional differences and commonalities, highlighting the economic power of the oil states, the historical connection among the countries of the Maghreb, and the strong connections that can and sometimes do take place among the countries of the Eastern Mediterranean Crescent. The map also illustrates those countries that remain relatively distinct within the region (such as Sudan and Yemen), as well as depicting the Western Sahara, which remains in a state of civil conflict.

▲ **FIGURE 4.16 Nationalization of Oil Industries in Middle East** This image of the Iranian flag next to an oil refinery illustrates the important link between national identity and oil production in the region. While Iran briefly nationalized its oil industry in the 1950s, it was not until 1979 that the new Islamic Republic nationalized the oil industry in the country for the long-term. The nationalization of oil production in Iran led to the end of British and U.S. corporate control of the oil industry in that country.

Despite attempts to nationalize key industries, many states in the Middle East and North Africa began to turn away from nationalization as their economies stagnated, standards of living declined, and national debt skyrocketed in the 1960s and 1970s. Pressured by the International Monetary Fund (IMF), the World Bank, and the U.S. Agency for International Development, who were lending these countries money (see Chapter 1), many states were forced to privatize multiple sectors of their economies and open up their markets to free trade. Eventually, nationalized economic systems and import substitution policies were dismantled in favor of global free trade. These changes resulted in increasing costs of food and basic necessities, cuts in spending on social programs, and reductions in public sector investment. Urban workers, government bureaucrats, and people on fixed incomes strongly felt the impact of these changes.

In theory, the rural peasantry was to benefit most from these policies, especially the opportunity to market crops more freely. However, corporate farmers became the main beneficiaries of these economic changes, as they were able to buy up land and use mechanized agricultural practices to reduce labor costs. These changes put into motion a whole new set of forces that lowered the living standards of both urban workers and peasants. These processes also pushed rural people to move into the cities to find employment. When they arrived, they were often confronted with decreased public services, not only in terms of schools and health care but also in terms of the most basic necessities, such as housing and clean water. As a result, many people, who continue to migrate to cities today, are forced to live in squatter settlements without sanitation and eke out a living in the **informal economy**—economic activities that take place beyond official record and are not subject to formalized systems of regulation or remuneration.

The Oil States A number of countries in this region derived much of their wealth from oil, including Bahrain, Iran, Iraq, Kuwait, Oman, Qatar, Saudi Arabia, and the United Arab Emirates. These so-called oil states are pivotal in the global economy for at least two reasons. First, much of the rest of the world is highly dependent on them for oil. Without oil from the oil states, the ability of a large portion of the globe to maintain productive economies would be severely hampered. Second, the impact of **petrodollars**—revenues generated by the sale of oil—is especially significant for the global economy, particularly the economies of Europe, North America, and East Asia, where they are spent, invested, and banked. The oil states of the Middle East are likely to continue to occupy a central role in the affairs of the region as well as the world, where guaranteeing a secure supply of oil is absolutely central to the smooth functioning of the global economy.

However, a nearly exclusive dependence on oil production in the oil states, particularly in the Persian Gulf region, leaves their economies highly vulnerable to fluctuations in the demand for oil. Recognizing this, some of the oil states have begun to diversify their economies through tourism, trade, and urban development. Saudi Arabia, for example, established a plan to build four new "Economic Cities" to attract foreign investment into the country. One of the planned cities is named "Knowledge Economic City," and aims to attract information industries to the country. These projects are a long way from completion, but they suggest how the region might respond to the changing nature of the energy economy over the next 20 years (**Figure 4.17**). Some countries are also attempting

▲ **FIGURE 4.17 Knowledge Economic City in Saudi Arabia** Countries throughout the region are trying to diversify their economies to lessen their reliance on oil revenues. The government of Saudi Arabia has increased investment in a number of planned cities, in hopes of luring international businesses and providing job opportunities for the local population.

to introduce new industries, such as textile production and food processing plants. Some are developing port facilities. Still others are resuscitating or introducing agriculture, though the scarcity of water makes irrigation an enormous technological challenge for all of these countries except for Iraq and Iran.

The Eastern Mediterranean Crescent The Eastern Mediterranean Crescent, made up of Egypt, Turkey, Lebanon, and Israel, has the potential to be a different kind of economic success story. Similarities are striking among the four states. For instance, all four possess more of a European orientation than many of their neighbors, certainly more so than the oil states. All four have a sizable middle class. And all four have the potential, because of their resource endowments, to continue building a diverse economic base.

There are also, of course, differences among these four countries; the most striking exist between Israel and the other three states. For example, Egypt, Turkey, and Lebanon have had strong agricultural bases for hundreds, if not thousands, of years (**Figure 4.18**). Israel's commitment to agriculture is more recent. The former three use agriculture as a major source of export revenues, but for Israel, agricultural production is more about achieving national food security (although

▲ **FIGURE 4.18 Harvesting Cotton in Southeastern Turkey** Much of Turkey's climate is conducive to agriculture, and the country contains numerous farming regions. Cotton is a major export crop; Turkey is also the world's largest exporter of sultana raisins and hazelnuts. Other crops are tobacco, wheat, sunflower seeds, sesame and linseed oils, and cotton-oil seeds. Opium was once a major crop, but its exportation was banned by the government in 1972. The ban was lifted two years later as poppy farmers were unable to adapt their land to other crops. The government now controls the production and sale of opium.

some Israeli foodstuffs are also exported). And, while Egypt, Turkey, and Lebanon are just beginning to encourage more industrial development, Israel already possesses a strong industrial base that is fairly diverse, allowing it to generate the second highest GDP per capita in the region after the United Arab Emirates. Turkey also possesses a fairly diversified economy, with a strong agricultural sector and substantial mineral wealth.

In contrast, national agriculture in Egypt is built on a fairly narrow base, largely because of the environmental constraints of the desert. But, like Turkey, Egypt has significant manufacturing capacity across a range of products from food processing to heavy machinery. Lebanon also possesses a strong agricultural base and, for many years, was a banking and financial center, connecting the region to the core of Europe and North America. However, civil war in 1975–1990 limited Lebanon's regional and global authority in the area of banking. A 2006 war with Israel further derailed Lebanon's political and economic stability. The conflict has since ended, and Israel has retreated from southern Lebanon, but tensions along the border remain and serve to destabilize Lebanon. As a result of these conflicts and the recent economic stressors associated with the global economy, Lebanon's economy is weak and the country is still deeply fragmented politically.

Strong political and cultural differences—in part driven by religious differences between the Jewish state of Israel and the predominantly Islamic states of Lebanon, Egypt, and Turkey—have prevented these countries from acting in concert, even though there is much to suggest that cooperation would be mutually beneficial. That said, these four countries appear to possess the necessary ingredients to participate in the world economy thanks to their histories, their economies, and their roles in regional politics.

The Maghreb The Maghreb is the region of northwest Africa that contains the coastlands and Atlas Mountains of Morocco, Algeria, and Tunisia and the mostly desert state of Libya (see Figure 4.15). The Maghreb region in general has a relatively strong economy based largely on oil and mineral exploitation, agriculture, and tourism. Algeria's oil industry provides nearly 90% of its export revenues. Libya, too, has substantial, high-quality petroleum reserves. Both Tunisia and Morocco are significant globally for their phosphate industries. All the Maghreb countries are also agricultural producers, though none is self-sufficient. The most important agricultural products of the region are wheat, barley, olives, dates, citrus fruits, almonds, peanuts, beef and poultry, and vegetables.

Hugging the southern coastline of the Mediterranean, with rugged mountains rising up from the coastal plains and then trailing off to the desert, the Maghreb is a spectacularly beautiful setting. The region offers a range of tourist experiences, in both luxury and economy style, from beautiful beaches to trekking areas in the Atlas Mountains or the Sahara to ancient archaeological ruins (**Figure 4.19**). Europeans have been frequent visitors to the Maghreb; a short flight brings them to warm temperatures, exotic landscapes, and inexpensive and sumptuous food. Algeria and Morocco are two of the fastest-growing economies in North Africa because of tourism. And Libya and Tunisia were making substantial strides as well until recent political events curtailed economic development. If stability returns to these countries, links across the Mediterranean with the European Union through

▲ **FIGURE 4.19 Tuaregs Crossing the Sahara** A Tuareg man and his children, dressed in traditional blue robes, cross the Sahara area with camels.

the Euro-Mediterranean Partnership, recently relaunched as the **Union for the Mediterranean**, may boost all sectors of the Maghreb, from resources to tourism.

Social Inequalities across the Region The Middle East and North Africa is a region of extreme contrasts of wealth and poverty. Despite the phenomenal wealth generated from oil production for some parts of the region, most of the region remains poor and highly dependent on an increasingly marginalized agricultural sector. In 2011, the UN Human Development Report listed Qatar as having the highest GNI per capita PPP in the region (and the world) at U.S. $107,721 and Yemen with the lowest in the region at U.S. $2,231. Not surprisingly, national statistics like per capita income hide all sorts of variation. For instance, within the same area, dramatic variation can occur between one urban neighborhood and the next, and there are also understandable differences between urban and rural spaces. Most of the extreme wealth of the region comes from oil-based revenues in Saudi Arabia, Kuwait, Iran, Iraq, Oman, Qatar, Libya, and the United Arab Emirates (**Table 4.1**). States with the largest populations tend to have the lowest per capita wealth. As Table 4.1 shows, the variation in the overall Human Development Index (HDI), as measured by life expectancy, educational attainment, and income, also varies dramatically across the region. Moreover, even though Qatar has the highest GNI per capita PPP in the world, it is only ranked 30th overall on the global HDI.

APPLY YOUR KNOWLEDGE Compare and contrast the economies of the oil the states and the states that do not produce large quantities of oil in the Middle East and North Africa. In cases where oil is not available, what sorts of economies have developed over time? Which countries have been most successful at diversifying their economies?

Territory and Politics

The countries of the Middle East and North Africa emerged from their colonial and mandate status burdened with a range of political challenges. Although the region has experienced wars and conflict

TABLE 4.1 Gross National Income (GNI) and Human Development Index (2011)

Country	Gross national income (GNI) per capita (Constant 2005 PPP$)	Human Development Index (HDI)*
Algeria	7,658	L (96)
Bahrain	28,169	VH (42)
Egypt	5,269	M (113)
Iran, Islamic Republic of	10,164	H (88)
Iraq	3,177	M (132)
Israel	25,849	VH (17)
Jordan	5,300	M (95)
Kuwait	47,926	H (63)
Lebanon	13,076	H (71)
Libya	12,637	H (64)
Morocco	4,196	M (130)
Oman	22,841	H (89)
Qatar	107,721	VH (37)
Saudi Arabia	23,274	H (56)
Sudan	1,894	L (169)
Syrian Arab Republic	4,243	M (119)
Tunisia	7,281	H (94)
Turkey	12,246	H (92)
United Arab Emirates	59,993	VH (3)
Yemen, Republic of	2,231	L (154)

SOURCE: United Nations Human Development Reports 2011; accessed April 2012 from http://hdr.undp.org/en/statistics/data/.
*Human Development Index is a combined measurement which includes life expectancy, educational attainment, and income. See Chapter 1, p. 29 for further explanation. Each country is given a general rating as Very High (VH), High (H), Medium (M), or Low (L). In the bracket is the ranking of the country overall in global terms. In 2011, there were 187 countries reported.

VISUALIZING GEOGRAPHY
Political Protest, Change, and Repression

In 2009, protests erupted in Iran as charges of election fraud were made against the president, Mahmoud Ahmadinejad. Although this uprising, known as the **Iranian Green Revolution**, took place in a Persian-dominated country, its occurrence was echoed in similar protests that began taking place two years later across the Arab-dominated parts of the region. These protests are led by a new generation of younger people who are calling for more open and transparent governments and political reform. The **Arab Spring** is a broad term used to refer to the numerous recent protests against dictatorships across the region, as people motivated by economic inequalities and lack of access to basic resources have taken to the streets to voice their concerns. The discontent that motivated Tunisia's **Jasmine Revolution**, for example, at the beginning of 2011, quickly found voice in many other countries in the region, including Algeria, Jordan, Saudi Arabia, Bahrain, Oman, Egypt, Libya, Yemen, Morocco, and Syria (**Figure 4.2.1**).

The levels and success of the protests have varied: Egypt has undergone a regime change; there has been "constitutional reform" in Morocco; and Libya, Yemen, and Syria have experienced violent insurgency. Insurgencies led to the successful removal of the leaders in both Libya and Yemen, while in Syria violence continues as protesters and the government fight to control the future of the country's political process. The Syrian conflict is motivated in part by high rates of unemployment among younger Syrians and human rights abuses by the Syrian government.

Protests in the region have been fueled by the global connectivity of people and through Internet sites. Viral videos—often home-made videos seen by millions of viewers—have been seen around the world on YouTube, and millions more people shared their experiences of the protests on Facebook. One of the most famous of these videos shows a woman being killed during the Green Revolution protests in Iran. Over 2 million people watched footage of the incident on YouTube.

In response to these protests and the role that social media is playing, governments in the region, such as Iran, have blocked websites, cell phone calls, and text messages. They have also banned rallies throughout the region, and some gatherings have been met with violent reprisals by state-supported militaries, as was the case in Bahrain. Casualty numbers are difficult to measure, but some estimates have suggested that by 2012 thousands of people had lost their lives in these conflicts (see Figure 4.2.1). Governments that have tried to crack down on their citizens have come under international pressure. For example, NATO forces directly assaulted positions of the then Libyan president, Muammar Qaddafi, in an attempt to aid resistance fighters. And the U.N. and other international organizations have made efforts to pressure the Syrian government to limit its repression and open the country to democratic and economic reform. ∎

---

for hundreds, if not thousands, of years, and certainly well before the Europeans arrived, it is generally agreed that most of the present conflicts stem from the colonial period. The colonial imposition of borders united peoples within the region who were antagonistic to each other or divided peoples who were once united. The strategic importance of the Middle East and North Africa to the political and economic interests of the world economy has also meant that the region has been subject to much contestation over its rich natural resources. At the same time that the region has been the site of bitter and, in some instances, seemingly irresolvable conflicts, it has also been the site of broad and sustained cooperation. The most significant unifying forces have been the religion of Islam and, for many of the countries, the Arabic language. These unifying forces of religion and language have helped many of the peoples of the region recognize that they have common political and economic goals. Recently, however, economic and class differences as well as a growing interest in democratic reform have shown that these unifying forces may not be enough to overcome some of the region's deep social divisions (see "Visualizing Geography: Political Protest, Change, and Represssion").

Ethnic Conflict There are various ethnic conflicts in the region, each with a unique origin but often stemming from age-old conflicts

and the legacies of colonialism. For example, the tensions that exist and the conflicts that have erupted between Iran and Iraq over the last 30-plus years are the result of several factors. One factor is the cultural differences between Persians (Iranians) and Arabs (Iraqis). Though the majority of both Persians and Arabs are Muslim, the ethnic origins, languages, geographies, and histories of the two groups are distinctly different. The differences and the conflicts between the two groups have very old beginnings. The Persians were conquered by the Arabs in the 7th century and converted to Islam beginning in the 9th century. Since then, there has been animosity between the peoples of these two countries; their relationship has more recently been complicated by their respective dealings with Britain and the United States. Iran's relationship with Arabs in the region has been tested in the last few years, as Iran has been seen to be supporting conflict in its neighbor Iraq and also in Syria. Some studies show that Iran's favorability among Arabs in the region has been declining for the last six years.

Other points of tension have occurred because of ethnic Kurdish groups who have been fighting the Turkish government for autonomous control over their shared Kurdish mountain region. Kurds have also been struggling against the Iranian and Iraqi governments for greater regional autonomy. About 20 million **Kurds**, a non-Arabic people who are mostly Sunni Muslims, live

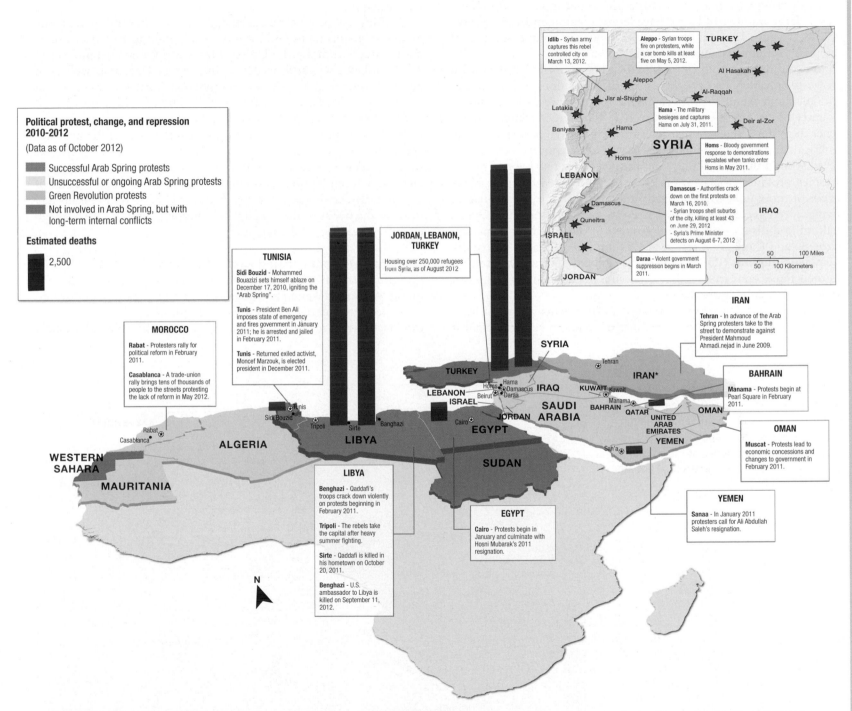

Political protest, change, and repression 2010-2012

(Data as of October 2012)

- Successful Arab Spring protests
- Unsuccessful or ongoing Arab Spring protests
- Green Revolution protests
- Not involved in Arab Spring, but with long-term internal conflicts

Estimated deaths

2,500

TUNISIA

Sidi Bouzid - Mohammed Bouazizi sets himself ablaze on December 17, 2010, igniting the "Arab Spring".

Tunis - President Ben Ali imposes state of emergency and fires government in January 2011; he is arrested and jailed in February 2011.

Tunis - Returned exiled activist, Moncef Marzouk, is elected president in December 2011.

MOROCCO

Rabat - Protesters rally for political reform in February 2011.

Casablanca - A trade-union rally brings tens of thousands of people to the streets protesting the lack of reform in May 2012.

JORDAN, LEBANON, TURKEY

Housing over 250,000 refugees from Syria, as of August 2012

IRAN

Tehran - In advance of the Arab Spring protesters take to the street to demonstrate against President Mahmoud Ahmadi.nejad in June 2009.

BAHRAIN

Manama - Protests begin at Pearl Square in February 2011.

OMAN

Muscat - Protests lead to economic concessions and changes to government in February 2011.

LIBYA

Benghazi - Qaddafi's troops crack down violently on protests beginning in February 2011.

Tripoli - The rebels take the capital after heavy summer fighting.

Sirte - Qaddafi is killed in his hometown on October 20, 2011.

Benghazi - U.S. ambassador to Libya is killed on September 11, 2012.

EGYPT

Cairo - Protests begin in January and culminate with Hosni Mubarak's 2011 resignation.

YEMEN

Sanaa - In January 2011 protesters call for Ali Abdullah Saleh's resignation.

Idlib - Syrian army captures this rebel controlled city on March 13, 2012.

Aleppo - Syrian troops fire on protesters, while a car bomb kills at least five on May 5, 2012.

Hama - The military besieges and captures Hama on July 31, 2011.

Homs - Bloody government response to demonstrations escalates when tanks enter Homs in May 2011.

Damascus - Authorities crack down on the first protests on March 16, 2010.
- Syrian troops shell suburbs of the city, killing at least 43 on June 29, 2012
- Syria's Prime Minister defects on August 6-7, 2012

Daraa - Violent government suppression begins in March 2011.

▲ **FIGURE 4.2.1** **Map of Protest, Change, and Repression in the Middle East and North Africa** This map shows the varying levels of success related to protests in the region between January 2010 and October 2012.

in the mountainous areas along the borders of Iran, Iraq, Syria, Turkey, and a small area in Armenia, called Kurdistan by Kurds (**Figure 4.20**). Of these, about 8 million Kurds live in southeastern Turkey. Throughout the 20th century and into the present one, the Kurds as an ethnic minority have faced repression and discrimination. Many Kurds have agitated peacefully as well as taken up arms in an attempt to gain outright independence or autonomy. Recent Turkish incursions into northern Iraq have been aimed at stopping Kurdish groups from agitating for regional autonomy within Turkey. A large number of Kurds have left the region entirely and now live in Western Europe, providing an important source of financial support for the resistance movement.

The ethnic and political conflict that began in the Darfur region of western Sudan also continues to unfold, as does wider conflict in Sudan and the new South Sudan, which became an independent country in 2011 (see Chapter 5, p. 174). The wider context of violence in western Sudan, in particular, has its roots in persistent social, political, and economic inequality between the country's core, centered around the Nile Valley, and the periphery represented by the western province of Darfur (an area about the size of Texas). Hundreds of thousands are now dead, and millions more are displaced, living in refugee camps, or have fled the country altogether. Over a half million displaced Sudanese people live outside their home territories, and there are a reported 1.62 million Sudanese **internally displaced persons** (IDPs), individuals who are uprooted but remain in their home country as refugees because of civil conflict or human rights violations (**Figure 4.21**).

The brutal violence in Darfur began in 2003, when native Africans in the province looking for a measure of freedom revolted against Sudan's authoritarian Islamic government. Media reports indicate that the Sudanese government aimed to end the rebellion by eliminating all the tribal Africans in the area so that Arabs could take over the land. In pursuit of this aim, the government provided support for a militia of African Arabs, who call themselves the **Janjaweed**, to undertake one of the most brutal campaigns of ethnic cleansing that Africa has ever seen. Although the Bush administration called the government's action genocide, the U.S. government took no protective action.

In May 2006, a peace agreement was brokered between the Sudanese government and local African communities after diplomatic intervention by the United States and Great Britain. Yet in 2012, violence was still occurring in parts of Darfur as well as in South Kordofan, along the new international border between Sudan and South Sudan (see Chapter 5, p. 174). The Blue Nile province of Sudan along the Sudan–South Sudan border has also seen violence in 2012, as people of South Sudanese origin pour out of the province and into the newly established country of South Sudan. It is clear that the challenges facing Sudan will not dissipate any time soon. In fact, the United Nations continues to consider the region a site of humanitarian crisis and estimates that over U.S. $1 billion is needed to serve the needs of the roughly 4.2 million people who have been affected by conflict in the region.

The Israeli–Palestinian and Israeli–Lebanese Conflicts The history of the Israeli–Palestinian conflict is complex. The situation is highly volatile, despite persistent local and international efforts to bring peace to the region. As with other conflicts in the region, the chief factors that inflame this seemingly intractable political problem

▶ **FIGURE 4.20 Kurdistan, "Land of the Kurds"** This map shows the area where the Kurdish population is dominant. Although Kurdistan is not an independent state, the map does provide a sense of the boundaries that (from the Kurdish perspective) it might ideally encompass.

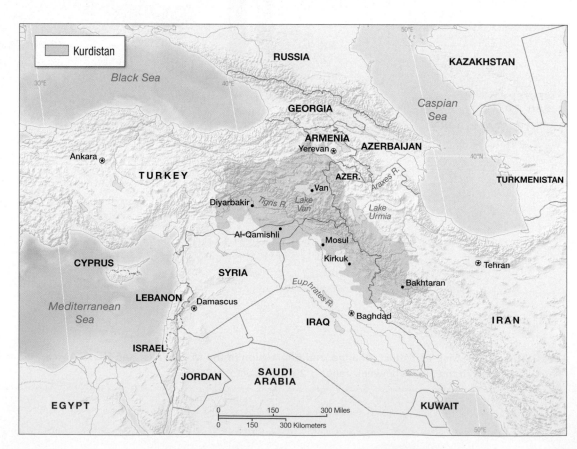

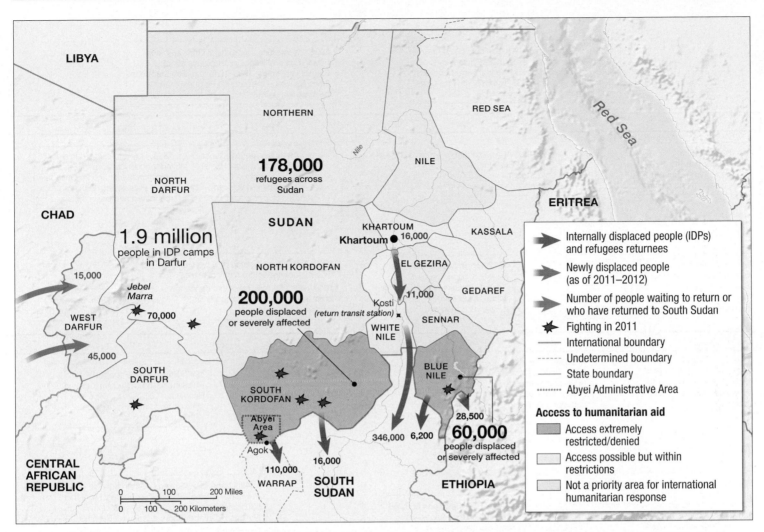

▲ **FIGURE 4.21 Civil Conflict and Human Rights Issues in Sudan** Even though a referendum created a new country in 2011, South Sudan, the conflict in Sudan continues, with many people displaced or without access to international humanitarian support. Some people are fleeing Sudan for the new South Sudan, while others are displaced by conflicts that remain internal to Sudan itself. The boundaries and names shown and the designations used on this map do not imply official endorsement or acceptance by the United Nations.

were exacerbated by British partitioning of the region during the colonial period. The official Jewish state of Israel is a mid–20th-century construction that has its roots in the emergence of **Zionism**, a late 19th-century movement in Europe. Zionism defined the need for an independent homeland in Palestine for the Jewish people, and thousands of European Jews, inspired by the early Zionist movement, began migrating to Palestine at the turn of the 19th century. When the Ottoman Empire was defeated in 1917, the British gained control over Palestine and the Jordan area and issued the **Balfour Declaration**, which supported the legal migration of Jews to Palestine.

The Balfour Declaration was highly problematic because the area was already populated by Palestinians. Some Palestinians viewed the arrival of increasing numbers of Jews and European sympathy for the establishment of a Jewish homeland as an incursion into the sacred lands of Islam. In response to increasing Arab–Jewish tensions as a result of mass migration of Jews into the area, the British decided to limit Jewish immigration to Palestine in the late 1930s through the end of World War II. In 1947, with conflict continuing between Jews and Palestinians, Britain an-

nounced that it would withdraw from Palestine in 1948 and turn it over to the United Nations. The United Nations, under heavy pressure from the United States, responded by voting to partition Palestine into Arab and Jewish states, with neither religious faction having exclusive control. The Jewish state would have 56% of the territory and an Arab state 43%; Jerusalem, a city sacred to Jews, Muslims, and Christians, was to be an "international city" administered by the United Nations. The proposed U.N. plan was accepted by the Jews and angrily rejected by the Arabs, who argued that a mandate territory could not legally be taken from an Indigenous population.

In 1948 Britain withdrew, Jews proclaimed the state of Israel, and war broke out. In an attempt to aid the militarily weaker Palestinians, combined forces from Egypt, Jordan, and Lebanon, as well as smaller units from Syria, Iraq, and Saudi Arabia, confronted the recently established Israeli state. The goal of the Arab forces was not only to prevent the Israelis from gaining control over additional Palestinian territory but to wipe out the newly formed Jewish state altogether. This war, known as **The First Arab–Israeli War**, ended in the defeat of Arab forces in 1949. Subsequent

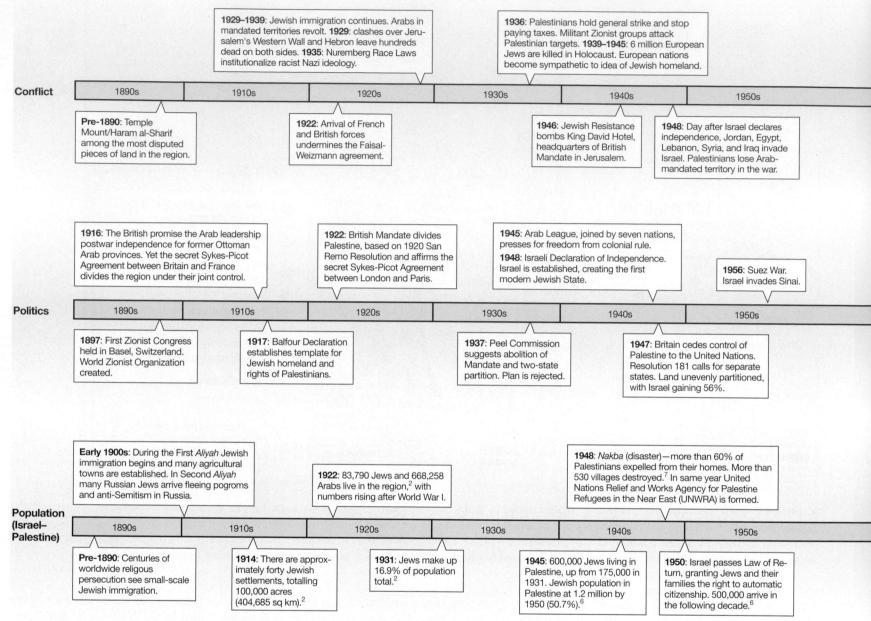

Conflict

1890s **1910s** **1920s** **1930s** **1940s** **1950s**

1929–1939: Jewish immigration continues. Arabs in mandated territories revolt. **1929:** clashes over Jerusalem's Western Wall and Hebron leave hundreds dead on both sides. **1935:** Nuremberg Race Laws institutionalize racist Nazi ideology.

1936: Palestinians hold general strike and stop paying taxes. Militant Zionist groups attack Palestinian targets. **1939–1945:** 6 million European Jews are killed in Holocaust. European nations become sympathetic to idea of Jewish homeland.

Pre-1890: Temple Mount/Haram al-Sharif among the most disputed pieces of land in the region.

1922: Arrival of French and British forces undermines the Faisal-Weizmann agreement.

1946: Jewish Resistance bombs King David Hotel, headquarters of British Mandate in Jerusalem.

1948: Day after Israel declares independence, Jordan, Egypt, Lebanon, Syria, and Iraq invade Israel. Palestinians lose Arab-mandated territory in the war.

Politics

1890s **1910s** **1920s** **1930s** **1940s** **1950s**

1916: The British promise the Arab leadership postwar independence for former Ottoman Arab provinces. Yet the secret Sykes-Picot Agreement between Britain and France divides the region under their joint control.

1922: British Mandate divides Palestine, based on 1920 San Remo Resolution and affirms the secret Sykes-Picot Agreement between London and Paris.

1945: Arab League, joined by seven nations, presses for freedom from colonial rule.
1948: Israeli Declaration of Independence. Israel is established, creating the first modern Jewish State.

1956: Suez War. Israel invades Sinai.

1897: First Zionist Congress held in Basel, Switzerland. World Zionist Organization created.

1917: Balfour Declaration establishes template for Jewish homeland and rights of Palestinians.

1937: Peel Commission suggests abolition of Mandate and two-state partition. Plan is rejected.

1947: Britain cedes control of Palestine to the United Nations. Resolution 181 calls for separate states. Land unevenly partitioned, with Israel gaining 56%.

Population (Israel–Palestine)

1890s **1910s** **1920s** **1930s** **1940s** **1950s**

Early 1900s: During the First *Aliyah* Jewish immigration begins and many agricultural towns are established. In Second *Aliyah* many Russian Jews arrive fleeing pogroms and anti-Semitism in Russia.

1922: 83,790 Jews and 668,258 Arabs live in the region,[2] with numbers rising after World War I.

1948: *Nakba* (disaster)—more than 60% of Palestinians expelled from their homes. More than 530 villages destroyed.[7] In same year United Nations Relief and Works Agency for Palestine Refugees in the Near East (UNWRA) is formed.

Pre-1890: Centuries of worldwide religous persecution see small-scale Jewish immigration.

1914: There are approximately forty Jewish settlements, totalling 100,000 acres (404,685 sq km).[2]

1931: Jews make up 16.9% of population total.[2]

1945: 600,000 Jews living in Palestine, up from 175,000 in 1931. Jewish population in Palestine at 1.2 million by 1950 (50.7%).[6]

1950: Israel passes Law of Return, granting Jews and their families the right to automatic citizenship. 500,000 arrive in the following decade.[6]

▲ **FIGURE 4.22 Timeline of Israeli–Palestinian Conflict and Change Since 1890** The history of modern Israel/Palestine is one of enduring conflict as Israel increasingly occupies Palestinian territory and Palestinians continue to fight to keep their homelands. This timeline shows the different conflicts as well as the larger political and population issues around which these conflicts have been played.

agreements then enabled Israel to expand beyond the U.N. plan into the western sector of Jerusalem. In 1950, Israel declared Jerusalem its national capital, though very few countries recognize it as such. **Figure 4.22** provides a time line of key events, and **Figure 4.23** (p. 154) depicts the changing political geography resulting from the conflict.

The territorial expansion of Israel resulted in hundreds of thousands of Palestinians being driven from their homeland, often inhabited by family lineages for many hundreds of years. (**Figure 4.24**, p. 155). Today, Palestinians live as refugees either in other Arab countries in the region, abroad, or under Israeli occupation in the West Bank, the Golan Heights, and the **Gaza Strip** (also known as the "**Occupied Territories**"). By the late 1980s, Palestinians who had remained in their homeland rose up in rebellion against 50 years of Israeli occupation. This violent uprising of Palestinians against the rule of Israel in the Occupied Territories, known as the **intifada**, often involved frequent clashes between fully armed Israeli soldiers and rock-throwing Palestinian young men.

In summer 2005, Israel began to cede territory back to Palestine under pressure from the U.N. As Israeli settlers moved from homes they had inhabited, in most cases for decades, critics argued that

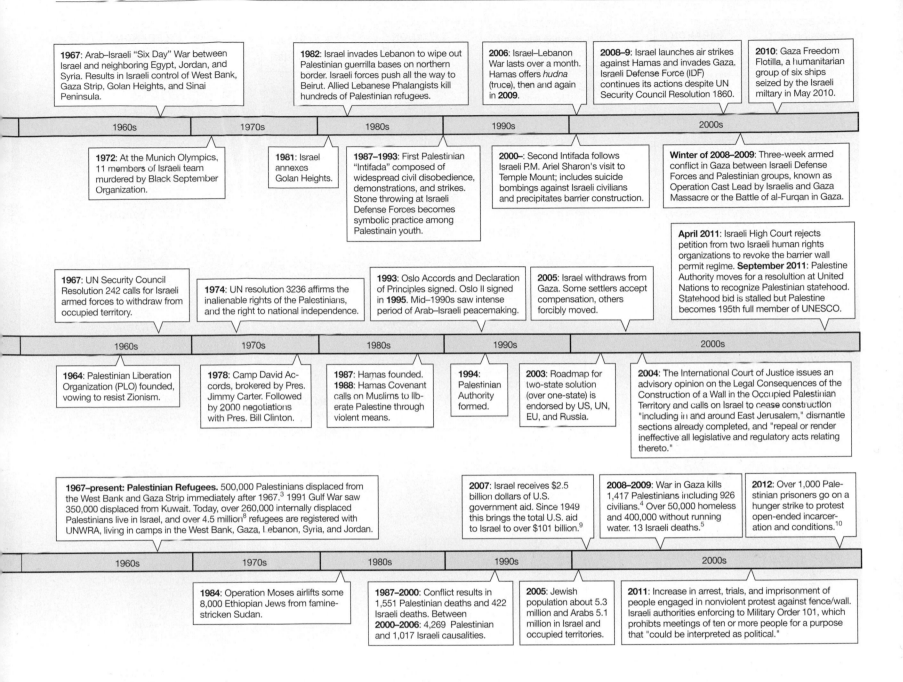

1967: Arab–Israeli "Six Day" War between Israel and neighboring Egypt, Jordan, and Syria. Results in Israeli control of West Bank, Gaza Strip, Golan Heights, and Sinai Peninsula.

1982: Israel invades Lebanon to wipe out Palestinian guerrilla bases on northern border. Israeli forces push all the way to Beirut. Allied Lebanese Phalangists kill hundreds of Palestinian refugees.

2006: Israel–Lebanon War lasts over a month. Hamas offers *hudna* (truce), then and again in **2009**.

2008–9: Israel launches air strikes against Hamas and invades Gaza. Israeli Defense Force (IDF) continues its actions despite UN Security Council Resolution 1860.

2010: Gaza Freedom Flotilla, a humanitarian group of six ships seized by the Israeli military in May 2010.

1960s | 1970s | 1980s | 1990s | 2000s

1972: At the Munich Olympics, 11 members of Israeli team murdered by Black September Organization.

1981: Israel annexes Golan Heights.

1987–1993: First Palestinian "Intifada" composed of widespread civil disobedience, demonstrations, and strikes. Stone throwing at Israeli Defense Forces becomes symbolic practice among Palestinain youth.

2000–: Second Intifada follows Israeli P.M. Ariel Sharon's visit to Temple Mount; includes suicide bombings against Israeli civilians and precipitates barrier construction.

Winter of 2008–2009: Three-week armed conflict in Gaza between Israeli Defense Forces and Palestinian groups, known as Operation Cast Lead by Israelis and Gaza Massacre or the Battle of al-Furqan in Gaza.

April 2011: Israeli High Court rejects petition from two Israeli human rights organizations to revoke the barrier wall permit regime. **September 2011**: Palestine Authority moves for a resolution at United Nations to recognize Palestinian statehood. Statehood bid is stalled but Palestine becomes 195th full member of UNESCO.

1967: UN Security Council Resolution 242 calls for Israeli armed forces to withdraw from occupied territory.

1974: UN resolution 3236 affirms the inalienable rights of the Palestinians, and the right to national independence.

1993: Oslo Accords and Declaration of Principles signed. Oslo II signed in **1995**. Mid–1990s saw intense period of Arab–Israeli peacemaking.

2005: Israel withdraws from Gaza. Some settlers accept compensation, others forcibly moved.

1960s | 1970s | 1980s | 1990s | 2000s

1964: Palestinian Liberation Organization (PLO) founded, vowing to resist Zionism.

1978: Camp David Accords, brokered by Pres. Jimmy Carter. Followed by 2000 negotiations with Pres. Bill Clinton.

1987: Hamas founded. **1988**: Hamas Covenant calls on Muslims to liberate Palestine through violent means.

1994: Palestinian Authority formed.

2003: Roadmap for two-state solution (over one-state) is endorsed by US, UN, EU, and Russia.

2004: The International Court of Justice issues an advisory opinion on the Legal Consequences of the Construction of a Wall in the Occupied Palestinian Territory and calls on Israel to cease construction "including in and around East Jerusalem," dismantle sections already completed, and "repeal or render ineffective all legislative and regulatory acts relating thereto."

1967–present: Palestinian Refugees. 500,000 Palestinians displaced from the West Bank and Gaza Strip immediately after 1967.[3] 1991 Gulf War saw 350,000 displaced from Kuwait. Today, over 260,000 internally displaced Palestinians live in Israel, and over 4.5 million[8] refugees are registered with UNWRA, living in camps in the West Bank, Gaza, Lebanon, Syria, and Jordan.

2007: Israel receives $2.5 billion dollars of U.S. government aid. Since 1949 this brings the total U.S. aid to Israel to over $101 billion.[9]

2008–2009: War in Gaza kills 1,417 Palestinians including 926 civilians.[4] Over 50,000 homeless and 400,000 without running water. 13 Israeli deaths.[5]

2012: Over 1,000 Palestinian prisoners go on a hunger strike to protest open-ended incarceration and conditions.[10]

1960s | 1970s | 1980s | 1990s | 2000s

1984: Operation Moses airlifts some 8,000 Ethiopian Jews from famine-stricken Sudan.

1987–2000: Conflict results in 1,551 Palestinian deaths and 422 Israeli deaths. Between 2000–2006: 4,269 Palestinian and 1,017 Israeli causalities.

2005: Jewish population about 5.3 million and Arabs 5.1 million in Israel and occupied territories.

2011: Increase in arrest, trials, and imprisonment of people engaged in nonviolent protest against fence/wall. Israeli authorities enforcing to Military Order 101, which prohibts meetings of ten or more people for a purpose that "could be interpreted as political."

the return of land was a hollow gesture as Israel continued the construction of physical barriers between Israelis and Palestinians (**Figure 4.25**, p. 155). Chief among these is the Gaza Strip barrier, which consists of 52 kilometers (30 miles) of mainly wire fence with posts, sensors, and buffer zones. Israel argues that the barrier is essential to protect the security of its citizens from Palestinian terrorism. Palestinians and other opponents of the barrier contend that its purpose is geographical containment of the Palestinians; they assert it is being erected to pave the way for an expansion of Israeli sovereignty and to preclude any negotiated border agreements in the future. But Israel argues it is purely for security not a part of a future border. The Gaza Strip barrier, completed in 2002, marked the beginning of the West Bank wall. When completed, the wall

will seal off another portion of Palestinian territories from Israel (**Figure 4.26**, p. 156).

These barriers continue to uproot and destroy Palestinian settlements and separate them from their livelihoods. In October 2003, the U.N. General Assembly voted 144–4 that the West Bank barrier was "in contradiction to international law" and therefore illegal. Israel called the resolution a "farce." Recently, the completion of the West Bank wall has been slowed by court challenges to its development and budgetary constraints. The future of the Palestinian–Israeli peace process remains in doubt today, as tensions remain between the Israel government and the Palestinian Authority, which currently governs the Palestinian territories in the West Bank. Complicating matters is the fact that the Gaza Strip is ruled by **Hamas** (an Islamist Palestinian

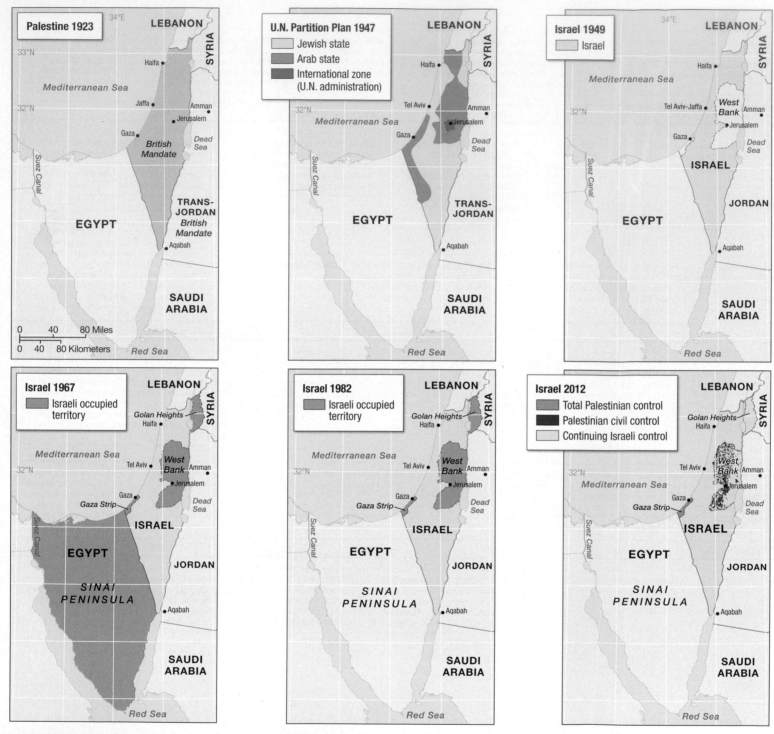

▲ **FIGURE 4.23 Changing Geography of Israel and Palestine, 1923–2012** Since the creation of Israel from part of the former Palestine in 1947, the regional political geography has undergone significant modifications.

political party). Israel refuses to recognize or negotiate with Hamas, which remains committed to the abolition of the Israeli state.

The political instability in Israel and Palestine affects other parts of the region. Lebanon's stability is very much tied to the situation in Israel and Palestine. Instigated by Israel's previous occupation of Lebanon (1982–2000), **Hezbollah** (a political and paramilitary organization) militias state as one of their goals the abolition of the state of Israel. In late July 2006, in response to the kidnapping of

two Israeli soldiers by Hezbollah militias, the Israeli army began sea and air strikes and ground incursions into southern Lebanon. Another conflict took place in 2010, when Lebanese and Israeli soldiers exchanged fire across the border. The tension between Israel and Lebanon can be traced in part to the end of the 1967 Arab–Israeli War. During that war, Palestinians began to use Lebanon as a base from which to launch attacks on Israel. Since then, despite periods of relative calm, Lebanon—especially the south, as

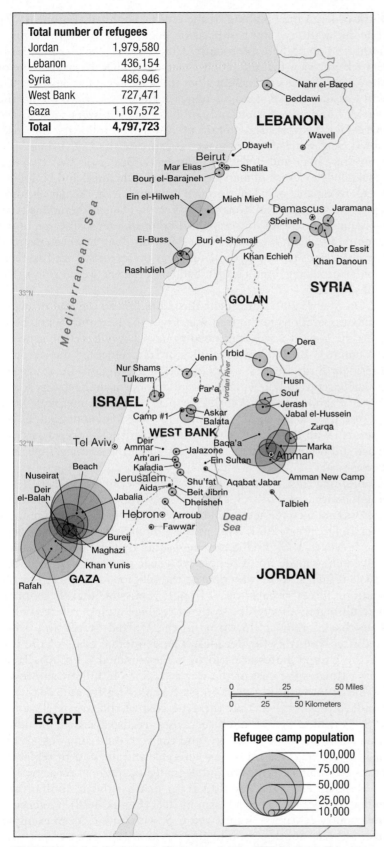

Total number of refugees	
Jordan	1,979,580
Lebanon	436,154
Syria	486,946
West Bank	727,471
Gaza	1,167,572
Total	**4,797,723**

Refugee camp population
- 100,000
- 75,000
- 50,000
- 25,000
- 10,000

▲ **FIGURE 4.24 Palestinian Refugees in the Middle East** This map shows the dispersion of Palestinian refugees, in camps and elsewhere, in the states around Israel and in the West Bank, Gaza, and the Occupied Territory. One of the biggest obstacles in the Israeli–Palestinian peace talks has been the question of refugee return and where Palestinians will be allowed to settle.

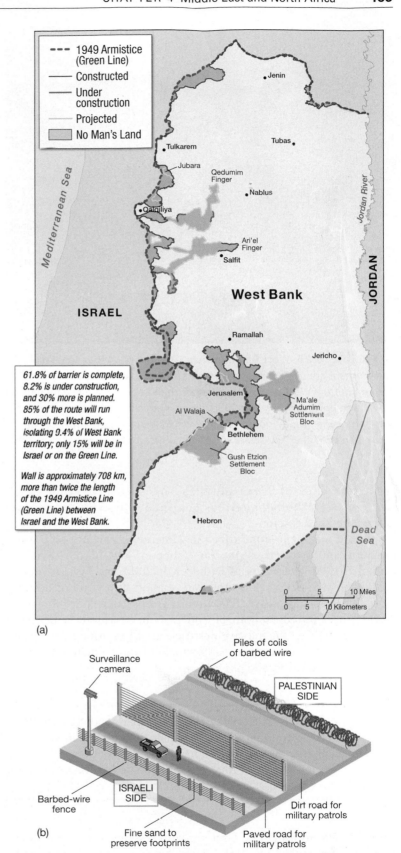

Legend:
- - - - 1949 Armistice (Green Line)
──── Constructed
──── Under construction
──── Projected
▓▓▓▓ No Man's Land

61.8% of barrier is complete, 8.2% is under construction, and 30% more is planned. 85% of the route will run through the West Bank, isolating 9.4% of West Bank territory; only 15% will be in Israel or on the Green Line.

Wall is approximately 708 km, more than twice the length of the 1949 Armistice Line (Green Line) between Israel and the West Bank.

(a)

(b)

Labels: Surveillance camera; Piles of coils of barbed wire; PALESTINIAN SIDE; Barbed-wire fence; ISRAELI SIDE; Fine sand to preserve footprints; Paved road for military patrols; Dirt road for military patrols

▲ **FIGURE 4.25 Israeli Barriers, 2012** (a). Shown are the planned and completed portions of the structures referred to as "security fences" by the Israelis, which are called "the wall" or the "apartheid wall" by Palestinians and other opponents. (b). The physical barriers consist of a network of fences, walls, and trenches equipped with monitoring and surveillance devices.

▲ **FIGURE 4.26 West Bank Barriers** The fences built by the Israeli government are designed to isolate and protect Jewish settlements throughout the West Bank. These fences mark more than just a military boundary, however; they also mark an economic boundary between wealthier Israelis and poorer Palestinians. Palestinians are often cut off from jobs and other key social services by the policed fences.

well as the cities of Beirut and Biqa—has been a target for Israeli attacks, as Hezbollah guerillas continue to operate from bases established in Lebanon.

There was some hope that the 21st century would be a more peaceful and prosperous one for Lebanon. The bloody civil war ended in 1990, Israeli troops withdrew in 2000, and Syria withdrew its troops in 2005, marking the end of almost 30 years of occupation. But Israel's attacks in 2006 against Hezbollah helped plunge the country back into open political and economic turmoil. Widespread civilian casualties, the massive destruction of key infrastructure and thousands of homes, and the displacement of approximately 1 million people derailed the fragile peace.

APPLY YOUR KNOWLEDGE Compare and contrast the role that ethnic and religious differences have played in various conflicts of the Middle East and North Africa. How do they complicate the image of this region as unified by one religion (Islam) and one ethnic identity (Arab)?

The U.S. War in Iraq The United States responded to the terrorist attacks of September 11, 2001, by declaring a global war against terrorism. First Afghanistan and then Iraq were identified as the greatest threat to U.S. security. Although the evidence of involvement in the attacks by Iraq and its leader, Saddam Hussein, was highly questionable, on March 19, 2003, after amassing more than 200,000 U.S. troops in the Persian Gulf region, U.S. President George W.

Bush ordered the bombing of the city of Baghdad (**Figure 4.27**). The declaration of war and invasion occurred without the explicit authorization of the U.N. Security Council, and some legal authorities take the view that the action violated the U.N. Charter. Some of the United States's staunchest allies (for example, Germany, France, and Canada) as well as Russia opposed the attack. Throughout the world, hundreds of thousands of antiwar protesters took to the streets for the weeks and months preceding and following the onset of war. The motivation for the war, as expressed by U.K. Prime Minister Tony Blair and President George W. Bush, was that Iraq had stockpiled "weapons of mass destruction"—chemical and biological weapons capable of massive human destruction. In the days leading up to the war, the U.N. weapons inspector, Hans Blix, and his team were unable to locate any weapons despite an intensive search of the country. President Bush and Prime Minister Blair, however, justified the invasion as part of an escalated "war on terrorism" and asserted that "neutralizing" Iraq's leader, Saddam Hussein, was necessary for global security.

On May 1, 2003, President Bush announced the end of major combat operations in the Iraq war. Over the next four years, however, the United States continued to have a very heavy presence in the country. By 2007, the stability of Iraq remained unclear, and President Bush ordered, over Congressional opposition, a surge of 140,000 troops to Iraq with the intention of providing security to the city of Baghdad and Al-Anbar Province. The surge met with some success, particularly in Al-Anbar, where some local communities cooperated with the U.S. Army to fight terrorism. But sectarian violence still occurs on a daily basis throughout the country, especially in Baghdad.

In spring 2009, President Barack Obama set an August 2010 deadline for withdrawal of U.S. combat forces from Iraq. The ambitious goal was not met; the actual pullout was completed almost 18 months later, in December 2011. Even as the United States pulled its troops out, however, U.S. intelligence suggested that the global **jihadist** movement—made up of people who seek to wage war on behalf of Islam against those who oppose the religion—was fueled by the coalition forces' occupation of Iraq. It is possible that these forces may ultimately carry out a cycle of revenge attacks referred to as "blowback," that could threaten the United States and other coalition countries for decades. The economic costs of the war for the United States are still to be determined. The U.S. Treasury estimates the cost of the war at U.S. $845 billion. However, Nobel Prize–winning economist Joseph Stiglitz and coauthor Linda Bilmes argued that the true cost of the war is closer to U.S. $3 trillion when the indirect costs are also assessed, including interest on the debt raised to fund the war, the rising cost of oil, health care costs for returning veterans, and the cost of replacing destroyed military hardware and degraded operational capacity.

Reports estimate 116,000 Iraqi deaths—both civilian and military—since the conflict began in 2003 (**Figure 4.28**). Violence in the country continues despite the U.S. withdrawal. As an example, over a three-month period beginning in 2012 there were an estimated 8,000 deaths. People are continuing to be displaced. The United Nations High Commission for Refugees indicated that as of January 2011, a total of 3.4 million Iraqis were living outside of Iraq, and another 1.8 million were displaced within Iraq. In 2012, the political situation in Iraq remained uncertain. (In one example of this

▲ FIGURE 4.27 Bomb Damage to Hussein's Government Infrastructure A member of the Combined Weapons Effectiveness Team assesses the impact point of a precision-guided bomb that ripped a hole through one of Saddam Hussein's government buildings in 2003.

lack of stability, a warrant was issued for the vice president on alleged terrorism charges.)

Nuclear Tensions In 2008, Iran alarmed the world by declaring its inalienable right to pursue uranium enrichment. In that same year, the United States, Russia, China, the United Kingdom, France, and Germany put forward an offer to deter Iran from continuing its nuclear weapons program while affirming Iran's right to pursue a peaceful nuclear energy program. Iranian President Mahmoud Ahmadinejad refused the offer. This refusal caused tensions to escalate between Iran and Israel, in particular. Both countries made a show of displaying their military capabilities. Iran, for example, tested its advanced missile and satellite technologies.

Iran's nuclear program development precipitated a strong response from many members of the international community, including the United Nations, which does not support Iran having nuclear weapons. As Iran refused to cooperate with the U.N., the United Nations established trade sanctions in an attempt to limit Iran's ability to benefit from its oil reserves. Despite U.N. sanctions, in February 2012, Iran declared that it had developed its own uranium to fuel a nuclear reactor. Iran's first nuclear power plant at Bushehr was reported to be operating at 75% of its full capacity. It is unclear if negotiations will yield a compromise between Iran and those who wish to see the country cease production of material that might potentially be used in a nuclear weapon. It is also unclear if international trade sanctions will be effective, as several countries, including China, continue to trade with Iran and purchase their oil.

APPLY YOUR KNOWLEDGE Briefly explain the role that colonialism has played in modern-day political and ethnic conflict in the Middle East and North Africa region. List three examples of current political and territorial issues that have their roots in colonial history.

Regional Alliances Despite all the conflict in the region, many political, economic, and cultural cooperative organizations are operating in the region today. **The Arab League**, for example, is a voluntary association of Arab states whose peoples speak mainly Arabic. Founded in 1945 by Egypt, Iraq, Lebanon, Saudi Arabia,

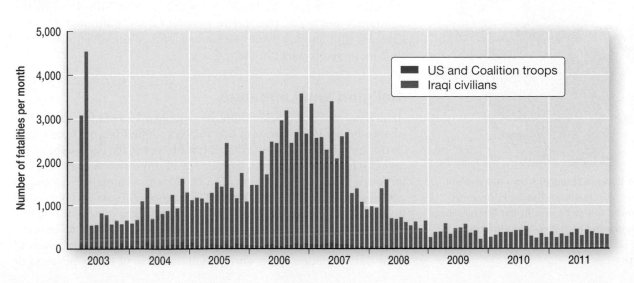

◄ FIGURE 4.28 Fatalities in Iraq War Counting the number of people who have died during this war is always a challenge. In this graph, data for civilian fatalities were compiled by "Iraq Body Count" (www.iraqbodycount.org), which is an independent organization that uses news reports to calculate the total loss of civilian life since the start of Operation Iraqi Freedom in 2003. U.S. and Coalition troop fatalities were gathered from "Iraq Coalition Casualty Count" (icasualties. org), which bases its numbers on published government reports.

Syria, Transjordan (Jordan, as of 1948), and Yemen, the Arab League (formally known as the League of Arab States), is the most unifying of all the Middle Eastern and North African regional organizations. Many other countries joined the League since its inception, including Algeria (1962), Bahrain (1971), Kuwait (1961), Morocco (1958), Oman (1971), Qatar (1971), Sudan (1956), Tunisia (1958), and the United Arab Emirates (1971). The Palestine Liberation Organization (now the Palestinian Authority) was admitted to the League in 1974 (**Figure 4.29**). In 1979, Egypt's membership was suspended after it signed a peace treaty with Israel, but Egypt was readmitted ten years later. The stated purposes of the Arab League are to strengthen ties among member states, coordinate policies, and promote common interests. The league is involved in various economic, cultural, and social programs, including literacy campaigns and programs addressing labor issues. It is also a high-profile political organization that acts as a sounding board on conflicts in the region, such as the 1991 Persian Gulf War and the Arab–Israeli conflict.

Another central and widely known regional organization, this one based on economic interests, is the Organization of the Petroleum Exporting Countries (OPEC) (see "Geographies of Indulgence, Desire, and Addiction: Petroleum," p. 138). Founded in 1960, OPEC has 12 members—Algeria, Angola, Ecuador, Iran, Iraq, Kuwait, Libya, Nigeria, Qatar, Saudi Arabia, United Arab Emirates, and Venezuela— four of which (Angola, Ecuador, Nigeria, Venezuela) are not part of the Middle Eastern and North African region. As is clear from the list, Middle Eastern Arab states dominate the membership.

Complementing as well as contrasting with the goals and objectives of OPEC is the **Gulf Cooperation Council (GCC)**. The GCC coordinates political, economic, and cultural issues of concern to its six member states—Saudi Arabia, Kuwait, Bahrain, Qatar, the United Arab Emirates, and Oman. The members of the GCC joined to coordinate the management of their substantial income from oil reserves and address problems of economic development as well as discuss social, trade, and security issues. All six of the states in the GCC are politically conservative monarchies wary of the revolutionary urges that swept the region during the Arab Spring movements and have transformed the monarchies of Egypt, Iran, and Iraq. The GCC has made very large sums of money available to all Arab countries for economic development as well as military protection during political crises.

Many other regional and international organizations have been established in the Middle East and North Africa, and new regional alliances are being proposed. The Arab Common Market is one such proposed alliance that has strong support in the region. Additionally, some individual states have made connections with organizations beyond the region, tying the region more securely to the rest of the globe. For instance, Turkey, already a member of NATO, applied for full membership in the European Union (see Chapter 2, p. 65), despite significant political barriers. Morocco, Tunisia, Jordan, and Israel signed agreements with the European Union that are leading to increased transnational integration beyond the region.

CULTURE AND POPULATIONS

In the Middle East and North Africa, issues of religion are uniquely tied to daily life as well as to formal political structures. In many of the region's countries, religion and politics are conjoined in states where their government is based on **Shari'a**. In these countries (Saudi Arabia, Iran, Oman, and Yemen), Islamic canonical law is the ideological foundation of the political institutions. Further, all the other countries of the region except for Israel, Syria, Turkey, and Sudan (which are either undeclared or **secular**—nonreligious—states) identify Islam as their state religion. It would be a mistake, however, to assume that the strong presence of Islam in this region translates into homogenous culture and politics. For example, although Saudi Arabia is an Islamic state, it is also an absolute monarchy in contrast to Yemen, which is a presidential republic with a two-chambered legislature (**Figure 4.30**).

The demography and distribution of populations in the region are the result of long-term historical processes as well as current environmental, political, and socioeconomic conditions. Demographic and urbanization patterns in the region were affected historically by trade (from the ancient Silk Road to the container port facilities in the Persian Gulf), conquest and war (the Muslim and Christian Crusades), and colonialism (and other European involvement in the region).

Religion and Language

If a reader were to quickly skim the news in a U.S. daily paper, he or she would get the impression that Iran is a unified Islamic state where everyone adheres strictly to Shari'a law. However, the geography of religion is not uniform in Iran. Similar significant differences in religious practice occur across the region at varying scales. For instance, although Iran is widely recognized as a state passionately committed to Islam, in Tehran, the capital city of more than 7 million inhabitants, many upper- and middle-class households are likely to be secular and more aligned with Western values than with the teachings of Islam. Iran is not unique; although about 96% of the population of

▲ **FIGURE 4.29 Protests at Arab League Headquarters in London, 1978** Members of the General Union of Palestinian Students organization participate in a "sit in" to demand a full-scale war in the Middle East. The protests came at a time when the Arab League was beginning to fracture; Egypt's membership in the League had been suspended after Egypt signed a peace agreement with Israel.

▲ **FIGURE 4.30 King of Saudi Arabia Meets with German Foreign Minister, 2012** The alliance between the Saudi Arabian monarchy and European and North American powers remains strong, despite the fact that Saudi Arabia has few democratic institutions. Saudi Arabia is a major global political player because of its vast oil wealth.

the Middle East and North Africa is Islamic, the way people practice Islam in the region is quite varied (**Figure 4.31**). And, naturally, other religions exist in the region. Christianity makes up about 3% of the population and Judaism ranks third, with less than 1% of the total population. There are also other religions with much smaller numbers of adherents. What this suggests is that the Middle East and North Africa is infused with especially active religious practices and belief systems. And, unlike in many parts of the globe where societies have become increasingly secular, religion in this complicated region is a central feature of everyday life for the vast majority of the inhabitants.

Islam **Islam** is an Arabic term that means "submission to God's will." As a religion founded primarily on the revelations of God to humankind, Islam recognizes the prophets of the Hebrew and Christian Testaments of the Bible, but considers Muhammad to be the last prophet and God's messenger on Earth. **Muslims** hold that Muhammed received the word of God from the Angel Gabriel in about 610 C.E.

There are two fundamental sources of Islamic doctrine and practice: the **Qur'an** and the **Sunna**. Muslims regard the contents of the Qur'an, the Islamic sacred book, as directly spoken by God to Muhammad. The Sunna is not a written document but a set of practical guidelines for behavior, the body of traditions derived from the words and actions of the prophet Muhammad. Islam holds that God has four fundamental functions—creation, sustenance, guidance, and judgment—and the purpose of people is to serve God by worshipping him alone and adhering to an ethical social order. In Islam, it is held that the actions of the individual should serve the ultimate benefit

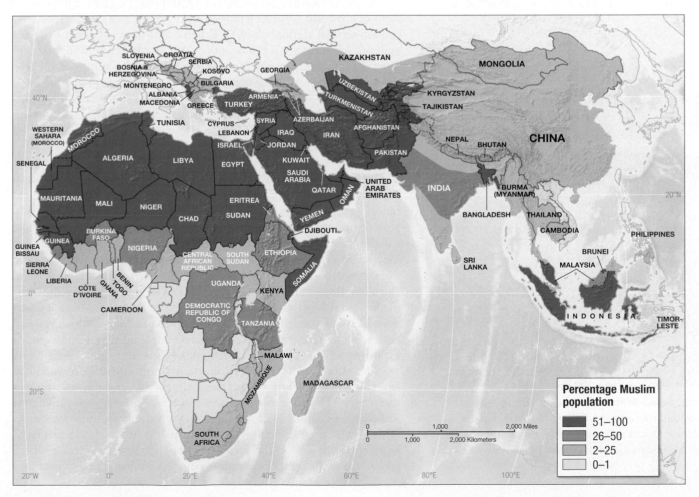

◀ **FIGURE 4.31 The Islamic World** The distribution of Islam in Africa, Southeast Asia, and South Asia testifies to the broad reach of Muslim cultural, colonial, and trade activities.

Percentage Muslim population
- 51–100
- 26–50
- 2–25
- 0–1

of humanity, not the immediate pleasures or ambitions of the self. A Muslim must fulfill five primary obligations, known as the five pillars of Islam: repeating the profession of the faith ("There is no god but God; Muhammad is the messenger of God"); praying five times a day facing **Mecca** (the city where Muhammed was born in 570 C.E.); giving alms (charitable giving); fasting from sunup until sundown during the holy month of Ramadan; and making at least one pilgrimage, called a **hajj**, to Mecca if financially and physically able (**Figure 4.32**). Islam's simple moral and ethical code, like Christianity's Ten Commandments, makes it a universalizing religion to its almost 1 billion adherents across the globe today.

The geographical origin of Islam is Mecca, in present-day Saudi Arabia. When Islam first emerged, Mecca was a node in the trade routes that at first connected Yemen and Syria and eventually linked the region to Europe and all of Asia. Because of Mecca's central location, Islam spread quickly through the interconnected trade networks of the region. Today, Mecca is still the most important sacred city in the Islamic world; it also continues to be a commercial center. Eventually, Medina also became a sacred city because Muhammad fled there when he was driven out of Mecca by angry merchants who felt his religious beliefs were a threat to their commercial practices.

Disagreement over the line of succession from the prophet Muhammad occurred shortly after his death in 632 C.E. and resulted in the split of Islam into two main sects, the Sunni and the Shi'a (sometimes known as Shi'ite). The central difference between these two sects revolves around the question of who should hold the *political* leadership of the Islamic community and what the *religious* dimensions of the leadership should be. The Shi'a contend that political leadership must be divine and must therefore derive from descendants

of the Prophet. The Sunni faction, which argues that the clergy (with no divine power) should succeed Muhammad, gained the upper hand and became dominant. As a result, Islam is practiced differently in many parts of the region.

The long-term global networks of Islam allowed Muslims to migrate out of the region—to Europe and the United States, for instance. This movement resulted in new ideas for how Islam might be practiced, and these ideas have, in turn, helped shape the practice of Islam globally. At the same time, Islam has been a counterforce to globalization. This is seen in its extreme form in the rise of **Islamism**, which is more popularly, although incorrectly, referred to as Islamic fundamentalism. Whereas *fundamentalism* is a general term that describes the desire to return to strict adherence to the fundamentals of a religious system, Islamism is an anti-Western and anti-imperial political movement. Islamists resist Western forces of globalization—namely modernization and secularization. But most Muslims are *not* Islamists; Islamism is the most militant movement within Islam today.

The basic intent of Islamism is to create a model of society that protects the purity and centrality of Islamic precepts through the return to a universal Islamic state—a state that is religiously and politically unified by incorporating principles from the sacred law of Islam into state constitutions. Most Islamists object to secularization and view the popularity of Western ideas as a move away from religion to a more secular society. Another aspect of the Islamist movement is the concept of **jihad**, a complex term derived from the Arabic root meaning "to strive." Current use of the term connotes both an inward spiritual struggle to attain perfect faith as well as an outward material struggle to promote justice and the Islamic social system. **Quital** (fighting or warfare) is one form of jihad and, according to the Qur'an, is a war of conquest or conversion against all nonbelievers. When a war directed against the enemies of Islam occurs, it can be interpreted as a holy war. But, for many, jihad is the peaceful struggle to establish Islam as a universal religion through the conversion of nonbelievers. One example of jihad today is the struggle of Shi'a Muslims for social, political, and economic rights within Sunni-dominated Islamic states. This is one of the core tensions that animates the politics of Iraq today, where a Sunni minority ruled for several decades at the expense of Shi'ite people.

Christianity, Judaism, and Other Middle Eastern and North African Religions Although Islam is the most widely practiced religion in the region, as noted earlier, it is by no means the only religion of political, cultural, or social significance. There are more than a dozen Christian sects—among them Coptic Christians in Egypt, Maronites in Lebanon, the Chaldean Catholic Church in Syria, and various orthodox affiliations, including Armenian, Greek, Ethiopian, and some Protestant faiths (**Figure 4.33**). Generally, Jews in the Middle East and North Africa are secular, observing some Jewish traditions. A small percentage practice **Orthodox Judaism**, which is based on a strict adherence to the religious texts of the Old Testament, or **Hasidim**, which is a mystical offshoot of Judaism.

The three regionally predominant religions—Islam, Judaism, and Christianity—have helped shape the people and the landscape of the region. The most obvious and enduring influences on the landscape have been places of worship and sacred spaces, more generally. Nowhere is the enduring interrelationship of the three religions more apparent than in the ancient city of Jerusalem. The centrality of Jerusalem as an ancient religious space, as well as its contemporary

▲ **FIGURE 4.32 Mecca, Saudi Arabia** Every year, during the last month of the Islamic calendar, more than 1 million Muslims make a pilgrimage, or hajj, to Mecca. In addition to the required pilgrimage, Islamic traditions require Muslims around the world to face Mecca during their daily prayers.

▲ **FIGURE 4.33 Coptic Church of St. George (Mari Girgis) in Cairo, Egypt** The Coptic Church serves between 6 and 11 million Egyptians today. The Church was established around 50 B.C.E. when the apostle Mark came to Egypt. Over its history, the Church has played an important role. It established one of the earliest schools of the Christian religion and was one of the founding members of the World Council on Churches. The Church is led by the Pope of Alexandria. Its membership includes not just Egyptians but about 1 million followers outside of Egypt.

significance as a place of pilgrimage for Jews and Christians, is very much tied up with Arab–Israeli conflicts.

APPLY YOUR KNOWLEDGE Briefly compare and contrast the different religious and social systems found in this region. How do these systems unify and divide the region socially? In what ways has the region's commercial history affected its current religious and political makeup?

Regional Languages Three major language families dominate in the Middle East: Semitic (including Arabic, Hebrew, and Aramaic); Indo-European (Kurdish, Persian, and Armenian); and Turkic (Turkish and Azeri). The regional language diversity is the result of successive migrations of different people. These different languages have influenced each other. For example, Persian is written in Arabic script, and Turkish incorporates vocabulary words from Persian and Arabic. All three are spoken in regional dialectics that are not always mutually understood. In addition to the prominent languages, some ethnic and religious communities have preserved their own languages for religious use, such as Coptic and Greek. Berber, an Afro-Asiatic language, is spoken in the region from Morocco to Egypt. Nubian, a Nilo-Saharan language, is spoken by Egyptians and Sudanese. And there are nearly 1 million Zazaki (an Indo-European language) speakers in southern Turkey. The most common foreign language in the Middle East is English. In North Africa, the colonial languages of French (Morocco, Algeria, and Tunisia), Spanish (Western Sahara), and Italian (Libya) are also spoken, as is English.

Cultural Practices, Social Differences, and Identity

The historical social issues in the Middle East and North Africa are as complex as that of any other region of the globe. As global media technologies such as satellite television and the Internet increasingly penetrate the region, the potential for new social and cultural forms to emerge and old ones to be reconfigured is increasing. The resulting cultural practices and social issues in the Middle East and North Africa reflect a continually changing mix of both regional and global influences.

Cultural Practice The cultures of the Middle East and North Africa have transformed the landscapes of many other parts of the world. For example, the cuisine of the eastern Mediterranean, especially that of Lebanon and Syria, as well as of Turkey, Egypt, Iran, Morocco, Tunisia, and Algeria, is available in many large cities in most of the world's regions. In the United States and Europe, it is also often available in smaller urban places; here, many young men and women from the Middle East and North Africa are studying at universities, and cafés offering strongly brewed coffee and regional cuisine have sprung up to serve them.

The types of food available from this region include *meze* dishes, predominantly subtly spiced appetizers or small plates. The range of mezes reflects the tastes and ingredients of a particular country or subregion within that country. Some of the most popular meze dishes include *baba ganoush*, a puree of toasted eggplant, sesame seeds, and garlic; *falafel*, a mixture of spicy chickpeas rolled into balls that are deep fried; *fuul*, brown broad beans seasoned with olive oil, lemon juice, and garlic; and *tabbouleh*, a salad of bulgur wheat, parsley, mint, tomato, and onion. Other regional specialties include grilled meats, especially lamb and chicken, often served with rice.

Other important contributions of the Middle East and North Africa to the world cultural scene have been made in the realms of dance, music, film, and architecture. The most widespread of the region's dances is the traditional belly dance, a women's erotic solo dance done for entertainment. The dance is characterized by undulating movements of the abdomen and hips and by graceful arm movements. Belly dancing is believed to have originated in medieval Islamic culture, though some theories link it to prehistoric religious fertility rites. Middle Eastern and North African music has become a staple of the contemporary world music scene also.

Film industries in the region are also finding their voice, producing some of the most provocative and engaging cinema in the world today. In 2012, the film *A Separation*, directed by Asghar Farhadi, won the Academy Award for Best Foreign Language Film (**Figure 4.34**, p. 162). The film traces the conflict that arises when a married woman desires to leave Iran and her husband does not. At the heart of their conflict is the need to care for an aging parent in Iran versus the desire to find better opportunities for their child outside Iran. The film illustrates the ongoing tensions and anxieties found in everyday Iranian society, as people struggle to find their voice in a society with a highly regulated cultural production system. It is clear that the film's message resonates with people well beyond Iran and the region. Other Iranian films are also garnering global attention while undergoing increasing scrutiny in their home country. In 2010, for example, two well-known Iranian directors—Mohammad Rasoulof and Jafar Panahi—were arrested for their work. In 2012, the Iranian government closed down Iran's House of Cinema, a 20-year old

▲ **FIGURE 4.34 Asghar Farhadi** In 2012, *Time Magazine* named Asghar Farhadi one of the world's most influential people. After graduating from Tehran and Tarbiat Modarres Universities in Iran with degrees in Theatre, he has become one of the most well-reviewed directors active in cinema today. His films are known not for their overt political criticism, but for sensitively documenting the human experience in Iran. As a result, while other directors have faced government repression for their work, Asghar Farhadi has thus far avoided government sanction.

independent organization that promotes Iranian independent filmmaking. Despite the crackdown within Iran, the Iranian film industry continues to thrive and produce art that gains global attention.

While film is one of the more popular forms of Middle Eastern and North African global cultural expressions, mosques may be the most prevalent evidence of the globalization of Middle Eastern and North African culture in the world today. Mosques serve as the main place of worship for Muslims but also serve many social and political needs. Mosques function as law courts, schools, and assembly halls; adjoining chambers often house libraries, hospitals, or treasuries. In the post-9/11 period, mosques have also been sites of conflict and tension. Recently, a proposal to place a mosque in downtown New York City met with strong resistance by some who viewed the placement of the mosque near the site of the former Twin Towers to be inappropriate. But mosques, like all religious spaces, provide important social services to their communities and play an important

role in meeting the needs of people who may not always find support from local or national governments.

Kinship, Family, and Social Order To understand societies in this region, it is important to understand concepts of kinship, family, and other personal relationships. **Kinship** is typically defined as a relationship based on blood, marriage, or adoption. However, in the Middle East and North Africa, the meaning of kinship is often expanded to include a shared notion of a relationship among members of a group; neighbors, friends, and even individuals with common economic or political interests can be considered kin. Kinship also often determines how people interact with each other. In some societies in the region, the relationships between men and women are determined by kinship identity. It is important to remember, of course, that at the local level, notions of kinship vary. Who is considered kin and what that means depends on local practice.

The concept of **tribe** is also central to the sociopolitical organization of the region, although the term *tribe* has not always been used appropriately by those outside the region. For instance, colonizers negatively associated the concept of tribe with primitive social organization. Generally speaking, a tribe is a form of social identity created by group members who share a common set of ideas about loyalty and political action. Tribes may be grounded in shared social, political, and/or cultural identities. A shared tribal identity leads to the formation of collective loyalties and a primary allegiance to the tribe over other social groups—nations, for example. Many pastoral communities in the Middle East and North Africa, who subsist on the breeding and herding of animals, are proudly tribal. The strong sense of tribal identity is linked to their experiences of living in mostly rural areas and among the closely knit communities of either sedentary or nomadic pastoral peoples. While most pastoralists practice **transhumance**, the movement of herds according to seasonal rhythms, others live in sedentary communities, where they herd close to home (**Figure 4.35**).

In most places in the region where the concept of tribe is adopted, it is not seen as a primitive form of social organization but rather a valuable element of local society. At the same time, tribes can be pressured to exert a stronger sense of national identity. For example, a state government might seek to destabilize tribal leadership in order to increase control over a particular people or community. Tribal communities also exist at border regions, where tribes have historical connections strongly aligned with people in another country. In this case, tribal loyalties might destabilize national governments and serve as spaces for transnational connection at the expense of national unity. This may very well be the case in some areas that remain unstable in the Middle East and North Africa, such as the Kurdish region along the Turkey, Iraq, Iran, and Syria borders and in the border region of Afghanistan and Pakistan.

Gender Many people harbor stereotypes about the restricted lives of the Middle Eastern and North African women who adhere to rigid Islamic traditions in this region. In reality, there is no single Islamic, Christian, or Jewish notion of gender in the region. Stereotypes of gender relations in the Middle East and North Africa do not capture the variation of gendered identities across lines of class, religion, generation, level of education, and geography (urban versus rural origins, for instance). As is true in many societies throughout the world, the ideological assumption that women should be subordinate to men is held by both men and women in many Middle Eastern and North African communities. Interestingly, it seems that many men in the region regard women's subordination as something natural, something that is effectively determined by biology. In contrast, most women in the region tend to regard their subordination as something that is the product of the society in which they live and operate and therefore something that can be negotiated and manipulated. The difference in how men and women view gender in the Middle East and North Africa is thus based on ideas that similarly undergird women's subordination in other parts of the world. What makes the Middle East and North Africa somewhat different is how control over women is exercised by restricting female access to public space and secluding women within private spaces (**Figure 4.36**).

National politics related to gender also affect how men and women see and represent themselves publicly in the region. Some societies, such as in Yemen and Bahrain, exercise very strict control over women's public movements: women are expected to cover themselves with veils and long, dark clothing when out on the streets. In some of the more secular societies of the region, such as Turkey and Egypt, women's public movements are less strictly regulated. In some countries, women's rights have recently been expanded (**Table 4.2**, pp. 164–165). The veil—from the all-encompassing full body garment, known as a **chador**, to a simpler head covering—is the means by which some women are able to effectively operate in public and yet remain in their personal space. Body coverings actually allow women physical and social mobility in societies that otherwise want to restrict women's movements to very private (often home) spaces.

Even as national policies regulate women's mobility, urban and rural practices often differ, as do practices between classes within urban areas. Generally, urban women's public movements tend to be

▲ **FIGURE 4.36 Gendered Architecture** Islamic architecture reflects gender differences within the culture. In different places these differences can be either strictly or more loosely observed. A classic aspect of Islamic architecture is the screen placed across windows in the women's parts of houses and in the interiors of some public buildings. In 1799, the Hawa Mahal in Jaipur, India, was designed to allow the women of the court to watch the activities in the street from behind stone-carved screens without being seen. Although the building is not located in the Middle East and North Africa, this regional architectural feature was incorporated to the Hawa Mahal because of the common practices among Hindus and Muslims as well as the significant Muslim population in Jaipur.

▼ **FIGURE 4.35 Berber Shepherd Campsite** The Berbers have lived in North Africa for thousands of years. *Berber* is the name applied to the language and people belonging to many of the tribes who currently inhabit large sections of North Africa. Pictured here is a Berber shepherd campsite in the High Atlas Mountains, where great herds of sheep are tended. Increasing numbers of Berbers are raising crops, a practice that signals the erosion of their nomadic practices.

TABLE 4.2 Measuring Gender Inequality and Women's Rights in the Middle East and North Africa

Country	Gender Inequality Index (world rank), 2011[1]	Seats in parliament (%, women)	Population with at least a secondary education, age 25 and older, W%/M%	Recent advances in women's rights[2]	Continuing challenges for women[3]
Algeria	0.412 (83rd)	7.0	36.3/49.3	Women can transfer their citizenship to their children and foreign husbands.	Men can divorce their wives without cause; women do not have the same right.
Bahrain	0.288 (42nd)	15.0	74.4/80.4	Sunni women have rights pertaining to marriage, divorce, guardianship, child custody, and inheritance.	Shi'a women do not have the same rights as Sunni women; marital rape of women is not a crime.
Egypt	No data	2.4[4]	43.4/59.3	Children of Egyptian mothers and foreign fathers can now have Egyptian citizenship.	A recent Political Rights Law canceled the previously established quota of 64 women in the People's Assembly, reducing the number of women who will likely serve in government.
Iran	0.485 (88th)	2.8	39.0/57.2	Women were visible participants in the recent political uprising that took place in the country.	Women are discriminated against in the areas of marriage, divorce, inheritance, and child custody; women must have male guardians' approval to marry; women must obtain men's permission to travel outside the country.
Iraq	0.579 (132nd)	25.2	22.0/42.7	Kurdistan's autonomous parliament in the north of the country passed the Family Violence Bill, which criminalizes, among other things, female genital mutilation.	The Iraqi parliament has reversed laws that gave women equal rights to men in the areas to acquire and change their nationality or pass on their citizenship to their children; everyday violence against women and girls remains a concern.
Israel	0.145 (22nd)	19.2	78.9/77.2	Women have full access to political participation.	Ultra-Orthodox Jews continue to segregate men and women; recent reports suggest that discrimination against women and girls is becoming a larger public issue as a result of Orthodox religious practices.
Jordan	0.456 (95th)	12.2	57.1/74.2	A special court to adjudicate "honor crimes" (violence against a family member) was formed in 2009.	Muslim women cannot marry non-Muslim men; non-Muslim women forfeit their rights to their children when the children turn seven; a recent constitutional change protects rights on race, language, and religion, but not gender.
Kuwait	0.229 (37th)	7.7	52.2/43.9	Kuwaiti women were granted voting rights in 2005; women were elected to parliament for the first time in 2009.	Kuwaiti women cannot marry non-Kuwaiti men; Kuwaiti family law for both Sunni and Shi'a women is adjudicated through Islamic law; women cannot pursue civil law cases against men.
Lebanon	0.440 (76th)	3.0	32.4/33.3	A 2010 bill was sent to parliament that would protect women against domestic violence (it has still not passed as of 2012); women's rights organizations have increased their efforts to encourage reforms in recent years.	Women continue to face gender-based injustices, such as the inability to pass citizenship on to their children and a penal code provision that offers reduced sentences for perpetrators of "honor crimes" (violence against a family member).
Libya	0.314 (51st)	7.7[5]	55.6/44.0	Women have had the right to vote since 1964 and enjoy legal protections in many areas, including property ownership.	Violence against women during the revolution was documented, including gang rape of women by Qaddafi-backed government forces. It is unclear how changes in the government as a result of the recent uprising will impact the status of women.
Morocco	0.510 (130th)	6.7	20.1/48.1	The new constitution (2012) guarantees equality for women; family law reforms from 2004 increased age of marriage and gave women more rights in divorce and custody.	Family law codes discriminate against women in the areas of inheritance and men's rights to repudiate their wives; a 2009 survey found that 55% of women between 18 and 64 had experienced some form of domestic violence.

[1]The Gender Inequality Index measures inequality in three areas: reproductive health, empowerment, and labor market accessibility (the ability of women to find employment). A score of zero (0) means total equality in these areas for men and women, while a score of one (1) means that women fare poorly in all these areas. Source: http://hdr.undp.org/en/statistics/gii/ (accessed May 14, 2012).

[2]Human Rights Watch, 2012, http://www.hrw.org/world-report-2012/world-report-2012-tunisia; Sanja Kelly, "Hard-Won Progress and Long Road Ahead: Women's Rights in the Middle East and North Africa." Freedom House, 2009, http://www.freedomhouse.org/sites/default/files/270.pdf; *Women's Rights in the Middle East and North Africa*, 2010 Edition. Freedom House, http://old.freedomhouse.org/template.cfm?page=444 (accessed May 14, 2012).

[3]Human Rights Watch, 2012 (http://www.hrw.org/world-report-2012/world-report-2012-tunisia); Sanja, Kelly, "Hard-Won Progress and a Long Road Ahead: Women's Rights in the Middle East and North Africa." Freedom House, 2009. http://www.freedomhouse.org/sites/default/files/270.pdf; *Women's Rights in the Middle East and North Africa*, 2010 Edition. Freedom House, http://old.freedomhouse.org/template.cfm?page=444 (accessed May 14, 2012).

[4]http://egyptelections.carnegieendowment.org/2012/01/25/results-of-egypt's-people's-assembly-elections (accessed May 14, 2012).

[5]These data were gathered before the establishment of a transitional government in Libya.

TABLE 4.2 (Continued)					
Country	Gender Inequality Index (world rank), 2011	Seats in parliament (%, women)	Population with at least a secondary education, age 25 and older, W%/M%	Recent advances in women's rights	Continuing challenges for women
Oman	0.309 (49th)	9.0	26.7/28.1	Oman's "Basic Law" bans discrimination on the basis of gender; women are increasingly participating in politics by voting and running for office.	Oman still adjudicates family matters through Islamic courts, which tend to discriminate against women in areas of divorce, inheritance, and child custody.
Palestine[6]	No data	No data	36.5/29.0[7]	Electoral laws were amended in 2005 to ensure greater political participation for women; women are extremely active in their communities and in civil society.	Internal political tensions coupled with Israeli military restrictions have seriously affected the health, employment opportunities, access to education, and political and civil liberties of Palestinian women.
Qatar	0.549 (111th)	0	93.2/96.7	Women have voting rights in municipal elections; Qatar's first female judge was appointed in 2010.	Family and personal law is adjudicated in Islamic courts, limiting women's rights; Qatar publishes no data on domestic violence.
Saudi Arabia	0.646 (135th)	0	50.3/57.9	Women gain right to vote in 2015 municipal elections; demonstrations have been successful in loosening cultural restrictions that ban women from driving (there is no national law against women driving); women no longer need a male representative to conduct government business.	Women are treated as minors under Saudi law; women cannot travel, study, or work without the permission of male guardians.
Sudan[8]	0.611 (169th)	24.2	12.8/18.2	Sudan is the lowest ranked country in the region at 169 of the total 187 countries surveyed; it is unclear if the creation of South Sudan will advance women's rights in either country.	Violence in the country against women is still rampant, particular sexual violence against internally displaced women in the Darfur region.
Syria	0.474 (86th)	25.0	24.7/24.1	The constitution guarantees gender equality; Syrian women have some of the highest rates of participation in government in the region.	Personal law and penal code discriminate against women and girls, particularly in terms of marriage, divorce, child custody, and inheritance; judges have the right to rule that "honor crimes" (violence against a family member) have been committed with "honorable intent."
Tunisia	0.293 (45th)	18.1[9]	33.5/48.0	"Convention on the Elimination of All Forms of Discrimination Against Women" was adopted in August 2011.	Discrimination exists in the areas of inheritance and child custody; rural women are less likely to understand their right to equal treatment under the law.
Turkey	0.443 (77th)	9.1	27.1/46.7	In May 2012, Turkey signed the Council of Europe Convention Against Domestic Violence and Violence Against Women.	Domestic violence rates remain high; spousal and familial killing of women rose in 2011; the courts fail to fully protect women from violence in the home, which remains endemic.
U.A.E.	0.234 (38th)	22.5	76.9/77.3	First female judge appointed in 2008; women can ask for divorce.	Family and personal law is adjudicated through Islamic courts; Emirati men can have up to four polygamous marriages while women cannot; Emirati women can have a no-fault divorce, but they lose their financial rights in the process.
Yemen	0.769 (146th)	0.7	19.4/64.8	Women recently participated in political protests against the government.	Domestic abuse, deprivation of education, early or forced marriage, restrictions on freedom of movement, exclusion from decision-making roles and processes, denial of inheritance, deprivation of health services, and female genital mutilation continue to plague Yemeni women.
Arab States[10]	0.563	12.0	32.9/46.2	N/A	N/A
Middle East and North Africa	0.431	10.88	43.45/52.4	N/A	N/A
World (187 countries)	0.492	17.7	50.8/61.7	N/A	N/A

[6]These data are for the "Occupied Palestinian Territory."
[7]These data come from UNESCO Institute for Statistics (2011) and refer to an earlier year than that specified.
[8]These data are for Sudan before the division of the country between Sudan and South Sudan.
[9]This number represents the most recent election in 2012, in which 49 women were elected to a possible 217 parliamentary seats. Source: http://www.hrw.org/world-report-2012/world-report-2012-tunisia.
[10]The Arab States include the following countries: Algeria, Bahrain, Djibouti, Egypt, Iraq, Jordan, Kuwait, Lebanon, Libya, Morocco, Occupied Palestinian Territory, Oman, Qatar, Saudi Arabia, Somalia, Sudan, Syrian Arab Republic, Tunisia, United Arab Emirates, and Yemen. Note that not all of these countries are in the Middle East and North Africa.

more restricted than those of rural women. This is largely because rural village life is typically defined by kinship relations, and women are allowed to operate relatively more freely among kin. In contrast, women in urban areas must move about in a world of both kin and strangers. This means that they must remain covered more often and spend less time in public areas. Social reality is not always fixed, however, and cultural assumptions and practices around gender are subject to negotiation and change. Although the predominant gender theme in the Middle East and North Africa is that women are subordinate to men, women can and do exercise significant household as well as political influence and independence across a range of societies in this region. Examples of this include women's control over family resources and their participation in national politics in the region.

Demography and Urbanization

For thousands of years, populations have moved within the region, at times as refugees, at other times voluntarily. Emigrants have also moved out of the region, settling all over the world and shaping the landscapes of cities and regions everywhere. Generally speaking, migration into and out of the region since the end of the colonial period has largely been related to factors that have pulled some immigrants to the region and pushed others to leave. Internal regional and national migration has also been significant and almost always related to the draw of urban economic opportunity.

The distinctive pattern of population distribution in the Middle East and North Africa also reflects the influences of environment, history, and culture. As noted earlier in the chapter, the mountains provide homes for many people in the region. Though mountain environments can present substantial challenges to human habitation, the availability of moisture means that these environments can support agriculture over a somewhat shortened growing season. Historically, the mountains provided safe havens for minority populations fleeing persecution and discrimination. The Druze in Syria and the Zayidis in Yemen are two groups that sought mountain refuge from their oppressors.

Whereas the highland areas are home to a small portion of the people of the region, the coastal areas, floodplains, and plateaus are the most densely populated landscapes of the region (**Figure 4.37**). The clustering of populations in these landscapes is hardly surprising, given that they are the ones where water is most abundant and environments the least harsh. The Iranian and Anatolian Plateaus are the most obvious examples of these landforms, but the highland plateau of Yemen also contains a sizable population. The coastal areas and floodplains, excluding the coastal areas of the Persian Gulf, are equally attractive to human habitation and constitute some of the most remarkable, highly engineered, and scenic of the region's landscapes. Other population clusters are found in and around the region's cities, which have been well established for centuries but have grown especially rapidly since the independence period of the 1950s. Even though the Middle East and North Africa is more urbanized than is popularly assumed, many of the people of the region still live in rural villages.

The total population of the 21 countries that make up the Middle East and North Africa is over 500 million, but given the inaccuracy, infrequency, and inconsistency of national censuses in the region, this number is only an approximation. Although most states have recently conducted censuses considered by population experts to be accurate and reliable, others, such as Turkey, tend to underestimate their minority populations (the Kurds in particular) for political reasons.

Recent U.N. data indicates that the decades-long population boom in the region appears to be coming to an end. Although this means that fertility rates have fallen and are now stabilizing, there are still challenges in providing for the health, education, and welfare of both children and the elderly as well as providing jobs for those in between. For instance, in several of the most populated countries in the region, including Egypt, Algeria, and Turkey, a large part of the population is younger than 15 years of age; the populations of these countries will continue to grow as these individuals reach reproductive age.

The global economy has increased the levels of inequality in the region, and other forces limit the life chances and standards of living of the region's population. For example, refugees and many migrant workers face very difficult economic circumstances. Since the first refugees left Palestine in 1948, millions of Palestinians have been displaced to refugee camps in Lebanon, Syria, Jordan, the West Bank, and the Gaza Strip. Many people are born and grow up in these camps—originally intended as temporary settlements—where basic provisions are poor. Other large refugee populations live in camps in Iran, which shelters over 1 million Afghan refugees and nearly 41,000 Iraqis as well as recent refugees out of Syria (**Figure 4.38**). Within Iraq there is also a significant problem associated with internally displaced persons (IDPs); it is estimated that over 1.3 million Iraqi citizens are living outside their homes. The plight of IDPs is actually worse than refugees because the IDPs' own governments are either unable or unwilling to provide them with the protection or assistance they have a right to expect. (See the discussion of ethnic conflicts earlier in the chapter.)

Pull Factors A number of so-called "pull" factors have drawn people to this region. For example, the founding of the state of Israel at midcentury drew large numbers of European Jews, particularly during and after World War II. Russian Jews arrived in Israel as a result of the end of the Cold War. More recently, Ethiopian Jews have migrated to Israel because of civil war. The strongest force pulling migrants to the Middle East and North Africa in the last 50 years, however, has been the oil economy. In the states of the Arabian Peninsula, several factors—small local populations, lack of skill, lack of interest, and cultural resistance to the kinds of jobs made available through the oil economy—have meant that workers had to be imported as **guest workers**. Guest workers have been brought in to work in all aspects of oil production, from exploration and well development to drilling, refining, and shipping (**Figure 4.39**, p. 168). And because oil revenues are increasingly reinvested in economic development projects in the region, even more regional jobs are being created outside the petroleum industry, ranging from service-sector positions to jobs in the building and construction industry.

To lessen the potentially dislocating impact of foreign workers on local social and cultural systems, immigration policy among the oil-producing states of the Arabian Peninsula favor

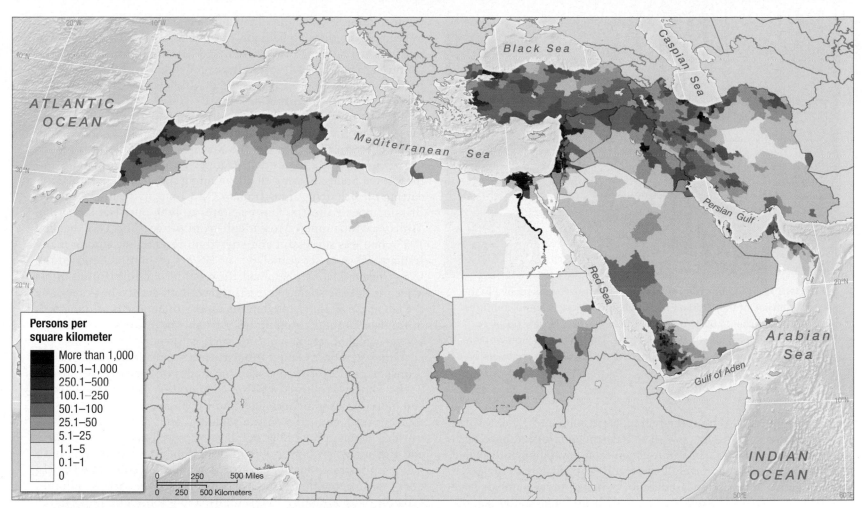

Persons per
square kilometer

More than 1,000
500.1–1,000
250.1–500
100.1–250
50.1–100
25.1–50
5.1–25
1.1–5
0.1–1
0

▲ **FIGURE 4.37 Population distribution in the Middle East and North Africa, 2012** Population distribution is heavily influenced by the availability of water. The intensity and unusual linear pattern of population concentration along the Nile River valley is a perfect illustration of this point. Other concentrations, such as along the eastern end of the Mediterranean Sea, the northern edge of Algeria, and the northern parts of Iran, Iraq, and Turkey, also reflect higher availability of freshwater.

◄ **FIGURE 4.38 Syrian Refugees in Turkey** Syrian refugees arrive at a camp in Islahiye, Turkey, about 100 km (60 miles) north of the Syrian border on April 6, 2012. The Turkish government reported that over 2,300 refugees arrived at a number of camps along the border within this 24-hour period.

▲ **FIGURE 4.39 Labor Migrants** Many immigrant workers live in the oil-rich countries of the Middle East and North Africa. In some countries, these workers make up 80% to 90% of the workforce, largely because there are so few local people to fill the jobs. These construction workers from India are at the site of an apartment complex of 40 skyscrapers on the outskirts of Dubai, U.A.E.

Muslim applicants. Within the region, large numbers of guest workers from Syria and Egypt, as well as Palestinian refugees, have been participants in the Arabian Peninsula oil economy, filling both skilled and unskilled positions. Overall guest workers constitute a rich global mix of nationalities. In the last decade or so, significant numbers have come from outside the region, especially from India, Indonesia, the Philippines, and Pakistan. Most of the workers who have come from other Middle Eastern and North African countries have been male; labor migrants from Indonesia and the Philippines have been female (see Chapter 10, p. 382).

Economic growth outside the oil economy of the region has also fostered migration to places like Beirut in Lebanon, Cairo in Egypt, and Istanbul in Turkey. These cities have been seen as important to the wider regional economy, although recent events in Egypt and Lebanon may be reducing that trend. Many sub-Saharan Africans migrated from their countries of origin using North Africa as a point of transit into Europe. These migrants are often relatively well educated and from moderate socioeconomic backgrounds. They move because of a general lack of opportunities, fear of persecution and violence, or a combination of both. Those who are unable to cross into Europe join growing immigrant communities in North Africa in Algeria, Libya, Egypt, Tunisia, and Morocco.

Push Factors The most consistent forces pushing migrants out of the Middle East and North Africa have been war, civil unrest, and the lack of economic opportunity (**Figure 4.40**). Often the latter two have been fostered and/or exacerbated by the imposition of the colonial political system. The case of Lebanon is illustrative. At the beginning of the period of European imperialism in the region, Lebanon became a French mandate under the League of Nations. Instead of promoting national unity among the many ethnic and religious groups of the Greater Lebanese mandate, France created a political administrative system of divide-and-rule that promoted

fragmentation. Predictably, after independence, Lebanon was beset by sectarian conflict. Rebellions, external attacks by Israel, and civil war between factions of Christian and Muslim militias ripped the country apart. This civil instability compelled tens of thousands of Lebanese people—both Christians and Muslims—to flee the country to escape violence (**Figure 4.41**).

British and U.S. political involvement in Iran similarly has caused many Iranians to flee. At the beginning of the Cold War between Western countries and the Soviet Union, Iran's oil reserves were considered to be of great strategic importance to Britain and the United States. Fearing that Iran might nationalize the industry, cutting off Britain's Anglo-Iranian Oil Company from the supply, Britain appealed to the United States to help oust Iranian Prime Minister Mohammed Mossadegh from office. The U.S.-backed 1953 coup was successful and stirred up resentment among many Iranians. Twenty-six years later, in 1979, a revolution deposed the U.S.-backed Shah Mohammad Reza Pahlavi and instituted in his place the Islamic state that exists today. Both events, but especially the fall of the Shah, resulted in the exodus of hundreds of thousands of Iranians to the United States, Britain, and Europe as well as to several Middle Eastern countries, such as Egypt.

Other significant emigration streams are made up of Algerians, Tunisians, and Moroccans, who have migrated mostly to Europe; Turks, who have followed a historical migration route to the Balkans and more recently to Germany; and Egyptians, who have migrated to other Arab countries in the region as well as elsewhere in the world. The migration of better-educated as well as low-skilled Turks and Egyptians has both pull and push dimensions to it. The push factor is illustrated by the departure of educated and skilled workers because economic growth has not kept pace with population growth; there are not enough high-paying jobs. The pull factors include European policies that enable temporary workers to take up low-paying, low-skill jobs that are not economically attractive to Europeans.

Cities and Human Settlement Although the predominant pattern of settlement in the region is a relatively small number of very large cities, a substantial number of medium-sized cities, and a very great number of small rural settlements, dramatic variations exist among countries. Israel is mostly an urban country, with almost 92% of its population living in cities. Sudan is largely a rural country, with only about 33% of the population living in cities. Cities in the region are growing dramatically each year. In fact, as of 2012, approximately 76% of the region's population was urban. Only about 50 years ago, most people in the region lived in small scattered rural settlements; however, since then, political independence and the development of the oil economy have been underlying factors in the increasing urbanization of the population.

The Middle East and North Africa has a long and distinguished urban history. Beginning with early empires, cities have been centers of religious authority, played pivotal roles in trade networks that extended into Europe and Asia, and reflected complex cultures. Today, cities in the region continue to play central administrative roles, and their political significance is as strong, if not stronger, than their religious significance. What has been most remarkable about contemporary urbanization in the region is that its rapid pace has led to the emergence of one or two very large cities in each country that contain a large proportion of the country's

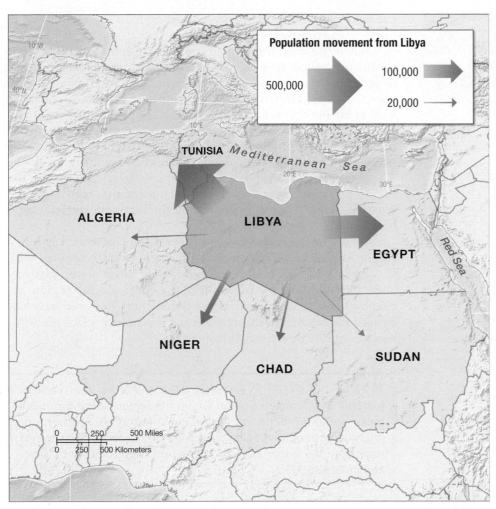

◀ **FIGURE 4.40 Internal Migration in the Middle East and North Africa** Conflict in Libya escalated with the start of a civil war in 2011, which resulted in the overthrow of long-time Libyan leader Muammar Qaddafi. During the conflict, over a million people fled Libya for neighboring countries. Those fleeing the violence in Libya included Libyans as well as citizens of Egypt, Nigeria, and Tunisia, who were living and working in Libya.

▼ **FIGURE 4.41 Lebanese Diaspora** This map shows the scattering of the Lebanese peoples around the world. Over the last century, Lebanon has experienced various waves of diasporic migration largely because of war and the tensions within this multiethnic, multireligious society.

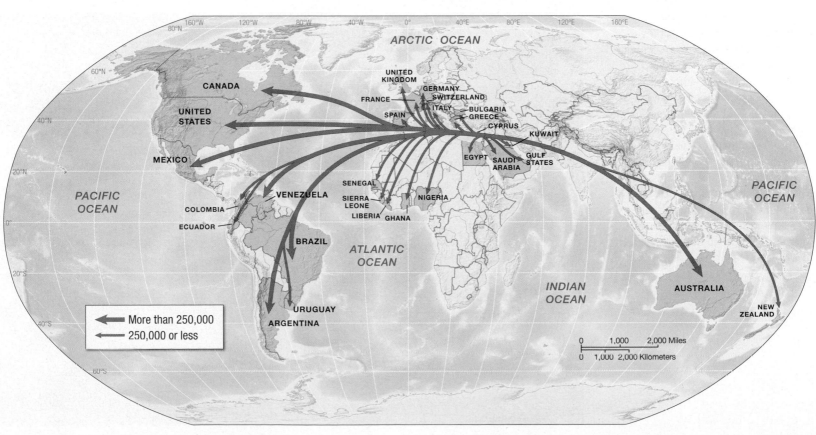

TABLE 4.3 Ten Largest Cities in the Middle East and North Africa, 2011

City	Population (millions)
Cairo, Egypt	11.2
Istanbul, Turkey	11.3
Tehran, Iran	7.3
Baghdad, Iraq	7.1
Khartoum, Sudan	5.2
Riyadh, Saudi Arabia	4.8
Alexandria, Egypt	4.3
Ankara, Turkey	4.2
Algiers, Algeria	3.9
Casablanca, Morocco	3.2

SOURCE: UNPD, World Urbanization Prospects: The 2011 Revision Population Database, http://esa.un.org/unup/; UN Inter-Agency Information and Analysis Unit, www. lauiraq.org/reports/GP-Baghdad.pdf (accessed April 25, 2012).

the urban system of the world economy, where capital, ideas, and people come together.

One of the most widespread problems of rapid urbanization is the inability of governments to meet the service and housing needs of growing urban populations. Inadequate and often poor-quality water supplies, electricity, sewer systems, clinics, and schools as well as air pollution and severe traffic congestion plague many cities in the region. Most critically, governments seem unable to provide housing for all who need it, and squatter settlements have sprung up on unclaimed or unoccupied urban land (**Figure 4.42**). These problems are compounded as the very largest cities in the region continue to attract even more migrants, who view the most well-known places as possessing the best opportunities for a better life.

In a few of the very wealthy oil-producing cities, such as Jubail (Saudi Arabia) and Doha (Qatar), enormous wealth coupled with very low populations has made the growth of urban places relatively uncomplicated. Many other cities of the oil-producing region, including Jeddah (Saudi Arabia) and Basra (Iraq), however, have not escaped the erection of shantytowns and the difficult social problems that accompany this type of urban change.

APPLY YOUR KNOWLEDGE Describe the various patterns of internal migration within the region. What produces these patterns? Similarly, what are some of the patterns of migration out of this region, and what produces these patterns? How have the urban areas of the region been affected by push and pull factors?

population and wield disproportionate political and economic influence. The brisk pace and extreme degree of urbanization in the region can be traced to the migration of rural people in search of economic opportunity in the cities as well as to the natural increase of urban residents. **Table 4.3** lists the major cities of the Middle East and North Africa. Many of these are critical nodes in

▶ **FIGURE 4.42 Squatter Settlements in Istanbul, Turkey** In Istanbul, squatter settlements are known as *gecekondu,* a Turkish word meaning that the settlements were built after dusk and before dawn. Geographer Paul Kaldjian's research in Turkey has shown that many residents of these settlements actually own land in the countryside, where they grow some of the food that sustains their lives in the city. Because there are no employment opportunities in rural areas, these individuals have been forced to migrate to the city in search of work.

FUTURE GEOGRAPHIES

Since the decline of de facto colonialism in the Middle East and North Africa in the 20th century, two related concerns have continued to attract worldwide economic and political investment there: oil and regional conflict. A third significant concern is internal: water.

Oil With oil as the foundation of the current global economic system, the region will continue to be an important economic factor in the global economy for many decades. And because tensions among different cultural and religious groups can complicate global access to oil, the Middle East and North Africa continues to draw the attention of the world's largest and most powerful states, especially the United States and Russia, as well as China and India. What is not so clear is how climate change and diminishing oil supplies will affect the region. One thing seems clear: OPEC's role as a self-regulating player in the global economy is likely to be challenged. For instance, climate scientists have proposed that OPEC establish a production quota system in response to global warming in order to reduce CO_2 emissions. So far, OPEC has resisted this type of external pressure on limiting production, but that does not mean

this proposal might not become appealing in the future. Moreover, as supplies become limited, OPEC's internal cohesion is showing signs of strain and members are becoming more autonomous to some extent.

Water Increasingly the resources of this region—where water is so scarce—are being invested in desalinization of seawater. Although seawater is abundant, the cost of removing its salt and minerals are exorbitant, and the process requires massive amounts of power. Once a region with a population that was well-adapted to the extremes of the environment, the nations in the Middle East and North Africa are increasingly home to modern, urban societies that have sought to overcome their environmental limitations through extreme exploitation of resources (**Figure 4.43**). If, instead of consuming water and related resources like citizens in less arid regions, the people in the Middle East and North Africa were once again motivated to adapt to their unique local environment, they could potentially develop sustainability practices and innovation that could reshape the region and be applied elsewhere in the world. Experiments like Masdar City in Abu Dhabi indicate that there is a growing understanding of the need for more sustainable practices.

Because of its scarcity, water is also a politically charged topic in the region, and access rights to water are the source of much conflict.

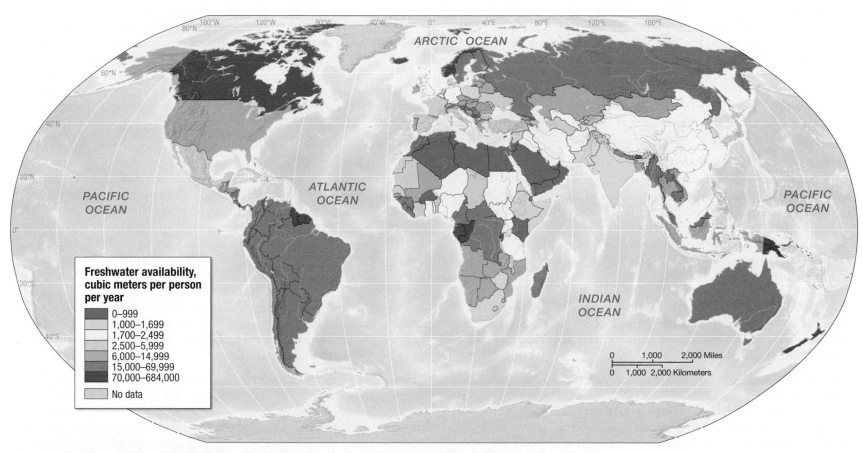

Freshwater availability, cubic meters per person per year

- 0–999
- 1,000–1,699
- 1,700–2,499
- 2,500–5,999
- 6,000–14,999
- 15,000–69,999
- 70,000–684,000
- No data

▲ **FIGURE 4.43 Map of World Water Scarcity** Even though water scarcity is a global issue, it is clear that the Middle East and North Africa is one of the most immediately vulnerable regions in the world. Adapting to the conditions of extreme water scarcity is a must for many of the countries in this world region.

Egypt has threatened military intervention in the Sudan should that country attempt to limit the downstream flow of the Nile. Another conflict involves Israel and Jordan, both of which inhabit the Jordan River valley and have access to its waters. The Jordan River is highly engineered through dams and diversions and is affected by intense groundwater pumping in the region, and at times its flow is little more than a trickle. The use and control of the Jordan River's water resources is also contested between Israeli and Palestinian, who each seek to use the river's resources for its own economic development.

Peace and Stability   There are a number of challenges to long-lasting peace and stability in this region. Recent events, such as the Arab Spring, suggest that many people are dissatisfied with the status quo in the region. The protests that brought down the leadership of Egypt, Tunisia, and Libya have opened up opportunities for new forms of government in these tightly regulated countries. But it is still unclear what will come next. In the context of the ongoing conflict between the Israelis and the Palestinians, there are still many challenges to establishing a brokered peace. The current prime minister of Israel is facing a tough internal political battle, and the Palestinians remain divided between Gaza and the West Bank. The continued expansion of the security fence through Palestinian territory also foments conflict. In addition, there remains uncertainty about the direction of Iran and Syria. In Iran, the government continues to push forward with a nuclear agenda that increasingly puts it at odds with major global powers. Syria also faces the ire of the international community for its treatment of its own citizens in the wake of protests that have recently rocked that country. Given all this, it is clear that the Middle East and North Africa will remain an important site for global and regional political negotiations for some time to come.

LEARNING OUTCOMES *REVISITED*

■ **Recognize the diversity of physical geographic environments in the Middle East and North Africa as well as the importance of key resources, such as oil and water.**

While aridity is the defining feature of the region's climate, coastal and mountain environments do exist. The region is resource-rich, particularly as a site of oil production. Other precious resources, however, such as water, remain scarce. Innovations are being developed to tap water resources and produce drinkable water from the vast oceans and seas that surround the region.

■ **Account for the diffusion of cultural and environmental practices, such as religion and agriculture, throughout the region.**

This region is home to three major world religions: Islam, Judaism, and Christianity. These religions have their foundation in the urban environments that emerged over time in the region, themselves a product of the region's ability to domesticate plants and animals. The Middle East and North Africa has long been a site of innovation, both technological and cultural. From this region, the knowledge of agriculture spread as did other cultural practices, such as writing and mathematics.

■ **Explain the natural and historical factors that contribute to the diverse economies of the Middle East and North Africa, how they vary from country to country, and how they are tied to global and regional processes.**

The Middle East and North Africa is home to some of the world's wealthiest and poorest economies. These stark contrasts have developed over time in relation to the availability of oil. As a rule, the wealthiest states are in the Persian Gulf region; poorer states are outside the belt of oil production. That said, several economies in the region have successfully diversified by exploiting agricultural, manufacturing, and service industries. Most notable of these is Israel, which has strong connections to Europe and a very diverse economy.

■ **Understand the geopolitical conflicts and tensions that bring instability to the Middle East and North Africa and the relationship between these conflicts and the political history of the region.**

Conflict and tension have their roots in the geopolitical organization of the region, a product of the colonial experience. In a number of areas, borders imposed under European rule brought different ethnic groups together under larger national banners. The ethnic and religious diversity of the region continues to challenge the stability of the region.

■ **Appreciate the differences in practices of kinship, gender, and religion in the region and how these factors affect the lives of women.**

Despite images of this region as a uniform space, the practices of kinship, gender relations, and religion vary. For example, women's mobility is quite diverse across different countries in the region. Women's mobility also varies between urban and rural spaces. Kinship—loosely defined by familial and neighborly relations—has a strong effect on the lives of many people in the Middle East and North Africa. Even though the region is predominantly Islamic (almost 96%), other religions, such as Christianity and Judaism, flourish throughout the region.

■ **Describe and account for the patterns of migration in the Middle East and North Africa, the push and pull factors that are involved in migration patterns, and the effects these patterns have on the region's demography.**

Many regional patterns of migration are tied to economical disparities. People migrate from the poorer countries to the wealthier countries in the region seeking work in the oil industries. Other migrants move to avoid conflict; refugee populations are increasing in the wake of the violence that has plagued the region. Some people with higher levels of education leave poorer countries to seek jobs in places such as Europe. These patterns of migration impact the region's overall demography. Urban areas, in particular, face challenges brought on by rapid and intense population growth.

KEY TERMS

afforestation (p. 137)

The Arab League (p. 157)

Arab Spring (p. 148)

aridity (p. 131)

Balfour Declaration (p. 151)

carbon market (p. 133)

chador (p. 163)

desalinization (p. 135)

desertification (p. 137)

dry farming (p. 132)

Fertile Crescent (p. 140)

The First Arab–Israeli War (p. 151)

fossil fuels (p. 136)

Gaza Strip (p. 152)

gravity system (p. 135)

guest worker (p. 166)

Gulf Cooperation Council GCC (p. 158)

hajj (p. 160)

Hamas (p. 153)

Hasidim (p. 160)

Hezbollah (p. 154)

import substitution (p. 144)

informal economy (p. 145)

internally displaced person (p. 150)

intifada (p. 152)

Iranian Green Revolution (p. 148)

Islam (p. 159)

Islamism (p. 160)

Janjaweed (p. 150)

Jasmine Revolution (p. 148)

jihad (p. 160)

jihadist (p. 156)

kinship (p. 162)

Kurds (p. 148)

mandate (p. 142)

Mecca (p. 160)

Muslim (p. 159)

nationalist movement (p. 142)

nationalization (p. 144)

oases (p. 135)

Occupied Territories (p. 152)

Organization of Petroleum
Exporting Countries (OPEC) (p. 138)

Orthodox Judaism (p. 160)

petrodollar (p. 145)

petroleum (p. 138)

Quital (p. 160)

Qu'ran (p. 159)

rain shadow (p. 132)

secular (p. 158)

Semitic (p. 141)

Shari'a (p. 158)

Shi'a (p. 141)

Sunna (p. 159)

Sunni (p. 141)

transhumance (p. 162)

tribe (p. 162)

Union for the Mediterranean (p. 147)

wet farming (p. 140)

world religion (p. 141)

Zionism (p. 151)

THINKING GEOGRAPHICALLY

1. How do countries in the Middle East and North Africa manage their water resources? List an example of an area where water has become a source of conflict in the region.

2. What are some of the factors that limit the possibility of a broader peace in the region? How have recent technological changes, such as the Internet, facilitated dialogue and resistance in the region?

3. How do regional views about gender affect the use of public and private space? Are rules concerning the veiling of women uniform throughout the region? If not, how do they vary geographically?

4. Describe the geographic distribution of refugees and guest workers within the Middle East. What are some current push and pull factors that motivate people to leave one part of the region and enter another?

5. The Middle East and North Africa is becoming highly urbanized. What factors are driving urban growth in this region? How are cities with considerable oil wealth handling their rapid urbanization?

6. Describe the spatial distribution of oil and natural gas production in the region. What are the consequences of the uneven distribution of these resources on this region's politics and economy? What impact has oil and natural gas production had on the local and global environment?

MasteringGeography™

Looking for additional review and test prep materials? Visit the Study Area in MasteringGeography™ to enhance your geographic literacy, spatial reasoning skills, and understanding of this chapter's content by accessing a variety of resources, including **MapMaster** interactive maps, videos, RSS feeds, flashcards, web links, self-study quizzes, and an eText version of *World Regions in Global Context*.

5 | Sub-Saharan Africa

Women play important roles in African economies and are seen as critical to the region's future development. In the village market in Senegal shown here, women sell crops they have grown as well as colorful clothing and household goods.

IN WEST AFRICA, LOCAL MARKETS ARE FULL OF WOMEN SELLING FOOD, textiles, and household goods. For example, in Ghana, the Kuapa Kokoo cooperative led by women cocoa farmers set up a chocolate company and, with some development assistance, succeeded in bringing its "Divine Chocolate" to a global market. In Ethiopia, Bethlehem Tilahun Alemu founded Sole Rebels, an environmentally friendly shoe company that sells its products worldwide. Some of these entrepreneurs and vendors are doing so well that they can afford luxury cars, such as Mercedes Benzes. In some parts of sub-Saharan Africa such as Togo, if a woman reaches this level of success, she may earn the nickname "Mama Benz."

In the last decade, some of the fastest-growing economies in the world emerged in sub-Saharan Africa, including those of Ethiopia, Nigeria, and Rwanda. Peace has come to many regions, and health indicators are slowly starting to improve. Behind this turnaround are many stories of individuals who have started businesses, campaigned to end civil wars, and labored in remote regions to improve health care and education. Many of these individuals are women.

- Compare and contrast the physical geographies of the diverse regions of sub-Saharan Africa.

- Explain the climate and vegetation of the Sahel and the Congo Basin, how desertification and deforestation occur and are managed, and why sub-Saharan Africa is vulnerable to climate change.

- Evaluate the role of physical geography, disease, and mineral resources in shaping contemporary conflicts, crises, landscapes, and economies in the region.

- Understand the legacies of European colonialism and the Cold War in the region.

- Explain how debt relief and the end of several conflicts may increase the chance of positive economic and political futures in sub-Saharan Africa.

- Describe the main causes of low agricultural production in sub-Saharan Africa and the role women are playing to help address agricultural and other problems in the region.

- Describe tensions that are emerging around conservation efforts, land grabs, and foreign investment in sub-Saharan Africa.

- Provide examples of how African peoples, goods, and culture have spread around the world as a result of slavery, colonialism, and globalization.

- Describe the spread of HIV/AIDS in the region, its impact, and the efforts made to end the epidemic.

When Ellen Johnson Sirleaf became president of Liberia in 2006, she became the first elected female head of state in Africa. Trained as an economist, Sirleaf worked for the World Bank and the United Nations before becoming involved in the politics of her country, which had been torn apart by civil war and violence. Sirleaf's administration sustained peace and reduced debt and corruption. As a result, she shared the 2011 Nobel Peace Prize for her "non-violent struggle for the safety of women and for women's rights to full participation in peace-building work."

Success stories such as this have led development agencies, charities, and governments to realize that supporting women can be the fastest way to reduce poverty, disease, and conflict and to foster an entrepreneurial spirit. Women often exceed expectations of productivity when they have access to cooperatives, small loans, and markets. And when women make a good living and can keep their families healthy, they often choose to have fewer children, thus reducing population growth rates that can stress resources. ∎

▼ **FIGURE 5.1**

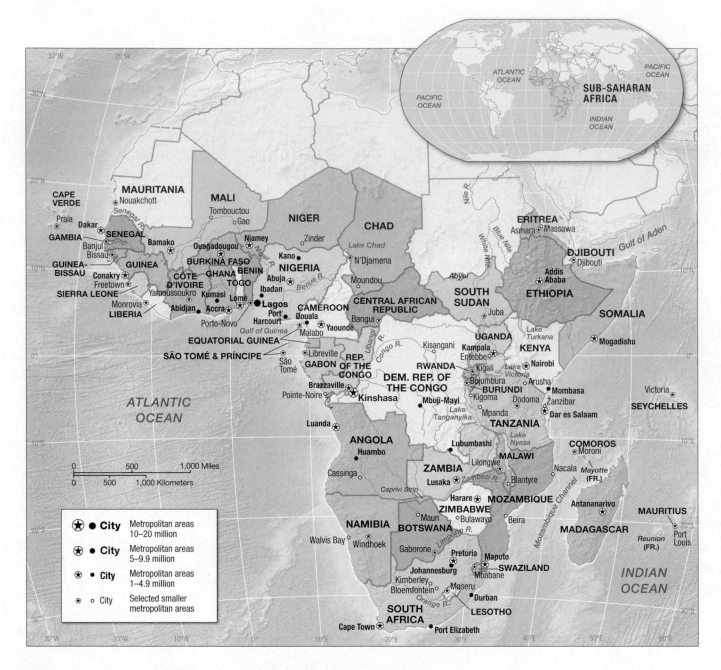

Sub-Saharan Africa Key Facts

- Major Subregions: West Africa, East Africa, Southern Africa, Equatorial Africa, Horn of Africa

- Major Physiogeographic Features: Sahara Desert, Cameroon Mountains, Fouta Diallon Highlands, Victoria Falls, Lake Victoria, Congo River, East African Highlands, African Rift Valley. Tropical climate is dominated by intertropical convergence zone, and the subtropical high leads to frequent drought in region.

- Major Religions: In addition to Christianity and Islam, sub-Saharan Africa is dominated by a range of traditional religions, particularly in the southern and eastern regions

- Major Language Families: Afro-Asiatic, Niger-Congo, Nilo-Saharan, French, English

- Total Area (total sq km): 22.64 million

- Population (2011): 883 million; Population under Age 15 (%): 43; Population over Age 65 (%): 3

- Population Density (per sq km) (2011): 36

- Urbanization (%) (2011): 37

- Average Life Expectancy at Birth (2011): Overall: 55; Women: 56; Men: 53

- Total Fertility Rate (2011): 5.2

- GNI PPP per Capita (current U.S. $) (2009): 2,251

- Population in Poverty (%, < $2/day) (2009): 72

- Internet Users (2011): 93,950,481; Population with Access to the Internet (%) (2011): 8; Growth of Internet Use (%) (2000–2011): 2,389

- Access to Improved Drinking Water Sources (%) (2011): Urban: 82; Rural: 47

- Energy Use (kg of oil equivalent per capita) (2009): 649

- Ecological Footprint (hectares per capita consumed/hectares per capita available, global scale) (2011): 1.5/1.8

ENVIRONMENT AND SOCIETY

Africa is a large, complex, and often misunderstood continent. Some people think of the region as a fertile tropical forest rife with exotic diseases or as an idyllic game reserve. Others perceive a harsh landscape devastated by war and drought or a magical place where rich cultural traditions reach back to the dawn of humanity. But sub-Saharan Africa is not neatly encompassed by any of these limited descriptions. The continental landmass called Africa straddles the Equator, stretching from the Mediterranean Sea to the southern tip in South Africa and from Senegal on the Atlantic to Somalia on the Indian Ocean (**Figure 5.1**). The continent's total area is about 30.4 million square kilometers (11.7 million square miles)—three times larger than the United States or Brazil. The sub-Saharan region is 22.64 million square kilometers (8.75 million square miles). The physical geography of this region has shaped ecosystems, the use of land, and the distribution of key mineral and water resources. The geography has, in turn, been transformed by human activities within and beyond the region. The satellite image (**Figure 5.2**) shows many of the key physical features of sub-Saharan Africa, including the Sahara Desert and the Rift Valley.

Sub-Saharan Africa has been defined and divided from North Africa based on historical, physical, and social characteristics that include a legacy of European colonialism and slavery, a mostly tropical climate, and the darker skin of many inhabitants. Although the race-based definition of sub-Saharan Africa is controversial, it is still used by both Africans and non-Africans to identify the region.

In this text, we discuss North Africa with the Middle East (see Chapter 4, p. 128) because of their shared characteristics (including similar physical environments of dry climates and the prevalence of Arabic language and ethnicity and Islamic religion). However, the geographical, racial, ethnic, and religious basis for dividing Africa between two world regions is artificial. The division oversimplifies both the cultural and historical distinctiveness of the two regions, the overlap between them, and the great variety they contain. For example, the Sahara Desert includes territory from both North and sub-Saharan African countries, and the Nile River links a long fertile corridor that stretches from Egypt in North Africa to South Sudan—the newest country in sub-Saharan Africa as of 2011—and then south into Ethiopia and Uganda. Because there are so many physical and human links across the continent of Africa, several sections of this chapter discuss broader patterns across the whole continent.

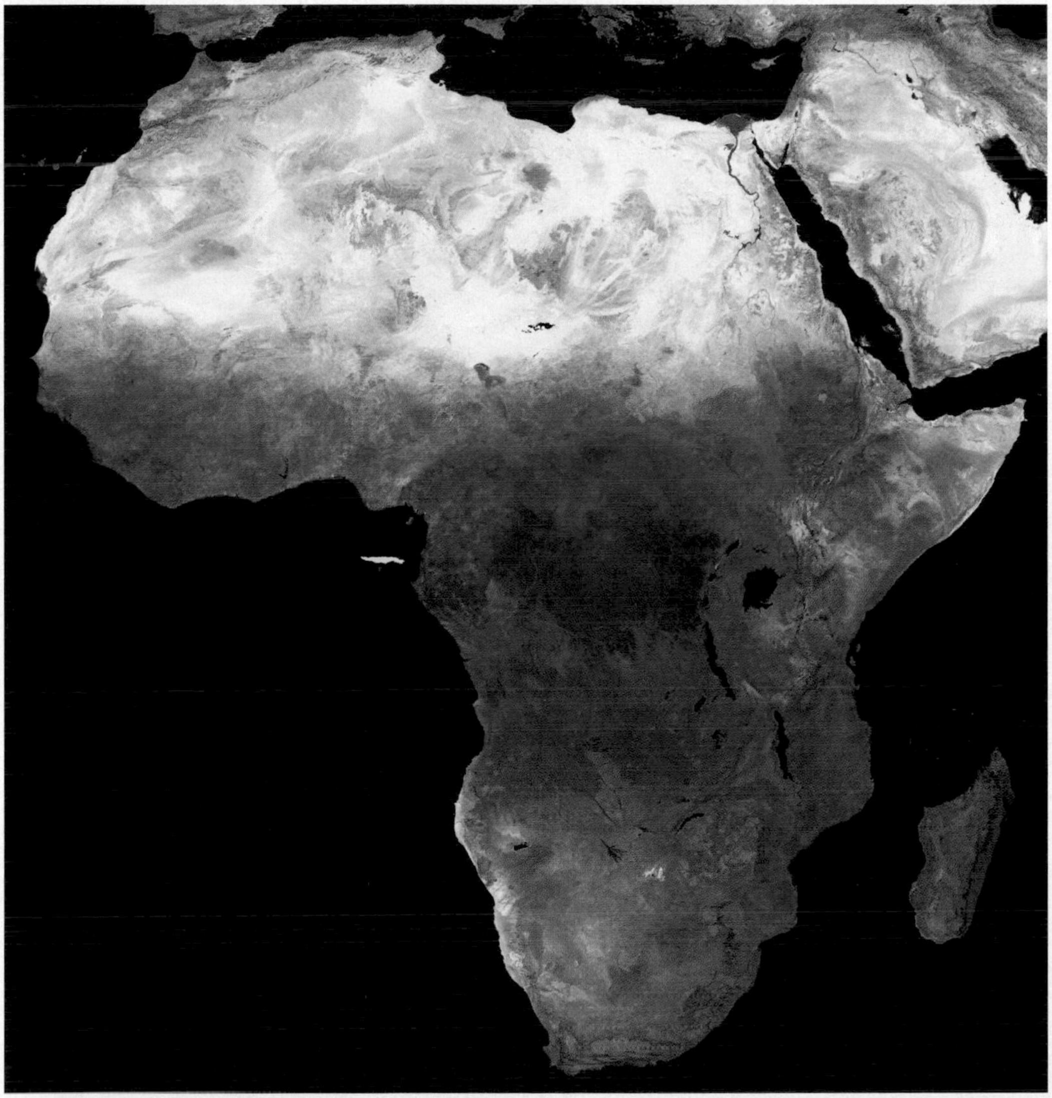

Climate, Adaptation, and Global Change

Most of sub-Saharan Africa has a tropical climate with temperatures that are higher than 20°C (70°F) and little frost except in highland areas (**Figure 5.3**, p. 178). The climate is dominated by two major features of the atmospheric circulation—the **intertropical convergence zone (ITCZ)** and the **subtropical high** (see Chapter 1, p. 9). The ITCZ is a low-pressure region where air flows together and rises vertically as a result of intense solar heating at the Equator, producing heavy rainfall of more than 1,500 millimeters (60 inches) a year over the Congo Basin. The subtropical high is a zone of descending air, which results in dry, stable air that causes desert conditions over the Sahara and Kalahari (Figure 5.6, p. 180). The areas between the ITCZ and the subtropical high, such as the West African coast and East Africa, experience seasonal rainfall as these global circulations shift northward in April and southward in October. Where seasonal rainfall is modest, semiarid conditions predominate, and variations in circulation bring unreliable rainfall and frequent droughts that result in agriculture and water resources management challenges. During December, very dry winds called the **harmattan** blow out of inland Africa.

▲ **FIGURE 5.2 Africa from Space** This satellite image shows the major landform subregions of the Sahara Desert (in light yellow and beige), the Congo rain forests (dark green), and the line of lakes along the East African Rift Valley, including the largest, Lake Victoria. The large island of Madagascar is also clearly visible.

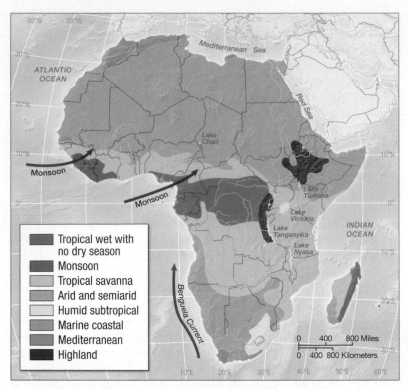

▲ **FIGURE 5.3 Climates of Sub-Saharan Africa** A climate classification of Africa shows several major climate zones spanning the Equator, including tropical wet and monsoon regions with higher rainfall, arid and semiarid desert, and savanna with low seasonal rains. In southern Africa, the climate is milder, with a Mediterranean climate on the west coast and a marine climate on the east coast.

The harmattan carry large amounts of dust and stress humans and animals. In July, southwestern trade winds blow onto the coast of West Africa, bringing seasonal monsoon rains (see Chapter 1, p. 9) to inland countries such as Mali.

These general patterns are modified by the effects of mountains, lakes, and ocean currents. The cold Benguela Current creates cool, dry conditions along the coasts of Angola and Namibia and intensifies the desert conditions of the Kalahari. High-altitude areas such as the East African highlands have higher rainfall and more moderate temperatures, which make them more favorable to agriculture and human settlement. South Africa is located in more temperate, cooler latitudes than the rest of the continent and experiences a mild Mediterranean climate.

Desertification and Famine The people of sub-Saharan Africa have adapted to the climate of the continent in various ways. Farmers work floodplains along seasonally variable rivers and

implement traditional irrigation and rainwater harvesting schemes in dry areas. Herders move their livestock to follow the rains. In spite of these adaptations, droughts resulting from variable rainfall—combined with political unrest, the growth of human and animal populations, and changes in land access and land use—threaten food security and trigger famine in the region. Land degradation and persistent drought can lead to a process called **desertification** in which arid and semiarid lands become degraded and less productive and desert-like conditions result. The main culprits are seen as climate change, overgrazing, overcultivation, deforestation, and unskilled irrigation.

The **Sahel** of West Africa is the southern border of the Sahara Desert and has highly variable rainfall (**Figure 5.4**). The human population in the Sahel are dependent on **pastoralism**—a way of life that relies on livestock—with some crop production along rivers and at oases. Beginning in about 1968, rains failed and drought covered most parts of the Sahel for seven years. Between 1968 and 1973, as many as 3.5 million cattle died, and 15 million farmers lost more than half their harvests in this area. A quarter of a million people died from famine before food relief could reach them. More recent famines precipitated by drought and conflict include those in Ethiopia (1984–1985, over 900,000 dead), Somalia (1991–1992, over 300,000 dead), and Sudan (1985, over 500,000 dead).

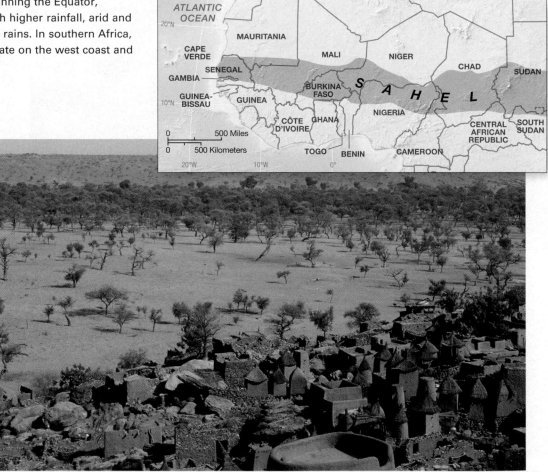

▲ **FIGURE 5.4 The Sahel** The map above highlights the Sahel region of West Africa. The photo shows a Dogon village in the Sahel area in Mali where trees (in the background) are planted to shade crops and keep back desert dunes.

During these crises, images of the drought and starving refugees in Africa began to appear in the international media, resulting in a relief effort and anguished debates among researchers and policymakers about what had gone wrong and what could be done to prevent future tragedy. Some researchers blamed the buildup of herds in the face of regular drought cycles in the Sahel. Sahelian pastoralists tend to be opportunistic, building up their herds in good years because livestock accumulation is the best way of amassing wealth and investing capital in this area. Other researchers argued that the roots of the crisis lay in changes in politics, economics, and land use stemming from colonial structures and overreliance on crops grown for export, such as peanuts.

Unfortunately, some of the international efforts to respond to drought backfired. For example, the deep wells that were drilled for cattle herds resulted in so many thirsty cattle gathered around the wells that all nearby vegetation was consumed, and the herds ultimately starved. Similarly, when food aid arrived in communities where some farmers still had crops to sell, food prices dropped as supply outstripped demand, and farmers could not then make a living. The U.N. Convention to Combat Desertification seeks to understand and reverse the processes that have led to so much devastation in this region. In recent years, Sahelian farmers in Niger and Burkina Faso have started to "regreen'" degraded areas through traditional techniques, such as planting crops in pits fertilized with manure, capturing rainwater, preventing erosion with rows of stones, and shading crops with planted trees.

Vulnerability to Climate Change Africa has been identified as extremely vulnerable to climate change: warmer temperatures and changes in ecosystems have already been observed in southern Africa (**Figure 5.5**). In terms of future climate, computer models are uncertain about how rainfall will change, but models show that temperatures will warm almost everywhere and that southern Africa will experience more intense droughts. Africa's vulnerability to climate change is caused as much by poverty as by climate itself. Many Africans are dependent on rain-fed agriculture and livestock, which will suffer greatly from the effects of drought and heat waves. Many people also lack access to safe water supplies and health care. Scientists forecast that by the year 2020, crop yields could fall by as much as half, the population experiencing water security problems could grow by 250 million, and warming temperatures may result in the increased spread of diseases such as malaria, especially in highland areas. Although the effects of climate change are predicted to be severe in the region, sub-Saharan countries have some of the lowest per capita greenhouse gas

emissions in the world. This illustrates the inequities of climate change, where those least to blame are often most affected. In recent years, African countries have demanded that the industrial world reduce emissions and assist the less industrial countries to adapt to climate change.

APPLY YOUR KNOWLEDGE Many scholars assert that the region of the United States and Canada (see Chapter 6, p. 222), which has very large per capita greenhouse gas emissions, is obligated to help sub-Saharan Africa, which has with low emissions, cope with the impacts of climate change. Conduct research on the Internet to learn more about what this would involve and what forms assistance could take.

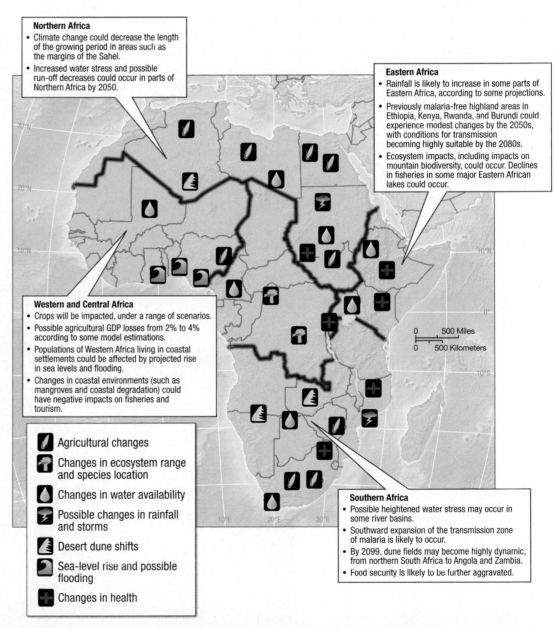

Northern Africa
- Climate change could decrease the length of the growing period in areas such as the margins of the Sahel.
- Increased water stress and possible run-off decreases could occur in parts of Northern Africa by 2050.

Eastern Africa
- Rainfall is likely to increase in some parts of Eastern Africa, according to some projections.
- Previously malaria-free highland areas in Ethiopia, Kenya, Rwanda, and Burundi could experience modest changes by the 2050s, with conditions for transmission becoming highly suitable by the 2080s.
- Ecosystem impacts, including impacts on mountain biodiversity, could occur. Declines in fisheries in some major Eastern African lakes could occur.

Western and Central Africa
- Crops will be impacted, under a range of scenarios.
- Possible agricultural GDP losses from 2% to 4% according to some model estimations.
- Populations of Western Africa living in coastal settlements could be affected by projected rise in sea levels and flooding.
- Changes in coastal environments (such as mangroves and coastal degradation) could have negative impacts on fisheries and tourism.

Agricultural changes

Changes in ecosystem range and species location

Changes in water availability

Possible changes in rainfall and storms

Desert dune shifts

Sea-level rise and possible flooding

Changes in health

Southern Africa
- Possible heightened water stress may occur in some river basins.
- Southward expansion of the transmission zone of malaria is likely to occur.
- By 2099, dune fields may become highly dynamic, from northern South Africa to Angola and Zambia.
- Food security is likely to be further aggravated.

▲ **FIGURE 5.5 Climate Change Vulnerability in Africa** Multiple stresses make most of Africa highly vulnerable to climate change. This map illustrates examples of climate change impacts on agriculture and food security, ecosystems, water, and health as well as risks that may occur as a result of sea-level rise in West Africa's coastal cities.

Geological Resources, Risk, and Water Management

The continent of Africa is the heart of the ancient supercontinent called **Pangaea**, the southern part of which broke off to form **Gondwanaland** about 200 million years ago. The landmasses that become Latin America and Asia then broke away from Gondwanaland, and the high plateau that remained became the continent of Africa.

Plate Tectonics Africa is still mainly a plateau continent, with elevations ranging from about 300 meters (1,000 feet) in the west, tilting up to more than 1,500 meters (5,000 feet) in the eastern part of the continent (**Figure 5.6**). Significant mountain ranges in western Africa include the Mount Cameroon and Fouta Diallon Highlands. Steep slopes, especially on the western edge of the plateau, drop to narrow coastal plains.

The higher areas of the plateau, where cooler temperatures and higher rainfall are hospitable to humans, include the Veld of southern Africa, the East African highlands of Kenya and Ethiopia, and the Jos Plateau of West Africa. The Sossusvlei sand dunes in Namib-Naukluft National Park are some of the highest in the world (**Figure 5.7a**). Volcanic peaks such as Kilimanjaro

(5,895 meters, or 19,340 feet), Kenya/Kirinyaga (5,200 meters, or 17,058 feet), and the Virungas (4,507 meters, or 14,787 feet) rise from the eastern plateau, which is split by a deep trough, where tectonic processes continue to pull the eastern edge of Africa away from the rest of the continent.

This trough is a **rift valley**—a large and long depression between steep walls formed by the downward displacement of a block of the earth's surface between tectonic faults. The East African Rift Valley runs more than 9,600 kilometers (6,000 miles) from the Red Sea in the north to Mozambique in the south and is from 50 to 100 kilometers (30 to 60 miles) wide. It has two major branches and is home to deep lakes, including Lake Tanganyika at 1,473 meters deep (4,832 feet). Lake Victoria, the third-largest lake in the world, lies between the two branches. The age, size, and depth of these lakes make them diverse freshwater ecosystems with important fisheries (**Figure 5.7b**).

Soils, Minerals, and Mining African soils tend to not be very fertile because of the great age of the underlying geology and because of high levels of rainfall that wash out nutrients from exposed soils. Soil fertility tends to be higher where volcanic activity has deposited ash, such as the East African highlands, and in wider river valleys,

▶ **FIGURE 5.6 The Physical Geography of Sub-Saharan Africa** Africa is a plateau continent, with rivers that often flow through inland deltas on the plateau or drop over waterfalls. One of the most significant features is the African Rift Valley, filled with elongated lakes and several active volcanoes.

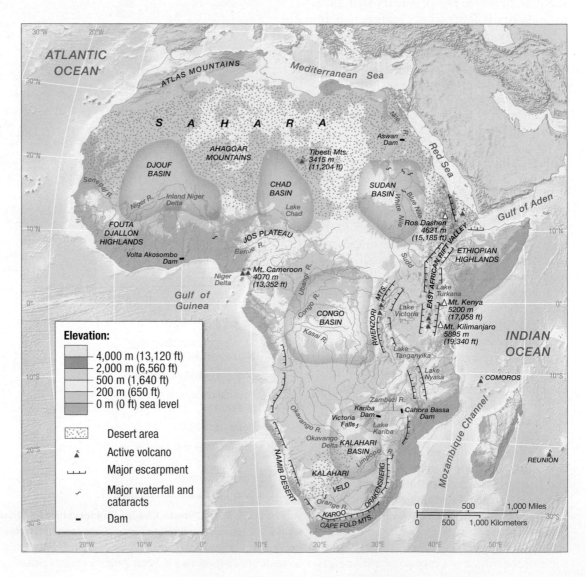

where sediment from floods creates richer soils. Tropical soils in wetter zones, such as central Africa, lose their fertility rapidly once the forest is cleared and the soil is exposed to rain, sun, and wind. Desert regions can have salty or alkaline soils that are toxic to crops. High iron and aluminum soils are also poisonous to plants and crops in some locations in the region.

Half the African continent is composed of very old crystalline rocks of volcanic origin that hold the key to Africa's mineral wealth. Ancient tropical swamps formed sedimentary rocks containing oil and other fossil fuels. These include coal in southern Africa and oil and gas in West Africa, particularly Nigeria and Gabon. Iron and manganese, important to manufacturing, are found in western and southern Africa, and most of the world's known chromium is found in southern Africa, especially in Zimbabwe and South Africa. Vast copper reserves are located in the southern Congo and in the copper belt of Zambia, where cobalt is also found. Bauxite, which is used in making aluminum, is found in a belt across West Africa, and uranium is found in Niger. Coltan, used in electronic devices such as mobile phones, is another key African mineral that is critical to industrial production elsewhere in the world.

There is gold in Ghana and Zimbabwe, and as much as half the world's gold reserves lie in the area around Johannesburg in South Africa. South Africa is also famous for diamonds; they

◀ **FIGURE 5.7 Physical Landscapes** (a) In southern Africa, the descending air of the subtropical high and the presence of the cold Benguela Current offshore create dry conditions. The spectacular Sossusvlei sand dunes are responsible for the spectacular scenery in this Namibia desert landscape. (b) Elongated lakes line the bottom of the African Rift Valley. Lake Bogoria, Kenya has several hot springs that result from tectonic activity and a large population of flamingos whose pink color comes from eating the organisms that live in the lake.

(a)

(b)

are found in Botswana and Namibia in southern Africa, at the edges of the Congo Basin, and in Sierra Leone in West Africa. Although these resources bring billions of dollars into Africa, they also make national and regional economies vulnerable to fluctuations in world market prices, especially in areas where minerals dominate exports. In 2008, African exports in physical terms were dominated by fossil fuels (75%), followed by metals (11%) and minerals (7%). The distribution of mineral wealth is uneven between and within countries in the region (**Table 5.1**).

Mineral resources have played an important role in African history. Salt was a key commodity in trans-Saharan trade from the 10th century to the present day. Gold was valued in West Africa from early times; Mansu Musa, the emperor of Mali, carried and traded so much gold on a pilgrimage to Mecca in 1324 that his actions depressed gold prices worldwide. Gold and diamonds spurred European colonial grabs for Africa as well as conflicts between colonial powers. The discovery of diamonds in 1867 at Kimberley and gold in 1886 in South Africa resulted in conflicts between colonizers and Indigenous groups. Gold and diamonds, together with oil, continue to incite conflict within Africa and to amplify interest in African economies on the part of other states and multinational corporations (see "Geographies of Indulgence, Desire, and Addiction: Diamonds," p. 184).

> **APPLY YOUR KNOWLEDGE** Do some research to determine which two or three products commonly and frequently used by you or by your friends and family contain minerals produced in sub-Saharan Africa.

Water Resources and Dams The routes of many of Africa's major rivers flow away from the coast and into inland wetlands and deltas before shifting back toward the ocean. For example, the immense Congo River, second only to the Amazon in terms of overall discharge, flows north across the Congo Basin before turning west toward the rapids that bring it down to the Atlantic. The Niger River flows north toward the Sahara into a large inland delta, before turning south toward its exit to the Atlantic in Nigeria. The Nile (discussed in more detail in Chapter 4) flows into the vast wetland known as the **Sudd**. These inland deltas create some of the richest ecosystems in the sub-Sahara African region and provide habitat for wildlife, fisheries, and grazing and irrigation opportunities.

Where rivers descend from the high plateaus to the coast in this region, they often cut deep valleys back into the plateaus and drop over rapids and waterfalls, such as Victoria Falls on the Zambezi River in southern Africa. This poses a serious problem for navigation by boat into the continent (**Figure 5.8a**). However, the rivers of Africa also provide considerable potential for hydroelectric development. Several large dam projects were initiated in the 1950s to harness the energy of the rivers, to provide electricity to industry and cities, and to irrigate agricultural fields. For example, the Federation of Rhodesia and Nyasaland (now the countries of Zimbabwe and Zambia) completed the Kariba Dam on the Zambezi at Kariba Gorge in 1959 (**Figure 5.8b**). This dam, which produces inexpensive electric power for this region of southern Africa, created Lake Kariba. The project required the resettlement of 57,000 people. Because of poor planning, some people ended up in places resembling refugee camps, where hygiene was very poor, and epidemics flourished. Thousands of wild animals, isolated as the waters rose behind the dam, were also relocated through "Operation Noah." On a positive note, some of the unanticipated benefits of the dam included development of a tourist industry, new lush animal habitats around the new lake, and a productive fishery. The electricity produced by the dam supports the copper-mining industry in Zambia.

The Akosombo Dam on the Volta River in Ghana required the relocation of 78,000 people. Its impacts were similar and included the loss of fertile sediment and fisheries below the dam. Both Kariba and Akosombo have been affected by drought in recent years, and the amount of electricity supplied has decreased as a result. The slower flow of the rivers and the resulting stagnant water behind the dams have increased the incidence of several diseases, including schistosomiasis and malaria. The successes and failures of these projects illustrate the importance of understanding both physical and social factors in development projects.

TABLE 5.1 Minerals in Sub-Saharan Africa Percent share of global reserves and production for selected minerals

Mineral	Share of world reserves (%)	Share of world production (%)	Main African producers
Aluminum	3	4	Mozambique, Egypt, South Africa
Chromites	12	37	South Africa, Zimbabwe, Madagascar, Sudan
Coal	4	3	South Africa, Zimbabwe
Cobalt	41	60	Democratic Republic of the Congo, South Africa, Zambia
Copper	4	7	Zambia, South Africa, Democratic Republic of the Congo
Diamonds	56	49	South Africa, Botswana, Democratic Republic of the Congo
Gold	34	18	South Africa, Ghana, Mali
Iron ore	1	3	South Africa, Algeria, Mauritania
Manganese	Less than 0.1%	23	South Africa, Ghana, Gabon
Natural gas	8	6	Algeria, Egypt, Libya
Oil	10	12	Nigeria, Angola, Algeria, Libya
Phosphate rock	53	25	Morocco, Tunisia, Egypt
Uranium	15	17	South Africa, Niger, Namibia

SOURCE: Calculated on the basis of data from the U.S. Geological Survey, British Petroleum, and OECD.

(a)

(b)

◀ **FIGURE 5.8 Victoria Falls and the Kariba Dam** (a) The Zambezi River cascades over Victoria Falls at the frontier between Zimbabwe and Zambia. In addition to viewing the spectacular waterfalls, tourists bungee jump from the suspension bridge, watch wildlife along the river above the falls, and raft down the river below the falls. (b) A few hundred miles downstream, the Kariba Dam generates electricity for the two countries.

Ecology, Land, and Environmental Management

Sub-Saharan Africa, where wildlife roam across grasslands and great apes and lemurs occupy forests, hosts some of the world's most fascinating ecosystems. But these ecosystems are under serious pressure from threats such as deforestation and poaching. African ecologies also include many pests and diseases that pose serious health threats to humans and agricultural systems. Low agricultural yields across the region undermine food security and health and increase pressures for land that result in encroachment on wildlife habitat and deforestation. Solutions include efforts to eradicate and cure disease, to increase food production on land that is already being farmed, and to establish parks and conservation programs.

Major Ecosystems African ecosystems are closely tied to climate conditions, as are ecosystems in the rest of the world. But African ecosystems also reflect a complex evolutionary history and physical

geography that have produced great diversity, unique plants, and the world's most charismatic community of animal species. The Congo Basin hosts Earth's second-largest area of rain forest (after the Amazon), which covers almost 2.2 million square kilometers (780,000 square miles). Other forests are found along the West African coasts, along the coast of Kenya, and on the island of Madagascar. Forests make up about 20% of African land area. These forests boast impressive biodiversity. They are home to monkeys and apes, such as chimpanzees and gorillas, as well as tropical hardwoods of significant economic value, such as mahogany. Like many forests around the world, African forests are threatened by demands for timber and firewood, by poaching and conflict, and by conversion to cropland. In addition, recent peace agreements in central Africa may open up the region to forest exploitation.

The island of Madagascar has been a special focus of international conservation concern. Less than 15% of the original forests

GEOGRAPHIES OF INDULGENCE, DESIRE, AND ADDICTION

Diamonds

Diamonds are associated with love and luxury for consumers around the world. Larger diamonds are graded, cut, polished, and set in gold or other metals and then sold for high prices in jewelry stores. About half of all diamond purchases are made in the United States. These glittering stones also have considerable industrial value and are valued for their hardness and strong, sharp cutting edge.

More than half of the world's diamonds originate from African countries, with a total annual value of roughly U.S. $8 billion. The diamond industry employs 10 million people globally, directly and indirectly. About two-thirds of the global diamond trade is controlled by a South African conglomerate, De Beers, which manages markets to ensure that prices remain high and the supply stable.

Most African diamond production takes place in South Africa, Botswana, and Namibia and contributes significantly to export revenue and local employment. In Angola, the Democratic Republic of the Congo, and Sierra Leone (**Figure 5.1.1**) mines

▲ **FIGURE 5.1.1 Diamond Mining** Miners dig for diamonds in Zimbabwe, where the government has been accused of human rights violations and where working conditions are very difficult.

and miners are often under the protection of armed guards or military forces. From the 1960s to the 1990s, diamonds became associated with corruption, violence, and warfare in these countries. Because of their immense value and their small size, which makes them easy to transport, diamonds were smuggled and used to purchase weapons. These weapons were ultimately used in brutal local conflicts and wars, resulting in countless civilian deaths. Some analysts suggest that these so-called **conflict diamonds** (also called blood diamonds) may have accounted for 10% to 15% of all global trade in diamonds during peak 1990s trade.

For example, diamonds funded a brutal civil war in Sierra Leone from 1991 to 2002, where rebels chopped off people's limbs with machetes to intimidate residents into leaving the country's eastern diamond zones. Conflict

diamonds from Liberia were being smuggled into neighboring countries for export, and diamonds from the Côte d'Ivoire found their way to the British and European markets. The Democratic Republic of the Congo (formerly Zaire) became a pawn in the struggle for diamonds in the 1990s, with Angola, Namibia, and Zimbabwe sending troops to protect the Kabila government, and Burundi, Rwanda, and Uganda assisting rebels.

In contrast, Botswana produces diamonds under peaceful conditions with the guidance of traditional leaders. The mines in Botswana employ more than 25% of the population and are responsible for 33% of the country's gross domestic product. Botswana is home to Jwaneng, the richest diamond mine in the world, whose name means "a place of small stones." The open-pit mine is owned by Debswana, a partnership between De Beers

and the government of Botswana. As a result, Botswana enjoys a standard of living that is much higher than the average for sub-Saharan Africa.

The Kimberley Process, which involves over 75 governments partnered with industry and civil society, was established in 2003 to prevent conflict diamonds from entering the mainstream rough diamond market. The process seeks to eliminate the use of diamonds to finance armed conflict. Today, the industry reports that less than 1% of diamonds is illegally sourced, and retailers often mention that their diamonds are conflict-free in advertisements. However, in 2011, the human rights watchdog group Global Witness left the Kimberley Process in protest when members signed an agreement to allow Zimbabwe to trade U.S. $2 billion in diamonds from fields linked to human rights abuses. ■

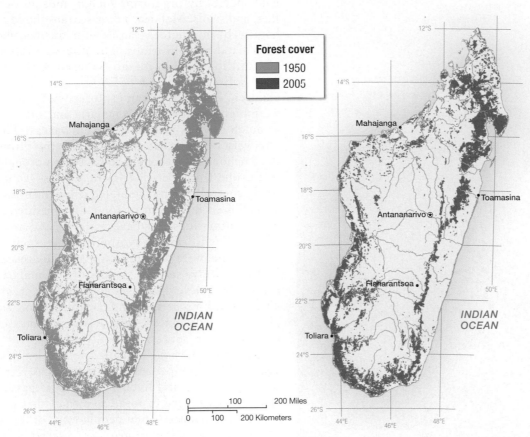

▲ **FIGURE 5.9 Deforestation of Madagascar** These maps, which compare forest cover in 1950 and 2000, illustrate the loss of the many important forest areas that protect lemurs and other rare species.

fuels and therefore rely on wood or charcoal for fuel. The increasing wood and charcoal demands of cities like Nairobi, the capital of Kenya, have had a tremendous impact on local forests, with serious deforestation tied to the city's energy needs occurring as far as 200 kilometers (124 miles) away. Projects to reduce energy demands by using scrap metal to make more efficient stoves have complemented the efforts of female-led nongovernmental organizations to protect trees in and around Nairobi. The best-known social movement of this kind is the **Green Belt Movement**, which counts 50,000 women as members. Led by Nobel Prize–winning environmental and political activist Wangari Maathai, Green Belt planted thousands of trees around Nairobi and has been the model for similar groups elsewhere in Africa and the world (**Figure 5.10**).

Drier areas in the region have mixed woodlands and grasslands. These **savanna** grassland areas cover about two-fifths of Africa. Savannas have open stands of trees interspersed with shrubs and grasses, vegetation typically found in tropical climates that have a pronounced dry season and experience periodic fires. The baobab tree is a symbol of this landscape, which is found in West Africa and in southern

of Madagascar remain (**Figure 5.9**). The island has one of the world's richest ecologies. Twenty-five percent of all the flowering plants in Africa (including the rosy periwinkle, which is used to treat the disease leukemia) are found here. It is also home to unique fauna, including perhaps 100 species of lemurs (see Table 5.2, p. 189), 800 species of butterflies, and numerous chameleons, cacti, and corals. Deforestation in Madagascar has a long history and is linked to clearing land for rice production, sugar plantations, and cattle ranches and cutting trees to export tropical hardwoods. About 80% of the population on the island consists of subsistence farmers who use a slash-and-burn technique called **tavy** to clear forests. Protected areas have been established to protect forests and foster ecotourism with the support of international conservation organizations, but poverty and mining still place surviving forests at risk.

In East Africa, the loss of forests has disproportionately affected women who are primarily responsible for heating and cooking and have limited access to electricity, gas, or petroleum

▲ **FIGURE 5.10 Wangari Maathai** Maathai, the founder of the Green Belt forest protection movement, led an initiative to protect the forests of the Congo Basin prior to her death in 2011.

▲ **FIGURE 5.11 Savanna** Grassland plains, such as the Serengeti of Tanzania and Maasai Mara in Kenya, have some of the densest concentrations of wild, hoofed, grazing mammals in the world. These animals coexist in the region with predators, such as the big cats—lions, leopards, and cheetahs such as the one shown here in the Maasai Mara. Larger herbivores in the region include elephants, giraffes, zebras, and rhinoceroses.

Africa near the Zambezi. Savannas provide extended grazing areas for both wildlife and livestock (**Figure 5.11**).

Deserts cover another two-fifths of sub-Saharan Africa. Deserts have very sparse and seasonal vegetation, for the most part, and feature drought-resistant plants, such as acacias and woody scrub. The Mediterranean climates of South Africa have produced a unique ecosystem dominated by vegetation (called fynbos) that is characterized by waxy or needlelike leaves and long roots that help plants survive long dry periods. The recent international popularity of the use of protea in flower arrangements and of rooibos tea has increased interest in this ecosystem, which is threatened by climate change.

Diseases and Pests African pests and diseases can have a devastating impact on human populations (**Figure 5.12**). Several of these diseases can be transferred to humans or domesticated animals from wild species by organisms such as mosquitoes, flies, and snails. **Malaria**, a disease transmitted to humans by mosquitoes, causes fever, anemia, and often-fatal complications. The disease affected 174 million people in sub-Saharan Africa in 2010, killing more than 655,000, many of them children. Early European explorers and settlers were highly vulnerable to malaria and suffered as much as 75% mortality in some areas. The discovery in 1820 that quinine—an extract from the cinchona tree, thought to have been brought to Europe from Peru by Jesuit priests—could partly control malaria facilitated colonialism and also allowed treatment of some local residents. But quinine does not cure malaria, and after World War II, several synthetic drugs, such as chloroquine, were developed to fight the disease. Unfortunately, several strains of the malaria parasite have developed resistance to these drugs. There is still no certain and cheap cure for the disease, although inexpensive mosquito nets can help prevent its spread.

Schistosomiasis (also called bilharziasis) is associated with a parasite that causes gastrointestinal diseases and liver damage. It is passed to humans who are exposed to a snail that is a host for the parasite when people work or bathe in slow-moving water in marshes, reservoirs, or irrigation canals. The disease is not fatal, but reduces general health and energy levels. The **U.N. World Health Organization (WHO)** estimates that 200 million people are currently infected by schistosomiasis in sub-Saharan Africa. River blindness (onchocerciasis) is transmitted by the bite of a black fly, which passes on small worms whose larvae disintegrate in the human eye and cause blindness. The eradication of river blindness by controlling fly populations with pesticides and treating victims with drugs has been relatively successful in West Africa. WHO estimates that 14.4 million live in high-risk areas, and 250,000 are blind as a result of this disease in sub-Saharan Africa. Other serious diseases include yellow fever, illnesses caused by rotaviruses, and sleeping sickness associated with the tsetse fly.

The **tsetse fly**, which lives in African woodland and scrub regions, is associated with both human and livestock diseases. In humans, the fly's bite causes sleeping sickness (trypanosomiasis). The symptoms of sleeping sickness include fever and infection of the brain that causes extreme lethargy and may end with death of the victim. Half a million people are infected with sleeping sickness in sub-Saharan Africa. It can be treated in early stages and prevented by a variety of pest-control measures. In domestic animals such as cattle and horses, the tsetse fly causes a disease called nagana, which is similar to sleeping sickness and causes fever and paralysis. This disease prevented the introduction of livestock into many parts of Africa and, as a result, preserved habitats for wild species in an unexpected benefit for conservation. Areas with an

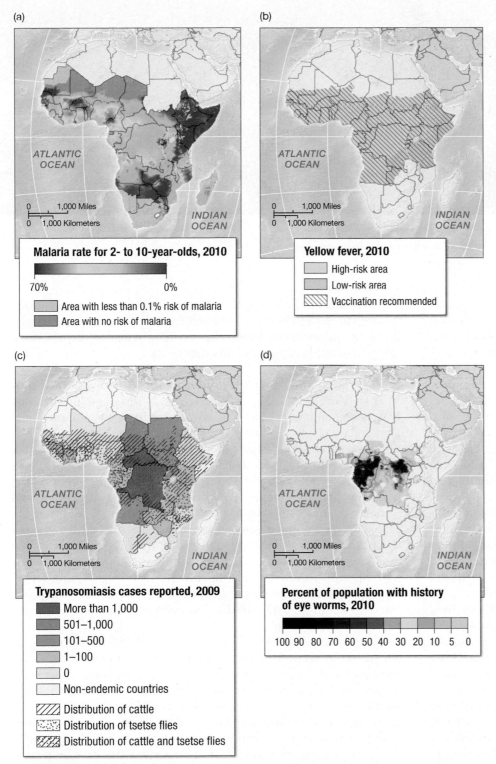

Many of these debilitating diseases are associated with the tropical climate and diverse ecologies of Africa. Their spread may have been facilitated by the expansion of human populations and the transformation of natural environments through deforestation and irrigation. Reducing the human toll from these diseases is a major challenge for scientific research, African governments, charitable organizations, and the WHO, which has targeted Africa for extra funds and programs. In 2011, the Bill and Melinda Gates Foundation committed over U.S. $1 billion to fighting malaria. The foundation's efforts have included the development of a malaria vaccine and distribution of mosquito nets and have spurred a multibillion-dollar global health initiative aimed at achieving major reductions in disease in poor countries, especially in Africa.

In addition to carrying disease, insects cause other problems for the human inhabitants of Africa. Locusts, a type of grasshopper, swarm in the millions after rains and can move across long distances, consuming all crops in their path and undermining farm livelihoods and food security. International efforts to control the devastation from periodic locust plagues include weather prediction, insect surveys, and insecticide application. Agriculture in the region is also vulnerable to large swarms of small birds, called Quelea, that eat vast amounts of grain crops.

APPLY YOUR KNOWLEDGE How does Africa's physical environment—including its climate, soils, pests—pose challenges for humans? What regional and international steps have been taken to overcome these challenges? How effective have these efforts been?

Land Use and Agriculture Less than 30% of the soil in sub-Saharan Africa is suitable for agriculture. In addition, agriculture is hindered by unsuitable climates and pests and diseases. However, Africa, the birthplace of the human species, is also the region where humans first adapted to the constraints of the physical environment. Early humans survived by hunting wild animals, fishing, gathering plants, and domesticating a number of crop and livestock species.

There is some disagreement about whether cattle were domesticated in Africa or introduced from the Middle East and Asia about 8,000 years ago. Archaeological sites from this period have provided evidence of domesticated livestock and of domesticated and cultivated grains, including sorghum. The highlands of Ethiopia are considered one of the early centers of **domestication**—the adaptation of wild plants and animals through selective breeding for preferred characteristics into cultivated or tamed forms. There, inhabitants still produce coffee, millet, and an important local cereal called teff used to make injera bread. Other crops domesticated in Africa include yams, oil palm, cowpea, and African rice. The introduction of maize from Latin America into Africa transformed land use and diets.

▲ **FIGURE 5.12 Tropical Infectious Diseases and Pests** These maps show the distribution of some of the more serious tropical diseases and pests in Africa. (a) Malaria is indicated by the proportion of 2- to 10-year-olds that were infected in 2010. (b) Yellow fever is indicated where vaccinations were recommended for international travelers in 2010. (c) Diseases associated with the tsetse fly are indicated by the distribution of cases of sleeping sickness (trypanosomiasis) in 2009. (d) River blindness is indicated by the prevalence of eye worm diseases in 2010.

elevation above 150 meters (480 feet), those with a long dry season, and those with sparse or no woodland are free from tsetse flies and became key areas for settlement.

▲ **FIGURE 5.13 African Pastoralism** The Masai of Kenya and northern Tanzania are herders—relying on cattle for meat, milk, and blood and moving their animals as the rains shift. In this photo, young men perform a traditional warrior jumping dance next to their cattle.

Traditional strategies for adapting to low soil fertility include **shifting cultivation**, which involves moving crops from one plot to another. One form of shifting cultivation is **slash-and-burn** agriculture. In this method, after an area is burned, the ash is used to fertilize crops. After a few years, when the nutrients are exhausted, farmers move on to a new area and leave the previous plot to return to forest or other vegetation. After a long fallow (rest) period, farmers return to clear and burn the land again. A modification of shifting cultivation is **bush fallow**, in which crops are planted around a village, and plots are left fallow for shorter periods than in the slash-and-burn system. Soil fertility is maintained through fallow periods and by applying household waste to the fields. The use of household compost to grow crops within a village is called "compound farming" and is popular in forest environments as well as in some urban areas. **Intercropping**—planting several crops together in a single field—is a technique for keeping the soil covered to reduce erosion, evaporation, and nutrient loss and improve soil fertility. Floodplain farming is an important technique in locations where there are seasonal floods, such as the inland delta of the Niger River and the Sudd wetlands along the Nile.

As noted earlier, pastoralism—a way of life that relies on livestock—is the agricultural activity best adapted to drier regions of Africa. Nomads, such as the Bedouin, migrate with their animals in search of pastures in the arid landscapes of the Sahel and North Africa. Other groups, such as the Fulani of West Africa, practice a system of seasonal herd movements called **transhumance**. They move their herds southward to wells and rivers in the dry season and drive them northward to take advantage of new pastures in the wet season. In some areas, local farmers let pastoralists graze their herds on harvested fields in the dry season. The famers' land is fertilized with animal manure in a mutually beneficial relationship with the pastoralists. In other areas, pastoralists are also farmers. Cattle are traded at regional markets and are a family investment for the future in places where there are few secure ways of accumulating capital (**Figure 5.13**).

Some explanations of land degradation in Africa, including desertification, blame Africans for overusing resources, especially through human and animal overpopulation, and for poor management of forests and soils. However, detailed case studies by researchers such as Michael Mortimore and Melissa Leach show that there are many examples of careful community management of forests and wildlife based on traditional rules and a deep knowledge of ecology.

For example, in Machakos, Kenya, a fivefold increase in population resulted in less rather than more soil erosion when members of the growing communities, especially women, began to terrace steep slopes, manage livestock, and diversify crops.

Conservation and Africa's Wildlife The rich biodiversity of sub-Saharan Africa is valued by local residents, tourists, and international environmental groups alike. However, differing views about its protection have resulted in many controversies about conservation. Traditional African societies hunted and gathered wild species for food; wildlife have long played a role in their spiritual beliefs. While human populations were low and hunting technologies were less effective, wildlife populations ranged naturally where climate, vegetation, and terrain were most suitable. As population, technology, and land use changed, especially after colonialism, human activity began to modify their habitats, and wildlife populations shifted. Europeans contributed to the decimation of African wildlife through indiscriminant hunting expeditions, eliminating animals near railroads and farms, and forcing Indigenous people off their lands into regions where they came into conflict with wildlife that encroached on the herds and fields of the local people.

Currently about 100 million hectares (386,000 square miles) or 5% of sub-Saharan Africa enjoy some sort of protected status, and there are more than 1,000 protected areas. Over half of these areas are in southern Africa. The major parks in East Africa and southern Africa, such as Serengeti in Tanzania and Kruger in South Africa, have become high-profile international tourist destinations, bringing millions of dollars to national economies and employing many local people. The parks are not without problems, however. Parks have been criticized for providing inadequate benefits to local people who may have been displaced and have lost traditional grazing and hunting rights, or who have had crops destroyed by marauding wildlife. In some parks, too much tourism and high animal densities have destroyed fragile habitats, and poaching has pushed some species close to extinction. Some parks are crowded and zoo-like, with dozens of tourist vehicles clustering around a group of lions, a cheetah, or a leopard.

Elephants, rhino, and gorillas are examples of charismatic **megafauna**, large species with popular appeal that are often used by conservationists as symbols for activism in worldwide campaigns (**Table 5.2**). For example, elephants are appealing because of their size and intelligence and the value of their ivory tusks. Elephants have been hunted for their tusks—known as "white gold"—for centuries. Pressure on herds in East and southern Africa grew with late 19th-century colonial demand for ivory décor, including piano keys, from Victorian England. Demand from Asia for decorative ivory also became a factor in the 1970s. A precipitous decline in African elephant populations from 2.5 million in 1970 to fewer than 500,000 in 1995 was a direct consequence of illegal killing and poaching, fueled by the ivory trade. The situation was aggravated by competition between people and elephants for land and by war and civil unrest, especially when firearms became available to unpaid soldiers and desperate refugees in regions where herds were unprotected.

Mounting international pressure resulted in a general ban on international sales of African ivory in 1989. In 1997, elephants were

TABLE 5.2 Charismatic Wildlife in Sub-Saharan Africa

Animal	Status	Geography	Threats	Solutions
African Elephant	Threatened; 470,000 to 690,000 individuals	Eastern, southern, and West Africa	• Poaching for bushmeat (meat from wild animals hunted for human consumption) and ivory • Human–elephant conflict • Changes in land use	• Reduce loss of natural habitat by protecting forests • Strengthen activities against poachers and the illegal ivory trade • Support and extend local wildlife authorities (training, funding) • Allow legal culling and ivory sales where elephant populations are high
Black Rhino	Critically endangered; 5,000 individuals	Cameroon, Kenya, Ethiopia, Namibia, South Africa, Rwanda, Swaziland, Tanzania, Zimbabwe, Zambia (reintroduced), Botswana (reintroduced)	• Hunting • Poaching for rhino horn trade • Habitat loss	• Institute antipoaching projects, including guards for individual rhinos • Establish and expand protected areas • Institute captive breeding programs • Preemptively remove horns so animals won't be hunted
White Rhino	Near threatened; 20,000 individuals	Botswana (reintroduced), Côte d'Ivoire (introduced), Democratic Republic of the Congo (DRC), Kenya (introduced), Namibia (reintroduced), South Africa, Swaziland (reintroduced), Zambia (introduced), Zimbabwe (reintroduced)		
Mountain Gorilla	Critically endangered; 786 individuals	Uganda, Rwanda, DRC	• Poaching for bushmeat • Habitat loss • War • Wildlife trade • Infectious diseases	• Prevent and reduce disease transmission through vaccination • Practice sustainable use of the species' forest habitats • Safeguard national parks and their resident gorillas • Increase fees for tourism • Reduce bushmeat hunting through education and introduction of alternative foods
Western Lowland Gorilla	Critically endangered; 150,000 individuals	Angola, Cameroon, Central African Republic, Congo, DRC, Equatorial Guinea, Gabon		
Lemurs	Of 100 species, 16% are critically endangered; most others are threatened; estimated at 750,000 individuals in 2008 (50% of 1950)	Madagascar	• Deforestation • Poaching for bushmeat • Climate change impact on forests	• Protect habitat • Institute captive breeding programs • Education to prevent hunting and consumption

listed under the Convention on International Trade in Endangered Species (CITES) (see Chapter 8, p. 316) as a species in need of protection. The ban on ivory sales was opposed by countries in southern Africa, which had seen a less serious decline in elephant populations and had some parks where elephants were destroying the habitats of other species. These countries were funding parks and conservation from the money earned from legal ivory and hide sales. In 2002, the United Nations granted permission to South Africa, Botswana, and Namibia to sell 60 tons of ivory. China was given permission by the United Nations to purchase ivory in 2008 and is now the global center of trading.

The rhino is under much greater threats, especially from poachers who hope to sell their horns. Rhino horn is sold in Yemen and other countries in the Middle East where it is made into dagger handles or in Asia where it is ground into a highly valued medicinal powder. Protecting the rhino from poachers who can make thousands of U.S. dollars from selling a single horn is a full-time and costly enterprise with individual rhino assigned their own armed guards. In central Africa, populations of gorillas and chimpanzees are hunted for trophies and **bushmeat** (meat from wild animals hunted for human consumption). Their habitats are being destroyed by deforestation despite the work of conservationists, such as Dian Fossey and Jane Goodall, who have studied their fascinating social behaviors and campaigned for their protection.

APPLY YOUR KNOWLEDGE Use the Internet to identify two groups campaigning on behalf of African wildlife. What arguments and data do they use in favor of protection and how do they take the needs of local people into account? What measures do they propose to protect endangered or threatened species?

HISTORY, ECONOMY, AND TERRITORY

Sub-Saharan Africa saw the dawn of humanity on the continent more than 2 million years ago. Its global role continued with the development of major trading societies about 5,000 years ago and the advent of the colonial system about 500 years ago. Most of sub-Saharan Africa was under European colonial domination by 1900. Independence movements, which began around 1950, coincided with Cold War tensions between the United States and the former Soviet Union, and these superpowers intervened in African political struggles and civil wars as they vied for global domination.

Since the beginning of the 21st century, much of sub-Saharan Africa has continued to struggle with the transition to independent and democratic government. Regional economies still continue to rely on a narrow set of exports to other world regions. Populations are still growing relatively fast, and many Africans are moving to cities. Although parts of Africa are becoming highly connected to other world regions through migration, telecommunication, and trade, other places are still isolated and rely on subsistence agriculture.

Historical Legacies and Landscapes

Contemporary African geographies—landscapes, livelihoods, and culture and politics—bear the imprint of prior periods of African history and of Africa's shifting connections to different regions of the world. These connections in turn have shaped other regions of the world. For example, colonialism spurred the worldwide trade in African slaves, resulting in a **diaspora** (the dispersion of people from their original homeland; see Chapter 1, p. 35) of African people that continues to influence the culture and societies of other world regions to this day. Colonialism also resulted in regional political boundaries that split ethnic groups across territories or clustered enemies within one territory.

Human Origins and Early African History Africa is often called the "cradle of humankind" because the earliest evidence of the human species (*Homo sapiens*) was unearthed there. Fossilized footprints of *Homo sapiens*'s earlier ancestor, the hominid (humanlike) *Australopithecus*, were found by archaeologist Mary Leakey at Laetoli in Tanzania and dated to 3.7 million years ago. Two-million-year-old stone tools have been found at several sites in Ethiopia and East Africa, including the famous site at Olduvai Gorge in Tanzania (**Figure 5.14**). Anatomically modern humans, who walked upright and had larger brains, lived at least 100,000 years ago in sites in southern Africa and along the Rift Valley. Genetic evidence shows that these humans are the ancestors of all modern humans—the most basic link between Africa and the world.

The complex societies in the Middle East and North Africa (see Chapter 4, p. 128), with their organization, systems, hieroglyphic writing, and hierarchical social organization, started to influence other parts of Africa around 5,000 years ago. By 500 B.C.E., some Indonesians had settled on Madagascar, introducing yams and bananas to mainland Africa. At about the same time, a strong kingdom had emerged at Aksum in Ethiopia and had adopted Christianity. Explorations, military campaigns, and European trading expanded into Africa from bases in the Nile Valley and North African Mediterranean coast. From about the 8th century, West African empires were associated with cities that were great centers of trade, scholarship, and power such as Timbuktu and Djenné in the Sahel (**Figure 5.15a**). By 1000 B.C.E., Zanzibar and the islands off the coast of East Africa were familiar to the Egyptians, Phoenicians, Greeks, and Romans (**Figure 5.15b**). As the Mediterranean empires extended their trade routes to the south and east, Zanzibar became one of several major commercial ports along the East African coast. In about the 3rd century B.C.E., merchants from southwestern Arabia also began trading with the island residents, bringing weapons, wine, and wheat to barter for ivory and other luxury goods. The Arabs also brought the religious traditions of Islam. The local language, Swahili, has many Arabic words and eventually became the trade language for eastern Africa.

The Advent of the Colonial Period in Africa With the development of faster and larger ships in the 15th century, contacts with Spain, Portugal, and China became part of the regular interaction between the Middle East and Africa. The Portuguese traded to get gold from coastal settlements in West Africa, and in 1497, the Portuguese explorer Vasco da Gama rounded the Cape of Good Hope at the southern tip of the African continent, initiating trade with the southeast coasts of Africa en route to India. In return for salt, horses, cloth, and glass, sub-Saharan Africa provided gold, ivory, and slaves to the world via Portuguese and Arab traders. For centuries, African slaves were in demand among Arabs, who used the slaves as servants, soldiers, courtiers, and concubines.

European colonialists took some time to establish control in Africa, and for many years only the coastal ports and trading posts were

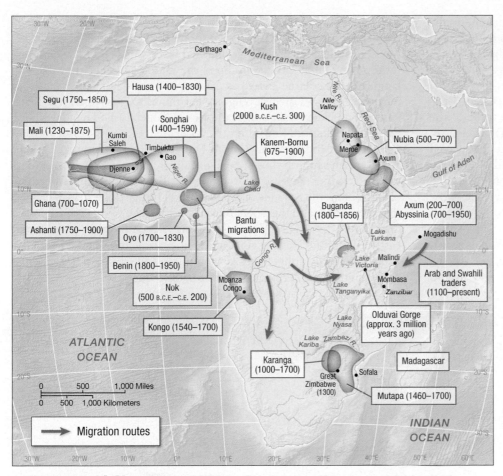

▲ **FIGURE 5.14 Map of African History** Sub-Saharan Africa had a rich historical heritage prior to the arrival of Europeans. The oldest human remains were found at Olduvai Gorge in present-day Tanzania. This map shows some of the locations and dates of the great kingdoms as well as the early migrations of Bantu people and Arab traders.

under European command. The European names for coastal regions along the west coast of Africa clearly indicate the commodities that they provided, from the Ivory Coast (Côte d'Ivoire) in the west to the Gold Coast (now Ghana) and Slave Coast (Nigeria and Benin) to the east. One of the main reasons for European reluctance to move inland was the reputation of Africa as the "White Man's Grave." The continent earned this nickname because so many Europeans died from diseases such as malaria, yellow fever, and sleeping sickness, against which they had no natural immunity. In addition, African armies attacked ports and fiercely resisted European attempts to move inland.

Slavery The coastal regions of Africa generated enormous profits for European traders mainly through slavery. The Portuguese took slaves for their own use on new sugarcane plantations on the Atlantic islands of Madeira and Cape Verde. In 1530, the first slaves were shipped to the Americas to work on plantations in Brazil. By 1700, 50,000 slaves were shipped each year to the Americas to provide labor on colonized lands and new plantations. The potential Indigenous labor supply in the Americas had been decimated by European diseases (see Chapter 6, p. 239), leaving plantation owners desperate for workers. Slavery was an important income source for some African coastal kingdoms, such as Dahomey and Benin, that captured enemies or residents of inland villages and sold them to the slave traders. It is estimated that more than 9 million slaves, who were mostly male, were shipped to the Americas from Africa between 1600 and 1870. At least 1.5 million slaves died

▼ **FIGURE 5.15 Early Centers of Commerce and Learning** (a) The Sankoré madrassah (school) includes three mosques that are ancient centers of learning and are now part of the University of Timbuktu. The city was a center for learning from at least 1000 C.E., and by the 1300s, Sankoré had 25,000 students and more than half a million books and manuscripts. It was also one of the key trading centers for West Africa. (b) The island of Zanzibar, just off the coast of Tanzania, was a key port on the Indian Ocean trading with the Middle East and Africa. The ship shown here is a traditional *dhow* used for trading and transport along the coast.

(a)

(b)

during the journey (**Figure 5.16**). The conditions of capture and transport were horrific. Hundreds of slaves were packed into the holds of ships with little food and water and were brutally abused by traders. In the Americas, many slaves introduced their own traditions—including ways of cultivating rice, religion, and music—that endure today.

By the end of the 18th century, social movements led to the banning of slavery in Britain in 1772, the end of slave trade in the British colonies in 1807, and the emancipation (freeing) of slaves in the British Caribbean in 1834. Slavery was a divisive issue that played a role in the U.S. Civil War; it was eventually abolished in 1863 by the Emancipation Proclamation in the United States. Some liberated slaves returned to Africa and became elites in the countries now known as Sierra Leone and Liberia.

European Settlement in Southern Africa European settlement was encouraged in southern Africa by the more temperate climate and the strategic significance of the trading routes from Europe to Asia around the southern tip of the continent. In 1652, the Dutch established a community at Cape Town. Cape Town soon became surrounded with small cattle and wheat farms, which supplied trade ships and local communities. The Dutch settlers modified the Dutch language into the new language that became known as Afrikaans. They belonged to the strict puritan Christian Calvinist religion, saw themselves as superior to black Africans, and became known as the **Boers** (which is the Dutch word for farmer). **Afrikaners** (including the *Boer*) are a Germanic ethnic group in Southern Africa descended from Dutch (including Flemish), French, and German settlers whose native tongue is Afrikaans. As British military and trading power grew in the 1800s, British immigrants were encouraged to settle around Cape Town and Durban. The British banned slavery in 1834, and the Boers moved north of the Orange River in the Great Trek, settling on the high pastures called the *veld*, or migrated eastward into the Natal region, where they came into conflict with the Zulus. The geography, territory wrangling, and racism of this colonial period framed the 20th-century politics of southern Africa.

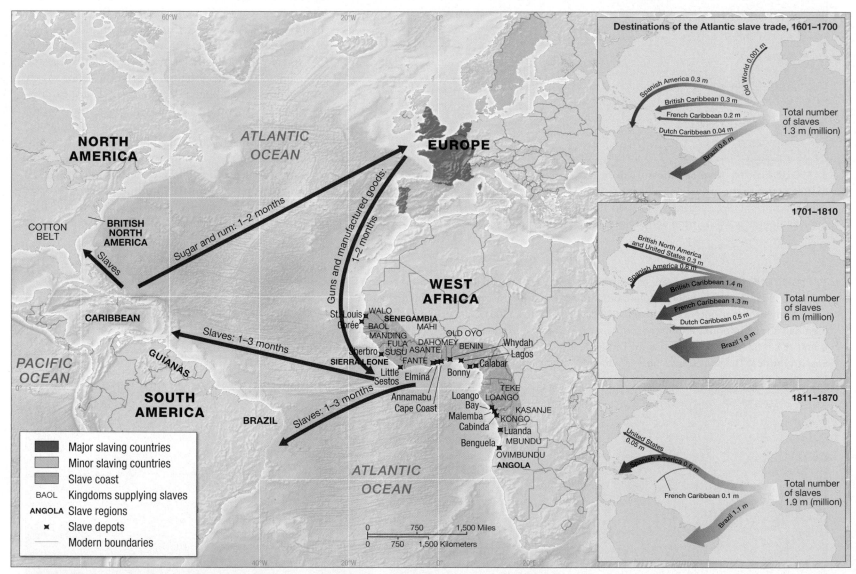

▲ **FIGURE 5.16 The Slave Trade** Millions of slaves were exported from Africa between 1600 and 1870, mainly from the West African coast. Some local leaders acted as suppliers in return for guns and manufactured goods. Slaves were sent to work on plantations in the Americas. The plantations sent sugar, rum, and other products back to Europe in a triangular trade.

(a)

(b)

▲ **FIGURE 5.17 Colonial Explorers** (a) David Livingstone is best known for his explorations of the Zambezi and his encounter with the magnificent waterfalls that he named after Queen Victoria. (b) Henry Stanley, a journalist sent to find Livingstone, poses as the conqueror of a hostile continent with his servant boy, Kalula.

European Exploration and the Scramble for Africa International interest in Africa increased dramatically after the 1850 discovery that quinine could suppress malaria. Explorers, traders, and missionaries moved to the interior of the continent seeking new territories, the source of the Nile, commodities, and souls to convert. Some of the most famous explorers were associated with the British **Royal Geographical Society (RGS)**, which supported and awarded medals of honor to many explorers, including David Livingstone, Henry Stanley, Richard Burton, and John Speke (**Figure 5.17**).

These Victorian explorers added to geographic knowledge of Africa, and their reports fueled colonial interest in the continent's resources and people. Their lectures at the Royal Geographical Society and elsewhere also increased interest in the discipline of geography and geography's role in Britain's colonial enterprise. Unfortunately, their books and those of other explorers contained many Victorian prejudices and racist and paternalistic attitudes that fostered the popular image of Africa as a barbarous and exotic continent in need of civilization and colonial supervision.

The competition among European powers, the reduced risk of African diseases, and the discovery of gold and diamond reserves further increased interest in the continent, especially from the British (Egypt, East Africa, and the West African coast), the French (West Africa), the Belgian King Leopold (the Congo), the Portuguese (southern Africa), and the Germans (**Figure 5.18**, p. 194). Pressure from commercial companies and even missionaries drove imperial ambitions and led to the incorporation of Africa into the emerging global capitalist economy. European governments granted exclusive concessions for trade and resource exploitation in Africa to private companies, who played a major role in the colonization of Africa. These companies, which were often given the right to police, to conscript, and to tax local populations, included the Royal Niger Company and British South Africa Company, both of which received royal charters in the late 1800s.

Between 1880 and 1914, European powers, especially the British, French, and Belgian governments, aggressively moved to colonize Africa. The British claimed the areas that are now known as Gambia, Ghana, Nigeria, and Sierra Leone in West Africa; Kenya, Uganda, Sudan, and part of Somalia in East Africa; and southern Africa, except for German South West Africa (Namibia), Portuguese Mozambique, the Cape Verde Islands, Angola, and the independent Boer region of South Africa. Germany claimed German East Africa (Tanzania) and Cameroon; Portugal claimed the Cape Verde Islands; and France and Belgium split the Congo. Italy took Somalia, Djibouti, Eritrea and coastal regions of Libya, and harbored ambitions to take over Abyssinia (now Ethiopia). The Spanish obtained a small coastal region of northwest Africa and Equatorial Guinea. Most of the remaining territory of West and North Africa was allocated to or taken by the French.

This hasty scramble for Africa culminated in the **Berlin Conference** of 1884–1885, a meeting convened by German Chancellor Otto von Bismarck to divide Africa among European colonial powers. Thirteen countries were represented at this conference, but it did not include a single African representative from sub-Saharan Africa, even though more than 80% of Africa was at that time still under African rule. The Berlin Conference allocated African territory among the colonial powers according to prior claims and laid down a set of arbitrary boundaries that paid little respect to existing cultural, ethnic, political, religious, or linguistic regions.

The next 20 years saw some rearrangement and consolidation of the European colonies as the British created regions of control called protectorates in southern Africa and expanded its control over the Sudan. In southern Africa, British entrepreneurs, including the ambitious Cecil Rhodes, responded to the discovery of gold and diamonds between 1867 and 1886 by acquiring the mines at Kimberley and the Rand and sparking a gold and diamond rush. Growing tensions between the British and Afrikaners resulted in the Boer War (1899–1902), which gave control of much of southern Africa to the British. The French eventually took control of many regions along the Niger River and Italy briefly occupied Abyssinia from 1889 to 1896. By 1914, almost all of Africa was under European colonial control except for Abyssinia (now Ethiopia), Liberia, and some interior regions of the Sahara Desert. A number of battles were fought in Africa during World War I. Germany's eventual loss in that war redistributed the German colonies to Britain, France, and Belgium.

The Impact and Legacy of Colonialism All these details about the reshuffling of African territory among European states can overshadow the everyday impact of colonial rule on African landscapes and peoples. The most general and enduring effects of colonialism include the establishment of political boundaries that disregarded traditional territories; a reorientation of economies, transport routes, and land use toward the export of commodities; improved medical care; and the introduction of European languages, land tenure systems, taxation, education, and governance. Many of the

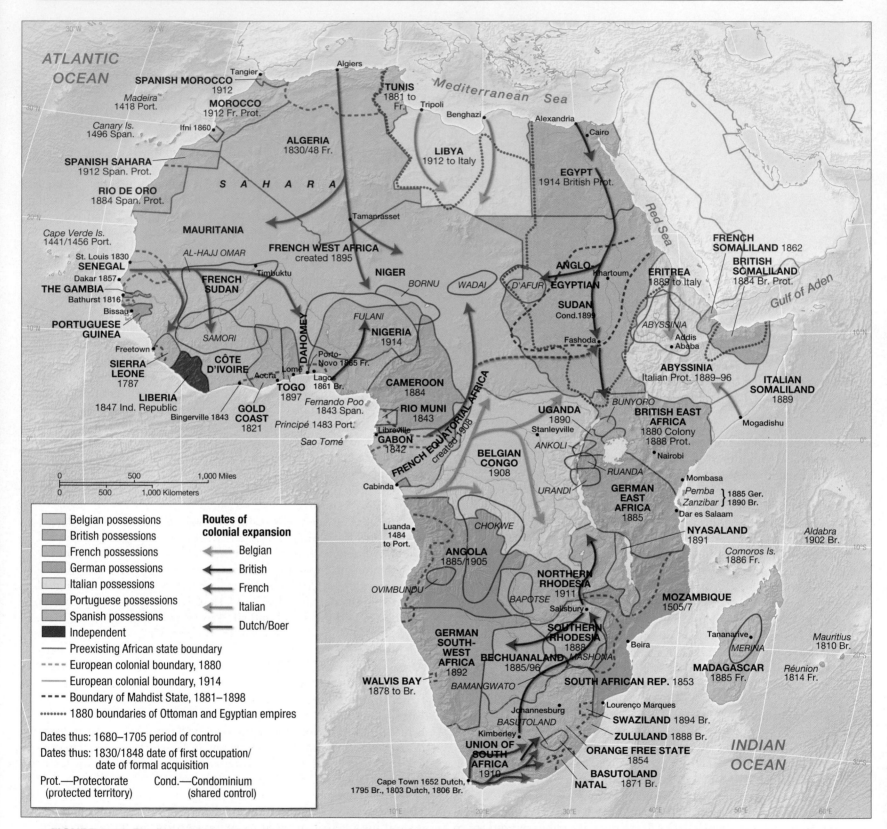

▲ **FIGURE 5.18 The Scramble for Africa** This map shows the routes and dates of the European colonization of Africa from 1880 to 1914.

new colonial boundaries divided Indigenous cultural groups and in some cases placed traditional enemies within the same country. For example, the Yoruba were divided between Nigeria and Benin. Nigeria comprised several competing groups, including the Yoruba in the southwest, the Ibo in the southeast, and the Hausa in the north. These groups still struggle over political and cultural differences today.

Colonial mines extracted large amounts of gold, diamonds, and copper for export to Europe. New roads and railways were constructed from inland to the coasts to speed the export of crops and minerals, but few efforts were made to link regions within Africa. The resulting infrastructure still facilitates trade beyond but not within Africa. Colonists also established plantations to produce crops, such as rubber, and used a variety of means, including taxation and intimidation, to persuade peasant farmers to produce peanuts, coffee, cocoa, or cotton for export. In the temperate climates of the East African highlands and southern Africa—areas that were more attractive to European immigrants—the best land was taken by white settlers for tea and tobacco plantations, livestock ranches, and other farming activities. By 1950, the geography of African agriculture clearly reflected this export orientation. Vast rubber plantations were owned by the Firestone Corporation in Liberia. Cocoa dominated the cropland of Ghana and the Côte d'Ivoire. Other dominant crops included cotton in Sudan, peanuts in French West Africa, and tea and coffee in East Africa (**Figure 5.19**). Traditional African land-tenure systems of communal land and flexible boundaries were often subsumed into privately owned and bounded plots, and traditional agriculture, decision-making processes, and legal systems were often replaced with European science, managers, and courts.

The effects and process of colonial rule varied. The British chose a paternalistic indirect rule for most of their African colonies. Preexisting power structures and leaders were made responsible to the British Crown and colonial administrators in a decentralized and flexible administrative structure. For example, local leaders were required to collect taxes—sometimes a hut tax based on the number of dwellings in a community, sometimes a poll tax based on the number of residents. To obtain money to pay taxes, people had to produce crops for sale to the Europeans. This was an indirect but effective way of transforming economies and land use to commodity production. Foreign ownership of land was prohibited in some cases, and traditional legal systems were used to resolve local conflicts. The British, preceded by missionaries, also introduced some European-style schools, and by the 1940s, a select group of Africans were attending overseas universities and given posts in government administration. The English language was common in many countries by the 1950s.

The French colonial policy was one of assimilation, encouraging local elites to evolve into French provincial citizens with allegiance to France. Agriculture and mining, however, were under close scrutiny from the French capital in Paris. By 1946, there were about 20 Africans, elected from West Africa, in the French parliament. The Belgian and Portuguese modes of colonialism in

(a)

(b)

▲ **FIGURE 5.19 Export Crops in Africa** (a) A tea plantation thrives below Mount Kenya. Tea was introduced into the Kenya highlands from India in 1903 and grown as a plantation crop on colonial landholdings. There are still 150,000 hectares (580 square miles) under production, providing about 10% of the world total. (b) A child carries a peanut plant near Djiffer, Senegal. Peanuts were introduced as a cash crop in French West Africa during the colonial period and are still important in Nigeria and Senegal.

central and southern Africa were much harsher, with direct rule and often ruthless control of land and labor. These authoritarian forms of control—which allowed for little local political participation, featured dominating official ideology, and involved the frequent use of armed force—provided an unfortunate model for leadership in independent Africa. Although the impact of the colonial period on Africa was dramatic and continues to this day, it is important to note that in most of Africa, formal colonialism only lasted 80 years, from about 1880 to 1960.

Independence The transition from colonies to independent states in sub-Saharan Africa was rapid and diverse. There were some relatively peaceful handovers to well-prepared African leadership. There were also some more violent transitions of power to divided or unprepared

TABLE 5.3	African Independence	
Decade	Colonial power	Independence granted to
1910	Britain	South Africa
1950	Britain	Ghana, Sudan
	France	Guinea
1960	Britain	Botswana, Gambia, Kenya, Lesotho, Malawi, Nigeria, Sierra Leone, Somalia, Swaziland, Tanzania, Uganda, Zambia
	France	Benin, Burkina Faso, Cameroon, Central African Republic, Chad, Congo, Côte d'Ivoire, Gabon, Mali, Mauritania, Niger, Senegal, Togo
	Belgium	Burundi, Rwanda
	Spain	Equatorial Guinea
1970	Portugal	Cape Verde, Mozambique, Sao Tome, Angola
1980	Britain	Zimbabwe
1990	South Africa	Namibia
	Additionally, the following two new independent countries seceded from other countries:	
1993	Ethiopia	Eritrea
2011	Sudan	South Sudan

local elites and militaries (**Table 5.3**). Independence movements were inspired by a variety of events, including Indian and Pakistani independence in 1947, by the half-million Africans who fought with the allies in World War II, by several foreign-educated activists, such as Kwame Nkrumah of Ghana and Jomo Kenyatta of Kenya, and by a Pan-African movement, led by black activists—including W. E. B. DuBois and Marcus Garvey—in the United States.

South Africa became independent in 1910, but most other British colonies did not gain independence until the 1950s. Although many British handovers were relatively peaceful, countries with significant white populations endured more violent transitions. In Kenya, white settlers controlled the best land, dominated the government, and set policy in the interests of the white minority, which consisted of only about 60,000 white residents. Many of the white settlers opposed independence, but Kenya eventually gained independence in 1963 in the wake of the violent Mau-Mau rebellion. In Southern Rhodesia (now Zimbabwe), the population of about 250,000 white (mostly British) settlers declared independence from the United Kingdom in 1965 rather than consider the possibility of rule by the majority of 6 million black Africans. Only after 15 years of conflict and international trade embargoes did an independent Zimbabwe finally emerge in 1980, with a mostly black government and a legacy of resentment against white residents.

In West and Equatorial Africa, transitions occurred dramatically in 1960 when France suddenly recognized a large group of independent countries. In most cases, strong economic and cultural ties were maintained with France, the franc remained the currency, and French troops were stationed in most countries. When Belgium abandoned its African colonies in the early 1960s, the countries of

Zaire, Rwanda, and Burundi were left with internal ethnic and political conflicts that manifested in later problems. Portugal hung onto its colonies of Angola and Mozambique until 1974, despite independence movements sponsored by the Soviet Union and Cuba, who supported rebel movements as part of the Cold War.

APPLY YOUR KNOWLEDGE What are the legacies of colonialism for African landscapes and politics? Choose two or three nations that are former colonies and research evidence of these legacies online.

Economy, Accumulation, and the Production of Inequality

Sub-Saharan Africa scores very low on many economic development measures compared with other world regions. The region had an average GNI PPP per capita of U.S. $2,251 in 2011, compared to the world average per capita of U.S. $11,574. Many national economies in the region are dependent on just a few low-priced exports, and large numbers of people make a living as subsistence farmers or lack formal employment. Africa is singled out for attention by international agencies and receives the highest amount of development assistance per capita of any world region. The World Bank has identified sub-Saharan Africa as "the most important development challenge of the 21st century."

Dependency, Debt, and African Economies Some blame sub-Saharan Africa's low levels of economic development on the harsh environment or the legacy of colonialism, arguing that colonial powers transformed the political and economic structures of Africa to serve their own interests. The colonial focus on obtaining cheap raw material to export undermined local agriculture and social development. Many African countries emerged from colonialism with their economies and trade dependent on just a few products (such as minerals or cocoa). The export value of these products declined over time in relation to the price of imports, especially manufactured imports. Because of these deteriorating **terms of trade**, small farmers had to produce twice as much coffee in 1990 as in 1960 to earn the money needed to purchase a bag of fertilizer. Some experts argued that sub-Saharan Africa (like other world regions, such as South America) needed to modernize and recommended development programs that included technology transfer, training, and large infrastructure projects, such as hydroelectric dams and roads. Other experts suggested that these actions would reinforce the dependency of the region on high-priced imports and outside expertise. This group argued that African countries needed instead to substitute local goods for costly imports by subsidizing local industry and setting up trade barriers to foreign imports.

These proposed development programs of modernization or import substitution required funds that were not easily available in Africa, and many countries looked outside the region to borrow money. Because many African countries had poor credit ratings with commercial banks and could not get loans on the private market, most of the loans that African nations were able to obtain were from other governments such as the United States or through international banks, such as the World Bank. Although some of these borrowed funds were certainly invested in infrastructural, industrial, and agricultural development, there is evidence that considerable sums were used to purchase arms or were diverted by ruling elites, who increased their own personal fortunes as over-all debt increased.

As in other regions, the multilateral agencies such as the International Monetary Fund (IMF) and the World Bank responded to the debt crisis in sub-Saharan Africa in the 1980s first by controlling the growing size of the loans through extending payment periods and adjusting interest rates and then by demanding that African governments cut budgets and remove subsidies and trade barriers in return for debt relief. In many parts of Africa, this required reduced public spending and the privatization of government-held companies. In many countries, the impact was severe, sending food prices soaring and increasing unemployment. The destitution created in Africa by economic crises and war prompted international agencies and others to try to cushion the impact by providing programs for alleviating poverty. Africa is the largest recipient of foreign aid, and receives one-third of the U.S. $131 billion that the United States gives globally. This is equivalent to about 4.2% of the region's total GNI and averages about U.S. $53 per person.

In many countries, the annual external debt was more than 50% of GNP in the 1980s and 1990s, and several countries were paying more than five times the value of their exports to service their debt each year. Official recognition that many of the countries had debt burdens that would permanently cripple development and destabilize governments led to several debt-relief programs. The Heavily Indebted Poor Countries (HIPC) initiative was adopted by the World Bank and the IMF in 1996 to restructure and forgive part of the debt of poor countries, including 33 in sub-Saharan Africa that showed a willingness to pursue policies of reduced government spending and free trade and to develop poverty reduction strategies. Much of the debt was controlled by the wealthy **G8** countries (the Group of Eight countries that include Canada, France, Germany, Italy, Japan, Russia, the United Kingdom, and the United States). Campaigns for debt relief included "Make Poverty History" in 2005. This was supported by many young people and humanitarian organizations across the world, as well as celebrities such as Bono of the band U2, who marched and pressured world leaders to act. In 2005, a G8 summit agreed to double the annual aid to the developing world by U.S. $50 billion and to write off the debts of the world's 18 poorest countries, 12 of them in Africa. Between 2001 and 2011, the total external debt of sub-Saharan Africa fell from U.S. $148 billion to U.S. $116 billion (**Figure 5.20**). By 2011, 26 African countries had their external debt cancelled in full at a cost in excess of U.S. $50 billion, which was funded by the World Bank, the IMF, and creditor countries. Debt relief has helped many countries, but critics point out that many of these countries are still disadvantaged by unequal terms of trade and that the cuts and privatization of public services have increased inequality and well-being.

Contemporary African Agriculture Minerals and petroleum (discussed earlier in the chapter) dominate the overall economy of sub-Saharan Africa in terms of GNI contribution. However, in terms of employment, 61% of the adult population works in agriculture. Most farmers in this world region have small plots and grow just

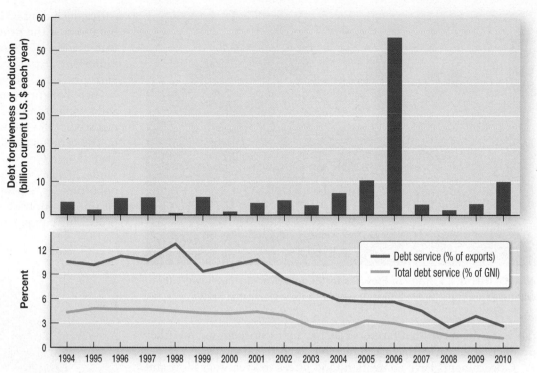

▲ **FIGURE 5.20 Declining African Debt** As a result of debt restructuring and forgiveness as well as improved economic conditions, debt across sub-Saharan Africa has declined in the last decade.

enough food to feed their families at a subsistence level. But commercial cash and export crops are also very important in some areas where export patterns for cocoa (Ghana, Côte d'Ivoire), cotton (Benin, Burkina Faso), nuts (Gambia), rubber (Liberia), tea (Kenya), and coffee (Burundi, Ethiopia, Uganda) were established in the colonial period. In some countries, there are also new agricultural export sectors. In Kenya, the most rapidly growing export sector is fresh vegetables and cut flowers, grown for export to Europe. Reliant on refrigerated air transport out of Nairobi airport, Kenya's horticulture industry is worth more than U.S. $3 billion a year. Kenya is now the number one exporter of flowers to Europe, providing more than half of all roses exported to the European Union (**Figure 5.21**, p. 198). Although these new industries provide employment and higher wages than some other sectors, the strict quality standards, perishable nature of the product, and reliance on air transport make it difficult for small producers to compete. There are also concerns about pesticide risks to workers. These and other related concerns have engendered a growing **fair trade movement** that is concerned with ensuring that producers are paid a reasonable wage and that crops, including flowers and coffee, are produced with more sustainable methods.

Per capita agricultural production has fallen or been stagnant in the region for many years because growth in agricultural production has not kept up with the growth in population and increased demand from urban populations. Agricultural progress has not raised the incomes or improved the nutrition of many of Africa's residents. Many areas in the region have become dependent on food aid, and exports have declined relative to imports. Agricultural problems in sub-Saharan Africa have been attributed to environmental degradation, lack of infrastructure, government policy, and unfair international market structures. Region-wide challenges, identified by geographer Godson Obia, include improving infrastructure for roads and storage and providing adequate incentives and rewards to local producers. The difficulties of getting crops to the market on Africa's dirt roads and tracks, especially in the

▲ **FIGURE 5.21 Kenyan Flowers for Export** Workers cut flowers that will be exported at the Wildfire flower farm near Naivasha, Kenya.

rainy season, mean that farmers risk having to store grain and other products in granaries that are vulnerable to pests and molds.

Governments have controlled food prices to keep wages low and quell unrest in urban areas, and this has in turn reduced the prices paid to farmers. International market volatility and a lack of information have made it difficult for farmers to move into new types of crops, and the general decline in the price of agricultural products in comparison to imports has also been problematic. The global rise in food prices in the wake of the 2008 global economic crisis did help some African farmers get more money for their crops, but was devastating for many poor consumers who could not afford to buy the higher-priced food.

One often-overlooked African success story is urban agriculture. Over 30 million city dwellers in the region cultivate crops. In cities such as Nairobi (Kenya), Lusaka (Zambia), Kano (Nigeria), and Kinshasa (Democratic Republic of the Congo), more than half the residents cultivate gardens, either at their homes or on unused land in the city. Kinshasa has been described as a giant garden plot, with crops growing at every roadside, on traffic islands, and on the airport perimeter. Crops from these urban plots contribute to urban food security and local incomes, as surplus is sold at local markets.

There is some concern that recent large-scale land purchases, often referred to as **land grabs**, may undermine African food security. This is especially a concern when purchasers plan to use the land for biofuels, mineral or water rights, or private export crop production. As of 2011, an estimated 50 million hectares (193,000 square miles), an area the size of Kenya, had been set aside in land deals in sub-Saharan Africa. Investors and purchasers include China and the Persian Gulf States, major Western banks, and individual U.S. investors. Large tracts of irrigated land that have been important to local food production in Ethiopia and Tanzania have been purchased. Contracts are often for very low cost or cheap leases and tend to generate only a few jobs for local people.

Manufacturing and Services Sub-Saharan Africa is less industrialized than many other world regions, although African manufacturing output has doubled in the last ten years. South Africa dominates the economy of the region with a substantial manufacturing sector that includes iron and steel, automobiles, chemicals, and food processing. Several major automobile companies produce their cars in South

Africa, including BMW and Toyota. One of the major goals is keeping more of the processing and finishing of goods in Africa rather than exclusively exporting raw materials. A number of countries have important textile and clothing industries often managed by female entrepreneurs. Some small businesses, including some owned by women, benefit from **microfinance** programs, which provide credit and savings to the self-employed poor, including those in the informal sector who cannot borrow money from commercial banks. Based on the demonstrated success of the Grameen Bank program in Bangladesh (see Chapter 9, p. 365), which provided small loans to thousands, African microfinance projects offer loans and secure savings opportunities to people who want to start or expand their businesses. Examples include loans to purchase sewing machines, food-processing equipment, agricultural supplies, and shop inventories.

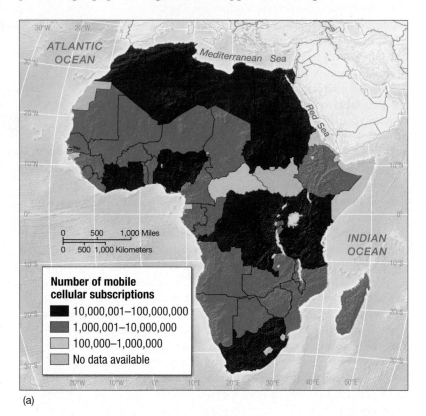

(a)

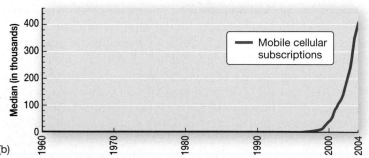

(b)

▲ **FIGURE 5.22 Connected Africa** (a) This map shows the rapidly expanding mobile phone network in Africa in 2010 with many subscriptions in southern, eastern, and western African coastal regions. (b) The graph shown here clearly indicates how recently and rapidly cell phone use has grown in Africa. Cell phones are used in many areas where there are no landline phones. Cell phones are charged by car batteries and solar energy and often shared in a community. Like people in all world regions, Africans use thier phones to obtain information about crop prices, call for medical help, access banking services, and keep in touch with friends and family.

The telecommunications sector has grown explosively with the rapid adoption of mobile phone technology in Africa. Mobile phones and the Internet are transforming the economy and culture, even in remote areas, with half a billion mobile phones in use across the region (compared to 4 million in 1998). New initiatives allow people to use their phones for banking transactions and for receiving weather and market-price information. Rwanda has set out to establish itself as a major node for Internet communication, although only just over 5% of Africans have Internet access. New undersea cables are planned to increase bandwidth and connection speeds (**Figure 5.22**).

The service sector is large in most countries in the region, with many people working as civil servants in government as well as in the tourist sector (especially in southern and eastern Africa). There are also two "hidden" contributions to economic development in most of Africa. The first is the **informal economy**; millions of people work as street vendors and maids, for example, and are neither taxed nor monitored by a government. In countries such as Zimbabwe, Tanzania, and Zambia, as much as half the GNI and more than three-fourths of the jobs are associated with the informal economy. The second contribution is from Africans working abroad who sent **remittances** of more than U.S. $21 billion home in 2010.

Since 2000, sub-Saharan Africa has experienced quite strong economic growth, above the world average at 5% per year, and with parallel growth in foreign direct investment from about U.S. $11 billion to U.S. $35 billion in 2010. Higher commodity prices contributed to growth for both mineral and agricultural products, with South Africa and Botswana especially competitive. Growth slowed with the global economic downturn in 2008, as export demand and foreign investment dipped but strong growth (of 4.7%) had returned by 2011 (**Figure 5.23**).

China and the African Economy China has become a major new economic partner for many countries in sub-Saharan Africa and, since 2009, has been the region's largest trading partner. Chinese exports to

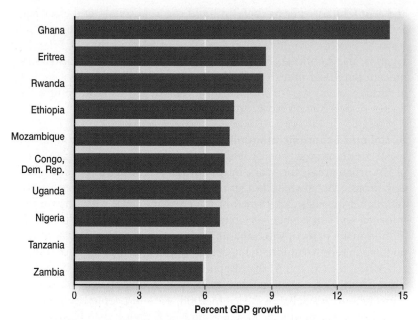

▲ **FIGURE 5.23 Economic Growth Rates in Sub-Saharan Africa** This graph shows the fastest growing economies in Sub-Saharan Africa in 2011.

the region include many low-cost goods, and cheap Chinese textiles and clothes have invaded markets, damaging local producers in areas such as West Africa. Access to raw materials drives Chinese investment and interest in Africa and, as a result, exports of wood, minerals, and foodstuffs to China are growing (**Figure 5.24**). China imports more than a quarter of its oil from Equatorial Guinea, Republic of the Congo, Gabon, Cameroon, and Nigeria as well as timber from the Congo Basin and huge quantities of critical minerals, such as cobalt, manganese, copper, and iron ore. China has also made massive investments in infrastructure in Africa. For example, in return for oil, China loaned money to Angola

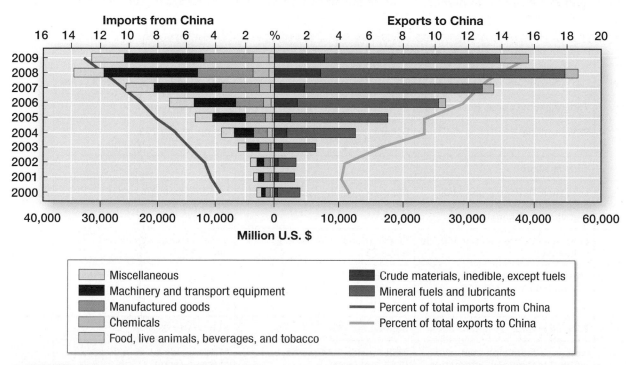

▲ **FIGURE 5.24 China and Africa** Trade between sub-Saharan Africa and China has grown dramatically since 2000. Food and raw materials travel from Africa to China and manufactured goods and equipment from China to sub-Saharan Africa.

and sent Chinese experts there to build roads, hospitals, and electrical and telecommunication systems. China has negotiated special economic zones to procure favorable tax and tariff treatment in Ethiopia, Nigeria, Uganda, and Zambia. As noted earlier in the chapter, China is also a significant purchaser of African land. Critics see the Chinese involvement in Africa as a new type of colonialism and claim that Chinese companies do not respect human rights or the environment.

Social and Economic Inequality and Development Economic and social conditions vary greatly within and among African countries. People in urban areas who work in the formal sector or are producers of cash crops generally have longer life expectancies, better service access, and higher incomes than those in rural areas and those who work in the informal and subsistence agricultural sectors. Income concentration is high in many parts of sub-Saharan Africa, with the richest 20% of the population receiving more than 60% of overall income in most of southern Africa, the Central African Republic, and Sierra Leone. In Namibia, a staggering 79% of national income is held by the top 20% of society. Throughout sub-Saharan Africa in 2008, 47% of people lived on less than the equivalent of U.S. $1 a day and 69.2% on less than U.S. $2.

Life expectancy in the region averages 14 years below the world average of 69 years and is lower than in any other world region. Conditions in Africa are particularly difficult for children. The infant mortality rate is 76 deaths per 1,000 children born (almost double the world average) and nutrition and immunization are at very low levels. In 2010, 19% of preschool children in Africa were underweight.

Within Africa, these generally gloomy average statistics do not reflect the situation in some regions where residents enjoy much better conditions. The United Nations ranks Cape Verde, Gabon, South Africa, Sao Tome and Principe, and Botswana as well as the Seychelles and the Mauritius higher on the Human Development Index (life expectancy, literacy, education, and GDP per capita) than other countries of Africa. Life expectancies in these countries are generally above 50. Levels of literacy, education, and income are also higher

than the regional averages in these areas, which tend to have better provision of basic services. For example, in South Africa, Gabon, and Botswana, 80% or more of the population has access to safe drinking water. In almost all countries in the region, life expectancy, incomes, and access to services are higher in urban areas than in rural ones.

The lowest 22 countries on the Human Development Index are all in Africa. Sierra Leone, for example, has a life expectancy of only 47 years and an annual GNI per capita averaging U.S. $324; these figures reflect the loss of life and economic collapse associated with civil war. Low life expectancies reflect some of the deficiencies in service provision in Africa. Two-thirds of the population lack access to safe drinking water in Ethiopia, Angola, the Democratic Republic of the Congo, and Sierra Leone.

Generally across the region, however, conditions have improved over the last 25 years. From 1970 to 2010, life expectancy has increased from 45 to 55 and infant mortality has dropped from 138 to 76 deaths per 1,000 children born. Literacy has also shown dramatic changes, increasing from 38% in 1980 to 62% in 2010. Improvements in conditions in specific countries are reflected in life expectancy and infant mortality changes. In Ghana, for example, life expectancy increased from 45 to 63 years since 1970, and infant mortality decreased from 131 to 50 deaths per 1,000 children. But war and AIDS have also affected Africa. Life expectancy in Botswana and Zambia started to drop in the 1990s as a result of AIDS and in Rwanda as a result of war and genocide. HIV/AIDS was also responsible for infant mortality increases in Botswana, Kenya, Zambia, and Zimbabwe in the late 1990s.

African development is increasingly driven by a new set of targets established by the United Nations called the **Millennium Development Goals (MDGs);** these aim to eradicate extreme poverty and hunger; achieve universal primary education; promote gender equality and empower women; reduce child mortality; improve maternal health; combat HIV/AIDS, malaria, and other diseases; ensure environmental sustainability; and develop a global partnership for development (**Table 5.4** and **Figure 5.25**). Sub-Saharan Africa has

TABLE 5.4 Millennium Development Goals: Results for 11 Villages after Three Years			
Millienium development goal	Millennium village goal	Measured indicator at start of program	Measured indicator after 3 years
Eradicate poverty and hunger	75% of people above poverty line		
Provide universal primary education	90% in school	School meals 26%	School meals 75%
Empower women	Girls in school		
Reduce child mortality	Less than 30 deaths per 1,000, 90% with measles vaccination and vitamin A supplement	Children sleeping under treated mosquito nets 7%	Children sleeping under treated mosquito nets 51%
Improve maternal health	Less than 150 deaths per 100,000 deliveries	Births attended by skilled nurse 31%	Births attended by skilled nurse 48%
Combat disease	Mother>child HIV/AIDS less than 5%, malaria less than 5%	Malaria 25%	Malaria 7%
Environmental sustainability	90% safe water, 75% sanitation	Safe water 17%	Safe water 68%
Global partnerships	Access to mobile phones, agricultural technology, alternative energy, and life-saving drugs	Maize yields 1.3 tons/hectare (0.53 tons/acre)	Maize yields 4.6 tons/hectare (1.9 tons/acre)

▲ **FIGURE 5.25 Millennium Villages** Women fetch water at a new borehole in Dertu, Kenya, one of 14 Millennium Villages where special investments have been made to help the community meet the Millennium Development Goals (MDGs).

traditionally ranked low on most of the MDG indicators. But some of the MDG indicators have shown progress since 1990. Millennium Villages have been established to end extreme poverty by providing affordable and science-based solutions to help people out of extreme poverty across Africa.

Many regional organizations are working to improve conditions across Africa. **ECOWAS**, (Economic Community of West African States) established in 1975 to promote trade and cooperation with West Africa, is an example of a program that promotes economic integration and political cooperation in sub-Saharan Africa. ECOWAS includes the countries of Benin, Burkina Faso, Cape Verde, Côte d'Ivoire, Gambia, Ghana, Guinea, Guinea-Bissau, Liberia, Mali, Mauritania, Niger, Nigeria, Senegal, Sierra Leone, and Togo. African integration was the dream of several independence leaders, most notably Kwame Nkrumah of Ghana, who called for a Union of African States in his 1961 speech, "I Speak of Freedom." Nkrumah believed that only by joining together could independent Africa reach its full potential. Nkrumah paved the way for the establishment of the Organization of African Unity (OAU) in 1963. The OAU, now called the **African Union** and based in Addis Ababa, Ethiopia, promotes solidarity among African states, the elimination of colonialism, and cooperative development efforts. The African Union was successful in mediating boundary disputes between Ethiopia and Somalia in the 1960s and in pressing for the end of the apartheid regime in South Africa.

APPLY YOUR KNOWLEDGE Using information from the Internet determine how three major indicators of development and equality have changed in one sub-Saharan country since 2000. Have conditions improved or declined? Can you suggest why this may be?

Territory and Politics

Peace and democratic politics in Africa have been hampered by the legacies of colonialism and the Cold War, ethnic rivalries, and the special interests of powerful individuals and sectors. Although some countries in the region were able to create or reestablish a sense of national identity following independence, others have experienced internal struggles or made claims on land beyond their current borders. The economic, human, and ecological cost of wars in sub-Saharan Africa has been a hindrance to investments in development. The Stockholm International Peace Research Institute reported that in the 1990s military expenditures by governments in sub-Saharan Africa ranged between U.S. $6.6 and U.S. $9.5 billion and that many governments were spending more on arms and the military than on education or health.

Geographers Samuel Aryeetey-Attoh and Ian Yeboah have identified multiple causes for continuing political instability in Africa, including ethnic conflict, poor leadership, outside interference, and the legacies of recent independence struggles and racist government. For example, Ghana, Nigeria, and Uganda all have experienced at least five coups since independence about 60 years ago. There is also concern about the number of elected leaders who have drifted toward one-party states and dictatorships, imposing accompanying repression and restrictions on freedom of speech. However, there are some signs of optimism in the more recent political geography of Africa. A number of countries in the region have moved toward democratic elections, and political and ethnic tensions were reduced in the 1990s as a result of the end of the Cold War and the cessation of apartheid in South Africa.

Apartheid Although South Africa is now seen as an international economic and political leader within and beyond sub-Saharan Africa, there was a period when it was an international outcast because of **apartheid**, its policy of racial separation. Under apartheid, black, white, and so-called colored (mixed-race) populations were kept apart and the South African government controlled the movement, employment, and residences of blacks. The history of racial segregation in the area dates to the establishment of a supply station by the Dutch East India Company in Cape Town in 1652. The Dutch were segregationists and attempted to prevent contact between whites and native people and held Africans as slaves.

Recall from earlier in the chapter that when Britain seized political control over the Cape in 1806, the Afrikaans-speaking Dutch settlers migrated north and became known as the Boers and later Afrikaners. The Boers imposed policies of strict racial segregation between blacks and whites that included the establishment of native reserves and the mandate that blacks needed permission to enter or live in white areas. These restrictions were known as the **pass laws**. Following a British military victory in 1902 over the Boers, the Union of South Africa was established under British dominion. The Union of South Africa included the Cape and the Boer territories to the north, and perpetuated many of the Boer attitudes toward the black population. Following complete independence from Britain in 1931, the Afrikaner-led South African governments imposed strict racial separation policies and transformed apartheid from practice to rule.

The Land Acts of 1954 and 1955 effectively set aside more than 80% of South Africa's land for the white minority. In addition, laws that required nonwhites to carry permits when in white areas were reinforced. Segregation was enforced through regulations that prevented social contact and marriage between races, established separate

(a)

(b)

▲ **FIGURE 5.26 Apartheid in South Africa** (a) Signs such as this one banning all non- white people from a beach were a common sight during apartheid. (b) After 40 years of apartheid, Nelson Mandela was freed from jail, President F. W. de Klerk agreed to share power, and Mandela was elected as the first black president of South Africa in 1994. This photo is of de Klerk and Mandela after Mandela's inauguration in 1994.

education standards and job categories, and enforced separate entrances to public facilities, such as stations and hotels (**Figure 5.26a**). Large-scale segregation was established in 1959 through the creation of ten **homelands** in South Africa. The homelands were set aside as tribal territories where black residents were given limited self-government but no vote and only limited rights in the general politics of South Africa.

For nearly 40 years, the white minority controlled the black majority. Protests against apartheid were quickly and ruthlessly repressed. Anti-apartheid movement leader **Nelson Mandela** was jailed, and activists such as Steven Biko were killed. International objections to apartheid demanded South Africa's withdrawal from the British Commonwealth. Economic and trade sanctions were imposed and voluntary investment and participation bans were instituted by some major international corporations and sports competitions. A number of white South Africans were also vocal in their opposition to the system, and in the 1990s, after years of both domestic and international protest, Mandela was freed from jail and South African President F. W. de Klerk's government agreed to extend political power to black citizens (**Figure 5.26b**). In 1994, South Africa held the first election in its history in which blacks were allowed to vote, and Mandela was elected the first black president of the country as leader of the African National Congress (ANC). The 1997 postapartheid constitution in South Africa includes one of the world's most comprehensive bills of rights and prohibits discrimination based on race, gender, pregnancy, marital status, ethnic or social origin, color, sexual orientation, age, disability, religion, conscience, belief, culture, language, and birth.

APPLY YOUR KNOWLEDGE Visit the website of the Apartheid Museum in South Africa (http://www.apartheidmuseum .org/) and find three examples of how apartheid was applied or experienced.

The Cold War and Africa Independence movements and transitions in Africa coincided with the global tensions associated with the Cold War between the United States and the Soviet Union (**Figure 5.27**). Many Africans found communist and socialist ideas of equity and state ownership appealing after the repression, foreign domination, and inequality of colonial rule. For example, in Tanzania in 1967, President Julius Nyerere developed the concept of an African socialism based on the traditional values of communal ownership and kinship ties to extended family expressed as *ujamaa* (familyhood). Nyerere believed that a socialist system of cooperative production would be more compatible with African traditions than individualistic capitalism.

Africa provided fertile soil for Cold War rivalry as newly independent nations searched for political models, struggled with civil wars and incursions from their neighbors, and sought assistance to develop their economies. For example, in the southern African regions of Angola and Mozambique, where the Portuguese had fiercely repressed independence movements, revolutionary movements espousing leftist ideals attracted the interest of the Soviet Union, China, and Cuba. These then communist powers provided military and economic assistance and trained young Africans in their universities. Thousands of Cuban advisors and large amounts of Soviet military aid were sent

The challenges of reconciliation for these two countries are enormous. The memories of violence are so fresh, the divisions are so deep, and the civil wars are now viewed as efforts at **genocide**. Genocide is defined as an effort to destroy an ethnic, tribal, racial, religious, or national group. Since 2000, considerable energy has been focused on reconciliation and peace, and respected leaders such as Nelson Mandela of South Africa and Julius Nyerere of Tanzania have been involved in the mediation. In Rwanda, there has been an effort to bring to justice those most responsible, to encourage forgiveness, and to de-emphasize ethnicity in politics. Serious efforts have also been made to rebuild the economy, tourism, and agriculture. In 2008, Rwanda elected a legislature in which the majority of politicians were women.

In West Africa, the worst civil conflicts were in Liberia and Sierra Leone in the 1990s. Many factors contributed to the unrest that led to conflict. The economies of these two countries were dependent on a very narrow range of exports that have volatile prices—rubber in Liberia and cocoa in Sierra Leone. Resettlement of freed slaves from other regions, who saw themselves as elites, caused resentment among Indigenous residents, and there was also considerable resentment among rural residents of urban wealth. In the 1990s, struggles for power between opposition groups within the countries were fueled by arms acquired through the sale of diamonds. Many civilians fled as refugees, and a series of military dictatorships took control. In Sierra Leone, civil war between rival paramilitaries killed more than 50,000 people and displaced 2 million people. Liberian civil wars killed at least 400,000 people between 1989 and 2003. These wars were notorious for atrocities that included rape and the cutting off of people's limbs and for the forced recruitment of child soldiers. **Child soldiers** are children under 18 years of age who are forced or recruited to join armed groups (**Figure 5.28**). In 2012,

▲ **FIGURE 5.27 Legacy of the Cold War in Africa** A Soviet tank lies abandoned in Aksum, Ethiopia.

to Angola and Namibia in the 1970s as Namibian rebels sought independence from South Africa. The United States and South Africa supported pro-Western movements in turn, which opposed communist governments and movements. In many countries, millions of dollars were expended on arms and other military assistance, thousands were killed, and rural areas were abandoned because of land mines.

Somalia sought Soviet aid as early as 1962 to support its efforts to annex adjacent territories with large Somali population such as the Ogaden area of Ethiopia. In Ethiopia, the mid 20th-century rule of Emperor Haile Selassie, supported by the United States and others, unfairly concentrated land and wealth, which contributed to periodic famines. In 1974, a military takeover in Ethiopia resulted in the promotion of socialist policies of self-reliance and widespread land reform in favor of peasants and workers. However, internal struggle continued in the country as rebel Eritrean forces pushed for independence. It was not until 1991, after 30 years of war, that Eritrea gained independence from Ethiopia.

Civil Wars and Internal Conflicts The two decades since the end of the Cold War in 1991 have seen some terrible conflicts in sub-Saharan Africa. Some of these conflicts have crossed international borders, but many are associated with internal civil war and unrest. For example, civil wars broke out in 1993–1994 in Rwanda and Burundi when long-standing tensions between the Hutu and Tutsi tribes within each country erupted into violence. In Burundi, more than 200,000 were killed and 800,000 refugees fled into neighboring countries. In Rwanda, more than 500,000 were massacred, and many Tutsis fled to neighboring countries where rebel Tutsi forces were organized. When these Tutsi rebels won control of Rwanda and Burundi, thousands of Hutus fled in turn to avoid retribution. Two million refugees ended up in the Democratic Republic of the Congo, with thousands more in Uganda, Kenya, and Tanzania.

▲ **FIGURE 5.28 Child Soldiers in Sierra Leone** Thousands of children were forced or recruited to join revolutionary forces during the civil war.

former Liberian president Charles Taylor was convicted by the International Criminal Court in The Hague for crimes against humanity including murder, rape and "conscripting or enlisting children under the age of 15 years into armed forces or groups, or using them to participate actively in hostilities."

Peace was established in Sierra Leone in 2002 and in Liberia in 2003 when U.N. forces disarmed rebels and militias. Programs of reconciliation were put into place, including efforts to rehabilitate child soldiers and those who had been injured during the conflict. In 2006, U.S.-educated economist Ellen Johnson Sirleaf won the Liberian presidential election and vowed to sustain democracy and rebuild the economy. Sirleaf became Africa's first elected woman head of state (see chapter opener, p. 175).

Some of the most difficult problems in sub-Saharan African are in the failed state of Somalia, one of the poorest countries in the world. A **failed state** is one where government is so weak that it lacks legitimacy and control over its territory, cannot provide public services, and corruption and crime dominate. The boundaries of the state of Somalia, drawn by Italy and Britain in 1960, left out some traditional Somali cultures and territories, which were instead granted to Kenya, Djibouti, and Ethiopia. Military governments have failed to regain these areas and lost internal legitimacy as a result. Within Somalia, rivalries between different clans continue to undermine stability, and Somalia has been in a state of civil war for much of the time since 1990. Clans in the northeast and southwest have declared autonomy and are being ruled by warlords. Hundreds of thousands of people have been killed. The combination of natural disasters (droughts, tsunamis, and floods) and conflict has resulted in thousands of Somalis becoming internal or international refugees.

This continuing civil unrest, the collapse of fisheries, high unemployment, and the country's location on the Gulf of Aden have provided opportunities for the young men of Somalia to engage in piracy (**Figure 5.29**). Arguing that they are protecting fishing grounds, pirates seize ships and demand ransoms. Ransoms totaled more than U.S. $200 million in 2010 and the pirates have rejuvenated some local economies with their spending. The ships that have been attacked or captured include oil and coal tankers, food aid shipments, tourist cruise ships, and private yachts. The U.N. Security Council has introduced sanctions against Somalia for its failure to control piracy. Other countries have offered antipiracy assistance and formed a maritime patrol in the Gulf of Aden. As of 2012, the attacks seem to be declining and many Somali pirates are on trial in courts around the world.

Zimbabwe in southern Africa has an economy that spiraled downward over the last decade as GNI declined, inflation soared into the thousands of percent, and unemployment reached 90%. Eighty percent of the population in Zimbabwe is below the poverty line and at risk of hunger, cholera, and other disease outbreaks. The problems in Zimbabwe partly stem from a difficult transition from British colonial rule. A white apartheid government declared independence in 1965 and took control of most of the good land; it was opposed by black guerilla movements. Although title to the lands had been given to white settlers by the colonial governments, and the farms were often well run, many black Africans believe that these lands were seized unfairly and should be returned to the Indigenous owners. In 1980, white rule ended when President Robert Mugabe was elected and promised to redistribute land. In 2000, noting that the

▲ **FIGURE 5.29 Somali Piracy** The British Royal Navy arrests a "mother boat" of Somali pirates in the Indian Ocean in 2012. The pirates use small boats to attack ships passing through the Gulf of Aden.

1% of the population that was white owned 70% of the productive agricultural land, Mugabe began a compulsory land distribution to blacks. The redistribution involved squatters moving onto white-owned farms, sometimes violently. Critics of Mugabe argue that land redistribution failed because peasant farmers did not have the knowledge or resources to succeed in commercial and export-oriented farming. Zimbabwe's problems have escalated beyond land conflicts, with rigged elections, human rights abuses, a lack of health care, the suppression of the media, and corruption. Life expectancy in Zimbabwe is low, infant mortality is high, and there is widespread poverty. Three million refugees have fled from Zimbabwe to South Africa. Although elections in 2008 were apparently won by the opposition, Mugabe has retained power. There are some small indications, however, that living conditions in Zimbabwe are slowly improving.

With extensive oil reserves and a population of over 160 million, Nigeria is one of the most powerful countries in Africa. Nigeria was formed when the British brought together several ethnic groups, including the Yoruba, Igbo, and Fulani, into a large colony, which eventually became independent in 1960. Since independence, Nigeria's government has been dominated by military dictatorships that have often brutally suppressed opposition and done little to address highly unequal distributions of wealth. In the 1960s, the Niger delta became the core of an oil-producing region. Conflict over boundaries and resources triggered the secessionist Biafra War in that decade, which resulted in more than a million dead. The economy and politics are now dominated by oil, which makes up more than 90% of all Nigerian exports by value. Political conflict continues due to festering ethnic rivalries and resentment that the oil wealth is not more fairly distributed.

Tensions in Nigeria have also arisen in relation to environmental pollution of the lands and waters of the delta (**Figure 5.30**). In 1995, the Nigerian government's hanging of Ken Saro-Iwa—a novelist and activist who protested oil spills in the area occupied by 500,000 Ogoni people—for his alleged role in political unrest provoked international outrage and sanctions. The conflicts in the oil region have had global implications for international oil companies, such as Shell, who have been accused of complicity with the government and subjected to consumer boycotts. Civilian rule and democratic elections have reduced tensions in Nigeria since 1999, and the economy has been growing fast. Notably, the manufacturing and service sectors, including banks and telecommunications, car manufacturing, film production, and textiles, have expanded. Crime and corruption, including international Internet and phone scams, are still, however, a problem in the area.

Other conflicts in the region include the war in the Democratic Republic of the Congo (DRC), which killed several million people between 1998 and 2008 both directly and as a result of famine and disease. There has also been conflict in Uganda, where the Lord's Resistance Army, led by Joseph Kony, uses child soldiers to fight the Ugandan government. Across the Sahel, religious differences between Muslims and Christians and unrest by Tuareg rebels have fueled conflict in Chad, Mali, and Niger. Sudan was divided into Sudan and South Sudan in 2011 and the birth of the new nation has resulted in some conflict as South Sudan identifies with sub-Saharan Africa (see "Visualizing Geography: South Sudan," p. 206).

Peacekeeping The end of many recent wars in Africa and subsequent peace settlements have been brokered and monitored by the U.N. and regional security forces. The U.N. Peacekeeping Forces operate under the authority of the U.N. Security Council and are intended to establish and maintain peace in areas of armed conflict with the permission of disputing parties. In Africa, U.N. forces, with their distinctive pale blue helmets, have been deployed in Angola, the Democratic Republic of the Congo, Sudan, South Sudan, Ethiopia, Eritrea, Rwanda, Burundi, Chad, Namibia, Somalia, Sierra Leone, Liberia, Mozambique, and Western Sahara (**Figure 5.31**).

Although there was some initial success in monitoring transitions to peace in Mozambique and Namibia, the success of these missions has ultimately been

▲ **FIGURE 5.30 Conflict in Nigeria** A resident of Lagos washes himself after trying to put out a fire caused when armed gangs tapped an oil pipeline. The fire killed 500 people in 2006.

mixed. U.N. forces failed to prevent massacres in Rwanda and Sierra Leone, were initially unable to establish peace in Somalia and the Democratic Republic of the Congo, and have had inadequate human or financial resources to sustain several operations. African leadership in promoting peace within the region is growing in importance.

▲ **FIGURE 5.31 U.N. Peacekeeping** The U.N. peacekeeping forces in sub-Saharan Africa include many soldiers from countries in the region, including South Africa, Nigeria, and Ethiopia. This photo shows Ethiopian soldiers prior to deployment to Liberia in 2003, wearing distinctive blue U.N. helmets.

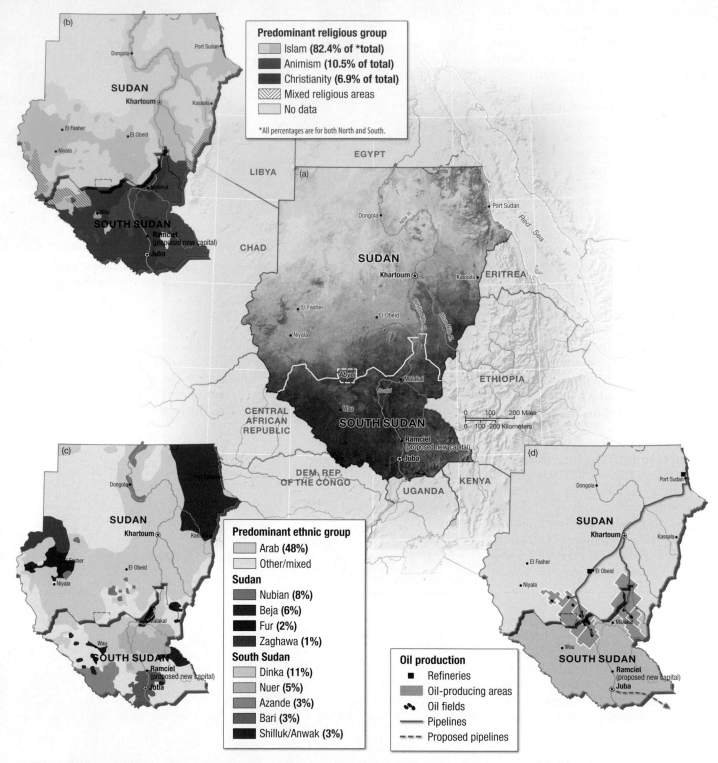

Predominant religious group
- Islam (82.4% of *total)
- Animism (10.5% of total)
- Christianity (6.9% of total)
- Mixed religious areas
- No data

*All percentages are for both North and South.

Predominant ethnic group
- Arab (48%)
- Other/mixed

Sudan
- Nubian (8%)
- Beja (6%)
- Fur (2%)
- Zaghawa (1%)

South Sudan
- Dinka (11%)
- Nuer (5%)
- Azande (3%)
- Bari (3%)
- Shilluk/Anwak (3%)

Oil production
- ■ Refineries
- Oil-producing areas
- Oil fields
- Pipelines
- Proposed pipelines

▲ **FIGURE 5.2.1 Maps of Sudan and South Sudan** These maps of Sudan and South Sudan illustrate some of the contrasts between the two new countries. (a) This satellite image shows the contrast between northern desert in yellow colors and the southern forests and swamps (such as the Sudd) in greener colors. (b) The north is dominated by Arab ethnicity and the south is home to mostly black African groups such as the Dinka and Nuer. (c) A map of the religious affiliations in different regions demonstrates the dominance of Islam in the north and Christianity and animism in the south. (d) This map pinpoints the location of oil producing areas that cross the boundary between Sudan and South Sudan. Also shown are existing pipelines crossing from South Sudan north to Sudan and a proposed new independent pipeline that will go south through Kenya to the coast.

How does a new country emerge, and how are its new boundaries determined? South Sudan became an independent state with full membership in the United Nations in July 2011. Historically a region of several tribes and empires, the area came under British colonial control when the Juba conference of 1947 created a single state of Sudan. Sudan was the largest nation on the African continent, encompassing more than 2.5 million square kilometers (almost a million square miles). Ruled mostly by military regimes that had close ties with the Arab and Islamic world, which repressed and discriminated against southern Sudan, civil war divided the country from 1955 to 1972 and 1983 to 1995. As many as 2.5 million died in these wars, and during this period, the southern part of Sudan suffered terrible famines and poverty. But crisis has not been limited to southern Sudan. In Darfur, which is located in western Sudan, conflict between the Sudanese government and non-Arab Indigenous populations has resulted in more than half a million deaths, and more than 2 million people have been displaced into refugee camps.

It took six years to create the new country of South Sudan following a peace agreement between the Sudanese government and southern liberation movements in 2005. The Republic of South Sudan became a reality as a result of a 2011 referendum, in which 99% of the population of southern Sudan voted for independence. The new country has a population of about 8.25 million people and an area of over 600,000 square kilometers (240,000 square miles). South Sudan's current capital is in Juba, but there are plans to create a new capital at Ramciel. The nascent Republic of South Sudan is not without serious problems, including tensions with Sudan over oil revenue and borders, intertribal clashes, destroyed or absent infrastructure, desperate poverty, and poor health.

The reasoning behind the boundaries of the new country can be seen in maps and statistics and political and religious inclinations (**Figure 5.2.1** and **Table 5.2.1**). The physical geography of South Sudan differs from the north, which features desertlike conditions. In contrast, South Sudan is dominated by a massive swamp called the Sudd, and also features tropical forests and grasslands. The predominant religions and races are also different in the two countries. Most of South Sudan identifies as Indigenous black Africans with traditional animistic or Christian religious beliefs and pastoral traditions. In contrast, the north is dominated by Arab and Islamic ethnicity and religious affiliations. South Sudan's orientation toward sub-Saharan Africa is manifested in its plans to introduce Swahili language (as an alternate to Arabic) and to build oil pipelines and rail routes south through Kenya and Uganda. ∎

TABLE 5.2.1 Sudan and South Sudan: A Comparison of Socioeconomic Indicators		
	South(ern) Sudan	**(Northern) Sudan**
Population	10.6 million	26 million
Area	644,329 sq km (250,000 sq miles)	1,886,968 sq km (730,000 sq miles)
GDP	U.S. $13.2 billion (2011 estimate) U.S. $1,456 per capita	U.S. $89 billion (2011 estimate) U.S. $2,495 per capita
Language	English and Arabic official languages, over 60 Indigenous languages	Arabic and English official
Religion	Traditional Indigenous animist religions, Christian	Sunni Muslim, small Christian minority
Exports	Oil (oil revenues form 98% of government budget), lumber	Oil, cotton, sesame, livestock, groundnuts, gum arabic, sugar
Literacy	27% of adult population	61% of adult population
Infant mortality	102 per 1,000 births	55 per 1,000 births
Life expectancy	62 years	63 years
Occupation	78% agriculture	80% agriculture
Sanitation	20%	35%
Poverty rate	51%	40%

These leadership efforts include negotiations led by former South African president Nelson Mandela and a West African peacekeeping force and monitoring group, called Economic Community of West African States Monitoring Group (ECOMOG), that is led by Nigeria under the auspices of the Economic Community of West African States (ECOWAS).

APPLY YOUR KNOWLEDGE Consult two or three recent U.N. or media reports on conflict, peacekeeping, or refugees in sub-Saharan Africa. Do these reports seem optimistic or pessimistic in tone? What factors contribute to reasons for hope or greater concern?

CULTURE AND POPULATIONS

Sub-Saharan Africa is a large and extremely diverse region, with hundreds of different ethnic groups who have their own traditional cultures, religions, and languages. In this region, a wide range of political systems and social organizations also persist. In addition to strong and unique cultural legacies, the region has experienced dynamic shifts in culture and attitudes associated with more recent global interconnections. It is difficult to draw generalities about a continent as large and diverse as Africa, but those who do generalize tend to highlight the importance of the extended family, ties to the land, oral tradition, village life, and music in traditional African culture.

The multiplicity of religions, languages, and dialects in Africa reflects the large number of distinct cultural or ethnic groups in Africa. Some writers use the term **tribe** to define these groupings and describe Africa as a *tribalist* society. The term tribe describes a form of social identity created by groups who share a collective set of ideas about loyalty and political action. In tribes, group affiliation is often based on shared kinship, language, and territory. Although it is used by many groups to identify themselves, some see the term tribe as negative (related to colonial perceptions of savagery) and now prefer to use the term **ethnic group**.

The largest ethnic groups in Africa are associated with the dominance of certain languages, such as Hausa, Yoruba, and Zulu, but almost all groups were either split geographically by colonial national boundaries or grouped together with their neighbors, enemies, or others with whom they shared no affinity. Attempts to consolidate ethnic groups across boundaries and struggles for power between groups within countries are a major cause of conflict in contemporary Africa.

Religion and Language

Traditional African religions have been described generally as animist (worship of nature and spirits; see Chapter 8, p. 304), but this description simplifies the wide variety of local religious beliefs in Africa. Although natural symbols, sacred groves of trees, and landforms often have significance in local religions, many African religions also include ancestor worship and feature a belief in a supreme being, several secondary gods or guardians, and good and evil spirits. In many of these belief systems, ancestors, priests, or witch doctors mediate and interpret the wishes of the gods and spirits, and rituals ensure the stability of society and relations with the natural world. More than 70 million people (about 10% of the total population) are reported to practice traditional religions on the continent of Africa. As in Latin America, traditional religion has blended with Christianity and Islam to create forms in which local traditional rituals are incorporated into religious services.

In about 300 B.C.E., Christianity began to spread into sub-Saharan Africa via North Africa and Ethiopia, but the pace of conversion accelerated rapidly under European colonial rule and European missionaries. Dutch Calvinism in southern Africa; Catholicism in French, Spanish, and Portuguese colonies; and Anglican beliefs in the British colonies all had strong influences. Of the 360 million estimated Christians in Africa, there are about 125 million Roman Catholics and 114 million Protestants. Evangelical Christianity is growing, attracting a following that equals more than 10% of the population across Africa. The vibrancy of Christianity in many countries draws more young people to careers in the church than in many other world regions, and Africa-trained pastors are taking leadership positions in European and North American churches.

Islam is another major religion in Africa, where it has 308 million adherents. Islam is important in the Sahel, North Africa, and in parts of East Africa, such as Somalia, Tanzania, and Ethiopia. Nigeria is now almost 50% Islamic. Spread by traders, Islam attracted believers from some West African pastoralists, such as the Fulani, who then went to war to eradicate animistic beliefs.

Religious differences have fueled numerous political conflicts in some regions of Africa, most notably where Muslim and Christian groups have been forced together by imposed colonial national boundaries. In West African countries bordering the Sahel, such as Nigeria, tensions exist between northern Islamic groups, such as the Hausa and southern Christians, such as the Igbo. As the "Visualizing Geography" box (p. 206) illustrates, the division between a mostly Islamic population in the north and mostly Christian one in the south was one reason for the division of Sudan into two countries in 2011.

The geography of languages in Africa is incredibly complex, with more than 1,000 living languages, 40 of them spoken by more than 1 million people (**Figure 5.32**). The dominant Indigenous languages, spoken by 10 million or more, include Hausa (the Sahel), Yoruba and Ibo (Nigeria), Swahili (East Africa), and Zulu (southern Africa). Hausa and Swahili are examples of a **lingua franca** (see Chapter 9, p. 344) and spoken as second languages by many groups to facilitate trade. English, French, Portuguese, and Afrikaans are also spoken in areas that were under colonial control and had European education systems or settlements. Arabic is common in areas near northern Africa and has strongly influenced Swahili along the east coast of Africa. Because most countries have no dominant Indigenous African language, some of these use a European language when they conduct official business and in their school systems. The countries with the most coherent overlap between territory and a dominant African language are Somalia (Somali), Botswana (Tswana), and Ethiopia (Amharic). In southern Africa, the Khoisan languages (spoken by the Bushmen, for example) are notable for their use of "clicking" sounds.

APPLY YOUR KNOWLEDGE How does the geography of language and religion in sub-Saharan Africa reflect the legacies of colonialism and the ways in which colonial boundaries cut across traditional ethnic groups?

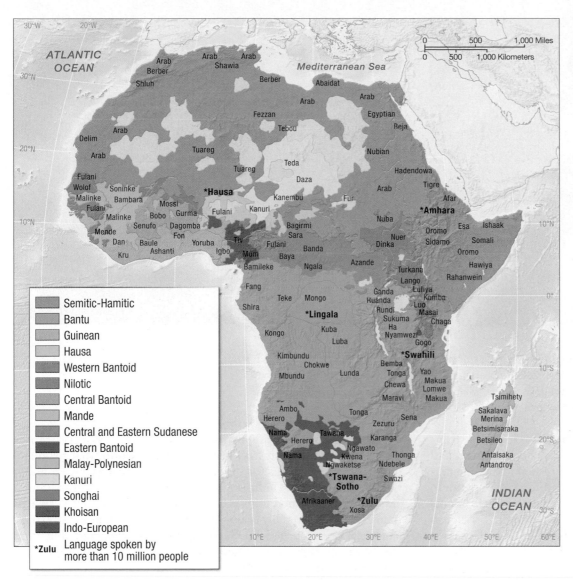

▲ **FIGURE 5.32 African Languages** The cultural complexity of Africa is demonstrated by the variety of Indigenous languages shown on this map. The Indigenous languages of Africa can be grouped into larger language families, including the Afro-Asiatic languages of North Africa, such as Somali, Amharic, and Tuareg; the Nilo-Saharan languages, such as Dinka, Turkana, and Nuer in East Africa; and the largest Niger-Congo group, which includes Hausa, Yoruba, Zulu, Swahili, and Kikuyu. Swahili and Hausa are spoken by millions as the trade languages of East and West Africa, respectively.

Cultural Practices, Social Differences, and Identity

Culture in Africa is as varied as religion and language and is changing rapidly as a result of interactions across and beyond the region. Through the slave trade, migration, and global communications, African traditions have influenced the Americas and other regions. In many areas, cultural traditions include the importance of the extended family and the kinship ties in social relations and obligations and the widespread respect for elders as sources of wisdom. Kinship ties going back multiple generations define "clan" allegiances in some subregions and sometimes drive primary loyalties. This is the case, for example, in contemporary Somalia, where interclan conflict has dominated recent political events. Different extended families are often linked through intermarriage. A transfer of wealth, sometimes

in the form of cattle, from the husband to the wife's family, may be offered as a mark of respect and an indication of the value of the woman's labor and companionship.

The tie to the land is connected to traditional forms of land tenure in Africa, where, in many areas, land was viewed as given by the spirits or held in trust for ancestors and future generations. The Elesi of Odogbolu (a traditional leader) in Nigeria has expressed this view in these words: "The land belongs to a vast family of which many are dead, few are living, and countless members are still unborn." This view endows the land with communal nature such that it cannot be bought or sold by individuals. In some cases, land rights are held by extended families or the community, rather than by individuals, and in other societies the chief or king controls land.

Traditions of reciprocity, where a gift is given to obtain a favor, and of helping family members can be a major source of cultural confusion in the region, according to African historian Ali Mazrui. He suggests that these traditions provide an explanation for the way in which some regional leaders have favored family members with jobs in their administrations and for the role bribes play in requests to government officials. Mazrui also notes that under colonialism, local residents viewed stealing from the government as a legitimate form of resistance because they felt that foreigners were robbing Africa of resources and funds through taxation.

Music, Art, and Film Africa has a rich tradition of music and the arts that have had an enormous impact on other regions of the world. For example, traditional music of the West African Sahel influenced the development of American blues, and West African coastal traditions influenced Afro-Caribbean music styles. Slavery was one way in which African musical, artistic, and food customs spread around the world, especially to the Americas. The arts are dynamic in much of Africa, often reflecting international influences and new youth identities.

Africa is often associated with music from percussion instruments, especially drums. Other instruments with metal keys that are plucked or tapped, as well as flutes and harp-like stringed instruments, are also commonly used in traditional and popular music. Such musical traditions vary widely across the continent and include the complex rhythms of women drummers in Tanzania and xylophone players in Uganda and the chanting of the Zulu of southern Africa. In West Africa, oral traditions are associated with singers and storytellers, some of whom receive the respected name of **griots**.

African popular music mixes Indigenous influences with those of the West. For example, the highlife music of West Africa is derived

(a)

(b)

▲ **FIGURE 5.33 African Music** (a) Senegalese musician Youssou N'Dour met with U.S. President George Bush and musicians Bono and Bob Geldof in 2007 to pressure for assistance in fighting disease and for debt forgiveness. (b) The Festival au Desert in Essakane, Mali showcases traditional music and popular singers. It is attended by tourists as well as local nomads.

from Caribbean calypso and military brass bands; it includes stronger percussion, soul influences, and exchanges between lead and background singers. Juju, or Afrobeat, which is related to highlife music, was made popular by Nigeria's King Sunny Adé and Fela Kuti. In French West Africa, singers such as Youssou N'Dour blend traditional African beats with powerful vocals, and in South Africa, singers such as Miriam Makeba received international recognition in the 1950s, presaging the popularity of the *a cappella* style of South African black musicians such as Ladysmith Black Mambazo. Sub-Saharan African music sometimes mixes with politics, as it does in many world regions. Popular musicians have often expressed political opinions against apartheid or corruption, and Senegalese musician Youssou N'Dour campaigns to combat malaria and is minister of tourism and culture for his country (**Figure 5.33a**). The Festival in the Desert, held in Mali each year, brings together musicians from across West Africa with local pastoralists, tourists, and international media (**Figure 5.33b**).

African art is also richly varied. In traditional African cultures, artists were valued specialists, often working under the patronage of kings and creating works of spiritual value. The masks and wood sculptures of the Dogon and Bambara people of West Africa are collected around the world and still maintain their cultural significance within the region. Kente cloth designs from northern Ghana have become meaningful in African-American identity. As interest in travel and world culture has grown, local artists have started to produce items for sale to tourists and to international distributors. Some organizations have sprung up to try and ensure fair trading principles and return as much value as possible to local people. But of course not all African art is traditional, and galleries in major cities as well as websites promote young modern artists to a global market. African literature has produced several Nobel Prize winners and internationally acclaimed novels. Millions have read the novels of Chinua Achebe from Nigeria, Doris Lessing of Zimbabwe, and Nadine Gordimer of South Africa.

Film is also a growing art form in the region. Nigeria is home to a booming film industry, known as **Nollywood**, that produces hundreds of recordings for the home video market each year and is the third-largest film industry in the world (after the United States and India). Nollywood films are often shot on location for very little money but are incredibly popular, often dealing with moral questions and social issues, such as AIDS, corruption, and witchcraft. The films are shown throughout Africa and also in diaspora communities in the United Kingdom and United States.

APPLY YOUR KNOWLEDGE Read a novel or view a recent film (or read a review of a book or a film) that is set in sub-Saharan Africa. How does the theme of the work you have chosen reflect the contemporary geography, problems, or culture of the region?

Sport As is the case across the world, for many young Africans, heroes are sports stars, especially athletes who make it on the international playing field. Even in remote villages, children play with improvised soccer balls, hoping to be spotted and eventually recruited to play for a European team. Superstars, such as Didier Drogba from the Côte d'Ivoire who plays for the English football club Chelsea (**Figure 5.34**), have become very rich and enormously popular. Another group of African sports stars are the long-distance runners of East Africa. Ethiopia and Kenya that have produced numerous Olympic gold medalists in both men's and women's events.

The Changing Roles of Women Women in Africa tend to have less education and lower incomes than men, but in most sub-Saharan African countries, they live slightly longer. Gender differences in Africa reflect the feminization of poverty; more than two-thirds of

▲ **FIGURE 5.34 African Sports Stars** Footballer Didier Drogba from Côte D'Ivoire (also known as the Ivory Coast) has gained international fame as a player for the English team Chelsea. He began his career in France, was named African footballer of the year in 2006, and is known for his charity work as a U.N. goodwill ambassador. In this photo Drogba is in his Chelsea colors holding the 2010 U.K. FA championship cup.

Africans who join the ranks of the poor are women. The average female annual income (PPP) in sub-Saharan Africa is about U.S. $2,000 less than that of men, and literacy rates are 56% for women compared to 74% for men. African women are more likely to be poor, malnourished, and otherwise disadvantaged because of inequalities within the household, the community, and the country. Women are less likely to receive an education, and overall, pay rates are lower in the workplace.

Patriarchy and cultural traditions mean that women may be required to eat less than men and eat only after the men in the family

have eaten. As noted earlier in the chapter, women bear a disproportionate responsibility for heavy household work, such as collecting fuel wood and water (**Figure 5.35**). The tradition of female circumcision has become a controversial struggle between those who see it as a human rights abuse involving brutal genital mutilation and health risks and others who see it as an important religious and symbolic experience. More than 100 million women in Africa are estimated to have undergone female circumcision.

Women are also disadvantaged by many traditional and modern institutions that define property rights. Land in some places may be passed on only to male children, and new land titles are often granted to male heads of household. Women perform critical functions in rural areas, meeting the basic needs of their families by preparing food and collecting wood and water, generating income with crafts and community work, and as acting as agricultural producers in both subsistence and commercial sectors.

Women control distinctly gendered spaces in the African landscape; parts of the home, the market, and the cultivation of certain trees and crops are reserved primarily for women. The U.N. Food and Agriculture Organization (FAO) reports that in 2010 women made up half or more of the population working in agriculture in sub-Saharan Africa and headed a quarter of all rural households. But women generally have much less access to agricultural inputs such as training, fertilizer, livestock, and land, and have less time to farm given their household responsibilities. The overall productivity is thus limited when development policies for agricultural training and technology to make physical labor easier are directed at men, and new projects for tree planting or cash crops focus on men. Women are often also disproportionately affected by environmental degradation, as deforestation and drought made more difficult the work of collecting wood, water, and food. Although some African governments and development agencies have recognized these disparities and established programs meant to improve conditions for women, poverty reduction among women has been patchy.

Those who understand the role of women's work in African communities and economies have criticized development policies that ignore, undervalue, or displace women. These policies gradually have

◄ **FIGURE 5.35 Women Collecting Firewood and Water** Women carry dung and firewood, which they have collected to use as fuel, in Chandeba, Ethiopia. Women often must walk long distances with heavy loads.

begun to change to include projects that provide technology, credit, and training to women and that link women's productive and reproductive roles and understand the gender-related differences and barriers to improving the lives of both men and women. Development agencies such as the World Bank now incorporate these approaches into many programs that support education, credit, and land-titling programs for women, women's organizations, and recognition of women's work.

Demography and Urbanization

Sub-Saharan Africa is still growing rapidly compared to other world regions. Its population was 883 million in 2011 and is projected to reach 1.5 to 2 billion by 2050. Although fertility has fallen from 5.5 in 2001 to 4.9 in 2011, the growth rate is 2.3% in the region with a doubling time of about 29 years. Fertility rates are lower in southern Africa. Approximately 43% of the African population is under age 15. This has major implications for future population growth. When this group starts to have children, demands on education systems, health care, jobs and resources may increase. The largest numbers of people in the region live in Nigeria, which has a population that is estimated to be over 160 million. Ethiopia, the Democratic Republic of the Congo (DRC), and South Africa have more than 50 million residents each. Some populations in the region, such as Nigeria, have not been reliably censused for many years, and population estimates are approximate. Overall population density in the region is relatively low, at just over 14 people per square km (36 per square mile) in 2010, compared to the global average of 20 (51). The only countries in the sub-Saharan Africa world region with population densities higher than the global average are the Indian Ocean islands and the Central African countries of Rwanda and Burundi (**Figure 5.36**).

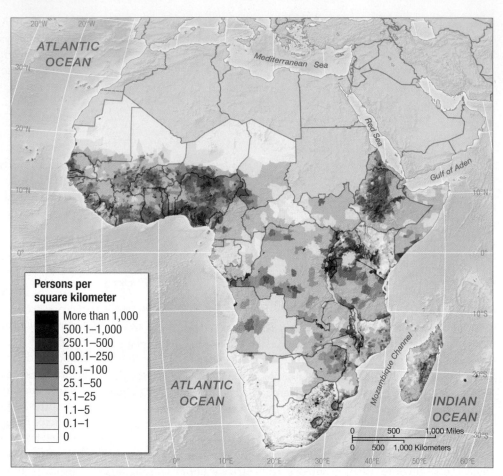

▲ **FIGURE 5.36 Africa Population Density** Africa is home to a mostly scattered rural population. The greatest density is along the West African coast, the southeast coast of South Africa, the highlands of East Africa, and along major rivers. Population concentrations tend to be associated with better soils and climate, with colonial centers for mining and export crops, and with coastal ports.

Population and Fertility What are the reasons for high fertility and birthrates in Africa, and what are the prospects for slowing population growth? Population geographers have found that although religious prohibitions of contraception and lack of access to contraceptive devices may play a small role, other factors are much more important. Children are the main source of security for elderly people in countries where there are few pensions or public services for the aged, and it is traditional for younger generations to respect and care for their elders. Children are valued in Africa for many reasons, including their ability to work in agricultural fields and as herders, to help with household work, and to care for younger siblings. Children are also a possible source of financial or other gain when they marry. Large families are also often perceived as prestigious, a spiritual link between the past and the present, and a way of ensuring family lineage. Even though infant mortality rates have improved with better nutrition and health care in much of Africa, many African families have internalized the need to have a large number of children to ensure that some survive to adulthood.

Many studies have also shown that conditions for women have a strong influence on fertility rates; younger marital ages, minimal

female education and literacy rates, and fewer opportunities for female employment all contribute to higher fertility rates. Fertility rates tend to be lower in urban areas with high rates of female education and employment, where women marry when they are older. In Ghana, for example, women with secondary or higher education have a total fertility rate of only 2.1 compared to a rate of 6 for women with little or no schooling. In Ghana, rural women have higher rates of fertility than urban women, and richer women have fewer children than poor women (**Figure 5.37**). Statistics such as these indicate that future population growth rates in the sub-Saharan African region and declines in fertility will depend on improved conditions for women.

HIV/AIDS and Other Health-Care Concerns In sub-Saharan Africa, more people are affected with the **human immunodeficiency virus (HIV)** that causes **acquired immunodeficiency syndrome (AIDS)** than any other part of the world. This region is home to two-thirds of all HIV cases diagnosed. In 2009, 22.5 million people had HIV in the region; the highest rates were in Botswana, Lesotho, and Swaziland (where 25% of the population was infected) and in South Africa (where 18% was infected). More than 15 million Africans have lost their lives to AIDS since the disease was identified in 1981. It has become the leading cause of early death in Africa, killing more people than malaria and warfare combined (**Figure 5.38**). Because of the

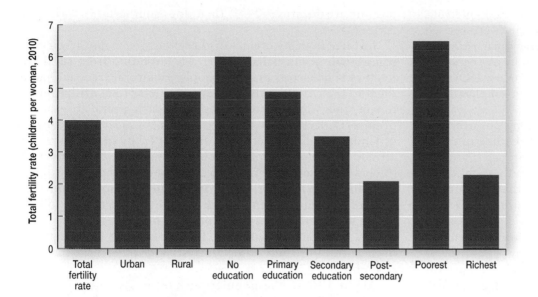

◄ **FIGURE 5.37 Status of Women and Fertility in Africa** In Ghana, women who are urban, better off, and better educated have lower fertility rates.

▼ **FIGURE 5.38 AIDS in Sub-Saharan Africa** These maps of the progression of the HIV epidemic in Africa from 1990 to 2009 show the infection rates among adults spreading from eastern to southern Africa in terms of percent of population with AIDS. The 2009 map and the graph of AIDS-related deaths from 1990 to 2009 indicate a slight slowing of infection and a decline in death rate as more people get treatment.

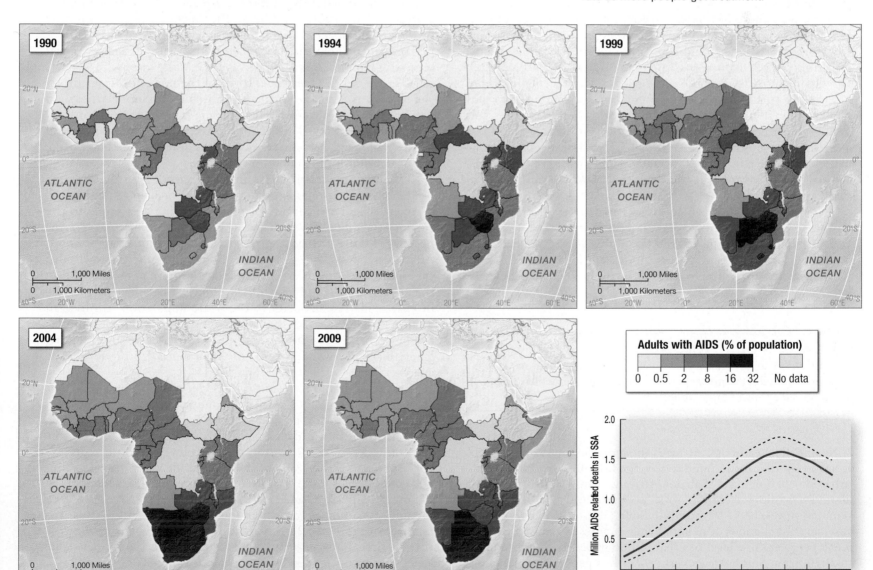

high mortality and infection rates associated with the AIDS epidemic, population projections for several African countries were revised downward in recent years. In Zimbabwe and Botswana, estimates of average life expectancy have been revised down by 20 years, and population growth rates have been reduced or even reversed. Overall, population projections for Africa for 2025 have been adjusted down by 200 million people.

The geography of AIDS in Africa varies by country, by regions within countries, and by social groups. Unlike in other regions, more women than men have AIDS in sub-Saharan Africa, and mothers often transmit the disease to their children. Frequently, married couples are both infected with AIDS and die from it. As a result, there were more than 14.8 million AIDS orphans in Africa in 2009. Poverty exacerbates the AIDS problem in sub-Saharan Africa because many cannot afford prevention (for example, through the use of condoms), testing, or antiretroviral medicines that can prolong the lives of people who are infected. Some governments have low health-care budgets and do not admit the severity of the AIDS epidemic. Few people have health insurance, diagnosis and treatment are often delayed, and interaction with other diseases, such as tuberculosis, increases mortality rates.

However, there has been considerable success in combating HIV/AIDS in the last ten years. A combination of domestic policies, foreign assistance, and subsidized treatment helped reduce infection rates by 25% across the region between 2001 and 2009 and deaths by 20% over the same period. Unprecedented international agreements with drug companies in combination with new assistance programs from the World Bank, charities, and donor countries are lowering the cost of AIDS drugs. More than a third of those infected now get antiretroviral medicines compared to only 2% in 2001. But many people with HIV and AIDS are still unable to obtain the drug therapies that will prolong their lives.

There is great concern about the potential impact of other emerging viruses in central Africa, specifically Ebola fever, which causes severe bleeding and kills more than 50% of its victims. So far, Ebola outbreaks, such as the ones in central Africa in 1995 and 2007 and in Uganda in 2000, have been contained, but only after 200 people died in each country.

APPLY YOUR KNOWLEDGE Which factors have contributed to the spread of AIDS/HIV in sub-Saharan Africa, what have been the impacts of the epidemic, and which policies are helping control the epidemic?

African Cities Although Africa is the most rural of world regions, it has been urbanizing rapidly over the last 40 years. In 1960, the urban population of sub-Saharan Africa was only 17 million people, about 20% of the overall regional population. By 2010, the urban population reached 319 million people, 37% of the total population, and the number of people living in cities is growing at about 3.6% per year. The level of urbanization varied greatly by country in 2010, from South Africa with 61% of its people in urban areas, to Ethiopia at 18% and Uganda at 13%. East Africa has a lower level of urbanization (24%) than does southern (59%) or West Africa (45%).

Just like in many other world regions, the major driver of urban growth in sub-Saharan Africa is migration from rural areas to cities. The factors pushing people from rural areas and pulling them to the cities are also somewhat similar to contributing factors in other world regions. People are leaving rural areas because of poverty, lack of services or support for agriculture, scarcity of land, natural disasters, and civil wars. Urban areas are more attractive because they offer jobs, higher wages, better services (including education and electricity), and entertainment. Urban areas have benefited from the **urban bias** of both colonial and independent governments in Africa, which have tended to invest disproportionately in capital cities that house centralized administrative functions. In addition, food prices have been kept down in the cities to reduce wage demands and to decrease the risk of civil unrest.

About 30% of the population lives in its country's largest city. Lagos (Nigeria), Kinshasa (the Democratic Republic of the Congo), and Johannesburg (South Africa) are the largest cities in the region, with populations of about 10.5 million, 8 million, and 3.6 million, respectively, and there are about 40 cities with more than 1 million inhabitants. Life in African cities has great contrasts of wealth and poverty. Many urban inhabitants live in the informal settlements that surround major cities, with inadequate social services, sanitation, water supply, energy, waste disposal, and shelter (**Figure 5.39**). Over time, some of the most notorious urban slums, such as Soweto, outside Johannesburg in South Africa, and Kibera in Nairobi, Kenya, have become important to the overall functioning of cities. These slums house many of the lower-paid workers and have become centers of social movements that make political claims for the poor.

The most powerful cities in the region in terms of economy and international presence are Johannesburg, Lagos, and Nairobi, which are centers of manufacturing and home to the regional headquarters of major international companies. Johannesburg (**Figure 5.40a**, p. 216) sits on the high plateau of South Africa, surrounded by gold and diamond mines and the townships that housed mineworkers and others who have come to work in the city. The geography of Johannesburg still reflects the legacy of apartheid. The majority of the white population still lives in wealthy residential areas, such as Hyde Park. In contrast, millions of black workers live in poor conditions in Soweto and Lenasia, where black people and people of Asian descent lived during the apartheid era.

Lagos—the colonial capital of Nigeria until the capital was relocated to Abuja in 1992—has a metropolitan population of more than 13 million people (**Figure 5.40c**, p. 216). Sited on a natural harbor, it was developed by the British as a rail terminus beginning about 1880. It became a leading cargo port, industrial area, and center for production of consumer goods. About 80% of Nigeria's trade goes through Lagos, although some trade is now shifting to the oil regions to the east. The Lagos area includes 53% of Nigeria's manufacturing, 62% of the gross industrial output, and 22 industrial estates. Lagos is also a cultural center, home to many well-known writers and musicians, and Nollywood, the West African film industry (see p. 210). Lagos is infamous for its traffic and crime problems. The average

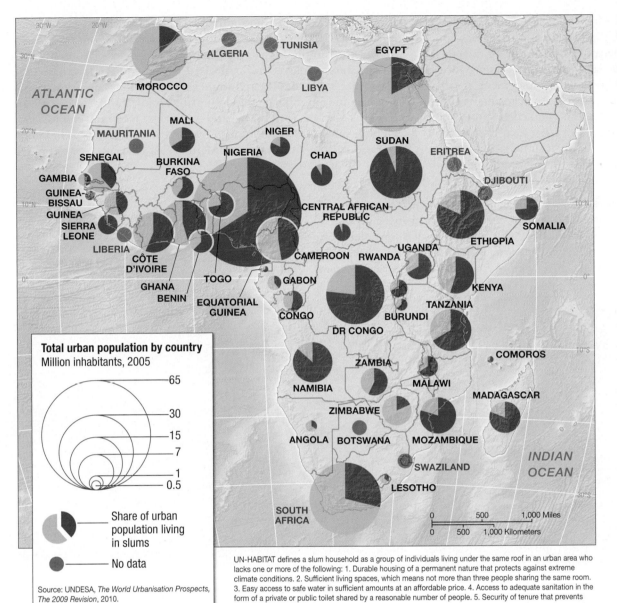

◄ FIGURE 5.39 Urbanization and Slums in Sub-Saharan Africa This graphic shows the total urban population in countries across Africa and the proportion of the people who live in slums.

Total urban population by country
Million inhabitants, 2005

- 65
- 30
- 15
- 7
- 1
- 0.5

Share of urban population living in slums

No data

Source: UNDESA, *The World Urbanisation Prospects, The 2009 Revision*, 2010.

UN-HABITAT defines a slum household as a group of individuals living under the same roof in an urban area who lacks one or more of the following: 1. Durable housing of a permanent nature that protects against extreme climate conditions. 2. Sufficient living spaces, which means not more than three people sharing the same room. 3. Easy access to safe water in sufficient amounts at an affordable price. 4. Access to adequate sanitation in the form of a private or public toilet shared by a reasonable number of people. 5. Security of tenure that prevents forced evictions.

commute to work is more than 90 minutes in polluted air and tangled traffic jams, made worse by inadequate bridges between islands. When food prices and unemployment increased in 1989, Lagos residents expressed their frustrations by rioting. Electricity and other services are insufficient; as a result of interruptions and lack of service, the National Electric Power Administration (NEPA) has been given the nickname of "Not Expecting Power Anytime."

Nairobi, Kenya (**Figure 5.40b**, p. 216), located in the cooler highlands where English settlers preferred to live, still displays its role as a center for colonization in East Africa. It is a major center of commerce and corporate headquarters and hosts several U.N. agencies, including the headquarters of the U.N. Environment Program. The Nairobi urban landscape shows some striking contrasts. The central business district is adjacent to Nairobi National Park, which is home to both wildlife and vast slums such as Kibera, where people live in dense, poor quality housing without adequate access to energy, clean water, or sanitation.

The Sub-Saharan African Diaspora and Migration within Africa Migration into Africa from other regions is overshadowed by the immense African diaspora and emigration from Africa to other regions. The descendants of slaves captured and sent away from the region represent a high percentage of the populations in Brazil, the Caribbean, and the United States. A second wave of emigration coincided with the aftermath of colonialism, a time when many Africans who retained British Commonwealth passports or French citizenship moved to Britain and France (or other Commonwealth countries such as Canada) in search of work. This second wave of emigration included many people living in the Caribbean (often the descendants of slaves) who became part of this secondary migration to England, Canada, or the United States. Other recent emigrations from Africa include the movements of white populations from South Africa and other countries to Europe, North America, and Australia. There has also been a "brain drain" of thousands of African students and professionals, especially doctors, per year who move to study or work in universities or companies in Europe and the Americas.

Contemporary migration within Africa is mainly associated with the search for employment and with flight from famine, floods, and violent conflict. Labor migrations became a factor during the colonial period, when the loss of traditional land to colonists and high taxes forced people to move in search of work, and employment became available in mines and on plantations. For example, residents from the Sahel migrated to work in peanut-, cotton-, and cocoa-producing areas in Senegal, Côte d'Ivoire, Nigeria, and Ghana; in East Africa, Hutus and Kikuyus migrated to work on Kenyan and Ugandan coffee and tea plantations and farms. The most significant labor migration of the last 100 years is from southern Africa, especially Botswana and Zimbabwe. Laborers from there have moved into South Africa to work in the mining industry. By 1960, more than 350,000 foreign workers were employed in South Africa. These migrations have disrupted family life, but the remittances that are sent back by workers have become an important contribution to local and, in the case of countries such as Botswana and Lesotho, national economies. Migrants now work in the South African service sector as well. Labor migration also continues from inland West Africa to coastal cities such as Abidjan, Accra, and Lagos.

▼ **FIGURE 5.40 Africa's Major Cities** (a) This view encompasses the poverty of the Alexandra slum township as well as the expensive business and residential area of Sandton in Johannesburg, South Africa. (b) Zebra roam in Nairobi National Park, which is adjacent to the business and tourist center of Nairobi home to headquarters of international organizations and international hotels. (c) In this poor and swampy neighborhood of Lagos, Nigeria, garbage and sewage create serious health hazards.

(a)

(b)

(c)

Refugees The U.N. High Commission on Refugees (UNHCR) estimated that there were more than 2.7 million refugees in sub-Saharan Africa in 2011 (down from 4.5 million in 2002). This number represents more than 20% of the world total. UNHCR reported that the "total population of concern," which takes into account refugees, stateless persons, asylum seekers, and internally displaced persons, was 10.5 million throughout the region. Refugees fleeing armed conflict, human rights violations, and drought in sub-Saharan Africa primarily originated from the Democratic Republic of the Congo (457,900), Somalia (760,800), and Sudan (462,100).

Refugee populations place serious burdens on neighboring countries that lack the resources to feed and resettle the impoverished starving arrivals. Guinea, for example, absorbed almost a half million refugees from Liberia and Sierra Leone, and Tanzania took in a similar number from Burundi. Most international refugees are housed in camps that are supported by international organizations and charities (**Figure 5.41**). Disease spreads rapidly in the crowded conditions of the camps, and food supplies are sometimes interrupted or diverted by military groups or governments. Refugees are often accused of spreading HIV and other diseases but are excluded from HIV/AIDS programs at the same time. Long-standing conflicts and loss of livelihoods in their own countries mean that many refugees spend extended periods in the camps with little hope of returning home. However, more peaceful conditions and carefully monitored repatriation have resulted in the return of refugee populations to countries such as Rwanda and Mozambique.

People forced to move within their own countries are some of the world's most desperate migrants because they may not fall under international definitions of refugees or qualify for assistance. The UNHCR estimates that there are 1 million internally displaced people in the Democratic Republic of the Congo, most of them inaccessible to relief organizations.

APPLY YOUR KNOWLEDGE How have HIV/AIDS, urbanization, and changes in the status of women altered population projections for countries in sub-Saharan Africa?

FUTURE GEOGRAPHIES

Sub-Saharan Africa experiences many connections to other world regions and to global systems through trade, migration, and the diffusion of rich and varied cultures, yet it is often overlooked by the media and the dominant global economies. It is often portrayed with negative images of poverty, war, and disease or as a vast nature reserve. In many ways, the prospects for sub-Saharan Africa are improving. Although some Africans maintain rural subsistence lives, disconnected from the world economy, others are working in transnational corporations or producing new exports for the global market. Many countries are experiencing successes as they work to discontinue reliance on foreign assistance, improve living standards, establish democratic governments, and increase food and economic production. Looking to the future, there are several key challenges and exciting opportunities for sub-Saharan Africa and several rather different possible outcomes in the region.

▲ **FIGURE 5.41 Refugees** Millions of Rwandans fled the country in the 1990s to camps such as this one in Katale, Democratic Republic of the Congo. Many of the refugees have returned to Rwanda now that conditions are more peaceful there.

Sustaining Representative and Effective Governments The transition to peace and democratic governments has occurred in many countries in sub-Saharan Africa, but several governments are still fragile and struggling with internal ethnic tensions, long-term corruption, great poverty, and high unemployment. Creating effective governments will require mechanisms for broad-based participation and reconciliation; transparent elections; a professional, reasonably paid, and incorruptible civil service; and reliable public services. Rwanda and Liberia are two countries that are making rapid turnarounds following devastating civil wars, but many others, such as South Sudan and Somalia, still suffer with weak or failed governance. In addition to undercurrents of ethnic tension in many areas in the region, there is also potential for conflict over religion—for example, between Islamic populations and others in the Sahel—and over valuable resources such as gold, oil, and diamonds.

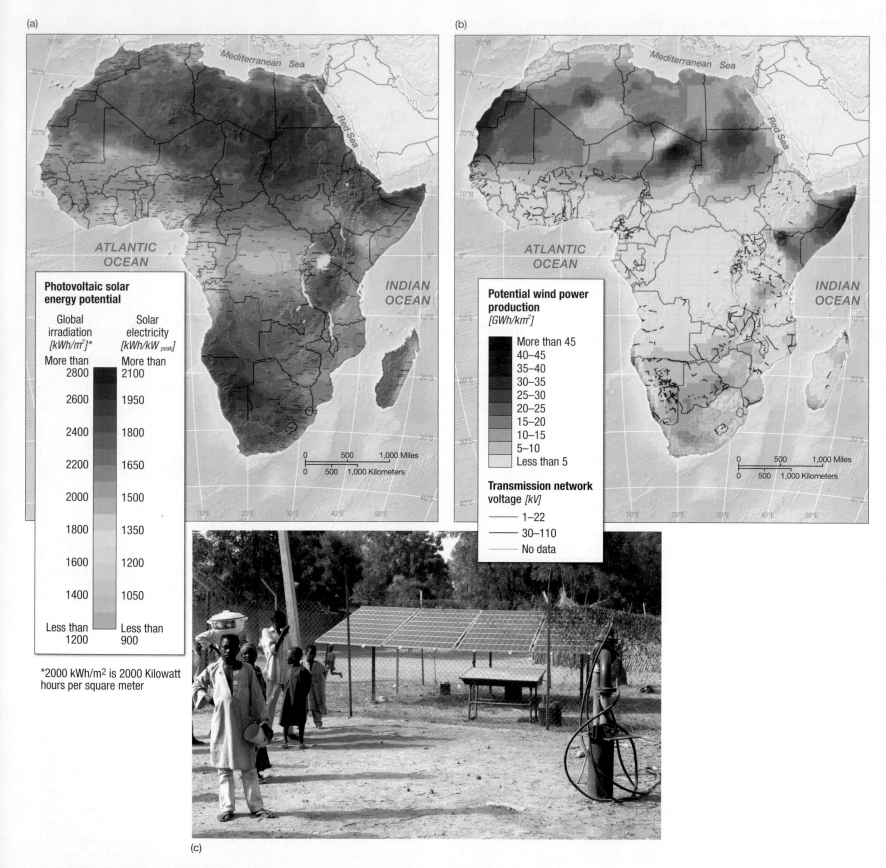

(a)

Photovoltaic solar energy potential

Global irradiation [kWh/m²]*

More than 2800

2600

2400

2200

2000

1800

1600

1400

Less than 1200

Solar electricity [kWh/kW peak]

More than 2100

1950

1800

1650

1500

1350

1200

1050

Less than 900

*2000 kWh/m² is 2000 Kilowatt hours per square meter

(b)

Potential wind power production [GWh/km²]

More than 45
40–45
35–40
30–35
25–30
20–25
15–20
10–15
5–10
Less than 5

Transmission network voltage [kV]

1–22

30–110

No data

(c)

▲ **FIGURE 5.42 Renewable Energy Potential in Africa** (a) This map shows the potential for solar photovoltaic energy in Africa. (b) Wind power potential is also great in Africa as illustrated by this map but is limited by the availability of electric grid connection. (c) Solar panels are being installed in the village of Ahoto in northern Nigeria.

Economic Opportunities Debt forgiveness, foreign investment, and local entrepreneurship have led to improvements in economic indicators for many African countries in the last decade. This region would benefit from economic diversification—more manufacturing and services—and from fairer trading and investment from a broader set of partners. Some regional economies—such as those in Botswana or Nigeria—are very dependent on world prices for precious minerals or oil—and others are vulnerable to fluctuations in the prices of food. If sub-Saharan Africa can reduce the gap between actual and potential yields, add value to resources through processing or manufacturing sectors, and build an educated workforce that can be employed in national and international services, economic growth can continue, ensuring food security and creating a middle class that can drive domestic demand. Countries such as Ghana and South Africa are showing how this can be done.

Health, Energy and the Environment Health prevention programs, child vaccination, and the increased availability of affordable medicine have already reduced infection and death rates from diseases such as yellow fever, malaria, and HIV/AIDS in sub-Saharan Africa. The continued success of these efforts, supported by international assistance, private philanthropy, and national governments, may include the development of malaria vaccines, greater attention to childhood and maternal health, and achievement of the Millennium Development Goals.

Africa is a critical region for the conservation of biodiversity, but efforts to protect forests and wildlife need to take into account the needs of local residents as well as the latest scientific research. Protecting biodiversity and forests may reduce the chances of new diseases that can emerge where humans eat or encounter forest animals and may prevent the loss of species. To make water and energy available to everyone in the region, there needs to be investment in water efficiency and treatment, especially in urban areas, and in energy efficiency, renewable energy, and electric grids in the region. Africa has enormous renewable energy potential that can reduce already low carbon emissions, bring electricity to remote areas, and fuel industrial development (**Figure 5.42**).

LEARNING OUTCOMES *REVISITED*

■ **Compare and contrast the physical geographies of the diverse regions of sub-Saharan Africa.**

Simple stereotypes that portray this world region as poor, diseased, disaster prone, economically underdeveloped, and conflict ridden overlook the great variations between and within countries. There is a great range of landscapes and livelihoods in the 49 countries in sub-Saharan Africa, and it is important to recognize differences in environment, wealth, culture, economy, and political dynamics.

■ **Explain the climate and vegetation of the Sahel and the Congo Basin, how desertification and deforestation occur and are managed, and why sub-Saharan Africa is vulnerable to climate change.**

The Sahel is the southern boundary of the Sahara desert, where descending air creates dry conditions and seasonal rains sometimes fail, bringing drought. Combined with overgrazing, overharvesting of wood, and overcultivation, drought can cause desertification. Careful farming and tree planting can prevent or reverse these effects. The Congo Basin is a vast river basin, where high rainfall associated with rising air at the Equator supports dense tropical forests that are being cut for timber and agriculture. Efforts are being made to protect the forests. Computer models are uncertain about how rainfall in the region will change in the future, but temperatures are predicted to warm almost everywhere and droughts will be more severe.

■ **Evaluate the role of physical geography, disease, and mineral resources in shaping contemporary conflicts, crises, landscapes, and economies in the region.**

Tropical climates and poor soils limit Africa's agricultural potential as do drought, tropical diseases, and pests. Food crises contribute to unrest and create refugees. Millions of people suffer from tropical diseases, such as malaria, yellow fever, sleeping sickness, and river blindness. Mineral resources such as oil and diamonds are important exports but have fueled serious conflicts across the region.

■ **Understand the legacies of European colonialism and the Cold War in the region.**

The legacies of European colonialism include trade patterns oriented to the export of primary products, unequal land distribution and apartheid, and political instability associated with contested boundaries and Cold War alignments. Colonialism also resulted in political boundaries that split ethnic groups across territories or clustered enemies within one territory.

■ **Explain how debt relief and the end of several conflicts may increase the chance of positive economic and political futures in sub-Saharan Africa.**

For many years, high rates of foreign debt and poor terms of trade for exports crippled the economies and well-being of residents in many African countries. Campaigns to end poverty contributed to policies that reduced and forgave debt across the poorest countries. This has stimulated higher levels of growth and investment. A number of serious conflicts—mainly over ethnic differences and resources—have ended across the region bringing hopes for peace, poverty alleviation, and better governance.

■ **Describe the main causes of low agricultural production in sub-Saharan Africa and the role women are playing to help address agricultural and other problems in the region.**

Agricultural production in the region is limited by harsh climates, poor soils, pests, and a lack of fertilizer and other inputs. Food distribution has also been limited by inadequate roads and conflicts. Many development experts believe women, who do much of the farming, could improve production if given access to inputs, training, and land. Women have been shown to earn money to contribute to household incomes when offered small loans and opportunities for business development.

■ **Describe tensions that are emerging around conservation efforts, land grabs, and foreign investment in sub-Saharan Africa.**

Efforts to protect ecosystems and wildlife such as elephants can create tensions between local people, conservationists, and tourists over land use, the protection of species that can harm humans and crops, bushmeat consumption, and park crowding. Global concern and higher prices for food and mineral resources are increasing international interest in purchasing large areas of land in sub-Saharan Africa. Chinese investment, in particular, has increased across much of the region, raising local and geopolitical concerns about Africans losing control of their land and economies.

■ **Provide examples of how African peoples, goods, and culture have spread around the world as a result of slavery, colonialism, and globalization.**

Slavery forcibly distributed people of African descent across many regions of the world, especially in the Americas. This process shaped the contemporary racial composition and culture of countries such as the United States and Brazil. Migration from the region continues as people flee conflict and poverty. African exports of minerals and agricultural products fuel industrial economies and are important in the production of manufacturing and consumer goods such as cell phones and jewelry. Globalization has brought African music, art, literature, and athletes to the world stage.

■ **Describe the spread of HIV/AIDS in the region, its impact, and the efforts made to end the epidemic.**

HIV/AIDS spread rapidly through sub-Saharan Africa since it emerged in 1981 with more than 15 million deaths and 20 million alive but infected by 2009, the majority of whom are women. The impacts include reduced life expectancy in countries such as Zimbabwe and Botswana, almost 15 million orphans, vulnerability to other diseases such as tuberculosis, and the loss of a generation of young professionals. Education on prevention and testing and improved health care and access to drugs, with the help of international charities and assistance, have reduced death and infection rates by 25%.

KEY TERMS

acquired immunodeficiency syndrome (AIDS) (p. 212)

African union (p. 201)

Afrikaners (p. 192)

apartheid (p. 201)

Berlin Conference (p. 193)

Boers (p. 192)

bush fallow (p. 188)

bushmeat (p. 190)

child soldier (p. 203)

conflict diamonds (p. 184)

desertification (p. 178)

diaspora (p. 190)

domestication (p. 187)

ECOWAS (p. 201)

ethnic group (p. 208)

failed state (p. 204)

fair trade movement (p. 197)

G8 (p. 197)

genocide (p. 203)

Gondwanaland (p. 180)

Green Belt Movement (p. 185)

griot (p. 209)

harmattan (p. 177)

homelands (p. 202)

human immunodeficiency virus (HIV) (p. 212)

informal economy (p. 199)

intercropping (p. 188)

intertropical convergence zone (ITCZ) (p. 177)

land grabs (p. 198)

lingua franca (p. 208)

malaria (p. 186)

megafauna (p. 188)

microfinance (p. 198)

Millennium Development Goals (MDGs) (p. 200)

Nelson Mandela (p. 202)

Nollywood (p. 210)

Pangaea (p. 180)

pass laws (p. 201)

pastoralism (p. 178)

remittances (p. 199)

rift valley (p. 180)

Royal Geographical Society (RGS) (p. 193)

Sahel (p. 178)

savanna (p. 185)

shifting cultivation (p. 188)

slash-and-burn (p. 188)

subtropical high (p. 177)

Sudd (p. 182)

tavy (p. 185)

terms of trade (p. 196)

transhumance (p. 188)

tribe (p. 208)

tsetse fly (p. 186)

U.N. World Health Organization (WHO) (p. 186)

urban bias (p. 214)

THINKING GEOGRAPHICALLY

1. Why has wildlife conservation become a priority in some regions of sub-Saharan Africa, and what successful strategies have been used to protect wildlife?

2. What are the current major patterns of migration between and within African countries and beyond, and how do they relate to employment opportunities, natural disasters, and conflict?

3. How has the geographic location of sub-Saharan Africa in tropical latitudes led to problems for agriculture and health and how have people adapted to these challenges?

4. To what extent has sub-Saharan Africa escaped the legacies of colonialism and the Cold War? In what places and sectors are the legacies still significant?

5. What are some of the positive trends in countries of the region and what is the role of debt relief, foreign aid, the end of conflict, and the work of women in these trends?

6. What has sub-Saharan Africa contributed to the demography and culture of other regions?

MasteringGeography™

Looking for additional review and test prep materials? Visit the Study Area in MasteringGeography™ to enhance your geographic literacy, spatial reasoning skills, and understanding of this chapter's content by accessing a variety of resources, including **MapMaster** interactive maps, videos, RSS feeds, flashcards, web links, self-study quizzes, and an eText version of *World Regions in Global Context*.

6 | The United States and Canada

Porcupine River caribou are swimming in a river in the Arctic National Wildlife Refuge, Alaska.

IN 2010, THE GWICH'IN NATION, A NATIVE POPULATION who have lived in the Arctic for 20,000 years, assembled in Fort Yukon, Alaska, for the biennial Gwich'in Gathering. Once, the ancestral lands of the Gwich'in people stretched from the Peel and Arctic Red Rivers in the south to the Mackenzie Delta in the north, the Anderson River to the east, and the Richardson Mountains to the west. The Gwich'in became a geographically divided nation, however, in the 19th century, when the United States and Canada laid claim to various parts of the Arctic. For the Gwich'in, who understand sovereignty not as a set of practices that are contained within political boundaries but as a way of being with one another and the natural world, multiple and sometimes conflicting external governmental systems are challenging their very way of life.

For thousands of years, the Gwich'in have met their subsistence needs by hunting the Porcupine River caribou herd, a massive assemblage of approximately 129,000 animals that provide food, clothing, tools, and a source of spiritual guidance. Each spring, the herd migrates from its winter range north to its spring calving grounds on the coastal plain

LEARNING OUTCOMES

- Describe the distinctive landscapes of the United States and Canada with respect to landforms and climate.

- Summarize the impact of the United States and Canada on global climate change, particularly in terms of their production of greenhouse gases.

- Explain the rise of the United States and Canada as key forces in the global economy.

- Describe how the global economic crisis of 2008 has affected the two countries as well as the impact it has had on economic inequality in the region.

- Compare and contrast the forms and practice of government of the United States and Canada.

- Summarize the Indigenous and immigrant histories of the United States and of Canada.

- Describe the geography of religion and language across the United States and Canada.

- Compare and contrast the urbanization processes that have shaped the United States and Canada over the last 100 years.

of northeastern Alaska and Yukon, where the Arctic National Wildlife Refuge (ANWR) is located. They return in the fall to ANWR.

ANWR is made up of 77,000 square kilometers or 7,689,027 hectares (over 29,700 square miles or almost 19 million acres) of the North Alaskan coast. It is the largest protected wilderness area in the United States, created by Congress under the Alaskan National Interests Lands Conservation Act in 1980. Importantly, Section 1002 of the act deferred a decision on whether to allow drilling for oil and gas exploration and development in 607,000 hectares (1,500,000 acres) of ANWR known as "1002 Area." The threat of oil drilling and gas exploration in 1002 Area has been a rallying issue for the Gwich'in since the 1990s, when elders expressed a need for the scattered communities of the Gwich'in nation to come together and develop collective strategies on the Porcupine River caribou and Yukon River salmon habitat protection. Because of the complexity of the nation's geography, the Gwich'in have presented their concerns not only to the U.S. and Canadian governments, but also to the United Nations. ■

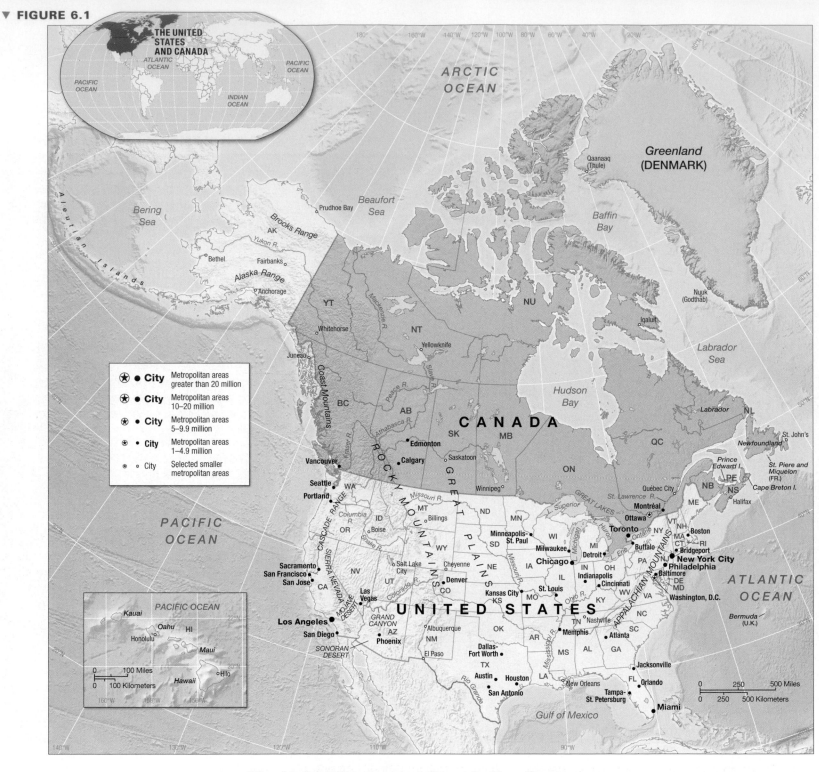

The United States and Canada Key Facts

- Major Subregions: United States of America, Canada

- Major Physiographic Features: Great Lakes, Rocky Mountains, Great Plains, Appalachian Mountains, Sierra Nevada, Grand Canyon, southwestern deserts

- Major Religions: Christian (Protestant and Roman Catholic)

- Major Language Families: English, Spanish, and French, from the Indo-European family which includes Germanic and Romance branches

- Total Area (total sq km): 19.6 million

- Population (2011): 346 million; Population under Age 15 (%): 19; Population over Age 65 (%): 13

- Population Density (per sq km) (2011): 16

- Urbanization (%) (2011): 79

- Average Life Expectancy at Birth (2011): Overall: 78; Women: 81; Men: 76

- Total Fertility Rate (2011): 1.9

- GNI PPP per Capita (current U.S. $) (2011): 44,790

- Population in Poverty (%, < $2/day) (2000–2009): 0

- Internet Users (2011): 272,960,859; Population with Access to the Internet (%) (2008): 79.9; Growth of Internet Use (%) (2000–2011): 99.6

- Access to Improved Drinking Water Sources (%) (2011): Urban: 100; Rural: 94

- Energy Use (kg of oil equivalent per capita) (2009): 7288

- Ecological Footprint (hectares per capita consumed/hectares per capita available, global scale) (2011): United States: 8.0/1.8; Canada: 7.01/1.8

ENVIRONMENT AND SOCIETY

The United States and Canada occupy the landmass of North America (**Figure 6.1**). This region has played a key role in the unfolding of global environmental processes largely because of its position with respect to climate patterns, plate tectonics, and ocean currents and the adaptations humans have made in response to these factors. People have occupied the region for millennia, though the most profound global environmental impacts and innovations have come about in the last 200 years. The region is a highly prosperous one, and the advanced standard of living enjoyed by people in the United States and Canada has had dramatic effects on the region's landscapes, from water environments to land uses to air quality.

Climate, Adaptation, and Global Change

Complex processes characterize the region and shape environment and society relations and practices domestically and across the globe. Because temperature, precipitation, and terrain patterns combine to influence vegetation, and to some extent soil, it is important to understand the role that climate plays in this region and the ways people in the region have adapted their practices to climate.

Climate Patterns The United States and Canada contain nearly every type of climate condition possible, with temperatures varying quite dramatically on any one day of the year from north to south. In the Canadian Arctic and Alaska in the north, it is very cold for most of the year; in the southern United States, it is warm for most of the year. Because of the moderating influence of the oceans on three sides, the Gulf of Mexico, and the interior Great Lakes, as well as the presence of mountains and plateaus, within-region differentials also occur (**Figure 6.2**). Central United States and Canada experience

the most dramatic temperature ranges; the coasts have a far narrower range because of the moderating influence of the oceans.

Variations in amounts of precipitation also have a profound effect on climate (**Figure 6.3**, p. 226). The east and west coastal areas tend to be mild and moderately wet, and the interior is largely arid because the north–south mountain chains prevent moisture-bearing clouds from moving inland to drop their moisture. Because of this, a moisture gradient exists that declines slowly but continuously from east to west as far as the three significant mountain ranges—the Rockies, Sierra Nevada, and Cascades. Once beyond these mountains, the moisture gradient rises dramatically toward the Pacific.

In the southeastern part of the United States, where no significant coastal ranges exist, moisture-bearing clouds more readily condense into rain. Here, the warm Gulf of Mexico is an important source of moisture and storms, including hurricanes. In the Arctic north, annual precipitation approximates desert conditions because of the dominance of very stable air masses with low moisture content and the dryness of cold air. The jet stream (see Chapter 1, p. 10) brings precipitation to most of the continent in the winter months. The warmer parts of the region—in the southern United States and Hawaii—experience this precipitation as rain, and the colder, more northerly parts experience snow. In areas around the Great Lakes, the warming effects of these large bodies of water add even more moisture to the mix, bringing especially heavy snowstorms (called "lake-effect" snow) to places like Buffalo, New York, on the northeastern tip of Lake Erie; and to Sault St. Marie, Ontario, on the channel between Lake Superior and Lake Huron.

Environmental Modification and Impacts The United States and Canada possess a bounty of resources: a range of minerals; vast forests; fertile, highly productive land; extensive fisheries; varied and abundant wildlife; and magnificent physical beauty. In addition to this physical wealth, this region has the technological capability and the drive to exploit resources to an extent achieved by few other regions on Earth. Consequently, this region experiences an extraordinarily high economic quality of life. The region also is home to material consumption that results in elevated levels of air and water pollution, soil contamination, solid waste disposal challenges, an ongoing problem of nuclear waste disposal, acid rain, extinct and endangered species, and insect and animal genetic mutations.

Certainly the most significant and pressing environmental problem in the United States and Canada today (as well as in many other places around the globe) is global climate change (see Chapter 1, p. 12). The United States and Canada play a major role in global climate change. Both countries produce high total and per capita greenhouse gas emissions resulting from production and consumption patterns. These emissions stem from the region's heavy reliance on fossil fuels, including high gasoline use and coal-fired electricity, as well as agricultural and industrial production that generate not only massive amounts of carbon dioxide

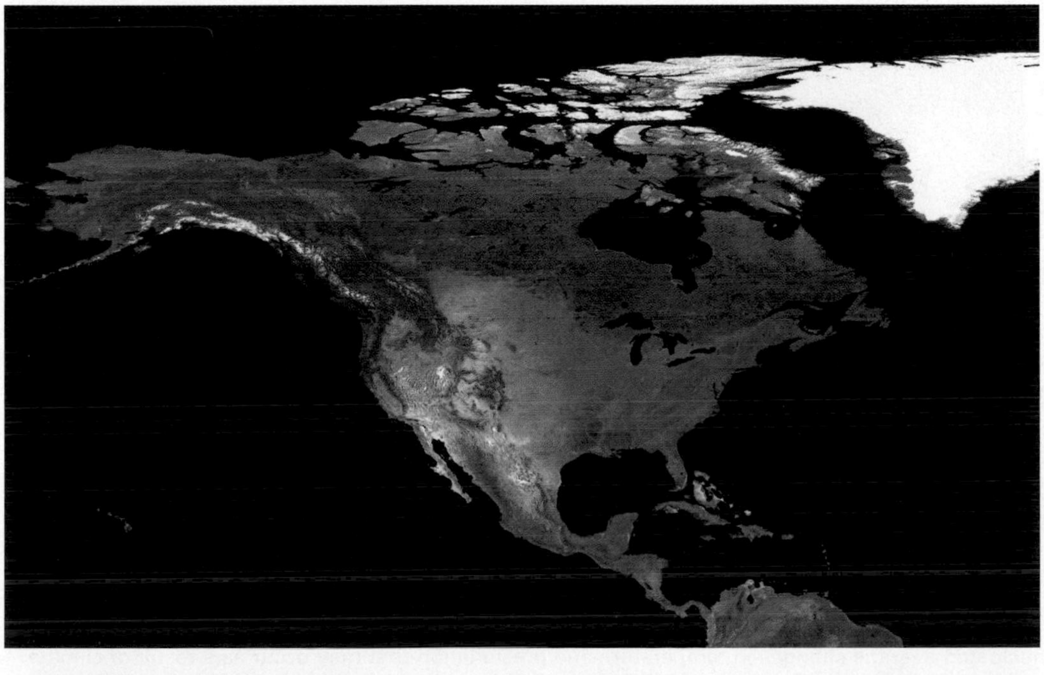

▲ **FIGURE 6.2 The United States and Canada from Space** The region is surrounded by oceans, gulfs, and bays with vast interior plains and high mountains on both the East and West Coasts. The United States and Canada contain vast mineral wealth, extensive forests, and supplies of clean water and enjoy high agricultural productivity.

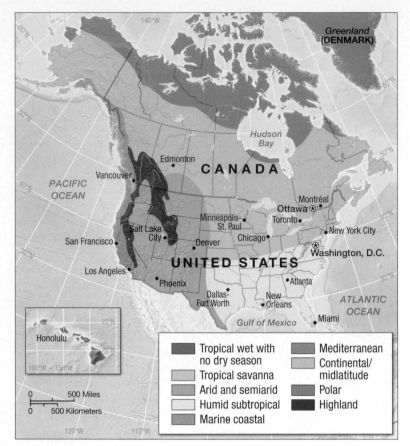

▲ **FIGURE 6.3 Climate of the United States and Canada** This is the only region in the world that contains the full range of climate types that occur globally, stretching from Alaska and the Canadian Arctic to the tropical Hawaiian Islands.

such as Miami (see "Visualizing Geography: Sea-Level Rise" in Chapter 1, p. 14) and other coastal communities where uncontrolled growth has made homes vulnerable and increased insurance costs.

■ Temperature increases could cause adverse health impacts, such as heat-related mortality and mosquito-borne diseases, and stress agricultural crops and livestock. In more northerly regions, warming might increase agricultural productivity (**Figure 6.4**).

■ More severe droughts, especially in the western United States, would increase competition among agricultural, municipal, and industrial users and ecosystems over the region's overallocated water resources and could increase risks of wildfire.

■ Changes in natural ecosystems could alter the current configuration of landscapes and valuable species; as a result, key species and areas would no longer be contained within parks and conservation areas.

APPLY YOUR KNOWLEDGE Search the Internet and find a map of the effects of global climate change on the United States and Canada. Identify two areas of the region with which you are familiar and determine how they will be affected by climate change.

Geological Resources, Risk, and Water Management

The United States and Canada feature a vast central lowland that includes the Canadian Shield, the Interior Lowlands, and the Great Plains. To the east are the Appalachian Mountains, which descend

but also other **greenhouse gases (GHGs)**, such as methane and nitrous oxide. Knowing the **carbon footprint**—the total set of GHG emissions caused directly and indirectly by an individual, organization, event, product, or place—is a way of understanding how GHGs are contributing to climate change and where their impacts might be reduced. The carbon footprint of both the United States and Canada is much larger than most other nations.

Climate change is indisputably a global problem with serious projected impacts and its effects are already being felt locally in North America. For instance, changes have already been detected in terrestrial and marine biological systems. Scientists using complex models that estimate changes in temperature, precipitation, and sea levels at the regional scale have predicted the following likely impacts of climate change on the United States and Canada:

■ An increase in sea levels along much of the coast would exacerbate storm-surge flooding and shoreline erosion, placing at risk those low-lying cities

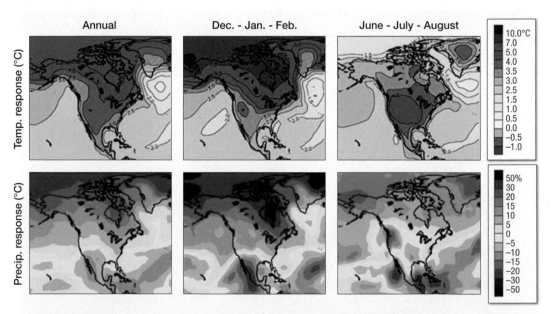

▲ **FIGURE 6.4 Temperature and Precipitation Due to Climate Change** These maps show the anticipated average changes in temperature and precipitation that may occur as a result of climate change in the region. Minimum winter temperatures and maximum summer temperatures are likely to increase more than the global annual mean in the southwestern United States. Projected warming is expected to be accompanied by a general increase in precipitation over most of the United States and Canada except the most southwesterly part.

gradually to the Gulf–Atlantic Coastal Plain and become broader as one travels farther south and southwest. To the west are three distinct topographical regions: moving from west to east are the mountains and valleys of the Pacific coastal ranges; then an **intermontane** set of basins, plateaus, and smaller ranges that lie between mountains; and finally the great Rocky Mountain range, which rises steeply and imposingly at the western edge of the central lowlands.

Physiographic Regions **Figure 6.5** provides an overview of the different physiographic regions of the United States and Canada. (Physiography is another term for physical geography, which is the branch of geography dealing with natural features and processes.) The Canadian Shield, shown in light brown, is a geologically very old region and very rich in minerals. The Interior Lowlands, in light green, is a glaciated landscape with fertile soils and abundant lakes

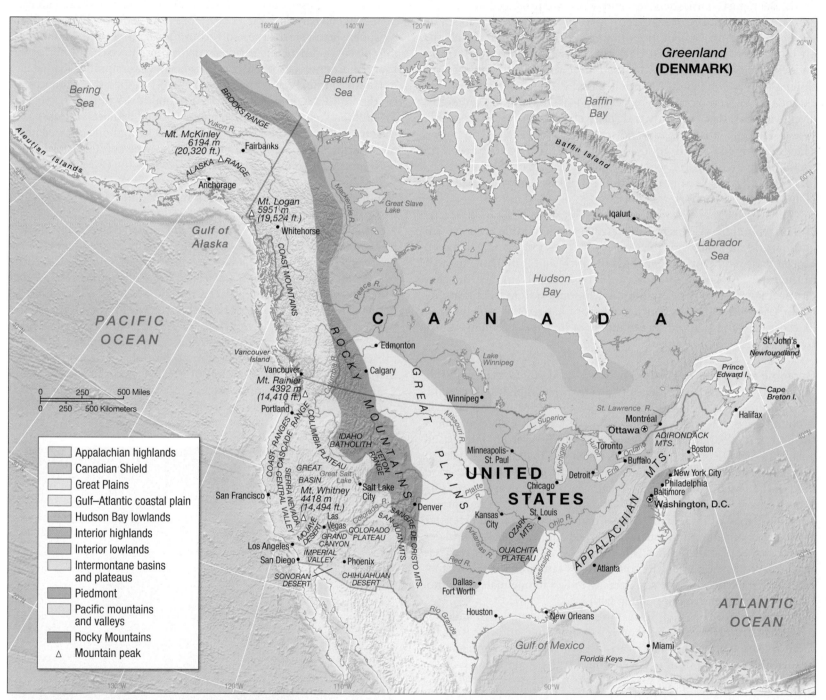

▲ **FIGURE 6.5 Physiographic Regions of the United States and Canada** The Interior Lowlands and Great Plains are central to agricultural production. The mountain ranges in the west are significantly higher and steeper because of their relatively recent emergence, whereas the Appalachian Mountains in the east are lower and more sloping because of their greater age and the effects of erosion. The eastern coastal region slopes toward the sea, but the West Coast is characterized by steep cliffs and fairly narrow beaches. Waves are also more dramatic in the west, and the western seashore is more dangerous and prone to riptides and deadly undertows.

and rivers. This part of the continental landmass is devoted mostly to agriculture, with some industry as well. The third primary physiographic subdivision is the Great Plains, shown in beige, an area that slopes gradually upward as you move west toward the Rocky Mountains. The Great Plains region experiences more rainfall than the surrounding regions. This is an area of extensive gently rolling and flat terrain with excellent soils and some of the world's most productive farms. Beef, pork, corn, soybeans, and wheat are produced here.

The major landform feature of the eastern United States and Canada is the Appalachian mountain chain. Surrounding the chain, on its eastern and southern flanks, is a coastal plain that can be divided into two subregions, the Piedmont area of hills and easterly sloping land and the Gulf–Atlantic Coastal Plain, a lowland area that extends from New York to Texas. The coastal plains are generally level, with soils that are sandy and relatively infertile. Where agriculture does occur, it is intensive and scattered. For instance, the lower Mississippi is an area of good soils and intensive agriculture. The Piedmont is the historic region of plantation agriculture—cotton, tobacco, and corn—but much of the soil in this region has been depleted or eroded from overfarming.

The Mississippi–Missouri river system is at the heart of U.S. economic geography. As the most extensive and navigable river system in the world, it, along with the five Great Lakes (Superior, Michigan, Huron, Erie, and Ontario), enabled the early and formidable growth of the interior of the United States. However, the engineering that has been applied to the Mississippi–Missouri system (including its major tributaries) over a century or more has irretrievably altered its ecosystem and exposed vast numbers of human settlements to extreme danger from flooding.

Geologic and Other Hazards The western coastal formation in this region sits along the fault line of two active crustal plates, the Pacific Plate and the Juan de Fuca Plate. Both plates are moving northward, rubbing against the more stationary North American Plate on which the continental landmass sits. As a result of this friction, the coastal area from San Diego through British Columbia to Alaska is subject to frequent tremors. Extreme and devastating earthquakes have occurred here in the past and are likely to occur again in the near future. The extraordinary views of the Pacific Ocean provided by the mountainous topography of the Pacific coastal region have attracted extensive home development on mountain slopes, and past earthquakes have destroyed many of these homes. The extensive rainfall in this area has also wreaked havoc. When the soil becomes saturated with rainwater, liquefaction occurs, and the soil literally moves in one massive slump, carrying very large structures along with it.

In the western United States, the intermontane basin and plateau formation between the Pacific coastal range and the Rockies includes the Columbia Plateau to the north, which begins at the headlands for the Columbia River, and the Colorado Plateau in the south, which includes the Grand Canyon. In between these two plateaus are the Great Basin in Nevada and Utah, which includes extinct lakes as well as the Great Salt Lake, and the southwestern deserts (Mojave, Sonoran, and Chihuahuan). The region's landscape features many contrasts and spectacular scenery that includes deep canyons, majestic mountains, and unique deserts. It occupies an area of dry climates, thin vegetation, and a general absence of many perennial surface streams. Wildfires, droughts, floods, and landslides are continuing threats in the western United States.

The physical geography of the U.S./Canada world region is the context for, and an inescapable reminder of, the enormous benefits and the sometime devastating costs of the "American way of life." There is no better example of those costs than the accident that occurred when British Petroleum's Deepwater Horizon oil rig exploded and caught fire on April 20, 2010, in the Gulf of Mexico, 400 kilometers (250 miles) offshore of Houston. The explosion killed 11 people and injured a least two dozen more. The well leaked an estimated 40,000 barrels of oil a day before it was effectively capped, and oil slicks spread across the Gulf from Florida to Texas, contaminating the environment and seriously damaging the coastal fisheries and tourism economy, especially in Lousiana, Mississippi, Alabama, and northern Florida. When the well was finally capped three months after the explosion, it is estimated that a total of 4.9 million barrels, or 780 million liters (205.8 million gallons), of oil had been released into the environment. As the world's largest accidental oil spill to date, it will continue to contaminate or damage marine and terrestrial life—both human and nonhuman—for decades to come (**Figure 6.6**).

Mineral, Energy, and Water Resources The United States and Canada possess great energy and mineral resources (see "Visualizing Geography: The Resource Bounty of the United States and Canada," p. 230). The two countries are among the top 20 producers of natural gas in the world. Canada is second only to Saudi Arabia in its proven oil reserves (proven sources are those where there is a reasonable certainty of being recoverable and produced), and although China is the largest producer of coal in the world, the United States possesses the largest proven coal reserves globally. Canada's oil is mostly located in the province of Alberta, in sand deposits that pose difficulties for extraction. The United States's oil deposits are primarily located in Texas, the Gulf of Mexico, Louisiana, Alaska, and California. Importantly, in addition to being a leading energy producer, the United States is the leading consumer of energy resources

▲ **FIGURE 6.6 Deepwater Horizon Explosion, April 2010** A blowout caused the Deepwater Horizon well to burn and sink, launching a massive offshore oil spill 60 kilometers (40 miles) southeast of the Louisiana coast in the Gulf of Mexico. Pictured here are fireboats attempting to suppress the blaze with high-pressure water spraying.

(though China is fast catching up). The United States relies mostly on Canadian imports of natural gas, coal, and oil to supplement this thirst for energy.

In addition to these nonrenewable sources of energy, in the western United States and Canada renewable hydropower is available, and solar and wind power are also becoming increasingly prominent on the energy landscape. Both Canada and the United States are also rich in a wide array of mineral and precious metal resources, many of which are integral to new technologies of production from copper to silicon.

APPLY YOUR KNOWLEDGE Search the Internet for information on the distribution of five global resources (such as oil, water, coal, a major foodstuff, and a mineral). Compare the United States and Canada with two other countries rich in similar resources and two others that are resource-poor. What generalizations can you make about the importance of resources to a country's overall wealth?

Ecology, Land, and Environmental Management

The Europeans who arrived on the Atlantic Coast of the United States and Canada beginning in the late 15th century encountered an environmentally diverse landscape thinly populated by Indigenous peoples. It is estimated that more than 30,000 years ago, the ancestors of these Indigenous peoples began the process of populating the continent and altering (and permanently changing) the environments they encountered. When European explorers arrived, they did not discover a pristine land, but one that had already been transformed by tens of thousands of years of human settlement.

Indigenous Land Use The first inhabitants of Canada and the United States came across the Bering Land Bridge, a series of landforms that once existed between northeastern Asia and northwestern North America. Once part of the seafloor, these land corridors were exposed during periods of worldwide glaciation and subsequent lowering of sea levels (**Figure 6.7**). Such areas began appearing about 70 million years ago, linking what is now northwestern Canada and

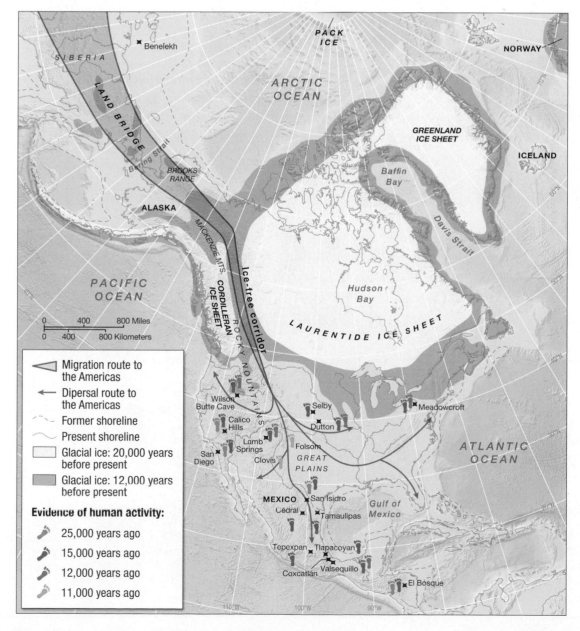

◀ **FIGURE 6.7 Migration of Neolithic Hunters into the Americas** It is believed that the first humans to enter the North American continent by way of the Bering Land Bridge were hunters pursuing large mammals and mastodons. When these animals grew scarce in one area, the small bands of nomadic hunters moved on, looking for new prey. They also fed themselves with roots, plants, and berries and learned to fashion clothing, weave baskets, and construct fishing nets. Because of the limits of the environment, the bands tended to remain limited in size and, as the population grew, small groups would break off and move on to increase their chances of survival in more plentiful surroundings.

VISUALIZING GEOGRAPHY
The Resource Bounty of the United States and Canada

The forests of the United States and Canada cover 719 million hectares (1.7 billion acres), about five and one-half times the size of England or 1,343 U.S. football fields. The United States and Canada also possess excellent soils and abundant water resources as well as extensive mineral wealth and energy rivaling, and in many cases surpassing, the resource wealth of other large, resource rich countries such as Russia, China, and Brazil. Indeed, much of this region's global economic success lies in its natural endowments and ability to exploit them. For instance, the United States produces twice as much food as is needed per person with 40% of food going to waste—enough to fill the Rose Bowl every three days! **Figure 6.1.1** and

Figure 6.1.2 provide insight into the mineral and energy resources possessed by the United States and Canada as well as the region's food production prowess.

Economic geographers, however, recognize that resource endowments—sometimes referred to as "natural capital"—are not enough to enable a region to be economically successful. It is also important to have a population that is educated and healthy and institutions that ensure economic growth and stability. For resource endowments to be realized and efficiently exploited, it is necessary to have not only a strong workforce—"human capital"—but institutions—"social capital"—that are high quality, trustworthy, and transparent.

A region can have all the natural capital that the United States and Canada possess and still not be prosperous if it lacks human and social capital as well as the capital (accumulated assets) to exploit those resources. Moreover, a country can have negligible natural capital and still be successful, as long as it has the human, social, and accumulated capital to secure what it needs through trade and labor. (Singapore is an excellent example of the latter; it is one of the richest countries in the world and has very few natural resources.) Both Canada and the United States enjoy vast natural capital. The following additional resources have allowed them to effectively exploit this natural capital:

▶ **FIGURE 6.1.1**
Mineral Wealth in the United States and Canada In addition to traditional energy resources such as oil and gas, both Canada and the United States possess a wealth of alternative energy sources such as wind and solar power. These energies riches, combined with abundant mineral deposits such as iron, copper, and gold, make the two countries among the richest in the world.

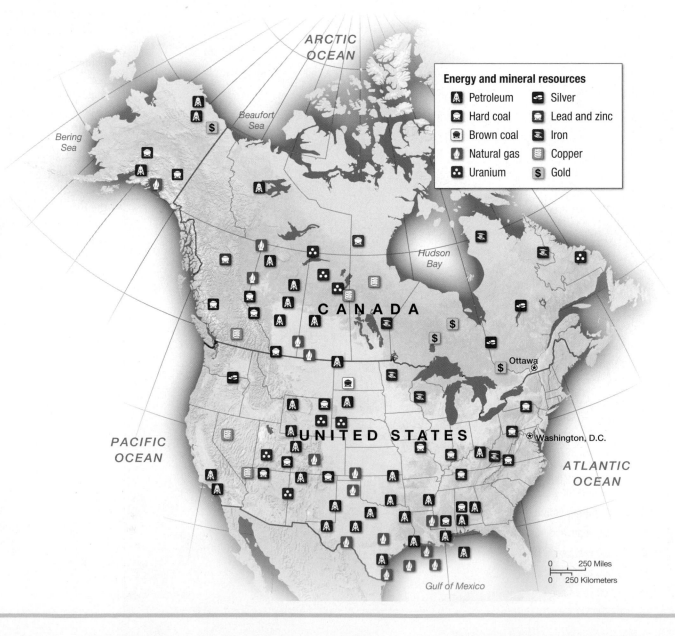

1. Savings and investment that allow the building of roads, factories, and the development of machinery
2. High-quality education, health care, and social security
3. The capacity to export and import goods and services so they can build up foreign capital and supplement domestic capital
4. Democratic forms of government, premised on honesty and openness, that are intended to support freedom and equality and a commitment to the nation
5. General economic stability that supports investment

What is missing from this list is a commitment to the sustainable exploitation of resources. While people in the United States and Canada enjoy a high material standard of living, their impact on the environment is commensurate with this high standard of living and their way of life is not environmentally sustainable. **Figure 6.1.3** shows the ecological footprint of the world, with the inset showing the United States and Canada, whose footprint is especially dramatic. Ecological footprints typically increase with levels of wealth and, as this figure makes clear, in order to maintain a healthy planet, the correspondence between wealth and ecological impact needs to be addressed in the region. ■

CANADA
Rank in the world by commodity (production value)

First Place
Mustard seed
Linseed
Lentils
Peas, dry
Canary seed
Mixed grain
Oats

Second Place
Rapeseed
Blueberries
Cranberries

Fourth Place
Fibre Crops

Sixth Place
Indigenous turkey meat

Seventh Place
Indigenous pig meat
Soybeans
Mushrooms and truffles
Wheat

Eighth Place
Raspberries
Rye

Ninth Place
Chickpeas
Indigenous cattle meat
Buckwheat

Tenth Place
Safflower seed

UNITED STATES
Rank in the world by commodity (production value)

First Place
Soybeans
Game meat
Indigenous chicken meat
Indigenous turkey meat
Cow milk, whole, fresh
Nuts
Maize
Almonds, with shell
String beans
Blueberries
Cranberries
Maize, green
Strawberries
Indigenous cattle meat

Second Place
Grapefruit (incl. pomelos)
Mushrooms and truffles
Carrots and turnips
Apples
Plums and sloes
Sugar beet
Hen eggs, in shell
Indigenous pig meat
Oranges
Tomatoes
Peas, dry
Pistachios
Sorghum
Walnuts, with shell
Safflower seed
Lettuce and chicory
Spinach
Cherries

Third Place
Hazelnuts, with shell
Wheat
Cotton lint
Lentils
Linseed
Potatoes
Pears
Grapes
Hops
Honey, natural
Onions, dry
Indigenous horse meat

Fourth Place
Raspberries
Pumpkins, squash, and gourds
Buckwheat
Groundnuts, with shell
Barley
Peaches and nectarines

▲ **FIGURE 6.1.2 Commodity Wealth**
It is one thing to have the energy and mineral wealth to sustain and develop economically, but add to that abundant commodity wealth and it's clear that the United States and Canada are able to be highly independent with respect to these key resources.

▶ **FIGURE 6.1.3**
Ecological Footprint of the World This figure shows the human ecological footprint index of the world, with the inset showing the region of the United States and Canada. Ecological footprint index is calculated based on the amount of productive land and sea area necessary to supply the resources a population consumes.

Human footprint index
0.0–1.0
1.1–10.0
10.1–20.0
20.1–30.0
30.1–40.0
40.1–60.0
60.1–80.0
80.1–100.0

northern and western Alaska with northeastern Siberia, Russia. Based on strong fossil evidence, it is believed that the various land bridges allowed plants and animals to move between the two continents and that the native peoples living in the region now are descendants of long-ago ancestors from Asia. At the end of the Pleistocene Epoch (about 2,600,000 to 11,700 years ago) the present-day Bering Strait between Alaska and Siberia opened and severed the intercontinental land connection.

These first human inhabitants moved into the United States and Canada, traveling southward along the western edge of the continental icecap. These Neolithic, or Stone Age, hunters probably originally came from northern China and Siberia. Descendants of these first hunters gradually moved farther and farther southward into the continental landmass, advancing eventually into Mexico. Abundant evidence suggests that they eventually spread throughout North America (as well as Central and South America) and adapted their ways of life to the conditions they encountered. As they began to settle, different groups introduced agriculture in different places. With game, fish, and wild and cultivated foodstuffs available, an economic system based on subsistence production and trade emerged. The cultivation of maize spread to wherever it could be grown, and new wild foods, like potatoes and tomatoes, were eventually domesticated.

In New England, prior to European contact, hunting, gathering, and some shifting cultivation existed among the Indigenous peoples. Hunter–gatherers were mobile, moving with the seasons to obtain fish, migrating birds, deer, and wild berries and plants. Shifting agriculture was organized around planting and harvesting maize, squash, beans, and tobacco. Hunter–gatherers and shifting cultivators all identified and used a wide range of resources. The economy was based on need, which was met by planting or foraging or through barter (for example, trading corn for fish). The prevailing practice was to take only what was needed to survive. Native people appeared to have no concept of private property or land ownership, and there is no evidence that a profit motive existed before contact with Europeans. Land and resources were shared. However, substantial vegetation change did occur as a result of New England native people's settlement and hunting activities, which resulted in some species depletion before the arrival of the Europeans.

Colonial Land Use Europeans saw the natural world they encountered in the United States and Canada much differently from how native peoples saw it. Most important, Europeans viewed resources as commodities to be accumulated, not necessarily for personal use but to be sold for profit or export. The arrival of the Europeans in the region meant that pressures on natural resources (especially wood, furs, and minerals) were hugely accelerated. In New England, where European settlement first occurred, there was extreme exploitation of white pine, hemlock, yellow birch, beaver, and whales that led ultimately to deforestation and extinctions.

The arrival of Europeans also meant a dramatic change in prevailing social understandings of the nature of land in the region. Native perspectives about the communality and flexibility of land were replaced by European views of land as private and as having fixed boundaries. European settlers wanted to own and fence a plot of land, which led eventually to the concentration of land in large private farms, plantations, and estates. Increasingly, the native people of the United States and Canada were forced onto less-productive land or reservations and were prohibited from hunting or gathering on private lands or from moving with the seasons as they had before. A **reservation**, called a **reserve** in Canada, is an area of land managed by an Indigenous tribe under the U.S. Department of the Interior's Bureau of Indian Affairs or, in Canada, under the Minister of Indian Affairs.

Contemporary Agricultural Landscapes Overall in Canada and the U.S. region, conditions are very good for agriculture as one moves from east to west until the precipitation gradient drops (**Figure 6.8**). Most of the agricultural productivity of the United States and Canada is concentrated in the Interior Lowland and Great Plains regions. Outside these subregions, soil fertility is often low, and rainfall is limited and infrequent although conditions become favorable again in the valleys along the Pacific Coast. Although natural conditions favor certain areas for the highest agricultural productivity, other parts of the region are also important agriculturally, largely because farmers have overcome the natural barriers to production through fertilizers, irrigation, pesticides, and other technological applications. This is the case in the Pacific valley areas, for instance, where irrigation water drawn from the Colorado River enables agriculture to flourish, and in the Southwest, where intensive irrigation supports significant cotton and citrus production. In Arizona and New Mexico, northward-flowing air masses bring summer monsoonal precipitation patterns

▲ **FIGURE 6.8 Winter Wheat in Canada** Pictured here are both fields and storage silos on a large farm on the Canadian prairie in Manitoba. Canada is one of the world's largest wheat producers.

▲ **FIGURE 6.9 Center Pivot Irrigation** The agricultural landscapes of the U.S. West, such as Farmington, New Mexico shown here.

that support unique desert vegetation. The summer and winter rains there allowed the ancestors of contemporary Native Americans to cultivate beans, squash, and corn using sophisticated irrigation systems in a landscape that would otherwise seem inhospitable to subsistence agriculture.

The agriculture of the Great Plains U.S. states and Canadian prairie provinces is large-scale and machine-intensive, dominated by a few crops, the most important of which is wheat. Winter wheat is planted in the fall and harvested in late May and June. Winter wheat is grown across much of the United States, but it predominates in the southern Plains from northern Texas to southern Nebraska. Spring wheat is grown primarily from central South Dakota northward into Canada and is planted in early spring and harvested in late summer or fall. It is suited to areas where winters are too severe for winter wheat. Most of the wheat grown in these former grassland ecologies use dry farming techniques, meaning the crop is not irrigated.

The interior valleys along the western coast are key areas of fruit and vegetable production (**Figure 6.9**). Rainfall is inadequate to sustain the crops so they are irrigated. More than half of the fruits, vegetables, and nuts consumed in the United States are produced in California. Washington State accounts for 50% of the U.S. apple supply. Fruit and vegetable production in Canada occurs largely in British Columbia, Quebec, and Ontario, with the latter being the leader in vegetable crops and the former dominating the export of cherry plantings around the world. The west coasts of both the United States and Canada are also where marijuana, an increasingly important though still largely illegal cash crop, is produced (see "Geographies of Indulgence, Desire, and Addiction: Marijuana" p. 234). Many new food movements have begun and disseminated from the west coastal area of these two countries. These movements, which are discussed in the following section champion organic production, slow food, and local food.

Alternative Food Movements **Organic farming** describes any farming or animal husbandry that occurs without commercial fertilizers, synthetic pesticides, or growth hormones. This type of

farming can be contrasted with conventional farming, which uses chemicals in the form of insecticides, herbicides, and plant fertilizers and relies on intensive hormone-based practices for breeding and raising animals. Organic production can be seen as a rejection of industrial farming practice and the increasing use of **genetically modified organisms (GMOs)** on U.S. and Canadian conventional farms. A GMO is an organism that has had its DNA modified in a controlled environment rather than through cross-pollination or other forms of evolution. Examples of GMOs include a bell pepper with DNA from a fish added to make the pepper more drought tolerant, a potato that releases its own pesticide, a soybean that resists fungus, and rice that includes more nutrients than non-GMO rice (**Figure 6.10**). Although conventional farming practices continue to dominate the region's agricultural sector, organic practices are a growing force not only for small farmers but also among larger corporate entities. Big-box retailer Walmart has even introduced organic foods into its superstores.

Local food is usually organically grown, but the designation "local" means that it is also produced within a fairly limited distance from where it is consumed. Most interpretations of local food establish a 160 km (100 mile) radius as the limit of what is truly local. In other words, a "locavore" is a person who consumes food that is produced on farms that are no farther than 160 km (100 miles) from the point of distribution. Local food movements have resulted in the proliferation of communities of individuals who have joined together to support the growth of new farms. **Slow food** is devoted to a less hurried pace of life and to the true

▲ **FIGURE 6.10 Protests against GMO Food in Mexico** Greenpeace activists in Mexico City protest the importation of GMO rice from the United States.

GEOGRAPHIES OF INDULGENCE, DESIRE, AND ADDICTION

Marijuana

By Conor Cash

According to the United Nations Office on Drugs and Crime, cannabis is the most commonly consumed illicit substance in the world, with an estimated 125 to 203 million users in 2009 (**Figure 6.2.1**). As with other illicit drugs, core countries constitute the lion's share of global demand and the United States and Canada represent the highest per capita use. Perhaps surprisingly, given the current emphasis on global commodity flows, the vast majority of marijuana that is produced worldwide is consumed locally. The United States and Canada are not exceptions to this, although marijuana produced in both countries travels extensively within national boundaries, accruing additional value as it moves across state and provincial lines.[1]

A number of diverse factors are involved in marijuana production in the United States and Canada, all operating within the broader context of large consumer markets and attendant high profits. Marijuana production occurs in a wide variety of locales, and there are a variety of vastly different systems of production for the drug. California produces approximately 75% of the marijuana consumed in the United States. In this state, cannabis is grown on large, illegal plantations on public lands (presumed to be financed by Mexican **cartels**—a collection of independent businesses formed to regulate production, prices, and marketing); medical grow operations, where plants are cultivated specifically for patients in legal medical marijuana buying clubs; and by home cultivators.

Marijuana production in Canada is substantial enough to supply the entirety of the country's domestic market (**Figure 6.2.2**). This is no small feat if one considers that some 10.6% of Canadians reported having used the drug in 2009. While British Columbia remains the largest producer of marijuana in Canada, both Ontario and Quebec have become sites of production in recent years.[2]

In many senses, the marijuana economy is one that is in transition. A patchwork of

---

[1]United Nations Office on Drugs and Crime (2011), World Drug Report, pp. 175–193.
[2]Bureau for International Narcotics and Law Enforcement Affairs, U.S. Department of State (2011), *International Narcotics Control Strategy Report, Vol. 1, Drug and Chemical Control*, pp. 174–78.

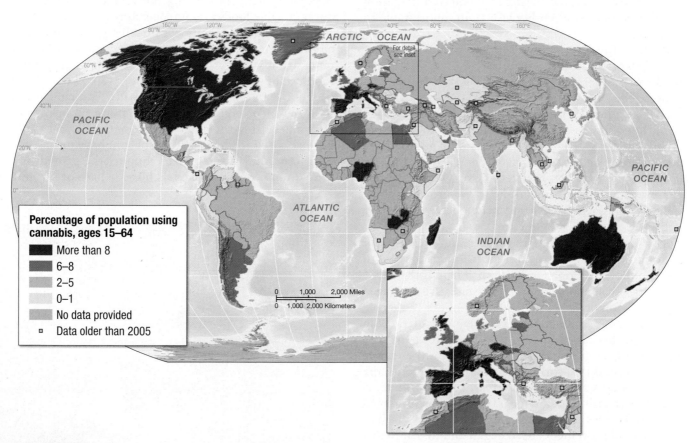

Percentage of population using cannabis, ages 15–64
- More than 8
- 6–8
- 2–5
- 0–1
- No data provided
- □ Data older than 2005

▲ **FIGURE 6.2.1 Global Cannabis Use** This 2011 map shows marijuana to be far more popular in wealthy countries than in poorer ones. The exception is Japan where no data are available but where it is widely recognized that marijuana use is not prevalent.

▲ **FIGURE 6.2.2 A Grow Operation in Canada** Grow operations are quite varied across the region. They can be legal ones that are monitored by government agencies like the one shown here at Island Harvest Farms in Duncan, B.C., or they can be illegal operations hidden within the walls of ordinary houses on an ordinary street.

medical marijuana laws and variations in criminal penalties among different administrative units of government make for complex and regionally distinct marijuana markets. Despite state and municipal initiatives to decriminalize marijuana possession and state medical marijuana laws that legalize some production, the U.S. and Canadian federal governments both maintain staunchly prohibitionist policies against the use of the plant. Both governments cite it as a "gateway drug" and avow that it leads to more harmful substance abuse, along with a litany of other public health and safety concerns. And both governments maintain large programs for eradication and interdiction of marijuana. In the United States, federal prosecutors have

made renewed efforts under the Obama administration to shut down the production and distribution of medical cannabis in California and 16 other states that have passed medical marijuana laws decriminalizing personal possession and use. However, public opinion seems to favor a relaxation of prohibition policies. A public opinion poll reported in the *Vancouver Sun* in February of 2012 stated that 66% of Canadians support legalization initiatives, while a 2012 Rasmussen poll assessed 56% support for the legalization and regulation of the plant among residents of the United States.

Ultimately, for all the profits generated by marijuana cultivation and all the funds expended by law enforcement on crop eradi-

cation and supply interdiction, there is very little that one can say about the industry that is not an estimate or a guess, as it remains an activity that is necessarily clandestine. It appears that prices have fallen in recent years due to large harvests and a reduction in legal distribution points stemming from federal crackdowns on dispensaries. Both prices and wages have fallen in California's "Emerald Triangle," according to a 2012 *Sacramento Bee*[3] article. ∎

[3]Hecht, Peter (May 5, 2012). California's 'Emerald Triangle' pot market is hitting bottom. The Sacramento Bee.

tastes, aromas, and diversity of good food. The movement also serves as a rallying point against globalization, mass production, and the kind of generic food represented by U.S.-based franchised restaurants. Its campaigns cover a range of specific causes, from protecting the integrity of chocolate to promoting the cultivation of traditional crop varieties and livestock breeds and opposing genetically engineered foods.

Although slow food, local food, and organic agricultural practices and the small farming landscapes they produce are proliferating across Canada and the United States, they in no way challenge the dominance of more conventionally produced, distributed, marketed, and consumed food. Moreover, as several critics of the movements have pointed out, these alternative practices are largely organized and promoted by white, middle-class people and exclude, often simply through cost and associated accessibility, poor people of color everywhere. In the United States and Canada, the result is that poor people turn to cheap, easily accessible food, also known as fast food, more than members of any other economic classes do.

APPLY YOUR KNOWLEDGE With a small group of your classmates, research online the ecological benefits of slow food. Provide examples of how local food and organic food production affect soil, water, and energy consumption.

Environmental Challenges Decades of federal environmental protection legislation have forced U.S. industries to curtail much of their polluting processes or move to manufacturing sites outside the United States. Nonetheless, various regions of the country still face serious and persistent environmental challenges. **Acid rain**, rainfall made acidic by atmospheric pollution that causes environmental harm, typically to forests and lakes, generated by industrial processes and automobile emissions, continues to pose a challenge in the traditional industrial areas on both sides of the U.S.–Canada border. Along the U.S.–Mexico border, air and water pollution are also problems and the legacy of past pollution-generating industrial practices lingers. The U.S. government currently is overseeing the cleanup of hundreds of **superfund sites**, locations officially deemed by the federal government to be extremely polluted and require extensive, supervised, and subsidized cleanup.

The most serious environmental challenge Canadians and Americans face today is their seemingly insatiable appetite for resources, especially energy resources. The impact of the high level of energy consumption not only seriously challenges Earth's supply of renewable energy resources, but is also leading to global climate change, as discussed earlier in this chapter. An illustration of this need for energy is the likelihood that Canada's **oil sands** reserves will be exploited for the fuel oil they contain. Oil sands are deposits of sand, clay, other minerals, and water that are saturated with bitumen, which is oil or petroleum in a solid or extremely viscous state. Although oil sands are found elsewhere in the world, the Athabasca deposit in Alberta, Canada, is the largest and already most developed in the world.

The refinement of oil sands would extend the burning of fossil fuel beyond current global known oil reserves, increasing the emission of greenhouse gases and accelerating the speed and severity of climate change. In a recent editorial in the *New York Times*, James Hansen, a highly respected physicist and director of the NASA Goddard Institute for Space Studies, stated:

"If we were to fully exploit this new oil source, and continue to burn our conventional oil, gas and coal supplies, concentrations of carbon dioxide in the atmosphere eventually would reach levels higher than in the Pliocene era, more than 2.5 million years ago, when sea level was at least 50 feet higher than it is now. That level of heat-trapping gases would assure that the disintegration of the ice sheets would accelerate out of control. Sea levels would rise and destroy coastal cities. Global temperatures would become intolerable. Twenty to 50 percent of the planet's species would be driven to extinction. Civilization would be at risk."[4]

Popular movements are in place that direct their efforts at reducing energy consumption in the region, and a growing consumer interest in sustainability has created a large and growing market for environmentally friendly—also known as "green"—goods and services in the United States and Canada. Leading corporations, such as the large automobile manufacturers, and energy companies, such as General Electric and Shell Oil, are making substantial investments in energy efficiency and energy technologies, and individual consumers and businesses are responding positively. Many greenhouse gas reductions, especially those associated with energy efficiency, such as home insulation and fuel-efficient transport, can be made at little or no cost and may even save people money. Individuals can also limit the effects of climate change in numerous ways, such as buying local foods (which reduces the energy expended in transportation and refrigeration), using less energy to heat and cool homes, and using reusable water bottles instead of disposable ones (which minimizes waste and the costs of disposal). In the absence of serious federal effort to reduce emissions, many cities and businesses across the United States and Canada have made their own commitments to mitigate climate change.

Several regional alliances have also been created to reduce greenhouse gas emissions. The alliances include the Regional GHG Initiative (RGGI) among the northeastern states in the United States and the Western Climate Initiative between California and British Columbia, Manitoba, Ontario, and Quebec. These regional alliances are taking on commitments to reduce their emissions. To make it easier to meet their goals, they have instigated **cap-and-trade programs.** In a cap-and-trade program, the "cap" is a government-imposed limit on carbon emissions. The "trade" occurs when unused emission permits or quotas from states (or firms) are sold by those who have been able to reduce their emissions beyond their quota to those who are unable to meet theirs. Many of these states are also working together to plan on how to adapt to climate changes that are already occurring. These plans include coastal protection as well as water management and conservation efforts. In addition, there is now a well-established and increasingly municipally organized structure for recycling materials and items from aluminum, paper, and glass to appliances, oil, and electronic goods (**Figure 6.11**).

---

[4]Hansen, J. 2012. "Game Over for Climate Change," *New York Times*, May 9).

▲ **FIGURE 6.11 Organic Waste Collection Bins** Many municipalities in both the United States and Canada have, in response to pressure from citizens, begun to collect organic waste that can be then composted and used for residential and government purposes or even sold to commercial users. Municipal composting programs, such as the one in Vancouver, Canada, helps to reduce the material sent to local and regional landfills.

HISTORY, ECONOMY, AND TERRITORY

Today, the United States and Canada are established democracies modeled on European political traditions. Both consolidated their leadership roles in the global economy early in the 20th century through an effective strategy of economic development and global geopolitical maneuvering. Much of the recent history of these two countries is the result of European colonization (**Figure 6.12**). More recent geographies of this region tie into wider movements of global immigration as well as internal migration (discussed in a later section of the chapter). Because of their comparable economic status, academics as well as policymakers and government agencies treat the two countries as a coherent region, although there are certainly differences between these two national spaces. That said, the commonalities provide a framework for thinking about this space as a world region. There is no doubt that the United States and Canada have had and continue to have an important effect on global relations. This region is also subject to dynamic change as it faces the changing geopolitical and sociocultural realties of our globalized world.

Historical Legacies and Landscapes

During the Age of Exploration in the 15th century, European views of the world did not include the existence of the North American landmass. And although 16th-century Spanish missionaries and explorers identified the southernmost section of present-day United States and Mexico to be of interest to their exploration and missionary efforts, most of the rest of North America was considered of little consequence because it presented none of the appearances of grandeur and resource potential that the Aztec, Maya, and Inca empires of Latin America did (see Chapter 7, p. 262). This lack of interest in the northern reaches of the continent changed dramatically in the 17th century, with consequences that continue to reverberate today.

Indigenous Histories The distribution and subsistence practices of the Indigenous people of the United States and Canada in about 1600 reflected the great diversity of cultural groups that occupied the continent. When Christopher Columbus arrived in the Caribbean, he assumed he had reached the Far East, or the Indies, and he called the aboriginal people he encountered *los indios*, or indians. As thousands of native languages existed at the time of European contact, there was no single word of self-description common to the diverse people who occupied the continent. As a result, the word *Indian* has endured as an extremely misleading and sometimes derogatory term for describing a wide range of regional cultural groups.

Estimates of the size of the native population of the region at the point of European contact range widely, between 1 million and 18 million. More than anything, these differing estimates indicate how little is known about the people who were thriving in the United States and Canada before the Europeans arrived. What is well known is that there was no common culture, particularly no common language, among the groups who first populated the region. Native American scholar Jay Miller estimates that in 1492, the Indigenous

▲ **FIGURE 6.12 New France in Quebec** The neighborhood of historic Old Quebec clearly shows the influence of French architecture in its buildings as illustrated in these early 18th-century residences with steep gable roofs. French is spoken widely in this neighborhood and in most of Quebec, where English is a second language.

The manner of their fishing.

▲ **FIGURE 6.13 Native Peoples Fishing in Virginia Colony, 16th Century** In addition to growing food, Indigenous people also trapped small game, such as beaver, for their fur. They fished the rivers and the coastal areas and harvested wild fruits and berries. At the time that Europeans arrived they had a strong sense of the location and quantity of natural resources to sustain their daily lives.

Lumping the settlers into one category of Europeans vastly simplifies the very complicated process of settlement that occurred when different people from different parts of Europe came to the United States and Canada. Although the occupation by missionaries and settlers who came to the region in the 16th and 17th centuries is widely known as the period of **Europeanization**, the process was actually highly selective and did not involve all of Europe. A few Western European countries—France, Spain, the Netherlands, and Great Britain—dominated colonization in the United States. In Canada, the European colonizers were predominantly Britain and France. In both Canada and the United States, Great Britain was by far the most influential of the four, though others did have substantial impacts, particularly the French in Canada and the Spanish in the United States.

Different groups from different regions of the four European countries settled in different parts of the United States and Canada. And given the hardships involved in immigrating, the unusual individuals and groups who assumed great risks in coming to the region represented only a small sample of their national cultures. As a result, the Europeanization process, as it unfolded along the Atlantic seaboard of the United States and Canada, created distinct colonial cultures and societies in different places.

Even though the settlers were not a homogenous group, European settlement consistently and routinely resulted in the exploitation and abuse of the native people. The history of colonial settlements, although at first peaceful, over time erupted into disputes over land claims that ended in violence and often outright massacres on both sides. After a time, moreover, not only were there conflicts between tribal people and colonists, but direct conflict also emerged among the various European groups that vied for control over land. In addition, colonists fanned the flames of rivalry between opposing tribal groups by providing them with arms, thus elevating the level of technology and intensifying the degree of violence in armed conflict. Finally, exposure to the diseases that the colonists brought with them had a devastating impact on native populations (see Chapter 7, p. 262). As native populations were decimated, defeated, or pushed onto reservations farther into the interior of the continent, the various European groups increasingly came into direct conflict, leading eventually to the **Seven Years' War** (1756–1763), the U.S. phase of which is known as the **French and Indian War** (1754–1763). This war left the British more or less triumphant over the whole of the European-inhabited territory of the United States and Canada.

inhabitants of the United States and Canada spoke 2,200 different languages, with many regional variations as well. Tribal culture and local environmental conditions were the frameworks within which daily life was governed and lived (**Figure 6.13**).

Europeans originally made contact with various individuals and tribes to solicit help with extracting resources, including animal furs; naval stores, such as tar and turpentine for shipbuilding; fish; and other primary-sector products exported back to Europe. The experiences of the Dutch, French, and English in North America differed substantially from that of the Spanish in Latin America (see Chapter 7, p. 262). The Spanish conquistadors vanquished sophisticated civilizations to plunder their gold and silver treasuries and make themselves rich in the process. The Dutch, French, and English were also interested in improving their financial situations, but they encountered a very different set of cultures with no centralized system of social control that they could exploit as the Spanish did with the Aztecs in Mexico. Instead, the eastern tribes that Europeans first encountered in the United States and Canada were small, autonomous groups, possessing an active sense of rivalry and competition with their neighbors. The Spanish were interested in massive occupation and exploitation, whereas for the Dutch, French, and English, exploration and colonization were commercial ventures.

Colonization and Independence No other landmass already occupied by a diverse range of complex and widely distributed societies, with the possible exception of Australia, has undergone such a dramatic transformation in such a short period of time.

In the two decades following the French and Indian War, residents of the original 13 colonies of the United States became disillusioned with their administrators in Britain—who were taxing them

▲ **FIGURE 6.14 German Immigrants in the United States** Frankenmuth, Michigan is a town in the east central part of the state with a history of German immigrant occupation. Shown here is the Bavarian Inn of Frankenmuth, also known locally as Little Bavaria. The city's name is a combination of two words. "Franken" represents the Province of Franconia in the Kingdom of Bavaria (in southern Germany), home of the Franks where the original settlers were from. The German word "Mut" means courage; thus, the name Frankenmuth means "courage of the Franconians."

the native populations that would have been an additional source of laborers.

African slaves had been a well-established commercial staple of the Mediterranean well before the Spanish and Portuguese introduced them to their newly captured territories in Latin America and the Caribbean, thereby establishing the Atlantic slave trade. Notably, Canada never participated to any significant degree in the trade in African slaves. By the early 17th century, England became the dominant slaving nation. As a result, slaves were a part of the social and economic system of the American colonies beginning with their founding. As an institution of formal social and economic organization, slavery endured in the South for more than 250 years, ending officially in 1870, following the end of the **U.S. Civil War** (1861–1865). Its legacy, however, continues to shape the landscape and identity of the region.

European Settlement and Industrialization

With the creation of new nations and the transformation of colonies into states, the relentless European settlement of the North American continent accelerated, more so in the United States than in Canada. By the middle of the 19th century, settlement in the United States had pushed beyond the Appalachian Mountains into the Interior Lowlands, including the upper Ohio and Tennessee River valleys and the interior southern states. France lost control of Canada to Britain in 1763, which had little impact on new settlement; however, by the end of the century, southern Ontario, in and around present-day Toronto, became attractive to settlers.

Historians argue that frontier settlement involved a continual process of national and personal reappraisal as well increasing geographical divergence, as new settlers encountered new landscapes. It was also a process of sustained mobility, so much so that mobility has come to be seen as characteristic of the region's inhabitants, especially in the United States. In the United States, the movement of the frontier was continuous *and* mostly contiguous, at least until settlers reached the Great Plains in the middle of the 19th century and confronted significant mountain ranges at its western edge. In Canada, westward expansion was interrupted early on by the vast, generally infertile, though heavily forested, Canadian Shield, which separated Ontario from the prairies. Many Canadian settlers leapfrogged across the Canadian Shield to the northern midsection of the country to acquire suitable farmland.

In the United States, the pace of westward expansion was accelerated by the federal government's decision in the late 1780s to sell public lands cheaply to citizens. By 1850, the development of the railroads reoriented the pace and direction of continental settlement—eastward from the Pacific Coast to the interior west rather than from the East Coast westward—at the same time that it diminished the previous isolation of pioneer settlements. By the close of the 19th century, the frontier process had resulted in a set of rural and farming areas that stretched across the continent. Each area was defined by its own experiences of the history and particular conditions of settlement and by its distinctive economic development.

to help recoup the high cost of the war—and launched their own war, the **American Revolution** (1775–1783), which led to the creation of a new, independent nation in the late 18th century. Even before the revolution, however, a process of **Americanization** had begun, as a generation of individuals of European parentage born in the U.S. colonies felt less loyalty and fewer cultural ties to the mother country (**Figure 6.14**). As a result, a new ethos of liberalism, individualism, capitalism, and Protestantism emerged, gained currency, and ultimately came to define a U.S. national character. The successful outcome of the revolution left the continent with a robust new nation dominated by Anglo-American institutions, which included slavery. Canada remained a colony, under British control, composed of both French- and English-speaking settlers. In Canada, a bloodless separation from Great Britain did not occur until well into the 19th century.

The Legacy of Slavery in the United States Although the impact of European colonization in the 16th and 17th centuries was felt all along the Atlantic seaboard of the United States and Canada, development of the southern United States following the end of the revolutionary period differed dramatically from that of its northern neighbors. Before the arrival of the Europeans, the area that now forms the southeastern United States was inhabited by a wide range of native tribes, among them the Cherokee, Choctaw, Chickasaw, Creek, and Seminole people. Early on, the region was occupied by European military personnel living in scattered outposts like Jamestown in Virginia. However, by the mid-17th century, the military outposts had given way to tobacco farms. At first, **indentured servants**—individuals bound by contract to the service of another for a specific term—from Britain were the primary source of labor on the tobacco and later indigo and cotton plantations. Increasingly, however, servants earned their freedom and were replaced by slaves from Africa. At the same time, disease, armed conflict, and demoralization reduced

▲ **FIGURE 6.15 Water-Powered Mill in New England** Early manufacturing sites sprung up along waterways where the power of rapidly running water or waterfalls provided energy to run the machines. This mill in Derry, New Hampshire, demonstrates the classic elements of early water power with the large water wheel shown being driven by the rushing falls.

The regional economy during this period was oriented to agriculture and trade activities. This means that trading agricultural crops and primary resources, such as fur, fish, timber, and minerals, provided an economic base for the expanding population. Yet by the mid-19th century, a new economy based on manufacturing was rapidly gaining momentum in the United States along the northern Atlantic seaboard, especially in and around southern New England and New York (**Figure 6.15**). At the same time, the rest of the United States and Canada was being settled by European immigrants or their descendants, making their livelihoods largely by farming.

As industrialization fueled the economies of the northern and midwestern United States, the differences between the northern and southern regions of the United States became increasingly pronounced. The commerce and industry fueling the national economy made the cities of the North the most important ones. The southern ports were less crucial as transfer points than were those in the North because the South had many navigable inland rivers. Ships could simply pick up agricultural products and raw materials or drop off their own cargoes by way of the rivers, thereby eliminating the need to stop at the coastal ports.

Whereas the North's economy was more diversified—based on commerce, agriculture, and industry—the South's was more simply tied to staple crop agriculture. Because the southern plantations produced much of the food and clothing for the region, and because most of the capital was invested in slave labor (which was unpaid and thus had no income to stimulate commercial exchange), the economy, as well as society and culture, made the South very different from the North and Canada. All these differences—especially slavery and the economic issues that surrounded it—divided the southern states from the northern ones. By 1861, the division was so deep that the South formed a new government, the **Confederate States of America**, and attempted to secede from the United States to protect slavery and the South's agricultural economy (see the previous section). Civil war ensued.

With the South in defeat following the end of the Civil War in 1865, territories in the West joined the United States as free states, contributing people, resources, and capital to the burgeoning U.S. economy. Fearing that the United States, emboldened by its Civil War victory, might launch an invasion of Canada and responding to agitation among the French-speaking minority, the British Parliament passed the **North America Act of 1867**. This act created the Dominion of Canada, dissolving its colonial status and effectively establishing it as an autonomous state with its own constitution and parliament. All the existing Canadian colonies joined the new Canadian confederation except British Columbia, which waited until 1871; Prince Edward Island, which joined the dominion in 1873; and Newfoundland, which remained independent until 1949.

By the early 20th century, the United States and Canada were occupied from east to west mostly by Europeans and Euro-Americans, and the industrialization of the U.S. economy was well on its way to transforming the landscape from one of rural agricultural settlement to one of urbanization and industrialization. The 1920 census documented for the first time in U.S. history that there were as many people living in cities as there were in rural areas. From that point onward, the United States and, soon after, Canada, became increasingly urbanized. Presently, over 80% of the U.S. and Canadian population lives in cities, up from 25% in the mid-19th century.

APPLY YOUR KNOWLEDGE Using national census data on any three ethnic groups in either the United States or Canada (http://www.census.gov or http://www.statcan.gc.ca), construct a map that shows their geographic distribution across the national territory in 1900, 1950, and 2000. Provide an explanation for changes in distribution.

Economy, Accumulation, and the Production of Inequality

By the end of the 19th century, Canada and the United States were fast becoming key players in the global economy. The United States, politically independent just before the onset of the Industrial Revolution (the rapid development of mechanized manufacturing that gathered momentum in the early 19th century), was able to become economically competitive thanks to several favorable conditions. Vast natural resources of land and minerals provided the raw materials for a wide range of industries that could grow and organize without being hemmed in and fragmented by political boundaries. Populations, which were expanding quickly through immigration, provided an ever-increasing market and a cheap and industrious labor force. Cultural and trading links with Europe provided business contacts, technological know-how, and access to markets and capital (especially British capital) for investment in a basic infrastructure of canals, railways, docks, warehouses, and factories. Because of these riches and infrastructural investments, the two national economies in the region took off and expanded.

The Two Economies Canada's path to global economic competitiveness was distinctive. Although most of its population enjoys a high quality of life and high levels of economic productivity, Canada is certainly an atypical economy largely because it was, until very recently, never highly industrialized. The primary sector (see Chapter 1, Figure 1.1, p. 4) was the major pillar of Canada's economic prowess. Recall from Chapter 1 that the primary sector includes primary or extractive industries such as mining and agriculture. Although this sector has declined in centrality, it continues to play an important role in Canada's economic structure. When industry did begin to grow and flourish in Canada after World War I, most of it occurred in the midsection of the country along a swath

of land at the U.S.–Canada border. A substantial proportion of the industries built there were branch plants of U.S. manufacturers. Canada's major trading partner is the United States, which imports more than half of all Canadian exports. What has been most remarkable about Canada's place in the world economy is that it became so successful as a **staples economy**—one based on natural resources that are unprocessed or only minimally processed before they are exported to other areas where they are manufactured into end products.

Over the last several decades of the 20th century, Canada's economy shifted. Today, its largest and most dynamic sectors are first, by a large margin, real estate, finance, and insurance, and second, manufacturing. Although an important energy-trading partner to the United States, the energy sector constitutes only 6.2% of the Canadian GDP (see Chapter 1, p. 28). **Figure 6.16** shows that, by comparison, the U.S. economy is more concentrated

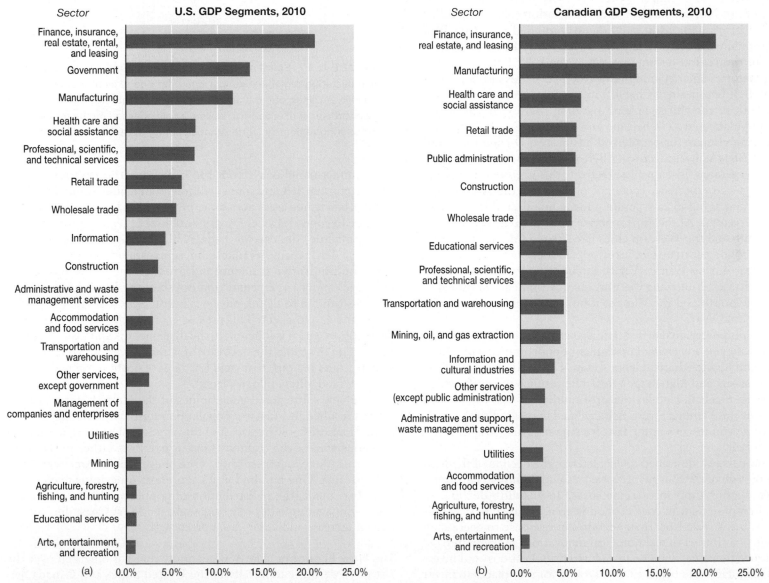

▲ **FIGURE 6.16 U.S. and Canadian GDP segments, 2010** (a) Trade and manufacturing dominate the U.S. economy. (b) In Canada, manufacturing and services dominate. Despite those differences, the economies of the two countries have much in common.

in trade and manufacturing. What the graphs do not show is that while the Canadian economy has increased its share in manufacturing, the United States has decreased its share. Both countries possess very large and powerful economies; the U.S. economy is dominant.

Today, the United States and Canada together produce more than one-quarter of the world's GNI (see Chapter 1, p. 28). The United States has the world's largest economy, and Canada is the ninth largest. As part of the recent restructuring of the global economy discussed in Chapter 1, pp. 6–7, the various regions of both the United States and Canada have experienced significant transformations in their economies, societies, political institutions, and even their physical environments.

Transforming Economies Political, economic, and social geographers would agree that the most important transformation in the structure of the global economy of the last 30 years has been the expansion of the service sector based on finance, real estate, and insurance. In the United States, the most recent wave of internal migration occurred as millions of U.S. residents moved from the historic core of North American industrialization (originally called the Manufacturing Belt but subsequently dubbed the **Rust Belt** and occasionally the **Snow Belt**) to the **Sun Belt**. The manufacturing area of the region became known as the Rust Belt because of its aging factories and declining infrastructure. It was also known as the Snow Belt because of its harsh winters. The area of the United States, south of the thirty-seventh parallel, that experienced this massive population growth became known as the Sun Belt because of its inviting climate for people and businesses. During the 1960s and the 1970s, the Manufacturing Belt in the Northeast and Midwest of the country (in cities like Detroit, Boston, New York, and Buffalo) began to experience economic problems in the form of high labor costs and aging infrastructure, mostly manifested in outdated technology systems. At this time, once-peripheral areas of the country, the South and the West (in cities like Atlanta, Miami, Houston, and Phoenix), began to attract investors. The military had invested in these areas during World War II, establishing bases and holding training exercises. Following the war, the government continued to invest in the South and the West as it built up its military capacity during the Cold War.

As the computer age dawned, numerous places in the South and the West were ripe for civilian investment opportunities with abundant land and highly educated labor forces. This labor was not accustomed to unions and high-wage rates. The result was a shift, which has since been rebalanced, as the profitability of old, established industries in the Manufacturing Belt declined compared to the profitability of new industries in the fast-growing new industrial districts of the Sun Belt.

The process just described to characterize the decline of the Rust Belt and the rise of the Sun Belt revolved around **deindustrialization**, a relative decline (and in extreme cases, an absolute decline) in industrial employment in core economic areas (**Figure 6.17**). This happens as firms scale back their activities in response to decreasing profitability. In this case, technological innovations in computerized production systems facilitated new industrial applications, and investors and manufacturers began to look around for new places to invest and build. Innovations in transport and communications technology, combined with these production innovations, created windows of opportunity. The result was the movement of capital investment

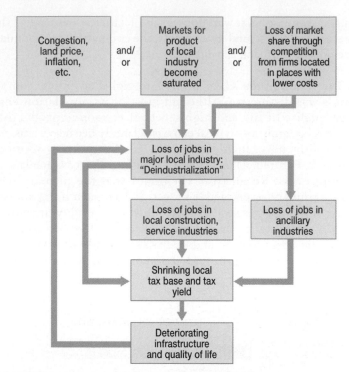

▲ **FIGURE 6.17 Spiral of Deindustrialization** When the advantages of manufacturing regions are undermined for one reason or another, profitability declines, and manufacturing employment falls. This can lead to a downward spiral of economic decline, as experienced by the traditional manufacturing regions of North America during the 1970s and 1980s.

in manufacturing away from the old industrial districts of the Manufacturing Belt and into small towns and cities in the Sun Belt, to suburban fringe areas near some of the old industrial districts, and offshore to countries where wages were lower.

Meanwhile, the capital made available from disinvestment in the Rust Belt became available for reinvestment in new ventures based on innovative products and production technologies. Old industries and a large proportion of established industrial regions were dismantled to help fund the creation of new centers of profitability and employment. This process is often referred to as **creative destruction**, and it is inherent to the dynamics of capitalism. Creative destruction describes the necessity of withdrawing investments from activities (and regions) that yield low rates of profit to reinvest in new activities (and, often, in new places).

The new subregional geographies of the United States just described resulted in a massive population redistribution. One in every three Americans now lives west of the Mississippi, making it the most populous area of the country. One in five lives either in Texas or California, the two largest states. One of every eight Americans lives in California. One hundred years ago, most Americans lived in the East. In Canada, the redistribution of population has been similar, with significant growth occurring along the West Coast due both to natural increase and immigration from Asia.

The New Economy The new economy that has emerged over the last three decades, especially in the United States and Canada, has fundamentally transformed industries and jobs through **information technologies (IT)**, the use of computer systems for storing, retrieving, and sending information. This shift to an IT-based economy has been facilitated by a high degree of entrepreneurialism and competition around the world. But the new economy was born in the United States, sired by the technological changes that emerged from Silicon Valley, California, nearly 50 years ago.

It is generally agreed that the previous economic order, the "old economy," lasted from 1938 to about 1974. The year 1974 was a critical year in global economic history: oil prices were skyrocketing and corporate profits were falling. The old economy's foundation was manufacturing geared toward standardized, mass-market production and run by stable, hierarchically organized firms focused on the U.S. market. Massive political and economic restructuring rocked the United States and Canada as well as Western Europe. Many regard 1975–1990 as the transitional period from the old economy to the "new economy."

The new economy is about more than just new technology. It is also about the application of new technologies to the organization of work—from the impact of biotechnology on farming to the impact of IT on management hierarchies. In short, the new economy is based on the application of IT to transform the organizational practices of firms and industries (see "Emerging Regions: The United States and the Caribbean," p. 244). Dynamism, innovation, and a high degree of risk are at the center of the new economy.

Canada has also been an active participant in the new economy, and part of the shift it has experienced into services and high-tech manufacturing is evidence of this (**Figure 6.18**). A particularly dynamic aspect of this change in Canada has been in science-based industries as well as research, development, and manufacturing in communications and transportation technologies. Quaternary sector transformations based on innovation and policy changes have also been important. (Recall from Chapter 1 that quaternary or information sector industries include information technologies, financial planning, and research and development.) The "new economy" that has emerged in the United States and Canada has helped increase wealth dramatically for some, but has also left many behind.

Wealth and Inequality Although globalization and the new economy helped improve the employment opportunities and level of wealth of many in this region, the global fiscal crisis that began in 2008 has seriously set back many in the United States and Canada (**Table 6.1**). Between 2009 and 2010, the real median income for whites and blacks in

▲ **FIGURE 6.18 Downtown Toronto** Toronto is Canada's largest employment hub, with one-sixth of the country's jobs. A world banking and finance leader, Toronto also supports a burgeoning variety of high-growth sectors, such as software and hardware design, biotechnology, pharmaceuticals, and telecommunications.

the United States declined at the same time that the poverty rate increased for non-Hispanic whites, blacks, and Hispanics. The nation's official poverty rate in 2010 was 15.1%, up from 14.3% in 2009—the third consecutive annual increase in the poverty rate. Real median income was U.S. $49,445 in 2010, a 2.3% decline from 2009.

In the United States, the poverty line for a family of two adults and two children is set at U.S. $23,050. In 2012, 20% of all families with two children had incomes below the federal poverty rate. Released in 2012, the most recent data brief from the *CIA World Fact Book* ranked the United States 50th globally in infant mortality, trailing countries such as Czechia,

TABLE 6.1	Income Disparity in the United States							
	Average After-Tax Income							
	Share of All Income			**Estimated**			**Change**	
Household Groups	1977	1999	2007	1977	1999	2007	1977–1999	1977–2007
One-fifth with lowest income	5.7%	4.2%	3.4%	U.S. $10,000	U.S. $8,800	U.S. $11,551	−12.0%	15.5%
Next lowest one-fifth	11.5	9.7	8.7	22,100	20,000	29,442	−9.5	33.2
Middle one-fifth	16.4	14.7	14.8	32,400	31,400	49,968	−3.1	54.2
Next highest one-fifth	22.8	21.3	23.4	42,600	45,100	79,111	5.9	85.7
One-fifth with highest income	44.2	50.4	49.7	74,000	102,300	167,971	38.2	127.0
1% with highest income	7.3	12.9	21.2 (2005)	234,700	515,600	868,000 (2004)	119.7	269.8 (1977–2004)

SOURCE: U.S. Census Bureau, "Income Data." 2012, http://www.census.gov/hhes/www/income/data/index.html, accessed July 14, 2012.

EMERGING REGIONS
The United States and the Caribbean

As IT improved the ability to move goods and connect information sources and services in an instant, the special importance of the Caribbean to the United States increased in the late 20th century. For example, over the last 20 years or more, U.S. companies have increasingly located many of their service jobs in the Caribbean due to the lower costs of labor there.

With more than 34 million people and 16 independent nations sharing a rich cultural and ethnic heritage, the Caribbean is a diverse region that includes some of the hemisphere's richest and poorest nations. The United States has a very long history of strong economic, strategic, and cultural ties with the Caribbean that stretches back to before U.S. independence (**Figure 6.3.1**). The triangular trans-Atlantic trade system linked the original U.S. colonies via the Caribbean to West Africa and Europe, which were an important source of slaves and a major market for the colonial farm surpluses, accounting for about a third of U.S. exports before 1815 (**Figure 6.3.2**).

Tourism is another economic activity that links the two regions. While traditional tourism opportunities in the Caribbean continue to be an important attraction for Americans, adventure and cultural tourism, eco-tourism, and upscale resorts increasingly draw U.S., Canadian, and European tourists to the region as well. Antigua, Belize, the Dominican Republic, and Jamaica offer Internet gaming—which is banned in the United States and Canada—drawing hundreds of thousands of online visitors from the United States.

Other examples illustrate the strategic importance of the Caribbean to U.S. interests. For instance, the 1895 Cuban revolt against Spain generated sympathy in the American public and Congress in favor of Cuban independence from Spain. In 1898, the battleship *Maine* mysteriously exploded in Havana Harbor. This event mobilized the

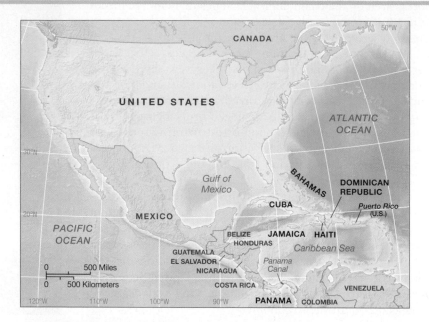

▲ **FIGURE 6.3.1 The United States and the Caribbean** Thirty-four countries officially constitute the Caribbean, ranging from Anguilla to Venezuela. The United States has economic relationships with all of them, including Cuba. As the countries of the Caribbean are the United States's nearest neighbors to the south, it is not surprising that the United States and the Caribbean are strongly linked through trade, culture, and politics.

▶ **FIGURE 6.3.2 Triangular Trade** Slave labor was fundamental to the plantations that grew colonial cash crops destined for European markets. European goods, in turn, were used to purchase African slaves, which were then brought from Africa to the Americas. Most of the residents of the Caribbean islands are the descendants of African slaves.

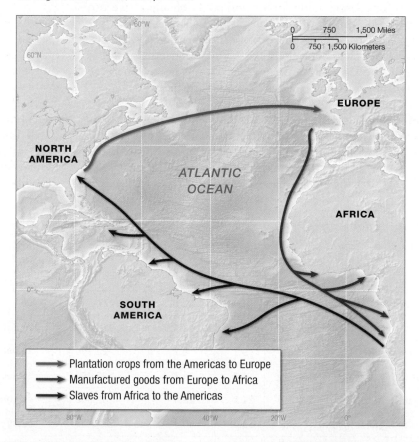

Plantation crops from the Americas to Europe
Manufactured goods from Europe to Africa
Slaves from Africa to the Americas

United States to go to war with Spain. During the short war, U.S. forces occupied Puerto Rico. The ensuing peace treaty gave Puerto Rico to the United States and left Cuba free of Spanish control and under American military occupation. A few years later, in 1904, U.S. president Theodore Roosevelt took over the stalled French construction of the Panama Canal in the Caribbean. Completed in 1914, the Panama Canal was a place of unique strategic importance. The canal significantly shortened the travel time of international maritime traffic, providing a safer and faster route to the U.S. West coast and the Pacific. After its completion, Washington viewed the entire Caribbean as a security zone. (Chapter 7, p. 262, provides a more comprehensive discussion of this history.)

For most of the 20th century, U.S. interests in the Caribbean were commercial and strategic—they still are today. Political unrest, discontent, and poverty have propelled Cuban, Haitian, Dominican, and Central American immigrants to the United States in unprecedented numbers. Cities along the Eastern Seaboard of the United States, such as New York, Washington, D.C., and Miami, contain large populations of Caribbean immigrants who have changed the businesses and labor landscapes as well as enlivened those cities with their culture, including language, food, music, and dance.

Throughout the history of their relationship, strategic issues have linked these two places. For example, narcotics trafficking and related violence have brought the two regions together in the Caribbean Basin Security Initiative (CBSI), which is intended to reduce illicit trafficking, increase security, and promote social justice. But the dominant role that the United States has played for so long in the region may now be eroding as a result of the increasing presence of China, especially in the Bahamas, Jamaica, Grenada, Trinidad and Tobago, Dominica, and Antigua and Barbuda. The government of China and private Chinese companies have been investing billions of dollars in the infrastructure of these countries in the form of schools, roads, port facilities, hospitals, sports stadiums, and resorts. China is also the sources of loans, investments, and gifts in the area (**Figure 6.3.3**).

Most analysts do not see a security threat in China's presence in the Caribbean, as the Chinese are not building bases or forging military ties. Some, however, do perceive a threat in the emerging superpower's efforts to forge economic inroads and popular support in this group of developing countries that once counted almost exclusively on the United States (and Canada) for help. ∎

▲ **FIGURE 6.3.3 China–Caribbean Economic Forum** While China's interest in the Caribbean is growing, as this recent conference on China–Caribbean economy and trade issues demonstrates, the United States remains one of the region's most significant partners along with the European Union.

Cuba, Portugal, and Canada. Child advocates argue that one of the reasons that the United States does so poorly when compared with other industrial countries is that it offers the lowest government benefits to families with poor children. Another reason is the stagnation of wages and high unemployment at the lower end of the wage spectrum.

In Canada, the percentage of families (with two adults and two children) living on less than U.S. $23,050 in 2012 was 9%—less that half of the U.S. poverty rate. Still, in 2012, one in ten Canadians lived in poverty. And poverty has been increasing for youth, young families, and immigrant and minority groups. Poverty among Indigenous groups is higher than among the rest of the Canadian population both on and off reserves. In fact, if the statistics for Indigenous people were viewed separately from those of the rest of the country, Canada's Indigenous population would slip to 78th on the UN Human Development Index—the ranking currently held by Kazakhstan. Over most of the 20th century, Canada provided a strong safety net for poor families, but in the last two to three decades, the poverty rate among children there has grown to the second highest (after the United States) among all developed countries.

The gap between rich and poor in the United States continues to widen. In 2012, the richest 10% of Americans controlled 75% of the wealth, leaving only 25% to the other 90% of Americans. In 2011, one worker making U.S. $10 an hour would have to labor for more than 10,000 years to earn what 1 of the 400 richest Americans pocketed in 2012. It is not surprising that in September 2011 **Occupy Wall Street** protestors began to appear in the United States. Occupy Wall Street is a popular social movement founded in Canada. With their slogan, "We are the 99%," the Occupy movement signals its dissatisfaction with the growing wealth gap between a small number of super-rich individuals and the other 99% of the population (**Figure 6.19**).

▲ **FIGURE 6.19 Occupy Movement Protest in Tucson, Arizona**
Different local groups may target different issues, but the central thrust of the Occupy movement is the prime concern that large corporations and the global financial system are shaping the world in a way that benefits a minority of wealthy people, undermines democracy, and is unsustainable overall. Here the protestors display their solidarity with the 99%.

APPLY YOUR KNOWLEDGE Choose one city in Canada and one in the United States and, referring to each city's website, compare and contrast the activities that form the basis of their economies. Then find data on median income levels and cost of living for both cities. How do the types of economic opportunities shape income in these two places? What factors shape the cost of living? The physical environment? Local demography?

The United States and Canada and the Global Economic Crisis

As discussed previously in this chapter, the global fiscal crisis that began in 2008 seriously set back the economies of the United States and Canada. When the failure of several major private financial institutions in October 2008 prompted panic among international financial markets, millions of households in these affluent countries faced the very real prospect of recession and loss of jobs, savings, and pension funds. In an attempt to prop up the international financial system and regional, national, and local economies, the governments of the United States and of other leading economies intervened with hundreds of billions of dollars of support for U.S. private financial institutions. In the United States, this involved the partial nationalization of private financial institutions: this was a dramatic reversal of the philosophy that had dominated the political economy over the previous 30 years.

How could this have happened? Part of the explanation lies in the steady increase in debt that had been fueling every aspect of the global economy, in particular the U.S. economy, since the late 1970s. Consumer spending had been financed increasingly by credit card debt. A housing boom had been financed by an expanded and aggressive mortgage market, and wars in Iraq and Afghanistan had been financed by U.S. government borrowing from overseas. By mid-2008, private debt in America had reached U.S. $41 trillion, almost three times the country's annual gross domestic product; the external debts of the United States had meanwhile reached U.S. $13.7 trillion. Everyone, it seemed, was borrowing from everyone else in an international financial system that had become extremely complex, increasingly leveraged, and decreasingly regulated.

The first signs of the crisis came in 2007, when credit losses associated with subprime mortgages—those given to a borrower with poor credit—led to difficulty for several large investment firms. Taking advantage of the relaxed controls on U.S. financial institutions, mortgage lenders had been selling their mortgages on bond markets and to investment banks to fund the soaring demand for housing. This led to abuses, as mortgage lenders no longer had an incentive to check carefully on borrowers. Many lenders began to offer mortgages to borrowers with poor credit histories and weak documentation of income. But few in the financial services industries fully understood the complexities of the booming mortgage market, and the various risk-assessment agencies underestimated the risks associated with these loans. Eventually, when interest rates increased, many households began to default on their payments, and as the bad loans added up, mortgage lenders found themselves in financial trouble. The impact of the mortgage market/housing crisis has been especially difficult for the states of Arizona, California, Nevada, and Florida, where speculation and overbuilding exceeded demand (**Figure 6.20**).

Because the entire world economy has become so interdependent, the problem in the United States quickly spread. Banks in other countries were caught up in the complex web of loans, and

▲ **FIGURE 6.20 Recession Ghost Town** The United States has a new kind of ghost town, such as this one, not part of the old West but the New West where the housing bust hit the hardest (Arizona, Nevada, and California). Developers caught up in the runaway housing boom overbuilt and oversold lots, houses, and condos, leaving neighborhoods barren with uninhabited model homes, eerily desolate luxury condos, and abandoned "McMansions" in the aftermath of the financial collapse that began in 2008. This faded billboard advertises the abandoned Desert Mesa development in North Las Vegas, Nevada.

by the fall of 2008, the banking industry was in crisis worldwide. As credit markets seized up, manufacturers and other businesses found it difficult to get the credit they needed to continue operating. Understandably, investors big and small were shaken; stock markets collapsed, and car manufacturers went bankrupt. Consumer confidence also plummeted, prompting retailers to cut back on orders. The sophisticated flexible production system of global commodity chains (see Chapter 7) meant that the impact across the globe was almost instantaneous.

Since 2008, some of the banks that survived have begun to recover. In late 2012, unemployment is still high and the stock market continues to remain volatile. But on the bright side, the economies of both countries in this region continue to grow, as does their manufacturing employment. At the same time that global crude oil prices are declining, U.S. household debt is shrinking, and though high, unemployment is dropping and the housing market appears to be stabilizing in both Canada and the United States.

The global financial crisis has generally been less pronounced in Canada, where the banks are more tightly regulated, more liquid, and less highly leveraged. Unlike risk-taking U.S.-based investment banks, Canadian banks tend to operate in a more traditional manner, with large numbers of loyal depositors and a more solid base of capital. But Canada's economy has not been entirely trouble-free. The Toronto Stock Exchange fell significantly in 2008 but has been showing strong signs of recovery. The increase in the value of the Canadian dollar has harmed its exports and the impact of the crisis on the United States—which takes the lion's share of Canada's exports—has had a direct impact on the Canadian economy. For example, the U.S. housing market collapse hurt Canada because much of the wood in new U.S. houses comes from Canada.

Territory and Politics

The United States and Canada are both **democracies**, an egalitarian form of government in which all citizens of a nation determine the laws and actions of their state. One of the chief requirements in a democracy is that citizens who meet certain qualifications have an equal opportunity to express their opinion. The United States became a federal republic founded on democratic principles in 1788 when it established its constitution. In the United States, the sovereign authority is the people. Canada, also a democracy, is a constitutional monarchy with a parliamentary type of government. The British sovereign is the foundation of Canada's judicial, legislative, and executive branches of government.

States and Government The United States and Canada are both federal states, which means that in each country, political authority is divided between autonomous sets of governments, one national and the others at lower levels, such a state/province, county, city, and town. A **state** is an independent political unit with recognized boundaries (though some of those boundaries may be in dispute). **Government**—one element of a state—is an entity that has the power to make and enforce laws. In a **federal system**, many political decisions are made at the local level. But there are differences even in federal states. Canadian federalism, for instance, allows more power to provinces than states have under U.S. federalism. The extremely close vote in Quebec in 1995—which would have enabled the province to become fully independent from Canada—is an illustration of the political power of provinces. Such an election or outcome would not be constitutionally possible in the United States.

During the first 100 years of the U.S. republic, the federal government spent most of its time regulating commerce. But beginning in the late 19th century, urged by constituents across the country, the federal government began to take an increasingly active and direct role in regulating and supporting all aspects of U.S. social and economic life, providing for social welfare; developing infrastructure, such as dams and highways; and transferring large amounts of tax dollars to contractors for the buildup of U.S. defense systems, especially during the Cold War.

Because the federal government was so heavily invested in all aspects of U.S. society, but especially in the economy, when the global economy experienced shock waves during the 1970s, the government was hit very hard. As corporations and businesses in the United States and elsewhere searched for remedies to their economic problems, the government did likewise and imposed dramatic restructuring on its own operations and programs.

The view that government's primary responsibility is as a guarantor of social welfare had dominated popular understanding in the United States since the 1930s. In the 1970s and 1980s, as local governments in the Rust Belt were declaring bankruptcy and the federal government was accumulating massive debt, popular opinion changed, and the role of government was reconfigured. Since the late 1980s, it has become routine for local governments in the United States to act more as entrepreneurs than as managers of the social welfare. For example, as deindustrialization accelerated in the Rust Belt, government agencies in the Sun Belt helped lure investment to the

▲ **FIGURE 6.21 Closed Mess Hall on Chanute Air Force Base in Rantoul, Illinois** Base decommissionings such as Chanute's had a significant effect on the region in terms of job loss when it occurred in 1993. More recent economic development initiatives have, however, looked toward turning problem sites such as this one around by repurposing them for new uses. This mess hall and the base headquarters are now on the Landmark Illinois list of endangered historic sites with an eye toward redevelopment.

region by offering tax breaks, creating needed infrastructure, and providing subsidies for private investment.

To reduce its mounting debt, the federal government in the United States shed its responsibilities for social welfare, passing these responsibilities on to state governments. The federal government also began to shut down military bases throughout the country as the fall of the Berlin Wall signaled the end of the Cold War (**Figure 6.21**). As it assumed less responsibility for social welfare and lowered military spending (until the recent military campaigns in Iraq and Afghanistan), the federal government oriented its role toward more actively facilitating the free flow of trade and the operation of transnational corporations.

Since independence, Canada has fostered a government that has been far more inclined to guarantee social welfare than that of the United States, though Canada also has a tradition of entrepreneurialism in government. In addition to continuing its tradition of providing social welfare, the state in Canada has accelerated its entrepreneurialism by directing support to expanding its tertiary sector (activities involving the sale and exchange of goods and services) and quaternary sector (activities involving the handling and processing of knowledge and information), particularly with respect to high-technology development.

APPLY YOUR KNOWLEDGE The United States and Canada are both democracies with federal systems, but they differ in many ways. Search the Internet for information about the electoral system in each country. How do they differ? How do these differences shape elections? How do they shape governing practices?

U.S. Political and Military Influence In combination with its economic strength, the United States achieved its place as the world's most powerful country through political and military strength. As the leading global military power, the United States significantly dwarfs Canada as well as Russia, the second ranked military power in the world. Because of its global dominance, we will focus on the United States in this section.

The first sign of the United States flexing its political and military muscle came with the Monroe Doctrine. Issued by President James Monroe in 1823, the Monroe Doctrine became the foundation of U.S. foreign policy in Latin America. The doctrine contended that European powers could no longer colonize the American continents and should not interfere with the newly independent Spanish–American republics. So long as Europe stayed out of the Americas, Monroe promised that the United States would not interfere with existing European colonies or with Europe itself.

U.S. militarism did not extend to an eagerness to get involved in European affairs, and the country only reluctantly entered World War I in 1917. Stunned by the nearly 5 million war casualties and the horror of the first highly technological engagement in history and eager to protect its growing economy, the United States entered a period of relative isolationism after World War I, rallying around the slogan of "America First." Although President Franklin Delano Roosevelt declared U.S. neutrality with respect to the European war in 1939, in 1942, the country entered World War II. The end of that war in 1945 marked a turning point in U.S. political and economic prowess as U.S. loans helped rebuild war-torn Europe and Japan. U.S. participation in the war effectively solidified the country's status as a world leader and led to the participation of the United States in several subsequent wars, including those fought in Korea and Vietnam.

All these wars were fought as part of the Cold War, which pitted the capitalist United States and Western Europe against the communist Soviet Union for the hearts, minds, and territories of peoples throughout the globe (see Chapters 1 and 4). The Cold War came to an end with the disintegration of the Soviet Union in 1991. A new era of global cooperation dawned, with the U.S. government and U.S. transnational corporations leading the way, though with markedly less militaristic fervor. The recent wars in Afghanistan and Iraq (see Chapters 4 and 9) have halted the brief hiatus in significant U.S. military involvement in global affairs, however (**Figure 6.22**).

U.S. War on Terror: Afghanistan, Iraq, and Pakistan The United States responded to the terrorist attacks of September 11, 2001, by declaring a global war against terrorism and identifying first Afghanistan and then Iraq as the greatest threats to U.S. security. With support from Canada as well as the United Kingdom and Australia, the United States invaded Afghanistan in October 2001 as part of its **War on Terror**. During this, the United States's longest-running war, over 100,000 troops have

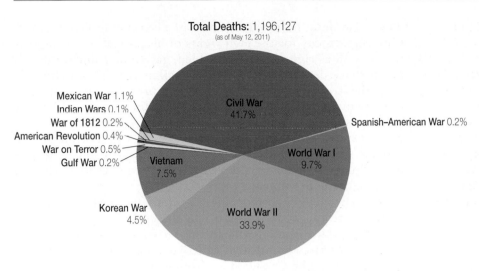

Total Deaths: 1,196,127
(as of May 12, 2011)

Civil War 41.7%
World War I 9.7%
Spanish–American War 0.2%
World War II 33.9%
Korean War 4.5%
Vietnam 7.5%
Gulf War 0.2%
War on Terror 0.5%
American Revolution 0.4%
War of 1812 0.2%
Indian Wars 0.1%
Mexican War 1.1%

▲ **FIGURE 6.22 U.S. War Dead** This graphic illustrates the dead among military personnel in all the wars in which the United States has been involved over the course of its history.

been deployed. The war costs some U.S. $2 billion a week (71% of the U.S. military budget in 2011). The number of troop fatalities (158) resulting from Canadian military activities in Afghanistan is the largest for any single Canadian military mission since the Korean War. Some 1,500 American lives have been lost, with troops routinely killed by roadside bombs, small-arms fire, and rocket attacks and in unspecified combat operations. On April 18, 2012, the United States and its allies finalized agreements to end the war in Afghanistan. A reduction of 33,000 troops began in summer 2012.

Although the evidence of involvement in the September 11, 2001, attacks by Iraq and its leader, Saddam Hussein, was highly questionable, on March 19, 2003, after amassing over 200,000 U.S. troops in the Persian Gulf region, President George W. Bush ordered the bombing of the city of Baghdad, Iraq. The declaration of war and invasion occurred without the explicit authorization of the UN Security Council, and some legal authorities take the view that the action violated the UN Charter. Some of the staunchest U.S. allies (Germany, France, and Canada) as well as Russia opposed the attack, and hundreds of thousands of antiwar protestors repeatedly took to the streets.

On December 15, 2011, the United States formally ended the war in Iraq. During more than eight years of war, nearly 4,500 U.S. service members were killed and 30,000 wounded. Tens of thousands of Iraqi troops and civilians died, as the United States deposed Saddam's regime and beat down an insurgency backed by al-Qaeda terrorists and sectarian revenge killings that threatened to destroy the country. **Al-Qaeda** is a global militant Islamist organization that operates as a network constituted by a multinational, stateless army and a radical Muslim movement. The United States and the United Kingdom, among other countries, as well as the United Nations Security Council, the European Union, and NATO, have identified it as a terrorist organization. Osama bin Laden, who was a member of a wealthy Saudi family, founded al-Qaeda, the organization responsible for the September 11, 2001, attacks on the United States.

While the United States has not formally declared war on Pakistan, the situation there is very much part of the War on Terror. With U.S. troops on the ground in Afghanistan, al-Qaeda moved operations to neighboring Pakistan, where Osama bin Laden was hiding. While the United States has not sent troops to Pakistan, it has been deploying drones—with the tacit agreement of the Pakistan government—in Pakistan's tribal areas against suspected al-Qaeda and Taliban targets since 2004 (**Figure 6.23**). A highly regarded study by the New America Foundation shows that there have been nearly 300 reported U.S. drone strikes in Pakistan (see Chapter 9). These strikes have involved the deaths of up to 2,680 individuals.

Tense relations now exist between Pakistan and the United States following the killing of bin Laden in 2011. That covert operation by U.S. Navy SEALS humiliated Pakistan, which has cut back on counterterrorism cooperation with the United States.

For the United States, the costs of the War on Terror are indeed still mounting. The U.S. Treasury estimates the cost of the war at U.S. $845 billion. Nobel Prize-winning economist Joseph Stiglitz and coauthor Linda Bilmes argue that the true cost of the war is closer to U.S. $3 trillion when the indirect costs are also assessed, including interest on the debt raised to fund the war, the rising cost of oil, health-care costs for returning veterans, and replacing the destroyed military hardware and degraded operational capacity caused by the war.[5]

---

[5]New America Foundation, 2011, "The Year of the Drone," www.Newamerica.net, accessed June 30, 2011.

▲ **FIGURE 6.23 U.S. Air Force MQ-1 Predator Drone** Unmanned aerial vehicles (drones) have been key weapons in anti-terrorist warfare in Pakistan. A drone can be as large as a passenger aircraft, as shown here, or as small as a hummingbird. Beginning in February 2012, when House Resolution #658 passed, the U.S. federal government as well as corporations and private individuals are authorized to use drones in U.S. airspace.

The assassination of Obama bin Laden on May 1, 2011, in Abbottabad, Pakistan, is seen by many as a symbolic end to the War on Terror. Bin Laden had been in hiding for a decade before he was killed in a covert operation ordered by President Obama and carried out by U.S. Navy SEALs and CIA operatives. Bin Laden believed that the United States and its political, economic, and cultural values, persecuted and oppressed Muslims around the world and that the only solution to ending this violence was to meet it with violence. His fundamental approach was to seduce the United States and other leading economic powers into endless war in Muslim countries. Bin Laden believed that endless war would lead to economic collapse for the United States.

APPLY YOUR KNOWLEDGE Research the impact of the costs of an ongoing war such as the War on Terror on a national economy. What costs, besides direct spending on troops and weapons, do governments incur when they wage war?

Social Movements Two of the most enduring, widespread, and effective social movements in the United States and Canada have been the women's movement and the environmental movement. The women's movement in the United States can be traced to the 1848 women's rights convention held in Seneca Falls, New York. From that point, inspired by women's movements in the United Kingdom, women agitated for the right to vote, to have the same civil rights as men, to have access to birth control, to improve working conditions for themselves, to earn equal pay for equal work, and to end discrimination in the workforce.

They also demonstrated in favor of disarmament and peace. Campaigns today for the civil rights of blacks and other minorities, as well as rights for the disabled, gay rights, and children's rights, have derived inspiration as well as tactics and rhetoric from the women's movement. The vice-presidential campaigns of Geraldine Ferraro (1984) and Sarah Palin (2008) and Hillary Clinton's (2008) presidential campaign are testimony to the effectiveness of the women's movement in the United States. Women's rights movements in Canada followed a similar trajectory. Women in Canada received the right to vote in 1918, two years before U.S. women.

The environmental movement in the United States can also be traced to the works of men like Henry David Thoreau and George Perkins Marsh, who in the 1850s and 1860s began to lecture and write about the destructive impact of humans on the natural world. The Canadian environmental movement emerged at roughly the same time and has continued to play an important role in Canadian politics since then. Contemporary environmental protection policy and sustainability movements echo the sentiments and commitments of these early activists. Today, citizens in both Canada and the United States are deeply engaged in a wide range of movements that are concerned with environmental issues and the cultural, social, and economic impacts of global warming (**Figure 6.24**).

CULTURE AND POPULATIONS

The U.S. and Canada region is culturally and politically diverse with influences that extend to remote corners of the globe. From jazz to hip-hop, horror movies to hamburgers, the influence of American culture on the rest of the world has been and continues to be dramatic. The influence of the rest of the world on the region has also been substantial. The region is notably adept at "sampling"—that is, taking what the world has to offer and translating it into something hybrid, popular, and desirable.

Religion and Language

The main language in the United States is English; Spanish is also widely spoken. Although the United States is popularly considered a Protestant country (including large numbers of Baptists, Methodists, Presbyterians, Lutherans, Pentecostals, and Episcopalians), Roman Catholics form the largest single religious group. The largest non-Christian religion is Judaism, and other non-Christian religions, such as Islam, Buddhism, and Hinduism, also have substantial followings.

In Canada, where there are two official languages (English and French), the situation for immigrants has always been somewhat different than in the United States. Instead of **assimilation**, Canadian popular opinion and public policy have advocated **multiculturalism**, the right of all ethnic groups to enjoy and protect their cultural heritage. Multiculturalism in practice includes protection and support of the right

▲ **FIGURE 6.24 Canadians Protest XL Pipeline** Activists protesting a proposed pipeline to bring oil sands oil to the United States from Canada march from the Canadian Embassy in Washington, D.C., to the White House. The Keystone XL pipeline, proposed by TransCanada, would carry 900,000 barrels of oil sands oil daily from Alberta, Canada, to Texas on the U.S. Gulf Coast.

to function in one's own language, both in the home as well as in official or public realms. Multiculturalism is premised on the belief that immigrants should not have to give up any of their original cultural attributes or practices.

A significant aspect of the geography of U.S. religion is its regional variation. For example, the Bible Belt, which stretches from Texas to Missouri, is dominated by Protestant denominations, many of them fundamentalist and evangelist. Mormons, or members of the Church of Jesus Christ of Latter-day Saints, are concentrated in Utah, where more than 75% of the population are adherents. Large numbers of Mormons also live in Nevada and Idaho. Large Catholic communities exist throughout the Southwest and in many of the large cities of the Northeast and upstate New York. Sizable concentrations of Jews occur in large U.S. urban centers like New York, Los Angeles, and Miami.

Cultural Practices, Social Differences, and Identity

Music, art, literature, dance, architecture, film, photography, sports, fashion, journalism, and cuisine, not to mention science, medicine, and technology, have all been shaped by the immigrants who have come to the United States and Canada. The influence of immigrants on music has been particularly impressive. Country, bluegrass, jazz, the blues, and rap all originated in the United States but have deep roots in the Old World. From jazz to rap, African Americans have been responsible for musical innovations that have been widely accepted, applauded, and imitated throughout the world.

Arts, Music, and Sports The early 20th-century origins and particularly U.S. expressions of jazz have been influential worldwide. West African folk music, brought by African slaves to the United States, forms one of the central foundations of jazz, but jazz was also influenced by European popular and light classical music of the 18th and 19th centuries. Dixieland jazz emerged from New Orleans and was played by white musicians who recorded the new music form on phonograph records. The spread of these recordings helped jazz become a sensation in the United States and Europe. African-American jazz groups—the originators of jazz styles such as ragtime, marches, hymns, spirituals, and the blues—were able to capitalize on the popularity of white Dixieland largely through the improvisational style of trumpeter Louis Armstrong. Armstrong migrated to Chicago in the 1920s, influencing local musicians and stimulating the evolution of the Chicago style (**Figure 6.25**).

About the same time that jazz caught on in Chicago, Harlem in New York emerged as a center for jazz, characterized by a highly technical, hard-driving piano style. Regional variations on the original Dixieland style emerged in other urban areas, where

▲ **FIGURE 6.25 Louis Armstrong** Jazz trumpeter Armstrong (1901–1971) was a foundational influence in jazz, shifting the music's focus from collective improvisation to solo performance. His style and virtuosity extended well beyond jazz to affect all forms of popular music in the United States and Europe.

significant populations of African Americans had settled. Jazz continued to flourish from the 1930s through the 1950s. In the 1960s, jazz began to lose popularity as audiences embraced mainstream rock and roll, which had itself been influenced by jazz and the blues. In the 1980s, jazz experienced a revival as a serious form of music, which it continues to enjoy today. Other distinctly U.S. musical and performance styles include rap, bluegrass, and musical theater.

Native populations in both the United States and Canada have also made significant cultural contributions to music, as well as handicrafts (especially basketry, rugs, jewelry, and pottery) and contemporary literature. People from all over the world travel to visit First Nations and Native American sites to view and collect their distinctive commercial products such as Navajo rugs from the U.S. Southwest and the wood carvings of the Haida people of Pacific Canada.

The game of baseball is another U.S. innovation, and it, too, has enjoyed widespread popularity beyond North America, especially in Caribbean countries like the Dominican Republic, Venezuela, and Cuba, but also in South Korea and Japan. The diverse nationalities of the players on the major league baseball teams in the United States and the Toronto Blue Jays in Canada demonstrate just how popular this sport has become worldwide. As a high-stakes commercial enterprise, baseball has traveled well. Players on the roster typically hail from Aruba, Australia, Colombia, Cuba, Curaçao, the Dominican Republic, Japan, South Korea,

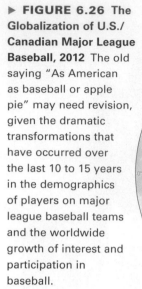

▶ **FIGURE 6.26** The Globalization of U.S./Canadian Major League Baseball, 2012 The old saying "As American as baseball or apple pie" may need revision, given the dramatic transformations that have occurred over the last 10 to 15 years in the demographics of players on major league baseball teams and the worldwide growth of interest and participation in baseball.

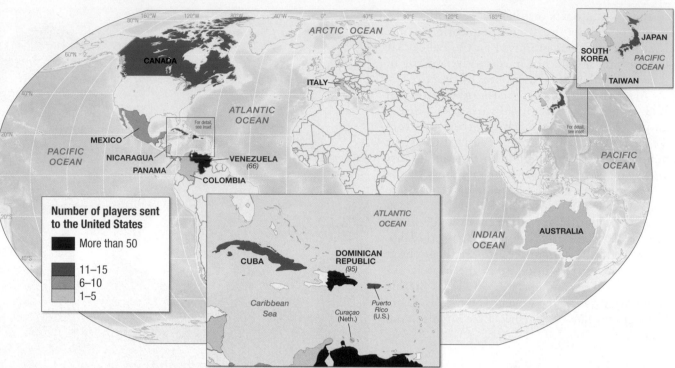

Mexico, Nicaragua, Panama, Puerto Rico, Taiwan, and Venezuela (**Figure 6.26**).

U.S. Cultural Imperialism Many scholars argue that "globalization" is really just a euphemism for "Americanization," and there is certainly some evidence to support this point of view. Consider the sheer numbers around the world who view *The Bachelorette*, drink Coca-Cola, and eat in McDonald's franchises. U.S. culture is embraced by local entrepreneurs around the world largely through consumer goods. It seems clear that U.S. products are consumed as much for their symbolism of a particular way of life as for their intrinsic value. Coca-Cola, Hollywood movies, rock and rap music, and NFL and NBA insignia have become associated with a lifestyle package that features luxury, youth, fitness, beauty, and freedom. Neither the widespread consumption of U.S. and U.S.-style products nor the increasing familiarity of people around the world with U.S. media and international brand names, however, necessarily adds up to the emergence of the Americanization of global culture. Instead, the processes of globalization are exposing the world's inhabitants to a common set of products, symbols, myths, memories, events, cult figures, landscapes, and traditions.

Canadian Cultural Nationalism Canadian culture tends mostly to represent a mix of immigrant and British settler influences. Canadian culture has had to battle the tremendous influence of commercialized U.S. culture, which has been enormously difficult to resist because of its geographical proximity. Canada has been very aggressive in its attempt to ward off the invasion of U.S. cultural products and has developed an extensive and very public policy of cultural protection against the onslaught of music, television, magazines, films, and other art and media forms.

Government bodies, such as the National Film Board of Canada and the Canadian Radio and Television Commission, actively monitor the media for the incursion of U.S. culture. For example, 30% of the music on Canadian radio must be Canadian. Nashville-based Country Music TV was discontinued from Canada's cable system in the early 1990s and replaced with a Canadian-owned country

music channel. Interestingly, some of the most famous U.S. pop culture icons were either invented by Canadians or are Canadian, such as William Shatner, *Star Trek*'s Captain Kirk, Superman (a concept invented by two Canadians), and popular singers and bands, such as Neil Young, Celine Dion, k.d. lang, Justin Bieber, and Broken Social Scene (**Figure 6.27**). Many Hollywood film and television actors, directors, producers and support personnel are also Canadian.

Sex, Gender, and Sexuality Gender is a category of identity, along with sex and sexuality, that has received a great deal of recent attention in the United States and Canada especially around issues of women's rights and gay marriage. An approach that has gained increasing acceptance is that these categories are not natural but are culturally constructed. That is, many people in both countries have begun to recognize that gendered behavior is not biologically determined, but that the practices that conform to notions of femininity and masculinity are imposed upon us by our place within larger social conventions. Just as baby boys are dressed in blue and baby girls in pink, an identity like male or female is learned and performed and not naturally given at birth. Instead, gender, sex, and sexuality are shaped by expectations that come through social conventions (boys wear trousers and girls wear dresses), the mass media (women who are powerful must also be feminine to be successful), religion (men are monks and women are nuns), and other powerful institutions that determine in advance what manifestations of sex, gender, and sexuality are socially permitted to appear as coherent or natural.

APPLY YOUR KNOWLEDGE Identify a cultural phenomenon (music, sports, language, sexual identity, or something else) and produce a sketch of the communities around the globe who participate in or appreciate it. For instance, soccer is an obvious option. Where is it played, what are the big rivalries, what sort of language connects those who watch it, how does its geography connect the community of fans?

▲ **FIGURE 6.27** **Broken Social Scene in Toronto** A Canadian band and musical collective, Broken Social Scene produces sounds from guitars, horns, woodwinds, violins, and experimental production techniques. Here they are performing at Canada Day, the country's national holiday that celebrates the uniting of three colonies into a single country in 1867.

Demography and Urbanization

Currently whites (of various European ancestry) constitute nearly three-quarters of the total U.S. population; African Americans, about 13%; Asians and Pacific Islanders, about 4%; and Native Americans, about 1%. Hispanics, who may also be counted among other groups, make up about 14% of the total U.S. population. As of May 2012, however, an important statistic was announced: more minority babies are being born in the United States than white babies. This statistic signals a far more multicultural population than the United States experienced in the 20th century and one that will be dominated by the demographics of minority populations.

Canada consists of primarily two founding ethnic communities, the British and the French. These two groups make up the largest piece of the demographic pie at over 50%. Other white European Canadians constitute about 15% of the population with Indigenous people contributing 2%; Asians, Africans, and Middle Easterners contributing 6%; and people of mixed background at 26%. Because the United States and Canada ask different questions about ethnicity on their national censuses, it is difficult to compare these numbers. Immigration history provides a more comprehensible picture.

Immigration The United States and Canada have varied and extensive immigration histories. U.S. immigration is frequently discussed in terms of waves, because the numbers and types of immigrants ebbed and flowed over time (**Figure 6.28**). At the beginning of the 19th century, the population of the new nation was largely dominated by English colonists and African slaves. There were also small numbers of Irish, Dutch, French, and Germans. The first large wave (1820–1870), in which overall immigration rose sharply to 2.8 million individuals, involved large numbers of Irish and German immigrants. The number of English immigrants declined. The newly arriving Irish were mostly peasants who fled the potato famine that had devastated the Irish economy and daily life. The Germans who came were mostly skilled craft workers seeking new opportunities.

In the second wave (1870–1920), in addition to the continuing stream of "old" immigrants from northern and Western Europe,

▶ **FIGURE 6.28** **U.S. Immigration, 1800–2010** This graph shows very dramatic peaks and troughs over the 200-plus year period. Some people object to the use of the term "waves" in reference to immigration, as it equates human lives to the flotsam and jetsam of ocean movements.

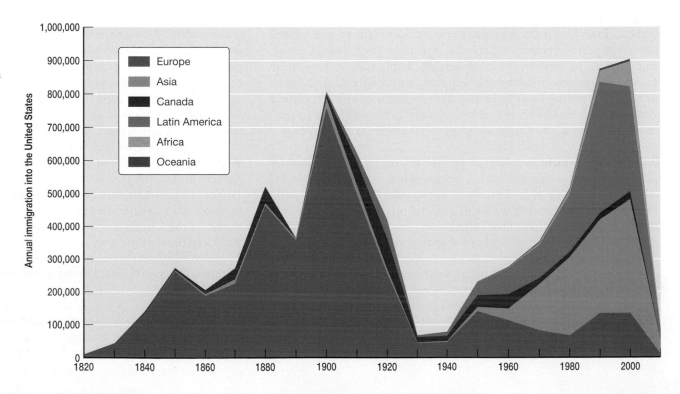

"new" immigrants from southern and Eastern Europe joined the flow into the United States. Between the 1870s and the 1880s, the absolute number of immigrants rose dramatically from 2.8 million to 5.2 million. Widespread economic depression in Europe and North America in the 1890s led to a decline in absolute numbers of immigrants (3.6 million). The numbers rose again in the first decade of the 20th century to an all-time high of 8.8 million. This wave included peasants, skilled workers, and successful merchants.

The third immigration wave (1970–present) is substantially different from the other two in that large numbers of more recent migrants have been from Asia and Latin America. Although Asians have been part of U.S. immigration history since the mid-19th century, Latin American migration to the United States is a 20th-century phenomenon that has increased since the changes brought about by the Hart-Cellar Act and the Immigration and Reform and Control Act of 1980. By 1990, Mexico had become (and still is) the largest source of immigrants to the United States. Most recent Mexican immigration into the United States has been to California, Texas, and Arizona, although there is a sizable Mexican population in Chicago and in many smaller U.S. cities where Mexican workers have been hired to harvest fruits and vegetables and to work in the meatpacking industry.

The immigration history of Canada is very similar to that of the United States, with one significant difference. French settlers dominated in Canada well into the 18th century, but by about 1750, other immigrant groups from Britain and Ireland joined the immigration stream. Canada also received some immigrants from the United States at least until 1810, when restrictive British policies made it difficult for Americans to immigrate. By the beginning of the 20th century, Canada and the United States had very similar experiences of immigration, including the restrictions that curtailed inflows of new migrants until the late 20th century. Today, Canada is a primary destination for Asian migrants, who make up nearly 50% of the immigrant stream.

Although over time the "new" immigrants of the second wave in the United States have largely been assimilated into mainstream U.S. life, experiencing increasing prosperity and social mobility in later generations, third-wave immigrants continue to confront racism and bigotry. They have been frequent victims of **hate crimes**, acts of violence committed because of prejudice against women; ethnic, racial, and religious minorities; and homosexuals (**Table 6.2**). Although the variety and range of national groups can potentially contribute to rich and interesting local and national expressions of culture, culture difference can also be seen as threatening to those who wish to protect a particular view of what it means to be American or Canadian.

Internal Migration In addition to foreign immigration, **internal migration**—the movement of populations within a national territory—has also played a role in both countries. In the United States, three overlapping waves of internal migration over the past two centuries have altered the population geography of the country. These three major migrations have been tied to broad-based political, economic, and social changes.

The first wave of internal migration began in the mid-19th century and increased steadily through the 20th century. This wave included both rural-to-urban migration associated with industrialization as well as movement of people from the settled eastern seaboard and Europe into the interior of the country. The latter is referred to as "westward expansion." Westward expansion took off in the early 19th century, when an official settlement policy was created (**Figure 6.29**, p. 256). Despite the emphasis on western expansion and rural settlement, between 1860 and 1920 the United States was transformed from a rural to an urban society. Industrialization created new jobs, and unneeded agricultural workers (along with foreign immigrants) moved to urban areas to work in the manufacturing sector.

The process has been quite similar for Canada, where urbanization has also been the dominant settlement process. In addition, there was also large-scale internal migration westward in Canada, mostly during the early decades of the 20th century. But there are also important differences in the immigration history of the United States and Canada. During the colonial period, most of the immigrants arriving in Canada were British and French, whereas the immigrant stream to the United States was more broadly based. Moreover, whereas immigrants from all parts of Europe migrated to Canada during the late 19th and early 20th centuries, the total number of people immigrating to Canada was smaller than the number immigrating to the United States. Finally, there has long been migration interaction between the two countries, with many Americans immigrating to Canada during the American Revolution. In the 20th century, Canadian immigration to the United States exceeded American immigration to Canada, as Canadians sought to secure a higher standard of living south of the international border.

The second wave of internal U.S. migration, which began early in the 20th century and continued through the 1950s, was the massive and very rapid movement of African Americans out of the rural South, where they had made livelihoods picking cotton, to cities in the South, North, and West. Although African Americans already formed considerable populations in cities such as Chicago and New York, large numbers of blacks moved out of the rural areas when mechanization of cotton picking reduced the number of jobs available. At the same time, pull factors attracted African Americans to the large cities. In the early 1940s, for example, large numbers of new jobs in the defense-oriented manufacturing sector became available when many urban workers left their jobs and entered the military. This second wave of migration can be seen as part of a wider pattern of rural-to-urban migration among agricultural workers as industrialization spread globally. After the war, a more important catalyst drove this migration: an increasing emphasis on high levels of mass consumption reoriented industry toward production of consumer goods and in turn stimulated large increases in the demand for unskilled and semiskilled labor. The impact on the geography of racial distribution in the United States was profound.

The third wave of internal migration began shortly after World War II ended in 1945 and continued into the 1990s (as was discussed earlier in this chapter). Between the end of the war and the early 1980s—and directly related to the impact of governmental defense policies and activities on the country's politics and economy—the region of the United States now commonly called the Sun Belt, and including most of the states of the U.S. South and Southwest, experienced a 97.9% increase in population. During the same period, the Midwest and Northeast, known as the Rust Belt, together grew by only 33.3%.

Urbanization, Industrialization, and New Growth Long before explorers and colonists arrived, Indigenous people built cities in the regions that would become the United States and Canada. A far more

TABLE 6.2	Hate Crimes
	Recent History of Hate Crime Laws and Incidents in the United States, 1971–2010
1971	Morris Dees and Joseph Levin, Jr., found the Southern Poverty Law Center in Montgomery, Alabama, seeking racial justice through the judicial system.*
1973	The National Gay and Lesbian Task Force founded to protect the rights of lesbians, gays, bisexuals, and transgender people to combat antigay violence and discrimination.*
1979	The Anti-Defamation League issues its first report of anti-Semitic events in *Audit of Anti-Semitic Incidents*.*
1981	Chip Berlet founds Political Research Associates, a nonprofit research center examining hate group activities targeting blacks, Jews, homosexuals, and other minorities.*
1988	Richard Lee Bednarski and his friends murder two gay men in Dallas, Texas. They each receive a lenient sentence for the two murders.*
1990	President George H. W. Bush enacts the Hate Crimes Statistics Act, which requires the *Federal Bureau of Investigation* to compile annual statistics on hate crimes.*
1991	Rodney King savagely beaten in Los Angeles by four police officers.* The National Asian Pacific American Legal Consortium founded with the goal of advancing human and civil rights for Asian Americans.*
1992	After a controversial ruling in the Rodney King case, the South Central section of Los Angeles is engulfed in rioting from April 30 to May 3. Fifty-eight people are killed.*
1993	The Hate Crimes Sentencing Enhancement Act added as an amendment to the Violent Crime and Law Enforcement Act of 1994.**
1994	The Violence Against Women Act passed, providing civil rights remedies for gender-motivated violence.*
1995–1996	A total of 318 church arsons occur over the course of two years; 130 people are arrested in conjunction with the arsons.*
1997	President Clinton devotes his weekly radio address to hate crimes, specifically citing bias crimes against lesbian, gay, bisexual, and transgender (LGBT) people. He asks Attorney General Janet Reno to review the laws concerning hate crimes and help the federal government develop a plan of action.** On November 13, the Hate Crimes Prevention Act is introduced in the House and the Senate (105th Congress). The bill extends the protection of the current federal hate crimes law to those who are victimized because of their sexual orientation, gender, or disability. It also strengthens extant law regarding hate crimes based on race, religion, and national origin.**
1998	On June 7, James Byrd, Jr., a black man in Jasper, Texas, is chained to a pickup truck and dragged along an asphalt road for two miles.* On October 6, two men in Laramie, Wyoming, believing Matthew Shepard to be a homosexual, tie him to a fence, and beat him. Shepard dies from injuries sustained in the beating.* In November, a bipartisan poll conducted for HRC finds that 56% of Americans support the Hate Crimes Prevention Act.**
1999	On July 1, Gary Matson and Winfield Scott Mowder, a homosexual couple, are shot and killed in their home near Redding, California.* On July 22, the Senate passes the Hate Crimes Prevention Act after it is incorporated as an amendment to the Commerce, Justice, and State appropriations bill.** On August 10, Buford Furrow, Jr., shoots five people at the North Valley Jewish Community Center in Los Angeles, later murdering Joseph Santos Ileto, a Filipino American postal worker.*
2001	On September 12, 300 men and women attempt to storm a mosque in a Chicago suburb a day after the terrorist attacks of 9/11.* Zainab al-Suwaij forms the American Islamic Congress to promote the diverse political and cultural interests of Muslim Americans.*
2007	In Jena, Louisiana, six black students known as the "Jena Six" are charged with attempted second-degree murder for attacking a white student, who was knocked unconscious. The attack comes after a series of racial incidents that began when white students dangled nooses from a tree at a local high school after black students sat underneath the tree, which had been traditionally deemed for white students only.* Over 10,000 people gather in an antiracism demonstration in Jena, Louisiana, to protest the treatment of six black students who had been arrested for beating a white student after numerous racially motivated incidents.* At Columbia University's Teachers College, a black professor finds a noose hanging from the door of her office, prompting an investigation by the New York Police Department's hate crime task force.* The Federal Bureau of Investigation reports that 7,722 hate crimes occurred in the United States in 2006—a 7.8% increase from the previous year.*
2008	West Virginia prosecutors file hate crime charges against Karen Burton, one of six white defendants charged with the kidnapping, torture, and sexual assault of Megan Williams, a 21-year-old black woman. Burton allegedly stabbed Williams in the ankle while using racial slurs. Pleading guilty to charges of assault and civil rights violations, Burton received the maximum sentence of 30 years in prison.*
2009	President Barack Obama signs the Matthew Shepard and James Byrd, Jr., Hate Crimes Prevention Act into law (as a provision of the National Defense Authorization Act).**
2010	FBI reports increase in hate crimes reported by law enforcement officers from previous year. The report, *Hate Crimes Statistics 2010*, found that of the 6,624 single-bias incidents in 2010, 47.3% were motivated by race, while 20% by religion, 19.3% by sexual orientation, 12.8% by ethnicity or national origin, and 0.6% by disability.***
2012	On August 5, at a Sikh temple in Oak Creek, Wisconsin, a single gunman, Wade Michael Page killed six people and wounded four others, including a responding police officer. Page, a white supremacist, later shot and killed himself. U.S. Attorney General Eric Holder called the mass murder a terrorist attack and a hate crime.

*ABC-CLIO, "Hate Crimes Timeline," http://www.historyandtheheadlines.abc-clio.com/ContentPages/ContentPage.aspx?entryId=1299856&currentSection=1296470&productid=21, accessed August 14, 2012.
**Human Rights Campaign,"Hate Crime Timeline," http://www.hrc.org/resources/entry/hate-crimes-timeline, accessed July 14, 2012.
***"FBI 2010 Report Indicates Slight Increase in Hate Crimes," The Leadership Conference on Civil and Human Rights, November 14, 2011, http://www.civilrights.org/archives/2011/11/2010-hatecrimes.html.

▶ **FIGURE 6.29** **U.S. Frontier Trails, 1834–1897** The western frontier is part of the history and mythology of the United States (and Canada). The various paths individuals traveled suggest the many ways they experienced the passage. The topography of the land west of the Mississippi was intimidating, with vast grasslands and plains, searing deserts, and unimaginable insects and animals.

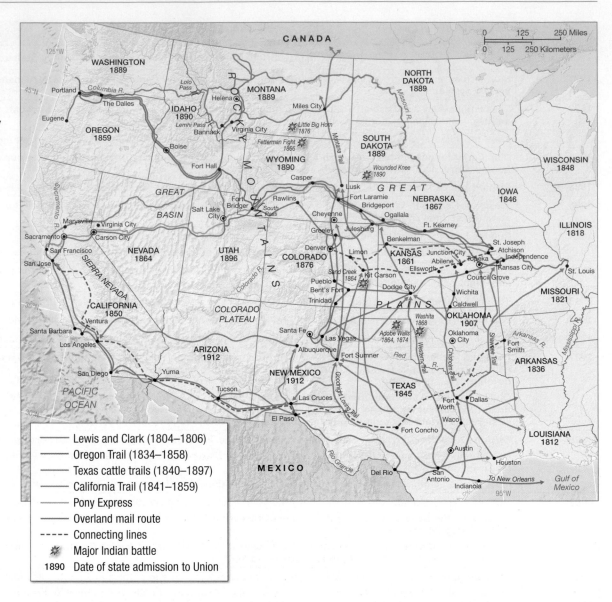

extensive and intensive urbanization process began during the colonial period. As Europeans colonized the new lands, they built cities as central places to organize commerce, defense, communication, and, later, administration and worship. In Florida and in the southwestern United States, the Spanish founded cities, which became symbols of their political and military authority (**Figure 6.30**). French explorers came not to settle but to reap commercial rewards, and they established urban centers to facilitate the exchange of goods. The Dutch also established urban settlements as trading centers for furs and slaves as well as other goods.

With few exceptions, the British played the largest role in shaping U.S. (and Canadian) urbanization and urban life. Sustained by trade as well as being administrative centers, Atlantic coastal cities such as Boston, Providence, Baltimore, Philadelphia, Charleston, and Savannah were also ports and key nodes in a globally expanding mercantile system. These urban centers enabled the transfer of resources, goods, and people, not only from the interior hinterland into the cities but also outward to Europe. At the same time, these cities received goods from England for U.S. consumer markets. Colonists saw their burgeoning cities not only as commercial centers but also as places where new ideas could be hatched and nurtured. They were

viewed as hearths of civilization, where European cultural practices confronted those of the new nation, creating in the process uniquely North American urban places.

As sites of innovation and cradles of culture, by the early 19th century, cities were also largely the places where a new economy based on manufacturing was born and flourished. By the mid-19th century, as industrialization disproportionately fueled the economies of the northern and midwestern United States, the differences between the northern and southern regions of the United States became increasingly pronounced persisting even into the 20th century.

Recall from earlier in the chapter, how this dynamic shifted in the 1970s between the North and South in the United States. Cities that had once been sleepy towns began to grow and by the 1980s and 1990s joined the list of some of the largest in the country. In Canada, this period saw the movement of the economy away from an overdependence on staples and toward more diversity, including finance, tourism, and high tech. Cities that had once been towns in Canada also began to grow and become more important to the national economy during this period, including Calgary, Edmonton, and Winnipeg.

▲ **FIGURE 6.30 Fort Matanzas in St. Augustine, Florida** Military defense sites were an important part of the early settlement of North America by Europeans, as they fought each other for control of territory. *Matanzas* means "slaughter" in Spanish and refers to the massacre of nearly 250 French Huguenots at the hands of the Spanish at this site, 175 years before the fort was constructed.

APPLY YOUR KNOWLEDGE Using the Internet, determine the top ten most populous Canadian and U.S. cities in each of the following years: 1900, 1930, 1960, and 2000. How has the list changed over time? Provide two reasons for any trends you observe.

Urban to Suburban Migration The first evidence of **suburbanization**—the growth of population along the fringes of large metropolitan areas—can be traced back to the late 18th and early 19th centuries, when real estate developers looked beyond the city for investment opportunities, and wealthy city dwellers began seeking more scenic residential locations. Later, residents fled to the suburbs to get away from the new immigrants and their increasing hold over urban machine politics. This process was rapidly accelerated with the introduction of new transportation technologies—first horse-drawn streetcars, then commuter rail services, and finally, automobiles. Each innovation in transportation allowed people to travel longer distances to and from work within the same or shorter time period.

North Americans chose to move to the suburbs in massive numbers, not in the least because the suburbs were considered by many to be more healthful places to raise a family. Suburbanization continues today in both the United States and Canada with a new wrinkle—retirees especially are searching out the good life on the far fringes of the metropolitan core in small towns like Bisbee, Arizona, and the Okanagan Valley in British Columbia (**Figure 6.31**, p. 258).

FUTURE GEOGRAPHIES

The United States continues to be the most powerful nation on Earth, and Canada is one of its staunchest allies. Both countries possess broad resource bases; a large, well-trained, and very sophisticated workforce; and a high level of technological sophistication. The United States has a domestic market that has greater purchasing power than any other single country as well as the most powerful and technologically sophisticated military apparatus. It also has the dominant voice and the last word in international economic and political affairs. The distinctive message that the United States promotes emphasizes free markets, personal liberty, private property, electoral democracy, and mass consumption. But, as discussed in Chapter 1, dramatic and rapid change is currently occurring as the world's political borders are being dismantled or rearranged around new economic relationships wrought by globalization.

U.S. Dominance and Its Challenge The future for the United States, at the moment, is unclear. Its economic dominance is no longer unquestioned in the way it was in the 1950s, 1960s, and 1970s. The 2008 collapse of the U.S. financial and housing industry left the country reeling. More important, the globalization of the world economy and ten years of war have severely constrained the ability of the United States to translate its economic resources into the firm control of international financial markets that it used to enjoy.

The al-Qaeda terrorist attacks of 2001 gave a new focus for U.S. geopolitical strategy. But the invasion of Iraq and Afghanistan, together with U.S. refusal to participate in high-profile international environmental agreements and its lack of cooperation with the United Nations, has caused concern among some about the future of the political, cultural, and moral leadership of the United States in global affairs. Many military thinkers, journalists, policymakers and scholars have warned that the formidable economic growth and increasing military buildup of the People's Republic of China could signal the relative demise of the United States. Canada, though never a world political power, still maintains substantial economic might. But like the United States, its past prominence is also likely to be challenged by up-and-coming economic powers like Brazil, Russia, India, and China.

It must, however, be recognized that the United States still maintains global economic and military prowess. Its GDP-PPP is the highest in the world, while the GDP-PPP of China, the next closest competitor, is nearly U.S. $5 trillion smaller. What's more, U.S. military might is formidable, with 500 bases around the world; an airforce that controls Earth's air and space; and a combined naval force that the remainder of the world combined doesn't come close to equaling. Those who advocate the position that China will eventually overtake the United States as the single global superpower also often fail to appreciate the interdependence of the Chinese and U.S. economies. Critically, China is the United States's number one trading partner, and vice versa (**Figure 6.32**, p. 258). The bottom line, argued by a number of well-respected observers, is that if the U.S. economy is destined for decline, then China's will likely go down with it. This is exemplified by the fact that the recession following the 2008 financial collapse slowed China's economic growth considerably. Perhaps the most central point to be gleaned from this

▲ **FIGURE 6.31 Okanagan Valley, British Columbia** The Okanagan Valley has become an attractive site for retirement living for Canadians because of its sunny climate and the wide range of available outdoor activities, including skiing, hiking, and fishing. Vineyards and orchards line many of the hills and the area is also a cultural attraction because of its First Nations history.

interdepence is to understand that we all sink or swim together, and that global political decisions should reflect that.[6]

Security, Terrorism, and War The enormous economic impact of ten years of war in Afghanistan and Iraq derives directly from security concerns brought on by the terrorist attacks of September 2001. While Osama bin Laden is dead, new security threats are anticipated, suggesting that security is likely to be a national concern and an economic priority for years to come. The focus on security by the U.S. federal government is reshaping governing structures as well as economic processes (creating new layers of rules, regulations, and policies as well as new ways of interacting politically with other states). The practices of daily life (as citizens deal with increased personal security measures and the transformation of human rights) are also changing around security concerns. In short, the War on Terror is not over. New challenges are likely to emerge from Asia. Canada, because of its proximity to the United States as well its dependence on the American economy for its own well-being, is likely to continue to be drawn into U.S. security concerns.

Environmental Change From observations drawn from the historical record as well as computer simulations, it is clear that the climates of the United States and Canada will continue to change into the 21st century and that this change will have significant effects. Models project warming of 4 to 9°F for the United States and Canada. Of course, the changes will vary across the subregions of the two countries. While effects vary based

on place, it is certain that animal and plant life will be affected, increased precipitation (rain and snow) can be expected in some places, there will be decreases in precipitation and drought in other places, and sea-surface temperatures in the Atlantic and Pacific and Bering Seas will increase. As the problems of climate change are becoming more clearly understood, a wide range of possible responses are being identified and developed that can begin to address the impacts. **Table 6.3** lists adaptation and mitigation measures to help governments and corporations address climate change.

▲ **FIGURE 6.32 Long Beach, California** More cargo and containers move through the Port of Long Beach than any other port in the United States. It is a major gateway for United States–Asian trade, especially for products that are imported by Walmart from contracted factories in China.

[6]This discussion is derived from the essay by Luke M. Herrington, 2011, "Why the Rise of China Will Not Lead to Global Hegemony," *e-International Relations*, http://www.e-ir.info/2011/07/15/why-the-precarious-rise-of-china-will-not-lead-to-global-hegemony/, accessed May 18, 2012.

TABLE 6.3 Corporate and Government Responses to Climate Change

Sector	Key Mitigation Technologies and Practices Currently Commercially Available	Policies, Measures, and Instruments Shown to Be Environmentally Effective	Key Constraints or Opportunities (Constraints Above; Opportunities Below)
Energy supply	Improved supply and distribution efficiency; fuel switching from coal to gas; nuclear power; renewable heat and power (hydropower, solar, wind, geothermal, and bioenergy); combined heat and power; early applications of carbon dioxide capture and storage (CCS) (e.g., storage of removed CO_2 from natural gas)	Reduction of fossil fuel subsidies; taxes or carbon charges on fossil fuels; feed-in tariffs for renewable energy technologies; renewable energy obligations; producer subsidies	Resistance by vested interests may make them difficult to implement
			May be appropriate to create markets for low-emissions technologies
Transport	More fuel-efficient vehicles; hybrid vehicles; cleaner diesel vehicles; biofuels; modal shifts from road transport to rail and public transport systems; non-motorized transport (cycling, walking); land-use and transport planning	Mandatory fuel economy; biofuel blending and CO_2 standards for road transport; taxes on vehicle purchase, registration, use, and motor fuels; road and parking pricing	Effectiveness may drop with higher incomes
			Particularly appropriate for countries that are building up their transportation systems
Buildings	Efficient lighting and daylighting; more efficient electrical appliances and heating and cooling devices; improved cook stoves, improved insulation; passive and active solar design for heating and cooling; alternative refrigeration fluids; recovery and recycling of fluorinated gases	Appliance standards and labeling; building codes and certification; public-sector leadership programs; including procurement	Periodic revision of standards needed, enforcement can be difficult
			Government purchasing can expand demand for energy-efficient products
Industry	More efficient end-use electrical equipment; heat and power recovery; material recycling and substitution; control of non-CO_2 gas emissions; and a wide array of process-specific technologies	Provision of benchmark information; performance standards; subsidies; tax credits, tradable permits, voluntary agreements	Predictable allocation mechanisms and stable price signals important for investments
			May be appropriate to stimulate technology uptake
Agriculture	Improved crop and grazing land management to increase soil carbon storage; restoration of cultivated peaty soils and degraded lands; improved rice cultivation techniques and livestock and manure management to reduce CH_4 emissions; improved nitrogen fertilizer application techniques to reduce N_2O emissions; dedicated energy crops to replace fossil fuel use; improved energy efficiency	Financial incentives and regulations for improved land management; maintaining soil carbon content; efficient use of fertilizers and irrigation	
			May encourage synergy with sustainable development and with reducing vulnerability to climate change
Forestry/forests	Afforestation; reforestation; forest management; reduced deforestation; harvested wood product management; use of forestry products for bioenergy to replace fossil fuel use	Financial incentives (national and international) to increase forest area, to reduce deforestation, and to maintain and manage forests; land-use regulation and enforcement	Constraints include lack of investment capital and land-tenure issues
			Can help poverty alleviation
Waste	Landfill CH_4 recovery; waste incineration with energy recovery; composting of organic waste; controlled wastewater treatment; recycling and waste minimization	Financial incentives for improved waste and wastewater management	Most effectively applied at national level with enforcement strategies
			May stimulate technology diffusion

SOURCE: IPCC, *Climate Change 2007: Synthesis Report*, p. 60; http://www/opcc.ch/pdf/assess,emt-report/ar4/syr/ar4_syr.pdf, accessed July 6, 2009.

LEARNING OUTCOMES *REVISITED*

■ **Describe the distinctive landscapes of the United States and Canada with respect to landforms and climate.**

The United States and Canada occupy an environmentally rich continent. Their mountain or coastal landscapes are highly scenic and their vast environmental diversity allows for highly productive agriculture as well as a wealth of resources, including energy in the form of oil, gas, coal, wind, water, and sun. The climates of the region encompass every possible variation from Arctic to tropical landscapes.

■ **Summarize the impact of the United States and Canada on global climate change, particularly in terms of their production of greenhouse gases.**

The United States and Canada play a major role in global climate change. Both countries, but especially the United States, are responsible for producing some of the highest total and per capita greenhouse gas emissions in the world as a result of their production and consumption patterns. These emissions stem from the region's heavy reliance on fossil fuels, including high gasoline use and coal-fired electricity, as well as agricultural and industrial production that generates not only massive amounts of carbon dioxide but also other greenhouse gases, such as methane and nitrous oxide.

■ **Explain the rise of the United States and Canada as key forces in the global economy.**

The United States's ascension to global dominance has been premised on the richness of its resource base, geopolitical prowess, and the complexity and diligence of its population. Canada has a similar history; however, it is the primary trading partner to the United States and at least some of its success has been due to that relationship.

■ **Describe how the global economic crisis of 2008 has affected the two countries as well as the impact it has had on economic inequality in the region.**

In recent years, both superrich and extremely poor populations have emerged in the United States and Canada. This inequality is due in part to the recent global financial collapse, during which people with low and modest incomes suffered enormously through the loss of jobs and housing. The crisis, in a sense, exacerbated a trend that was already in force, as the rich have gotten richer and the middle class and the poor have lost wealth.

■ **Compare and contrast the forms and practice of government of the United States and Canada.**

The United States and Canada are established democracies modeled on European political traditions. Both are federal states, which means that power is allocated to local government units (provinces in Canada and states, counties, cities, and towns in the United States). The United States is a republic, while Canada is a constitutional monarchy with a parliamentary type of government.

■ **Summarize the Indigenous and immigrant histories of the United States and of Canada.**

The Indigenous and immigrant histories of the United States and Canada are strikingly similar; this is hardly surprising since they were inhabited by similar Indigenous groups and settled by the same groups of Europeans at the same time. Canada, however, never participated in any significant trade in African slaves. Settlement of both places largely proceeded from east to west until the 20th century, when large numbers of immigrants arriving from Asia and other parts of the Pacific Rim began settling from west to east.

■ **Describe the geography of religion and language across the United States and Canada.**

The official language of the United States is English; in Canada, it is English and French. This simple statement, however, does not reflect the variations of language and religion that characterize both countries. With so many ethnic groups occupying these two countries over their relatively short histories, it is not surprising to find multiple languages spoken and multiple religions practiced on the streets of Toronto and New York and ethnic neighborhoods in large cities as well as in small towns and even in rural areas.

■ **Compare and contrast the urbanization processes that have shaped the United States and Canada over the last 100 years.**

In the United States, urbanization occurred early on in the north and midwestern part of the country, as the processing of agricultural goods and the rise of manufacturing drew immigrants in search of jobs. In Canada, the situation was similar. The most recent changes to cities in both countries have come from a period of deindustrialization of old manufacturing areas and urban and population growth in the Sun Belt in the United States and in the west of Canada. High technology growth has been the most recent influence on cities in both the United States and Canada.

KEY TERMS

acid rain (p. 236)

al-Qaeda (p. 249)

American Revolution (p. 239)

Americanization (p. 239)

assimilation (p. 250)

cap-and-trade programs (p. 236)

carbon footprint (p. 226)

cartel (p. 234)

Confederate States of America (p. 240)

creative destruction (p. 242)

deindustrialization (p. 242)

democracy (p. 247)

Europeanization (p. 238)

federal system (p. 247)

French and Indian War (p. 238)

genetically modified organisms (GMOs) (p. 233)

government (p. 247)

greenhouse gases (GHGs) (p. 226)

hate crime (p. 254)

indentured servant (p. 239)

information technology (IT) (p. 242)

intermontane (p. 227)

internal migration (p. 254)

local food (p. 233)

multiculturalism (p. 250)

North America Act of 1867 (p. 240)

Occupy Wall Street (p. 246)

oil sands (p. 236)

organic farming (p. 233)

reservation/reserve (p. 232)

Rust Belt (p. 242)

Seven Years' War (p. 238)

slow food (p. 233)

Snow Belt (p. 242)

staples economy (p. 241)

state (p. 247)

suburbanization (p. 257)

Sun Belt (p. 242)

superfund site (p. 236)

U.S. Civil War (p. 239)

War on Terror (p. 248)

THINKING GEOGRAPHICALLY

1. Find an image that shows an aspect of the physical geography of either Canada or the United States (a mountain range, a river system, the prairie, etc.) and discuss the way that this particular feature is relevant to agriculture, climate, adaptation, ecology, or resources.

2. The economies of Canada and the United States are critically important to the global economy. Referring to a specific commodity, mineral, or other natural resource, describe how the production of that item in either Canada or the United States shapes another country through its consumption of that particular commodity or resource.

3. Discuss the three main waves of immigration into the United States. When did they occur, and where did the immigrants come from? Discuss the three main waves of internal migration within the United States. How have African Americans, many of whose ancestors were brought to the United States long before other immigrant groups arrived, been affected by immigration and internal migration?

4. Choose a city in both the United States and Canada and describe how and why its fortunes have changed over the course of 100 years—from 1900 to 2000.

5. The future of U.S. global dominance is, of course, unknown. Construct a cogent argument for its continuation and one for its decline and then indicate which future you believe is the most likely to take place and why.

MasteringGeography™

Looking for additional review and test prep materials? Visit the Study Area in MasteringGeography™ to enhance your geographic literacy, spatial reasoning skills, and understanding of this chapter's content by accessing a variety of resources, including **MapMaster** interactive maps, videos, RSS feeds, flashcards, Web links, self-study quizzes, and an eText version of *World Regions in Global Context*.

7 | Latin America and the Caribbean

The Lonely Planet travel guide states: "Twenty-first century Cuba promises to be like nowhere else you've ever visited: economically poor, but culturally rich; visibly mildewed, but architecturally magnificent; infuriating, yet at the same time, strangely uplifting." This street scene from Havana, Cuba, shows one of the American classic cars that have been maintained since the 1950s.

CUBA, A LARGE ISLAND THE SIZE OF KENTUCKY located only approximately 100 miles from the U.S. mainland, has been a touchstone for relations between the United States and the Latin American and Caribbean region for more than a century. The United States granted Cuba independence after winning a war against Spain in 1898, but retained the right to intervene in Cuban affairs and to maintain a military base on the island. From the 1920s to the 1950s, U.S. companies controlled much of the Cuban economy—sugar, mining, ranching, and utilities—creating resentment in Cuba, and in 1959, a left-wing revolution led by Fidel Castro gave birth to a communist government. Castro's government nationalized private land, utilities, and the media and developed strong links with the Soviet Union. Cold War anticommunist sentiments prompted an unsuccessful U.S. invasion at the Bay of Pigs in 1961. Soon after, the Soviet Union placed nuclear missiles in Cuba, provoking the Cuban Missile Crisis confrontation between the United States and the Soviet Union. The United States imposed trade and travel embargoes on Cuba for decades after the crisis. Since the

LEARNING OUTCOMES

- Explain the physical origins and social impacts of natural and human-caused disasters in the region, including earthquakes, hurricanes, deforestation, and the effects of El Niño and climate change.

- Describe the ways in which the Maya, Aztecs, and Incas adapted to environmental challenges.

- Summarize the common experiences and legacies of colonialism in the region.

- Compare and contrast the economic policies and impacts of import substitution, structural adjustment programs, and free trade in countries across the region.

- Describe the historical sequence of U.S. intervention, the rise of authoritarianism, key revolutions, and the emergence of democracy across the region.

- Describe the history and current distribution of Indigenous peoples and cultures across the region, the ways in which the plants they domesticated have influenced global food and health, and how slavery and colonialism changed the demography of the region.

- Understand the geography and impacts of the drug trade and international tourism in the region.

- List the factors promoting urbanization in the region and provide examples of the social and environmental problems associated with urbanization.

1959 revolution, thousands of Cubans have fled to the United States, especially to Florida, partly because of political repression and dissatisfaction and economic problems that include limited supplies of food and consumer goods. Cuba's exports and economy were weakened by the U.S. embargo and the fall of the Soviet Union. However, today many social indicators are relatively positive in Cuba. Cuban agricultural practices are more sustainable than those of many other countries. In addition, Cubans enjoy a high life expectancy (78.3 years) and a low infant mortality rate (4.9); both of these factors are linked to Cuba's strong public health system. Tourism is thriving, especially at popular beach resorts; more than 2 million vacationers visit the island annually, mostly from Canada and Europe. The next decade is likely to bring many changes to Cuba: travel restrictions may be lifted, private enterprises may start to flourish, trade will expand, and the government is moving toward broader political participation. Cuba's size, resources, and connections to the world could make it a leader within the region. ■

Latin America and the Caribbean Key Facts

- Major Subregions: Central America, the Southern Cone, Caribbean Region, Gulf of Mexico

- Major Physiogeographic Features: Andes Mountains; Amazon Basin. Influenced by global atmospheric circulation, the climate is varied across the region and includes tropical, midlatitude, and Mediterranean systems

- Major Religions: Mainly Christian (primarily Roman Catholic); Indigenous religions are also widely practiced

- Major Language Families: Spanish and Portuguese, both of which descend from the Romance branch of Indo-European language; also Indigenous languages, such as Mayan and Quechua

- Total Area (total sq km): 20 million

- Population (2011): 596 million, including 196 million in Brazil, 114 million in Mexico, and 47 million in Columbia; Population under Age 15 (%): 28; Population over Age 65 (%): 7

- Population Density (per sq km) (2011): 29

- Urbanization (%) (2011): 80

- Average Life Expectancy at Birth (2011): Overall: 74; Women: 77; Men: 71

- Total Fertility Rate (2011): 2.2

- GNI PPP per Capita (current U.S. $) (2011): 10,130; Highest, Trinidad and Tobago: 23,439; Lowest, Haiti: 1,123

- Population in Poverty (%, < $2/day) (2008): 13

- Internet Users (2011): 235,874,427; Population with Access to the Internet (%) (2011): 46; Growth of Internet Use (%) (2000–2009): 1,205

- Access to Improved Drinking Water Sources (%) (2011): Urban: 97; Rural: 80

- Energy Use (kg of oil equivalent per capita) (2009): 1,277.8

- Ecological Footprint (hectares per capita consumed/hectares per capita available, global scale) (2011): 2.6/1.8

ENVIRONMENT AND SOCIETY

Latin America is the southern part of the large landmass of the Americas that lies between the Pacific and Atlantic Oceans (**Figure 7.1**). The Americas are often divided into the two continents of North America (Canada, the United States, and Mexico) and South America (all the countries south of Mexico). Latin America includes all the countries south of the United States from Mexico to the southern tip of South America in Chile and Argentina. Latin America is a world region that covers more than 20 million square kilometers (7.7 million square miles). "Latin" America, defined by the shared Latin-based languages of Spanish and Portuguese, may not be the most accurate name for this region, which includes countries where Indigenous languages and English are important. One alternate name for Latin America is Central and South America, where Central America includes Mexico, Belize, Guatemala, Nicaragua, Honduras, Costa Rica, and Panama. This chapter also includes the Caribbean region—the arc of islands that sweep from the U.S. state of Florida to Colombia across the Caribbean Sea (**Figure 7.2**). We include the Caribbean with Latin America because of shared histories of colonialism and common cultures and contemporary problems; however, in the material in this chapter, we also note some distinct differences.

Climate, Adaptation, and Global Change

Humans have transformed Latin American and Caribbean landscapes for centuries. Research has shown that many of the seemingly pristine forests were cleared centuries ago and have grown back, that grasslands have been repeatedly selectively burned or grazed, that mountains were carved into terraces, and that the waters of the deserts have been stored or diverted for hundreds of years (**Figure 7.3**, p. 266). Contemporary human settlements and agriculture are extremely vulnerable to natural disasters and epidemics, and the human geography of Latin America and the Caribbean continues to be influenced by environmental conditions and climate change. An understanding of the physical geography and the ways in which people have adapted to and modified their environmental surroundings is integral to appreciating the historical and contemporary human geography of the region and the challenges to its sustainable development.

The overall climates of Latin America and the Caribbean are determined by global atmospheric circulation, including the positions

◀ **FIGURE 7.2 Satellite Image of Latin America** Some key physical features of Central and South America are clearly visible in this satellite image, including the verdant green of the Amazon Basin and the mountain ridges of the Andes and Central America. The image also shows several lakes—Lake Nicaragua in Central America, Lake Maracaibo in northern Venezuela, and Lake Titicaca on the Andean Altiplano.

▲ **FIGURE 7.3 Andean Terraces and Alpaca** The Incas constructed terraces, not only to reduce soil erosion and provide a flat area for planting, but also to decrease frost risks by breaking up downhill flows of cold air and allow for irrigation canals to flow across the slopes in efficient ways. Here, domesticated alpacas graze near the archaeological site of Machu Pichu.

of the Equatorial low-pressure and subtropical high-pressure zones and the major global wind belts (see Figure 1.10, p. 11). Climatic patterns in the region provide good general examples of how global circulation affects regional climate, vegetation, and human activity (**Figure 7.4**).

Recall from Chapter 1 that the general circulation of the atmosphere is driven by the differential heating and rotation of Earth, with warm air rising at the Equator at the intertropical convergence zone (ITCZ), cooling as it rises. The warm air produces high rainfall and then flows poleward, finally sinking and drying around the Tropics of Capricorn and Cancer. The Equatorial zones of high temperatures and rainfall provide conditions for the rapid growth of vegetation. For example, in the rainforests of the Amazon, annual rainfall ranges from 1.5 to 2 meters (60 to 80 inches). Where the air sinks over the tropics, it holds so little moisture by the time it reaches ground level that these regions are characterized by the very low rainfall, sparse vegetation, and dry conditions of deserts. Latin America includes the Sonoran and Chihuahuan Deserts of Mexico and the Atacama Desert of Chile.

As Earth spins, it drags air flowing from the tropics to the Equator into an east–west flow called the **trade winds**, and air flowing poleward from the tropics into a west–east flow called the **westerlies**. The trade winds flowing across the Atlantic north of the Equator frequently produce rain on the Caribbean Islands and east coasts of Central America. In the Southern Hemisphere, they bring rain to the east coast of Brazil. Easterly winds moving across the warm Caribbean absorb moisture, especially during the fall, when the sea surface is warmest. When storms start to circulate, the warm sea fuels

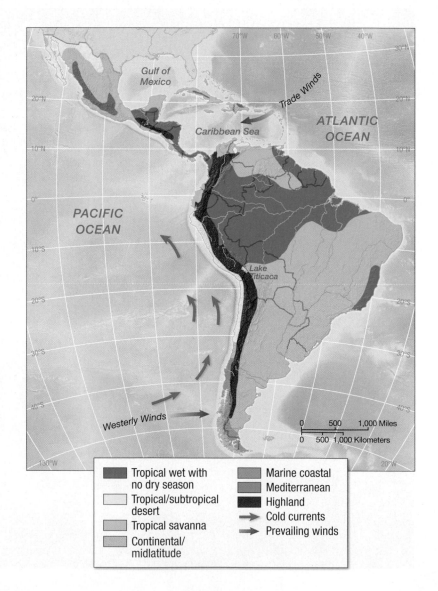

▶ **FIGURE 7.4 Climate Regions of Latin America** Latin America's climate is influenced by major wind and pressure belts and the configuration of land and oceans. The average pattern shown in this map tends to shift northward in June and southward in December, bringing seasonal changes to many regions.

both the moisture and energy of the storms, producing the hurricanes that regularly cross the Atlantic coast of Latin America.

The areas in this region on the margins of the trade winds and at the edges of the Equatorial rainfall zone have highly seasonal climates with a distinct rainy season. Latin America's extensive grasslands occur where seasonal shifts in wind and pressure belts bring moderate rains. The westerlies bring heavy rains to the southern country of Chile. When the global circulation shifts southward in December, storms spinning out of the Northern Hemisphere westerlies also bring rain to northern Mexico.

The coastal mountain chains, especially the Andes, clearly illustrate the role of topography in regional climate, as ocean winds that encounter coastal mountains are forced to rise even higher, cooling to the point that they release most of their moisture in the form of rain and snow. The high precipitation over the Andes feeds the rivers that water the lowlands east and west of the mountains, most notably the Amazon. However, mountains also create a **rain shadow** effect. When this happens, winds passing over mountains from the coast to the interior lose their moisture and become warmer and drier as they descend to the interior, creating arid conditions to the leeward of mountain ranges (see Figure 1.9, p. 11). This is the phenomenon that creates the dry region of southern Argentina, known as Patagonia, to the east of the Andes; creates the drier regions of Chihuahua in northern Mexico in the lee of the Mexican Sierra Madre; and affects the leeward side of some Caribbean islands. Latin America also provides a classic example of how the temperatures of the ocean can influence the climate of adjacent landmasses. Winds flowing across the very cold ocean that moves northward off the coast of Peru and Chile pick up very little moisture. This exacerbates the already dry conditions promoted by descending air over the tropics and forms the Atacama Desert, one of the driest spots on Earth.

Altitudinal zonation is a classification of environment and land use according to elevation based mainly on changes in climate and vegetation from lower (warmer) to higher (cooler) elevations. In Latin America, these zones are defined in a classification of mountain environments: *tierra caliente* (warm land); *tierra templada* (moderate land); and higher *tierra fría* (cold land). The very high altitudes are called the *tierra helada* (icy land). Each of these zones is associated with characteristic vegetation types and different agricultural activities (**Figure 7.5**). Communities locate fields at different elevations to adapt to and to take advantage of specific climatic and soil conditions. At higher altitudes, potatoes grow and animals graze; at lower altitudes, grains such as wheat and corn grow; and finally, vegetables and fruit are found at lower levels with more tropical climates. Global warming is shifting these zones upward in many parts of the tropics.

Climatic Hazards and Patterns Storms are a formidable force in this region. In October 1998, Hurricane Mitch dumped a year's worth of rain (about 1.2 meters, or nearly 4 feet) on Central America in 48 hours (**Figure 7.6a**, p. 268). Flash floods and mudslides on

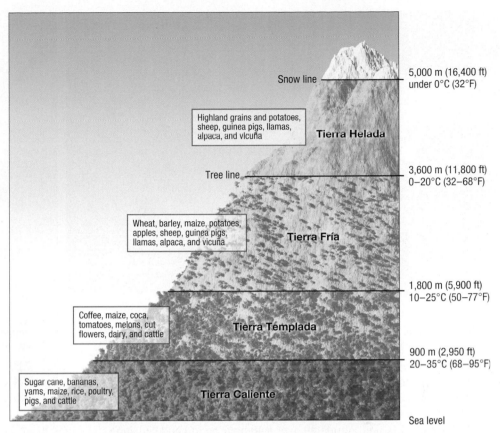

Highland grains and potatoes, sheep, guinea pigs, llamas, alpaca, and vicuña

Tierra Helada

Snow line — 5,000 m (16,400 ft) under 0°C (32°F)

Wheat, barley, maize, potatoes, apples, sheep, guinea pigs, llamas, alpaca, and vicuña

Tierra Fría

Tree line — 3,600 m (11,800 ft) 0–20°C (32–68°F)

Coffee, maize, coca, tomatoes, melons, cut flowers, dairy, and cattle

Tierra Templada

1,800 m (5,900 ft) 10–25°C (50–77°F)

Sugar cane, bananas, yams, maize, rice, poultry, pigs, and cattle

Tierra Caliente

900 m (2,950 ft) 20–35°C (68–95°F)

Sea level

▲ **FIGURE 7.5 Altitudinal Zonation** The climate and vegetation in mountainous regions such as the Andes creates vertical bands of ecosystems and provides a range of environments for agricultural production. The zones shown here are typical of mountain zones near the Equator, but cooler zones start lower down the mountains at latitudes further from the Equator.

deforested slopes left nearly 10,000 people dead, almost 20,000 missing, and more than 2.5 million temporarily dependent on emergency aid. Honduras, the second-poorest nation in the Western Hemisphere, was the hardest hit. Of the 6 million people living in Honduras, nearly 2 million were affected by the storm, 1 million lost their homes, and 70% of the country's productive infrastructure was damaged or destroyed. This disaster created immediate food shortages and decimated vital export crops such as bananas, coffee, and shrimp, which made up half the country's annual export revenue.

Nature creates hazards, but it is humans who create *vulnerability* by creating social inequality, environmental degradation, and unequal access to resources. In countries like Honduras and Haiti, which have high levels of poverty and where many people live and farm on steep slopes or in flood-prone areas, it is difficult for the poor to recover from disasters. These vulnerable populations lack the money and resources to rebuild their homes and livelihoods. In 2008, a series of hurricanes destroyed 70% of the crops and killed more than 800 people in Haiti. Several other hurricanes have caused serious damage in the region since Mitch, including Hurricanes Stan and Wilma in 2005.

The climate of the region does not remain constant from year to year. One of the most important causes of climate variability is **El Niño**—a periodic warming of sea-surface temperatures in the tropical Pacific off the coast of Peru. El Niño results in worldwide changes in climate, including droughts and floods. It brings warmer and wetter winds to the coasts of Peru and Ecuador, with

(a)

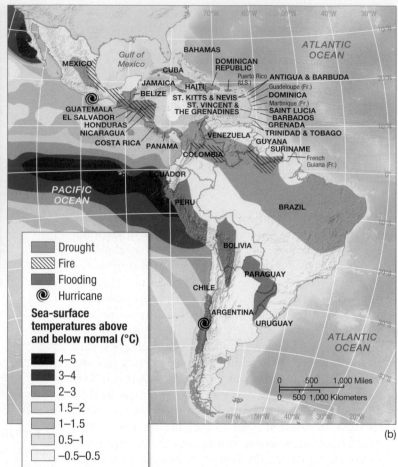

Drought
Fire
Flooding
Hurricane
Sea-surface temperatures above and below normal (°C)
4–5
3–4
2–3
1.5–2
1–1.5
0.5–1
−0.5–0.5

(b)

▲ **FIGURE 7.6 Climate Hazards** (a) Hurricane Mitch was one of the strongest hurricanes on record when it hit Central America after moving across the Caribbean in October 1998. With winds of more than 290 km/hour (180 mph), rains intensified over Central America's coastal mountains, causing severe flooding, especially below deforested hill slopes. Hurricane Mitch inundated vast areas of Honduras, dropping about 1.2 meters (4 feet) of rainfall over a 48-hour period and causing tremendous damage in the capital Tegucigalpa, shown in this photo. (b) When an El Niño occurs, the ocean warms off the coast of Peru and causes heavy rain and flooding along the usually dry Peruvian coast. The Pacific warming is linked to other changes in atmospheric circulation that produce drought in northeast Brazil, the Caribbean, and the altiplano; floods on the Parana River; and droughts and fires in Mexico and Central America.

high rainfall and flooding. When this happens, the sensitivity of the global atmospheric circulation is such that the links between Pacific Ocean temperatures and conditions elsewhere produce droughts in northeast Brazil, floods in southern Brazil and northern Mexico, and fewer Caribbean hurricanes as well as droughts in southern Africa, Australia, and Indonesia (**Figure 7.6b**). In some years, the ocean off Peru gets colder than normal, producing a contrasting global pattern called **La Niña** (the periodic abnormal cooling of sea-surface temperatures in the tropical Pacific off the coast of Peru). La Niña causes floods in northeast Brazil, drought in northern Mexico, and more intense Pacific hurricanes.

Mexico City's most infamous environmental problem is air pollution, which is intensified by the climate of the high basin where the city is located. Polluted air is often trapped in this basin by the surrounding mountains and by inversions (where warm air traps cold air near the ground). Mexico City is at more than 2,000 meters (6,000 feet) and has a dry climate for much of the year with rains only washing out the build up of pollutants in the rainy season. At this altitude, fuel burns less efficiently and humans must breathe more air because of the lower oxygen levels. All of these factors combine to result in pollution levels that are dangerous to human health more than 100 days a year (see Figure 7.31, p. 299). Thousands of automobiles, trucks, and buses, many with inadequate emission controls, are

responsible for about 75% of the air pollution. Dust, fires, industrial plants, and miscellaneous energy create the remainder. Although the government has implemented air pollution controls, the continuing growth of the city and of car ownership has prevented any significant decline in pollution levels. Other cities in the region, such as Santiago, Chile, have similar air pollution problems.

Climate Change Latin America and the Caribbean are already experiencing the effects of global warming, especially in the Andes, where glaciers and snowfields are shrinking (**Figure 7.7**). This retreat is creating serious problems for communities that rely on water from ice and snow for irrigation and drinking water and for larger-scale hydroelectric schemes, where dry conditions mean electricity cutoffs in major cities. Climate change may also cause more intense hurricanes. The Amazon forest is an important "sink" for greenhouse gases because the vegetation takes up the carbon dioxide that would otherwise warm the planet. The Amazon is also a critical rainfall source, generating rain from evaporated moisture. Unfortunately, there are some indications that higher levels of carbon dioxide and droughts associated with global warming, combined with deforestation, are drying the Amazon and reducing its ability to absorb carbon. If this continues, the Amazon could become a globally significant so-called **tipping point**. Tipping points areas where climate could shift suddenly as a result of climate and environmental change. In the Amazon, drying, fires, and associated carbon emissions could flip the vegetation from forest to grasslands and make the area a source of emissions rather than sink for carbon.

▲ **FIGURE 7.7 Andean Glaciers at Risk** Climate changes caused by increasing levels of greenhouse gases from fossil fuels and deforestation have already produced higher average temperatures in mountain areas. Andean glaciers and snow cover are shrinking, and the rivers that depend on snow and ice melt are drying up, creating problems for irrigated agriculture, urban drinking water, and hydroelectric power generation. Here the city of La Paz, Bolivia, is overlooked by the icy peaks and glaciers of Illimani, a mountain reaching 6,438 m (21,122 ft)

▲ **FIGURE 7.8 Carbon Credits for Forests** This tree nursery in Chiapas, Mexico supplies a forest carbon project as part of an international program to reduce greenhouse gas emissions. The trees take up carbon dioxide, compensating for emissions and generating credits that countries or companies can buy to meet their emission reduction responsibilities.

Although there are many traditional adaptations to climate variability in Latin America—including complex irrigation systems developed by the Aztecs, Maya, and Incas (that will be discussed later in the chapter)—global warming may test the limits of this region if it brings drier and warmer conditions. The islands of the Caribbean are also at risk from climate change and warmer temperatures because of the potential increase in the intensity of hurricanes and the impact of warmer oceans on coral reefs. In addition, in the Caribbean, sea-level rise threatens coastal ecosystems and settlements (see "Emerging Regions: AOSIS" in Chapter 11, p. 444).

In Latin America, efforts are being made to prevent dangerous climate change by reducing greenhouse gas emissions from fossil-fuel burning and from deforestation. For example, Mexico and Brazil have both promised to cut their greenhouse gases by one-third by 2020 by switching energy sources to gas, biofuels, and renewables and by protecting forests. Latin American countries also receive financial assistance for reducing emissions through the **international carbon markets**. In Europe, countries and companies are required to reduce emissions and can earn "carbon credits" toward meeting their reduction obligations by investing in reducing emissions in developing countries. For example, the Norwegian and British governments have provided funding to replant and protect forests in Latin America as a way to earn credits toward their international promises to reduce emissions (**Figure 7.8**). **REDD**—Reducing Emissions from Deforestation and forest Degradation—programs allow countries and companies with high emissions to get credit for emission reductions by providing financial and other incentives for forest protection in the developing world. For example, California is allowing companies in that state to meet emission reduction targets by participating in forest protection programs in the Amazon.

APPLY YOUR KNOWLEDGE If you hear on the news that an intense El Niño or bad hurricane season has been predicted, what areas and people of Latin America and the Caribbean would you expect to be affected, and how?

Geological Resources, Risk, and Water Management

The physical landscape of Latin America varies widely and includes striking mountain ranges, high plateaus, and enormous river networks (**Figure 7.9**, p. 270). The two largest-scale physical features in Latin America are the Andes Mountains and the Amazon Basin, both easily seen from space (see Figure 7.2, p. 265). The Andes are an 8,000-kilometer-long (5,000-mile-long) chain of high-altitude peaks and valleys that for the most part run parallel to the west coast of South America. Their highest peak, Aconcagua, is 6,960 meters (22,830 feet). Major South American rivers such as the Plata and the Orinoco provide transport routes into the interior as well as water resources for agriculture and hydroelectricity generation. The two major deserts—the Sonoran and the Atacama—are located along the Pacific coasts of Mexico and Chile.

Other important physical features include the mountainous spines of Mexico and Central America and the high-altitude plateaus that lie between or next to the mountain ridges. In the Andes, the plateau is called the **altiplano**. This subregion, as well as the Mexican Plateau, or Mesa Central, is an important area of human occupation because it provides flatter, cooler, and wetter environments for agriculture and settlement than do the adjacent steep-sloped mountains, dry lowland deserts, and humid lowlands.

The Caribbean Islands and the Caribbean coast of Central America have large areas of limestone geology, where water tends to flow underground and create large cave systems, such as those in the Yucatán Peninsula of Mexico and in Puerto Rico. Coral reefs, a key feature of the Caribbean landscape, are created when living coral organisms build colonies in warm, shallow oceans. These reefs, hosts to myriad other marine animals, are fragile ecosystems that are easily damaged by boats, divers, pollution, and environmental change.

▲ **FIGURE 7.9 Physical Regions and Landforms of Latin America and the Caribbean** The Amazon Basin and the Andes Mountains are the two largest physical features in Latin America.

▲ **FIGURE 7.10 Natural Disasters** The 2010 earthquake in Haiti killed 300,000 people and destroyed buildings and infrastructure in Haiti's capital, Port au Prince. Haiti's population was especially vulnerable to the earthquake because of widespread poverty.

Earthquake and Volcanic Hazards Many of the mountain areas and island chains in the region are the result of a long history of tectonic activity in Latin America and the Caribbean, which are situated at the intersection of several major tectonic plates (see Figure 1.14, p. 16). Earthquakes and volcanoes occur near the plate boundaries in parts of the Caribbean, Central America, the Andes, and Mexico.

Volcanic activity poses threats to human activity when eruptions and ash destroy crops and lives. Shifting tectonic plates have produced devastating earthquakes that have ravaged the capital cities of Mexico City, Mexico; Managua, Nicaragua; Guatemala City, Guatemala; and Santiago, Chile. Such natural disasters cannot be blamed solely on geophysical conditions but also on vulnerability. The greatest damages occur when people live in unsafe houses or on unstable slopes because they lack the money or power to live in safer places, cannot afford insurance, or are unable to obtain warnings of impending natural disasters, such as volcanic eruptions and hurricanes. The Haitian earthquake of 2010 was one of the most devastating in recent history, killing more than 300,000 people and displacing over 1 million residents. It was followed by a massive worldwide rescue and relief operation, which, as of 2012, had still not rehoused many victims or cleared the rubble of collapsed buildings (**Figure 7.10**).

Mineral Resources The mineral wealth of Latin America is typically found where crustal folding brings older rocks near the surface. Major precious-metal mining districts include the Peruvian and Bolivian Andes, where mountains of silver were excavated in the Spanish colonial period at Cerro de Pasco and Potosí, and where lead, zinc, and tin are still important; the silver region of the Mexican highlands; and the gold and iron mines at Carajas on the edge of the Brazilian Plateau. World-class iron deposits are found on the southern edge of the Brazilian Highlands at Itabira, on the northern edge of the Guiana Highlands at Cerro Bolívar, and in northern Mexico. Copper is the geological treasure of the southern Andes, especially northern Chile (**Figure 7.11a**), and is also important in northern Mexico. The shores of the Caribbean, including the Guianas and Jamaica, have deposits

of bauxite (a mineral used in the aluminum industry). These minerals, especially gold and silver, were the foundation of the European colonial economies and currently dominate the export economies of countries such as Chile and Bolivia. They are a focus of foreign interference and ownership and have often transformed local labor and environmental conditions.

The other critical resources associated with Latin America's geology are fossil fuels. Coal is found in northern Mexico, Colombia, Brazil, and Venezuela. The earliest oil booms and later gas developments occurred in Venezuela around Lake Maracaibo and on Mexico's Gulf Coast (**Figure 7.11b**). In all of Latin America's oil

(a)

(b)

▲ **FIGURE 7.11 Minerals in Latin America** (a) The world's attention focused on copper mining in Chile in 2010 when an accident at the Copiapó mine trapped 33 miners underground for more than 2 months before their eventual rescue. In this photo one of the miners celebrates after reaching the surface. (b) Oil development has been associated with serious pollution. The Ixtoc well spilled 3 million barrels into the sea off the coast of Campeche, Mexico, in 1979. In the Americas, the Campeche spill disaster has been equalled only by the Deepwater Horizon spill in the Gulf of Mexico in 2010.

regions, environmental pollution has been a serious problem, leading to waterways contaminated with waste oil, widespread ecosystem damage, and serious health problems among local residents. The most recent developments are in the Amazon, where oil and gas were discovered in 1967. The Amazonian oil deposits are found mostly in remote forest areas, where land rights of Indigenous people are not secure, and as a result, conflicts have erupted between Peru and Ecuador and among governments, corporations, and Indigenous groups.

Water Resources The three largest river basins in Latin America are the Amazon, the Plata, and the Orinoco, which all flow to the Atlantic Ocean (see Figure 7.1, p. 264). The rivers in the Plata Basin (including the Paraná, Paraguay, and Uruguay Rivers) originate in the Andes and the Brazilian Highlands. The Plata system has become a major source of energy through large hydroelectric dams such as the Itaipu Dam on the Paraná, and there are controversial plans to dam and divert the water resources of the Pantanal—a vast and ecologically diverse wetland east of the Andes that lies mostly in Brazil south of the Amazon Basin. The Orinoco drains the *llanos*, the grasslands of Colombia and Venezuela. The Amazon tributaries flow downward and eastward from the Andes into an enormous river network that covers a basin of more than 6 million square kilometers (2.3 million square miles) and includes vast rainforests. The Amazon Basin includes the river itself and the surrounding landscape, about two-thirds in Brazil and parts of Peru, Ecuador, Bolivia, Colombia, Venezuela, Guyana, Suriname, and French Guiana. The Amazon River and its tributaries carry 20% of the world's freshwater and provide transport, sediment, and fish that support the agriculture, diets, and mobility of the people of the basin.

Latin America has several large freshwater lakes, among them Lake Nicaragua in Nicaragua and Lake Titicaca at the border of Bolivia and Peru. Major waterfalls such as Iguaçu Falls—where Brazil, Argentina, and Paraguay meet (**Figure 7.12**)—and Angel Falls, Venezuela—the tallest waterfall in the world at a height of 985 meters (3,230 feet), which falls off the flat-topped mountain of Auyantepui—have become increasingly popular tourist destinations.

Ecology, Land, and Environmental Management

The diversity of Latin America's physical environments has produced astounding **biodiversity**—a large number of different species. Latin America's biodiversity is substantial because of the size of the continent, its range in climates from north to south, the altitudinal variations within short distances, and its comparatively long history of fairly stable climates and isolation from other world regions. Many tourists are attracted to the colorful birds, interesting animals, and verdant plants of the tropical environment (**Figure 7.13a**).

Desert ecosystems, such as in the Atacama Desert of northern Chile and Peru, are associated with drier climates (**Figure 7.13b**). Between the moist forests and dry deserts lie ecosystems where alternating wet and dry seasons produce vegetation ranging from scattered woodlands to dry grasslands. Grasslands are also found at higher altitudes, where there is not enough precipitation or temperatures are too low to support highland forests. In Argentina, the *pampas* grasslands cover more than 750,000 square kilometers (300,000 square miles) and have become important to the cattle economy. Other large grassland ecosystems include the *llanos* of Colombia and Venezuela and the *cerrados* of Brazil. The high grasslands of the Andean altiplano provide habitat for grazing animals such as the llama, wild guanaco, and vicuña. The long coasts of Latin America and the islands of the Caribbean include about 50,000 square kilometers (about 19,000 square miles) of mangrove ecosystems, or about 25% of the world's total (**Figure 7.13c**).

The wetter climates of Latin America are associated with magnificent forest ecosystems, including the tropical rainforests of the Amazon, Central America, and southern Mexico and the temperate rainforests of southern Chile. The Amazon forest ecosystems are

▼ **FIGURE 7.12 Iguaçu Falls** Iguaçu Falls, on the Paraná River, where Brazil, Argentina, and Paraguay meet, has become a major tourist destination.

(a)

(b)

(c)

◄ **FIGURE 7.13 Latin American and Caribbean Ecosystems** Ecosystems in this region range from forests and grasslands to deserts and coastal mangroves. (a) The Pantanal wetlands and grasslands of Brazil support rich biodiversity that includes the capybara—the world's largest rodent. (b) The Atacama Desert of Peru and northern Chile is one of the driest locations on Earth. Here the desert meets the sea in the Paracas National Reserve in Peru. (c) Mangrove ecosystems border many coasts in tropical Latin America and the Caribbean, providing protection from storms and supporting important fisheries.

notable for the sheer number of species found within small areas of forest. The warm and wet climates of much of Latin America and the large diversity and prevalence of pests and diseases have limited development in these areas. For example, malaria (see Chapter 5) is endemic in much of the Amazon Basin and lowland Central America.

The most dramatic transformation of nature by early people in Latin America was plant and animal domestication. Starting more than 10,000 years ago, wild plants and animals in the region were domesticated into cultivated or tamed forms through selective breeding for preferred characteristics. Many of the world's major food crops were domesticated by Indigenous Latin Americans. These include the staples of maize (corn), manioc (cassava), beans, and potatoes as well as vegetables and fruits, such as tomatoes, peppers, squash, avocados, and pineapples (**Figure 7.14**, p. 274).

Tobacco, cacao (chocolate), vanilla, peanuts, and coca (cocaine) were also domesticated in Latin America. In dry areas, people tried to ensure water supplies by building small dams and channels to bring water to crops. Latin America has very few Indigenous domesticated animals; however, the camelids (llama and alpaca) were tamed and bred for wool, meat, and transport, and dogs (similar to the Chihuahua) and guinea pigs were also bred for pets and meat.

APPLY YOUR KNOWLEDGE List the main ingredients in your typical weekly diet. How many of the products that you regularly eat are based on crops originally domesticated in Latin America? Did you eat a meal in the course of the past week that *did not* contain foods originally from Latin America?

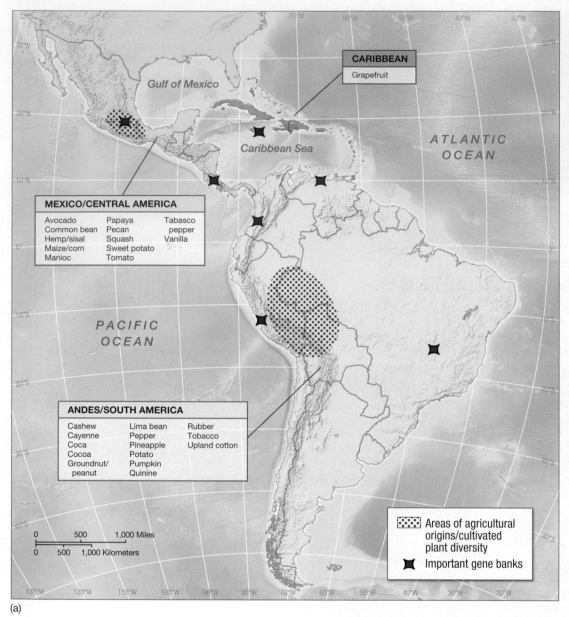

(a)

CARIBBEAN

Grapefruit

MEXICO/CENTRAL AMERICA

Avocado	Papaya	Tabasco
Common bean	Pecan	pepper
Hemp/sisal	Squash	Vanilla
Maize/corn	Sweet potato	
Manioc	Tomato	

ANDES/SOUTH AMERICA

Cashew	Lima bean	Rubber
Cayenne	Pepper	Tobacco
Coca	Pineapple	Upland cotton
Cocoa	Potato	
Groundnut/	Pumpkin	
peanut	Quinine	

Areas of agricultural origins/cultivated plant diversity

Important gene banks

(b)

(c)

◀ **FIGURE 7.14 Domestication of Food Crops in Latin America** (a) Latin America has two important centers of domestication in Mexico and the Andes, which are indicated on the map by the small red dots. The red bursts on the map denote research centers, where attempts are being made to preserve genetic diversity in crop gene banks. (b) Maize (corn) was domesticated in Mexico from a wild grain called teosinte. Bred for a variety of microenvironments and tastes, traditional maize has many shapes and colors. (c) The various colors and shapes of potatoes grown in the Andes illustrate the diversity of types domesticated in these regions.

Maya, Incan, and Aztec Adaptations to the Environment As in other regions of the world, increased yields from domesticated crops in ancient Latin America created a surplus that permitted the specialization of tasks, the growth of settlements, and ultimately the development of highly complex societies and cultures. Complex societies in this region included the great Maya, Inca, and Aztec Empires (**Figure 7.15**). These groups all modified their environments to increase agricultural production and to exploit water, wood, and mineral resources. In some cases—most notoriously that of the Maya civilization—people placed so much pressure on regional landscapes that environmental degradation may have precipitated social collapse. The Maya occupied the Yucatán Peninsula as well as a considerable portion of Guatemala and Honduras from about 1800 B.C.E. and reached a peak of population and political development in about 250 C.E. But in about 800 C.E., the great Maya cities such as Copán, Palenque, Tulum, and Tikal were abandoned, and overall population declined dramatically.

Many scholars believe that one reason for the Maya collapse was their overuse of the soils. Faced with rapid declines in the fertility of soils after clearing the rainforest, the Maya adapted by burning the forest to capture the nutrients in the trees through the ash. They moved on to clear another patch and then another once the declining fertility of each cleared area resulted in reduced yields. This adaptation to rainforest environments mirrors farming methods in other parts of the world and is called slash-and-burn (see Chapter 5, p. 188), or swidden, agriculture. The Maya also developed methods for growing crops in wetland areas by building raised fields that lifted plants above flooding, but at the same time took advantage of the rich soils, frost protection, and reliable moisture of wetland environments.

There is evidence that the Maya cleared vast areas of forest. Large-scale forest clearing has been linked to regional changes in climate, with increases in temperature and decreases in rainfall that threaten agriculture. Soil erosion, droughts, and declining soil fertility in Maya times, are believed to have contributed to a decline in the

▲ **FIGURE 7.15 The Maya, Inca, and Aztec Empires** This map shows the extent of pre-European empires in Latin America.

amount of food available to feed the large population in the latter years of the empire, especially when rulers were demanding food as tributes. Signs of nutritional stresses have been detected in human skeletons from the period of collapse.

The Inca Empire, with lands stretching from northern Ecuador to central Chile, dominated the Andean area from about 1400 C.E. The Incas responded to the difficulties of living in a mountain environment in a variety of ways, most notably by constructing many miles of agricultural terraces so they could farm the steep slopes of the Andes (see Figure 7.3, p. 266). The Aztecs, who settled in central Mexico in the 1300s, constructed an extensive network of dams, irrigation systems, and drainage canals in the basin of Mexico to cope with the highly seasonal and variable rainfall pattern. They also developed the *chinampa* agricultural system, which involves building islands of soil and vegetation in lake and wetland environments to allow for cultivation. Evidence suggests that the Aztecs cleared the forests in the basin of Mexico, which may have contributed to a drop in the water table and to a resulting water crisis that led to the abandonment of some settlements.

The widespread evidence of these modifications from Latin American history has been used to debunk what geographer William Denevan has called the **pristine myth**. This is the erroneous belief that prior to the Europeans' arrival in 1492, the Americas were mostly wild and unmodified by the human inhabitants of the region. In fact, large areas were cultivated and deforested by Indigenous populations. The environmental adaptations and impacts of the Maya, Incas, and Aztecs still echo in the traditional technologies used in some regions of Latin America and in the continual efforts to use technology to benefit from the physical environment and avoid its hazards. But the previous inhabitants' overuse of their environment may also serve as a warning about the implications of current patterns of widespread deforestation, overuse of the land, and depletion of water resources.

APPLY YOUR KNOWLEDGE List three ways that the impacts that the Maya, Incas, and Aztecs made on their environment might serve as warnings or offer solutions for current environmental issues in the region.

The Fate of the Forests The forests of Latin America and the Caribbean host a significant share of the world's biodiversity, provide food and shelter for forest dwellers, and provide **ecosystem services**, such as water and soil protection, as well as the removal of greenhouse gases from the atmosphere. These services benefit countries in the region and the planet as a whole.

Covering more than a half-billion hectares (about 1.2 billion acres), the forest of the Amazon Basin contains water, forest, mineral, and other resources of great value, yet has had relatively low population density until recent years (see "Visualizing Geography: Amazon Deforestation," p. 276). The colonial image of the Amazon Basin varied. Sometimes it was viewed as a tropical Eden with untapped resources; at other times it was seen as an impenetrable, disease-ridden, jungle of savage tribes. The region was of botanical interest and held little economic attraction until the late 19th century, when development of the automobile industry in the United States and Europe caused an explosion in the demand for rubber. Rubber is obtained by tapping the latex sap of rubber trees, which at the time were found only in South America. Local rubber tappers, or *seringueiros*, sold the rubber to middlemen to meet the needs of the industrialized nations. The middlemen traded in turn with the "rubber barons," who constructed enormous mansions and a magnificent opera house in the Amazonian port of Manaus. The end of the rubber boom in Latin America is said to have occurred when Henry Wickham shipped thousands of rubber tree seeds from Brazil to Kew Gardens, in England. The seeds were cultured and the seedlings exported to Southeast Asia, where they became a successful cash crop. The success of the more efficient Asian plantations, especially in Malaysia, drove the Amazon into decline because Brazilian trees were too susceptible to disease when grown on plantations.

Chile has about one-third of the world's remaining areas of relatively undisturbed temperate forests. The forests in Chile are home to towering old-growth *alerce* trees, which can live for up to 3,000 years, and are similar to the redwoods in the western United States (both species rely on the heavy seasonal rains from westerly storms). But these forests are threatened by the international demand for wood and paper that supports a Chilean forest industry. Exports have expanded and are important to the national economy. The industry logs ancient forests and replaces them with plantations, where fast-growing, nonnative species are grown. Conservationists have been able to obtain some protection for the old-growth forests by purchasing land and supporting environmental legislation that establishes protected areas. Moves to promote sustainable forestry and certify

VISUALIZING GEOGRAPHY
Amazon Deforestation

Understanding the rate and causes of Amazonian deforestation is a priority for many scientists and for the Brazilian government. Development in the Amazon frontier became a focus of Brazilian government policy in the 1970s. At that time, settling the Amazon was seen as a way to reduce the pressure posed by impoverished peasants as well as a way to secure national territory through settlement and colonization. Several highways were built across the Amazon, including the Trans-Amazon from Recife to the Peruvian border and the Polonoreste from Brasilia to Belém (**Figure 7.1.1**). Brazil gave landless peasants title to plots of land if they promised to develop them productively, and peasants migrated in thousands along the

new roads. In addition, tax breaks for large landholders encouraged ranching.

Estimates of the rate of Amazon Basin deforestation do not always agree because of differences in the way in which forests are defined and satellite images are analyzed. Clouds and smoke prevent accurate assessments in some regions. But the general consensus is that perhaps 15% of the Amazon forest has been cleared since 1900. In satellite images of the region, the process of Amazonian deforestation can be clearly seen in the networks of new roads and associated forest clearance (**Figure 7.1.2**). Satellite images also show the thousands of fires that are set each year to clear land. These fires produce a dense layer of smoke that closes airports and chokes local residents.

Photos and images also illustrate how the pattern of development and deforestation varies spatially, with some remote areas still relatively untouched and others along roads and around cities almost completely transformed to ranching and crop production At the southern margins of the Amazon Basin, global and national markets for soybeans are driving land use changes but cattle ranching is still the largest overall cause of deforestation. Other threats to the forest include the construction of dams that flood extensive areas of forests and mining, which uses mercury, a hazardous chemical element that pollutes ecosystems and causes health problems.

The fate of the Amazon has attracted global attention, drawn by scientists and

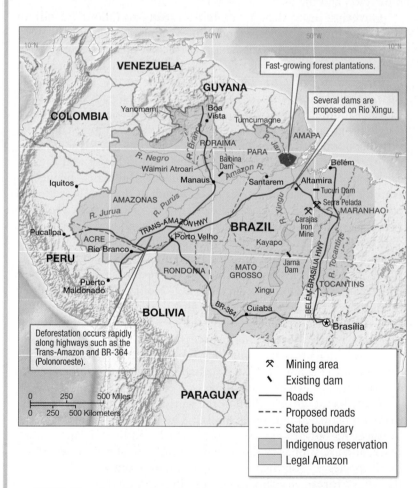

▲ **FIGURE 7.1.1 Development in the Amazon** Development projects that are causing deforestation and other environmental problems in the Amazon Basin include major roads, such as the Trans-Amazon Highway; dams, such as Tucuri; and agricultural development around cities. The Amazon Basin has many Indigenous groups, some of whom have gained reserve status for their lands.

▲ **FIGURE 7.1.2 Satellite Image of Deforestation** A satellite image from the southern region of the Brazilian Amazon in the state of Rondônia shows how the forest is cleared as roads and people move into the area. The cleared areas are lighter brown, and the forests are green.

environmental organizations concerned about the impacts of such large-scale forest loss on global biodiversity and climate. The Brazilian government has responded by removing some of the tax breaks for development, intensifying monitoring and control of deforestation, and establishing parks and reserves to protect the forest. One of the best-known reserves (west of Rio Branco in the state of Acre) is named for Chico Mendes, a rubber tapper who organized resistance to deforestation by large ranchers and was murdered in 1988. He pushed for the establishment of areas that were protected for appropriate extractive uses called **extractive reserves**.

The Brazilian Remote Sensing Agency's (INPE) official estimates of annual forest loss record a slowing of deforestation since 2005 (**Figure 7.1.3**). This is partly due to the creation of Indigenous reserves, parks, and protected areas; extractive reserves; and sustainable use areas (**Table 7.1.1**). ■

TABLE 7.1.1 Forest Protection		
Type of protection	Percent of Brazilian Amazon	Reduced deforestation between 1997 and 2008 by %
Indigenous reserve	23.6%	31%
Park	9.2%	57%
Sustainable use	12.2%	14%
Not protected	55%	0%

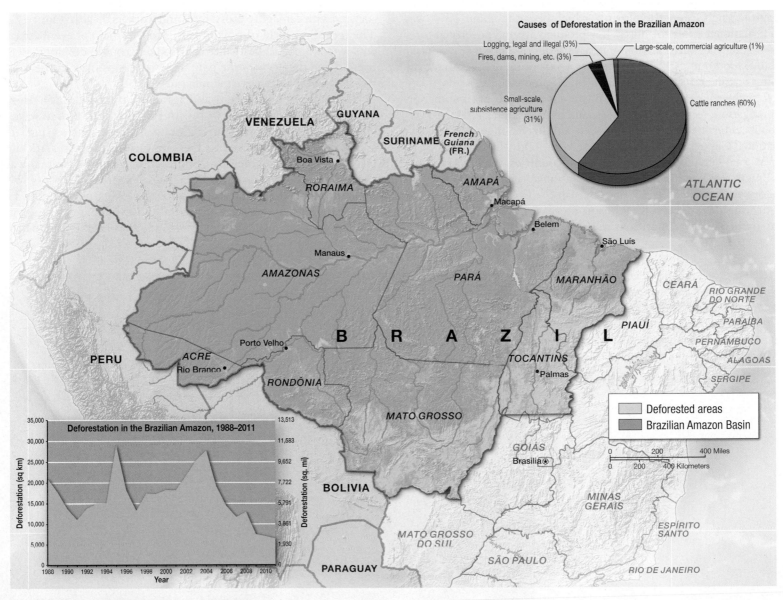

▲ **FIGURE 7.1.3** Deforestation in Brazilian Amazon

and label Chilean exports as sustainably produced may also help protect the remaining ancient rainforests.

Many of the Caribbean Islands were also once covered by tropical forests in precolonial times, but today, many are left with only small patches of forest. For example, Cuba, Haiti, and the Dominican Republic have lost most of their original forests. Clearing for export plantations, timber, and fuel wood has caused deforestation and biodiversity loss and increased flood and landslide risks on the steep slopes of many islands. Tree planting and abandonment of agricultural land has restored some forests, especially in Cuba.

The Green Revolution and Agriculture Increasing yields on existing land rather than clearing new fields helps prevent deforestation and food insecurity. For the first part of the 20th century, the yields of most agricultural crops in Latin America and the Caribbean were very low. Farmers with small plots of land in this region could not produce enough to feed themselves, let alone to sell in the market. As population and urban-consumption demands increased, countries such as Mexico and Brazil had to import basic food crops, such as wheat and corn. The **Green Revolution**, which began in about 1950, was an international effort to address low productivity and poverty in rural areas through the introduction of new varieties of crops that produced far more food but also required higher levels of industrial inputs, including fertilizer, water, and pesticides. Higher-yielding seeds of key crops, such as wheat, rice, and corn, were introduced in combination with irrigation, fertilizers, pesticides, and farm machinery. Mexico was a global center for Green Revolution technology and home to the International Center for Improvement of Maize and Wheat (CIMMYT) near Mexico City. Other Latin American countries, such as Argentina and Brazil, also promoted Green Revolution agricultural modernization for key crops, such as rice and soybeans.

Although the Green Revolution increased crop production in many parts of Latin America, it was not an unqualified success. The Green Revolution increased dependence on imports of chemicals and machines from foreign companies and thus contributed to the debt problem. Its benefits accrued to wealthy farmers in irrigated regions who could afford the new inputs, while poorer farmers whose land was watered only by rainfall fell behind or sold their land. In some cases, such as in southeastern Brazil, machines replaced workers, thus leading to unemployment. Green Revolution technology and training also tended to exclude women, who play important roles in food production. The most serious criticism of the Green Revolution was that it contributed to the worldwide loss of genetic diversity: a wide range of local crops and varieties were replaced with a narrow range of high-yield varieties of just a few crops. Planting single varieties over large areas (monocultures) also made agriculture vulnerable to disease and pests.

The new agricultural chemicals introduced in the Green Revolution, especially pesticides, contributed to ecosystem pollution and worker poisonings, and the more intensive use of irrigation created problems of water scarcity and salinization (the buildup of salt deposits in soil). The increased use of imported pesticides on export crops is associated with damage to ecosystems and workers' health (**Figure 7.16**). The term the **circle of poison** refers to the chain of events that occurs when imported pesticides are used on crops in developing countries, which then export the contaminated crops back to the regions where the pesticides were manufactured.

Increasingly, both forests and agricultural land in Latin America today are being converted to **biofuel** production. Biofuels are energy sources derived from living matter. Crops such as sugar cane and corn can be converted to fuels by various processes, including fermentation. Vegetable oils (biodiesel) or crop and animal wastes can also be used as fuels. Brazil is a major exporter of ethanol, which has been commercially produced since the 1980s, mostly from sugar cane. Although biofuels can potentially slow climate change by reducing fossil fuel use, there are concerns that biofuel production is

▶ **FIGURE 7.16 Pesticide Use** A farmer spraying pesticides on tomatoes near Casapamba, Ecuador. Vegetables growing export are often given heavy doses of pesticides to ensure the produce looks of high quality, despite the risks to the local farmworkers.

driving deforestation, diverting land from food production, and consuming limited water and fertilizer resources.

Ecotourism and Conservation Latin America and the Caribbean have seen a boom in tourism that is often focused on the natural attractions of the region, such as coasts and rainforests. Environmentally oriented tourism, or **ecotourism**, aims to protect the environment and provide employment opportunities for local people. Costa Rica, for example, has won high praise for protecting 30% of its territory by setting it aside in biosphere and wildlife preserves (**Figure 7.17**). Costa Rica has more bird species (850+) than are found in the United States and Canada combined and more varieties of butterflies than all of Africa. Twelve distinct ecosystems contain more than 6,000 kinds of flowering plants and more than 200 species of mammals, 200 species of reptiles, and 35,000 species of insects. The payoff for Costa Rica is the escalating number of tourists who come to visit its active volcanoes, palm-lined beaches, cloud forests, and tropical parks. Ecotourism is the country's main source of foreign exchange, and over 2.2 million tourists visited in 2011 alone. Ecotourism has been a mixed blessing for some rural areas of Costa Rica. The benefits are not shared equally among residents, and some regions are becoming so crowded that environmental degradation is occurring.

The ecological diversity of Latin America also makes biological prospecting—**bioprospecting**—for new medicines and products with commercial uses a viable endeavor. For example, Costa Rica has signed agreements with multinational pharmaceutical companies, which give the companies rights to prospect and develop in return for turning over a share of profits to the Costa Rican government and to local people.

▲ **FIGURE 7.17 Conservation in Costa Rica** Costa Rica protects a high percentage of its land in parks.

Other new approaches to conservation include Payments for Environmental Services (PES), a program that pays residents to protect their local environment because preservation of that environment brings value to others. For example, upstream communities are paid to protect their forests because of the role the forests play in sustaining the flow of rivers for downstream water users, such as plantations and hydroelectric plants.

> **APPLY YOUR KNOWLEDGE** Consider the ways that the Green Revolution, biofuels, ecotourism, and payments for environmental services have changed environmental and social conditions in Latin America and the Caribbean. Keeping in mind the positive and negative impacts, do you think these programs and policies should be expanded across the region and to other world regions?

HISTORY, ECONOMY, AND TERRITORY

Much of Latin America shared the common experience of Spanish and Portuguese colonialism that was characterized by the dominance of the Spanish and Portuguese languages, religion (Roman Catholicism), and legal and political institutions. The Caribbean coast of Latin America and the Caribbean Islands had a more diverse colonial experience as a result of British, Dutch, and French takeovers. Colonialism in the region also resulted in the collapse of Indigenous populations; European migration and settlement; and European control of resource extraction, trade links, and other economic activity. Most of the mainland portion of the region became independent in the 19th century and was drawn into global trade relations, especially with Britain and the United States. In the 20th century, the Latin American region experienced rapid integration into global markets and the transition from revolutionary and military governments to democratic ones. Contemporary Latin America has some thriving economies and industrial regions, which serve as international centers of commerce.

Historical Legacies and Landscapes

Archaeological evidence from sites in Chile, Mexico, Peru, and Ecuador suggests a human presence in Latin America from at least 10,000 years ago. Early inhabitants were the descendants of migrants who came from Siberia into North America across the Bering Straight. The Indigenous people of Latin America lived by hunting, fishing, and gathering but also eventually domesticated some of the world's most important agricultural crops such as corn and potatoes. Complex societies emerged in the Andes (the Inca, from about 1200 C.E.), Mexico, and Central America (the Maya, from about 1800 B.C.E., and the Aztec, from about 1200 C.E.) as discussed earlier in this chapter. The Maya, for example, constructed monumental cities and had a written language, mathematics, and astronomical observatories.

The Colonial Experience in Latin America The integration of Latin America into the global system of political, economic, ecological, and social relationships began more than 500 years ago with the arrival of Spanish and Portuguese explorers at the end of the 15th century. The 15th and 16th centuries were a period of innovation in Europe, with changes in manufacturing technology and economic policy. Improvements in shipbuilding and navigation allowed Europe to explore and trade with other regions of the world, including Asia to the east and Africa to the south. Europeans also sailed west in search of new routes to Asia.

The most famous of these European explorers was Christopher Columbus, an Italian from the port city of Genoa. Under the sponsorship of Queen Isabella of Spain, Columbus was commissioned to search for new territory and trading opportunities on a western route to the Indies (as Asia was then called). Having set sail from southern Spain on August 3, 1492, with three small sailing ships—the *Santa María*, the *Pinta*, and the *Niña*—Columbus arrived in the Caribbean in October and landed on a small island in the Bahamas, to which he gave the name San Salvador (**Figure 7.18**). On his first voyage, Columbus also visited Cuba and another island that he called Hispaniola (now Haiti and the Dominican Republic). When the *Santa María* was wrecked on the north coast of Hispaniola, Columbus left behind 21 volunteers to found a colony and returned to Spain on the *Niña* with six local people, several parrots, and some gold ornaments.

Because he intended to establish permanent settlements, Columbus's second voyage in 1493 was much larger, with 17 ships and 1,500 men. This second excursion was met by hostility from Indigenous residents. On his third and fourth voyages, Columbus explored the island of Trinidad and the coasts of Venezuela and Central America. The Spanish wanted the new lands to be assigned to Spain rather than to its rival, Portugal, and negotiated the **Treaty of Tordesillas** (1494) to divide the world between Spain and Portugal along a north–south line about 1,800 kilometers (1,100 miles) west of the Cape Verde Islands. With the approval of Pope Julius II, Portugal received the area east of the line, including much of what is now Brazil and parts of Africa, and Spain received the areas to the west (see Figure 7.18).

Columbus was followed in subsequent decades by others seeking gold, territory, and additional resources. His successors included notable explorers (called *conquistadors*) such as Hernán Cortés, who landed in Mexico in 1519 and went on to conquer the Aztec Empire; and Francisco Pizarro, who seized control of the Inca Empire centered in Cuzco, Peru, in 1533. The Portuguese began their colonization with the landing of Pedro Álvares Cabral in 1500 at Porto Seguro in southeast Brazil.

Mexico City and Lima became the headquarters of the main Spanish administrative units (or viceroyalties) of New Spain (Mexico and Central America) and Peru (Andean and southern South America).

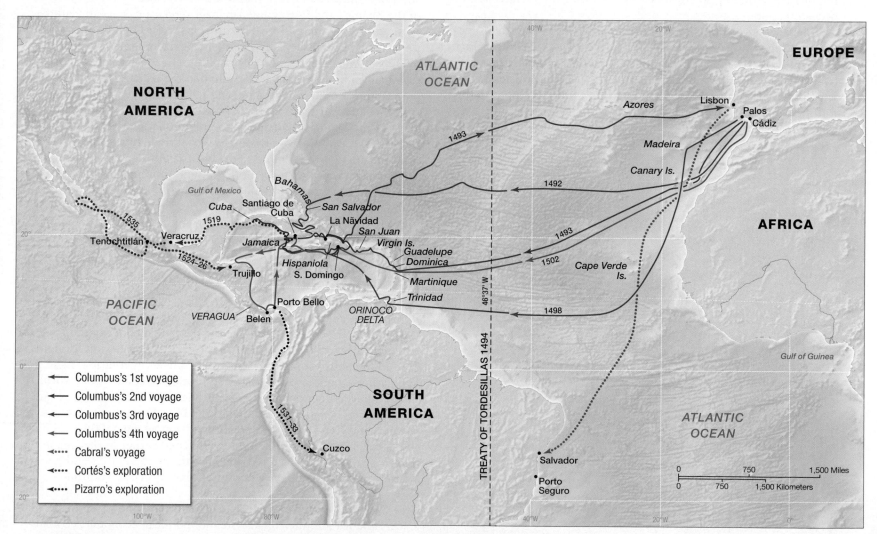

▲ **FIGURE 7.18 Colonial Voyages and the Treaty of Tordesillas** The map shows the major voyages and missions of Columbus, Pizarro, Cabral, and Cortés and the division of Latin America between Spain and Portugal under the Treaty of Tordesillas in 1494. The initial line set in 1493 was contested by Portugal and shifted farther west.

The colonial push into Latin America took place over at least two centuries, with some places incorporated earlier than others and some regions never really coming under complete colonial control because of their remoteness (the Amazon) and/or local resistance (parts of the Andes). Spain demanded that their colonies provide gold and silver for the Spanish crown, convert the Indigenous people to the Catholic religion, and become as self-sufficient as possible through the use of local land and labor. The Spanish crown demanded 20% of all mining profits—the so-called *Quinto Real*, or royal fifth.

The search for local labor to work in the mines and fields of the Spanish colonizers was frustrated by one of the most immediate and significant impacts of the European arrival in Latin America—the **demographic collapse**. After about 1500, the Indigenous populations of the Americas began to rapidly die off as a result of diseases introduced by the Europeans to which residents of the Americas had no immunity. Because of the long isolation of the Americas from other continents, Indigenous people lacked resistance and immunity to European diseases, such as smallpox, influenza, and measles. When they caught these diseases from Europeans and then from each other, mortality rates were very high, and up to 75% of the population of Latin America died in epidemics in the century following contact with the Europeans. This massive mortality demoralized local people and led to the abandonment of their settlements and fields. It also resulted in a scarcity of labor to work in the colonial mines, missions, and farms.

The introduction of European diseases into the Americas is just one example of the interaction between the two continents that historian Alfred Crosby has called the **Columbian Exchange**. The Columbian Exchange refers to the interchange of crops, animals, people, and diseases between the Old World of Europe and Africa and the New World of the Americas that began with the voyages of Christopher Columbus in 1492 (**Figure 7.19**). When the Spanish and other colonial powers arrived in new lands, they brought with them favorite plants and animals from their homelands or other areas that they planned to introduce into the new colonies. They also, unintentionally, brought diseases, weeds, and pests (such as rats) that stowed away on their ships. For their return voyages, the explorers and colonists collected species that they hoped could be sold or traded back in Europe and elsewhere. The colonists introduced crops and domesticated animals, such as wheat, cattle, fruit and olive trees, horses, sheep, and pigs as well as sugar, rice, citrus, coffee, cotton, and bananas from North Africa and the Middle East (see "Geographies of Indulgence, Desire, and Addiction: Coffee," p. 282). The colonizers took corn, potatoes, tomatoes, tobacco, and possibly syphilis back to Europe.

Over longer periods, these exchanges had additional effects. The clearing of land for European crops, such as wheat and sugar, and the overgrazing by cattle and sheep contributed to soil erosion and deforestation in Latin America and the Caribbean. Newly introduced

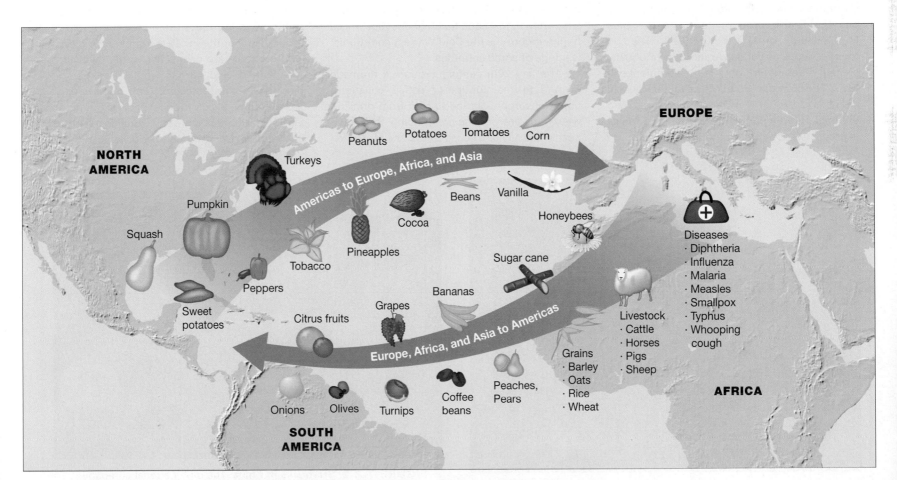

▲ **FIGURE 7.19 The Columbian Exchange** The arrival of Europeans in the Americas initiated the extensive exchange of crops, animals, and diseases between continents, including the introduction of European diseases to the Americas, which resulted in the deaths of millions of Indigenous people. The introduction of wheat, sugar, and livestock transformed the landscapes of Latin America, and European and African diets (and health) were altered forever with the arrival of corn, potatoes, and tobacco.

GEOGRAPHIES OF INDULGENCE, DESIRE, AND ADDICTION
Coffee

Coffee is one of the world's most popular drinks, stimulants, cultural traditions, and traded commodities. Over 2 billion cups of coffee are consumed daily, and coffee consumption has risen by 37% since 2000. Coffee was one of the most important crops introduced into Latin America and the Caribbean by colonial powers and today the region produces half of the world's coffee.

Coffee comes from the twin seeds, known as beans, inside the fruit of the coffee plant. Roasting and brewing of coffee spread from Ethiopia, where it was first domesticated, to Yemen, and, by the 16th century, across the Middle East. It then spread to Venice, Italy, and the rest of Europe, fostered by the coffee houses where men met to socialize and do business. During the 18th century, the Dutch planted coffee in their tropical colonies, especially Indonesia and Surinam; the French brought production to what is now Haiti; the Portuguese introduced its cultivation in Brazil; and the British brought coffee to the Blue Mountains of Jamaica (**Figure 7.2.1**). The two main varieties of coffee, Arabica and Robusta, contain different amounts of the stimulant caffeine. The stronger (up to 4% caffeine) Robusta coffee is grown at warmer temperatures, and the milder, more popular (1% caffeine) Arabica is grown in cooler, often mountainous areas.

Coffee has shaped politics and economies in Latin America and the Caribbean. Coffee is a labor-intensive crop, and slaves produced much of the region's coffee under brutal working conditions, which prompted a successful slave revolt in Haiti in 1791. In 19th-century Brazil, the expansion of coffee production, which relied on slave labor, caused deforestation but made coffee a cheaper beverage for workers in other parts of the world. In Guatemala and El Salvador, the growth of coffee production in the 19th century drove Indigenous peoples from their lands and into forced labor on coffee plantations. By the 1870s, politics in El Salvador became controlled by powerful coffee farming families who controlled the country under authoritarian rule and violent repression, fueling civil wars and human rights violations throughout the 20th century. In the 1950s, Colombia entered coffee markets and led efforts to maintain prices through an international agreement, which the United States supported in the hope that higher coffee prices might undermine the Cold War era communist leanings of some countries.

The late 20th century brought a dramatic change to the culture of coffee consumption. Consumers started to shift from drinking weak "instant" coffee to grinding their own beans and visiting local coffee shops. Coffee purveyors, such as Peet's and Zabar's, introduced free-roasted, carefully sourced specialty beans to the market, and three friends opened the first Starbucks outlet in Seattle in 1971. By 2011, there were more than 17,000 Starbucks stores worldwide (**Figure 7.2.2**).

The coffee industry in Latin America and the Caribbean is now at risk from climate change and the globalization of coffee production. Frosts and hurricanes have the capacity to devastate coffee crops, and climate change may alter conditions for production. The collapse of the International Coffee Agreement in 1989 caused prices to drop from U.S. $1.50 a pound to less than 50 cents, undermining the livelihoods of small farmers. Compounding the problems, in the late 1980s, development banks encouraged new countries, especially Vietnam, to enter coffee production. This led to a glut in production and another drop in coffee prices. Concerns about justice and environment also began to influence the coffee economy in the 1990s. Some consumers want assurance that their coffee comes from places where people are paid a fair wage, pesticides are banned, and human rights, forests, and biodiversity are protected. "Fair trade coffee" has become a common pronouncement on labels that also boast about qualities such as shade-grown, bird-friendly, or organic. ∎

▲ **FIGURE 7.2.1 A Coffee Plantation in Jamaica** A worker picks coffee in the Blue Mountains of Jamaica.

▲ **FIGURE 7.2.2 Starbucks in China** The multinational company Starbucks owns coffee shops around the world, including this one in Beijing, China.

rats, pigs, and cats ate the food that traditionally supported local species and consumed the local fauna, especially ground-dwelling birds. One of many species of Andean potatoes carried back to Europe became the foundation of the Irish diet; this was the potato that became blighted in the mid-19th century, resulting in famine and Irish migration to the Americas. Corn and manioc (cassava) were introduced into Africa and became new staples, whereas peanuts and cacao became the basis of new African export economies. Cotton was introduced from Latin America to India; pineapple was distributed to the Pacific, including Hawaii; and tobacco became an addictive habit across the world.

The colonial powers introduced several new forms of land tenure and labor relations that still influence contemporary landscapes in Latin America. In areas where Spain wished to directly control the land, they granted land rights over large areas to colonists, often military leaders, and to the Catholic Church, ignoring traditional local uses and establishing fixed property boundaries. These *latifundia* (large rural landholdings or agricultural estates) typically occupied the best land. Other farmers were forced onto small plots of land, known as *minifundia*. **Haciendas** were established to grow crops such as olives and wheat, mainly for local consumption in mines, missions, and cities, rather than for export.

Plantations are large agricultural estates established in the colonial period that grow crops such as sugar or tobacco for export. Labor for the Spanish haciendas, plantations, and mines was obtained initially through the institution of **encomienda**, a system by which groups of Indigenous people were "entrusted" to Spanish colonists. The colonists demanded tribute in the form of labor, crops, or goods and were assigned the responsibility of converting Indigenous groups to the Catholic faith and teaching them Spanish. These forms of labor control did not produce a large enough workforce, especially in areas where the demographic collapse had devastated local populations. To make up for the labor shortage, colonial trading routes were established to import slaves, mainly from Africa, to the Caribbean, Central America, and Brazilian plantations (see Chapter 5). Slaves worked in the production of sugar and other export crops. In many parts of Latin America, the colonial legacy of land grabs, labor exploitation, and racism still frames contemporary attitudes toward Indigenous people, who still struggle to regain the land and dignity lost during the colonial period.

The most important export commodities in Spanish colonial America were silver, produced mainly from mines in Mexico and Bolivia; sugar, grown on plantations in Cuba and southern Mexico; tobacco from Cuba; gold from Colombia; cacao (for chocolate) from Venezuela and Guatemala; and indigo, a deep blue dye, from Central America. Spain derived enormous wealth from the bonanza of the silver mines at Potosí (Bolivia), which produced half of the world's silver in the 16th century. This rapid influx of money led to inflation in Europe, and Spanish industry suffered as upper classes in both Spain and the colonies chose to purchase luxuries from other parts of Europe.

APPLY YOUR KNOWLEDGE Visit a Website that features tourist or economic information for a Latin American or Caribbean country. In what ways are the legacies of colonialism apparent in the landscapes, cities, or economy of the country?

Independence Over time, merchants and landowners in the Spanish colonies started to resent the strict control of trade and taxation by Spain and began scheming to keep the revenue from mines and crops for themselves. Spain was weakened by wars in Europe, and revolutionary movements in the United States and France provided inspiration. Pressure built for independence in the colonies, and by the beginning of the 1800s, regional independence movements were calling for liberation from Spanish and Portuguese rule.

Between September 16, 1810, when priest and peasant leader Miguel Hidalgo called for Mexican independence, and 1824, when Simon Bolívar finally led northern South America to independence, a series of regional revolts led to the formation of independent republics in Mexico, Argentina, Peru, Colombia, Chile, and Brazil. Borders were still not settled, however, and in 1848, Mexico was defeated in its war with the United States and forced to cede large portions of its territory to the state of Texas and to what would become the states of Arizona, California, Colorado, Nevada, and New Mexico.

In the Caribbean, after several rebellions against France (inspired by the French Revolution), former slaves declared Haitian independence in 1804 and occupied the rest of the island of Hispaniola until the Dominican Republic gained independence in 1844. Cuba remained under Spanish control until after the Spanish–American War in 1898, when limited independence was granted under U.S. influence and frequent intervention. Puerto Rico also shifted from Spanish control to a U.S. territory after 1899; it is still currently part of the United States, with some residents desiring independence and others demanding the country be granted full status as a U.S. state.

Independence came late to most of the British Caribbean. Jamaica, Barbados, and Trinidad and Tobago became independent in the 1960s. Other British colonies, such as the islands of Antigua, Barbados, St. Lucia, St. Kitts, and Nevis, and Belize in Central America, did not become fully independent until the beginning of the 1980s.

Six islands in the region remain as Dutch protectorates with full autonomy for internal affairs (Aruba, Bonaire, Curaçao, Sint Maarten, Saba, and St. Eustatius), and Martinique, Guadeloupe, St. Martin, and St. Barthélemy are overseas departments of France. The British Virgin Islands, Turks and Caicos, Montserrat, Anguilla, and the Cayman Islands are still colonies of Britain, and the U.S. Virgin Islands are under the control of the United States.

Economy, Accumulation, and the Production of Inequality

After independence was gained, the loss of trade with Europe and internal struggles left Latin America economically unstable for the first half of the 19th century. In about 1850, as conditions stabilized and industrialization in Europe and North America created new demands, new consumers, and new profits to invest, money became available to the Latin American economies that were open to free trade and foreign investment.

Foreign capital helped develop export economies for nitrate (used to make fertilizer) and copper in Chile; livestock in Argentina; coffee in Brazil, Colombia, and Central America; bananas in Central America and Ecuador; tin in Bolivia; and silver and henequen (a fiber used in making sacks and matting) in Mexico. Foreign-owned companies, which were mostly British in the 19th century, ran many of the new export activities and developed railroads and banks. Countries that produced these exports developed economies that were highly dependent on, and closely linked to, a volatile world market. This condition is still evident today in the vulnerabilities of Chile (38% of the country's total export value in 2010 was from copper), Paraguay (39% from soya beans in 2010), Ecuador (13% from bananas in 2010), and Venezuela (93% from crude petroleum in 2010).

Economic Shifts In Latin America, the aftermath of the 1929 stock market crash demonstrated the extent to which the region had become integrated into the global economy. Throughout the region, declines in exports, restrictions on investment, and a general economic crisis ensued. This, together with a general awareness that foreign ownership and low prices for unprocessed exports made Latin American economies vulnerable to world conditions, led to the development of the new economic strategy of **import substitution**. Import substitution is a trade and economic policy that supports the development of domestic industries with government protection. In essence, domestic producers are incentivized to produce goods for the domestic market that were formerly purchased from foreign producers.

Governments in Mexico, Brazil, and Argentina moved aggressively to implement import substitution policies from the 1930s to the 1960s. They protected domestic industries through import tariffs (taxes) and quotas. Government nationalization (the process of converting key industries from private to governmental organization and control) and investment in new manufacturing industries fostered production of chemicals, steel, automobiles, and electrical goods in regions such as northeastern Mexico (steel), Mexico's Gulf Coast (petrochemicals), and São Paulo, Brazil (automobiles). But growing criticisms of import substitution emphasized the oversized government bureaucracy and the high costs of subsidizing industries that were inefficient and produced goods of poor quality.

After 1960, economic growth in Latin America began attracting international investors and banks. The governments of Mexico, Argentina, Brazil, and Chile were offered the largest loans, but almost all Latin American governments took advantage of the initially low-interest loans to support economic development and other projects. When interest rates rose and debt payments soared in the early 1980s, Latin American governments were unwilling to cut back on popular subsidies and social programs and instead borrowed more money, ran budget deficits, and overvalued their currencies. The resulting runaway inflation and debt reached unprecedented levels. Mexico and Brazil owed over U.S. $100 billion and were paying enormous amounts in interest each year. In 1982, Mexico threatened to default on its loans. Fears that debt default across the region would destabilize the international financial system prompted financial institutions such as the International Monetary Fund (IMF) and the U.S. government to find a solution to Latin America's debt crisis. The United States extended the repayment period for debts and lent more money, while the International Monetary Fund moved to restructure loans on the condition that governments initiate new economic policies. Mexico got a U.S. $48 billion bailout in 1994. But the crisis continued across Latin America; the 1980s have been called Latin America's "lost decade" because of the slowdown in growth and deterioration in living standards.

The new economic policies that were the condition for debt restructuring included requirements that countries curb inflation by cutting public spending on government jobs and services, increase interest rates, control wages, and devalue currencies to increase exports. These **structural adjustment programs (SAPs)** required

▲ **FIGURE 7.20 Maquiladora** The North American Free Trade Agreement increased the number of Mexican export assembly plants, called *maquiladoras*, which employ thousands of workers, especially women, along the border and elsewhere in Mexico. This plant in Tijuana makes implantable electronic medical devices.

the removal of subsidies and trade barriers, the privatization of government-owned enterprises such as telephone and oil companies, reductions in the power of unions, and an overall focus on export expansion. These policies, while reducing inflation and debt, had very negative effects on some people and sectors, especially the poor. Food prices increased as subsidies were cut. Health services and education were reduced, and unemployment increased as government jobs were eliminated. As a result of structural adjustment programs in Peru in 1990, gasoline prices went from 10 cents to U.S. $2 per gallon. In the Caribbean, Jamaica, Barbados, and Trinidad and Tobago cut their public sectors and privatized public enterprises.

Free Trade In addition to the structural adjustment policies (SAPs) that involved government cutback and privatization of key sectors such as utilities, governments in the region have also been encouraged to expand free trade through regional agreements to remove barriers among trading partners. The most dramatic step in this direction was taken by Mexico, which in 1994 joined the **North American Free Trade Agreement (NAFTA)** with the United States and Canada.

NAFTA set out to reduce barriers to trade among its three member countries, through reducing customs tariffs and quotas. Advocates of NAFTA argued that free trade would create thousands of jobs in Mexico with higher wages and that these opportunities would reduce migration to the United States. Mexican agriculture would shift to growing high-value fruits and vegetables, where it had a comparative advantage during the winter, and Mexico would be able to reduce food prices by importing low-cost grain from the United States and Canada. Free trade was also linked to financial stabilization and to promises of more democratic government in Mexico.

Studies suggest that NAFTA has led to the creation of thousands of new jobs in Mexico and to increased wages in some industries. The *maquiladora* industry, in particular, has generated more than

1.5 million jobs in Mexican cities. *Maquiladoras* or *maquilas* are manufacturing facilities where components can be imported duty-free and assembled for export (**Figure 7.20**). NAFTA increased the growth of the *maquiladoras* at and beyond the border, including factories producing electronic equipment such as computers, clothing, and appliances. However, in recent years, some facilities have closed as a result of competition from China. There have been concerns about pollution from the factories and about poor workplace conditions. Nongovernmental organizations (NGOs) and community groups have demanded improved environmental protection and safety on both sides of the border. As a result, weak amendments have been added to the agreement to protect the environment and working conditions.

Unfortunately, many of the hoped-for benefits of NAFTA were frustrated by the economic crises that followed a currency devaluation in 1994 in Mexico and by continuing inequality in both urban and rural areas. The 1994 Zapatista rebellion in Chiapas was partly driven by opposition to NAFTA. Other regional trade agreements include MERCOSUR (Spanish acronym for "southern common market"), initiated in 1991 that links Chile, Argentina, Brazil, Paraguay, and Uruguay in a trade agreement; and CARICOM (Caribbean Community), formed in 1973 to create a trade zone in the Caribbean. The United States also has free trade agreements with Central American countries (CAFTA-Central America Free Trade Agreement) and with Chile, Colombia, and Panama. The Andean Community trade agreement links Peru, Ecuador, Colombia, and Bolivia.

APPLY YOUR KNOWLEDGE What were some of the negative consequences of the economic policies that were associated with debt restructuring in the region? Why did some people in Latin America oppose structural adjustment and free trade policies?

Contemporary Economic Conditions The share of GNI (gross national income) paid as interest on debt has dropped from more than 5% to less than 1% in the last 25 years, and economies in Latin America and the Caribbean have grown and diversified considerably in the last decade (**Figure 7.21**). For example, the region's largest economies, Brazil and Mexico, grew rapidly from 2000 to 2011. Brazil is now the world's sixth-largest economy, and the region is home to major manufacturing zones in cities such as São Paulo (Brazil) and Monterrey (Mexico).

The World Bank estimated the overall GNI (gross national income) of the Latin America and Caribbean region in 2010 at U.S. $5 trillion and per capita GNI at U.S. $7,821. These figures show that this world region is much more prosperous overall than sub-Saharan Africa, South Asia, or Southeast Asia. Total exports in 2011 were valued at about U.S. $1.3 trillion and were balanced by imports. The Latin American region also received more foreign investment in private capital than any other low- or middle-income region. This all suggests that overall economic conditions in Latin America and the Caribbean are positive and have improved in the last decade, with the overall picture heavily influenced by the success of Brazil. The growth of Brazil's economy merits its inclusion in a new association of emerging economies with Russia, India, China, and South Africa, which are collectively referred to as BRICS (see "Emerging Regions: BRICS," p. 286).

In contrast to Brazil, other countries in Latin America and the Caribbean still suffer from producing limited commodities for export. As a result they have difficulty producing jobs, food, or goods for their domestic populations. Economies that are dominated by just a few exports can be very vulnerable to global economic fluctuations. Sustained demand for oil and minerals is important to the economies of Venezuela, Ecuador, and Trinidad and Tobago (oil), and Chile and Peru (copper and gold). The mining industry is still important in many parts of the Andes; however, despite some unionization and attempts to improve technology and working conditions, many miners still endure high levels of respiratory diseases and accidents (see Figure 7.10, p. 271). The profits from mining have tended to concentrate in multinational corporations or in a few families who control the major companies.

The agricultural economy of many parts of Latin America has changed considerably in recent years. Some countries, such as Argentina, have maintained a traditional focus on grains and livestock; others, such as Guatemala, still export colonial crops of bananas, coffee, or sugar cane. Some countries in the region prioritize food self-sufficiency, while others have shifted to crops that are apparently more competitive in international trade, such as fruit, vegetables, and flowers. These **nontraditional agricultural exports (NTAEs)** have become increasingly significant in areas of Mexico, Central America, Colombia, Ecuador, and Chile, where they have replaced grain and traditional exports such as coffee and cotton

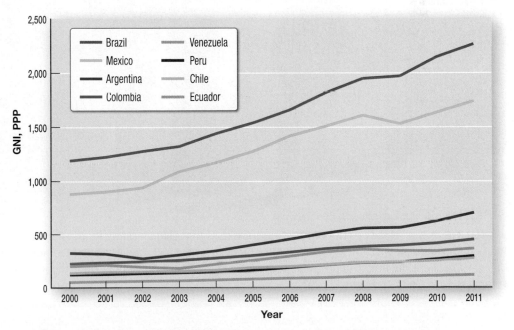

▲ **FIGURE 7.21 Economic Growth** This graph shows the growth of major Latin American economies since 2000 in U.S. $ billions.

EMERGING REGIONS
BRICS

The World Bank and other U.N. agencies have often clustered countries in Latin America and the Caribbean with those in Africa and Asia into the category of less economically well off "developing" countries. But the utility of this label is breaking down as key countries in Asia, Latin America, and Africa emerge as major economic and political powers. The emerging economies of Brazil, India, and China joined with Russia to form an alliance of countries referred to as BRIC. The BRIC countries are home to 3 billion people and massive economies that rank at the top of the world on several major indicators (**Table 7.3.1**). The first summit of BRIC countries in 2009 focused on the topics of global economy, financial reform, and strategies for cooperation in the global arena.

Although South Africa's economy and population are smaller than the other four countries, it was invited to join the association in 2010 as a strong representative and economic leader on the African continent. When South Africa joined BRIC, the group adopted a new acronym: BRICS (**Figure 7.3.1**).

Brazil is the economic leader in Latin America. Brazil benefits from abundant and diverse natural resources, including agricultural land and minerals, and from a manufacturing sector supported by private sector entrepreneurs and the government. A global boom in prices for agricultural commodities

TABLE 7.3.1 Rankings of Brazil, Russia, India, and China				
	Brazil	**Russia**	**India**	**China**
Area	5	1	7	3
Population	5	9	2	1
GDP	7	6	3	2
Exports	18	9	14	1
Foreign exchange reserves	7	3	6	1
Mobile phones	5	4	2	1

▲ **FIGURE 7.3.1 Heads of the BRICS Countries** (L to R) President Dilma Rousseff of Brazil, Russian President (now Prime Minister) Dimitry Medvedev, Indian Prime Minister Manmohan Singh, Chinese President Hu Jintao, and President Jacob Zuma of South Africa pose prior to the BRICS summit in New Delhi on March 29, 2012.

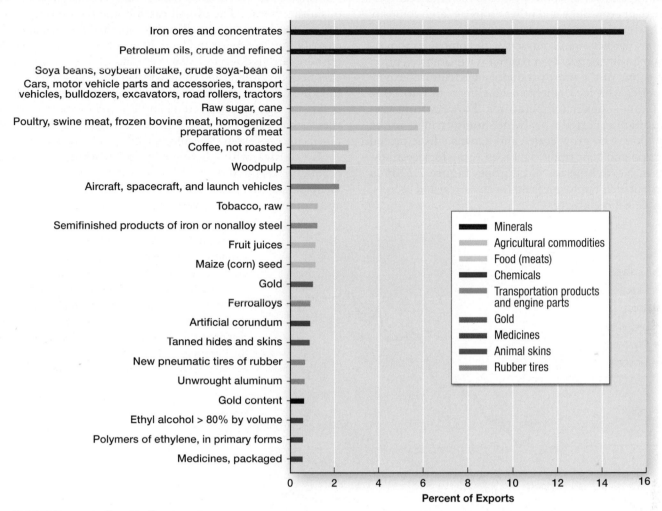

▲ **FIGURE 7.3.2 Brazil's Exports** This diagram of Brazil's exports illustrates the importance of agricultural commodities such as sugar, soybeans, and coffee and of mineral exports, including iron and oil, but also the growing importance of manufacturing (cars, aircraft).

produced by Brazil, such as beef and soybeans, has contributed to economic success. Investments in aircraft (such as the Embraer commercial plane), automobiles, textiles, and iron and steel have also paid off in terms of exports (**Figure 7.3.2**). Other features of Brazil's economic success include government support for science and technology and social policies that have sustained domestic consumption and a growing middle class.

The BRICS countries represent a shift in global economic power away from North America, Europe, and Japan. Members of the new group have joined to challenge the United States on several issues. Goldman Sachs, a global investment firm, predicts that these countries will become dominant suppliers of manufactured goods, raw materials, and even services. Although they all are nurturing a growing middle class and increasing consumer demand, per capita income levels in all the BRICS countries except Russia are still relatively low. The political power of Russia and China is already represented in their

permanent membership of the U.N. Security Council, and Brazil and India are leading possibilities should U.N. membership be expanded. Brazil is taking big steps in extending its own geopolitical power, including assuming leadership roles in U.N. peacekeeping operations in Haiti and Lebanon. While some critics see the BRICS grouping as a convenient label for a set of otherwise disconnected emerging economies, many international companies consider this coalition as key to their expansion strategies. ■

(**Figure 7.22a**). The NTAE crops are vulnerable to climatic variation and to the vagaries of the international market, including trending tastes for foods and health scares about pesticide or biological contamination. Rather than grow these and other crops on large company landholdings, the current strategy involves *contract farming* for multinational corporations; farmers sign contracts with companies to produce crops to certain quality standards in return for a guaranteed price.

Central Chile is one of the most productive agricultural export zones in Latin America with its Mediterranean climate of warm, wet winters, and moderate summer temperatures. Flowers from Colombia and Ecuador are exported worldwide. Some farmers in the region, especially those who grow coffee and bananas, have turned to organic production and "fair trade" labels as a way to access new consumer markets in North America and Europe (**Figure 7.22b**); in Brazil, several nutrient-rich rainforest fruits, such as açai and acerola, have gained international popularity.

Fisheries are another critical component of Latin American and Caribbean food and export systems, and activities range from subsistence fisheries in small coastal villages to large-scale commercial offshore fisheries. The overall catch in the region was more than 12 million metric tons (13.2 million US tons) in 2009, making a significant contribution to exports in Chile, Ecuador, Peru, and Costa Rica. But offshore catch is down by half in Peru and Chile since the early 1990s as a result of overfishing and climate variability. Aquaculture and mariculture (the cultivation of fish and shellfish under controlled conditions) in coastal lagoons has become an important export sector in countries such as Chile (salmon), Ecuador (shrimp), and Honduras, but has raised concerns about the ecological effects of intensive fish farming.

Costa Rica, with a better-educated workforce and more stable economy, has lured high-technology companies such as Microsoft, General Electric, and Intel to build factories near San Jose. Textiles and clothing are another very important component of the economy

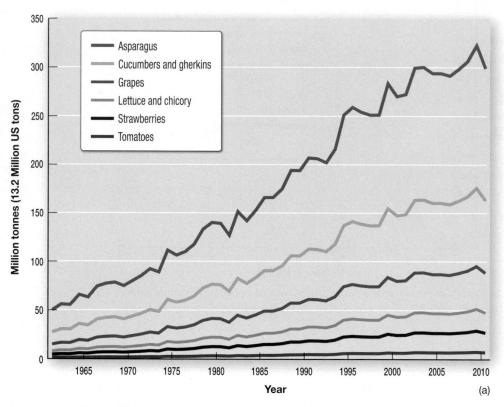

◄ **FIGURE 7.22 Nontraditional Agricultural Exports** (a) Production of fruit and vegetables has increased in Latin America, partly as a result of export of nontraditional crops. Tomato, cucumber, strawberry, and lettuce production has grown rapidly in response to demand from North America and Europe, and grapes in response to the growth of the Chilean wine industry. (b) Fair trade and organic certification provide new markets for sustainable products from Latin America.

(a)

(b)

in the region. Countries such as Haiti, El Salvador, and Honduras produce millions of low-cost T-shirts and other garments. In the Caribbean, tourism is the cornerstone of many economies. Almost 24 million international tourists visited in 2011, generating revenues of more than U.S. $150 billion. Mexico's tourist industry, focused on North America, almost matches the Caribbean with more than 20 million visitors. Many come to the Caribbean coast resorts around Cancun. Ecotourism, as discussed earlier in the chapter, is also a growing industry, especially in Costa Rica.

The **informal economy** includes economic activities that take place beyond the official record and are not subject to formalized systems of regulation or remuneration. In Latin America and the Caribbean, the informal sector comprises a variety of income-generating activities of the self-employed that do not appear in standard economic accounts. These include street vending, shoe shining, garbage picking, street entertainment, prostitution, crime, begging, and guarding or cleaning cars.

Inequality Despite many economic successes, inequality is an issue in many countries of the region. Wealth and land are often concentrated in the hands of the elite, and many people remain poor and landless. The highest average annual incomes of U.S. $10,000 to U.S. $20,000 per person are reported in Brazil, the Bahamas, Trinidad and Tobago, Venezuela, Uruguay, and Chile. In the poorest countries, such as Haiti, Bolivia, Honduras, and Nicaragua, annual incomes average less than U.S. $2,000 per capita. Many of these countries have unequal distribution of incomes within their populations, with the richest 20% of the population receiving more than 10 times the wealth of the poorest 20%. In most of Europe, there is a 4 to 1 ratio; in the United States the figure is 8 to 1 (the wealthiest 20% receive eight times the amount that the poorest 20% do).

However, both aggregate poverty and inequality have declined across the region as a result of stronger economies, more women in the work force, and policies that raised minimum wages and provided support for the poor. In Brazil and Mexico, government programs now focus on reducing poverty. Brazil's internationally praised *Bolsa Familia* (family allowance) program gives money to poor families as long as their children attend school and are vaccinated. The funds are usually spent on food, school supplies, and clothing and are estimated to have been responsible for 20% of the country's decline in poverty since 2001. Mexico has a similar cash transfer program, called *Oportunidades*, which is conditional on school and clinic attendance, and helps the poorest 30% of the population.

Social and health conditions are often considered a better measure of overall inequality within and between countries than economic measures. National improvements in life expectancy, infant mortality, and literacy, for example, reflect improvements at the lower end of the scale rather than for the better-off segments of the population. Latin America and the Caribbean tend to compare more favorably to the rest of the world on social and health indicators than on measures of income and income inequality. For example, life expectancy averages 74 years and adult literacy rates are also relatively high, averaging 91%.

There are wide gaps in social and health conditions within Latin America and the Caribbean. Haiti, Central America, and the Andes tend to have much worse conditions than do southern South America, northern South America, Costa Rica, Mexico, and

the English-speaking Caribbean. For example, in 2010, the average Haitian lived only 62 years, and the literacy rate was only 48%, compared to a life expectancy of 79 years in Costa Rica where literacy reaches 96%.

National indicators also hide large variations in economic and social conditions within Latin American and Caribbean countries. In Mexico, the southern regions of the country have lower incomes and life expectancies than do the northern and central areas of the country. In Brazil, the northeastern and Amazon zones have higher infant mortality, lower life expectancy, and lower average monthly incomes than the southern parts of the country. Each Latin American and Caribbean country has its own geography of inequality, with the more rural regions generally experiencing lower social and economic conditions.

APPLY YOUR KNOWLEDGE What evidence could you use to argue that economic conditions in Latin America and the Caribbean have improved significantly in recent years? Are there counterarguments?

Territory and Politics

Latin America and the Caribbean is a dynamic world region. Economic, political, cultural, and social changes have been rapid in the 20th century and have varied in their nature and impact among and within countries. Latin American and Caribbean countries have taken divergent political paths. The region has included socialist and military governments; authoritarian, single-party, and multiparty systems; and highly centralized and very localized administrations. The challenges of creating functioning national governments and promoting economic growth dominated the postindependence period in the 19th century. The 20th century saw regional factions, the working class and the poor demanding reform through revolution and populist movements, and threats to military and authoritarian rule. One of the most dramatic shifts in the region has been the transition from the military and authoritarian governments that predominated in the 1970s to the mostly democratic systems that prevailed by 2000.

U.S. Influence Latin America in the 20th century became caught up in Cold War politics between the United States and the Soviet Union. The United States has traditionally viewed outside intervention in the region by any country other than itself as threats to its own security. This stemmed from 1823, when U.S. President James Monroe issued the **Monroe Doctrine**, which stated that European military interference in the Western Hemisphere, including the Caribbean and Latin America, would be considered a threat to the peace and security of the United States, and treated as a hostile act. The doctrine also stated that, in return for European noninvolvement in the Western Hemisphere, the United States would not interfere in European affairs. This doctrine set the stage for subsequent U.S. involvement and intervention in Latin America and the growth of U.S. economic and political dominance in the region (**Figure 7.23**, p. 290). To that end, the United States intervened to maintain stability and economic access in Cuba (1896–1922), Haiti (1915–1934), and Nicaragua (1909–1933).

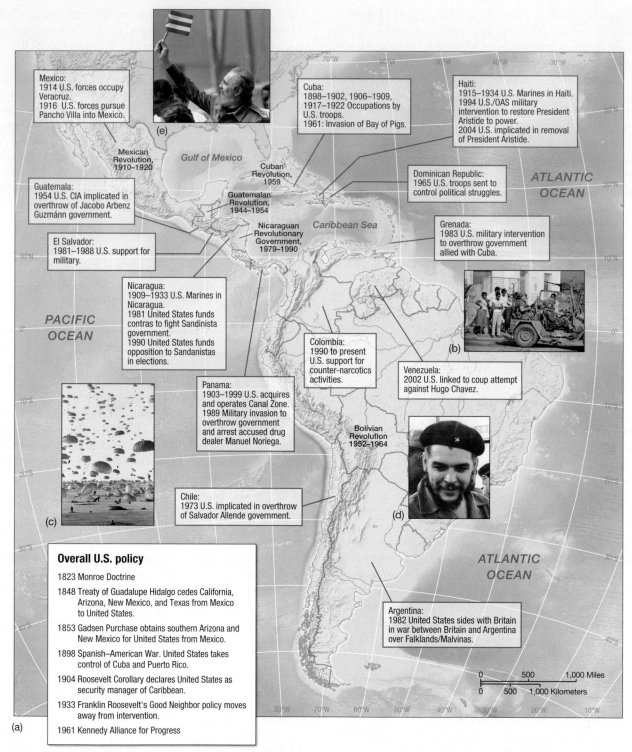

Mexico:
1914 U.S. forces occupy Veracruz.
1916 U.S. forces pursue Pancho Villa into Mexico.

Mexican Revolution, 1910–1920

Gulf of Mexico

Cuba:
1898–1902, 1906–1909, 1917–1922 Occupations by U.S. troops.
1961: Invasion of Bay of Pigs.

Cuban Revolution, 1959

Haiti:
1915–1934 U.S. Marines in Haiti.
1994 U.S./OAS military intervention to restore President Aristide to power.
2004 U.S. implicated in removal of President Aristide.

ATLANTIC OCEAN

Guatemala:
1954 U.S. CIA implicated in overthrow of Jacobo Arbenz Guzmánn government.

Guatemalan Revolution, 1944–1954

Dominican Republic:
1965 U.S. troops sent to control political struggles.

Caribbean Sea

El Salvador:
1981–1988 U.S. support for military.

Nicaraguan Revolutionary Government, 1979–1990

Grenada:
1983 U.S. military intervention to overthrow government allied with Cuba.

Nicaragua:
1909–1933 U.S. Marines in Nicaragua.
1981 United States funds contras to fight Sandinista government.
1990 United States funds opposition to Sandanistas in elections.

PACIFIC OCEAN

Colombia:
1990 to present U.S. support for counter-narcotics activities.

Venezuela:
2002 U.S. linked to coup attempt against Hugo Chavez.

Panama:
1903–1999 U.S. acquires and operates Canal Zone.
1989 Military invasion to overthrow government and arrest accused drug dealer Manuel Noriega.

Bolivian Revolution 1952–1964

Chile:
1973 U.S. implicated in overthrow of Salvador Allende government.

ATLANTIC OCEAN

Overall U.S. policy

1823 Monroe Doctrine

1848 Treaty of Guadalupe Hidalgo cedes California, Arizona, New Mexico, and Texas from Mexico to United States.

1853 Gadsden Purchase obtains southern Arizona and New Mexico for United States from Mexico.

1898 Spanish–American War. United States takes control of Cuba and Puerto Rico.

1904 Roosevelt Corollary declares United States as security manager of Caribbean.

1933 Franklin Roosevelt's Good Neighbor policy moves away from intervention.

1961 Kennedy Alliance for Progress

Argentina:
1982 United States sides with Britain in war between Britain and Argentina over Falklands/Malvinas.

0   500   1,000 Miles
0   500   1,000 Kilometers

▲ **FIGURE 7.23 U.S. Interventions in Latin America and Latin American and Caribbean Revolutions** (a) This map shows U.S. military intervention and those areas where there were major revolutions in the 20th century. In most cases, the United States intervened as a result of Cold War politics and in the interests of U.S. national security to prevent formation of governments allied with the Soviet Union or with socialist orientation. Most revolutionary movements have been inspired by calls for land reform and socialist policies. (b) U.S. marines intervened in Grenada in 1983. (c) U.S. paratroopers invaded Panama December 21, 1989. The invasion was ordered by U.S. President George H. W. Bush to seize Manuel Noriega to face charges on drug trafficking in the United States. (d) Guerilla leader Che Guevara fought in the Cuban and Bolivian revolutions. (e) Cuban leader Fidel Castro successfully led the Cuban Revolution in 1959.

▲ **FIGURE 7.24 Panama Canal** The Panama Canal cuts through the isthmus that joins North and South America and is a critical transport route for both cruise ships and cargo. Large ships must pass through several locks such as these at Miraflores, Panama.

An example of U.S. interests is the Panama Canal, which joins the world's two great oceans—the Atlantic and the Pacific—across the Isthmus of Panama in Central America and shortens the ocean trade route between them by weeks (**Figure 7.24**). In November 1903, Panama proclaimed its independence from Colombia and concluded the Hay/Bunau–Varilla Treaty with the United States. The treaty conceded rights to the United States "as if it were sovereign," in a zone roughly 16 kilometers (10 miles) wide and 83 kilometers (50 miles) long. The treaty specified that in that zone, the United States would build a canal, then administer, fortify, and defend it "in perpetuity." In 1914, the United States completed the 83-kilometer (50-mile) lock canal, dramatically cutting shipping travel time between the Pacific and Atlantic Oceans. The canal is one of the world's greatest engineering triumphs; it has three sets of locks that lift enormous ships up and across a lake 24 meters (80 feet) above sea level. More than 25,000 people lost their lives building the canal, mainly from tropical diseases such as malaria and yellow fever. The size of these locks has limited the size of many international ships to what is called the Panamax. Supertankers and some aircraft carriers cannot fit.

The early 1960s saw the beginning of sustained pressure within Panama for the renegotiation of the Hay/Bunau–Varilla Treaty. A new treaty, signed in 1977 handed control of the Canal Zone to Panama and established a joint U.S.–Panamanian Panama Canal Commission that was to administer the canal until the end of 1999, at which time the United States would withdraw forces from Panama. During the 1980s, Panama's military government was led by Manuel Noriega, who was implicated in drug trafficking and refused to recognize civilian elections supporting opposition candidates. After economic sanctions against Noriega's government failed, the U.S. military landed in Panama on December 20, 1989, citing the need to protect U.S. lives and property, to fulfill U.S. treaty responsibilities to operate and defend the canal, to assist the Panamanian people

in restoring democracy, and to bring Noriega to justice. Noriega was overthrown, imprisoned in France, and there was a return to democratic rule. When the canal handover took place in 1999, Panama celebrated its liberation from U.S. domination as well as its full control of a valuable economic asset. Tolls for the canal bring in more than U.S. $1.9 billion a year, and Panama is currently expanding the canal with new locks and a deeper channel for larger ships.

The Cold War and Revolution At the beginning of the 20th century, social and economic inequality produced frustration among the poor, the landless, and opposition or regional factions in several countries. Internal tensions between elites and other groups, especially landless peasants, complicated relationships between the Latin American and Caribbean region and the United States. As the Cold War between the democratic West and the Soviet Union intensified in the 1950s, a series of revolutions in Mexico, Guatemala, Cuba, Bolivia, and Nicaragua reverberated around the hemisphere and the world.

The Cuban Revolution, led by Fidel Castro in 1959, created a socialist state on the largest island in the Caribbean only 100 miles from the Florida Keys (see chapter opener). Since the revolution, the United States has taken an aggressive stance against the Cuban government. It mounted the Bay of Pigs invasion in 1961 and a series of embargoes on trade. The United States maintains a large military base at Guantánamo Bay in eastern Cuba, infamous as a terrorist holding center, where high-wire fences separate the base from the rest of communist Cuba.

The spread of socialist ideas about working-class activism and the need for land reform led to the election of socialist governments in Guatemala in 1954 and Chile in 1970. In both cases, redistribution of land and nationalization of key industries threatened the local elite and U.S. interests. The United States was implicated in assassinations and military coups that overthrew socialist leaders Jacobo Árbenz Guzmán in Guatemala in 1954 and Salvador Allende in Chile in 1973.

In the 1960s and 1970s, the dual threats of economic instability and communist ideas contributed to a rise in nondemocratic authoritarian and military governments in the region. Seeking financial order and control of socialist movements, the military took control of governments in Brazil in 1964, Chile and Uruguay in 1973, and Argentina in 1976. Although central authoritarian control certainly provided some degree of economic stability and growth, the military governments aggressively kept social order by repressing dissent, especially among students and workers who were branded as having socialist ideals. In Argentina, the military government's so-called Dirty War against dissenters in the 1970s is alleged to have killed 15,000 people and forced many others to leave the country. In Chile, the military government of General Augusto Pinochet was accused of orchestrating similar "disappearances" and human rights abuses from 1973 to 1990.

In Nicaragua, concentration of wealth and land under the Somoza dictatorship fostered rebellion that resulted in the establishment of the socialist Sandinista government in 1979. Again, Cold War anticommunist sentiments led the United States to support—covertly—a counterrevolutionary movement. Guerilla movements inspired by socialist and communist ideas that opposed powerful ruling elites and military governments also emerged in El Salvador (the FMLN from 1979), Colombia (the FARC from 1964), Peru (the Shining Path from 1980), and Bolivia (from 1962), and were severely and often violently repressed by ruling governments, sometimes with U.S. assistance.

In Guatemala, for 35 years beginning in about 1960, authoritarian governments suppressed leftist groups and forced Maya populations to flee into neighboring Mexico or farther north or to retreat into remote mountains, where the military, which did not differentiate between ordinary people and guerillas, annihilated whole communities. The difficulties faced by the Indigenous populations of countries such as Guatemala are characteristic of the discrimination that persists against Indigenous cultures in Latin America. The discrimination that stemmed from the colonial period has been reinforced by elites, who have promoted the image of Indigenous people as uncivilized, underdeveloped, and rebellious, in order to take their lands, undermine their religion and language, and force them into low-paid occupations as farm workers, miners, or servants.

APPLY YOUR KNOWLEDGE Use the Internet to research an example of an authoritarian/military government, leftist revolution, or U.S. intervention in Latin America or the Caribbean. (See the examples in Figure 7.23, p. 290.) List three ways the memories and repercussions of the event you research are still influencing domestic politics or international relations in the area.

Democracy Public and foreign outrage at authoritarian repression and human rights violations, the inability of military governments to solve economic problems, the end of the Cold War, and international and internal pressures that linked economic globalization to democratic governance resulted in gradual transitions to democratic governments in most Latin American countries. For example, transitions took place in Argentina in 1983, Brazil in 1985, and Chile in 1989. In Argentina, the departure of the military government was hastened by the loss of a war with Britain when Argentina invaded the Falkland Islands (called the Islas Malvinas in Latin America) in 1982. Most of Latin America is now under democratic rule, although

democracy has been fragile. Economic and other crises have toppled governments in countries such as Bolivia and Argentina. In other countries, heavy-handed tactics and questionable elections have maintained one-party systems.

Political opposition and activism in the region often take the form of **social movements** that organize against cuts in government services, for land reform, and for specific resources and issues, such as housing, water, human rights, or environmental protection. New Indigenous social movements have organized to promote language, culture, and land rights. In Brazil, the Rural Landless Workers movement (*MST* in Portuguese), which is supported by an estimated 1.5 million people, organized land occupation, winning titles for 350,000 families, and achieving significant political change when their leaders eventually won elections. In Bolivia, coca farmers organized against the suppression of coca cultivation, and public-sector employees fought against job cuts and the privatization of water. The Bolivian city of Cochabamba became famous for a "water war" in 2000 when residents organized against the sale of the municipal water system to an international consortium, which had raised prices. A general strike and roadblocks added to pressure on the government, which revoked the private contract and turned control of the water system over to the locals. The leader of the coca farmers, Evo Morales, was elected as the first Indigenous president of Bolivia in 2006 (**Figure 7.25**). In Argentina, protests led to the ouster of a president in 2001 and several years of demonstrations against banks, rising food prices, and elites.

A number of these social movements have translated into changes in electoral politics. A backlash against budget cuts, privatization, and inequality brought more left wing, antiglobalization, and populist leaders to power after 2000. The new leaders include Presidents Lula of Brazil, Chávez of Venezuela, Morales of Bolivia, Ortega of Nicaragua, and Bachelet of Chile. President Hugo Chávez of Venezuela has led the **ALBA** group (*Alianza Bolivariana para los pueblos de nuestra América*, or Bolivarian Alliance for the Peoples of Our America) that includes Venezuela, Bolivia, Cuba, Ecuador, and Nicaragua and opposes free trade and U.S. influence in the region.

▲ **FIGURE 7.25 Bolivia** Conflicts over coca cultivation and water in Bolivia contributed to the election of left-leaning president Evo Morales. In this photo, Morales is shown in Indigenous clothing posing with a vicuna whose wool he has just sheared.

Drugs The drug economy has seriously destabilized politics in some regions of Latin America, especially in Bolivia, Colombia, Mexico, and Peru. Latin America produces drugs that are illegal in many countries, including cocaine (from the coca plant), heroin (from poppies), marijuana, and methamphetamine ("meth"). Latin American farmers grow drugs because of the high prices they yield in comparison to other agricultural products. In areas where crop yields are low, people have only small plots of land, and market prices for legal agricultural crops do not cover production costs, drug production is an attractive option or necessary evil. Most of the drugs produced in the region are exported to the United States and Europe. The farmers receive only a fraction of the street value of the drugs.

The bulk of drug exports are controlled by powerful families in Colombia and Mexico, who manage the transport systems from rural Latin America by land, air, and boat into the main distribution and consumption centers in the United States, such as Los Angeles and Miami. Analysts contend that farmers will continue to produce crops that can be processed into drugs until they can obtain a better living from other crops or other means of employment or until the demand is controlled in the United States. They argue that the United States should be focusing on controlling demand within its own borders or on limited legalization of consumption, rather than on fighting a "war on drugs" in Latin America.

In some areas, the drug trade has exacerbated political conflicts. For example, in Colombia, several guerilla movements with links to powerful drug lords controlled large areas despite opposition by government military units. The United States became involved in the strife and supported the Colombian government's antidrug activities. The United States sprayed pesticides on the illegal crops and provided military training and equipment to the Colombian army and police groups, who repressed rebel groups in the drug-producing regions. The army and the police were able to capture the leaders, extradite them to the United States for trial, and provide alternative livelihoods for farmers. By 2010, cocaine production was cut in half, and the area saw a corresponding drop in violence and the political corruption associated with drug production.

As the power of the drug economy declined in Colombia, it rose steeply in Mexico. Mexican **cartels**—organizations and networks that work together to control a product—now control most of the drug flow to the United States, including cocaine from Colombia and meth, heroin, and marijuana from Mexico (**Figure 7.26a**). Drug traffickers control production areas but also influence the police, the army, judges, and political leaders through intimidation and bribery. Power struggles between rival cartels and narcotic traffickers have escalated to the point where more than 11,000 people were estimated killed in drug-related violence in 2010. The Mexican army is struggling to control the situation, which is fuelled by corrupt police forces and a flow of guns from the United States (**Figure 7.26b**).

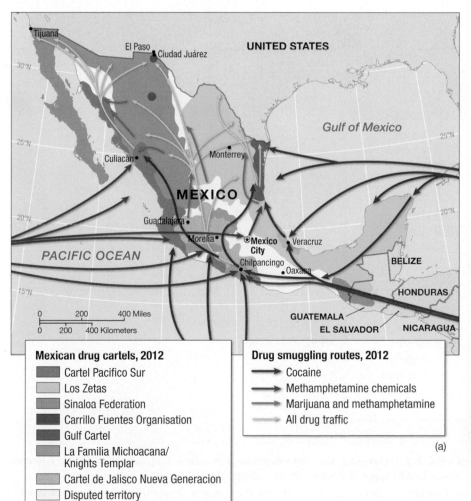

Mexican drug cartels, 2012

- Cartel Pacifico Sur
- Los Zetas
- Sinaloa Federation
- Carrillo Fuentes Organisation
- Gulf Cartel
- La Familia Michoacana/ Knights Templar
- Cartel de Jalisco Nueva Generacion
- Disputed territory

Drug smuggling routes, 2012

- → Cocaine
- → Methamphetamine chemicals
- → Marijuana and methamphetamine
- → All drug traffic

(a)

◀ **FIGURE 7.26 Drugs in Mexico.** (a) The influence of the main Mexican drug cartels is pervasive. (b) The Mexican army is shown here with its largest drug bust of marijuana in Tijuana, Baja California, on October 18, 2010.

(b)

CULTURE AND POPULATIONS

The mixed racial and ethnic composition of Latin America is echoed in many aspects of the cultural heritage and social practices in the region. Indigenous culture, including traditional dress, crafts, ceremonies, and religious beliefs, persists in regions such as highland Guatemala and Peru. This is partly a result of colonial policies that kept Indian communities separate while demanding tribute and labor from them, and partly because of resistance to the adoption of European culture by conservative Indigenous religious and political leaders. Cultural traditions such as these are now promoted to tourists and revalued through Indigenous social movements, which seek political rights and recognition. For example, Indigenous centers such as Quetzaltenango in Guatemala or Lake Titicaca in Bolivia and Peru are promoted as tourist destinations, where traditional crafts may be purchased and photos taken (often for a price) of women and children in traditional colorful woven garments.

Religion and Language

One of the main objectives of Spanish and Portuguese colonialism was the conversion of Indigenous peoples to Catholicism. Although some Indigenous people fiercely resisted missionary efforts, others found ways to blend their own traditions with those of the Catholic Church and create new **syncretic** religions. The term syncretic means that practices have co-evolved and merged with one another over the centuries. The process of conversion was facilitated by the reported appearance of the brown-skinned Virgin Mary of Guadalupe to an Indian convert in Mexico on December 9, 1531. Conversion rates also benefitted from the efforts of some priests to protect local communities from the Spanish efforts to obtain land, tribute, and labor by force. A 2010 report estimates that more than 432 million people in Latin America and the Caribbean are followers of Catholicism and 94 million are Protestant. Islam and Judaism draw about 1.7 million and 400,000 people respectively, and more than 10 million people have traditional Maya beliefs.

The slave trade brought African religious traditions to Latin America and the Caribbean, and these often merged with Indigenous and Catholic beliefs into syncretic beliefs that are followed by more than 30 million people. These belief systems include Candomble and Umbanda in Brazil, Voodoo in Haiti, and Santería in Cuba and other islands. Candomble and Umbanda are both sects of the Macumba religion, whose rituals involve dances, offerings of candles and flowers, sacrifices of animals such as chickens, and mediums and priests who use trances to communicate with spirits including several Catholic saints. Voodoo (also spelled Voudou) rituals include drumming, prayer, and animal sacrifice to important spirits, which are based on traditional African gods and Catholic saints. These rituals are led by priests who act as healers and protectors against witchcraft. Santería, which is closely connected to the Yoruba religion of West Africa, incorporates Catholic saints and African spirits associated with nature, and features rituals similar to other Latin American and Caribbean religions.

Recent decades have seen the emergence of a new form of Catholic practice, **liberation theology**, which focuses on needs of the poor and disadvantaged. Liberation theology is informed by the perceived preference of Jesus for the poor and helpless and by the writings of Karl Marx and other revolutionaries on inequality and oppression. This new orientation to the poor was espoused by the Second Vatican Council, called by Pope John XXIII in 1962. In response, priests preached grassroots self-help to organized Christian base communities and often spoke out against repression and authoritarianism. Also in recent decades, evangelical Protestant groups with fundamentalist Christian beliefs have grown and spread rapidly in Latin America and the Caribbean. Their message of literacy, education, sobriety, frugality, and personal salvation has become very popular in many rural areas. Estimates suggest that up to 40 million Latin Americans are now members of such churches.

Indigenous languages endure in several regions of Latin America (**Figure 7.27**). Widely spoken languages include Quechua in the

▲ **FIGURE 7.27 Languages of Latin America** The official language of most of mainland Latin America is Spanish, except for Brazil, whose official language is Portuguese; French Guiana, French; Suriname, Dutch; and Belize and Guyana, English. The dominant European languages of the Caribbean reflect the colonial histories of the islands.

Andean region (13 million people), English Creole and French Creole in the Caribbean (10 million), Guaraní in Paraguay (4.9 million), Aymara in the Andes (2.2 million), Mayan in Belize, Guatemala, Honduras, and southern Mexico (6 million), and Nahuatl in Mexico (1.5 million). Spanish is the dominant language across most of mainland Latin America, except for Brazil (Portuguese), Belize and Guyana (English), Suriname (Dutch), and French Guiana (French). In the Caribbean, the dominant languages are English, Spanish, French, and Dutch, although many people speak local versions of European languages, such as Haitian creole (a simplified version of French that is spoken by 12 million people) and Caribbean English.

Recent population censuses that have attempted to record race and ethnicity show some general patterns that correlate with the population history of the region. These include the African heritage from the slave trade and the mixing of Indigenous and European peoples to create mixed or mestizo ethnicity (a **mestizo** is a person of mixed white (European) and Indian ancestry; see the material on demography later in the chapter). In recent surveys, Latin Americans list themselves as about one-third mestizo, one-third white, 20% black, and 10% Indian. The largest percentages of Indigenous populations are in Guatemala, Bolivia, Peru, and Mexico; the largest black populations are in Brazil, Colombia, Cuba, Venezuela, Panama, and the Dominican Republic; and the largest numbers of self-identified and censused white populations are in Argentina, Chile, Costa Rica, and Uruguay. There are also several million people of Asian descent in Latin America. In particular, there are a large number of people of Chinese and Japanese descent in Brazil and Peru.

These numbers hide subtle differences in how different countries record, construct, and perceive race and ethnicity. For example, a tendency to identify with Europe may increase the proportion of those who report themselves as white in Argentina, whereas a national pride in mestizo heritage can increase the number of people who self-identify as being of mixed-race heritage in Mexico.

Brazil has promoted an image of Brazilian racial democracy and equality, where skin color is called "coffee," and musical, religious, and dietary traditions merge into a uniquely Brazilian culture. This myth of racial democracy is contradicted by evidence of continuing racism in Brazil and other Latin American countries. Studies show that race and class correlate strongly. Afro-Brazilians are on the whole poorer, less healthy, less educated, and more discriminated against in employment and housing. In Mexico, the media have tended to promote lighter skin as more desirable through the choice of more European-looking actors in commercials and other programs, and job advertisements still ask for "good appearance," hinting at a preference for non-Indigenous features.

APPLY YOUR KNOWLEDGE What evidence could you marshall to argue against stereotyping Latin America as a region that is Spanish-speaking, Catholic, and comprised of mostly Indigenous populations?

Cultural Practices, Social Differences, and Identity

For the most part, cultural practices in Latin America have been evolving toward global cultural trends promoted by formal education and popular media, especially television. Differences within the region have been partly erased by the common experience of popular television shows—especially *telenovelas* (soap operas)—and by education systems and media that promote modern urban lifestyles.

The foods of Latin America blend Indigenous crops, such as corn or potatoes, with European influences, especially from Spain. Although Mexico and Jamaica are associated with spicy dishes that include *chile*, the food is quite mild in the rest of Latin America. In livestock-producing areas, such as Argentina, grilled meat is extremely popular, but in much of Latin America, the poor eat simple meals of rice, corn, potatoes, and—for protein—beans. Modified versions of Mexican cuisine have diffused throughout North America and are the basis for many chain restaurants. Foods in the Caribbean reflect the medley of cultures in the region with African, Asian, and European influences combining in dishes of fish, chicken, pork, a range of vegetables, fruits, and starch crops (rice, potatoes, and yuca).

Music, Art, Film, and Sports Latin American and Caribbean art and literature display incredible variety and regional specialization. Traditional textiles, pottery, and folk art are sold to tourists and by import stores in North America and Europe. Literary traditions include magical realism and famous authors such as Gabriel García Márquez blend imaginary and mystical themes into their fiction. Latin American and Caribbean authors have won six Nobel Prizes for literature. Works of noted Mexican artists such as Frida Kahlo and Diego Rivera are numbered among the masterpieces of global 20th-century art.

Latin American and Caribbean music enjoys worldwide popularity. Caribbean global influences include the reggae of Jamaica and steel drum bands of Trinidad (**Figure 7.28**), which resonate with the

▲ **FIGURE 7.28 Steel Drums of Trinidad** A steel drum band performs on the main stage of the Carnival in Port of Spain. The Carnival is the city's main annual festival, attracting an audience from around the world.

rhythms of Africa. Latin music includes salsa, samba (Brazil), mambo (Cuba), tango (Argentina), and mariachi and ranchera (Mexico). Also very popular is Latin pop and rock music, often produced in Miami, Latin America's business capital in the United States, and featuring stars such as Jennifer Lopez, Shakira, and Enrique Iglesias. Mexico, Argentina, and Brazil have thriving film and TV industries.

Whereas the English-speaking Caribbean, especially the West Indies, produces some of the world's best cricket players, several Spanish-speaking islands, such as the Dominican Republic, are associated with famous baseball players, such as Alex Rodriguez, Pedro Martinez, and Albert Pujols, who gained fame and fortune in the U.S. leagues. Soccer (fútbol) is the most popular sport in mainland Latin America. Enormous stadiums host tens of thousands of fans in countries such as Argentina, Brazil, and Mexico and children in the poorest communities play in the streets with makeshift balls and goals.

Gender Relations   Certain cultural views of the family and gender roles have been characteristic of Latin America. Multiple generations often live and work together, individual interests are subordinated to those of the family, and the traditions of **machismo** and **marianismo** define gender roles within the family and the society. Machismo constructs the ideal Latin American man as fathering many children, dominant within the family, proud, and fearless. Marianismo constructs the ideal woman in the image of the Virgin Mary; she is chaste, submissive, maternal, dependent on men, and closeted within the family. Latin American society has been generally patriarchal, and institutions have prohibited or limited women's right to own land, vote, get a divorce, and secure a decent education.

These stereotypes are, of course, contradicted by many individual cases and are breaking down in the face of new geographies and global cultures. Family links are weakened through migration and the isolation of many living spaces—from each other and from those of other family members—in urban environments. Men's and women's roles are changing as fertility rates decline and women enter the workforce and politics. Latin American and Caribbean feminists have organized to obtain the right to vote; to effect changes in divorce, rape, and property laws; to gain access to education and jobs; and to elect women to political office.

Still, gender inequality is prevalent in Latin America. Female literacy, on average, is 2% to 15% lower than that of male populations. Women tend to earn much less money on average than men. In Latin America and the Caribbean, women earned between 39% and 61% of the male rate in 2011. In Chile, for example, female income was U.S. $8,845 (PPP) in 2011 compared to U.S. $19,897 for men. This inequality has been associated with systematic institutional biases that denied women in many countries the right to vote or the right to marital property until the 1950s; with cultural traditions that discourage more than a few years of education for women; and with employment structures that pay women less than men or pay less for traditionally female work, such as domestic service work and food processing.

Demography and Urbanization

Prior to the arrival of the Europeans in about 1500, Latin America is estimated to have had a population of approximately 50 million people, including large concentrations within the empires of the Aztecs and Incas and many smaller groups of hunters, gatherers, and agricultural communities. The demographic collapse discussed earlier in this chapter dramatically reduced Indigenous populations, but significant Indian populations remained in Mexico, northern Central America, and the Andes.

Colonialism also changed the demographic profile of Latin America through the intermixing of European and Indian peoples and the importation of slaves from Africa to the Americas. Few European women accompanied the early Spanish and Portuguese explorers and settlers, and many of the newcomers fathered children with Indian women through force, cohabitation, or marriage. Racial mixing occurred among European, Indian, and African populations, especially in Brazil. The resulting mixed-race populations were classified according to their racial mix. The most common category was that of mestizo; other categories included *mulatto* (Spanish/African) and *zambo* (African/Indian).

Slave imports to Latin America from Africa totaled more than 5 million people during the colonial period, including 3.5 million to Brazil and 750,000 to Cuba. Many of the Caribbean Islands, including Haiti and Trinidad, with very small Indigenous and European populations, had a large number of African slaves working on plantations, and African populations also settled along the plantation coasts of Mexico, Central America, northern South America, and Ecuador. Although slavery was not abolished until the mid-1800s (1888 in Brazil), escaped and freed slaves formed communities as early as 1605, most famously the African community of Palmares, which was an autonomous republic from 1630 to 1694 in the Brazilian interior. These settlements, also called **maroon communities**, were created by escaped and liberated slaves in other regions as well, such as Jamaica.

There is also a legacy of diasporas in contemporary Latin American populations. Asian immigration to the region began during the colonial period and picked up after the end of slavery when Chinese, Indian, and Japanese workers were brought over to work on plantations and in construction. The workers had to pay off the cost of their travel and sustenance. Europeans other than the Spanish and Portuguese settled in the more temperate climates in the region, especially in Argentina. There, many families have Italian, German, or British names. Six million Italians and Spanish also migrated to Argentina. And some areas, such as Patagonia, are associated with Welsh immigration and culture.

Population Growth   The overall population of contemporary Latin America totaled 596 million people in 2011. The distribution is clustered around the historical highland settlements of Central America and the Andes and in the coastal colonial ports and cities (**Figure 7.29**). The population has grown rapidly since 1900, when the regional total was 100 million, mainly as a result of high birthrates and improvements in health care. Brazil (196 million) and Mexico (114 million) have the largest populations today. Fertility rates reached 3.5 children per woman in Bolivia, Guatemala, and Haiti. But fertility rates have declined throughout much of the region. The 2005–2010 rate approached under two births per woman in Argentina, Brazil, Chile, Colombia, Cuba, Mexico, Uruguay, and many small islands of the Caribbean. Although this will eventually slow population growth considerably across the region, a large percentage of the population is under age 15, especially in Central America, and populations are likely to continue to grow as this cohort enters its reproductive years.

Higher fertility rates are characteristic of poorer, rural regions, where infant mortality is high, children can contribute labor in the

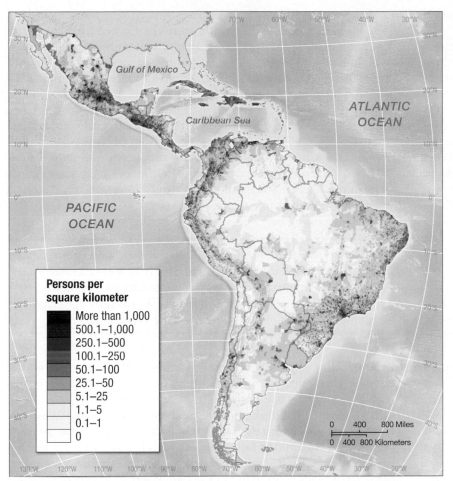

People leave rural areas because wages are low; because services such as safe drinking water, health care, and education are absent or limited; or because they do not have access to enough land to produce food for home consumption or for sale. Unemployment as a result of agricultural mechanization, price increases for agricultural inputs, and the loss of crop and food subsidies has also driven people from rural areas to the cities. Other push factors have been environmental degradation and natural disasters, such as Hurricane Mitch in Honduras, as well as long-running civil wars or military repression of rural people.

Cities pull migrants because they are perceived to offer high wages and more employment opportunities as well as access to education, health, housing, and a wider range of consumer goods. Governments often have an urban bias and provide services and investment to cities, which are seen as the engines of growth and the locus of social unrest. Social factors that encourage migration to the cities include the promotion of urban lifestyles and consumption habits through television and other media, and long-standing social networks of friends and families that link rural communities with

fields, and women do not have access to education, employment, or contraception. Fertility rates in the region have dropped as people move into the cities, health care improves, and more women work and are formally educated. Mexico illustrates this pattern with lower fertility rates in urban, industrial, and higher-income states near Mexico City and the U.S. border and higher fertility rates in the poorer, more rural southern states. Attitudes toward family size in Latin America are also affected by the Catholic Church's position against contraception and by the culture of machismo, which views high male fertility as a measure of status.

Migration within the Region More than 150 million people are estimated to have moved from rural areas to cities in Latin America and the Caribbean in the 20th century. The reasons for this massive rural–urban migration include factors that tend to push people out of the countryside and others that pull people to the cities (**Figure 7.30**).

▶ **FIGURE 7.30 Major Migration Streams in Latin America** There are major migration streams within countries and between countries in Latin America and to the rest of the world from Latin America. The most significant overall trend is rural-to-urban migration, but poor people are also moving into frontier regions in the Amazon and southern Mexico; from the Andes to work in Argentina and Venezuela; and away from political unrest and natural disasters. The Caribbean has flows to Europe, especially from the English-speaking islands, and to the United States, especially from Cuba, Haiti, the Dominican Republic, and Puerto Rico.

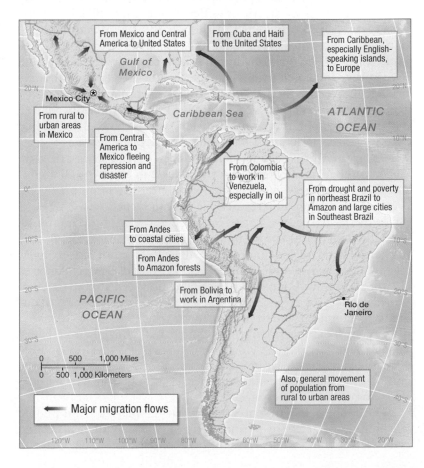

people in cities who can provide housing, contacts, and information to new migrants.

Although most people in the region have migrated to cities within their own countries, there are several other important migration flows within the Latin American region. Several countries have encouraged the colonization of remote frontier regions by providing cheap land and other incentives. For example, the building of roads and availability of land in the Amazon created a stream of migrants from coastal regions of Brazil to the interior, and the development of irrigation in Mexico and Chile attracted migrants to desert regions. People have moved out of the Andes to work in mining, agriculture, and oil in Argentina and Venezuela, and out of Central America to Mexico either as refugees or workers seeking higher wages. Some of the smaller migrant streams have included better-off sectors of society—for example, many intellectuals left Chile, Argentina, and Brazil for Mexico, Venezuela, and Costa Rica when leftists and students were being repressed by military governments in the region.

The Latin American and Caribbean Diaspora Latin American and Caribbean people have also left the region in considerable numbers, creating a global Latin American and Caribbean diaspora. The United States hosts the largest number of people outside the region who define themselves as being of Latin American or Hispanic heritage. Many Mexican families became part of the United States when the land they lived on became a U.S. territory following the U.S.–Mexican War in 1848. They use the phrase "the border crossed us, we didn't cross the border" to emphasize that they are not migrants but long-standing residents. Between 1900 and 1930, 1.5 million Mexicans (10% of the total population) migrated to the United States to escape the chaos of the Mexican Revolution and to fill labor shortages created by World War I. Although 400,000 Mexicans (some of them U.S. citizens) were deported during the Great Depression in the early 1930s, the growth of the U.S. economy from 1940 and World War II created such a demand for low-cost labor, especially in agriculture, that the U.S. and Mexican governments introduced a formal guest farm worker program. This program distributed 4.6 million temporary permits for Mexicans to work in the United States between 1942 and 1964. Many **braceros** (defined as a guest worker from Mexico given a temporary permit to work as a farm laborer in the United States) never returned to Mexico, and migration continued through social networks after the program ended, even as U.S. immigration restrictions were tightened.

In the past 50 years, Latin American and Caribbean migration to the United States has been dominated by Mexicans (about 40% of the total and 60% of those defined as illegal), but also includes large numbers of people from Cuba (15%) and Central America (10%). Significant Latin American populations can also be found in Canada and Europe (especially in Spain, where large numbers of migrants from Andean countries work in low-paid jobs). The Caribbean diaspora includes migration to the United States (mainly from Cuba, Jamaica, and Puerto Rico). However, because of colonial links to Britain, large numbers of Caribbean people have also migrated to Europe and British Commonwealth countries, especially from Jamaica and Barbados to Britain and Canada.

The money that is sent back to Latin America from people working temporarily or permanently in other countries is called **remittances** and can make a significant contribution to national and local economies. The World Bank estimated that more than U.S. $60 billion was sent to the region from the United States in 2011. Many communities in the Caribbean and Mexico rely on these funds to build houses, purchase agricultural inputs, or educate their children. This is one of the new informal flows of international financial capital in the global economy.

Migration—both with and without documents—from Latin America to the United States has dropped dramatically since 2005 and may have reached a net flow of zero from Mexico (with migration into the United States balanced by those returning home) after 2010. Explanations for the drop-off include recession and unemployment in the United States, improved conditions in Latin America, and stricter border enforcement.

APPLY YOUR KNOWLEDGE Immigration from Latin America and the Caribbean to the United States is a controversial issue in U.S. politics. Conduct some research and summarize the factors that promote this migration flow. Then come up with one argument in favor of this migration flow and one against.

Urbanization Eighty percent of people in Latin America and the Caribbean now live in cities, and the levels of urbanization in this region are among the highest in the world, ranging from about 71% in most of Central America to more than 90% in Argentina, Uruguay, and Venezuela. This compares to a regional average of only 10% in 1900. The region also hosts several of the world's largest metropolitan areas, sometimes called **megacities**, including Mexico City and São Paulo. These two urban areas number among the ten largest cities in the world, and each has more than 20 million people in its metropolitan areas. Not far behind are Buenos Aires at 14 million, and Rio de Janeiro at 13 million. A major cause of urban growth is migration, although the redefinition of city boundaries (to include metropolitan regions) and internal population growth have also played a role. In many countries, the population of the largest city in an urban system is disproportionately large in relation to the second- and third-largest cities in that system. This so-called **urban primacy** is characteristic of Argentina (Buenos Aires has 35% of the national population), Peru (Lima, 43%), Chile (Santiago, 39%), and Mexico (Mexico City, 22%). Concentration of population and development in one or two cities within a country can create problems when physical and human resources, political power, and pollution are all focused in one major settlement.

The megacities are global centers of commerce but also have many urban social and environmental problems. Mexico City is the economic, cultural, and political center of Mexico. It produces 40% of the gross national product and is home to the headquarters of leading Mexican companies and the regional offices of multinational corporations. However, its modern high-rise office buildings and elegant colonial plazas are clouded with a layer of pollution that obscures the snowy peaks of the volcanoes that ring the city (**Figure 7.31**). As in other large Latin American cities, many of the new migrants to Mexico City cannot afford to rent or purchase homes and settle in irregular settlements, or *barrios*, that surround the city. As much as 50% of Mexico City's housing is defined as "self-help construction" and ranges from cardboard and plastic shanties to sturdier wood and brick structures with aluminum or tile roofs. Many of these

▲ **FIGURE 7.31 Mexico City** Mexico City is located in a high basin surrounded by mountains, including the volcano of Popocatépetl. Air pollution obscures the view of the volcanoes many days of the year.

settlements occupy steep hill slopes, valley bottoms, and dry lakebeds that are vulnerable to flooding, landslides, and dust storms. One such example, Neza-Chalco-Itza, houses more than 4 million people on the shores of Lake Texcoco.

A second megacity, the city of São Paulo, Brazil, is located on a high plateau about 50 kilometers (30 miles) from the Atlantic coast, and has wide avenues and many skyscrapers around the central business district. But it is surrounded by neighborhoods of poorer and slum housing (**Figure 7.32**). São Paulo is the major financial center for Brazilian and international banks and has recently developed a large telecommunications and information sector. Brazilian geographer Milton Santos reports that São Paulo, which employs more than 2 million manufacturing workers and produces 30% of Brazil's gross national product, has morphed from a commercial center to a manufacturing hub to a service and information core for the global economy.

The cultural and media center of Brazil is the city of Rio de Janeiro, which has been overshadowed by the economic growth of its rival São Paulo (**Figure 7.33**, p. 300). Rio was the capital of Brazil from 1822 to 1960. Its urban structure includes an older city center with a wealthier residential zone and beaches such as Copacabana toward the south, and a poorer, more industrial zone to the north. Rio and São Paulo followed a similar pattern as other Latin American cities, attracting millions of migrants who have settled informally around the urban core. The **favelas**, a Brazilian term for the informal settlements or shanty towns that grow up around the urban core of Rio, lack good housing and services and are home to 11.4 million people. In São Paulo, 28% of residents have no drinking water, and 50% have no sanitation. The crowding, high land costs, violent crime, poverty, traffic problems, and pollution of the cities are starting to cause people and economic development to shift to smaller neighboring cities across Latin America.

FUTURE GEOGRAPHIES

The region of Latin America and the Caribbean has changed dramatically in recent years as the forces of market liberalization, economic integration, democratization, urbanization, and environmental degradation have transformed the region and changed its relationships with other world regions. Each of these processes has interacted with local conditions to produce a new mosaic of distinct geographies and produced new opportunities and challenges for people and policymakers throughout the region. Looking to the future, there are several important trends and challenges for Latin America and the Caribbean. Key challenges relate to climate change, economic forces, and political movements.

Sustainability in a Changing Climate The first challenge is that of sustainable development in a changing climate. The future of the region will depend on its ability to manage the resource demands and pollution emissions of new industrial development and farming technologies and the pressures of new consumption habits and growing populations. The region faces serious risks; climate change may degrade water supplies and ecosystems, increase disaster losses, and encroach on coastlines.

As discussed earlier in the chapter, many countries in Latin America and the Caribbean are moving to new low-carbon energy sources, and this, together with protecting forests, may reduce the region's contribution to greenhouse gas emissions. But even if the region limits or reduces emissions, it remains vulnerable to developments that would increase emissions in other regions and cause worldwide warming. For this reason, some countries are making efforts to adapt their agricultural areas, islands, and cities to warmer conditions and sea-level rise.

▲ **FIGURE 7.32 São Paulo** High-rise office buildings in this capital city are surrounded by slum housing called *favelas*.

▲ **FIGURE 7.33 Rio de Janeiro** Rio de Janeiro has a stunning location with a harbor overlooked by Sugar Loaf Mountain and the beaches of Copacabana and Ipanema. The city hosted the 2012 Rio+20 Earth Summit and will host the 2014 soccer World Cup and the 2016 Olympics.

Another sustainable development challenge is to reduce environmental problems in the region's cities. More and more people are likely to live in cities, and national and urban authorities need to rehabilitate older infrastructure and extend it into slums and suburbs to ensure clean air and water, waste disposal, energy supplies, effective transportation, and safe housing. However, as smaller cities become more attractive, companies and people may relocate to smaller regional settlements that are more livable.

Emerging Economies Serious challenges are also emerging from rapid economic growth in the region. Brazil in particular is becoming a world power, joining India and China as a major player in the global economy. Continued economic growth in the region will depend on increasing demand from a larger domestic middle class and further diversification of the economy. Other countries in the region, such as Mexico, Chile, and Costa Rica, are also well on their way to having more diversified, wealthier economies with larger middle classes, stronger international exports, and more robust domestic demand. Diversifying economies based on tourism or a narrow range of exports could help stabilize economies across the region, especially in the Caribbean. The region's economic future may ultimately depend as much on markets within the region as beyond, especially if the global economy remains unstable, and as the United States and China continue to negotiate their complex and dominant trade relationship.

Ensuring Democracy In addition to climate and economics, politics is always of interest in this region. Although most of the region has moved toward more democratic governments, given Latin

America and the Caribbean's diverse political parties and active citizen's movements, there are many undercurrents that could threaten fragile democracies. These include poverty and inequality, autocratic rule, powerful militaries, and the drug economy. Mexico's future in particular may depend on reducing the power of the drug economy that makes citizens and businesses feel insecure. In a number of countries, the regional backlash against government cuts has resulted in the election of governments that pay more attention to social safety nets but may still be influenced by powerful private sector interests and have an inadequate tax base. Marginalized populations, including Indigenous groups and the desperately poor of countries such as Haiti, still seek representation and fair treatment. Wide access to TV and other media means that the poor are more aware of inequalities.

Surveys of public attitudes across the region indicate many residents of Latin America and the Caribbean are positive about the future, but their greatest concerns are about economic problems and crime. They also, in general, support democracy and environmental protection. Brazilians were most positive, with more than 70% of those surveyed believing that the economic future will be better for their families, followed by Colombians (61%). But these numbers contrast with El Salvador, where less than a quarter of the respondents see a positive economic future. Overall, however, almost three-quarters of people in the region say they are satisfied with their lives. If Latin America and the Caribbean can continue to build the core values of democracy and grow their economies while protecting their environments and the poor, the region could have an increasingly prominent role in global affairs and a positive future.

LEARNING OUTCOMES REVISITED

■ **Explain the physical origins and social impacts of natural and human-caused disasters in the region, including earthquakes, hurricanes, deforestation, and the effects of El Niño and climate change.**

The region has mostly tropical climates ranging from rainforests to deserts, with cooler climates in highland regions and in southern South America. The climate and geology of the region produce many extremes; this includes devastating earthquakes, such as those that occurred in recent decades in Haiti and Chile, and hurricanes across the Caribbean and Central America. Changes in ocean currents and temperatures, called El Niño, periodically bring heavy rains and droughts to the Andes and other parts of the region. The impacts of these disasters are increased by the poverty and substandard housing that makes people vulnerable to the effects of natural disasters and climate change. The tropical forests of the region, most notably the Amazon, have been cut down for timber and agricultural land, causing environmental degradation such as drought, soil erosion, and species loss and disruption of forest communities. Efforts to stem deforestation have been successful in some countries: examples include Brazil, where reserves were set aside for rubber tappers, Indigenous groups, and conservation; and Costa Rica, where forests are seen to provide tourism and protect biodiversity and watersheds.

■ **Describe the ways in which the Maya, Aztecs, and Incas adapted to environmental challenges.**

The Maya, Aztecs, and Incas adapted to difficult environmental conditions by terracing hillsides to farm steep slopes, developing innovative irrigation techniques for drier areas, and using slash-and-burn farming to clear and fertilize soils in forest environments. They also used raised fields to farm flooded areas and domesticated wild animals. The case of the Maya illustrates how overfarming can lead to the collapse of a society when farming exhausts the soils and deforestation creates local climate changes.

■ **Summarize the common experiences and legacies of colonialism in the region.**

Latin America and the Caribbean share a history of European colonialism, which left a legacy of Spanish language and culture from Mexico to Chile and of Portuguese language in Brazil. Colonialism in the region also resulted in widespread practice of the Catholic religion. Other European countries, such as Britain and France, colonized the Caribbean and its coast. Colonial powers oriented regional economies to export minerals and crops and introduced slaves when European diseases decimated local Indigenous populations.

■ **Compare and contrast the economic policies and impacts of import substitution, structural adjustment programs, and free trade in countries across the region.**

As Latin American governments emerged as independent in the 20th century, concerns about low export prices and high import prices led many countries to prioritize domestic manufacturing and set barriers to imports through import substitution policies. When debt forced many countries to appeal for assistance, structural adjustment programs required countries to sell off publicly owned companies and utilities, cut government spending, and remove trade barriers. The opening of Latin American economies was encouraged by free trade agreements, such as NAFTA, which promoted hemispheric trade. As the region entered the 21st century, many economies were growing rapidly across the region. In Brazil, Mexico, and Chile, diverse economies, a growing middle class, declining poverty, and strong export and domestic demand have defined these countries as emerging industrial economies.

■ **Describe the historical sequence of U.S. intervention, the rise of authoritarianism, key revolutions, and the emergence of democracy across the region.**

U.S. influence in the region was formalized with the Monroe doctrine. The United States also had continued concerns about left-leaning governments in the region, especially during the Cold War. These concerns led to a series of U.S. interventions in countries such as Cuba, Chile, and Guatemala and also were the basis for the U.S. control of the Panama Canal. Powerful landholding elites and military governments dominated much of the region during the late 19th and 20th century, often violently repressing dissent and Indigenous peoples. Revolutionary movements succeeded in countries such as Mexico and Cuba and led to land reform and nationalization, but other such movements were eventually overthrown in countries such as Chile, Guatemala, and Nicaragua. In recent years, democratically elected governments have been established across the region, and social movements have been able to achieve some of their goals.

■ **Describe the history and current distribution of Indigenous peoples and cultures across the region, the ways in which the plants they domesticated have influenced global food and health, and how slavery and colonialism changed the demography of the region.**

The Indigenous populations of the Americas declined rapidly following European arrival in a demographic collapse. The Indigenous peoples of the region had domesticated plants that included maize and potatoes, which became global food staples, as well as tobacco, which became a worldwide health problem. African slaves were introduced as colonial labor across the region, especially in Brazil and the Caribbean, and racial mixing led to new ethnic categories, such as mestizo. Today significant Indigenous populations remain in Guatemala, Bolivia, and Mexico and a large percentage of people in countries such Brazil, Cuba, Jamaica, and Haiti identify themselves as black.

■ **Understand the geography and impacts of the drug trade and international tourism in the region.**

The region receives millions of international tourists seeking tropical beaches, especially in the Caribbean and Mexico as well as some tourists who are seeking ecotourist experiences in countries such as Costa Rica. Tourism supports economies, but can undermine local culture, overuse resources, and cause pollution. The drug trade, primarily driven by demand in the United States, has promoted the growth of powerful criminal groups known as drug cartels, which have destabilized politics and increased violence in Colombia and Mexico.

■ **List the factors promoting urbanization in the region and provide examples of the social and environmental problems associated with urbanization.**

In contemporary Latin America, migration flows are driven by economic conditions as people—pushed by unemployment and lack of health, education, energy, and water in rural areas—move in search of work and services to cities. Latin America's largest cities, such as Mexico City, Rio, Santiago, and São Paulo, have serious problems with pollution, congestion, and waste. They are often surrounded by slums that lack services but house millions of people who often work in informal economies.

KEY TERMS

ALBA (p. 292)

altiplano (p. 269)

altitudinal zonation (p. 267)

biodiversity (p. 272)

biofuel (p. 278)

bioprospecting (p. 279)

bracero (p. 298)

carbon market (p. 269)

cartel (p. 293)

circle of poison (p. 278)

Columbian Exchange (p. 281)

demographic collapse (p. 281)

ecosystem services (p. 275)

ecotourism (p. 279)

El Niño (p. 267)

encomienda (p. 283)

extractive reserves (p. 277)

favela (p. 299)

Green Revolution (p. 278)

hacienda (p. 283)

import substitution (p. 284)

informal economy (p. 289)

La Niña (p. 268)

liberation theology (p. 294)

machismo (p. 296)

maquiladora (p. 284)

marianismo (p. 296)

maroon communities (p. 296)

megacity (p. 298)

mestizo (p. 295)

Monroe Doctrine (p. 289)

nontraditional agricultural exports (NTAEs) (p. 285)

North American Free Trade Agreement (NAFTA) (p. 284)

plantation (p. 283)

pristine myth (p. 275)

rain shadow (p. 267)

REDD (p. 269)

remittances (p. 298)

social movements (p. 292)

structural adjustment programs (SAPs) (p. 284)

syncretic (p. 294)

tipping point (p. 268)

trade winds (p. 266)

Treaty of Tordesillas (p. 280)

urban primacy (p. 298)

westerlies (p. 266)

THINKING GEOGRAPHICALLY

1. How did colonization change the ecology and environment in the region through the Columbian Exchange and other processes?

2. Which regions of Latin America are particularly vulnerable to natural hazards, and why and how might climate change and alterations in land use, such as deforestation, increase these risks?

3. How did colonialism restructure the economies of Latin America and what impacts did that restructuring have?

4. Why has the United States played such a prominent role in Latin America and the Caribbean?

5. In addition to the United States, which other world regions have had interests in the region and why?

6. How do free trade, migration, and drugs link Mexico and the United States?

7. How have urban–rural contrasts driven the growth of major cities in Latin America, and what are the main environmental and social problems of these urban areas?

8. To what extent did Latin America shift from authoritarianism and revolution to democracy in the late 20th century? How has the rise of social movements influenced politics in the region?

9. What has contributed to the recent economic success of Brazil?

MasteringGeography™

Looking for additional review and test prep materials? Visit the Study Area in MasteringGeography™ to enhance your geographic literacy, spatial reasoning skills, and understanding of this chapter's content by accessing a variety of resources, including **MapMaster** interactive maps, videos, RSS feeds, flashcards, Web links, self-study quizzes, and an eText version of *World Regions in Global Context*.

8 | East Asia

This photo illustrates the terrible aftermath of the 2011 earthquake and tsunami off the coast of Japan. Situated along the western edge of the Pacific Ring of Fire, the region is under constant threat of such disasters. Though architecture and urban planning have been adapted to anticipate catastrophic events, the magnitude of the recent disaster was such that preparation was not enough to protect inhabitants.

IN THE SUMMER OF 2012, SOCCER BALLS, BOATS, ROOF BEAMS, and bicycles washed ashore on the West Coast of North America. These floating emblems of the wealth of Japanese society were swept out to sea by the tsunami that struck Tohoku on March 11 of the previous year. More than 15,000 people were killed, hundreds of thousands of structures collapsed, millions of households lost electricity and water, and hundreds of thousands of residents were evacuated in the wake of perhaps the most powerful earthquake in Japanese history. Destructive waves and seismic aftershocks led to meltdowns at the Fukushima Daiichi Nuclear Power Plant complex. Global nuclear experts, from the United States and elsewhere, were dispatched to control the growing threat. Ultimately all of Japan's nuclear power supply was taken offline. By May 2012, a country that had previously received 30% of its electricity from nuclear sources was nuclear-free.

The events surrounding the disaster illustrate several characteristics of the East Asian region. The region's eastern flank sits on an area of high seismic activity at the edge of the

LEARNING OUTCOMES

■ Explain how the landscape of the East Asian region has been modified by humans over the centuries, particularly in regards to water management, and how these adaptations may be impacted by climatic change.

■ List the main physical subregions of East Asia and describe their tectonic origins and related hazards.

■ Identify the adverse ecological effects of urbanization and modernization of agriculture.

■ Distinguish the centers of power in East Asia in the precolonial, colonial, and postcolonial periods and describe how and why they have changed over the course of history.

■ Compare the development of different economic experiments in the 20th and 21st centuries in China, South and North Korea, and Japan.

■ Identify central geopolitical tensions within East Asia and their historic and recent drivers.

■ Use specific examples to show how East Asian cultural practices and systems are the product of distinct regional traditions as well as linkages across and beyond the region.

■ Describe the various forces that have caused significant declines in population growth across East Asia.

■ Summarize and explain the major immigration trends in the region such as rapid urbanization.

Pacific plate. The very existence of the Japanese islands is a product of the same violent tectonic forces that set off the 2011 tsunami. The high level of dependence on nuclear power is also a geographical phenomenon. Limited domestic natural resources (such as coal or oil) led to Japan's adoption of nuclear power. The expansion of nuclear power was also a response to available human resources; technical expertise, education, and training are hallmarks of Japanese industrial development. In this way, distinctive regional conditions led to Japanese nuclear capacity and at the same time made the coastal region vulnerable to disaster.

The mass of garbage that traveled across the Pacific Ocean to come ashore in Alaska, British Columbia, and Washington was carried on currents driven by prevailing westerly winds that connect the weather patterns of East Asia and North America. The fate of these two regions is intertwined through climate just as it is connected by the shared technical capacity of Japanese and U.S. nuclear expertise. The "Great East Japan Earthquake" clearly demonstrates the high-stakes interplay of global forces and regional conditions. ■

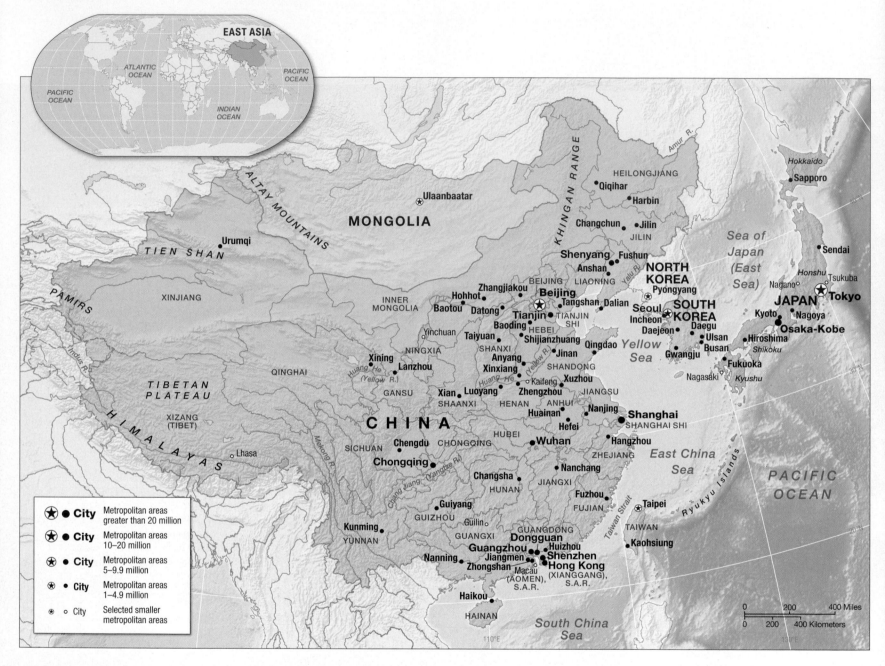

▼ FIGURE 8.1

EAST ASIA

City legend:
- (★) ● **City** Metropolitan areas greater than 20 million
- (★) ● **City** Metropolitan areas 10–20 million
- (★) ● **City** Metropolitan areas 5–9.9 million
- (★) • City Metropolitan areas 1–4.9 million
- (✦) ○ City Selected smaller metropolitan areas

East Asia Key Facts

- Major Subregions: Tibetan Plateau, Central Mountains, Continental Margin, Japanese archipelago

- Major Physiogeographic Features: The Tibetan Plateau includes the Himalayan Mountains (including Mount Everest) and is the source of the great Mekong and Indus Rivers. Cool summers and cold winters

- Major Religions: Buddhism, Daoism, Confucianism, and various folk religions

- Major Language Families: The dominant Sino-Tibetan family contains the various subbranches of the Chinese language. Japonic and Altaic families are also widely spoken.

- Total Area (total sq km): 9.6 million

- Population (2011): 1.581 billion, including 1.345 billion in China, 128.1 million in Japan, and 49 million in South Korea; Population under Age 15 (%): 16; Population over Age 65 (%): 10

- Population Density (per square km) (2011): 134

- Urbanization (%) (2011): 54

- Average Life Expectancy at Birth (2011): Overall: 75; Women: 78; Men: 73

- Total Fertility Rate (2011): 1.5

- GNI PPP per Capita (current U.S. $) (2009): 9,650

- Population in Poverty (%, < $2/day) (2000–2009): 36

- Internet Users (2011): 676,055,833; Population with Access to the Internet (%) (2011): 58.5; Growth of Internet Use (%) (2000–2011): 595.5

- Access to Improved Drinking Water Sources (%) (2011): Urban: 98; Rural: 83

- Energy Use (kg of oil equivalent per capita) (2009): 2,369

- Ecological Footprint (hectares per capita consumed/hectares per capita available, global scale) (2011): 3.7/1.8

ENVIRONMENT AND SOCIETY

The most striking and geographically significant environmental feature of East Asia (**Figure 8.1**) is its position on the east end of the vast Eurasian continent. This position affects regional climate, resource endowments, and historical flows of people, trade, and culture. Its formidably high interior plateaus and towering mountains are the engines that drive the lengthy river systems. Its peninsular and island coasts create a set of oceanic subregions abutting the continental mass (**Figure 8.2**). These characteristics are the product of a long geotectonic history that links the region to other world regions. The Tibetan Plateau and Himalayan ridge are the result of a collision with the South Asian Plate; the island boundaries (including all of contemporary Japan) are the result of tectonic activity. The boundary region is therefore part of the tectonically active so-called **Ring of Fire**, a circle of volcanoes and earthquake zones that surrounds the Pacific Plate, stretching from Southeast Asia through the Philippines, north across Japan and Kamchatka and then down the Pacific coast of North America. Together, these physical regions set the stage for the complex adaptation of regional cultures to climate, landscape, and a diverse set of ecosystems.

Climate, Adaptation, and Global Change

The geological uplift of the Tibetan Plateau that occurred between 65 million and 2 million years ago is the key to understanding certain aspects of the climate of East Asia (**Figure 8.3**, p. 308). When the uplift first occurred, the elevation of the newly formed Tibetan Plateau cut off the moisture formerly brought into the interior of East Asia from the Indian Ocean. This contributed to the gradual desiccation (drying) of the Tibetan Plateau, whose once numerous lakes are now much reduced in size. Because of its mountain barriers and sheer distance from the coast, much of western and northwestern East Asia today averages less than 125 millimeters (5 inches) of rain a year. On the Tibetan Plateau, high elevations make for cool summers and extremely cold winters. Farther north, in Xinjiang, Qinghai, Gansu, and Mongolia, summer temperatures can be extremely hot. The Turfan Depression, some 154 meters (505 feet) below sea level, is one of the hottest places in East Asia, with recorded temperatures in excess of 45°C (113°F).

To the east of this vast arid region are two distinctive continental climatic regimes. The northern regime is subhumid (slightly or moderately humid, with relatively low rainfall) and extends southward as far as a latitude of about 35°N, encompassing the Northeast China Plain, the North China Plain, the northern parts of the Korean peninsula, and the Japanese archipelago. Winters here are cold and very dry. Summers are warm, with moderate amounts of rain from the southeasterly monsoon winds. Rainfall, however, is extremely variable, so that both drought and flooding occur frequently. The southern regime is humid and subtropical. It extends west from the plains of the middle and lower Chang Jiang valley as far as the Sichuan basin, and south as far as the southernmost coastlands of South China. Winters here are mild and rainy, and summers are hot with heavy monsoonal rains. Overall, annual rainfall is 1,200 millimeters (47 inches), higher than in the north. In the arid and subhumid regions of East Asia, drought is a critical natural hazard, causing widespread famine as a result of crop failures in drought years. In addition, the subhumid parts of East Asia tend to be prone to flooding.

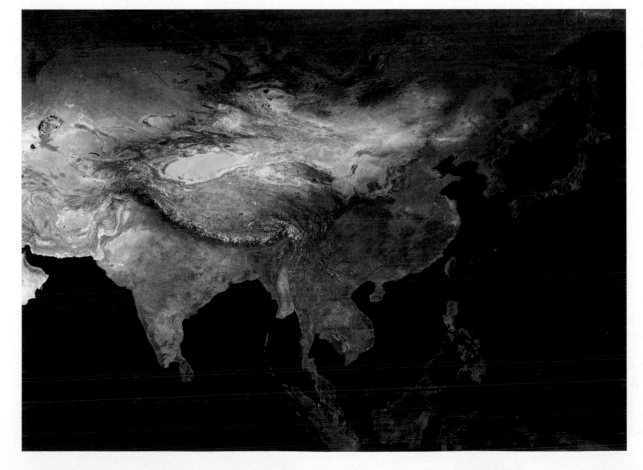

◀ **FIGURE 8.2 East Asia from Space** Much of East Asia consists of vast areas of upland plateaus and mountain ranges.

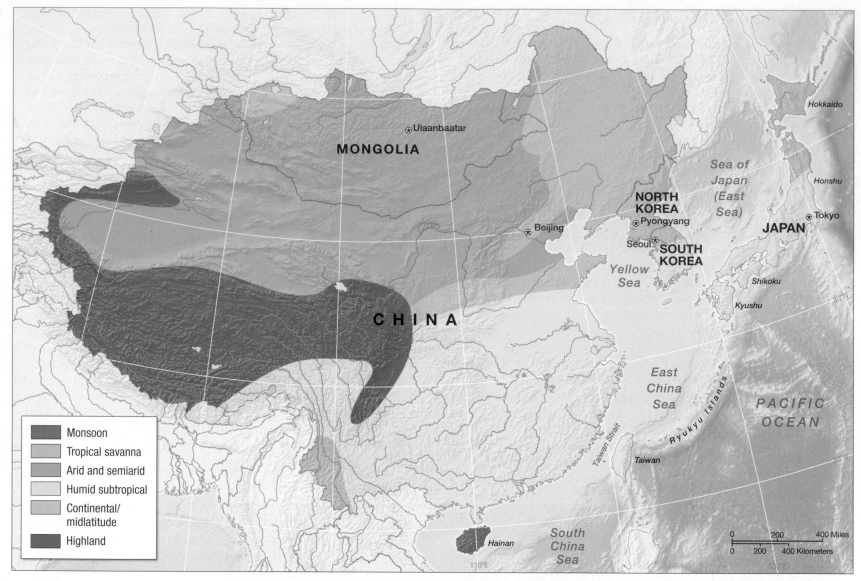

▲ **FIGURE 8.3 Climate Map of East Asia** East Asia is divided between four main climates, including the variable highland climate, the arid climates of the interior, the continental climates of the north, and the humid subtropical climates of coastal China and southern Korea and Japan.

Adapting to Semiarid and Subtropical Climates Each of these distinctive climates influences regional cultures, especially as humans adapt to very different rainfall patterns and temperature ranges. The western deserts and grasslands of Xinjiang and Mongolia, for example, have a long and well-developed herding tradition that mixes settled agriculture with herd mobility to best utilize infrequent and unevenly distributed rainfall.

The cultures of southeastern China and Japan, in contrast, have developed systems to capture and control the flow of water to maximize production year-round and to combine rice production with fish production. This rice–fish culture, which dates to at least as early as 300 C.E., allows the production of protein and rice. Locals raise fish in wet rice ponds, where rice plants provide shade and organic material for the fish and the fish eat the pests that might attack the rice plants, while also oxygenating the water and the soil. The system also produces its own fertilizer, often making supplementary chemical inputs

unnecessary. Water is harnessed through complex systems of carefully designed irrigation canals and levees (**Figure 8.4**). This technique likely originated in China but has since diffused across East Asia and is found as far away as the island of Java in Southeast Asia and in parts of India. Innovative adaptations to climate change have been, in this way, a cultural connection between East Asia and the rest of the tropical world.

Climate Change A warming trend over the last 50 years, especially pronounced during the winters, presents serious problems and challenges for this region. The impact of this overall warming is unclear, but it probably will mean decreases in catch in the Pacific fisheries on which most countries in the region heavily depend. It may also do serious damage to the delta ecosystems along the coasts of China and Japan. Rainfall regimes are less easy to predict because trends over the last century include drier average conditions in some

▲ **FIGURE 8.4 Dams and Levees of a Complex Rice System** The integration of fish farms and rice paddies, seen here on the banks of the Li River, China, is an ancient technique that maximizes production of both proteins and grains.

change, urban areas in Japan and China represent an enormous challenge for mitigating climate change. China's 2008 CO_2 emissions (7 billion tons) surpassed those of the United States (5.5 billion tons). This figure continues to rise, largely as a result of ongoing Chinese development, urbanization and the accompanying transportation, construction, and lifestyle demands. Japan is already a major emitter of greenhouse gases (emitting 1.32 billion tons in 2008). Japan's per capita rate far exceeds that of China due to its smaller population and energy-demanding lifestyle. Where each person in China emits approximately 6 tons per year, the per-person rate in Japan is 9 and in the United States more than 17.6 tons. Future rounds of negotiations over greenhouse emissions and climate change will increasingly focus on China, especially if urban areas continue to grow (**Figure 8.5**).

parts of the region (including northeast China) and wetter trends in others (including arid western China). Intense rainfall appears to be on the increase, with more dramatic single rain events causing increased flooding in western and southern China and Japan.

Flooding is being increasingly addressed by the Chinese government, which has invested in dams, flood control, and resettlement programs to reduce climate change risk. As recently as 2011, however, flooding in southwestern Sichuan Province in China destroyed crops and housing, causing the displacement of thousands in what was probably the worst flooding on record for the area. Droughts, meanwhile, have become more common in areas to the north and east, with associated dust storms and crop failures in northern China and livestock die-offs in Mongolia. In the same year as the Sichuan flooding, China's major agricultural regions confronted the most dramatic drought in more than 60 years. Despite efforts at engineered controls, these floods and droughts, and the prospect of increasing climate events like these, vex Chinese planners.

While problematic, these conditions may open onto new regional adaptations. Drought-tolerant crop options are part of the historical crop diversity of the region. Similarly, traditional livestock breeds—including hardy Mongolian cattle varieties and a range of Tibetan goats—are well-adapted for water scarcity and variability. Some of the secrets for adapting to climate change in the future likely lie in the adaptive history of East Asian producers of the past.

Given the critical influence of urban activities on the greenhouse gas emissions that drive climate

APPLY YOUR KNOWLEDGE Cite a specific example of a way people in East Asia have modified their physical environment to adapt to weather and climate patterns. To what degree might people need to modify this adaptation method in the face of climate change?

▲ **FIGURE 8.5 Chinese Industrial Emissions** China, whose greenhouse gas emissions are growing more rapidly than any other country's, is seeking ways to modernize its industry and energy production into cleaner forms. This photo shows emissions from a cement plant in Baokang.

Geological Resources, Risk, and Water Management

Much of East Asia consists of plateaus, basins, and plains separated by narrow, sharply demarcated mountain chains. These broad physical regions contain a great diversity of river systems and elevational gradients, providing water resources as well as hazards. The satellite photograph of East Asia in Figure 8.2 (p. 307) suggests a general, threefold physical division: the Tibetan Plateau; the central mountains and plateaus of China and Mongolia; and the continental margin of plains, hills, continental shelves, and islands (**Figure 8.6**).

The Tibetan Plateau In the southwest, the Tibetan Plateau, an uplifted **massif** (a mountainous block of Earth's crust) of about 2.5 million square kilometers (965,000 square miles), forms a plateau of several mountain ranges—including the Himalaya Mountains—with peaks of 7,000 meters (22,964 feet) or more. The Tibetan

Plateau is a unique physical environment, a vast area violently uplifted in relatively recent geological time to produce the youngest, highest plateau in the world. Its lofty mountains tower to heights of between 6,000 and 8,000 meters (19,684 to 26,245 feet): the Himalayas contain the world's highest peak, Mount Qomolangma (Mount Everest), at 8,848 meters (29,027 feet).

Plains, Hills, Shelves, and Islands The bulk of the population of East Asia lives on the continental margin, which includes the great plains of China: the Northeast China (Manchurian) Plain, the North China Plain, and the plains of the Middle and Lower Chang Jiang (Yangtze River) valley. Most of these plains lie below 200 meters (656 feet) in elevation. South of the Chang Jiang is hill country, with elevations of about 500 meters (1,640 feet); along the coast and the Korean peninsula are uplifted hills and mountains of 750 to 1,250 meters (2,460 to 4,100 feet). The Japanese archipelago of Hokkaido, Honshu, Shikoku, and Kyushu forms the outer arc of

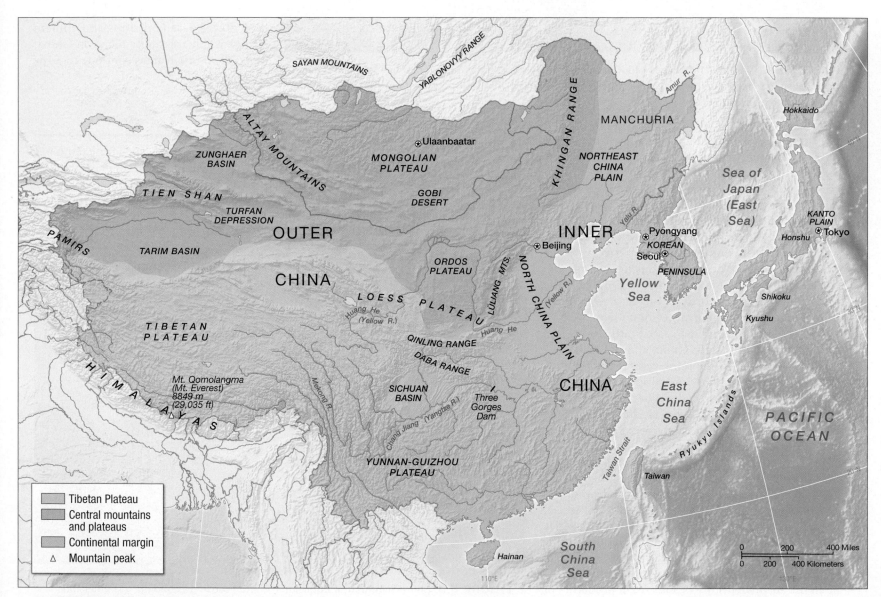

▲ **FIGURE 8.6** East Asia's Physiographic Regions East Asia features three broad physiographic divisions: the Tibetan Plateau; the central mountains and Outer China; and the continental margins in an arc spanning Inner China, the Japanese archipelago, the Korean peninsula, and Taiwan.

East Asia's continental margin. A backbone of unstable mountains and volcanic ranges projects from the shallow seafloor and extends to the island of Taiwan. This outer perimeter of islands and peninsulas is part of the Ring of Fire that girdles the Pacific Ocean.

Earthquakes in a Still-Forming Region

The key to understanding the topography of the region is the plate tectonics that have had a dramatic influence in recent geological times. The entire Tibetan Plateau was uplifted between 65 million and 2 million years ago. The Indian–Australian Plate moved northward and pushed up against the Eurasian Plate, not only uplifting the Tibetan Plateau, but also causing the mountain-building episode that resulted in the Himalayas. The weathering of the newly created mountains provided huge quantities of clay, silt, and other fluvial deposits that now blanket the plains. To the east, the movement of the Pacific Plate toward the Eurasian Plate caused the folding and faulting that resulted in the mountains of the peninsulas and islands

▲ **FIGURE 8.7 The 2011 Tsunami** The devastating wave that hit the Japanese coast in March 2011, seen here crashing over Miyako City, was set off by an offshore earthquake.

of the continental margin, including the Korean peninsula and the Japanese archipelago.

The geological uplift of the Tibetan Plateau also increased the gradient of the rivers that flow off the borders of the plateau, enabling them to incise (cut their way into) the plateau and produce a great number of gorges and canyons of considerable length and depth. The rivers are still incising rapidly, and the Tibetan Plateau itself is still not stable: seismic disturbances continually occur along the Himalayan foothills, often bringing disaster to settlements there.

The physical environments of East Asia are, in fact, relatively hazardous. In addition to regular earthquakes along the Himalayan foothills, much of the North China Plain is subject to seismic activity. In May 2008, a massive earthquake destroyed schools, houses, and factories across the Sichuan Province of central China, taking the lives of approximately 70,000 people. Much of Japan, too, is subject to earthquake hazards.

Most dramatic of all such recent tectonic events are those surrounding the Great East Japan Earthquake of 2011 discussed in the chapter-opening vignette. The Pacific Plate, which moves east at a rate of 8 to 9 centimeters (3.1 to 3.5 inches) per year, is subducting under the Eurasian Plate (see Chapter 1, p. 2). This violent process leads to the periodic release of massive amounts of energy. This is what happened in March 2011, when a 9.0-magnitude earthquake occurred just 70 kilometers (43 miles) east of the Peninsula of Tōhoku, sending shockwaves through the ocean and creating the resulting legendary tsunami. The enormous waves took between 10 and 30 minutes to reach the heavily populated shore of the country, causing widespread devastation (**Figure 8.7**).

Flooding, Flood Management, and Hydroengineering

Like earthquakes, inland flooding hazards are an ongoing problem in

this region. This is partly because the rivers that flow over long courses across the plains carry a tremendous amount of silt. This silt is deposited in more sluggish lower reaches, building up the height of riverbeds and making river courses unstable. The Huang He (Yellow River) is the largest and most notorious of these rivers, having changed its course several times in the last two centuries. The river's name derives from the color it is given by its heavy silt load: a distinctive brown-gold color. That color reflects both its productivity, which comes from the fresh nutrients and soil it delivers to the agricultural plains, as well as its danger, which results when it breaks out from its high riverbeds, flooding the farms and villages of the densely populated North China Plain. Three of the most devastating floods in worldwide historical record were along the Huang He: in 1938 (up to a million casualties), 1931 (more than 1 million casualties), and 1887 (as many as 2 million casualties).

The people of East Asia have over the millennia developed sophisticated techniques to modify the landscape for water control. Over the centuries, marshes were drained, irrigation systems constructed, lakes converted to reservoirs, and levees raised to guard against river floods. These traditions of water management are remarkable examples of coordinated action from strong imperial states taming highly irregular and unpredictable physical systems. This success has caused some researchers to term the regions' historic empires as **hydraulic civilizations**—civilizations that survive, thrive, and expand based on their capacity for controlling water.

The tradition continues into the present (as discussed in the earlier material on climate change). A massive public works program aimed at hydroelectric facilities, irrigation and drainage, and improved waterways has been the cornerstone of the Chinese central government approach to the economic development of the country's interior.

▲ **FIGURE 8.8 Three Gorges Dam** The Three Gorges Dam is one of the world's largest-ever engineering projects. The dam is located in central China where the Chang Jiang narrows to form the Xiling, Wu, and Qutang Gorges.

Ecology, Land, and Environmental Management

The ecologies of East Asia reflect the broad land and climate forms of the region. These natural divisions have been subject to dramatic transformation under successive civilizations that sculpted the physical environment over millennia. The principal human impact has been the clearance of land for farming. For several thousand years, as the population grew and premodern civilizations flourished, much of humid and subhumid East Asia was cleared of its forest cover. Even the topography was altered to suit the needs of growing and highly organized populations, as hills and mountainsides were sculpted into elaborate terraces to provide more cultivable land (**Figure 8.9**).

Several of the world's most important food crops and livestock species were domesticated by the peoples of East Asia, beginning around 6500 B.C.E. Millet, soybeans, peaches, and apricots were domesticated in the northerly subhumid regions. Rice, mandarin oranges, kumquats, water chestnuts, and tea were domesticated in the more humid regions to the south. Hemp was domesticated as a source of fiber and oil. Chickens and pigs were among the livestock species domesticated and mulberry bushes were domesticated as silkworm fodder. Only in the arid western parts of East Asia and the inhospitable Tibetan Plateau—a broad region often referred to as "Outer China"—do contemporary regional landscapes reflect large-scale natural environments relatively unmodified by human intervention. The scenery here is often spectacular, encompassing

The Three Gorges Dam on the Chang Jiang (**Figure 8.8**) is emblematic of this strategy and of China's ambitious determination to modernize. It has also, unfortunately, become emblematic of the problematic social and environmental consequences of large-scale infrastructure development. The dam spans 2 kilometers (1.25 miles) from bank to bank. The reservoir extends 650 kilometers (404 miles) upstream. Its construction submerged 19 cities, 150 towns, and 4,500 villages and hamlets, and displaced millions of residents. Final completion of the dam's 32 massive generators occurred in 2012; the dam will provide almost 10% of China's energy output, bringing electric power to millions of rural households and opening central China to a much broader range of industries. The increased water level in the enormous reservoir improves river navigation and helps flood control along the Chang Jiang. However, a significant potential problem is that the dam sits on an active fault line. The weight of the water in the reservoir could trigger an earthquake, with devastating consequences. Critics also claim the dam is threatening the region's ecosystem. The river blockage endangers several species indigenous to the region, including the Chinese alligator, the white crane, and the Chinese sturgeon. And conservationists have no hope of recovering more than a fraction of the historical artifacts and relics that have been permanently submerged.

APPLY YOUR KNOWLEDGE Describe two main physical features of the East Asia region. What tectonic forces shaped the landscape into these forms? List two hazards associated with these forces.

▲ **FIGURE 8.9 Agricultural Terracing, Yuanyang, China** This technique for carving productive farmland out of steep slopes is practiced across the world, but the elegantly worked hillsides of rural China are a testimony to the efficiency and productivity of the region's ancient rice culture.

snow-clad mountains, vast swamps, endless steppes, and fierce deserts.

Tibetan and Himalayan Highlands Not surprisingly, the rugged terrain of the Tibetan Plateau is quite sparsely inhabited. Outer China contains barely 4% of the population of East Asia. Large areas are effectively uninhabited, and in general population density does not exceed one person per square kilometer. Nomadic pastoralism remains the chief occupation within the region; the principal source of livelihood is the yak, along with sheep and goats. Yaks supply milk and wool as well as meat. Yak wool is the fabric used for the darkish, rectangular tent that is a distinctive component of the Tibetan landscape. In recent decades, Chinese reforms have brought an expansion of farming to the more sheltered valleys of the region, with wheat, highland barley, peanuts, and rapeseed as the principal crops. The mixed forests of Gaoshan pine, Manchurian oak, and Himalayan hemlock are increasingly being brought into timber production, accelerating a long history of deforestation in the region and extending the imprint of humankind on this wild region. At the eastern edge of the plateau, overgrazing, combined with a decade of hotter, drier weather, has destroyed the thin topsoil, with the result that this legendary country is now gradually being degraded.

Steppe and Desert Northwestern China (**Figure 8.10**) and much of Mongolia is a territory of temperate desert—grasslands, scrub-covered hills, and basins with plant communities tolerant of high levels of mineral salts in the soil—although dense forests of tall spruce grow on the cool, wet flanks of the Tien Shan and Altay Mountains. Inner Mongolia is mostly a mixture of steppe and semidesert.

Much of Mongolia is taken up by the Gobi Desert, where bare rock and extensive sand dunes dominate the landscape. These natural landscapes were impacted significantly from the 1950s to the 1980s by an extensive agricultural modernization program similar to that undertaken in the Soviet Union (see Chapter 3). Hundreds of thousands of acres of land were plowed, and large-scale irrigation systems were installed to allow for farming—especially wheat, root vegetables, and fodder crops. More recently, Mongolia's extensive mineral resources (including oil, coal, copper, lead, and uranium) have been exploited, and an industrial sector has emerged, stimulating urban growth. As of 2012, Mongolia's mining sector represents 70% of its exports. Coal exports reached a level of 25 million tons in 2011, at a value in excess of U.S. $2 billion. The benefits from this development are poorly distributed however. More than a third of Mongolians live on less than U.S. $1.25 a day.

Inner China, Korea, and the Japanese Archipelago The remainder of East Asia—"Inner China" plus the Korean peninsula, the Japanese archipelago, and Taiwan—can be divided broadly into the landscapes of the subhumid regions to the north and those of the humid and subtropical regions to the south. Mixed temperate broad-leaf and needle-leaf forests on the hills and mountains dominate the natural landscapes of the northern part of Inner China, along with the northern part of the Korean peninsula and the northern half of the Japanese archipelago. Higher elevations and more northerly latitudes are dominated by spruce, fir, and birch trees; lower elevations and more southerly latitudes are dominated by maple, basswood, and oak as well as areas of extensive and productive agriculture. In contrast, the winter landscape of the North China Plain is brown and parched, and on windy days the air is thick with the dust blowing from the silty soils.

The island of Taiwan, the southern part of the Korean peninsula, and the southern half of the Japanese archipelago, are dominated by low hills covered with a secondary growth of evergreen monsoon forest, leaving less than 10% of the land for agriculture. Nevertheless, the cultivable land is very productive, and the climate allows for double-cropping of paddy rice and for cash crops such as sugarcane, mulberry trees, and hemp.

Around Guilin and in the Zuo River area to the south, near Nanning, erosion has dissolved areas of soft limestone, leaving spectacular pinnacles several hundred feet tall. These improbably shaped hills contrast with intermediate areas of intensely cultivated flatlands and winding rivers, making for some strikingly beautiful landscapes

▲ **FIGURE 8.10 Northwestern China** Surrounded by mountain ranges, the vast plains and deserts of northwest China stretch for thousands of kilometers.

▲ **FIGURE 8.11 The Guangxi Basin** A unique combination of geomorphology and agricultural development has led to one of the world's most spectacular landscapes. From classic paintings to contemporary tourist posters, this dramatic karst (eroded limestone) scenery serves as an iconic representation of traditional China.

(**Figure 8.11**). These striking landforms, forests, and coasts have provided the inspiration for thousands of years of Chinese art.

Revolutions in Agriculture The land management systems of East Asia are thousands of years old, well established, and rooted in large-scale manipulation of both land and water. Technological innovations in production have swept the region since the 1960s, with significant implications for environmental quality and human health. These strategies are a part of the **Green Revolution**, an international effort to introduce and encourage the cultivation of new varieties of crops which produce far more food but also require higher levels of industrial inputs including fertilizer, water, and pesticides.

The environmental impacts of these changes are potentially enormous, especially considering the vast areas of East Asia under cultivation. In China, for example, 554 million hectares (1.4 billion acres) of land are under agricultural cultivation, almost 13% of the global total. In 2009, farmers in north China used roughly 590 kilograms of nitrogen fertilizer per hectare (526 pounds per acre) every year. This is six times more nitrogen fertilizer than farmers use in America on the same amount of land, and has resulted in the release of approximately 225 kilograms per hectare (200 pounds per acre) of excess nitrogen per acre into the environment. Released in waterways, fertilizers can radically damage biodiversity and the health of streams and rivers.

More dramatically, the heavy use of herbicides and insecticides, a crucial component of Green Revolution technology, have resulted in widespread illness in rural areas. Roughly 17,000 annual deaths in China are estimated to result from unintentional (nonsuicidal) exposure to pesticides. This high incidence rate has led in recent years to new provisions for rural health care in the country. It has also encouraged the adoption of genetically modified varieties of crops, modified to require less pesticide inputs. In a 2011 study, the planting of one type of genetically modified cotton was shown to reduce pesticide use by 50% in general and by 70% for the most

toxic chemicals. The result has been a decline in such deaths overall. Nonetheless, the continued high incidence of pesticide poisonings suggests that many other environmental effects from intensive agriculture may be unmeasured and unobserved in waterways and soils in China (**Figure 8.12**).

APPLY YOUR KNOWLEDGE Suggest two advantages and two disadvantages of Green Revolution technology for people and the environment in East Asia.

▲ **FIGURE 8.12 Pesticide and Fertilizer Application in China** Since the Green Revolution, East Asians have utilized large amounts of insecticides, herbicides, and fertilizers, which endanger farmers as well as environmental quality. This farmer is spraying pesticides on vegetables in Xinlou village, northeast of Guangzhou.

Conservation and Environmental Issues The dense settlement of East Asia holds implications for native fauna. Five important mammal species (including the wolf) have become extinct in Japan. In China, there are more than 385 threatened species, according to the IUCN (International Union for the Conservation of Nature). Tigers are all but gone in China; 90% of their population vanished in the last century and as few as 90 tigers may remain there in the wild today. Prominent and iconic among threatened species is the giant panda. These animals lost half their habitat (upland forest) in the 1970s and 1980s, bringing them to the edge of extinction. They present an especially difficult conservation challenge due to the difficulties breeding in captivity pose for these animals. The main threat to wild species in the region is habitat loss due to human activities, including intensive industrial logging and mining. China also hosts a large illegal trade in animal parts, used in medicines, from species like musk deer and bears (see "Geographies of Indulgence, Desire, and Addiction: Animals, Animal Parts, and Exotic Pets"). This trade is international in scope and impacts tiger populations as far away as India (see Chapter 9) as well as elephants in Africa.

There may be a location where East Asian biodiversity has thrived over the last few decades, though it remains hard to tell. This area is the so-called **DMZ** (or **demilitarized zone**), a no-man's-land located between North and South Korea and established at the end of hostilities between the two powers in the 1950s. (There is a detailed discussion of this conflict later in the chapter.) Since that time, the area has been entirely uninhabited by humans and has been heavily strewn with land mines. As a result, wildlife has come to thrive in the long corridor between the two states. Experts have identified 2,900 plant species, 70 mammal species, and 320 bird species in the DMZ, with many endangered species among them, including the red-crowned and white-naped crane, the Amur leopard, and the Asiatic black bear. There are even rumors that tigers survive there, though these reports are so far unsubstantiated. This accidental wilderness, a product of a century of conflict, is an example of an **Anthropocene** environment: it is heavily influenced by human activities and also a product of unpredictable natural forces (**Figure 8.13**).

Throughout East Asia, cities continue to grow at a rapid rate, stretching outward to devour not only agricultural land, but also natural areas. Emblematic of urban growth problems is the **Asian Brown Cloud**. This is a blanket of air pollution 3 kilometers (nearly 2 miles) thick that hovers over most of the tropical Indian Ocean and South, Southeast, and East Asia, stretching from the Arabian Peninsula across India, Southeast Asia, and China almost to Korea (see "Visualizing Geography: The Asian Brown Cloud").

The urban growth problem is acute in Taiwan, an island of merely 36,000 square kilometers (13,900 square miles), two-thirds of which is a mountainous and hilly region surrounded by burgeoning coastal urban development. Here, the increase in urbanized areas moves inevitably inward toward the island's natural core areas. Even so, Taiwan has a significant cultural advantage over some of its neighbors when it comes to preserving open lands. People in the region are more historically tolerant of high-density urban living, allowing

▲ **FIGURE 8.13 Wildlife in the DMZ** A water deer swims in a pond inside the heavily fortified demilitarized zone in Chulwon. Sixty years of conflict between North and South Korea have left the demilitarized zone largely untouched by humanity, allowing rare and endangered species to thrive.

more people to use less space. Vertical urban development in major Taiwanese cities (in high-rises and closely packed homes) helps offset the pressure of urban growth on the environment. Whether this culture will persist, on the other hand, as wealthier residents become capable of buying and developing larger properties, remains to be seen.

Elsewhere in East Asia, efforts to protect core areas from rapid urban growth take two forms. First, the establishment of national parks and protected areas in East Asia is increasing rapidly. China claims 145 areas larger than 100,000 hectares (247,105 acres) under protection, Japan claims 7, and South Korea 1. The total land area dedicated to these is inevitably limited, however, and rarely are they situated in the path of urban growth. Although six national parks have been established on Taiwan, they cover only 308,000 hectares (761,000 acres), less than 13% of the core nonurbanized area. The second mode of protection is deliberate planning efforts to manage density and capitalize on high-density residential and industrial development. Many East Asian cities are increasingly working to achieve urban economic growth without paving over adjacent natural areas.

HISTORY, ECONOMY, AND TERRITORY

East Asian civilizations are indisputably some of the oldest in the world, and the mark of these cultures is indelible on the contemporary landscape. Consider, for example, the Great Wall of China, which is 2,500 years old and still visible from space. In East Asia, sprawling civilizations were founded from large-scale agricultural

GEOGRAPHIES OF INDULGENCE, DESIRE, AND ADDICTION
Animals, Animal Parts, and Exotic Pets

Trade in ivory—a precious commodity hewn from the teeth and tusks of animals—dates back thousands of years. Ivory has unique structural and aesthetic qualities and can be carved into startling sculptures and made into artwork and furniture. People from many cultures in Asia, Africa, and the Americas have long hunted elephants, hippopotami, walruses, sperm whales, and narwhal for their ivory. With the rise of colonialism and widespread naval trade in the 18th century, regional trade systems in ivory became entangled in global trade and almost limitless demand. The results have been devastating. Where African and Asian elephants numbered in the millions at the turn of the 20th century, current estimates put their populations at 50,000 and 500,000, respectively. Whales and other ivory-producing species have also been ravaged by the trade (**Figure 8.1.1**).

A thriving global market exists for the parts of other animals as well. The bile of some bears is prized for medicinal purposes, as are parts of tigers, most especially skulls. The incredible demand for tiger parts in Chinese medicine, coupled with the size of the East Asian market and the growing affluence of Taiwanese, Chinese, Japanese, and Korean consumers, has meant an almost unstoppable poaching problem for the dwindling tiger populations in adjacent countries and regions. It is hard to know the size of the global market in these goods, but it is likely at least a U.S. $6 billion-a-year trade. A bowl of tiger penis soup sells for $320 in Taiwan. The humerus bone of a tiger retails for as much as U.S. $3,190 per kilogram in Korea.

Living wild animals are also valuable and have become part of a global market for exotic pets, with consumers in the United States making up the largest part of that market. Exotic pets in the United States range from Australian kangaroos and Southeast Asian tropical fish to African snakes and Asian and African wild cats (**Figure 8.1.2**). More than 650 million animals were imported legally into the United States in 2006, for example; millions more enter illegally. A raid on a single dealer's warehouse in Texas in 2010 resulted in the seizure of more than 27,000 animals, including hundreds of iguanas stuffed into shipping crates without food or water.

The negative consequences of this trade are numerous and diverse. Animals

▲ **FIGURE 8.1.1 Elephant Tusks** Customs officers in Osaka, Japan, display a record-setting seized cache of ivory which totalled 2.8 tons. Osaka is a top black market destination for elephant tusks.

themselves suffer from mishandling, abandonment, and abusive conditions. The over-harvesting of animals contributes to the destruction of their populations in the wild, a serious problem for the most endangered animals. The removal of these animals can also be extremely harmful to their native habitat. The harvesting of tropical fish frequently results in the destruction of coral reefs. Finally, many wild animals carry rare diseases, which owners frequently contract.

Dealing with this damaging trade has proven difficult for governments, since these luxury items are easy to smuggle and highly valuable. One effort has been the **Convention on International Trade in Endangered Species (CITES)**, a global treaty first signed in 1990 that restricts trade in rare or endangered animals and their parts between member countries. Many observers credit CITES with reducing the rates of decline for high-profile species, like elephants and tigers. Critics suggest that banning trade in objects like ivory altogether, however, results in artificially high prices, actually serving to *increase* demand, desirability, and the incentive to poach wild species. ■

◀ **FIGURE 8.1.2 Pet African Serval Kittens with Their Owners in the U.S.** International trade in exotic live animals—including reptiles, fish, and mammals—has had devastating effects on habitat and populations in the wild.

VISUALIZING GEOGRAPHY
The Asian Brown Cloud

The Asian Brown Cloud, which hovers across East and South Asia, was first identified by U.S. Air Force pilots and is now clearly visible in satellite photographs (**Figure 8.2.1**). The cloud consists of sulfates, nitrates, organic substances, black carbon, and fly ash, along with several other pollutants. This cocktail of contamination is the result of a dramatic increase in the burning of fossil fuels in vehicles, industries, and power stations in Asia's megacities. Forest fires used to clear land also contribute, as do emissions of millions of inefficient cookers burning wood or cow dung. A study of the Asian Brown Cloud sponsored by the UN Environment Programme involving more than 200 scientists suggests that the cloud not only influences local weather but also has worldwide consequences.

The smog of the Asian Brown Cloud reduces the amount of solar radiation reaching Earth's surface by 10% to 15%, with a consequent decline in the productivity of crops. But it can also trap heat, leading to warming of the lower atmosphere. It suppresses rainfall in some areas and increases it in others, while damaging forests and crops by contributing to acid rain. The haze is also believed to be responsible for hundreds of thousands of premature deaths from respiratory diseases. Because of the overall westerly drift of air over the Pacific, moreover, this cloud does not stay put. Particulate pollution of all kinds in Asia can drift across the Pacific and impact places as far away as North America (**Figure 8.2.2**).

Given that much of the source of the cloud comes from the direct burning of biomass (wood stoves, for example), modernization of heating and cooking technologies may lead to a diminution of the problem. Alternative energy sources, such as solar cook stoves used in parts of Asia or the use of new clean technologies to create cooking gas, could have a significant positive impact on the problem providing a double benefit for both regional and global health. ■

▲ **FIGURE 8.2.1 The Asian Brown Cloud** The Brown Cloud hovers in the left-hand side of this image, just west of the Korean peninsula, totally obscuring the landmass of China beneath. The pollution cloud can be distinguished as a think brown haze spread across the Asian landmass, distinct from the normal clouds, which are smaller and white. In this image, some brown haze can also be seen drifting east over the Pacific.

▲ **FIGURE 8.2.2 Westward Drift of Asian Pollution** Asian pollutants can drift east across the Pacific to the Americas. Notably, most pollutants are created in the production of consumer goods destined for North American markets.

Historical Legacies and Landscapes

State societies, with organized bureaucracies, consolidated militaries, and unified legal systems, appeared in East Asia centuries ago. The geographic footprints of these societies, in the form of canals, irrigation systems, road networks, and fortresses, remain visible in the landscape to this day.

Empires of China China has had a continuous agricultural civilization for more than 8,000 years. The first organized territorial state was that of the Xia dynasty, a Bronze Age state that occupied the eastern side of the Loess Plateau (in present-day Shanxi Province) and the western parts of the North China Plain (northwestern Henan Province) between 2206 and 1766 B.C.E. The Xia dynasty was succeeded by the Shang dynasty (1766–1126 B.C.E.). During this dynasty, walled cities appeared. The first *unified* Chinese empire was that of the Qin dynasty. Emperor Shih Huang-ti (221–209 B.C.E.) established an imperial system that lasted for 2,000 years. Shih Huang-ti established a centralized bureaucratic administration and had the Great Wall (**Figure 8.14**) built to protect China from "barbarian" nomads. The stability of the Qin dynasty was a precondition for the success of the **Silk Road** as an economic and cultural link among the civilizations of China, Central Asia, India, Rome, and, later, Byzantium (**Figure 8.15**). This trade route made East Asia the effective center of gravity for the global economy for centuries, a role the region is reestablishing in the present era.

The history of the Chinese empire is complex, with constantly shifting territorial boundaries. Successive dynasties tended to move from vigorous beginnings, with power concentrated around a strong

▲ **FIGURE 8.14 The Great Wall** The Qin dynasty began construction of the Great Wall from 316–209 B.C.E. The Wall that visitors see today dates from the Ming period and was built between the late 14th and mid-16th centuries.

production. High levels of social organization were needed to manage irrigation and taxation. This organization led to persistent state territories and widespread political consolidation in empires. East Asia also played an important role as the historical eastern anchor for a global trade system established long before the colonial era. During the postcolonial era, the region has been home to highly divergent economic experiments established throughout its subregions.

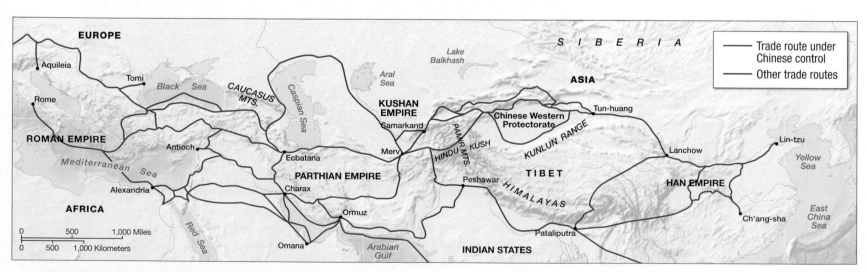

▲ **FIGURE 8.15 The Silk Road** The political and economic stability created by successive Chinese empires fostered a trade corridor connecting the region to North Africa and Europe and the rise of thriving trade cities along its length.

center. Typically, this period was followed by a slow loss of control as regional centers gained more power, which ultimately led to a final collapse as a forceful new dynasty was established. Sometimes the dominant empire was fragmented or subdivided for extended periods, but over two millennia the overall trend favored larger and more consolidated continental empires.

The first major dynasty to run through this cycle was the Han dynasty, from 206 B.C.E. to 220 C.E. Han emperors extended the empire westward, allowing the Chinese to control more trade routes along the Silk Road. Another important economic development took place early in the 7th century, under the Sui dynasty, when the northern and southern regions of Inner China were linked by the first of a series of **Grand Canals**. These artificial rivers are among the largest in the world, and the longest of these, stretching from Beijing to the city of Hangzhou, is now 1,794 kilometers (1,114 miles) in length. These canals connected the plentiful rice regions of the south to the Sui capital in the northwest (present-day Xi'an). By the 15th century, China's canal system was more than 1,000 kilometers (621 miles) in length. Comparable canals were not cut in Europe until the 18th century. In 1279, the Mongols, under Kublai Khan, conquered China, establishing imperial rule over most of both Inner and Outer China. The Mongol dynasty, known as the Yuan dynasty, was expelled after less than 100 years and was succeeded by the Ming (1368–1681) and the Qing (1681–1911) dynasties.

Japanese Civilizations and Empires The Chinese concept of the centralization of power in an imperial clan spread to Japan in the 6th century C.E. A succession of Japanese dynasties maintained a rigid system in which a subjugated peasantry sustained a relatively sophisticated but inward-looking civilization. Japan's distinctive civilization was largely a result of the introduction of Buddhism (covered in more detail later in this chapter), which arrived from India via China and Korea. Japanese civilization is also characterized by the legacy of feudal rule under a central family. In 794 C.E., the ruling elite established a capital in Kyoto, which became the residence of Japan's

imperial family for more than 1,000 years and the principal center of Japanese culture (**Figure 8.16**). As an industrial system developed in the western hemisphere, the Tokugawa dynasty (1603–1868) strove to sustain traditional Japanese society. To this end, the patriarchal government of the Tokugawa family excluded missionaries, banned Christianity, prohibited the construction of ships weighing more than 50 tons, closed Japanese ports to foreign vessels (Nagasaki was the single exception), and deliberately suppressed commercial enterprise.

At the top of the imperial social hierarchy were the nobility (the *shogunate*), the barons (*daimyos*), and the warriors (*samurai*). Farmers and artisans represented the productive base exploited by these ruling classes. In terms of spatial organization, the Japanese imperial economy was built around a closed hierarchy of castle towns, each the base of a local noble lord, called a **shogun**. The position of a town within this hierarchy depended on the status of the shogun, which in turn related to the productivity of the agricultural surroundings. As a result, the largest cities emerged among the rich alluvial plains and the reclaimed lakes and bay heads of southern Honshu. Largest of all was Edo (known today as Tokyo), which reached a population of around 1 million by the early 19th century under the Tokugawa Empire.

Imperial Decline The dynasties of Imperial China and Imperial Japan both eventually succumbed to a combination of internal and external problems. Internally, the administration and defense of growing populations began to drain the attention, energy, and wealth of the imperial regimes. The cultural and social elites tended to be focused on the arts, humanities, and self-promotion at the imperial court, rather than on economic or social development. As the economy stalled, peasants were required to pay increasingly heavy taxes, driving many into grinding poverty and thousands to banditry.

By the early 19th century, both China and Japan had moved into a phase of successive crises—famines and peasant uprisings. Both

◀ **FIGURE 8.16** Nijō **Castle** Built in 1601 by the founder of the Tokugawa shogunate, the castle is a classic example of high feudal Japanese architecture. The ornate interior gardens and courtyards are organized hierarchically, so visitors would have access only to what their status dictated.

countries were presided over by introverted and self-serving leadership. The peasantry began to flee the countryside in increasing numbers in response to a combination of rural hardship and the lure of the relative freedom and prosperity of the cities. The imperial courts of both China and Japan suppressed the spread of knowledge of modern weapons because they feared internal bandits and domestic uprisings. Though the crossbow, guns, and gunpowder were innovations of East Asia, the 19th century found the Japanese and Chinese empires relying on antiquated military technology compared to the Europeans and Americans.

External problems also undermined both imperial regimes. European traders had been a growing presence in East Asia in the 18th century but had been restricted to a few ports. Initially, Europeans bought agricultural produce (mainly tea) in exchange for hard cash (in the form of silver currency). Over time, this proved to be an unacceptable drain on European treasuries, and the British insisted on being able to trade opium (grown in India for export by the East India Company) for tea and other Chinese luxury goods. This policy eventually provoked China into a military response. In 1839, the Chinese, having prohibited the sale of opium several decades before, destroyed thousands of chests of opium aboard a British ship. This was just the excuse the British needed to exercise their superior weaponry. The so-called **Opium Wars** (1839–1842) ended with defeat for the Chinese and the signing of the **Treaty of Nanking**, which ceded the island of Hong Kong to the British and allowed European and American traders access to Chinese markets through a series of treaty ports (ports that were opened to foreign trade as a result of pressure from the major powers). After this, the opium trade to China expanded enormously; the British profited from Chinese opium addiction both through sales of opium in China and through its production in India (**Figure 8.17**).

Shortly afterward, in 1853, U.S. Admiral Matthew Perry anchored his flagship in Edo Bay (now Tokyo Bay) to "persuade" the Japanese to open their ports to trade with the United States and other foreign powers. The Japanese quickly complied in an acknowledgement of East Asia's abject weakness in the face of superior Western military technology. In both China and Japan, fear of the Western powers galvanized feelings of nationalism and fear of foreigners. This fear precipitated a period of civil war among the feudal warlords that culminated in revolutionary change.

Political and Industrial Revolutions Revolution came to China in 1911, when the Qing dynasty was overthrown and replaced by a republic under the leadership of Sun Yat-Sen's Nationalist Party (or Kuomintang). After a lengthy civil war and a struggle against the Japanese in World War II, Communist forces under **Mao Zedong**

▲ **FIGURE 8.17 British Opium Trade** This illustration from the 1880s shows a contemporary drying room in an opium factory, filled with thousands of vessels of the drug. At its height, the opium trade addicted untold tens of thousands of Chinese workers and turned a tidy profit for Britain.

unified the country. These forces came to control almost all of China except for the island of Taiwan, where the Nationalist leadership had retreated under U.S. protection. China would emerge from these struggles as an isolated nation, but one with enormous unrealized economic potential.

Japan also reacted to the humiliation of Western assertiveness, but in a remarkably different way. Japan's transition from feudalism to industrial capitalism can be pinpointed to a specific year—1868—when the Tokugawa dynasty was toppled by the Meiji imperial clan. In the Meiji clan, a clique of samurai and daimyo were convinced that Japan needed to modernize to maintain its national independence. Under the slogan "National Wealth and Military Strength," the new elite of ex-warriors set out to industrialize Japan as quickly as possible.

A distinctive feature of the entire modernization process in Japan was the very high degree of state involvement. Japanese governments intervened in the late 19th and early 20th centuries to promote industrial development by supporting capitalist monopolies and taxing the agricultural sector heavily. A strong feudal cultural order, a well-developed educational system, and a very lucrative sericulture (silk production) system also fostered economic development. Finally, and most important, were the spoils of military aggression (**Figure 8.18**). Japan's naval victories over China (1894–1895) and Russia (1904–1905) and the annexation of Taiwan (1895),

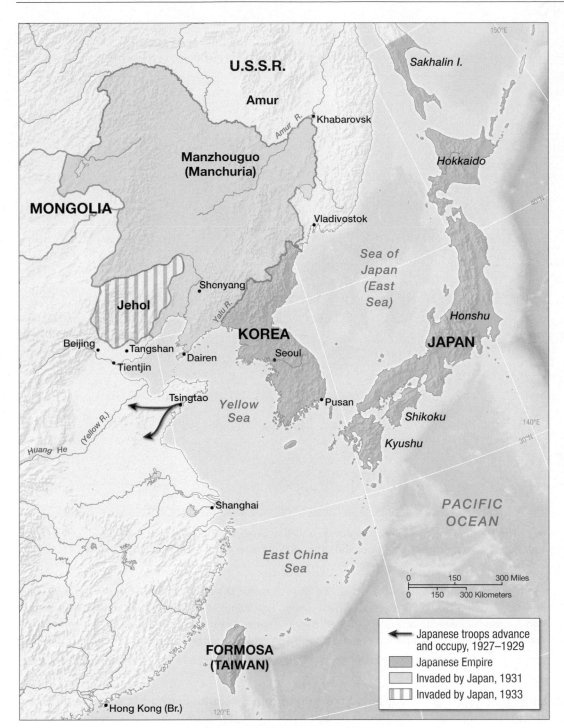

◄ **FIGURE 8.18 Japanese Expansionism** In the late 1920s and early 1930s, Japan pursued a policy of military aggression to secure a larger resource base for its growing military–industrial complex.

Korea (1910), and Manchuria (1931) provided expanded markets for Japanese goods in Asia. In 1931, the Japanese army advanced into Manchuria to create a puppet state and, in 1937, began a full-scale war with China. The Japanese military leadership ultimately overplayed its hand, however, by attacking the United States at Pearl Harbor in December 1941. The resulting sustained U.S. offensive against Japan, culminating in the dropping of atomic weapons on the cities of Hiroshima and Nagasaki (resulting in hundreds of thousands of deaths), led to Japan's defeat in World War II in 1945. Japan's industry lay in ruins. In 1946, output was only 30% of the prewar level.

Economy, Accumulation, and the Production of Inequality

Resistance to colonialism and the formation of new nation–states in East Asia during the postcolonial era brought with it turbulence and violence as well as economic experiments. There are several contrasting systems of economic management in the states of the region today, each with its own pattern of accumulation and development. Japan, China, and South Korea, most notably, all have large economies, but their road to economic development has led the three nations down three very different paths.

EMERGING REGIONS

The Pacific Rim

Between 1980 and 2012, the direction of flows in international trade reversed. Before the liberalization of the Chinese economy and the rise of the so-called Asian Tigers, the exports and imports of the United States typically looked *eastward* toward Europe and *northward* toward Canada. In 2012, Canada still remains the largest importer of U.S. goods and services, but China is a close second, and imports from China dominate all U.S. markets, as do those from Japan and other thriving Pacific economies. Trade among East Asian countries is also strong, including imports and exports that move between South Korea, Malaysia, China, and Japan. The resulting network of connections around the Pacific Ocean has drawn together the countries at its edges to form a new region: the **Pacific Rim**. If defined generously to include all 47 nations and territories bordering the Pacific, the emerging Pacific Rim region accounts for 48% of world trade and 58% of world GDP. Even if interpreted as consisting of just Japan, China, and the United States, the Pacific Rim region comprises the world's three largest economies (**Figure 8.3.1**).

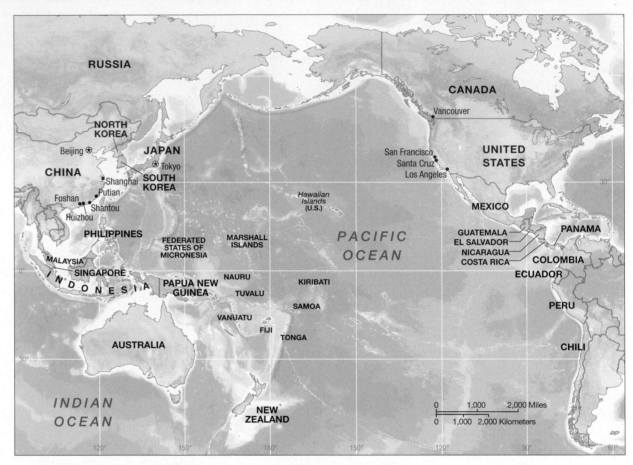

▲ **FIGURE 8.3.1 The Pacific Rim** Economic integration of urban centers between China, Canada, Taiwan, the United States, South Korea, Singapore, and other players map out the emergence of a new world region.

This emerging region is marked by its coastal urban economic centers, which are tied far more closely to one another than they are to their respective interior areas. Among these, the trade centers of China are the fastest growing. In 2010, four coastal Chinese export cities were numbered among the ten fastest-growing cities in the world: Shantou, Putian,

These very different paths to economic development have led to a current regional economic climate marked by interesting similarities. All the states in the region have geographically uneven patterns of development and accumulation. Agriculture and rural areas, most notably, have suffered in both Korea and China. All of these nations, moreover, are moving toward greater regional integration. Mutual investment and integration of firms between states are hallmarks of East Asian economies in the 21st century. This has led to the emergence of a new kind of economic region in East Asia. The new region is *externally* more competitive internationally and shifting toward increasing accumulation and centrality within the world system, but *internally* is increasingly competitive and fractured, with some subregions expanding while others contract.

Japan's Economy: From Postwar Economic Miracle to Stagnation Though WWII left Japan in ruins, incredibly, within five years, the Japanese economy recovered to its prewar levels of output. Having begun the postwar period at the bottom of the ladder of international manufacturing, the nation soon found itself at the top. By 1980, Japan had outstripped the major industrial core countries in the production of automobiles and television sets, for example. The Japanese, in short, not only achieved a unique transition directly from feudalism to industrial capitalism, but they also presided over a postwar "economic miracle" of impressive dimensions.

Several factors helped transform reconstruction into spectacular growth in Japan. These included exceptionally high levels of personal savings, rapid acquisition of new technology, and extensive levels of government support for industry. The "economic miracle" achieved

Huizhou, and Foshan. These municipalities are also becoming some of the most cosmopolitan. The financial and trade center of Shanghai, for example, is marked by the use of dozens of languages, and a fully eastern- and southern-looking economic orientation. Shanghai is as closely tied to Singapore as it is to Beijing.

Much the same can be said of cities on the eastern edge of the Pacific. Vancouver is home to Canada's largest port, ranks first in North America in total foreign exports, and trades more than U.S. $43 billion of goods every year, with most of these moving west to Asia. After English, the most commonly spoken languages in Vancouver are those of the Pacific Rim: Cantonese, Mandarin, Korean, and Tagalog.

Cultural, political, and social innovations have always followed the flow of economic goods. For this reason, the Pacific Rim is more than just an economic region. The rise of cross-Pacific popular culture can be seen in film, for example. Chinese action movies and Korean horror films are popular on all sides of the Pacific (**Figure 8.3.2**). The Pacific Rim Film Festival, held for the last 23 years in Santa Cruz, California, hosts movies from Japan, South Korea, China, New Zealand, the Philippines, India, Mexico, Australia, and the United States.

It may be premature to call the Pacific Rim a fully coherent region. Cities like Singapore, Tokyo, and Vancouver obviously have different cultures, have emerged from different histories, and sit within different political and economic spheres. Nonetheless, the tightening links between these sites are clearly creating something very new. ∎

◄ **FIGURE 8.3.2**
Global Reach of Pacific Rim Culture Hong Kong action star Jackie Chan poses in front of the poster for his movie *Medallion* in Taipei. Major films of this kind show to packed audiences all around the Pacific Rim and beyond.

was remarkable not only for its overall success in terms of economic performance, but also for the interdependence of government and industry, characterized by some as "Japan, Inc." Orchestrated by the Ministry of International Trade and Industry (MITI), the state bureaucracy guided and coordinated Japanese corporations—organized in business networks known as **keiretsu**. Favorable trade policies, technology policies, and fiscal policies helped Japanese industry compete successfully in the world economy. Today, Japan is a global economy with extensive linkages not only within East Asia but also throughout the **Pacific Rim**, a loosely defined region of countries that border the Pacific Ocean (see "Emerging Regions: The Pacific Rim").

Though in the late 1980s the success of the Japanese economy made Japan, Inc. the envy of global capitalism, the nation's economy suffered significantly in the intervening period (**Figure 8.19**, p. 324). Starting in the early 1990s, Japan entered a period of stagnant economic growth, which persists to the present. In the 1990s, GDP grew at approximately 1.5% annually (while it grew in excess of 4% each year in the 1980s). The previous guarantees of Japanese economic culture (including long-term employment) disappeared, suicide rates rose, and the government was unable to pull the economy from its nosedive. The strong role of banks in providing capital in the Japanese business model may have been a significant contributor to the crisis, in a period when global liquidity has declined around the world. It is possible, therefore, that the elements of the Japanese economy that allowed its miracle to occur may have contributed to the miracle's end. The most recent global economic recession only accentuated the crisis in Japan. Japan is particularly

▲ **FIGURE 8.19 Japan, Inc.** By the 1970s, the Japanese economy was a wonder of efficiency and productivity. In this photo, visionary Sony President Akio Morita holds one of the revolutionary products that catapulted Japan to its position as the world leader in consumer technology in the late 20th century.

dependent on export markets in the United States and elsewhere, where consumer spending has slowed dramatically. In 2012, the Japanese economy showed signs of turning around, but that year microchip producer Elpida Memory, Inc., filed for bankruptcy protection (its liabilities totaled U.S. $5.52 billion). This was the largest corporate failure in Japan since World War II and showed that the Japanese recovery still has a ways to go.

Perhaps more crucial, Japan has been forced to compete for foreign investment with China. And several Japanese electronics giants, including Toshiba Corp., Sony Corp., Matsushita Electric Industrial Co., and Canon, Inc., have expanded operations in China even as they have shed tens of thousands of workers at home. Olympus manufactures its digital cameras in Shenzhen and Guangzhou. Pioneer has moved its manufacture of DVD recorders to Shanghai and Dongguan. China's gains have often occurred at Japan's expense.

China's Economy: From Revolutionary Communism to Revolutionary Capitalism The economic story of China is the reverse of that of Japan, with a long period of postwar stagnation, followed by a recent meteoric rise. After World War II, Communist leader Mao Zedong faced the task of reshaping society after two millennia of imperial control. In 1958, in what became known as the **Great Leap Forward**, Mao launched a bold scheme to accelerate the pace of economic growth. Land was merged into huge communes, and an ambitious **Five-Year Plan** was implemented. The impact of the Five-Year Plan on the landscape was dramatic. Whereas pre-Communist China had an average farm size of 1.4 hectares (3.5 acres), the new agricultural communes averaged 19,000 hectares (46,949 acres) in size, with between 30,000 and 70,000 workers. Instead of a patchwork of fields, each with a different crop and presenting a rich palette of browns,

yellows, and shades of green, vast unbroken vistas now appeared, planted with crops dictated by the central planners.

China's planners were concerned only with increasing overall production. They paid little or no attention to whether a need for products existed, whether the products actually helped advance modernization, or whether local production targets were suited to the geography of the country. Several years of bad weather, combined with the rigid and misguided objectives of centralized agricultural planning, resulted in famine conditions throughout much of China. It is estimated that between 20 and 30 million people died from starvation and malnutrition-related diseases between 1959 and 1962.

Radical economic changes followed in the 1980s. Following a long era of economic stagnation, the Chinese economy has been growing at double-digit rates for much of the past 25 years. Under the leadership of **Deng Xiaoping** (1978–1997), China embarked on a thorough reorientation of the economy, dismantling central planning in favor of private entrepreneurship and market mechanisms and integrating itself into the world economy. Agriculture was decollectivized, state-owned industries were closed or privatized, and centralized state planning was dismantled.

The result is a unique economy. China is capitalist in the sense that privately owned manufacturing plants and business employ its millions of workers, but it is a planned economy as well, since state agencies and functionaries endorse, guide, and support certain sectors and markets. For example, in a few short years, China has become a major producer and exporter of solar panels. When confronted with a massive shortage of polycrystalline silicon in 2007 (the main raw material for these devices), the Chinese government directed investment into silicon production and procurement from foreign sources. This represents a kind of **state capitalism**, a market-based economy with private ownership and investment where the state continues to own some firms (so-called "national champions"), to seek and obtain technology, and to carefully control the value of its currency.

This model has shown some remarkable results. In recent years, China's manufacturing sector has grown by almost 15% annually. China has extended its trade policy, permitted foreign investment aimed at Chinese domestic markets, and normalized trading relationships with the United States and the European Union. In 2001, it joined the World Trade Organization. Growth has been further bolstered by the development of **Special Economic Zones (SEZs)** set up as carefully segregated export-processing areas that offer cheap labor and land, along with tax breaks, to transnational corporations (**Figure 8.20**).

In the 21st century, regional investors have responded positively and enthusiastically to China's openness to capitalist investment and trade. Networks of business connections sprang up quickly in the coastal cities and regions adjacent to Hong Kong and Taiwan. Once the business networks were in place, capital flowed in from all over the globe. A new web of interdependent trade and investment emerged, linking coastal China with Singapore, Bangkok, Penang, Kuala Lumpur, Jakarta, Los Angeles, Vancouver, New York, and

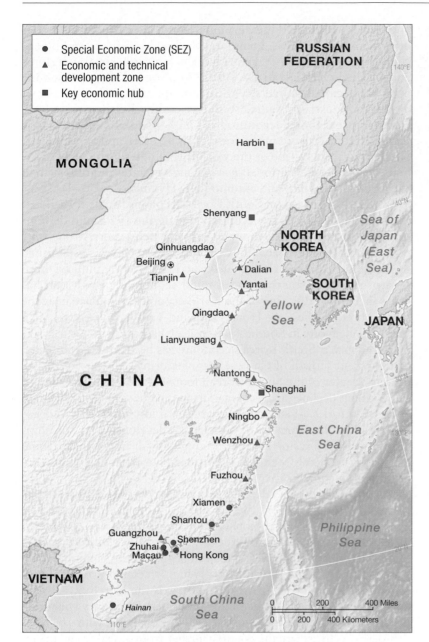

▲ **FIGURE 8.20 China's Special Economic Zones** Since the 1980s, China has been establishing areas where special tax exemptions and low export restrictions encourage foreign investment and technological development.

▲ **FIGURE 8.21 Walmart China** The sale of inexpensive consumer goods in the United States at stores like Walmart has long fueled the Chinese economic boom. Now, a growing consumer class makes China itself a target locale for big-box retailers.

for U.S. $30 for a 60-hour week. Studies of the industry show that some of the female workers in Chinese garment factories are as young as 12 years old. Girls from rural villages are sent to work as sewing machinists in city workshops. They sleep eight to a room and sew seven days a week from 8 A.M. to 11 P.M. As a result of these conditions, the retail margin (gross sales profit) on clothing made in workshops in China and sold in Europe and the United States is 200% to 300%, compared to a margin of only 70% or so on clothing made in the West. The success of these industries has resulted in rising wage demands by Chinese garment workers. Competition from new labor markets—like the Philippines—means that the continued growth of the garment industry in China is questionable; firms may outsource garment work to even poorer places in the future (**Figure 8.22**).

Sydney. In 2011, the United States bought more than U.S. $399 billion worth of goods made in China. Walmart was the single largest U.S. importer, according to a study by the Citizens Trade Campaign, procuring at bargain prices from China everything from T-shirts to car stereos. The success of the Chinese export boom has created a burgeoning consumer class in China and led to the emergence of big-box retail stores across the country (**Figure 8.21**).

The low cost of labor underpins the preponderance of "Made in China" apparel labels in U.S. stores. Whereas skilled clothing workers in Toronto's garment district, for example, are paid U.S. $350 to $400 a week, plus benefits, for a 33-hour week, their counterparts in Hong Kong are paid U.S. $250 for a 60-hour week, with no benefits; and in China, skilled garment workers can be hired

▲ **FIGURE 8.22 Textile and Garment Production in China** The growth of exports from China has depended on low wages for industrial labor, especially women's labor. Pictured here is a garment factory in Shenzhen.

This new pattern of accumulation in China occurs at a sacrifice. Environmental degradation and pollution are some of the consequences of rapid industrialization. And the success of industrial modernization has intensified the gap among different regions and between urban and rural areas. The countryside and interior have not benefited at the rate the coastal industrial belt has, and this disparity creates the potential for political tensions within China. Low wages and difficult working conditions have also created controversies for manufacturers in China. In a high-profile case, Apple Inc. was accused of poor working conditions in factories and employing large numbers of child workers. In 2012, the company called for a third-party audit of its supplier factories in China.

The current global economic crisis suggests several important things about the current trajectory of Chinese economic development. First, the increasingly globalized economy is dependent on consistent markets for Chinese industrial products (ranging from toys to shower curtains to processed food). This makes China somewhat vulnerable to downturns in consumer spending in places like the United States and Europe. Conversely, China's massive international mobilization of capital has included the acquisition of significant debt from other nations. As of 2011, China, which owns more than U.S. $1.2 trillion in United States bills, notes, and bonds, is the largest foreign holder of U.S. debt. This gives China great political power in negotiations with the United States, but again makes China vulnerable to the changing status of the U.S. economy. As the Chinese economy has become more globalized, it has become more exposed to distant regional economic patterns.

Two Koreas, Two Economies The end of World War II brought about an important geopolitical change to Korea as well. In 1949, under the sponsorship of Soviet troops that had moved south from Manchuria into the northern part of Korea, **Kim Il Sung**, an anti-Japanese Marxist–Leninist nationalist, came to power. Almost immediately, U.S. troops occupied the southern part of the country in response to the threat of communism from the north. A key site in the Cold War (see Chapter 3), Korea suddenly found itself divided into two parts along the 38th parallel (latitude). In June 1950, Kim Il Sung's troops carried out a major attack on South Korea. The United Nations intervened, and U.S. forces rapidly rolled back the North Korean forces all the way to the Chinese border. This prompted China to become involved on behalf of the North Koreans, and there followed a devastating war. Casualties were horrendous, with hundreds of thousands killed and much of the country destroyed. In 1953, the 38th parallel was finally restored as the border, but the two Koreas remain bitter rivals and the war has never officially ended.

The division resulted in two very different economic experiments on the Korean peninsula. In South Korea (the Republic of Korea), whose economy was crippled by lengthy wars between the 1930s and the 1950s, the government crafted a state-led development effort that carefully controlled imports, managed internal competition between firms, and took out enormous international loans to invest heavily in private industrial development. The initial emphasis was on **import substitution**, which involves the development of domestic industries with government protection to produce goods for the domestic market. Between the mid-1960s and mid-1970s, the focus changed to favor the growth of export-oriented, labor-intensive manufacturing. The South Korean government facilitated the development of these export industries by providing incentives, loans, and tax breaks to firms and by encouraging the growth of giant, interlocking industrial conglomerates called **chaebol**. In the late 1970s, South Korean economic planners decided to diversify the economy with greater emphasis on heavy industry, chemicals, and assembly. By the mid-1980s, manufactured goods accounted for 91% of total exports. More than half of these were in the form of ships, steel, and automobiles. South Korea is now one of the world's leading exporters and is the home of major international *chaebol* conglomerates in areas such as electronics (Samsung) and automobiles (Hyundai-Kia).

The radical success of South Korea's economy made it the leading so-called **Asian Tiger** economy during the 1990s and demonstrated how capitalist development could be tied to very strong state policies regulating the economy. In addition to Korea's established strength, East Asia has two other Asian Tigers, newly industrialized territories that have experienced rapid economic growth that has lifted them from the edge of the world economy to a place near the center of trade. These are Hong Kong and Taiwan (the fourth Asian Tiger is Singapore, in Southeast Asia; see Chapter 10).

Like Japan's, South Korea's economy has been adversely affected by the recent growth of Chinese manufacturing. The city of Busan, once the center of the South Korean footwear industry, experienced an enormous decline in sales of manufactured goods and is now full of deserted factories. Most of that export sector lost out to new Chinese factories. On the other hand, this city has developed tourism, shipping, and luxury residences, leveraging growth in the tertiary and quaternary economic sectors (see Chapter 1) to offset declines in secondary production.

North Korea (officially the Democratic People's Republic of Korea) has long followed the path of a closely planned, centrally guided economy, a model pioneered by the Soviet Union and China in the era before economic reform. This form of central planning often retards capitalist investment and adaptation to changing market conditions. North Korea remains highly economically isolated. A major economic barrier for the country is the incredibly disproportionate fraction of the economy dedicated to military spending, perhaps as much as 25% (**Figure 8.23**). This, coupled with sanctions that periodically isolate the country further from global trade, means a North Korean economy that is largely stagnant, small, and marked by shortages and hunger.

North Korea began a cautious transition of its economic policy during the first years of the 21st century toward a more competitive stance. Specifically, it began to experiment with Chinese-style Special Economic Zones to foster some foreign investment. The impact of this experiment is impossible to determine at this early stage. It certainly represents part of an ongoing trend, however, that makes East Asia more externally competitive as a world region for investment, while also becoming a place of more bare-knuckle internal competition, as subregions and nations compete for scarce investment capital.

APPLY YOUR KNOWLEDGE Compare and contrast the economic experiments in China, South and North Korea, and Japan. What have the costs and benefits of each strategy been for the economic prominence and power of each country? For their people?

◀ **FIGURE 8.23** **North Korean Military** A significant proportion of the North Korean gross domestic product is dedicated to supporting the military, rendering the economy anemic. This photograph shows North Korean female soldiers marching in a military parade in Pyongyang on April 15, 2012, to mark 100 years since the birth of the country's founder, Kim Il-Sung.

Territory and Politics

East Asian political geography is marked by competing historical trends. On the one hand, regional historical struggles, remnants of the 20th century, play an outsized role in state power and geopolitics. Specifically, the war in Korea remains, for all intents and purposes, unresolved. The fate of Taiwan also remains in question more than a half century after it was seized by Chinese Nationalists. On the other hand, the expansion of Chinese geopolitical power has escaped the bounds of the region. Chinese influence is now being felt not only across North America and Europe, but also in the resource frontiers of Africa, even as the Chinese government struggles to maintain its own internal control.

An Unfinished War in Korea The partition of North and South Korea in 1953 (described earlier in the chapter) left North Korea behind the world's most heavily militarized border. The first period of leadership in that country set the tone for politics and development for the decades that would follow. The nation's first leader, Kim Il Sung, governed the country according to a philosophy called *juche*: a mixture of centrally planned economics, self-reliant nationalism, and the cult of the personality. The people still refer to Kim Il Sung as the "Great Leader" although he died in 1994. From its inception, North Korea's dynastic leadership has retained tight relations with China and a posture of continued hostility toward South Korea.

Blessed with valuable natural resources, including significant reserves of coal and iron ore and good potential for hydropower, North Korea set out to become a major power within East Asia. For decades, more than one-quarter of the country's GDP has been expended on the military, including the development of nuclear power capability that could lead to nuclear weaponry. To carry out this strategy of national development, Kim Il Sung imposed an austere regime directed through central planning and a regimented way of life dominated by a totalitarian government.

In 1994, Kim Il Sung died and was succeeded by his son **Kim Jong Il**, who came to be referred to as the "Dear Leader." Estimates of the total number of deaths because of North Korea's food shortages between 1995 and 2005 range between 1 and 3 million. By 2011, after 58 years of sacrifice and austerity, personal incomes in North Korea were only one-tenth of those in South Korea, and infant mortality rates were five times higher. North Korea remains not only one of the world's most impoverished societies but also one of the most closed and rigid. Listening to foreign radio broadcasts is punishable by death, citizens can be detained arbitrarily, and there are an estimated 150,000 political prisoners.

In 2012, Kim Jong Il died and was succeeded by his own son, **Kim Jong-un**, an enigmatic figure and the youngest head of state in the world (he was born in 1983). Kim Jong-un governs one of the most highly militarized countries in the world. North Korea has the world's fifth-largest standing army (after China, the United States, Russia, and India), even though it has a relatively small population (an estimated 24 million in 2012). North Korea's weapons and missiles are sold indiscriminately in world markets to raise sorely needed foreign exchange. Its own arsenal—including possible nuclear capability—is pointed menacingly at South Korea and Japan. This threat represents a major geopolitical problem for the region, but, in general, North Korea has attempted to use this threat only as a periodic lever for economic aid and political concessions from both China and the United States. North Korean threats are typically followed by regional negotiations and concessions, a period of quiet, and then a return to a threatening posture.

Most recently, North Korea has made efforts to expand its missile capacity, testing increasingly long-range missiles that might in the future—in theory—carry nuclear payloads. The country may possess as many as 800 ballistic missiles, the longest-range of which reaches far into the Pacific Ocean, making it within potential firing range of Japan and, perhaps eventually, Europe and the United States. In April 2012, North Korea attempted a much-anticipated rocket launch, which most world leaders denounced. The result was a failure, as the rocket broke into pieces over the ocean only moments after takeoff. This event revealed that North Korean missile capacity is not yet well-developed and is still error-prone. However, the continuing threat of North Korean power, uneven at best, is still enough to consistently demand the attention of world leaders.

Unresolved Geopolitics in Taiwan When the Communist revolution created the People's Republic of China (PRC) in mainland China, the ousted Nationalist government of China established itself in Taiwan as the Republic of China in 1949. Granted diplomatic recognition by Western governments but not by the PRC, Taiwan immediately became a potential geopolitical flashpoint in the Cold War. Large amounts of economic aid from the United States helped prime Taiwan's economy. An authoritarian regime in Taiwan suppressed opposition to government policies. By the early 1960s, the nation's political stability and cheap labor force provided a very attractive environment for export processing industries.

Taiwan lost its full international diplomatic status in 1971, when U.S. President Richard Nixon's rapprochement with the PRC led to China's entrance into the United Nations. In 1987, Taiwan's government lifted martial law, began a phase of political liberalization, and relaxed its rules about contact with the PRC (also referred to as "mainland China"). Mainland China still claims Taiwan as a province of the PRC and has offered to set up a Special Administrative Region for Taiwan, with the sort of economic and democratic privileges that the PRC has given Hong Kong. Geopolitical tensions periodically flare up in conflicts over Chinese or U.S. naval exercises and other symbolic acts. So, while it is extremely unlikely that any real conflict will determine the future of Taiwan's status, meaningful diplomatic and trade relations have not emerged despite a long period of effort.

APPLY YOUR KNOWLEDGE Go to the home pages of the Democratic People's Republic of Korea (North Korea): http://www.korea-dpr.com/, the Republic of Korea: http://www.korea.net/, and Taiwan (Republic of China): http://www.gio.gov.tw/. How do the political visions of these states, as presented in the text on these sites, reflect their distinct geopolitical visions and histories?

The Contested Periphery of Tibet As early as the 7th and 10th centuries C.E., the expanding and unified Han Empire began to move into the margins of the Tibetan region. By that time, Tibet had developed a distinctive culture based on Tibetan Buddhism—developed from a fusion of the region's ancient religion (known as *Bon*) with Tantric Buddhism imported from India. During the Ming (1368–1644) and Qing (1644–1911) dynasties, the entire Tibetan Plateau became part of the Chinese empire. Far away from the center of power, Tibet was divided into numerous small tribes and districts, with feudal chieftains ruling repressively and autocratically, generation after generation. In 1950, China seized on this fact as justification for its invasion of the strategically important region.

▲ **FIGURE 8.24 The Dalai Lama** The spiritual leader of Tibetan Buddhism resides in Northern India but continues to press for Tibetan autonomy and reform of Chinese state control.

Between 1950 and 1970, the Chinese invaded Tibet, destroying much of Tibetan cultural heritage and causing the death of an estimated 1 million Tibetans. In 1959, the Tibetans rose in an attempted revolt, and their spiritual leader, the **Dalai Lama,** fled into exile. In 1965, Tibet was granted the status of an autonomous region by the PRC (**Figure 8.24**), called Xizang, but by that time, large numbers of Chinese had flooded in, often taking positions of authority and leaving Tibetans as disadvantaged, second-class citizens. There were serious anti-Chinese disturbances in 1988, and since then the volatile situation continues to simmer. Today in Xizang, all Buddhist monasteries are rigidly controlled by the police and Communist Party officials. Expressions of devotion to the Dalai Lama are banned. About 60% of the area's population is now of Han Chinese descent. These developments have given rise to significant ethnic tensions within the region. In the rest of the world, the exiled Dalai Lama has acquired global celebrity status as a result of his international tours promoting Buddhism and publicizing the "cultural genocide" committed by China in Tibet. The Dalai Lama does not seek national independzence for Tibet but greater autonomy. For their part, the Chinese cannot understand the ingratitude of the Tibetans. As the Chinese see it, they have saved the Tibetans from feudalism and built roads, schools, hospitals, and factories.

As the Dalai Lama ages, however, and Chinese authorities and the Chinese military exercise further control of Tibet, it is not clear what the future holds for the region. China has expressed some interest in naming the Dalai Lama's successor. Since that successor is considered by Tibetan Buddhists to be the reincarnation of the last Dalai Lama, this puts the Chinese Communist Party in the awkward position of interpreting, intervening, and tacitly acknowledging matters of Buddhist faith.

The Geopolitics of Globalized Chinese Investment The success of state capitalism in China has resulted in two trends that have put pressure on global geopolitics. First, the enormous production systems of the industrial zones of China have tested the limits of available raw materials and energy resources within the PRC. Petroleum and minerals essential to sustaining production in Chinese factories are in insufficient quantities given the current rate of use, even though China is well endowed with resources. This has led to a search for new sources of inputs, and has led China abroad into new resource frontiers, especially into Africa.

At the same time, the industrial growth of the PRC has made a handful of business operators (and government officials) enormously wealthy. As a result, there is significant capital available in China for investment abroad. Although some of this capital has been used to purchase the debt of other industrial nations (notably the United States), more speculative investment has also become attractive for the Chinese. Chinese investments overseas, especially in underdeveloped areas (again including Africa) have grown. As of 2012, China purchases more than one-third of its oil from Africa and is the continent's largest trade partner.

The convergence of these two trends makes China a large and quickly growing presence in the developed world, with implications for international conflict. In a notable case, Chinese development money has flowed heavily into many countries in Africa, including Sudan, a country under international scrutiny for high levels of uncontrolled and state-tolerated violence and genocide (see Chapter 5). In the Chinese point of view, its supply of development capital to countries like Sudan provides a net benefit to Sudan's citizens. The Chinese position is that the mode of governance a nation chooses is not the business of other members of the world community. China also points to the historical development of poor nations by wealthier ones in the past and stresses that international condemnation of such investment is merely a symptom of jealousy on the part of European nations, who have long colonial histories of their own. From the point of view of the United States and the European Union, on the other hand, Chinese support for the Sudanese regime subverts any efforts to bring the disastrous situation under control. The moral indifference of Chinese investment represents an impediment to peace and human rights.

Whatever the outcome in the case of Sudan, the rise of China as an international development and investment actor on a global scale is a key factor in geopolitics and the formation of new regions. China and Africa were linked as a part of the global trade system 1,000 years ago. This connection was only broken during the colonial era. With the rise of a powerful China, Africa and East Asia are becoming relinked, with transformative possibilities for each region. Changing global relations portend the creation of new regions.

APPLY YOUR KNOWLEDGE Consider the political changes of fortune for the great powers of East Asia (especially Japan and China) before colonialism, under colonialism, and in the last few decades. How has the center of power shifted across the region?

CULTURE AND POPULATIONS

Though urbanization has been an unquestionable recent theme in East Asia, traditional societies in the region revolve around family, kin networks, clan groups, and language groups developed in largely rural contexts, with a strong bureaucracy enforcing social order. In China and Japan, society was historically rigidly hierarchical, and individuals were subsumed within the family unit, the village, and the domain of the local lord. In these environments, important social values include humility, understatement, and refined obsequiousness, with particular deference being shown to older persons and those of superior social rank.

To a degree, traditional values persist as a distinctive dimension of the cultural geography of the region. Not surprising, in rural areas and smaller towns, and among older people in general, traditional cultural values remain. Rural communities with a great variety of distinctive ways of life still characterize much of the region. There are, however, marked regional variations in cultural traditions within East Asia. In addition to the variations in religious adherence associated with different ethnic groups, there are some striking regional differences in language, diet, dress, and ways of life. The changing relations of the region to other global cultures, moreover, has transformed local ways of living.

Religion and Language

The cultural traditions of East Asia, especially its religions and languages, show signs of merging and hybridizing with one another, while they remain diverse and independent. Japanese and Chinese languages share some symbols, for example, and Buddhism is found in many parts of the region. However, the languages are also entirely distinct, and Buddhist practices in Mongolia differ from those of Japan.

Religions Chinese culture found spiritual expression in the philosophy of **Confucianism**. Unlike formal religions, Confucianism has no place for gods or an afterlife. Confucianism's emphasis is on ethics and principles of good governance and on the importance of education as well as family and hard work. Confucianism proved to be ideally suited to Imperial China. It remains the most widely recognized belief system in China, even after several decades of discouragement by the Communist regime. Other formalized religions in China include Daoism (throughout much of the country), Tibetan Buddhism (in Xizang), and Islam (in large parts of Inner Mongolia and Xinjiang).

For a large number of Chinese, however, folk religions are more important than any organized religion. **Animism**—the belief that nonliving things have spirits that should be respected through worship—continues to be widely practiced. **Ancestor worship** is also observed. This practice is based on the belief that the living can communicate with the dead and that the dead spirits, to whom offerings are ritually made, have the ability to influence people's lives. The costs of ancestor worship can be financially burdensome. Offerings involve burning paper money and hiring shamans and priests to perform rituals to heal the sick, appease the ancestors, and exorcise ghosts at times of birth, marriage, and death.

Japanese Indigenous culture was expressed in **Shinto**. Shinto is not a distinctive philosophy but, rather, a belief in the nature of

In East Asia today, feng shui practitioners focus most on the layout and interior design of homes and offices. The principles and beliefs of feng shui are complex, but some of the key ideas are as follows: Water and mirrors enhance *qi* flow. Narrow openings or hallways cause *qi* to flow too quickly, negating its beneficial effects. Straight lines are a frequent cause of high-velocity *qi* that can be dangerous to health and well-being. The numbers eight (representing "prosperity") and nine (representing the fullness of heaven and Earth) are good, but four is bad ("death"). The placement of furniture and other objects, as well as the use of certain colors (purple is popular) and motifs (birds are favored), are all believed to enhance *qi* flow and, therefore, career success, strong family relationships, and good health. Ideally, each room should include some representation of each of the five fundamental elements (fire, metal, water, wood, and earth).

▲ **FIGURE 8.25 Shinto Gate** The classic archway of the Shinto shrine (Torii) in Miyajima, Japan, is emblematic of Japan's deepest religious traditions. The gate marks a location of passage between the profane and the sacred.

sacred powers that can be recognized in every individual existing thing (**Figure 8.25**). The traditions of Shinto can be thought of as the traditions of Japan itself. Seasonal festivals elicit widespread participation in present-day Japan, regardless of people's religious affiliation. These festivals usually entail ritual purification, the offering of food to sacred powers, sacred music and dance, solemn worship, and joyous celebration. Buddhism is also important in Japan. It was introduced to Japan in the 6th century c.e. from Korea. During the Nara period (710–784 c.e.), Buddhism was vigorously promoted and led to the birth of distinctive Japanese art and architecture. Later, between the 12th and 14th centuries, Zen Buddhism was introduced from China, along with new painting styles, new skills in ceramics, and tea-drinking customs.

Feng Shui Throughout East Asia there is widespread adherence to **feng shui** (pronounced "fung shway")—not a religion, but a belief that the physical attributes of places can be analyzed and manipulated to improve the flow of cosmic energy, or *qi* (pronounced "chi"), that binds all living things. Feng shui involves strategies of siting, landscaping, architectural design, and furniture placement to direct energy flows. The oldest school of feng shui, known as the Land Form (or simply Form) School, dates back to the Tang dynasty (618–907 c.e.). From its origin in the jagged mountains of southern China, the Form School used hills, mountains, rivers, and other geomorphological features to evaluate the quality of a location. Subsequent modifications of feng shui introduced the idea that specific points of the compass exert unique influences on various aspects of life. For example, the south, with its orientation toward the Sun's path and away from cold north winds, was declared a most auspicious direction, conducive to longevity, fame, and fortune. Influence over career and business success was attributed to the north.

Languages "Chinese" is linguistically a term that is somewhat imprecise. In the region of East Asia dominated by Chinese people, residents speak not only many Chinese dialects, but also an enormous range of diverse minority languages (like Uyghur in the west, Sibo in the northwest, and Chuang, Yi, Nakhi, and Miao in the south). The dominance of Han people within China is reflected in the geography of language, however. Mandarin, the language of the old imperial bureaucracy, is spoken by almost all Han people as well as Hui and Manchu people (and the people of Taiwan). Significant regional variations, however, still remain in Mandarin dialects. The other 53 ethnic minorities in China have their own languages.

Chinese writing dates from the time of the Shang dynasty (1766–1126 b.c.e.). It features tens of thousands of characters, or **ideographs**, each representing a picture or an idea. A single dictionary contains 50,000 distinct characters. Chinese children may learn a few thousand characters in their first ten years, but advanced reading or writing includes several times that number of characters. Ideographs are enormously complex, especially in their studiously artful calligraphic form; a single Chinese character might take as many as 33 brush strokes to complete (**Figure 8.26**).

More recently, the government of China has facilitated literacy by simplifying characters and reducing the number of characters in use. The PRC has also adopted a new system, **pinyin**, for spelling Chinese words and names using the Latin alphabet of 26 letters. Previously, Chinese words and names had been translated (usually by Westerners) using a Western system, known as the Wade-Giles system, with often misleading results and a proliferation of mispronunciations. Older Western textbooks, reference books, and atlases, for example, refer to Chongquing as Chunking, Mao Zedong as Mao Tse-Tung, and so on.

Though the Japanese adopted some Chinese pictographic characters in the 3rd century c.e., the two main languages are largely

▲ **FIGURE 8.26 Chinese Calligraphy** A Chinese calligrapher at work writing on a scroll.

unrelated. Japanese ideographs—symbols representing single ideas or objects (*kanji*)—number more than 10,000. These are joined by numerous other symbols (*kana*) that represent phonetic syllabus or word fragments. Collectively, the language is wholly unique.

The affiliation of Korean to other languages in the region is uncertain. Although it contains Chinese words, its structure and grammar is far closer to Japanese. Even so, the linkage between Korean and Japanese is unclear. The Korean alphabet, unlike Japanese and Chinese, is phonetic (where characters represent independent sounds), though the text is written in clustered groups, which are wholly unlike that of other languages in the region. The fundamental differences between the languages of East Asia suggest the independent development of diverse cultures within the region, each with its own linguistic and symbolic logic. Their clear divergences remind us that the region is a product of interactions, not an inevitable coherent whole. At the same time, the shared symbols and mutual influences illustrate a long history of strong regional ties.

Cultural Practices, Social Differences, and Identity

Culture in East Asia is a product of local evolution and development as well as intense interaction and struggles between distinct cultural groups and between advocates of tradition and those of modernity. This was most obviously seen in China during the **Cultural Revolution**. In 1966, in an attempt to restore revolutionary spirit and to reeducate the privileged and increasingly corrupt Communist Party officials, Mao Zedong launched what he called a "Great Proletarian Cultural Revolution." The Cultural Revolution brought a sustained attack on Chinese traditions and cultural practices. Millions of people were displaced, tens of thousands lost their lives, and much of urban China was plunged into a terrifying climate of suspicion and recrimination.

Only with the death of Mao Zedong in 1976 did the Cultural Revolution come to an end. Since that time, new museums, folk art celebrations, and espousal of commitments to traditional culture

have become more common throughout China (**Figure 8.27**). This may in part reflect an interest in selling a managed form of Chinese culture to foreign governments and tourists, or it may be an intentional effort to utilize certain art forms, language, and music in the service of Chinese nationalism. Whatever its motivation, the back-and-forth struggle over traditional practices is a key feature of the East Asian cultural landscape, as can be seen in the changing state of gender roles, conflict over ethnic identity, hybrid food traditions, and new cultural traditions that have come to travel worldwide.

Gender and Inequality Traditional East Asian cultures insist on distinct and hierarchically ordered relationships between men and women. There are numerous dramatic historical examples of patriarchy. The Chinese tradition of foot binding, practiced mainly by elites, involved the often painful disfiguring of women's feet. The Japanese Meiji Civil Code of 1898 denied women many basic legal rights. Neo-Confucian culture in Korea, as elsewhere in the region, focused on the woman's role of providing a male heir. Once married, women in Korea moved away from their families to assume the lowest position in their husbands' households.

Much of this tradition has been overturned since the end of World War II. Independent, working, professional women are ubiquitous in Japan since the time of its postwar economic miracle. In China, traditional gender roles were erased during the Communist period, where society was strictly gender-neutral, images of female

▲ **FIGURE 8.27 Traditional Opera Culture in China** This performance in the Beijing Opera includes an appearance by Monkey, a favorite character from Chinese classic literature.

beauty and ideas of romance were suppressed, and the participation of women in political and economic activities reached almost 100%.

Just as recent economic reform in the region has been accompanied by the reemergence of regional inequality, economic and social reforms led to changes in the status of women. In China, competitive markets in education since the 1990s have resulted in a fourfold expansion of women college students in the 1990s. These better-educated women have made their mark in China's labor markets, earning high salaries and challenging traditional gender boundaries for top jobs. Between 1991 and 2001, the number of marriages in rural areas arranged by parents in China fell from 36% to 16%. But at the same time, there has been an explosion of femininity and sexuality in the media, not always with positive results for women. Traditional gender stereotypes have resurfaced. This has played out in labor markets, as older and poorer women have borne the brunt of layoffs in the state sector and have been left with lower-paid occupations.

As a result, there is a widening income gap between men and women in China. Urban women who made, on average, 70% of men's wages in 1990 made about 67% of men's wages in 2011. Gender inequality is most marked in rural settings, where women's wages, on average, are only 56% of men's, and where gender roles have regressed farthest. Stereotypical roles of marriage and home ownership have resurfaced and affected women's access to real estate assets. In 2010, only 37.9% of women owned homes compared to 67.1% of men. Legally, women in China have equal rights to land ownership as men; however, a recent survey in 17 Chinese provinces, undertaken by the global land rights group Landesa, found that only 17.1% of existing land contracts and 38.2% of existing land certificates include women's names. Other countries in the region harbor similar disparity. In South Korea in 2007, the employment rate for women with college degrees was 61.2%, at the very bottom among advanced economic countries. Women in South Korea receive 66% of men's wages for the same work, according to the Korean Employment Information Service.

Ethnicity and Ethnic Conflict in China Paralleling changes in gender roles, many traditional ethnic divisions have been both challenged and reinforced in recent years. Although there are few immigrant populations of any significant size in the region, there are some important ethnic variations, especially in the People's Republic of China, which officially recognizes 56 ethnic groups. The dominant group in China is the Han, who make up more than 91% of the population. The Han people originally occupied the lower reaches of the Huang He and the surrounding North China Plain. They spread gradually inland along river valleys, into present-day Korea, and toward the humid and subtropical south. Han colonialism and imperialism established them as the dominant group throughout the jungle lands of the south and southeast (as far as present-day Vietnam) and in Taiwan. In the 19th century, the Han spread north into Manchuria.

Today there are more than a billion Han in China. The remaining 55 minority groups add up to less than 110 million people. They are mostly residual groups of Indigenous peoples, such as the Miao, the Dong, Li, Naxi, and Qiang, who live in the remote border regions, removed from central authority in Beijing and relatively economically disadvantaged.

Given these circumstances, it is not surprising to find that tensions exist between several of the larger minority groups and the dominant Han. Perhaps the best-known case is that of the Tibetans, whose troubled relationship with the People's Republic of China is discussed earlier in the chapter. Ethnic communities like the Tibetans, with their own highly distinctive cultural traditions and religions, face explicit state policies of "sinification"—the ostensible "modernization" of their economies and communities in a way that appears unmistakably colonial. The ethnic Uyghurs, who predominate in westernmost China (and in many post-Soviet central Asian states) have also faced suppression from the PRC. Starting in 2009, violence broke out between Uyghurs, ethnic Han people, and state police in the city of Ürümqi in far Western China. The Chinese state looks upon Uyghur activism as hostile to the interests of the PRC nation and has branded some Uyghurs as terrorists. When the United States sought to release four Uyghurs held at the U.S. prison for terrorist suspects in Guantánamo Bay (who, like many detainees, were never convicted of any crime) as part of their effort to close that facility, the Chinese government demanded their extradition back to China. Disquiet and riots have continued intermittently into 2012, with Uyghurs citing ongoing economic marginalization and restrictions on the practice of Islam (**Figure 8.28**).

▲ **FIGURE 8.28 Uyghurs in Western China** Police patrol past ethnic Uyghurs on a street in Urumqi, capital of China's Xinjiang region where bloody ethnic violence left almost 200 dead and 1,700 injured in 2009. A major ethnic minority group in China, the Muslim Uyghur population frequently faces discrimination.

Food For people outside East Asia, "Chinese food" or "Korean food" may conjure up images of an undifferentiated set of globalized generic dishes (from Kung Pao chicken to Mongolian barbecue). The diverse and nuanced food traditions in East Asia, however, emphasize the regionally unique character of food practices and their dependence on traditional food sources, as well as the complex influences from other regions.

In terms of regionally distinct foods, for example, within China there are fundamental differences between the cuisines of Inner and Outer China and between those of the northern and southern areas within Inner China. In Outer China, milk-based dishes—yogurt, curds, and so on—are common and herds of sheep, goats, cows, horses, camels, and yaks also provide meat dishes. Inner China, by contrast, has little stock-raising and, consequently, few milk-based dishes; its principal meat dishes feature ducks, poultry, and pigs. Within Inner China, the humid and subtropical south has developed a cuisine based on rice, while in the subhumid north, noodles are the staple.

More generally, the distinctive character of East Asian cuisine is heavily influenced by historical linkages with the rest of the globe. Many Chinese dishes, for example, are marked by the creative use of hot chilies. This Central American domestic plant only became part of Asian cuisine after 1492. The meat-filled dumpling, a standard fundamentally "Chinese" staple on Asian menus in Britain, France, and the United States, was likely introduced to China in the 3rd century c.e. from elsewhere in Asia (**Figure 8.29**). Similarly, the vegetarian traditions in East Asian cuisine thrived under the influence of imported Buddhist teaching (originally from South Asia—see Chapter 9), and had to compete against meat-loving habits of people from different regions and faiths. East Asian foods have become regionally distinct in part because of the peculiar characteristics of their local conditions, including diverse local plant and animal species and cultural habits and traditions. And they have also become distinctive through their unique interaction with other food traditions from around the world.

East Asian Culture and Globalization In addition to food, other East Asian cultural products have become popular beyond the region. East Asian cultures have long been part of the process of globalization. Chinese art and artifacts were popularized in Europe and North America in the 19th century. Chinese and Japanese cuisines have been introduced to cities throughout the rest of the world. Japanese architecture and interior design have influenced Modernist design. Religions such as Tibetan Buddhism and Zen Buddhism have a growing following in both Europe and North America, and feng shui has found adherents in many Western countries.

Two stylized artistic Asian entertainment forms that have swept the globe in recent years are manga and anime. **Manga** refers to Japanese print cartoons books (*komikku*), which date at least to the last century. Contemporary manga novels cover topics from fantasy to history, usually with an easily recognized and common style, typically involving characters with large eyes and childlike features (**Figure 8.30**). **Anime**, in a general sense, refers to all Japanese animated cartoons. More specifically, however, the term "anime" is associated with a distinctive style of animated film. Though stylistically anime can vary from highly realistic to more abstract, internationally it is most closely associated with the styles of manga. Chapters, settings, and stories draw heavily from manga print cartoons and feature human forms with large eyes and exaggerated expressions. Early popularity of anime film in the United States came in the form of cartoon television series. *Speed Racer* (originally Mach GoGoGo), a cartoon that ran in the late 1960s, made global audiences familiar with the styles, themes, and colorful palette of both Japanese *manga* and *anime*. Since then, this art form has exploded internationally. Nintendo's Pokémon cartoons (and their associated game products) are a multibillion-dollar global business. Of these newly globalizing East Asian cultures, Chinese culture in particular is entering a period of global ascendance. Chinese music and products are increasingly familiar throughout Asia and internationally.

▲ **FIGURE 8.29 Chinese Street *Bao Zi* (Dumplings) in Shanghai** *Bao zi* are dough dumplings filled with vegetables. This classic Chinese street food is common all over China.

▲ **FIGURE 8.30 Manga** These Manga books (Japanese comics), which are enjoyed by a global audience, are on sale at the annual Paris book fair.

▶ **FIGURE 8.31** Beijing National Stadium, "The Bird's Nest" Constructed for the 2008 Beijing Summer Olympics, the Bird's Nest cost approximately U.S. $423 million and seats more than 80,000 people.

China's global cultural profile is reinforced by its growing status as a major destination for tourism. In 2009, China received 51 million international visitors, following only France, the United States, and Spain. In 2011, foreign exchange revenue from tourism to China generated U.S. $35 billion. One of the major events to catalyze this increased traffic was the 2008 Beijing Summer Olympics, which was watched by 4.7 billion viewers worldwide (according to Nielsen Media Research). This event put the "new" China on the map as a global cultural leader, and its cities—Beijing first among them—as destinations on par with New York, Tokyo, Paris, or Sydney.

Despite problems with air quality, the Chinese Olympic games was seen as a tremendous success, involving spectacular architecture (**Figure 8.31**), well-coordinated events, and the hosting of tens of thousands of overseas guests. The hidden costs of that success and similar efforts to accommodate tourism and development are notable. Stringent police control was instituted, free expression was quashed, and many small traditional neighborhoods, or **hutongs**, were demolished to make way for construction (and ongoing urban development). Chinese culture, in this case, has been projected and marketed globally in part through the use of force, over which the Communist Party still maintains a monopoly.

APPLY YOUR KNOWLEDGE Identify some East Asian cultural traditions that have elements or ingredients introduced from other regions. Does describing these cultural phenomena as "traditional," "local," or "regional" have economic or political implications?

Demography and Urbanization

The most striking demographic characteristic of East Asia as a whole is the sheer size of its population. With a total population of some 1.56 billion in 2011 and almost 14% of Earth's land surface, East Asia contains about 22% of its population at an overall density of only 11 persons per square kilometer (28 per square mile). As **Figure 8.32** shows, the bulk of the population is distributed along coastal regions and in the more fertile valleys and plains of Inner China. Population densities in Outer China are very low, less than one person per square kilometer (between two and three per square mile) in most districts—the same as in the far north of Canada. In contrast, population densities throughout Inner China and in Japan, North and South Korea, and Taiwan are very high: between 200 and 500 per square kilometer (518 to 1,295 per square mile) on average.

Population Density High densities of population reflect very high levels of agricultural productivity as well as past population growth rates. High densities often translate to overcrowding, especially in areas where farmland is valuable and where topography restricts settlement. The most striking examples of crowding are in Japan, where mountainous regions preclude urban development on much of Honshu, the main island. The great majority of Japan's 127.5 million people live in dense conditions in the towns and cities of the Pacific Corridor, where space is so tight and expensive that millions of adults, unable to afford places of their own, live with their parents. The average dwelling in metropolitan Tokyo is about 60 square

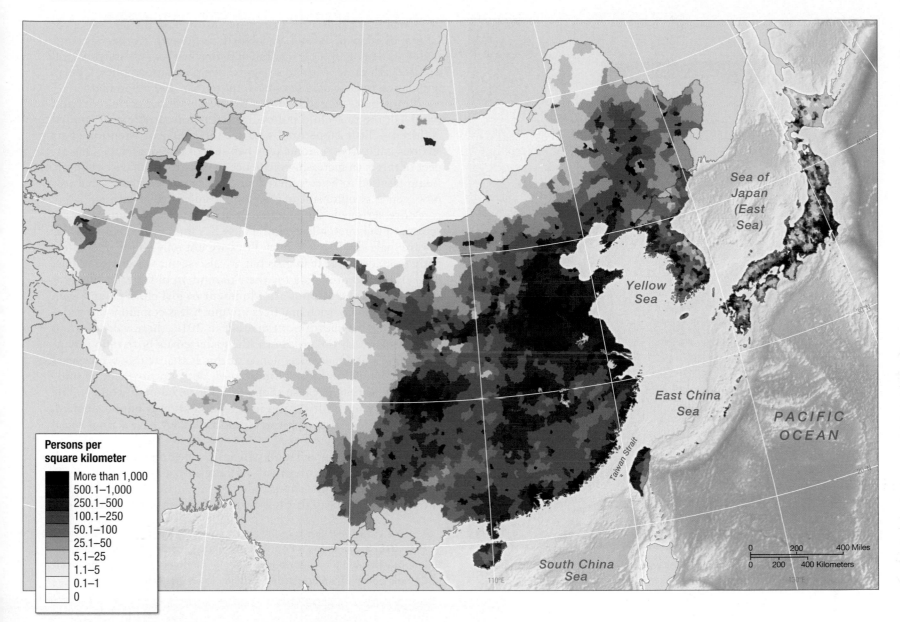

▲ **FIGURE 8.32 Population Density in China** The density of population is very high throughout most of the continental margins, as well as the fertile basin and river valleys. Arid and mountainous interiors are sparsely populated.

meters (646 square feet) (**Figure 8.33**, p. 336). (By comparison, the median dwelling size in metropolitan Los Angeles is just over 150 square meters, or 1,615 square feet.).

People's behavior and urban development patterns also follow from a need to use space efficiently. Japanese daily life is filled with rules, and many of them have evolved because of space shortages. In many Tokyo neighborhoods, for example, trash is picked up four times a week, mainly because people have no place to store it. Skyscrapers are the norm in urban development, with cities "sprawling" upwards by elevator instead of outwards by car. Japanese commuters are used to tight conditions in subway trains during rush hour and sharing public spaces such as parks with large numbers. As a result, transportation and daily life in Japan are relatively free of hassles, despite the high overall density of people.

Population Control In China, crowded towns and cities are more often a reflection of high rates of the rural-to-urban migration of low-income households than of a sheer lack of space. China certainly does not lack space, and until the last century China's population was officially regarded as being too small rather than too large. Between 1961 and 1972, China's population increased by 210 million people as a demographic transition took effect (see Chapter 1, p. 2). Improved medical care and public health reduced death rates, while birthrates remained high, encouraged by the political leadership. During the 1960s, China's population grew at a rate of 2.6% per year.

By the early 1970s, the average family size was 5.8 children, and it became clear that China's communal mode of production could not sustain such a large population. Increasingly, population growth was seen as a threat to the country's chances of economic

▲ **FIGURE 8.33 Small Home, Tokyo, Japan** Scarcity of space makes elegant but efficient home design an architectural goal in Japan.

2011 was 1.54 children/woman, whereas the rate in 1980 was 2.75. China, like Japan, Taiwan, and North and South Korea, has completed its demographic transition and now has low birthrates as well as low death rates. Without its aggressive population policy, China's population today would have been more than 300 million larger.

Nevertheless, in addition to the personal and social coercion involved, the one-child policy has led to the problem of spoiled children—"little emperors," who are the center of attention of six anxious adults (two parents plus two sets of grandparents). One aspect of this is an increasing incidence of obesity in Chinese children. Being fat used to be considered a hedge against bad times, but it is now seen as symptomatic of the cultural shift in child-raising.

More serious is the practice of aborting female fetuses in response to the one-child policy. There has long been a cultural bias toward male children in China. Boys are not only considered inherently more desirable than girls but are also seen as insurance against hard times and providers for their parents in old age. In the past, this bias took the form of abandonment of girl children and even infanticide. Today, such practices are much less common, but widespread selective abortion means that in 2010, there were close to 118 boys for each 100 girls born (the male/female birth ratio is typically 106/100). One consequence of this is that within a few years there will be 50 million Chinese men with no prospect of finding a wife, simply because there will not be enough women. The prospect of such a tilted ratio in the population is unprecedented, except in periods after major wars, and will likely have serious social implications.

Today, China's population policies are beginning to be relaxed. Not only has the rate of growth of China's population slowed satisfactorily, but also the sharp reduction in birthrates over the past two decades will mean that for the next three or four decades, there

▼ **FIGURE 8.34 China's Population Policy** This billboard in Chengdu, China, proclaiming the importance of family planning embodies the ideal of China's policy of limiting urban families to a single child.

development. In response, China's Communist Party instituted an aggressive program of population control. A sustained propaganda campaign (**Figure 8.34**) was reinforced in 1979 and strict birth quotas were imposed: one child for urban families; two for rural families; and up to three for families that belonged to ethnic minorities. The policy involved rewards for families giving birth to only one child, including work bonuses and priority in housing. In contrast, families with more than one child were fined and penalized by a 10% decrease in annual wages. Their children were not eligible for free education and health-care benefits.

In China's major cities, this **one-child policy** was rigorously enforced. Economic trends and health-care technologies along with policy efforts further thwarted population growth. Urbanized families found costs associated with childcare to be prohibitive and abortions were made freely available for unsanctioned and unwanted pregnancies. In terms of reducing population growth, the policy has been very successful. The total fertility rate in China in

will be a pronounced aging of the population, creating a top-heavy situation in which more and more elderly people will have to be supported by fewer and fewer younger workers.

The population policy is not the only force in the region leading to declining birthrates. Though East Asian culture has traditionally favored early marriages, the professionalization of women in many parts of the region has changed this dramatically. Women in East Asia, even when they are working full-time jobs at 40 hours or more per week, are expected to be the caregivers for their children, husbands, and aging parents, who often live with their extended families. This can amount to another 30 hours of housework per week for women, compared to an average of perhaps 3 hours for men, and lead to overall dissatisfaction with marriage for women. A 2011 survey in Japan revealed that fewer women felt positive about their marriage than men.

As a result, marriage has become a far less desirable situation for young women. In Japan in 2011, 22% of women between the ages of 35 to 39 remained single. This trend is also seen in other East Asian countries. In Taiwan, 21% of women in the 35–39 age bracket were single, 17% in Singapore, and 12.6% in South Korea. These statistics reflect new opportunities, identities, and ideas for young women in society as they become more independent and publically assertive (**Figure 8.35**).

> **APPLY YOUR KNOWLEDGE** What are the positive outcomes for China and the Chinese people of the one-child policy? What are the policy's negative effects? What other forces have led to declining birthrates in the region?

Migrations For decades, internal migration within East Asia has been dominated by the movement of people from peripheral, rural settings to the towns and cities of more prosperous regions. Until very recently, migration was strictly controlled in China. During the Communist era, urbanization was restricted and controlled, and many urban dwellers were returned to the countryside as part of reeducation polices during the Cultural Revolution. Other regulations aimed at restricting rural-to-urban migration included raising housing costs for migrants and fining employers who hired transient workers without permission. The gradual liberalization of the Chinese economy has inevitably led to increased rural-to-urban migration, however, and the Chinese government's explicit policy is now to urbanize its population. At the end of 2011, for the first time in the country's history, the urban population of China surpassed that of rural areas, with 691 million people in cities and only 657 million living in the countryside.

In Japan, levels of urbanization have increased dramatically since World War II. In 1950, almost half of Japan's population was dispersed throughout the country in farming households in rural areas. During the period of postwar economic recovery and growth, rural-to-urban migration occurred very rapidly as manufacturing industries expanded, mainly along the Pacific Corridor from Tokyo to northern Kyushu. There was a brief phenomenon of **counterurbanization** (the net loss of population from cities to smaller towns and rural areas) during the 1970s, as some businesses sought to escape the congestion and inflated land prices of metropolitan areas and as some people sought quieter and more traditional settings in which to pursue slower-paced lifestyles. This counterurbanization was selective, affecting only a few places and regions.

By 1990, some 46.8% of Japan's total land area was officially designated as "depopulated" and eligible for special funding. With less than 7% of Japan's population, these rural areas were left with declining economies and aging populations. Japan's cities, in contrast, grew at a

▲ **FIGURE 8.35 Japanese Professional Women** Urban, single women in Japan are marrying later and working longer.

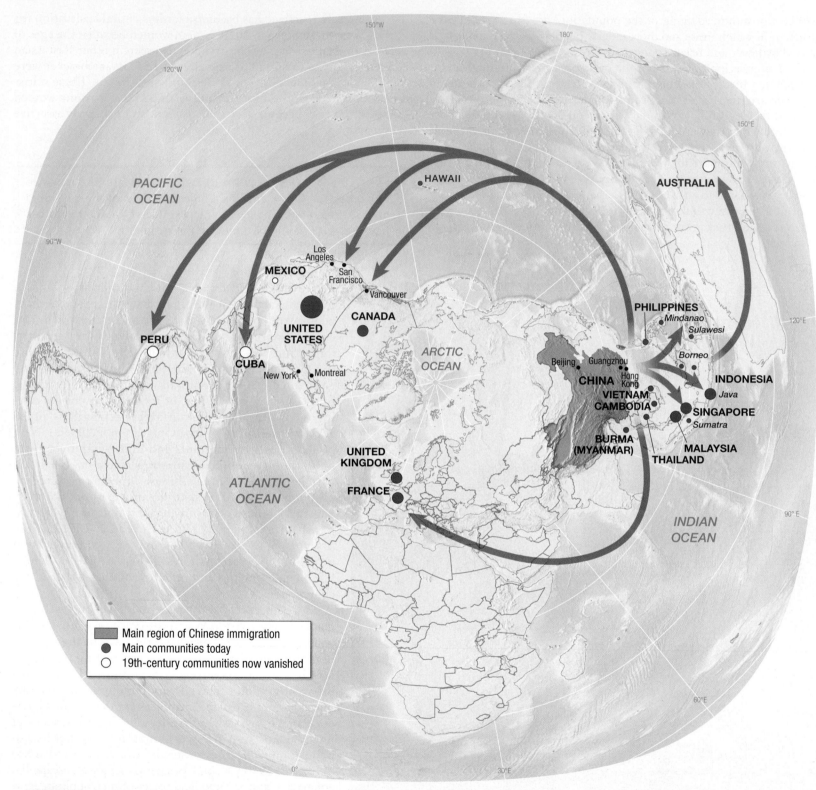

▲ **FIGURE 8.36 The Chinese Diaspora** Chinese emigration began on a large scale during the 19th-century Industrial Revolution, when Chinese laborers found opportunities for employment in newly colonized lands.

terrific pace—partly through migration and partly through natural increase of their younger populations. By 2010, nearly 70% of Japan's population was urbanized.

South Korea has experienced a similar pattern of rural-to-urban migration. Seoul has been the focus of a shift of overall levels of urbanization increasing from 21.4% in 1950 to 83% of total population in 2010. Even North Korea, with poor levels of productivity, experienced a steady increase in urbanization: from 31.0% in 1950 to 60% in 2010.

Diasporas East Asia is distinctive insofar as it has few immigrant populations of any significant size. Distrust of foreigners of all kinds has been long-standing within East Asia. The Communist regimes of China and North Korea contributed to this mindset as well by maintaining tightly closed borders for several decades. In contrast, there has been considerable emigration from East Asia. The Chinese diaspora (**Figure 8.36**) dates from the 13th century, after the conquest of China by the Mongols in 1279. Some Chinese took refuge in Japan, Cambodia, and Vietnam. The Yuan (Mongol) dynasty promoted the basis for a wider Chinese diaspora through its trading colonies in Cambodia, Java, Sumatra, Singapore, and Taiwan. Under the Ming and Qing dynasties, Chinese communities developed around additional trading colonies, including those in Indonesia, the Philippines, and Thailand.

After defeat in the Opium War (1839–1842) and the opening of trade with Western powers, many more Chinese emigrated, creating the basis for the modern Chinese diaspora. The Industrial Revolution and the opening of the world economy through imperialism and colonization provided overseas opportunities. Between 1845 and 1900, 400,000 Chinese are estimated to have emigrated to the United States, Canada, Australia, and New Zealand. Over the same period, an additional 1.5 million Chinese emigrated to Southeast Asia (Indonesia, Malaysia, Singapore, Thailand, and Vietnam), working in mines, on road- and railroad-building, and as farm laborers. Another 400,000 emigrated to the West Indies and Latin America, mainly Chile, Cuba (after 1847), and Peru (after 1848).

Chinese immigration to the United States began with the California gold rush of the 1850s. By 1860, there were 35,000 Chinese in the United States, concentrated in San Francisco, Los Angeles, Seattle, and Portland and scattered in the mines, railway construction projects, and ranches of California. Other significant concentrations of Chinese immigrants developed in Boston, New York, and Philadelphia. But intense discrimination against the Chinese—who were, as geographer Susan Craddock shows, wrongly blamed for everything from outbreaks of cholera to the moral corruption of 19th-century cities—led to a decade of U.S. federal restrictions on Chinese immigration, beginning in 1882. It was only after World War II that Chinese immigration to the United States once more became a significant flow. Today, the Chinese diaspora is the largest in the world and one of the most prosperous. The past quarter-century has seen a great increase in Chinese immigration to the United States, Canada, and Western Europe. In 2009, the Chinese population of the United States was 3.8 million.

The Korean and Japanese diasporas are much smaller but are important elements of the contemporary geography of both East Asia and North America. In Japan and elsewhere in Asia, Korean immigrant populations remain a discriminated-against minority, even after three generations. More recent Korean immigration was to Hawaii and North America, mainly Los Angeles and Vancouver. The strongest component of the Japanese diaspora is in North America. Today, the Japanese-American population numbers more than 1.3 million, with concentrations in Honolulu, San Francisco, and Los Angeles.

Great Asian Cities Japan and Korea have become highly urbanized in the years since World War II, and China is urbanizing quickly in the current era. The resulting cities have numerous environmental problems and stresses, but they have also radically propelled new forms of architecture, public transportation, and urban planning and design. Tokyo, for example, is a conglomeration of 23 wards, each with its own governing body, all linked together through a central metropolitan government. This innovation allows this city, with its 35 million people, to manage urban affairs with flexibility and local initiative, even while coordinating key functions. For example, 27 million people ride Tokyo's highly organized rapid public transportation system every day, traveling to distant parts of the city.

The new cities of China are all the more spectacular for the recentness of their growth. The city of Guangzhou, for example, recently grew to a population of more than 12 million people (**Figure 8.37**, p. 340). The bursting economic activity of the city is embodied in a construction boom, where innovative structure, curvilinear design, and expressive experimentation are creating a whole new skyline. As a reminder of the regional interconnectedness of rural and urban areas, however, it is critical to recall that the glistening new buildings of Guangzhou are, at least in part, constructed by impoverished migrants from the countryside. More than 150 million rural laborers live in China's cities as temporary, and typically second-class, citizens.

FUTURE GEOGRAPHIES

The future geographies of East Asia over the next 20 years rest on the trajectory of three issues with uncertain paths: (1) the slowing of the Chinese economy in the next few years, (2) the fate of the Korean peninsula, and (3) the role of China in international geopolitics.

A "Soft Landing" for the Chinese Economy? The Chinese economy is the world's second largest. At its peak, it grew at an annual rate of 14.2%. This incredible rate of growth, which was never expected to last forever, made possible a total transformation of Chinese society. People's employment expectations, personal wealth, and ability to find and maintain work catapulted upward after the stagnation of the late Communist period. At the start of 2012, Chinese economic growth was 8.1%, a figure many countries would envy, but one that was even lower than what pessimistic experts had predicted. This downturn was in part the result of a natural slowing of the economy,

▲ **FIGURE 8.37 A Construction Site in Guangzhou, Guangdong, China.** Hundreds of new skyscrapers now form the skyline of this bustling city at the mouth of the Pearl River in southern China.

coupled with the sharp downturns in Europe and the United States, the key consumers of Chinese goods.

This slower growth will likely continue in coming years. The question is how quickly the economy will slow and what effect the downturn will have on Chinese workers, consumers, and investors. In a country that has traded political freedoms for the promise of economic prosperity, a sharp downturn in the economy might lead to considerable unrest. On the other hand, a slow cooling, where reasonable growth in employment and GDP per capita can be maintained, might open the door to social and political stability. East Asia's fate, as well as that of the rest of the world, may lie in whether the Chinese economy comes in for a hard landing, or a soft one.

Emerging Korean Conflict or Reconciliation The ascendance of Kim Jong-un as supreme leader of North Korea in 2011 ushered in some intense uncertainties. Many speculate that this young unknown leader may attempt to make a bold military statement in the interest of asserting his authority. Conversely, his arrival may represent a possibility for new openness between the country and its

neighbors and the world. Early signs are not especially encouraging. A failed missile test almost immediately after Kim Jong-un's accession suggests the world can expect ongoing aggressiveness. The nontransparent nature of the government and the extreme political power of the single family of rulers also point to intransigence and unpredictability.

On the other hand change is still possible. Efforts at diplomacy between the Koreas have been extremely limited to date. But the Korean people continue to hold a tacit interest in unification. These trends make it possible to at least imagine a very different outcome, in which a unified Korea emerges in the 21st century, which would hasten regional disarmament and bring East Asia back from the brink of a conflict that has simmered for 50 years.

Emerging Chinese Hegemony As we have seen, the growth of the Chinese economy has placed the country in a new political position relative to the rest of the world. Will the 21st century be the "Chinese Century" or a century of diverse economic players, including China among many? How central and autonomous will the Chinese economy be as global trade sees a reduction in the absolute

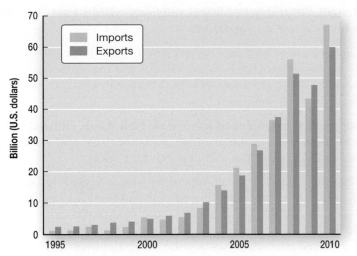

▲ FIGURE 8.38 Chinese Investment in Africa Chinese investment in Africa has risen dramatically in recent years, as China pursues energy resources and minerals to support its booming industrial economy.

dominance of historically critical players (such as the United States)? China is now the major trade partner with Africa (**Figure 8.38**) and Chinese development has come to compete with U.S. and European foreign aid in Africa (**Figure 8.39**).

An extremely strong political China signifies interesting possibilities, because that country is in a unique position to oversee negotiations with North Korea, for example, and to provide a strong counterweight to Russia and other global resource powers. As noted earlier, a strong political China is also in the position to support and extend the power of problematic and dangerous states. For example, in the wake of recent extreme violence on the part of the Syrian state against its citizens (resulting in as many as 8,000 civilian deaths), China joined Russia in vetoing two UN Security Council resolutions condemning the violence. How and where China extends its power within and beyond the region will determine a great deal about global economics and politics for the next 20 years.

◄ FIGURE 8.39 Chinese Development in Africa Chinese workers from the Zhongyuan Petroleum Exploration Bureau (ZPEB) celebrate the completion of the major part of an oil pipeline with Sudanese workers. China gives more loans to underdeveloped countries than the World Bank. Most of the money ends up in Africa. The influence of China over African politics parallels this rise.

LEARNING OUTCOMES *REVISITED*

■ **Explain how the landscape of the East Asian region has been modified by humans over the centuries, particularly in regards to water management, and how these adaptations may be impacted by climatic change.**

The management of waterscapes in the region has long involved large and sophisticated irrigation systems, often overseen by expert centralized authority. This continues to be the case, especially in China, where hydroelectric investment and water diversion are key aspects of development. Worsening flooding and droughts resulting from climate change are expected to test the capacity of these systems.

■ **List the main physical subregions of East Asia and describe their tectonic origins and related hazards.**

East Asia is distinguished by the vast continental interior of China and the peninsular and island coasts formed at the edge of the Ring of Fire. Risks associated with the region's physical features include flooding from the enormous drainages across China and earthquakes such as the one that triggered the recent 2011 Japanese disaster.

■ **Identify the adverse ecological effects of urbanization and modernization of agriculture.**

Widespread urban growth has led to a decline in wildlife and placed stress on open space, leading to "vertical" urban development in the region. Traffic and industry have led to serious air quality problems and a persistent "Asian Brown Cloud." Green Revolution agriculture has resulted in high levels of overfertilization, but new genetically modified crops may have caused a decrease in pesticide use and poisonings.

■ **Distinguish the centers of power in East Asia in the precolonial, colonial, and postcolonial periods and describe how and why they have changed over the course of history.**

The historical geography of the region is characterized by the core economic power of China, which was at the center of global trade networks for more than a thousand years prior to colonization and is becoming so again. The colonial era and the upheavals of World War II drove China from its historically dominant position in the region, allowing for a reconfiguration of the region dominated by trans-Pacific trade between Japan and the United States.

■ **Compare the development of different economic experiments in the 20th and 21st centuries in China, South and North Korea, and Japan.**

Japan's economy is conglomerate-oriented and based on banking-dependent industrialization. South Korea utilized import substitution to grow its industrial capacity. North Korea continues to feature an authoritarian and centralized planning approach to economic development. Chinese development, however, dramatically switched from central planning to state-led capitalism. All of these developments have led to spatially uneven patterns of development.

■ **Identify central geopolitical tensions within East Asia and their historic and recent drivers.**

The geopolitics of East Asia revolve around the shifting balance of economic power between Japan and China and the unresolved problem of the Korean peninsula, where the Korean War has never officially come to an end.

■ **Use specific examples to show how East Asian cultural practices and systems are the product of distinct regional traditions as well as linkages across and beyond the region.**

In this region, many of the dominant languages (including Japanese and Korean), religious and philosophical traditions (including Confucianism, Buddhism, and Shintoism), and food traditions (including iconic local dishes like dumplings) have distinct regional roots but display shared characteristics, feature some elements from outside the region, and are often the object of political contestation.

■ **Describe the various forces that have caused significant declines in population growth across East Asia.**

The one-child policy in China led to a major decline in the fertility rate, although with some social problems, including the imbalance in the populations of girls and boys. Cultural change and professionalization of women, in countries including Japan, Taiwan, and South Korea, have also contributed to later marriage ages and declining fertility rates.

■ **Summarize and explain the major immigration trends in the region such as rapid urbanization.**

East Asian populations have long been diasporic. The region has also seen a rapid pattern of urbanization as a result of industrialization, especially in trade centers along the Pacific Rim. Urbanization has led to a growing disparity of wealth and power between cities and rural areas.

KEY TERMS

ancestor worship (p. 329)

anime (p. 333)

animism (p. 329)

Anthropocene (p. 315)

Asian Brown Cloud (p. 315)

Asian Tiger (p. 326)

chaebol (p. 326)

Confucianism (p. 329)

Convention on International Trade in Endangered Species (CITES) (p. 316)

counterurbanization (p. 337)

Cultural Revolution (p. 331)

Dalai Lama (p. 328)

demilitarized zone (DMZ) (p. 315)

Deng Xiaoping (p. 324)

feng shui (p. 330)

Five-Year Plan (p. 324)

Grand Canals (p. 319)

Great Leap Forward (p. 324)

Green Revolution (p. 314)

hutongs (p. 334)

hydraulic civilizations (p. 311)

ideograph (p. 330)

import substitution (p. 326)

keiretsu (p. 323)

Kim Il Sung (p. 326)

Kim Jong Il (p. 327)

Kim Jong-un (p. 327)

manga (p. 333)

Mao Zedong (p. 320)

massif (p. 310)

one-child policy (p. 336)

Opium Wars (p. 320)

Pacific Rim (p. 323)

pinyin (p. 330)

Ring of Fire (p. 307)

Shinto (p. 329)

shogun (p. 319)

Silk Road (p. 318)

Special Economic Zones (SEZs) (p. 324)

state capitalism (p. 324)

Treaty of Nanking (p. 320)

THINKING GEOGRAPHICALLY

1. How has water management been an important part of the civilizations of East Asia? To what degree has water management continued to be a factor into the present?

2. How and to what degree was East Asia economically linked to the rest of the world in the precolonial era? What is the Silk Road?

3. Explain how the central government involves itself in corporate activities in China, Japan, and South Korea.

4. What are the driving forces creating the emerging Pacific Rim region? Is this region a fully coherent or unified one? Why or why not?

5. How and why is the status of women changing in East Asia?

6. What are some dominant East Asian cultural traditions? How have East Asian beliefs and traditions affected other world regions?

7. How has the state been involved in population control in East Asia? What other forces have influenced the region's birth rates?

MasteringGeography™

Looking for additional review and test prep materials? Visit the Study Area in MasteringGeography™ to enhance your geographic literacy, spatial reasoning skills, and understanding of this chapter's content by accessing a variety of resources, including **MapMaster** interactive maps, videos, RSS feeds, flashcards, Web links, self-study quizzes, and an eText version of *World Regions in Global Context*.

9 | South Asia

Lunchbox deliverymen pick up hot food from homes and deliver it to some 200,000 hungry Mumbai office workers in time for lunch. This army of *dabbawallas* has been studied by business schools around the world as a model of organization worth emulating.

A WOMAN IN THE SPRAWLING SUBURBS OF MUMBAI, INDIA, finishes cooking her husband's lunchtime meal in the middle of a busy workday morning. She places it, still piping hot, into small metal tiffin, a set of stacked lunch boxes for carrying food. From here, the tiffin begins an epic journey of dozens of miles, changes hands five times, and somehow, remarkably, winds up on the desk of her husband, a systems engineer working in a downtown Mumbai high-rise. Tens of thousands of lunches miraculously travel similar routes every day, by way of a highly organized system.[1]

The tiffin delivery system depends on a huge network of *dabbawallas*—food delivery men—who pick up tiffins from homes, deliver them by bicycle to trains, pick them up at local stations, carry them on their heads through crowded streets, and finally hand them off for final delivery. A system of complex colorful markings on every tiffin ensures that they arrive, still

[1]Saritha Rai, "In India, Grandma Cooks, They Deliver," *The New York Times*, May 27, 2007.

LEARNING OUTCOMES

- Describe the pattern of the monsoon climate, explain its origins, and identify the key adaptations people in the region use to cope with change and uncertainty.

- Distinguish the physiographic and engineered components of environmental hazards in the region, especially catastrophic flooding.

- Explain how some land uses contribute to environmental destruction and declines in South Asian biodiversity, while others promote the maintenance, protection, and restoration of wildlife.

- Provide examples of rulers, governments, and economic systems that historically put South Asia at the center of global trade, and explain how this position was later exploited and overridden by colonial governance.

- Identify the major engines of recent rapid South Asian economic expansion and discuss the uneven benefits of expansion.

- Describe the key relationships and borders that present the greatest current geopolitical challenges for South Asia.

- Summarize the characteristics and distributions of the region's major religions and language groups and identify their overlap, uniting features, and mutual influence.

- Distinguish key traditional cultures and emerging consumer cultures and cite examples of how each has impacted and been impacted by global connections.

- Explain where and under what conditions South Asian populations have begun to stabilize or decline and where they continue to grow relatively rapidly.

warm, in the hands of the right office worker. Internet communication has made the system even more efficient. The startling complexity of this network together with the remarkable simplicity of its core concept say a lot about the always innovative economies of South Asia.

First, although this system is old, it continues to innovate and expand. As Mumbai's labor force grows and its middle class thrives, the demand for hot food delivery from home has only increased. The global connections of South Asia to other world regions has created new economies in high-tech businesses, and also expanded traditional ones, like that of the dabbawallas.

Second, although the business thrives on information technology, both in the ancient form of colored markings and the expanding use of the Internet, it also depends heavily on labor. In a country of more than 1 billion people, high technology and working hands are *both* part of daily life. In South Asia, the emerging global, urban, modern, and high technology office world is not so much replacing the rural, traditional, and manual systems of the past as it is becoming firmly interlinked with them. ∎

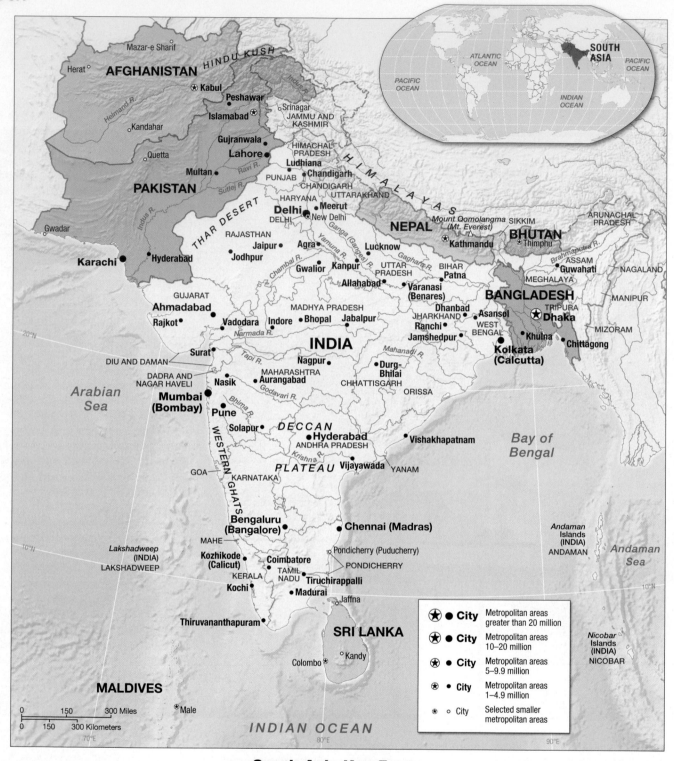

South Asia Key Facts

- Major Subregions: Deccan Plateau, Mountain Rim, Peninsular Highlands, Coastal Fringe
- Major Physiogeographic Features: Bay of Bengal; Himalayas (including Mount Qomolangma or Everest); Khyber Pass; Indus, Ganga (Ganges), and Brahmaputra Rivers; Indus Plains. Dominant aspect of climate is southwesterly summer monsoons
- Major Religions: Hinduism, Islam, Buddhism
- Major Language Families: Indo-Aryan (including Hindi, Bengali, and Urdu), Dravidian
- Total Area (total sq km): 5 million

- Population (2011): 1.65 billion, including 1.24 billion in India, 150.7 million in Bangladesh, 176.9 million in Pakistan; Population under Age 15 (%): 33; Population over Age 65 (%): 4
- Population Density (per sq km) (2011): 415
- Urbanization (%) (2011): 27
- Average Life Expectancy at Birth (2011): Overall: 65; Women: 66; Men: 64
- Total Fertility Rate (2011): 3.1
- GNI PPP per Capita (current U.S. $) (2009): 3,101; highest is Bhutan: 5,290; lowest is Afghanistan: 860

- Population in Poverty (%, <$2/day) (2000–2009): 55x
- Internet Users (2011): 161,634,316; Population with Access to the Internet (%) (2011): 17.3; Growth of Internet Use (%) (2000–2011): 2,886
- Access to Improved Drinking Water Sources (%) (2011): Urban: 92; Rural: 79
- Energy Use (kg of oil equivalent per capita) (2009): 415
- Ecological Footprint (hectares per capita consumed/hectares per capita available, global scale) (2011): 1.3/1.8

ENVIRONMENT AND SOCIETY

Two aspects of South Asia's physical geography help define it as a world region (**Figure 9.1**). Each of these features also highlights the region's connections to other regions. The first striking feature is the **monsoon**, seasonal torrents of rain on which the livelihoods of South Asia depend. The monsoon system is propelled by the annual heating and cooling of the landmass of Central Asia to the north and the constant feeding of moisture from the Indian Ocean to the south. This signature physical condition is a result of South Asia's position between other world regions. The second feature, as the satellite image reveals (**Figure 9.2**), is the forbidding mountain ranges that sets South Asia apart from the rest of Asia. This arc of mountain ranges serves as a porous boundary around the people of South Asia. Like the monsoon, this distinctive feature is born from the interaction of regions, as the South Asian Plate collides with the Asian Plate, creating massive uplift and ongoing mountain growth.

Climate, Adaptation, and Global Change

The climate of South Asia is distinguished by the dramatic effect of the southwesterly summer monsoon, a deluge of rain that sweeps from south to north across the region. The word *monsoon* derives from the Arabic word *mausim*, meaning "season." Although now widely used to describe any windy, rainy season, the term *monsoon* was originally applied only to the distinctive seasonal winds in the Indian Ocean. The name was given to the winds by Arab traders who relied on them to power their sailing ships on annual voyages in quest of spices, ivory, and fine fabrics.

The Monsoon In most of South Asia, the seasonal pattern of climate consists of a cool and mainly dry winter, a hot and mainly dry season from March into June, and a wet monsoon that bursts in June and lasts into September or later. The engine behind the monsoon is the heating and cooling of the Asian continent to the north of the region. During the early summer, the interior parts of Asia begin to heat more rapidly than the areas to the south, creating low-pressure convection as hot air rises all across Asia. By midsummer, drawn onward by this low pressure, the jet stream and the atmospheric circulatory system move north, carrying moisture from the ocean. This wet monsoon moves inland bringing drenching rains. The rains slowly progress from the island of Sri Lanka, across the southern coastal zones of India, across Bangladesh and the plains of India, and finally into the northwestern parts of India and Pakistan (**Figure 9.3**, p. 348). The northern hills and uplands exert a strong **orographic effect**, causing moist air from the sea to lift, condense, and produce heavy rainfall. The arrival of the wet monsoon season is announced by violent storms and torrential rain (**Figure 9.4**, p. 348). In winter, this system reverses, as relative low pressure over the Indian Ocean draws air outwards from the interior of Asia. The result is northeasterly winds, which blow from the interior toward the sea. These are the dry monsoon winds that typically do not bring rain.

One exception to this pattern occurs in parts of Afghanistan, Pakistan, and northwest India, where shallow low-pressure systems move through from the west, bringing light but useful rainfall in late winter. Another exception occurs in northern and eastern Sri Lanka, where trade winds bring some winter rains. The southern part of Sri Lanka, the Maldives, and the Nicobar Islands are far enough south to be affected by intertropical convergence and so these locations rarely experience a dry month all year. The result is the overall pattern of climate across the region, with tropical climates hugging the coasts and the islands, a broad belt of humid subtropical climate following the Ganga (Ganges) and abutting the Mountain Rim, and arid climates in the desert northwest of India and Pakistan in the interior of the Indian Plateau (**Figure 9.5**, p. 348).

Adapting to Rainfall and Drought For much of South Asia, there is a significant risk of drought and famine if there is a late or unusually dry monsoon season. In 2009, India experienced a massive drought that affected over half of its political districts. More than 700 million people were affected; food prices increased by more than 10%, and food had to be imported by the government to head off rampant speculative buying and hoarding. Even where the monsoon does not fail completely, it can be highly uneven, so that villages in one region may receive sufficient rainfall to plant and harvest, while nearby settlements do not.

The fierceness of the monsoon rains, as well as their unpredictability, spatial unevenness, and occasional failure, has led to a range of ingenious adaptations. Inventive cropping systems, agricultural biodiversity, social networks, and water storage ensure stability and survival. Systems for growing crops across South Asia have evolved

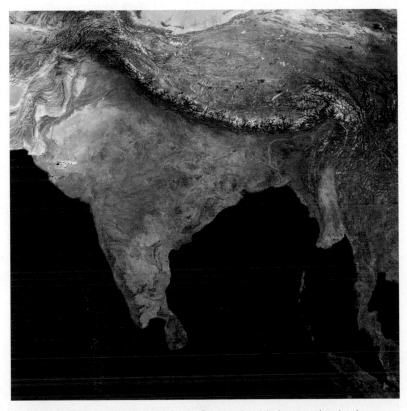

▲ **FIGURE 9.2 South Asia from Space** As this image clearly shows, South Asia is naturally bordered by mountains and oceans.

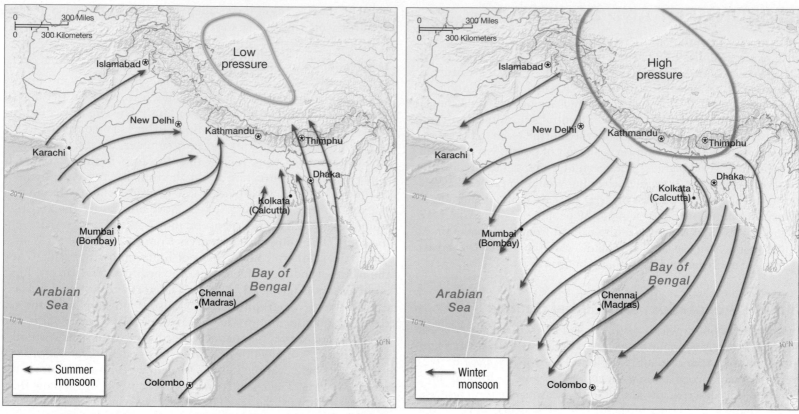

▲ **FIGURE 9.3 Summer and Winter Monsoons** The wet summer monsoon arrives in June and lasts into September or later. In winter, the prevailing dry monsoon winds are then northeasterly.

▲ **FIGURE 9.4 Monsoon in South Asia** An Indian hand rickshaw puller wades through waters after a heavy downpour in Kolkata during a fierce monsoon during 2007.

▶ **FIGURE 9.5 Climate Map of South Asia** South Asia is home to a variety of climate regions, with desert in the northwest and wet tropics in the far south.

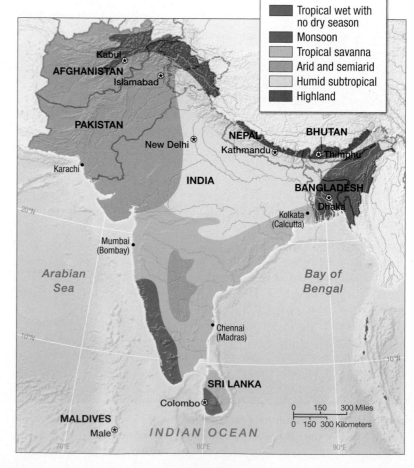

◀ **FIGURE 9.6 Itinerant Herding in India** The goatherds shown here are migrating with their flocks from arid Rajasthan, India. Livestock keepers adapt to the vagaries of climate by strategically moving animals to find forage.

to maximize efficiencies in the use of soil and water. In the mountain areas of the Himalayas and the Karakoram, enormous labor has gone into stabilizing the land for cropping in the face of potentially catastrophic soil erosion. **Terracing** of these slopes over generations has created a distinctive landscape and allowed sustained agricultural yields amid torrential rainfall and steep slopes.

In more arid regions, agricultural systems maximize **intercropping**, the mixing of different crop species that have varying degrees of productivity and drought tolerance. In bad rainfall years, farmers salvage a harvest by relying on the heartiest part of the crop. The agrodiversity of traditional South Asian farming is extremely high and includes grains like millet. That hardy crop can grow to harvest after soaking up only two or three light rainstorms.

Livestock production has also adapted to the region's climates. In Afghanistan, the Himalayan regions of Nepal and India, and the arid desert districts of India and Pakistan, local communities tend hardy breeds of goats and sheep, periodically moving the animals to different climate zones to take advantage of local pasture conditions. This practice, referred to as **transhumance**, involves a traditional social system that allows members of families to be away many months of the year tending animals (**Figure 9.6**). It also accommodates modern technology. Herders now transport their animals by truck on paved highways and use radios and computers to access up-to-date weather information.

During the 1960s, efforts were made to boost agricultural production through the use of modern technologies as part of the **Green Revolution**. During the Green Revolution, countries such as the United States, global agencies such as the United Nations Food and Agriculture Organization (FAO), and multilateral lending agencies such as the World Bank extended new technologies to the region. These innovations included high-yielding varieties of wheat and corn, which could double or triple production per hectare of land. These crops required more inputs than their traditional counterparts, however, including more water, fuel, fertilizer, and pesticides. The result was an increase in productivity, but also increased dependency on purchased farming products, decreased native seed diversity, and exhausted soils and water supplies. The food security benefits of the Green Revolution are undeniable. However, the negative impacts of the system, which undermined many of the traditional systems geared toward coping with uncertainty, are also apparent in the era of climate change.

Implications of Climate Change It is too soon to tell precisely what global climate change might portend for South Asia. Recent years have actually seen an increase in the intensity of rainfall, which has led to flooding. There have been record-breaking and recurring floods in Bangladesh, Nepal, and northeast India. Typhoons (hurricanes), which are endemic to the Bay of Bengal and a serious environmental hazard in Bangladesh, have grown less frequent in recent history but more intense. These storms are now harder to predict and have increasingly devastating impacts. Coastal areas are vulnerable to the sea-level rise that will accompany melting of polar ice as well. Should sea level rise by 1.5 meters (5 feet) in the next century, as many as 22,000 square kilometers (8,494 square miles) could be lost to the sea in Bangladesh, impacting at least 18 million people. Even where inundation is not complete, the seasonal storms that impact the Bay of Bengal will become all the more serious, as storm surge comes inland into low-lying areas (**Figure 9.7**, p. 350).

Conversely, climate change may also increase drought events in some areas; crop yields in many regions have already suffered in recent years from unexpected high temperatures and heat waves. More frequent and recurring failures of the monsoon are possible, especially for arid northern India and Pakistan, where the monsoon

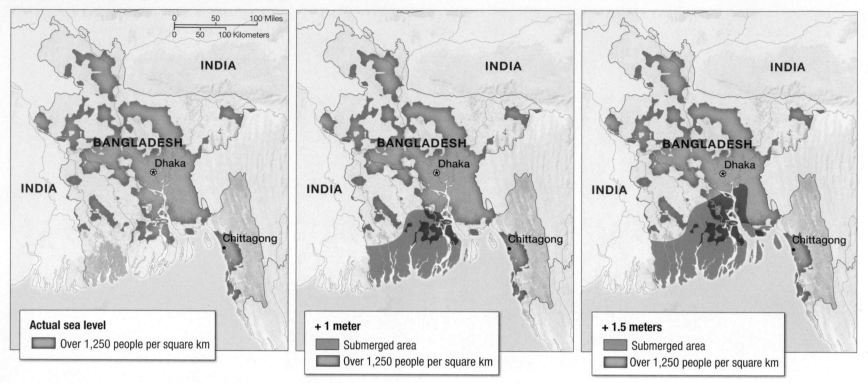

▲ **FIGURE 9.7 Forecasted Hazards of Sea-Level Rise for Bangladesh** With only 1.5 meters (5 feet) of sea-level rise, much of Bangladesh could be inundated. The blue areas on the second and third maps show the areas that will be underwater if sea level rises.

often fails to reach (**Figure 9.8**). Water scarcity in cities is also an increasing problem. Climate change will likely exacerbate this situation by making the monsoon less predictable.

One of the most uncertain impacts of global climate change on the region has to do with the status and condition of Himalayan glaciers. As the Intergovernmental Panel on Climate Change (IPCC) reports, these glaciers, which cover 17% of the mountain areas of the Himalayas and represent the largest body of ice apart from the polar

▲ **FIGURE 9.8 Drought in South Asia** A village boy leads his goat through a parched pond on the outskirts of the eastern Indian city of Bhubaneswar, in the month of May. Huge swathes of rural farmland turn dry during this period before the annual monsoon rains. If these rains do not arrive, difficult conditions will continue indefinitely.

ice caps, are receding faster than in any other area of the planet. Should these glaciers experience serious decline, the major rivers of the region (Ganga, Indus, and Brahmaputra) will lose all but their seasonal (monsoonal) water sources. This threatens agriculture as well as the drinking water supplies for more than a half-billion people, especially during the lengthy dry season. Future glacier status is difficult to predict, however; recent research shows that some glaciers in the Himalayas have benefited from increasing local precipitation and gained a small amount of mass between 1999 and 2008.

Another poorly understood but serious problem posed by climate change is disease ecology. Nearly the entire population of this region is vulnerable to certain diseases, especially those spread by mosquitoes. These include malaria, which a 2010 study suggests may kill between 125,000 and 277,000 people per year in India alone. With overall global warming, it is reasonable to anticipate that some cities will experience increasingly lengthy mosquito breeding seasons and a significant, if not exponential, escalation in rates of disease. Devastating viral fevers, such as Chikungunya and dengue, may even increasingly impact wealthy residents in cities like Delhi, where breeding of some mosquito species is dependent on clean standing water (commonly found in gardened suburbs). If climate change leads to increased precipitation, there will be a greater opportunity for standing water and mosquito breeding, a concern for both rich and poor in South Asia.

APPLY YOUR KNOWLEDGE How has adaptation to the monsoon rains prepared the people of South Asia for the vagaries of climate change? What new hazards may be too unprecedented to be readily addressed based on past experience and traditions?

Geological Resources, Risk, and Water Management

In geological terms, South Asia is a recent addition to the continental landmass of Asia. The greater part of what is now South Asia broke away from the coast of Africa about 100 million years ago. It drifted slowly on a separate geological plate for more than 70 million years until it collided with the southern edge of Asia. The slow but relentless impact crumpled the rocks on the south coast of Asia into a series of lofty mountain ranges and lifted the Tibetan Plateau more than 5 kilometers (3.1 miles) into the air. The Himalayas, which stand at the center of South Asia's Mountain Rim, are still rising (at a rate of about 25 centimeters—9.8 inches—per century).

The principal physiographic features of South Asia reflect this major geological event. They include the Peninsular Highlands, the Mountain Rim, the Plains, and the Coastal Fringe (**Figure 9.9**). The Peninsular Highlands of India form a broad plateau flanked by two chains of hills. The highlands rest on an ancient layer of volcanic rocks, together with some very old sedimentary rocks. This has remained a relatively stable landmass for much of the past 30 million years. However, between 65 and 55 million years ago, immense eruptions of lava buried parts of the peninsula beneath dense

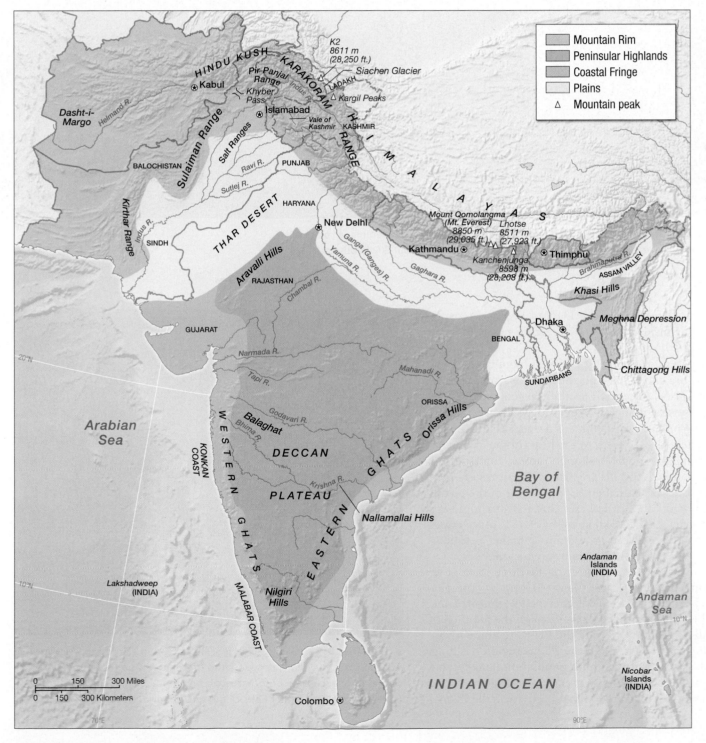

◀ **FIGURE 9.9**
South Asia's Physiographic Regions The physical geography of South Asia is framed by the highland plateaus of an ancient continental plate and a young mountain rim; these are separated by broad plains.

Legend:
- Mountain Rim
- Peninsular Highlands
- Coastal Fringe
- Plains
- △ Mountain peak

▲ **FIGURE 9.10 Deccan Plateau** The distinctive geology of the subcontinent's peninsular plateau is carved by erosion into layers of volcanic rock, shown here at Malshej ghat, about 200 kilometers (125 miles) from Mumbai.

volcanic rock. Today, this lava layer—the Deccan Plateau—covers about one-third of the Peninsular Highlands, creating a truly distinctive landscape (**Figure 9.10**).

The Mountain Rim is a vast region of spectacular mountain terrain, remote valleys, varied flora and fauna, ancient Buddhist monasteries, and fiercely independent tribal societies. The physical geography of the region is complex, with several mountain ranges together sweeping in a 2,500-kilometer (1,554-mile) arc that contains numerous high peaks, including K2 (8,611 meters; 28,250 feet). At the heart of the Mountain Rim are the young Himalayan and Karakoram ranges. Interspersed among the high peaks and the foothills are protected gorges and fertile valleys that sustain isolated settlements. The largest of the fertile valleys is the Vale of Kashmir, a fertile basin that is 130 kilometers (81 miles) long and between 30 and 40 kilometers (19 to 25 miles) wide. In this and other valleys, it is possible to grow grains at higher elevations and to maintain orchards of apricots and walnuts on the valley slopes.

These mountains also serve as the source of the three great river systems of the region—the Indus, the Ganga (Ganges), and the Brahmaputra—which begin within 1,600 kilometers (994 miles) of one another but flow in three different directions through the mountains and into the plains. All three rivers provide the Plains region with an uneven flow of melting snow. As a result, the Plains region has long been widely irrigated, has supported a high population density, and has been the site of many of the region's great cities (**Figure 9.11**). Cultivation of wheat and rice is the predominant activity in the Plains.

▲ **FIGURE 9.11 Ganga (Ganges) River Flowing Through Varanasi** A set of steps (or *ghats*) descend into the Ganga (Ganges) in the sacred city of Varanasi. Some of the greatest and most ancient cities of the region were built on the banks of the this important river.

▲ **FIGURE 9.12 The Coast of Sri Lanka** Traditional stilt fishermen practice their craft at Kogalla, Sri Lanka. The Coastal Fringe of South Asia is home to distinctive fishing cultures and rich trading civilizations.

The Bengal Delta, which covers a large portion of Bangladesh and extends into India, is the product of the Ganga and Brahmaputra Rivers and their **distributaries** (river branches that flow away from the main stream). Because these rivers lie in deep deposits of sand and clay and carry such enormous quantities of water, they are almost impossible to control with engineering works; as a result, flooding is normal. In the monsoon season, about 70% of the delta is flooded up to 1 to 2 meters (3 to 6 feet) deep.

The narrow Coastal Fringe is the product of marine erosion that has cut into the edge of the mountainous Peninsular Highlands and of alluvial deposits in the form of deltas and mudflats. Many of South Asia's largest and most prosperous cities developed from trading posts that were established along the Coastal Fringe in the 17th century. The largest among them—Chennai (formerly Madras), Colombo, Karachi, Kolkata (formerly Calcutta), and Mumbai (formerly Bombay)—remain important global cities to this day. During the rainy monsoon seasons, the Coastal Fringe is filled with luxuriant growth, especially along the southwest Malabar Coast of India and the southwest coast of Sri Lanka, where rich harvests of rice and fruit support dense rural populations (**Figure 9.12**).

The Hazard of Flooding Where the summer monsoon hits hills and mountains, the rains are especially heavy. The peninsula typically receives between 2,000 and 4,000 millimeters (79 to 158 inches) of rainfall where the southwest monsoon winds meet the steep slope at the edge of the Peninsular Highlands. Similar levels of annual rainfall fall in the central and eastern parts of the Mountain Rim. Cherrapunji, in the Khasi Hills south of the Assam Valley, boasts the world's highest average annual rainfall record of 11,437 millimeters (450.6 inches). Cherrapunji once recorded as much as 924 millimeters (36.4 inches) in one day at the onset of the monsoon season. The outflow from river flooding from these

highlands, coupled with heavy local rainfall, can have a doubly hazardous effect. In 2011, extremely heavy monsoon precipitation had devastating effects. Dams overflowed in the state of Gujarat, and more than 100 people died or went missing in India and Nepal. Over 5 million people were affected and almost 1 million were left homeless.

In Bangladesh, monsoon rains produce widespread flooding. Swollen by rains, the distributaries of the Ganga (Ganges), Brahmaputra, and Meghna regularly spill over into the low-lying delta areas. If rains are unusually heavy, flooding can be disastrous, inundating villages, drowning people and livestock, and ruining crops. Some geographers have suggested that annual flooding is becoming more pronounced and point to deforestation in India and Nepal as the cause of increased runoff.

The more arid regions of the northwest, including parts of both Pakistan and India, are also susceptible to periodic flooding. In Pakistan in 2010, heavy and unrelenting rainfall in the Indus River basin left approximately one-fifth of Pakistan's total land area underwater (**Figure 9.13**). The floods affected 20 million people, destroying homes and farmland, with a death toll of approximately 2,000 people. The economic impact likely exceeded U.S. $43 billion. The Pakistan floods were a natural event, insofar as they were the result of an unusually high and widespread rainy season. The impacts of the event were made far more devastating, however, because of the way the floodplain has been engineered and managed. Specifically, the Indus has been increasingly "straightjacketed" through the construction

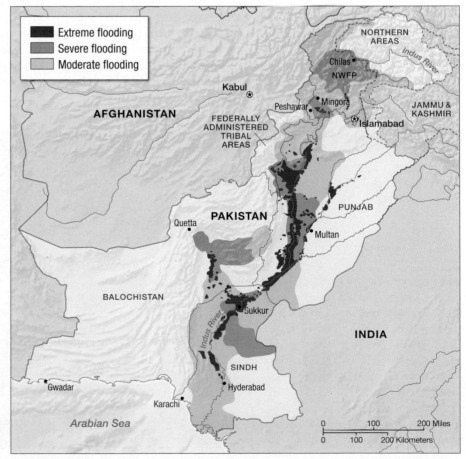

▲ **FIGURE 9.13 Flooding in Pakistan, 2010** The extent of this massive flood was national in scope, sweeping away whole districts and resulting in thousands of deaths.

of levees or embankments, especially by wealthier landholders, to allow for irrigation of their land without the risk of seasonal high waters. These infrastructural changes reduce the risk of typical, shorter-term, smaller floods and maximize producer profits. In the process, however, these alterations (along with widespread deforestation on hillsides) tend to accelerate and concentrate floodwaters during more extreme events. The geography of the Pakistan floods shows how climate events and human decision making can combine to lead to tragedy on a vast scale.

> **APPLY YOUR KNOWLEDGE** Why did landholders and government officials in Pakistan resort to building large-scale engineered structures (dams and dikes) in the Indus Valley? How did their efforts reduce or increase flooding risk?

Energy and Mineral Resources The geological resources of the South Asia region are modest by global standards, but there are several important deposits. The Peninsular Highlands region contains valuable deposits of iron ore, manganese, gold, copper, asbestos, and mica. Coal is found in workable quantities in parts of the Mountain Rim. Oil-bearing ranges have been located in the Assam Valley and Gujarat and natural gas is present in Bangladesh and Sindh. New oil deposits have recently been discovered in the Indian state of Rajasthan.

The geology of the region also yields large quantities of sandstone and marble, construction materials that have long been a fabric of the region's architecture, including monumental works like the Taj Mahal in Agra and numerous major structures in Delhi (**Figure 9.14**). Although these materials have proven incredibly valuable for construction and export on international markets, the labor

and environmental conditions at quarries are highly problematic. Workers, including young people, labor in conditions that expose them to fine particles leading to silicosis in their lungs. Pay and schools are also subpar in mining areas. Mining in these areas also damages watersheds, wildlife habitats, and forest cover.

Arsenic Contamination A dramatic case of geology's impact on humans in the region was the discovery in 2000 that millions of **tube wells** in Bangladesh were drawing arsenic-contaminated water. Tube wells are lined with a durable and stable material, usually cement. This makes it possible to sink them deeper than traditional water wells. These wells were installed in Bangladesh as a result of a campaign in the 1970s by the United Nations Children's Fund (UNICEF) intended to provide drinking water free of the bacterial contamination that was killing more than 250,000 children each year.

Unfortunately, the well water was never tested for arsenic contamination, which occurs naturally in the groundwater. By the 1990s, high rates of certain types of cancer throughout much of Bangladesh led researchers to investigate and identify the cause as arsenic-contaminated water. According to a 2010 study by the British medical journal *Lancet*, up to 77 million people in Bangladesh are exposed to toxic levels of arsenic, taking years off their lives. Medical statistics indicate that one in ten people who drink such water over a prolonged period will ultimately die of lung, bladder, or skin cancer. Tube wells that test positive for unsafe arsenic concentrations are now painted red as a warning to residents. Unfortunately, in many areas, these comprise most of the available water sources. New technologies have emerged in recent years to treat contaminated water, including the use of filters. Because some arsenic may be absorbed through eating crops irrigated with contaminated water, however, a large-scale and comprehensive solution to this problem remains elusive.

Ecology, Land, and Environmental Management

The ecologies and landscapes of South Asia are diverse and support a huge range of life. The ecosystems of the region include deep deserts, thick forests, isolated mountains, and a critical coastal zone. This great diversity of habitats has led to a wide-ranging diversity of species, many of which are at risk. The desert regions are home to several antelope and deer species and an impressive range of birds, including the Great Indian Bustard. Forests are home to monkeys, tigers, deer, and panthers, and also serve as the last remaining breeding sites of the Indian wolf. The mountains of the region, including the Hindu Kush of Pakistan and the Himalayas of Nepal, India, and Bhutan, sustain animals like the one-horned rhinoceros in the foothills and the rare clouded leopard in the high country abutting Tibet.

One of the most notable and endangered ecosystems in the world, the world's largest contiguous area of tidal mangrove forests stretches through the delta regions of the Ganga (Ganges)

▲ **FIGURE 9.14 Marble and Sandstone in Monumental Architecture** The stone quarried in India and Pakistan have been a crucial component in the construction of the great works of classical architecture, including the Jama Masjid and the Taj Mahal (shown here).

▲ **FIGURE 9.15 Cranes in India** The Sarus Crane, *Grus antigone*, is an all-year resident bird in northern Pakistan and India (especially Central India and the Ganga plains). Averaging an impressive 156 centimeters (5 feet) in length, this large endangered bird depends on conservation reserves and wetlands for habitat but also thrives in agricultural areas.

across the coastal edge of West Bengal in India and into Bangladesh. In the southernmost reaches of the delta, saltwater penetrates the channels. This creates a distinctive ecology of untouched tropical swamp forest—the **Sundarbans** ("beautiful forest")—that is home to crocodiles, Chital deer, and one of the largest single populations of Bengal tigers in the world.

These species and the landscapes, although distinctive to the region, are part of a larger distributive pattern of flora and fauna. This pattern reflects how the ecology of South Asia is deeply influenced by the region's connection to other parts of the world. Many of the key plant and animal species of the northwestern part of India and most of Pakistan are common to the region's arid and mountainous neighbors stretching west across Iran to Southwest Asia and North Africa, whereas the forests and mountains of the northeast are connected to those of Southeast Asia. A number of the remarkable winter bird species of the subcontinent migrate from their summer homes as far away as northern Siberia (**Figure 9.15**). The Siberian crane migrates 5,000 kilometers (3100 miles) from Russian to northern India every year, crossing Kazakhstan, Uzbekistan, Turkmenistan, Afghanistan, and Pakistan in the process.

Land Use Change Humans have inhabited this region for centuries and intensive development of the landscape has left its mark. For example, residents have cleared large areas of natural forest to cultivate land from earliest times. For the most part, however, subsistence farmers, herders, fisher folk, and artisans have historically drawn on local resources only for their food, traditional medicines, housing materials, and fuel. Their activities, although resulting in some **deforestation**, were managed in such a way that many critical habitats were preserved.

The arrival of European traders, the era of British rule, and the subsequent period of independence coincided with accelerated and persistent environmental transformation. In 1750, when the British were beginning their imperial conquests, more than 60% of South

Asia was still forested. British imperial rule brought systematic clearing of land for plantations and the methodical exploitation of valuable tropical hardwoods for export to Europe and North America. By 1900, only 40% of South Asia remained forested. The most rapid period of change has been the past 50 years, as the independent countries of South Asia have sought to modernize and expand their domestic economies. Only 21% of India remains forested today, and less than half of that is intact, natural forest—the rest consists of forest plantations. About one-third of the forest plantations in India consist of eucalyptus, a fast-growing, non-Indigenous species that is very demanding of soil moisture.

In a region so densely occupied by human beings, on the other hand, expecting primeval, dense, and pristine forest to persist is unrealistic. According to recent satellite surveys of the region, some large areas are still covered in mixed forest, with tree cover thriving in agricultural areas, wastelands, and even in and around cities. The proportion of land area that has at least a density of 10% forest cover on it is 44% in India, 71% in Nepal, 73% in Bhutan, and 91% in Sri Lanka. This compares favorably to the roughly 50% of the area worldwide that has at least some tree cover. While these fragments and patches may be no substitute for the original forest cover of 2,000 years ago, they do sustain some native biodiversity. A 2008 study by Stanford University scientists showed, for example, that intensively cultivated palm plantations in southern India harbored 114 native bird species, including rare Hornbill species. South Asia is ultimately a region with high levels of **anthropogenic** forest cover (tree cover created or retained by human beings), which can serve as wildlife habitat. Therefore, while stresses on the landscapes continue, there is also reason for optimism.

Agriculture and Resource Stress The rural population of South Asia varies greatly by country. Nepal (18% urban), Afghanistan (23% urban), and Bangladesh (23% urban), are less urban than India (29% urban), Bhutan (33% urban), and Pakistan (35% urban). Nevertheless, the region as a whole is still considered heavily rural, and demand for land-based resources is high. In rural areas, a large number of people live directly off the land, at subsistence levels, but in ways that have protected and preserved the environment. Increasing demands for rural resources in the regional and global economy, however, have begun to place serious stress on the land. Commercial forestry, mining, road and dam construction, and the spread of industrial agriculture have become far more common in recent decades. These land uses are themselves hard on the land and they also displace local farmers, forcing these poorer subsistence producers onto unproductive soils and arid hillsides. As environmental resources diminish, a destructive cycle is set in motion. Rural people, even those disconnected from the booming global economy, are forced to use their limited resources in increasingly unsustainable ways, depleting sources of fuel wood, exhausting soils, and draining water resources.

In parts of Punjab and Haryana—the "breadbasket" of India, where almost a third of the country's wheat is grown—73% of crops are dependent on groundwater. Unfortunately, the water table has fallen by an average of 55 centimeters (more than 2.5 feet) per year in the last decade. Because 86% of water resources are used for agriculture in India, cities must compete with farming for very scarce water resources. Chennai and New Delhi are just two examples of the many large cities in India that are heavily dependent on supplementary water supplies hauled in by tanker (**Figure 9.16**, p. 356).

▲ **FIGURE 9.16 Water Tanker Truck in India** Urban dwellers are shown collecting water from a tanker in New Delhi. An unprecedented water crisis looms large over this national capital. Though a municipal tanker is shown here, shortfalls are often met by water salesmen, who draw off the public water supplies to sell to water-starved local neighborhoods.

Environmental Pollution Water is often polluted in addition to being scarce. In India, about 200 million people do not have access to safe and clean water; about 690 million lack adequate sanitation, and an estimated 80% of the country's water sources are polluted with untreated industrial and domestic wastes. The Asian Development Bank has estimated that fewer than 1 in 10 of the industrial plants in South Asia comply with pollution-control guidelines. Only 10% of all sewage in South Asia is treated. In India, 1,769 tons of organic water pollutants are discharged *daily* into local waters. As of 2011, Indian cities alone were estimated to generate about 68.8 million tons of solid waste each year, most of which is disposed of in unsafe ways: burned, dumped into lakes or seas, or deposited into leaky landfills.

Air pollution has also become a serious issue. According to the Tata Research Institute in New Delhi, air pollution in India causes an estimated 2.5 million premature deaths each year. Motor vehicle emissions are a major contributor to urban air pollution. Approximately 2.6 million cars were sold in India in 2009, more than three times as many as were sold in 2003. The increasing congestion of city streets and highways has resulted in a corresponding rise in respiratory diseases. Studies show these effects are unequally distributed. A study comparing urban men showed that the poorest men, who work disproportionately outdoors, have far worse respiratory problems than wealthier ones, who tend to work in offices.

In the last decade, municipal rules, national statutes, and clever innovations have made headway into the air pollution crisis. The Supreme Court of India handed down a set of decisions in the early 2000s to address the issue and mandated that all public vehicles in New Delhi—including buses, taxis, and rickshaws—convert to compressed natural gas fuel. The court also has limited the passage of diesel trucks through the city during daytime and closed a number of polluting industries. These actions have had a remarkable effect. New Delhi's air was visibly cleaner by 2012 and residents experienced an overall improvement in respiratory health.

Disappearing Megafauna Global extinction of species is another ongoing crisis in South Asia. According to the chief global scientific conservation organization, the World Conservation Union, as of September 2011, there are 47 critically endangered species in India. These include many magnificent and charismatic large animals, such as rhinoceroses, elephants, and tigers. Cheetahs disappeared from the wild in India more than 50 years ago; the last sighting was in 1948, when three young males were killed by a hunting party in the jungles of Bastar in Madhya Pradesh, central India. The Bengal tiger (**Figure 9.17**) and the Asian elephant are emblematic of critically endangered species in South Asia. Tigers are threatened because of loss of habitat and remorseless poaching (see "Geographies of Indulgence, Desire, and Addiction: Animals, Animal Parts, and Exotic Pets" in Chapter 8). Pakistan is also home to a number of iconic species, including the snow leopard of the Himalayas, which are declining for similar reasons. In Bhutan, the golden leaf monkey and the red panda are endangered.

Equally dramatic, though regrettably greeted with less global concern, is the decline of a large number of vulture species. These enormous and adaptive birds of prey are crucial to the functioning of natural ecosystems, as they scavenge and recycle carcasses and waste. Nesting in high rocks and thriving on the enormous garbage dumps of India and Pakistan, vultures were found in huge numbers only a few years ago, but are now in fast decline. The Indian vulture (*Gyps indicus*) experienced a cataclysmic 97% population decrease between 2000–2007 (**Figure 9.18**). There are many reasons for their decline, including loss of habitat and pressure from human settlement. One surprising culprit is the veterinary drug *diclofenac*. This drug is widely administered to livestock to reduce pain and allow them to work harder and longer. When ingested by vultures, which scavenge the carcasses of

▲ **FIGURE 9.17 Bengal Tiger** The Bengal tiger is one of the most seriously endangered species: its habitat is disappearing and it is still hunted by poachers.

▲ **FIGURE 9.18 Indian Vulture** While less charismatic than other endangered species, the vulture is among the most ecologically important; the decline of numerous vulture species in Pakistan and India in the last decade is viewed as a conservation crisis.

dead cows and bulls, the drug causes kidney failure and death. A recent ban on the drug may contribute to recovery, though populations are so seriously reduced that only captive breeding may restore them.

APPLY YOUR KNOWLEDGE Visit the World Wildlife Fund (WWF) Web page, which details the organization's work in the Himalayas at http://www.worldwildlife.org/what/wherewework/easternhimalayas/projects.html. What conservation efforts and activities does the organization emphasize? To what degree do their programs address the full range of threats to wildlife in South Asia?

HISTORY, ECONOMY, AND TERRITORY

South Asia has long been at the center of global cultural exchange and commerce. Sophisticated cultures and powerful political empires have emerged and spread across the region. Its resources and geographic situation on sea-lanes between Europe and the East Indies made it especially attractive to European imperial powers, and in modern times it has become an emergent industrial and high-tech sector. Contemporary territorial and security issues in the region are the outcome of global engagements. Pakistan and India have internationalized their ongoing enmity by becoming nuclear states. In recent decades, the United States has also become more deeply invested in securing stability in the region in support of its war in Afghanistan.

Historical Legacies and Landscapes

South Asia appears to sit, in terms of physical geography, in an isolated position, yet nothing could be further from the truth. The seas to the west and east of the jutting peninsula connect the subcontinent by trade and invasion to Africa in the west and the great kingdoms of Southeast Asia to the east. Traders and invaders have constantly traversed the rugged mountains. These travelers connected the region to the Persian and Arab worlds to the northwest and Mongolian and Chinese empires to the north.

Harappan and Aryan Legacies The first extensive imprint of human occupation in the region dates from at least 5,300 years ago, when the people of the **Harappan Civilization** began to irrigate and cultivate large areas of the Indus Valley. Between 3,000 and 2,000 B.C.E. this Indus Valley culture utilized sophisticated agricultural techniques to produce enough surplus, primarily in cotton and grains, to sustain an urban civilization. The Harappans built enormous, organized cities at the sites of Mohenjo Daro and Harappa (located in present-day Pakistan) as well as great trading centers at Lothal (in contemporary India). At their cultural and economic height, they became the world's largest ancient civilization. Harappan cities had organized water and sewage systems, followed grid planning, and were built around docks and markets. Distinctively manufactured Harappan trade goods, including jewelry and fine crafts, are found in archaeological sites across Mesopotamia and as far as the Mediterranean (**Figure 9.19**). The success of the Harappans illustrates the degree of global integration that South Asia achieved from the very start.

▲ **FIGURE 9.19 The Indus Valley Civilization** This King Priest figure, carved from steatite (soapstone), demonstrates the enormous artistic talents of the Harappan Civilization, also known as the Indus Valley Civilization. Urban, organized, and with trade linkages across West Asia, the Harappans were the first to put South Asia at the center of global trade, more than 5,000 years ago.

The disappearance of the Harappans, likely following a shift in both climate and the course of the Indus River, was followed immediately by the rise of the **Aryan** civilization (1500 to 500 B.C.E.). This group of livestock-keeping peoples from the northwest populated the plains of the Ganga (Ganges) and developed a series of kingdoms. Their advent was accompanied by farming and new trade links, and they slowly transformed the plains of the northern part of the subcontinent from a green wilderness to a heavily agricultural zone with large dense settlements. The story of the Aryans (named after the language that they spoke) highlights the long history of invasions into the subcontinent by outside cultures and the debt South Asian culture owes to its many neighbors.

Early Empires The **Mauryan Empire** (320–125 B.C.E.) was the first to establish rule across greater South Asia. By 250 B.C.E., the emperor Ashoka had established control over all but present-day Sri Lanka and the southern tip of India. Securing control wreaked such havoc and destruction, that Ashoka renounced armed conquest and adopted a policy of "conquest by *dharma*," that is, through the example of spiritual rectitude and chivalrous obligations. **Dharma**, a key element of Buddhist teachings, is part of the legacy of Ashoka's reign, which also includes Buddhist principles of vegetarianism, kindness to animals, nonacquisitiveness, humility, and nonviolence. During this period, the culture of South Asia became closely interlinked with that of other regions thanks to the Buddhist missionaries that Ashoka sent throughout Asia. Ironically, this successful evangelization led to the survival of Buddhism throughout the rest of Asia after its disappearance in peninsular South Asia.

After Ashoka's death in 232 B.C.E., the Mauryan Empire fell into decline, and northern India soon succumbed to invaders from Central Asia. After more than four centuries of division and political confusion, the **Gupta Empire** (320–480 C.E.) united northern India. The Gupta period is generally regarded as a great classical period. It produced the decimal system of notation, the golden age of Sanskrit and Hindu art, and contributions to science, medicine, and trade. Other invasions and empires would follow, but these traditions would persist.

Mughal India Toward the end of the 15th century, a clan of Turks from Persia (now Iran), known as the **Mughals**, moved east into the region. Led by Babur, they conquered Kabul, in what is now Afghanistan, in 1504. By 1605, Babur's grandson, Akbar, had established control over most of the Plains, and in the next century Mughal rule extended to all but Sri Lanka and the southern tip of India from 1526–1707 (**Figure 9.20**).

Akbar's rule was an extraordinary time. He synthesized the best of the many traditions that fell within his domain. Traditional kingdoms and princely states were kept intact, but they were integrated in a highly organized administrative structure with an equitable taxation system and a new class of bureaucrats. Persian became the official language, yet Akbar abolished a tax on non-Muslims. Mughal rule did not seek to impose Islam on Indigenous populations but Mughal commitment to the religious precepts of Islam gave great stature to Islam. Over time, Islam proved attractive to many, especially in the northwest (the Punjab) and the northeast (Bengal). By 1700, mosques, daily calls to prayer, Muslim festivals, and Islamic law had become an integral part of the social fabric of South Asia. Spectacular architecture became a signature of Mughal rule; The Mughals built lavish mosques, palaces, forts, citadels, towers, and

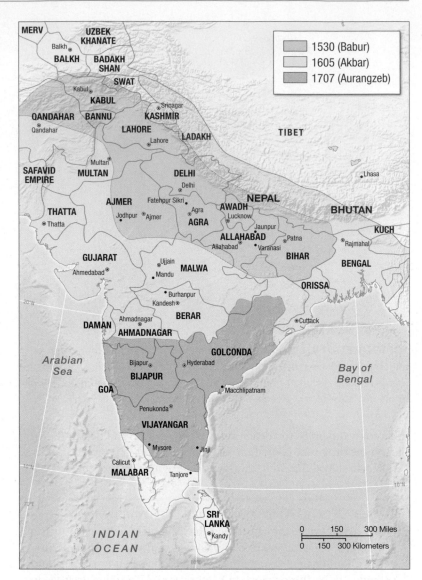

▲ **FIGURE 9.20 Mughal India** Two centuries of Mughal rule began in 1504, with the conquest of Kabul, in present-day Afghanistan. By 1707, Mughal rule extended almost to all the subcontinent.

gardens, including the iconic Taj Mahal. Internal collapse, coupled with conflict with neighboring Hindu kingdoms, left South Asia open to the increasing interest and influence of European traders and colonists.

British Colonialism and the Raj European traders were a regular presence along the coasts of South Asia long before the dissolution of the Mughal Empire. Portuguese Vasco da Gama arrived in India in 1498. Early in the 17th century the British East India Company (established in 1600) and the Dutch East India Company (established in 1602) followed the Portuguese. South Asia, situated as it was on the route between Europe and the East Indies, provided attractive intermediate stops for trade in textiles.

By the 1690s, European trading companies had established a permanent presence in several ports, though they had little interest in establishing settler colonies or exerting political authority at first. The British East India Company was the most successful of these, and in 1773 the British government transformed the company into an administrative agency. Soon afterward, the British pushed ahead with aggressive imperialist policies in South Asia, using a mixture of force, bribery, and political intrigue to gain exclusive control over the region.

Toward the end of the 1700s, the focus of the British shifted beyond trade to economic imperialism. The central goal of British policy was to harness the agricultural wealth of the region and to export raw materials, like tea, opium, and cotton, while dismantling native industry, creating a condition of **dependency** of the region on the import of finished goods like textiles (made with Indian cotton!) from industrializing England. One by one, the territories of native rulers were annexed or brought under British protection to affect this outcome (**Figure 9.21**). During this period, the region was ruled not by the British Crown, notably, but by the East India Company itself, which kept its own standing army. Western institutions of property and law were extended to secure and maintain rights and control,

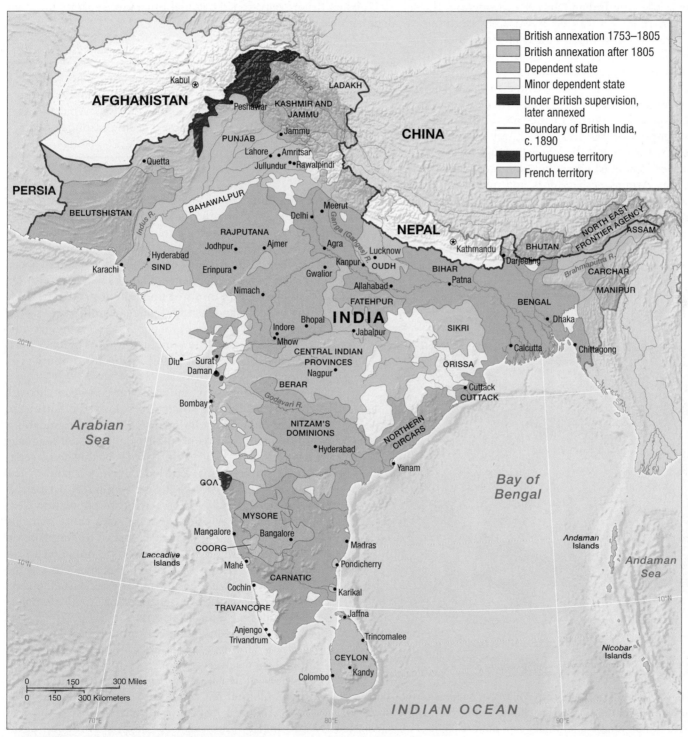

▲ **FIGURE 9.21 The British Conquest of India** British interest in India began as a consequence of merchant trade in the early 18th century. By 1890, the British had come to control, directly or indirectly, most of South Asia.

while modern industrial infrastructure—railroads, roads, bridges, and irrigation systems—led to increased efficiencies in production and transportation. In an effort to create and maintain a trained cadre of local managers and bureaucrats, Western educational curricula flourished and British-style public universities were established.

The inevitable anticolonial reaction came to a head in 1857, when an Indian Army unit rebelled purportedly because its soldiers were jailed for upholding their Hindu religious principles by refusing to use ammunition greased with animal fat. The incident quickly spread into a yearlong civil uprising—the Indian Mutiny—throughout the north–central region. The rebellion was put down with enormous and brutal force (mutineers were strapped to discharging cannons) and, in 1858, the British Crown assumed direct control over India. In 1876, Queen Victoria was declared empress of India.

The **Raj**, British rule over South Asia, extended to the border of present-day Afghanistan by 1890. Well into the 20th century, the British worked to extend political control further northward, dueling against the Russian Empire for political and military control beyond the Khyber Pass. The British and the Russians fought real and proxy wars over this border region, which includes both Afghanistan and parts of Western China. Armies, ambassadors, and spies were engaged in a conflict that became known as the **Great Game**. That conflict was never fully resolved, and the British were never able to maintain control of Afghanistan during their long reign in South Asia (**Figure 9.22**).

PUNCH, OR THE LONDON CHARIVARI.—November 30, 1878.

"SAVE ME FROM MY FRIENDS!"

"IF AT THIS MOMENT IT HAS BEEN DECIDED TO INVADE THE AMEER'S TERRITORY, WE ARE ACTING IN PURSUANCE OF A POLICY WHICH IN ITS INTENTION HAS BEEN UNIFORMLY *FRIENDLY* TO AFGHANISTAN."—*Times*, Nov. 21.

▲ **FIGURE 9.22 The Great Game in Afghanistan** This 1878 political cartoon shows Afghanistan asking to be saved from the colonial ambitions of its "friends," Russia (Bear) and Britain (Lion). Rivalry between Russia and Britain over political control of Afghanistan in the 1900s resulted in instability and warfare in the region for decades. This conflict has contemporary implications as well.

Where the British did rule, they brought dramatic change, including the development of plantation agriculture. They grew coconuts, coffee, cotton, jute, rubber, and tea (see "Geographies of Indulgence, Desire, and Addiction: Tea"). The Raj also introduced Western industrial development and technology to South Asia, displacing Indigenous crafts and industries. It fostered Western political ideas of national territorial sovereignty and the materialism that accompanies free markets in land, labor, and commerce. This was an explosive legacy that manifested itself at the conclusion of the Raj in 1947, when Britain partitioned colonial India into separate independent national states.

APPLY YOUR KNOWLEDGE Compare and contrast the global integration of the South Asian economy before colonialism to global integration under the Raj. Which trade connections and governance systems became stronger under British rule? Which became weaker?

Partition Long-standing grassroots resistance to British imperial rule gained momentum through the **Indian National Congress Party**, formed in 1887 to promote greater democracy and freedom. A leader and the inspirational figure of this movement was **Mohandas Gandhi**, who advanced a vision of social justice and independence. Gandhi and his followers emphasized methods of nonviolent protest, including boycotts and fasting. Under Gandhi's leadership, the case for national independence became irrefutable. Soon after the conclusion of World War II, the British set about withdrawing from South Asia altogether.

In creating new, independent countries through **partition**, Britain sought to follow the European model of building national states on the foundations of ethnicity, with a particular emphasis on language and religion. As a result, Britain established a separate Islamic country called Pakistan ("land of the pure"). Administrative districts under direct British control that had a majority Muslim population were assigned to Pakistan, together with those princely states whose ruling maharajas wished to join Pakistan rather than India, which was to be home to a predominantly Hindu population. Pakistan was created in two parts, East Pakistan and West Pakistan, one on each shoulder of India, separated by 1,600 kilometers (994 miles) of Indian territory.

In 1947, when Pakistan and India were officially granted independence, millions of Hindus found themselves as minorities in Pakistan, whereas millions of Muslims felt threatened as minorities in India. Communal violence erupted. In desperation, more than 12 million people fled across the new national boundaries—the largest refugee migration ever recorded in the world. Hindus and Sikhs moved toward India and Muslims moved toward

GEOGRAPHIES OF INDULGENCE, DESIRE, AND ADDICTION

Tea

Tea, a mild drug that makes a refreshing drink, was cultivated in China and Japan for centuries before becoming a commodity in international trade. Thereafter, tea became a catalyst of economic, social, and political change, and the social and economic fabric of large parts of South Asia was destroyed by colonial mercantile forces that transformed land into tea plantations.

Tea was first brought to Europe by Portuguese traders and marketed as a medicine. Tea drinking was adopted by the royal court of King Charles II in England (1660–1685) and quickly became an indulgence of the European bourgeoisie. By 1700, more than 8 million kilograms (17 million pounds) of tea were being exported to Europe each year, mostly to Holland, Portugal, and England. As East Asian trade came to be dominated over the next century by the British East India Company, the cost of tea fell and the taste for it spread. It soon became an addiction for the middle and working classes in Britain and a desirable indulgence for many in the American colonies. A significant market for tea also developed in Russia and in many countries of the Middle East and North Africa.

South Asia became a source of tea only in the 19th century, when the British sought to find commercial advantage in its newly acquired territories in Assam (northeast India). In 1848, the Assam Company hired botanist Robert Fortune to travel incognito to China to discover the secrets of cultivating and processing tea. After four expeditions, he brought to Assam not only information but also 12,000 seedlings, the specialized tools used in processing tea, and a skilled Chinese workforce. The British colonial government granted land on inexpensive leases to anyone who had the capital to establish a tea plantation. Inevitably, those with the capital were European settlers, and thousands of would-be planters flocked to Assam. Valuable timber was cut, forests were cleared, and the land not needed for tea plantations was rented to tribal people from Bihar and Orissa, who were contracted to work—for miserably little pay and in dreadful conditions—as laborers in the plantations.

The opening of the Suez Canal in 1869 cut transport costs, made tea even cheaper, and allowed producers a bigger profit margin. Tea plantations in Assam and neighboring Bengal increased sixfold in size between 1870 and 1900, by which time there were a half-million plantation workers in the region. In this same period, Ceylon—now Sri Lanka—emerged as a major tea-producing area. In the 1870s, 250,000 acres of coffee plantations in Ceylon were destroyed by blight, leaving coffee planters bankrupted and land extremely cheap. Thomas Lipton, a prosperous grocer from Glasgow, Scotland, arrived in Ceylon in 1871 while on a world cruise, bought dozens of the plantations at bargain prices, and turned them over to the production of tea.

Lipton was a pioneer of "vertical economic integration." His company controlled all the operations from growing tea to processing, management, transport, blending, packaging, and marketing. Through this system, Lipton was able to cut the cost of tea to European consumers by 35%. Lipton and other tea planters brought in low-caste Tamil Hindus from famine-stricken areas of south India to work as laborers in the Ceylonese plantations. The conditions in which they were forced to work were atrocious, as bad as those of their counterparts in Assam and Bengal. Being low caste, low paid, and isolated on estates in the highlands, and with no political voice, the "estate Tamils" were effectively enslaved on the plantations.

After independence in India, Bangladesh, and Sri Lanka, most European planters were forced to sell their tea plantations, either to the respective governments, under nationalization programs, or to Indigenous business interests. The conditions for plantation workers have remained relatively unchanged, however. Plantation workers are still an impoverished group, a fact that is often belied by the image of smiling tea-pickers dressed in traditional costume, adorned with jewelry, working in sunshine amid beautiful scenery (**Figure 9.1.1**). ■

▲ **FIGURE 9.1.1 Tea Pickers** At the beginning of the economic chain that links South Asian producer regions with consumers in Europe and North America, tea pickers work long hours in harsh conditions for meager pay.

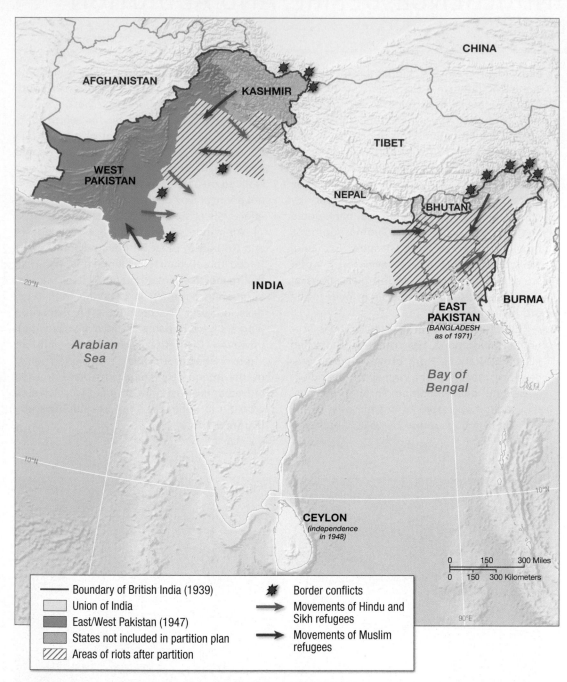

▲ **FIGURE 9.23 Partition of India and Pakistan, 1947** The decision to create two states from British India had both immediate and long-term consequences; the slaughter that followed independence resulted in the deaths of as many as 1 million people, and peace has never prevailed over the region.

West Pakistan quickly developed into regionalism, and East Pakistani leaders called for secession. As a result, the country was split into two independent states in 1971: West Pakistan became Pakistan, and East Pakistan became Bangladesh. Meanwhile, neither India nor Pakistan could agree on the status of Kashmir, and the two countries briefly went to war over the region in 1948, in 1965, and again in 1971.

Economy, Accumulation, and the Production of Inequality

In the decades since independence, South Asia—and India, in particular—has come to play a large role within the economic world system. India is the world's most populous democracy and has maintained stable parliamentary and local government through elections and rule of law since adopting its constitution in 1950. Between 1950 and the 1990s, many key institutions in India—including banks, utilities, airlines, railways, radio, and television—were government owned and operated. High tariffs, restrictions on foreign ownership, and a complex system of permits, licenses, quotas, and permissions all kept foreign investors away and suppressed the energy of Indian entrepreneurs. Since that time, however, India has charted a more open and entrepreneurial economy. Key institutions have been privatized, and foreign investment has flowed into the country, helping generate exceptionally high economic growth and marked economic inequality.

The New South Asian Economy Over the past 20 years, India has been the second-fastest-growing country in the world—after China—averaging well above 6% growth per year. This booming economy, moreover, has shown terrific resilience in the face of recent global financial turmoil. While most economies around the world shrank during 2008 and 2009, India's (almost alone) actually posted job growth across industrial and technical sectors. By 2012, India had become the 19th largest exporter and tenth largest importer in the world, a fully international economy linked closely to the growth of China and the large economies of Southeast Asia.

Pakistan (**Figure 9.23**). As many as 1 million people were killed in the resulting confusion and violence. In Kashmir, a Hindu maharaja had elected to join India, but Pakistani forces intervened to protect the majority Muslim population, splitting the territory in two, with implications that have never been resolved.

The British granted independence to the island of Ceylon as a Commonwealth dominion in 1948. In 1949, Britain handed to India formal control over the external affairs of the kingdom of Bhutan. In 1968, Britain granted independence to the Maldives. The Raj was finally over, but the legacy of partition remains. In some ways, the states of South Asia are still adjusting to the 1947 partition of India and Pakistan. In Pakistan, divergent regional interests in East and

India now has an affluent middle class estimated at about 200 million. This represents a huge, well-educated, and sophisticated consumer market, which has become an agent of globalization. Market reforms have triggered an associated cultural change. Flaunting success is no longer frowned upon, and India's expanding middle class is increasingly unabashed about its cars, mobile phones, and vacations in Dubai and Singapore. Broader elements of this growth

▲ **FIGURE 9.24 Consumer Culture in South Asia** Brigade Road in Bangalore, India, is one of the country's largest commercial and shopping centers. Elite restaurants, stores, and clubs line the road, representing the booming consumer culture and economy of an emerging Indian middle class.

can be seen across South Asia, especially in the rise of the tech sector, the expansion of global entrepreneurship, and the boom in tourism.

One of the most dramatic examples of growth is that of the software industry in south India. Bangalore, in particular, has become a thriving industrial and business center as a result of a tech boom in India. Long one of the premier science and technology centers of India, industries in Bangalore manufacture aircraft, telecommunications equipment, watches, radios, and televisions. But the city has become world famous for its software industry. As free-enterprise capitalism and direct foreign international investment have risen in India, a burgeoning pool of Indian programmers has emerged, who are experts at writing concise, elegant code. Altogether, Bangalore boasts more than 500 high-tech companies that employ 100,000 people, and the city is home to the Indian Institute of Science, a world-renowned technical school. These relatively affluent employees of the tech industry have contributed to the city's progressive and liberal atmosphere and its lively commercial centers, featuring fast-food restaurants, theme bars, and glitzy shopping malls (**Figure 9.24**). They have also brought new opportunities in the service and construction industries in their wake, along with unprecedented growth. Though the state of Karnataka, where Bangalore is located, actually has declining natural growth rates, the Bangalore urban district grew by 47% between 2001 and 2011, to around 9.59 million people.

Not all jobs made possible by high technology are those in software design. Another major growth industry is **call centers**, where workers respond to sales

and customer service phone calls from throughout the world, keeping long hours for only modest pay. Educated Indian workers are especially attractive to foreign companies for this work, since their typically excellent English skills suit the needs of phone conversation with customers in North America and the United Kingdom, two major markets (**Figure 9.25**).

The region's economic growth has not come exclusively through external foreign investment. Many homegrown South Asian industrial empires have emerged to compete and invest in industrial development abroad. Some of the most prominent household names in India are the heads of industrial families whose firms emerged before independence and are now known internationally. The Tata Group, for example, was originally a hotel company founded in the 1860s by entrepreneur Jamsetji Tata. It is now a multinational conglomerate company headquartered in Mumbai, whose operations extend to materials (Tata Steel); automobiles and trucks (Tata Motors); information (Tata Consultancy Services and Tata Technologies); as well as chemicals, energy, and communications. Another mogul is Lakshmi Mittal, the wealthiest man in India. His company, ArcelorMittal, is the world's largest steelmaker.

The reach of these South Asian conglomerates is international. Tata has acquired, for example, English tea companies, South Korean auto plants, and Anglo-Dutch steel firms. ArcelorMittal is listed on the New York Stock Exchange. It owned and finally announced the closing of the quintessentially American steel plant, Bethlehem Steel, in Lackawanna, New York, in 2008. Like their Chinese corporate rivals, these South Asian firms are also actively seeking energy and natural resources in Africa. In this sense, globalization has not only reached inward to South Asia, but has also been a vehicle for the expansion of the region's influence abroad.

▲ **FIGURE 9.25 Call Center in India** At 1 A.M. local time in Bangalore—peak workday hours in the United States—the noise inside a 24/7 customer call center crescendos to a climax as nearly 1,300 phone conversations are being held at the same time. Though the call-center industry is a successful signature component of the regional economic boom of the past 20 years, young workers in this industry work long hours for poor pay and benefits with little chance of advancement.

▲ **FIGURE 9.26 Trekking in Nepal** Tourism has become a critical component of the South Asian economy; in Nepal, foreigners pay considerable sums to hike the stunning terrain of the Himalayas.

The Tourism Boom Tourism is another growing sector in the South Asian economy. Tourism in India is the country's largest service industry, employing approximately 9% of the total workforce. But it is in the nearby country of Nepal where tourism truly rules the economy. Nepal's boundaries enclose 8 of the 10 highest mountains in the world, including Mount Everest, making the country a destination for mountain climbers, trekkers, and nature enthusiasts from around the world. In 2009, Nepal earned U.S. $852 billion from tourism which employs roughly 20% of the economically active population (**Figure 9.26**).

There are downsides to being tourism-dependent. In Nepal, the growth of visitors has also meant the growth of trash and the deforestation of hillsides to support campfire fuel. When a tourism-dependent country experiences apparent instability or even a minor disease outbreak, tourists are quick to stay away, making the sector vulnerable to dramatic and rapid downturns. Even so, much of the recent growth in South Asian tourism has been *internal*, with emerging wealthy classes in India and Pakistan beginning to tour their home region in large numbers. Though apt to spend less than foreign tourists, their numbers are far higher than that of foreigners (Indian tourists outnumbered international tourists by more than 41 times in 2010) and they are less easily frightened by news reports.

Inequality and Poverty Not all economic change in the region has been positive. The rapid growth of India's affluent middle class serves to highlight the desperate situation of a larger group: the extremely poor. According to a state-run Indian study in 2007, 77% of Indians, roughly 836 million people, live on less than 50¢ US per day. These workers exist in the margins of the new boom economy, in the informal labor sector in abject poverty. Twenty-six percent of the population lives on less than 12 rupees (25¢ US) per day, entirely

bypassed by the booming global economy of the country (**Figure 9.27**).

The liberalization of India's economy has also had less-predictable consequences. Lifting export controls has enabled farmers with access to large amounts of capital to reorganize their production toward lucrative overseas markets, with the result that domestic consumers have to pay more for traditional staples. For example, many farmers are switching from growing grains for local consumption to growing cash crops like cotton and tobacco. Others are turning to the cultivation of flowers and strawberries, which are airfreighted abroad. And now that a global market has become aware of high-quality local specialties, such as the fragrant basmati rice of the Himalayan foothills and the short-season Alphonso mangoes of Maharashtra, their prices within India has put them in the luxury class, out of reach of many of the consumers who have traditionally regarded them as staples.

In rural South Asia, illiteracy remains common, and even the most basic services and amenities are lacking. Life expectancy is low; hunger and malnutrition are constant facts of life. In urban areas, poverty is

▲ **FIGURE 9.27 Poverty in India** A homeless man in Kochi in the state of Kerala seeks goods for resale in the garbage. Poverty is crushing in the region, but drives people to creative and informal economic activities.

compounded by crowding and unsanitary conditions. In South Asia's largest cities, one-third or more of the population lives in slums and squatter settlements, and hundreds of thousands are homeless. Clean drinking water is limited, and most poor households do not have access to a latrine of any kind. Overcrowding, a lack of adequate sanitation, high levels of ill health and infant mortality, and rampant social pathologies characterize the worst concentrations of poverty. Much of this poverty results from the lack of employment opportunities in cities that are inundated with people. To survive, people who cannot find regularly paid work resort to various ways of gleaning a living. Some of these ways are imaginative, some desperate. Examples include street vending, shoe shining, craftwork, street-corner repairs, and scavenging on garbage dumps.

Life is especially difficult for those living as squatters, or slum-dwellers, in the growing cities of Pakistan and India. Typically they have no sewage, no formal road connection, and no running water. The result is a large range of diseases that persist in squatter communities, including malaria, tetanus, diarrhea, dysentery, and cholera, which remain the principal causes of death among children under age five.

Squatters and slum-dwellers, who have no formal rights to the land they may have occupied for years or decades, face a further challenge of poverty in that they have no legal standing and little political voice. Consider Dharavi, one of Asia's largest slums, located in Mumbai, a city where slum-dwellers constitute perhaps half of the 12 million citizens (**Figure 9.28**). Despite the hardships these residents face in living in this slum, its location between two suburban rail lines is convenient for local workers and rents are low enough (about 185 rupees or U.S. $4 per month) for poor families to afford.

Residents have rigged electricity and often have televisions. They have collectively created a range of small, informal industries, such as sewing, leatherwork, and pottery, whose products are sold throughout the city. The annual value of businesses here is hard to estimate, but it may be as high as U.S. $650 million annually. And yet the land these Dharavi squatters occupy, many for decades, lies in the near center of the city, on prime property for development. As a result, the citizens face constant threats of eviction because politicians and developers are keen to acquire the land. While residents have organized to oppose evictions, they possess little in the way of formal rights. Poverty in South Asia is an economic condition, but also a political one.

Among South Asia's poor, women bear the greatest burden and suffer most. South Asian societies are intensely patriarchal, though the form that patriarchy takes varies by region and class. The common denominator among the poor throughout South Asia is that women not only have the constant responsibilities of motherhood and domestic chores but also have to work long hours in informal-sector occupations. In many poor communities, 90% of all production is in the informal sector, and more than half of it is the result of women's efforts. In addition, women's property rights are curtailed, their public behavior is restricted, and their opportunities for education and participation in the waged labor force is severely limited.

Women's subservience to men is deeply ingrained within South Asian cultures, and it is manifest most clearly in the cultural practices attached to family life, such as the custom of providing a dowry where the family of the bride provides hefty payments to the family of the groom. The preference for male children is reflected in the widespread (but illegal) practice of selective abortion. Within marriages, many (but by no means all) poor women are routinely neglected and maltreated.

The picture is not entirely negative, however, and one of the most significant developments has been the emergence of women's self-help movements. The best known of these is the **Grameen Bank**, a grassroots organization formed to provide small loans—**microfinance**—to the rural poor in Bangladesh. The Grameen Bank, which claims to be a financially sustainable, profit-making venture, runs completely against the established principles of banking; it lends money to poor borrowers who have no credit. As of October 2007, 7.34 million people have borrowed from the Grameen Bank. Cumulatively, the bank has loaned more than U.S. $11.35 billion. U.S. $10.11 billion (97%) has been repaid. This rate is far higher than any conventional financial institution operating in the country. The average size of a Grameen loan is about U.S. $120, typically enough to purchase a cow, a sewing machine, or a silkworm shed. Studies have shown that the bank's operations have resulted in improvements in nutritional status, sanitation, and access to food, health care, clean drinking water, and housing, and that more than one-third

▲ **FIGURE 9.28 Dharavi in Mumbai** One of the largest slums in Asia, Dharavi teams with economic activity and energy as a population of more 1 million people live in dense, improvised housing.

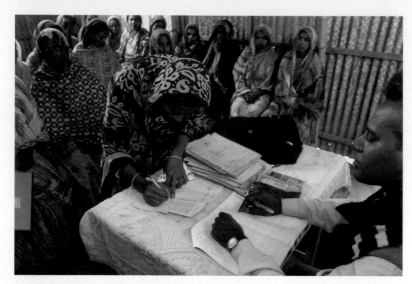

▲ **FIGURE 9.29 Rural Enterprise** Microcredit programs such as those pioneered by the Grameen Bank have enabled tens of thousands of rural women to begin small businesses. This photo shows women signing up for loans in Gazipur, Bariali, Bangladesh.

▲ **FIGURE 9.30 Opium Production in Afghanistan** Two men work in a field of opium poppies, which grow in abundance in Afghanistan. Despite efforts by U.S. and Afghani authorities to stamp out opium production, poppies remain among the most reliable source of revenue for poor farmers in the region.

of all borrowers have risen above the poverty line, with another third close to doing so. The most distinctive feature of the Grameen Bank is that 97% of its borrowers are women (**Figure 9.29**).

Children in impoverished settings are even more vulnerable than women. Throughout South Asia, the informal labor force includes children. In environments of extreme poverty, every family member must contribute something, and children are expected to do their share. Industries in the formal sector often take advantage of this situation. Many firms farm out their production under subcontracting schemes that are based not in factories but in home settings that use child workers. In these settings, labor standards are nearly impossible to enforce.

The International Labour Office has documented the extensive use of child labor in South Asia, showing that many children under ten years of age are involved in manual labor: weaving carpet, stitching soccer balls, making bricks, handling chemical dyes, mixing the chemicals for matches and fireworks, sewing, and sorting refuse. Most of them work at least 6 and as many as 12 hours a day. A particularly cruel type of exploitation of child labor is **bonded labor**. This kind of bondage occurs when persons needing a loan but having no security to back up the loan pledge their labor, or that of their children, as security for the loan.

Opium Production Although far from the poverty of urban areas, rural dwellers also face stark economic choices. Farmers are confronted with the fact that agricultural products—including cotton, coffee, rubber, wheat, cut flowers, and rice—are now produced and distributed worldwide, which makes global competition fierce and keeps commodity prices perilously low. At the same time, production costs have risen: fertilizer and pesticide costs are high; scarce water must be paid for or pumped; and new hybrid seeds represent an additional expense. For rural producers, finding a crop that can maintain its value amid this agroeconomic chaos is critical to survival.

In war-torn Afghanistan, the crop that most clearly fills this requirement is the opium poppy (see "Geographies of Indulgence, Desire, and Addiction: Opium and Methamphetamine" in Chapter 10).

Grown in South Asia for millennia, the poppy is the source of a resin rich in morphine, which is drawn from the pods of the flowering plants in the field by making a small incision, often using a traditional tool called a *nishtar*. Harvested in large quantities, the resin is refined into morphine and typically sold for export. Further refinement into heroin, happens in-country or after export. This means there is a steep markup in price, but farmers still make reliable profits growing opium. In 2002, one kilogram (2.2 pounds) of opium was worth U.S. $300 for the producer, a value far exceeding that of cut flowers. For this reason, as of 2012, Afghanistan is believed to be the largest illicit opium producer in the world, and efforts at eradication of poppy production by the government of Afghanistan and the U.S. military have been unsuccessful. Exports from Afghanistan amount to perhaps U.S. $4 billion, with farmers retaining a quarter of that value before officials, insurgents, and traffickers take their cut. Poverty in Afghanistan is the driver that maintains the global opium trade (**Figure 9.30**).

APPLY YOUR KNOWLEDGE Visit the Web page of *The Economist* (http://www.economist.com/content/indian-summary) that compares the strength of the economies of the states of India (such as Kerala or Rajasthan) with the strength of the economies of other countries (such as Tunisia or Singapore). How poor are the poorer states? How wealthy are the wealthier ones? What different economic activities are taking place to account for differences in these economies?

Territory and Politics

As it does in other world regions, the colonial imposition of arbitrary borders continues to have an enormous impact on governance and conflict in South Asia. Due to the region's tremendous cultural diversity, national boundaries tend to encompass diverse ethnic, linguistic, and religious groups. Boundaries also divide groups. The partition of British India in 1947 demonstrated this, as did the subsequent secession of Bangladesh from Pakistan. Cultural diversity and uneven economic development have also given rise to regionalism, and separatism. These movements contribute to political tension, social unrest, and, occasionally, outright rioting or armed conflict. Further geopolitical complications have arisen in the region as a result of the U.S.-led effort to drive the **Taliban** (a fundamentalist Muslim movement that ruled much of Afghanistan between 1996 and 2001) out of Afghanistan, following the September 11, 2001 terrorist attacks on the United States. This has exacerbated persistent tensions among the United States, Pakistan, India, and Afghanistan.

Indo-Pakistani Conflicts and Kashmir South Asia was of great interest during the Cold War, in part owing to its location south of the Soviet Union. During the late 20th century, both the Soviet Union and the United States vied for favor with South Asian states, leading to a long-standing alliance between Pakistan and the United States. South Asia has become even more of a geopolitical hot spot since the end of the Cold War.

Part of the reason for this is the unresolved question of Kashmir (introduced earlier in the chapter). Kashmir, whose predominantly Muslim population found itself isolated as a minority within India at partition, has three times been the cause of war between India and Pakistan (in 1948, 1965, and 1971). Its northern border is not an accepted international border—it is a "line of control" established after the 1971 war (**Figure 9.31**). Pakistan controls the northwestern portion of what India claims as Kashmir, and China controls the northeastern corner. In 1986, Muslim separatists began a renewed campaign of insurgency in the Indian-controlled portion of Kashmir. Since 1989, more than 30,000 people—separatist guerillas, policemen, Indian army troops, and civilians—have died in the

guerilla campaign aimed at incorporating Kashmir into Pakistan as part of a larger Islamic state. Pakistan sent its own forces across the border into the Kargil Peaks district in 1999. Pakistani troops were withdrawn after India launched a full-scale military offensive to evict them and the United States put pressure on the Pakistani government. The problem of Kashmir, however, remains unresolved.

This conflict is aggravated by the fact that both India and Pakistan have developed nuclear weapons. India announced in May 1998 that five nuclear tests had been carried out in the Thar Desert close to the Pakistani border while the territorial dispute over Kashmir was still simmering. The test triggered a fervent bout of national pride within India, but it prompted Pakistan to respond within a few weeks with its own series of nuclear tests. Tensions mounted again in November of 2008, when roughly a dozen carefully coordinated attacks were carried out in the dense business and tourist district of Mumbai, over the course of a few hours. More than 170 people were killed in the attacks, and hundreds more were wounded (**Figure 9.32**). In the period since, there has been some easing of hostility, with

▲ **FIGURE 9.32 Mumbai Terrorist Attacks** Firefighters try to douse the fire as smoke rises from the Taj Hotel building in Mumbai. The 2008 terrorist attack on luxury hotels and properties in Mumbai by armed gunman left 164 people dead and at least 300 wounded. India holds Pakistan responsible for the attacks.

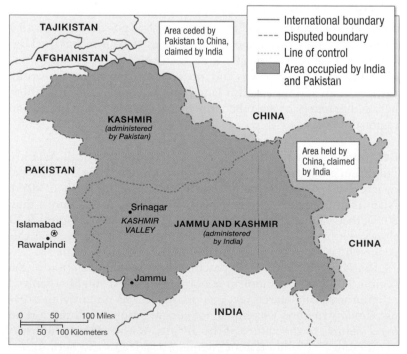

▲ **FIGURE 9.31 Kashmir and the "Line of Control"** Established by a brief war following partition, the unsettled boundary that splits the region of Kashmir has become a problem for both Pakistan and India.

the opening of borders and increased cultural exchange between the two nations.

Geopolitics of Afghanistan Indo-Pakistani tension is exacerbated by the international troubles surrounding Afghanistan. Afghanistan has long been of particular geopolitical significance because it is situated pivotally between Central and South Asia. Control of its mountainous terrain and mountain passes—such as the Khyber Pass—has frequently been disputed. For much of the latter part of the 20th century, Afghanistan had economic and cultural ties to the Soviet Union and began to pursue a Soviet-style program of modernization and industrialization. This provoked resistance from a zealous group of fundamentalist Islamic tribal leaders called the **mujahideen**, who were armed and trained by Pakistan. A total of more than 120,000 Soviet troops were sent to Afghanistan to confront the resistance; however, like British colonizers before them, they were unable to establish authority in Afghanistan outside the capital city of Kabul.

The guerilla war ended in 1989 with the withdrawal of the Soviet Union. With the demise of their common enemy, the mujahideen militias' ethnic, clan, religious, and personality divisions became problematic, and civil war ensued. Eventually, in 1996, the hard-line Islamist faction, the Taliban, gained control of Kabul and most of Afghanistan. The Taliban regime imposed harsh religious laws and social practices on the Afghan population. The regime also harbored an entirely new geopolitical force, whose actions have had worldwide implications: Osama bin Laden and his al-Qaeda terrorist network. Al-Qaeda was responsible for the September 11, 2001, attacks on the Pentagon and the World Trade Center in the United States. Consequently, as a core component of the **War on Terror**, Afghanistan became the focus of a U.S. military operation, Enduring Freedom, which resulted in the defeat of the Taliban and installation of a U.S.-backed government in Kabul in December 2001.

Military progress in Afghanistan, however, slowed during the subsequent years. NATO forces, along with the U.S. military, managed to maintain control of Kabul and some rural areas, but Taliban attacks intensified in outlying areas, preventing NATO and U.S. forces from fully controlling the country. The major goals of the U.S. operation, therefore, remain, although on April 18, 2012, the United States and its allies finalized agreements to end the war in Afghanistan. The Taliban has managed to survive in the border regions between Afghanistan and Pakistan, as have some dwindling representatives of al-Qaeda leadership.

The Afghan–Pakistan Frontier Between 2008 and 2012, the unresolved conflict in Afghanistan spilled over into Pakistan. Several of that country's frontier districts lie adjacent to the long, remote, mountainous border with Afghanistan. These are borders that tribal communities have freely traversed throughout history and where government control has been difficult to impose. Because Taliban forces operate in these areas and al-Qaeda may potentially be sheltered there, U.S. forces have conducted military operations in the area. These operations in the War on Terror include the use of **drones**, unmanned aerial vehicles flown out of bases in Afghanistan into Pakistani territory that are used to spy and strike enemy targets. Drone strikes have resulted in collateral civilian casualties, triggering both popular ire and official Pakistani outrage. Pakistan views these actions as transgressions of its territory.

Tensions increased in May of 2011, when U.S. Navy SEAL commandos made a raid on the hiding place of al-Qaeda leader Osama bin Laden in Abbottabad, Pakistan. Bin Laden had been hiding, essentially in plain sight, in a suburban area occupied mostly by retired Pakistani military officers. The fact that bin Laden was found and killed in Pakistan was an embarrassment and source of frustration for Pakistan's government, and Pakistan–U.S. relations deteriorated further as a result. The United States transports matériel and personnel into Afghanistan through Pakistan, and Pakistan is crucial to the maintenance of any peace in the future. As a result, problems along the border of Afghanistan–Pakistan translate into broader unsettled geopolitical relations.

> **APPLY YOUR KNOWLEDGE** Examine the map of South Asia in Figure 9.1. What areas and features on the map do you think are of most strategic concern from India's point of view? Now consider the map from Pakistan's point of view. What areas and features do you think Pakistan considers to be the most strategic?

CULTURE AND POPULATIONS

South Asia is a region with deep cultural roots that are tangled with those of neighbors and the broader world. Even where regional traditions have not been mixed or hybridized, there are differences in the extent traditional cultures have accommodated or resisted globalization. New South Asian cultural traditions are increasingly global. For example, Bollywood cinema and Punjabi food are now common on the streets of London and New York.

Religion and Language

Religious and linguistic traditions in South Asia possess elements that have evolved fully within the region, such as Hinduism and the Dravidian languages. At the same time, many elements are wholly imported products of invasions and interactions from across the world, such as Islam and the English language. The signature cultural beliefs and practices of the subcontinent are the product of global and regional synthesis.

Religion The two most important religions in South Asia are Hinduism and Islam. **Hinduism** is the dominant religion in Nepal (where about 90% of the population is Hindu) and India (about 80%). Islam is dominant in Afghanistan (99%), Bangladesh (more than 80%), the Maldives (100%), and Pakistan (about 80%). **Figure 9.33** shows the broad regional patterns of religion in South Asia; this geography is much more complex when considered in detail.

Roughly a billion people practice the dominant religion of the region, Hinduism. Hinduism is not considered a single organized religion with one sacred text or doctrine; it has no unifying organizational structure, worship is not congregational, and there is no agreement as to the nature of the divinity. Rather, Hinduism exists in different forms in different communities. It features both formal texts and major common beliefs, as well as diverse regional practices, idiosyncratic deities, and local legends. These traditions originally derive from the Vedas, a collection of oral poems that date from the

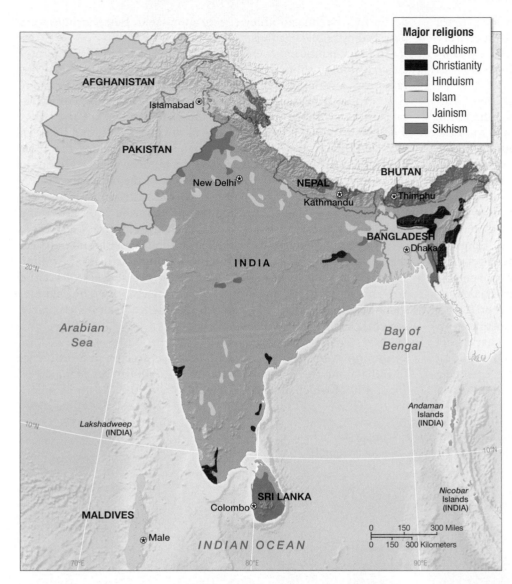

Major religions
- Buddhism
- Christianity
- Hinduism
- Islam
- Jainism
- Sikhism

◄ **FIGURE 9.33 The Geography of Religion in South Asia** Partition between India and Pakistan in 1947 resulted in mass migrations of Hindus from Pakistan to India and of Muslims from India to Pakistan. However, more than 123 million Muslims still live in India.

123 million Muslims still reside in India today. Islam is practiced in South Asia in a manner similar to its practice elsewhere in the world (see Chapter 4). Certain unique regional traditions also exist. South Asian mosque architecture can be distinctive, especially Mughal architecture, which fuses Hindu and Muslim styles. Many local traditions in Islam are **syncretic** with Hindu traditions, moreover, meaning that practices have co-evolved and merged with one another over the centuries. For example, a local forest may be protected by a Hindu goddess according to local Hindus and by the spirit of a local Muslim saint according to Muslims. Like their Hindu neighbors, Indian Muslims rarely eat beef.

Numerous other religions were founded in South Asia and persist. **Buddhism** originated in South Asia, although it is followed by only about 2% of the region's population (see Chapter 8). Buddhism is the predominant religion in Bhutan and Sri Lanka, and an enclave of Buddhism is found in Ladakh, the section of Kashmir closest to China. Buddhists trace their faith to Prince Siddhartha, a religious leader who lived in northern India prior to the 6th century B.C.E. and who came to be known as Buddha (the Enlightened One). Buddhism stresses nonviolence, moderation, and the cessation of suffering. *Jains* are another distinctive religious group whose origins are in South Asia. Jains, like Buddhists, stress self-control and nonviolence,

10th century B.C.E. There are several other important Hindu texts, including the epic tales of the *Ramayana* and the *Mahabharata*. The key common aspects that emerge from these are an acceptance of polytheism and many faces and versions of god(s); belief in **karma**, the idea of cosmic responsibilities for actions and deeds visited upon eternal souls throughout an endless cycle of reincarnated lifetimes; faith in *dharma*, the actions aligned with a right way of being and doing in the universe; and the practice of *puja*, worship and prayer, typically at a temple or shrine (**Figure 9.34**). Nonviolence is also an important tenet and many, though by no means all, Hindus are vegetarians. Geographically, Hinduism is written into the landscape of South Asia. Within a village or a city, certain sites, wells, trees, or stones may be sacred. And throughout India there are some especially sacred sites and cities which are the destination for pilgrims, including Varanasi (associated with the Hindu god Shiva), Haridwar (where the sacred Ganga River enters the plains from the Himalayas), and Ujjain (the site of the Kumbh Mela, a huge religious fair), among others.

A sizable minority population in the region adheres to religions other than Hinduism. The most significant of these other religions in terms of numbers is **Islam**. Although several million Muslims migrated from India to Pakistan at the time of partition, more than

▲ **FIGURE 9.34 Prayer at a Hindu Holy Site** Hindu worshippers are shown here immersed in the waters of the Ganga (Ganges) River at the sacred city of Varanasi, performing *puja* (prayer). Hindu rituals take place in temples, in homes, and outdoors at sacred sites such as this one.

and their roots also can be traced back to the 6th century B.C.E. Guru Nanak founded the monotheistic Sikh religion in the 16th century C.E. in Punjab. Sikhism remains largely associated with the people of the Punjab, where its sacred Golden Temple sits in the city of Amritsar (**Figure 9.35**).

There are also almost 20 million Christians in India. The Portuguese brought Roman Catholicism to the west coast of India in the late 1400s, and Protestant missions, under the protection of the British East India Company, began to work their way through the region in the 1800s. Christianity is most widespread in the state of Kerala, in southwest India, where nearly one-third of the population is Christian.

Religious Identity and Politics Historically, the religions of South Asia have displayed a high level of syncretism. Recent changes in identity and geopolitics, however, have exacerbated religious antagonisms and fostered social movements oriented around religion. In Afghanistan and Pakistan, notably, powerful Islamist movements have attempted to promote traditional culture as the basis of contemporary social order. There is strong adherence in both countries, for example, to traditional forms of dress and public comportment. In Afghanistan, the ultraorthodox Taliban rulers imposed a harsh version of Islamic law that followed a literal interpretation of the Muslim holy book, the Qu'ran. Under Taliban laws, murderers were publicly executed by the relatives of their victims, adulterers were stoned to death, and the limbs of thieves were amputated. Lesser crimes were punished by public beatings.

In India, political movements founded in a conservative interpretation of Hinduism have also been on the increase since the 1950s. **Hindutva**, a social and political movement that calls on India to unite as an explicitly Hindu nation, has given birth to a range of political parties over the years, including the Bharatiya Janata Party (BJP). The BJP attracted popular support in the late 1980s, and between 1999 and 2004 it was the key partner in a 24-party coalition that came to power in national elections. Given the number of non-Hindus in the country and the explicitly secular nature of the national constitution, the rise of Hindu nationalism is viewed with suspicion and fear by many. The BJP failed to capture a place in the 2009 governing coalition but remains active in state governance (**Figure 9.36**).

Islam and Hinduism both provide cultural symbols and practices that conservative and intolerant groups and parties have rallied around and used to organize. The continued success of secular government in India, however, and, to a lesser degree in Pakistan, shows that these parties and groups are neither universally embraced nor consistently successful.

Language A great diversity of languages is spoken in South Asia. In India alone there are approximately 1,600 different languages, about 400 of which are spoken by 200,000 or more people. There

▼ **FIGURE 9.35 The Sikh Golden Temple at Amritsar** The Golden Temple at Amritsar is the holiest site of the Sikh religion, where the sacred text of the faith is housed. The temple is a destination for hundreds of thousands of pilgrims each year.

different languages. There are thriving regional language presses; however, film and television are dominated by Hindi and Tamil with some Telegu programming.

English, the first language of fewer than 6% of the people, links India's states and regions; English is the **lingua franca** (a widely recognized common second language) throughout the region for commerce, government, and travel. As it is in other former British colonies in South Asia, in India, English is the language of higher education, the professions, and national business and government. South Asian literature includes numerous classic novels written in English.

Cultural Practices, Social Differences, and Identity

The cultural practices of South Asia are enormously diverse and represent a merging of Indigenous traditions with a range of customs from other cultures. Some of the traditions most widely associated with social identity in the region include caste and religious identity, but new cultural phenomena are also widespread, including emerging consumerism and anticonsumerism.

▲ **FIGURE 9.36 Indian Bharatiya Janata Party (BJP)** Supporters of the youth wing of the Bharatiya Janata Party were arrested by police as they broke barricades near the Indian Parliament during a 2012 protest. Characterized by its saffron colors and lotus leaf symbolism, BJP iconography merges religious identity and political action.

is, however, a broad regional grouping of four major language families:

1. Indo-European languages, introduced by the Aryan herdsmen who migrated from Central Asia between 1500 and 500 B.C.E., prevail in the northern plains region, Sri Lanka, and the Maldives. This language family includes Hindi, Bengali, Punjabi, Bihari, and Urdu.
2. Munda languages are spoken among tribal hill people who still inhabit the remote hill regions of peninsular India.
3. Dravidian languages (including Tamil, Telegu, Kanarese, and Malayalam) are spoken in southern India and the northern part of Sri Lanka.
4. Tibeto-Burmese languages are scattered across the Himalayan region.

Written languages also vary owing to the diverse populations that have settled in the region over millennia. Most notably, Hindi and Urdu are in the same language family and are extremely similar in structure and vocabulary, but while the former is written in the Devanagari script derived from ancient Sanskrit writing, the latter is written in Nastaliq, a modified version of the Persian script (**Figure 9.37**).

In India, the boundaries of many of the country's constituent states were established on the basis of language. This led to fracturing and splintering of groups and communities since no single language is spoken or understood by more than 40% of the people and all areas have dominant languages but many minority languages as well. There have been efforts to establish Hindi, the most prevalent language, as the Indian national language, but this has been resisted by many states whose political identities are closely aligned with

Caste, Marginality, and Resistance A very important—and often misunderstood—aspect of the region's cultural traditions is **caste**, which is a system of kinship groupings, or **jati**, that are reinforced by language, region, and occupation. Though associated with Hinduism, and therefore most common in India and Nepal, forms of caste distinctions are not unknown in Pakistan, showing that this is a broader cultural tradition of the region as a whole. There are several thousand separate jati in India, most of them confined to a single linguistic region. Many jati are identified by a traditional

▲ **FIGURE 9.37 Devanagari and Nastaliq Scripts** The word Hindi is written above in Devanagari and the word Urdu is written below in Nastilaq script. The two are closely related Indo-European languages and are often orally indistinguishable; however, their respective written scripts derive from entirely different cultural lineages.

occupation, from which each derives its name: *jat* (farmer), for example, or *mali* (gardener), or *kumbhar* (potter). Modern occupations such as assembly-line operators or computer programmers, of course, do not have a traditional jati and people doing these jobs do not cease to be members of the jati into which they were born. People within the same jati tend to sustain accepted norms of behavior, dress, and diet. They are also **endogamous**, which means that families are expected to find marriage partners for their children among other members of their jati.

In each village or region, jati exist within a locally understood social hierarchy—the caste system—that determines the norms of social interaction. Each individual person's jati is fixed by birth, but the position of the jati within the local caste system is not. Nevertheless, the broad structure of caste systems always places certain groups at the top and others at the bottom. Caste systems tend to hold in high esteem those who are religious and those who are especially learned. Those who pursue wealth or hold political power are typically less well regarded, and those who perform menial tasks are accorded the lowest status. Priestly jatis—known as Brahmins—are at the very top of the caste hierarchy. Brahmins are expected to lead ascetic lives and revere learning. At the opposite end of all caste systems are the so-called untouchables—whose members dispose of waste or dead animals. In an effort to eliminate the demeaning term *untouchable*, today most people in these communities prefer to be referred to as Dalits, meaning "the oppressed."

Traditionally, these lower-caste groups were forced to live outside the main community because they were deemed to be capable of contaminating food and water by their touch. They were denied access to water wells used by other communities, refused education,

banned from temples, and subjected to violence and abuse. Although these practices were outlawed by India's constitution in 1950, discrimination and violence against lower-caste communities is still routine in some areas. A number of more marginal castes have been legally assigned an official status by the Indian government. Members of these "scheduled castes" (**Figure 9.38**) sometimes receive set-asides or quotas for government positions.

In urban areas, practices relative to caste vary widely but are less regularly observed in general. Most urban Indian newspapers, however, still contain a lengthy "matrimonial" section, where advertisements are placed for "suitable" matches for arranged marriages, most often within the same specific caste group. Discrimination is also experienced by groups based on criteria other than caste. **Adivasi**, or tribal people, in India represent a significant part of the manual labor force in the regions where they predominate. Tribals tend to hold the least land, have the poorest job prospects, and live in the most difficult of conditions. Like lower-caste communities, these groups have received government assistance and infrastructural development, but their marginal social and political position persists.

Increasingly, marginal castes and adivasi communities have become more politically active, and some violent resistance movements are associated with these groups. So-called **naxalite** insurgencies, violent armed uprisings, have occured across India, especially in the poorest states: Chhattisgarh, Jharkand, and Bihar. Because they attack landlords and police stations, the naxalites are considered terrorists by the government of India, but they have a considerable following in rural areas and among students. Nepal has experienced a similar long-standing, rural insurgency.

▶ **FIGURE 9.38 Caste and Poverty** Slum-dwellers live under a bridge in Mumbai as the city prepares to host a World Social Forum, where one of the key debates addresses the problem of the untouchables (138 million people out of a total population of one billion people) who still live on the margins of society, excluded by the hierarchical Hindu caste system.

Within Pakistan, ethnic tensions have developed due to regional and linguistic differences, and many regional minorities have sought autonomy, including groups in Baluchistan, the rugged southwestern area of the country. A number of violent, albeit scattered, secessionist movements are active in Baluchistan, which has a long border in common with Afghanistan and a lengthy history of independence and resistance. Baluchistani society is heavily tribal and culturally conservative and has been isolated from the development and growth in the rest of the country. The situation in Baluchistan is made more complicated by its proximity to Afghanistan.

Popular Culture Traditional cultures are deep and abiding in South Asia, but there are very few places where the impact of contemporary global culture is not heavily felt. The growth of a large and affluent middle class in India has brought Western-style materialism to larger towns and cities. Fast-food outlets, ATMs, name-brand leisure wear, consumer appliances, video games, luxury cars, pubs, clubs, and shopping malls abound. The Indian credit card industry is growing quickly. As of 2011, 29% of total credit card holders in India have premium cards; this statistic reflects the importance of the very highest income earners to the credit industry. Despite recent global economic contraction, between 2009 and 2010, India's luxury goods market grew 20%, to U.S. $5.8 billion. Indian consumers have an increasing appetite for jewelry, electronics, and fine dining, wines, and spirits—markets that are entirely new to the economy.

The preeminent sport in both India and Pakistan is cricket, a legacy of British colonialism, which today generates huge sums in betting and supports a star system to rival that of baseball in the United States. More surprising, however, is the rising popularity of motor sports, especially Formula One racing. India staged its first Formula One race in 2011, which attracted thousands of fans. The sport now has its own homegrown superstars, like Indian driver Narain Karthikeyan.

Cable television arrived in India in the early 1990s and has since become ubiquitous. The number of homes with television in India grew from 120 million to 148 million between 2007 and 2011. Ninety-four million homes have cable. While foreign programming is popular, the top ten Indian television shows of 2011 (a mix of reality shows, game shows, and soap operas) did not include a single foreign program. Programs address very regional themes. One popular program, *Balika Vadhu*, set in the rural state of Rajasthan, follows the troubles of a young girl who was married at the age of eight and overcomes the challenges of sexism, discrimination, and economic hardship. In this way, South Asian television, though borrowing globalized themes and styles, is fully regionalized and represents a singular force for local change.

When Mumbai was still known as Bombay, the city developed a huge Hindi-language film industry that was eventually nicknamed **Bollywood**. The 1,000 films Bollywood releases each year reach an audience of more than 3 billion people, which includes domestic audiences and Indians abroad. Hollywood's audience is somewhat less than 3 billion viewers annually, by comparison, including important and lucrative overseas sales of CDs and DVDs. Bollywood revenues, however, are less than U.S. $2 billion annually, a fraction of the U.S. $70 billion made by Hollywood.

Bollywood movies thrive on a local markets and regional traditions (**Figure 9.39**). Originally drawing on classical Hindu mythology and traditional social values, the roots of Bollywood lie in traditions of folk theater and performance. Most modern Bollywood movies address contemporary themes and characters, however, in ways that relate directly to the lives of viewers. Although India also produces avant-garde movies that have been recognized for their artistic and

◀ **FIGURE 9.39** Bollywood
Bollywood superstar Shahrukh Khan dances in *Kal Ho Naa Ho* (*Tomorrow May Never Come*). The film, an international success, grossed millions of dollars worldwide.

VISUALIZING GEOGRAPHY
Gross National Happiness

Bhutan is the smallest and most isolated of the countries of South Asia, with a population of less than a million people, a low level of urbanization, and a highly traditional, rural, Buddhist society. It is also a country that has intentionally limited immigration and tourism and has been historically isolated from much of the rest of the world. The country's king began to open the country to international influence and trade in 1972, but chose to do so on local terms, pri-

oritizing developments and economic change that would bring serenity and satisfaction, rather than mere prosperity. In doing so, he coined the term "gross national happiness" (playing on the economic indicator of gross national product) and sought to devise a measure of the spiritual development and quality of life of the population as opposed to per capita economic activity. The government adopted the principle and built it into its planning process. Bhutan

began to survey people's happiness for planning purposes beginning in 2005. Whether the people in Bhutan are happier than others in South Asia is not entirely clear; however, this approach to culture contrasts dramatically with that of neighboring states, who have embraced the accumulation of material goods as the primary goal of public welfare.

The gross national happiness concept has since expanded to an international

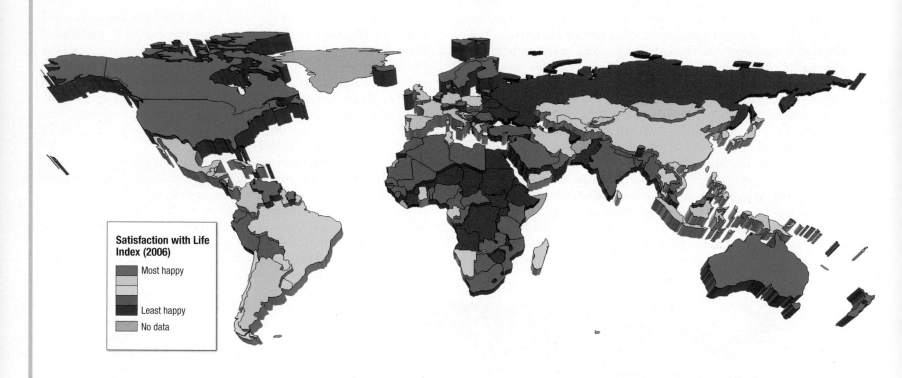

Satisfaction with Life Index (2006)

- Most happy
- Least happy
- No data

▲ **FIGURE 9.2.1 Satisfaction with Life Index** This map shows the relative self-reported happiness of people, on average, in countries around the world. The index used in the map is calculated from a direct survey of people's happiness.

dramatic content, most Bollywood products are exuberant, spectacle-driven entertainment: melodramatic fantasies that mix action, violence, romance, music, dance, and moralizing into a distinctive, formulaic form. Bollywood stars such as Amitabh Bachchan, Salman Khan, Katrina Kaif, Shahid Kapur, Aishwarya Rai, and Shahrukh Khan have their careers and private lives monitored by adoring fans with an intensity that Hollywood agents would envy. Most Bollywood films have some sort of musical content, and the songs dominate Indian pop charts. Soundtracks include sitars, synthesizers, pianos,

and violins; scores move effortlessly from classical Indian ragas to Mozart to hip-hop and rap music.

In sum, the sheer size and market power of growing middle classes has meant that an externally imposed global culture has not emerged in South Asia. Rather, luxury-good producers, television networks, and film studios all produce regional products with international styles in local languages (including Hindi, Tamil, Urdu, and Bengali). A distinct regional popular culture has emerged, albeit one marked heavily by global consumerism. These new forms

movement. Efforts have been made to survey and measure people's happiness from various countries. Data has been combined relating to people's health, experience at work, social stability, and so on. These kinds of measures were featured in the first Global Gross National Happiness Survey, as well as a number of related studies of national and global well-being. Happiness is hard to measure and map, but current assessments suggest that it is widely variable from country to country. **Figure 9.2.1** shows national happiness based on the "Satisfaction with Life Index", created by Adrian G. White, an analytic social psychologist. Compare this map to the map in **Figure 9.2.2**, which is based on the "Happy Planet Index" (HPI), a mix of average life satisfaction, life expectancy, and ecological footprint data. The inclusion of ecological information means that countries are scored in the HPI based on well-being relative to how their lifestyle choices impact the planet. Note how some countries score well on one index and not another. Notably, some countries are happy, but at high environmental cost. This raises the question not only of what happiness is, but what the acceptable environmental price of happiness is. ■

Happy Planet Index (2006)

Most happy

Least happy

No data

▲ **FIGURE 9.2.2 Happy Planet Index** This index combines people's well-being with their impact on the environment and consumption of resources. Essentially, this map measures how efficient people are at being happy with little. The countries that are colored green are happy *and* have smaller ecological footprints.

of popular consumer culture have not gone without contestation. Many local groups, nongovernmental organizations, blog authors, and cooperatives continue to draw upon the rich traditions of the region to criticize consumerism as the new predominant culture. Operating on a national scale, as a remarkable example, the government of Bhutan has worked to recognize **gross national happiness**, rather than economic output, as the measure of national success (**Figure 9.40**, p. 376). (See "Visualizing Geography: Gross National Happiness.")

The Globalization of South Asian Culture Culture also flows outward from South Asia. Mysticism, yoga, and meditation found their way into Western popular culture during the "flower power" era of the 1960s after the Beatles visited India. South Asian methods of nonviolent protest developed in the 20th century by Gandhi, such as boycotts and fasting, have spread around the world. Contemporary South Asian literature from writers such as R. K. Narayan, Aravind Adiga, Satyajit Ray, and Salman Rushdie has found a global readership. South Asian music is well-represented in the

▲ **FIGURE 9.40 Bhutanese Children** Junior high school students smile for the camera in Thimphu, Bhutan. Though one of the least developed and economically active countries in South Asia, tiny Bhutan has led a worldwide effort to revaluate development by stressing gross national happiness rather than material gain.

"international music" sections of Western record stores, and some artists, such as Sheila Chandra, have crossed over into a broader international audience.

One of the most visible global influences of South Asia is its food. Consider chicken *tikka masala*, a dish possibly conceived in Punjab, which has become global in scope owing solely to its popularity abroad (**Figure 9.41**). The dish has many fully regional South Asian components; roasted chicken (tikka) and spicy sauce (masala) are signatures of Pakistani and Punjabi cuisine. But the specific mix of yogurt, cream, and spices into a kind of gravy is novel, and it has become the most popular dish in British restaurants. Britain's foreign secretary, Robin Cook, declared the meal "a true British national dish" in 2001 precisely because it took the food components of South Asia and domesticated them to British tastes.[2]

APPLY YOUR KNOWLEDGE Describe two key cultural traditions from South Asia. To what degree have modern media and consumer culture accommodated and absorbed these traditions?

Demography and Urbanization

South Asia has the second-largest, fastest-growing population of all the world regions. The total population of South Asia in 2011 was 1.62 billion, with India accounting for just over 1.24 billion. The

rate of natural growth varies, with Afghanistan growing at a relatively high rate of 2.8%, while India is growing at a far lower 1.5%. Both these rates are high relative to China's modest 0.5%, making the region one of the fastest growing in Asia. There is very high density in most of the region (**Figure 9.42**). The overall density of population in India is 316 persons per square kilometer (819 per square mile), compared to 131 persons per square kilometer (338 per square mile) in China, and 29 persons per square kilometer (75 per square mile) in the United States. Patterns of population density reflect patterns of agricultural productivity. Rich soils support densities of more than 500 people per square kilometer (1,300 per square mile) in a belt extending from the upper Indus Plains and the Ganga (Ganges) plains through Bengal and the Assam Valley. Similar densities are found along much of the Coastal Fringe.

Declining Growth In many parts of the region, population growth is slowing or halting. Although some rural areas have fertility rates as high as six children per woman, a rate consistent with historic averages, others have declined to far lower figures, some to as few as two births per woman. Little of this change resulted from the explicit and harsh population policies favored by regional governments in the past. These efforts, which began in the 1950s and continued into the 1970s, were largely misguided. In the mid-1960s, for example, the government of India announced specific demographic targets and opened "camps" around the country for the mass insertion of intrauterine devices (IUDs). The program soon failed, mainly because of negative public reaction to the poor training of health workers and unsanitary conditions in the camps. Next came vasectomy camps. More than 10 million men were coerced into being sterilized in the 1970s in

▲ **FIGURE 9.41 Chicken Tikka Masala** This dish is by no means traditional, and may have been invented far from South Asia, but its unmistakable Pakistani and Punjabi elements are a testimony to the linkages and influences South Asia has across the globe.

[2]"Robin Cook's Chicken Tikka Masala Speech"—Extracts from a speech by the foreign secretary to the Social Market Foundation in London. http://guardian.co.uk, April 19, 2001, accessed May 16, 2012.

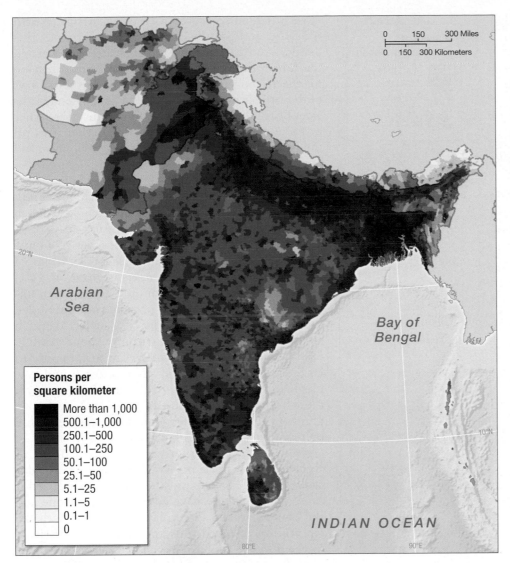

▲ **FIGURE 9.42 Population Density in South Asia, 2010** The density of population is very high throughout most of South Asia, but especially so in the plains and in subregions with good soils and humid climates.

Persons per square kilometer

More than 1,000
500.1–1,000
250.1–500
100.1–250
50.1–100
25.1–50
5.1–25
1.1–5
0.1–1
0

literacy in the population, especially of women, whose education is consistently correlated with lower fertility rates. The final factors include high levels of participation in the labor force by women in Kerala, as well as a local matrilineal tradition, which gives women significant control over property. Conversely, states in India where population growth remains high, like Rajasthan and Uttar Pradesh in north India, typically have low women's employment, low women's literacy, and a poorly developed health infrastructure. Several Indian states are now following Kerala's example of emphasizing women's education and better infant and maternal care.

The effect of women's education, employment and health is reproduced across the region. Pakistan and Bangladesh make a useful comparison. Although both countries have high levels of poverty and historically discriminatory cultures in terms of women's access to employment and education, they have startlingly different population profiles. In 2011, the total fertility rate in Bangladesh was 2.4, while in Pakistan the figure was a far higher 3.6. The differential availability of health care for women is certainly a factor; ongoing efforts in Bangladesh to provide clinical coverage across the country allow women to access immunizations for their infants and family planning for themselves, leading to dramatic and encouraging declines in birthrates. Contraceptive use among married women in Bangladesh in 2011 was 56%, compared to 27% in Pakistan. These statistics indicate that while large-scale cultural factors are important—both Pakistan and Bangladesh are Muslim majority nations—contextual health and employment factors matter far more. If places like Kerala and Bangladesh are indicators, it is possible that the boom in population growth across the subcontinent will halt within a few decades.

an "emergency drive." Not surprising, a popular backlash put an end to the sterilization program. These efforts did little to curtail population growth.

Instead, these demographic changes have come about as a result of infrastructural, educational, and economic changes. The unevenness of this effect can be seen across the region. **Figure 9.43**, p. 378 shows the average fertility rates by state in India. Notably, many of the states of southern India have rates of roughly two births per woman, effectively a condition of **zero population growth**, or ZPG. In the state of Kerala, often taken to be an example of demographic success, this change appears to have been the result of a combination of factors. First, rural health care in the state is well-developed and well-distributed. With access to good maternal and infant care, infant mortality rates are low, which discourages families from having more children to ensure against possible deaths. Good health care also means access to birth control, either in the form of prophylactics or sterilization. This is coupled with high levels of

APPLY YOUR KNOWLEDGE Which countries in South Asia have the lowest growth and fertility rates? Which states and regions within India have the lowest fertility rates? What do these states and regions have in common?

Urbanization South Asia is still very much a land of villages. Indeed, India has the largest rural population in the world. In 2011, approximately 65% of Pakistan's population and 71% of India's lived in rural settings; in Afghanistan, Bangladesh, and Sri Lanka, the rural population is about 77%; the tiny state of Bhutan is 67% rural, with Nepal at 82%. Rural-to-urban migration is shifting the balance toward towns and cities, however. In 2010, there were 55 cities of 1 million or more in the region, including 11 with 5 million or more residents. Between 1960 and 2010, Mumbai—the largest metropolis in South Asia—grew from 4.1 to 20 million; Dhaka, in Bangladesh, grew from

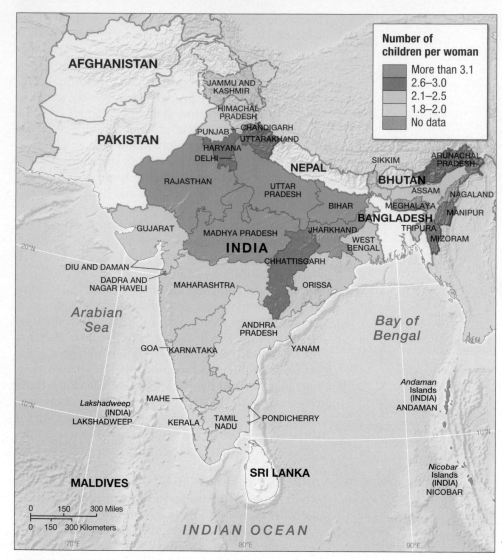

▲ **FIGURE 9.43 Fertility Rates across the States of India** Although high fertility and large family sizes are the norm in some parts of the country, many states, like Kerala, have achieved nearly zero population growth.

fewer than 650,000 to more than 14.8 million; and Karachi, Pakistan, grew from 1.8 to 13.1 million.

The explosion of cities, middle-sized urban areas, and large towns has fundamentally changed the character of South Asian politics, landscapes, and culture. Urban areas are characterized by intense mixing of ideas, dress, and traditions, including many religions, classes, and castes. The dynamism of cities extends beyond the signature urban areas of Mumbai and Bangalore to the tens of thousands of middle-sized towns, where housing construction is booming. The competition for land at the edge of cities is intense, as agricultural land prices have risen dramatically through speculation, and formerly quiet rural areas have been quickly incorporated into bustling metropolises.

The South Asian Diaspora The population dynamics of South Asia are highly influenced by its relationship to the rest of the world. The South Asian diaspora—the movement of Indians, Pakistanis, and Bangladeshis abroad—has involved between 5 and 6 million people,

most of whom relocated to Europe, Africa, North America, and Southeast Asia. The origins of this diaspora can be traced to the abolition of slavery in the British Empire in 1833. The consequent demand for cheap labor on the plantations and on the railways of the British Empire was filled in part by emigrants from British India. In the mid-19th century, thousands of Indians left for the plantations of Mauritius (in the Indian Ocean), East Africa, the West Indies, and South Africa.

After World War II, the pattern changed significantly. Former British colonies excluded South Asian immigrants; both Burma and Uganda expelled most of them. But a new destination for South Asian emigrants opened up as the postwar economic recovery in Europe resulted in a severe shortage of labor on assembly lines and in transportation. Britain received more than 1.5 million South Asian immigrants, whose permanent presence not only filled a gap in the labor force, but has also served to enrich and diversify British urban culture. About 800,000 South Asians moved to North America, mainly to larger metropolitan areas, where most found employment in service jobs. From the 1970s onward, there has been a steady stream of South Asian immigrants to the oil-rich Persian Gulf states, recruited on temporary visas to fill manual and skilled jobs.

South Asia has experienced a **brain drain** of significant proportions over the past several decades, as some of the most talented and well-educated young people have emigrated to Europe and the United States. The brain drain began with the emigration of physicians and scientists to Britain in the 1960s and accelerated as South Asian students, having completed their studies in British and American universities, stayed to take better-paying jobs rather than return to South Asia. The idea of living abroad gained popularity among India's cosmopolitan and materialist middle classes as newspaper and television features publicized the global successes of Indian emigrants.

In the last two decades, the most distinctive aspect of the brain drain from South Asia has been the emigration of computer scientists and software engineers from India to the United States and parts of Europe. Between 2000 and 2010, the Indian population in the United States grew by 69%, to total of almost 3 million. This population is very well-educated. Seventy-one percent hold doctorates, and the American Association of Physicians of Indian Origin reports there are approximately 35,000 Indian doctors in the United States. Similar statistics hold for Pakistani Americans, though they have faced greater anti-Islamic discrimination. India has worked to maintain relationships with its successful migrants abroad, designating them as "persons of Indian origin" and providing them with identity cards and privileges in India. This is done not so much in the hope that these second- and third-generation foreigners will eventually return to India, but instead so that they will financially invest in, and identify politically with, their ancestral homeland.

FUTURE GEOGRAPHIES

How the shape and character of South Asia will change over the next 50 years depends on the outcome of several tensions and trends that link the region to forces and places beyond its boundaries. First, the resolution of the war in Afghanistan will affect the future political conditions of the region with implications for the ongoing Indo–Pakistani conflict. Second, the speed at which birthrates in the region decline will determine the ultimate total population, labor force, and resource burdens. Finally, the future of food production technologies will determine whether South Asia will be a global breadbasket or the site of hunger.

The Fate of Afghanistan The bulk of U.S. troops are scheduled for withdrawal from Afghanistan by 2014. The stability and overall geopolitical position of Afghanistan after the United States' exit has a great bearing on the condition of the region as a whole. If an orderly exit is achieved and a semistable regime remains in place, with normalized relationships with Pakistan, it is far easier to imagine a peaceful outcome for the region. In that case, for example, India and Pakistan might resume their discussion about the state of Kashmir and other disputes, including communication about shared rivers and groundwater.

If, on the other hand, the situation deteriorates in Afghanistan, India might see a geopolitical advantage in intervening—directly or indirectly—in Afghanistan's situation. This could lead to redoubled tensions or all-out war. Pakistan's government already suspects India of interfering in Afghanistan's affairs. India, conversely, views Afghanistan as a possible training ground and launching pad for anti-Indian insurgents. Afghanistan likewise is concerned that Pakistan may support insurgents who could undermine the Afghani government. This spurred Afghanistan to sign a strategic partnership agreement with India in 2011, a move that dismayed Pakistan. The highest stakes of the settlement of the Afghanistan conflict are unquestionably for Afghanistan, Pakistan, and India, rather than the United States.

Toward Zero Population Growth South Asia is home to more than one-sixth of the world's people. But growth rates in many parts of the region have declined dramatically in recent years, with some areas actually achieving zero population growth. It is reasonable to assume that overall demographic growth in the region will halt before 2050; however, how much sooner growth ends depends on several factors, including women's education, health-care availability, women's participation in the labor force, and the degree of urbanization. If conditions in these areas improve even modestly, as they have in Bangladesh where the fertility rate has fallen from seven births per woman in 1975 to 2.5 births per woman in 2011, the prospect of a smaller final population are good. At the time population growth ends, however, the region will be composed predominantly of older people, presenting a serious economic puzzle for a generation of younger people who will have to support a large, dependent population.

New Food Technologies or Old South Asia is a food-exporting region and a breadbasket of rice and wheat. Much of its agricultural success has been leveraged on implementation of the Green Revolution. But with crop yields declining, soils becoming less vital, and water increasingly scarce, it is unclear how food production will be maintained. Some are calling for a new Green Revolution, which will involve the use of **genetically modified organisms (GMOs)**, specifically, genetically modified (GM) food crops to spur an increase in production. To date, no genetically modified food crops have been introduced in either Pakistan or India, where opposition to GM crops has been vehement (**Figure 9.44**). Both countries, however, do grow *Bt cotton*—a variety of the cotton modified through cross-breeding to fight pests without the use of pesticides. In 2012, Bangladesh began movement toward allowing the cultivation of GM eggplant, and Pakistan has begun to consider GM maize.

There are others who suggest the best solutions to the food problem lie in traditional water, soil, and crop management techniques, precisely those discarded during the Green Revolution. Advocates for traditional crops and techniques argue that these are adapted to uncertain rainfall conditions and so may contain the

▲ **FIGURE 9.44 Protesting Genetically Modified Crops** An activist dressed as an eggplant protests in Bangalore against *Bt brinjal*, or genetically modified (GM) eggplant crops. Though genetically modified cotton is grown in Pakistan and India and the Green Revolution introduced many cross-bred crops into the region, consumers have balked at the production of genetically modified food crops, especially eggplant.

genetic keys to surviving climate change. This group insists that the best resource for the next century is the vast knowledge of traditional farmers.

It is very likely that the answer to the question of which technology is best lies somewhere in between. South Asia remains characterized by its embrace and utilization of new ideas and technology in concert with traditional practices and knowledge. It is possible to imagine a situation in the near future where traditional rice production knowledge from Sri Lanka and Bangladesh, coupled with genetically modified varieties of the region's key crops, will make South Asia a resource for future global food solutions instead of the target for future development aid.

LEARNING OUTCOMES *REVISITED*

■ **Describe the pattern of the monsoon climate, explain its origins, and identify the key adaptations people in the region use to cope with change and uncertainty.**

The monsoon climate is caused by the relative pressure between the Asian interior and the Indian Ocean, which results in a band of intense summer rains and winter dry winds. People of the subcontinent have adapted to monsoon conditions with drought-tolerant cropping, transhumance, and the terracing of steep hill slopes. Green Revolution technology has put some of these traditional adaptations at risk.

■ **Distinguish the physiographic and engineered components of environmental hazards in the region, especially catastrophic flooding.**

The major rivers of the region flow from Himalayan mountains and glaciers. Periodic heavy rainfall and steep drainages can contribute to flooding. When these natural events are exacerbated by human engineering of rivers and deforestation of hill slopes, the landscape is vulnerable to rare but catastrophic flooding.

■ **Explain how some land uses contribute to environmental destruction and declines in South Asian biodiversity, while others promote the maintenance, protection, and restoration of wildlife.**

The history of human settlement of the region has been one of steady deforestation, which has been accelerated in recent years, especially during and after colonialism. This loss of forest results in a decline in habitat for native species, like tigers and rhinoceroses. However, human land uses like tree crop plantation (including coffee and rubber) can also produce habitats for wildlife.

■ **Provide examples of rulers, governments, and economic systems that historically put South Asia at the center of global trade, and explain how this position was later exploited and overridden by colonial governance.**

From the time of the Harappan Civilization and its extensive trade routes to the era of the Mughal Empire and its linkages between the Hindu and Muslim worlds, South Asia has been at the center of links that reach west to Africa and north toward China. Seeking to exploit the region's location, colonial powers, especially the British, slowly dismantled the productive capacity of South Asia's economy and increased its dependence.

■ **Identify the major engines of recent rapid South Asian economic expansion and discuss the uneven benefits of expansion.**

The new South Asian economic boom is based on growth in high technology (such as software design), service-sector employment (such as call centers), and exporter industrialists with a global reach. This combination has resulted in the rapid industrialization of many parts of the subcontinent and new consumer affluence for an emerging middle class. A significant majority of the region's population has not been touched by this economic growth, however, and many women, children, and rural workers still live in grinding poverty.

■ **Describe the key relationships and borders that present the greatest current geopolitical challenges for South Asia.**

Long-standing antagonism exists between Pakistan and India, who since the time of partition have contested the status of the region of Kashmir. The U.S.-led NATO war in Afghanistan has fueled further antagonisms between the two states. The outcome of that conflict will either lead to opportunities for further diplomacy or continued conflict between these two nuclear states.

■ **Summarize the characteristics and distributions of the region's major religions and language groups and identify their overlap, uniting features, and mutual influence.**

The dominant religions of the region are Hinduism and Islam, though many faiths with fewer followers are also practiced, including Buddhism and Sikhism. Similarly, popular languages, including Hindi, Urdu, and Bengali, have only subregional dominance, and countless other languages prevail in different areas. The result is a region of many languages that have mixed vocabularies and scripts and a number of religions that are syncretic.

■ **Distinguish key traditional cultures and emerging consumer cultures and cite examples of how each has impacted and been impacted by global connections.**

Key elements of traditional South Asian culture include caste, gender-specific social roles, and religious identity. As consumer culture has increased, television, luxury goods, and a booming film industry have taken center stage. Indian culture has greatly influenced global culture through cultural aspects such as food and music, which are so common now in North America and Britain as to be considered local.

■ **Explain where and under what conditions South Asian populations have begun to stabilize or decline and where they continue to grow relatively rapidly.**

South Asia has one of the largest and densest populations of all world regions, but growth rates are dropping rapidly, especially where health care, birth control, and education are available to women. Areas with high population growth (northwest India and Pakistan) contrast sharply with areas with far lower growth (southern India and Bangladesh) in these aspects.

KEY TERMS

adivasi (p. 372)

anthropogenic (p. 355)

Aryan (p. 358)

Bollywood (p. 373)

bonded labor (p. 366)

brain drain (p. 378)

Buddhism (p. 369)

call centers (p. 363)

caste (p. 371)

deforestation (p. 355)

dependency (p. 359)

dharma (p. 358)

distributary (p. 353)

drone (p. 368)

endogamous (p. 372)

genetically modified organisms
(GMOs) (p. 379)

Grameen Bank (p. 365)

Great Game (p. 360)

Green Revolution (p. 349)

gross national happiness (p. 375)

Gupta Empire (p. 358)

Harappan Civilization (p. 357)

Hinduism (p. 368)

Hindutva (p. 370)

Indian National Congress Party (p. 360)

intercropping (p. 349)

Islam (p. 369)

jati (p. 371)

karma (p. 369)

lingua franca (p. 371)

Mauryan Empire (p. 358)

microfinance (p. 365)

Mohandas Gandhi (p. 360)

monsoon (p. 347)

Mughals (p. 358)

mujahideen (p. 368)

naxalite (p. 372)

orographic effect (p. 347)

partition (p. 360)

Raj (p. 360)

Sundarbans (p. 355)

syncretic (p. 369)

Taliban (p. 367)

terracing (p. 349)

transhumance (p. 349)

tube well (p. 354)

War on Terror (p. 368)

zero population growth (p. 377)

THINKING GEOGRAPHICALLY

1. How do the Himalayan Mountains and the monsoons help define the character of South Asia?

2. How did Britain transform the physical and cultural geographies of South Asia in terms of agriculture, infrastructure, and economy?

3. Compare poverty in rural and urban South Asia. With malnutrition and illiteracy rates high, what strategies do the poor employ to survive in cities and in the countryside?

4. Which qualities of South Asian economies have made them resilient in the face of recent global economic crises?

5. As South Asians migrated worldwide, which jobs did they take? Which cultural traditions did they bring with them?

6. Why do India and Pakistan both make a claim to the Kashmir region?

7. How are key external political and military linkages between South Asia and the rest of the world affecting internal relationships within the subcontinent?

MasteringGeography™

Looking for additional review and test prep materials? Visit the Study Area in MasteringGeography™ to enhance your geographic literacy, spatial reasoning skills, and understanding of this chapter's content by accessing a variety of resources, including **MapMaster** interactive maps, videos, RSS feeds, flashcards, Web links, self-study quizzes, and an eText version of *World Regions in Global Context*.

10 | Southeast Asia

Thailand had some of its worst floods in decades in October 2011. On the day this photograph was taken, hundreds of laborers in Nawa Nakorn Industrial District, Bangkok, were working to stem the flow of water into the city's industrial zones.

IN THE FALL OF 2011, MASSIVE FLOODS SWEPT through central Thailand, bringing Bangkok, one of Southeast Asia's largest cities, to a grinding halt. By the time the floodwaters receded, it was estimated that almost 800 people had lost their lives in the city's worst flooding in almost 50 years. Like flooding in other parts of the world, the floods in Bangkok had a dramatic impact on the city's poorest residents, who often live in fairly unstable housing along waterways or other low-lying areas of the city, such as along rail lines and in between larger high-rise buildings.

Flooding is nothing new to the region or to Bangkok, whose low-lying lands have long been advantaged by the intense rains of the monsoon seasons. The region surrounding Bangkok benefits greatly from the rains that allow year-round rice cultivation along Thailand's largest river, the Chao Phraya. The 2011 floods, however, did more than bring irrigation water; they also wrought intense physical, social, and economic damage. Thailand serves as an important manufacturing site for producers from around the world.

- Assess the relationship between global patterns of climatic and geological change over time, the modern-day map of physical landscapes in the region, and how people have adapted to this region's physical geographies.

- Describe the historical and current relationship between local economic development and global economic issues and trends.

- Assess the broader history of political tension in the region, noting the role global conflicts have played.

- Trace the historical development of social life in the region, explaining how and why differences between places emerge over time.

- Identify the major social issues in the region today and describe how they are the result of historical, geographical, and social patterns.

- Explain the patterns of demographic change and migration in the region and their effects.

Japanese investment in Thailand has been particularly strong since the 1970s. But the recent floods are challenging the staying power of Japanese corporations, who have the choice to invest in other parts of the world, far from the Bangkok region. Even as workers cobbled together makeshift dams in an effort to stem the flooding in Japanese-owned businesses, as shown in the chapter-opening photo, decision makers at those same companies were considering pulling up roots and moving their business to higher and safer ground either in other parts of Thailand or in other countries, such as China.

As this story makes clear, Thailand is subject not only to global geographic processes of monsoon rains and flooding but also to the characteristics of global capitalist production. An event such as the 2011 floods in Thailand has not only local effects; it can also have wide-ranging global consequences, as local governments and global businesses struggle to respond. How residents and institutions cope with a crisis affects local as well as global players. ∎

▼ **FIGURE 10.1**

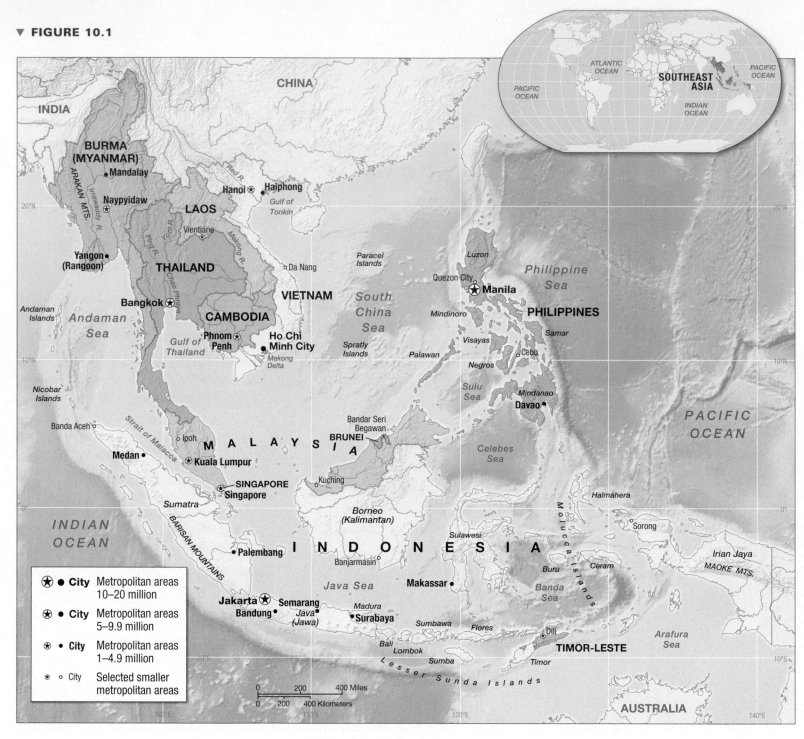

Southeast Asia Key Facts

- Major Subregions: Indochina, Malaya Archipelago, Mainland Southeast Asia, Insular Southeast Asia

- Major Physiogeographic Features: Mekong, Red, Chao Phraya, and Irrawaddy Rivers; more than 20,000 islands, including the world's third largest, Borneo; mostly monsoon climate

- Major Religions: Islam, Theravada Buddhism, Mahayana Buddhism, Catholicism, Hinduism

- Major Language Families: Austro-Asiatic, Malayo-Polynesian, Papuan, Tibeto-Burmese, Tai-Kadai with over 500+ distinct ethnic and linguistic groups

- Total Area (total sq km): 4 million

- Population (2011): 602 million; Population under Age 15 (%): 28; Population over Age 65 (%): 6

- Population Density (per square km) (2011): 134

- Urbanization (%) (2011): 42

- Average Life Expectancy at Birth (2011): Overall: 71; Women: 73; Men: 68

- Total Fertility Rate (2011): 2.3

- GNI PPP per Capita (current U.S. $) (2009): 4,490

- Population in Poverty (%, < $2/day) (2000–2009): 42

- Internet Users (2011): 156,315,808; Population with Access to the Internet (%) (2011): 31.1; Growth of Internet Use (%) (2000–2011): 92.6

- Access to Improved Drinking Water Sources (%) (2011): Urban: 92; Rural: 80

- Energy Use (kg of oil equivalent per capita) (2009): 2,030

- Ecological Footprint (hectares per capita consumed)/(hectares per capita available, global scale) (2011): 2.1/1.8

ENVIRONMENT AND SOCIETY

Geographers in Southeast Asia have long understood that the relationship between the region's physical geography and human geography is complex (**Figure 10.1**). On the one hand, this region's position in relation to global climatic patterns and large rivers and oceans has fostered rice cultivation and fishery development for over 5,000 years. On the other hand, the region's proximity to tectonic plate boundaries makes it susceptible to **tsunamis** (very large ocean waves caused by earthquakes or other tectonic disturbances under an ocean or sea), land-based earthquakes, and volcanic activity. These tectonic processes give this region its complex physical geography (**Figure 10.2**). The people of Southeast Asia have adapted to all of these environmental conditions, modifying both physical landscapes (developing agricultural practices that take advantage of annual heavy rains) and the biological and geographical world around them (creating new crops and agricultural practices and harnessing biodiversity for medicinal purposes).

Climate, Adaptation, and Global Change

Although most common images portray Southeast Asia as a uniform tropical region, the climatic map makes it clear that it has a variety of climatic subregions (**Figure 10.3**, p. 386). Southeast Asia's climatic diversity is a function of the region's relation to the wider global climatic processes tied to the intertropical convergence zone, or ITCZ (see Chapter 1, p. 9), which produces the **monsoon**—a wind system that reverses directions periodically and produces seasonal rain patterns throughout the region. In Southeast Asia, as the sun heats inland Asia from May to October, low pressure builds over the continental interior, and winds flow in from cooler high-pressure regions over the Indian and Pacific oceans. These monsoon winds are laden with moisture and produce heavy rain over land, particularly where air currents rise and cool to produce orographic precipitation on highlands and south-facing island slopes. Orographic effects—where mountains limit rainfall in interior parts of the region—means that parts of Southeast Asia receive much less rainfall than other places in the region. In parts of Thailand, Burma, and Laos, for example, rainfall may be limited to just a few months a year. In many parts of Indonesia, Malaysia, Singapore, and the Philippines, however, rain may fall every month of the year. In places where rainfall happens only part of the year, the onset of the monsoon is a momentous event. It brings some alleviation from the very high and oppressive temperatures as rains cool the air, and it brings water for crops and drinking. But, as the chapter-opening vignette illustrates, the monsoons can also cause severe floods, and the constant heavy rain can promote the growth of molds and funguses, which may have an impact on the health of people living in this region.

Today, this region is most commonly classified by its two major subregions: mainland and island (or insular). The mainland consists of Burma, Thailand, Laos, Cambodia, Vietnam, and the peninsula of Malaysia, while the island region consists of Singapore, Indonesia, Timor-Leste, and the Philippines. On mainland Southeast Asia, the November-to-March period brings cooler temperatures as air flows south from the large highlands and plateaus of East Asia. Ocean temperatures become warmer relative to land, and lower pressure prevails over the oceans. During this period, the monsoon winds reverse and blow out of the continental interiors and across the ocean. This period is marked by a drier climate on the mainland, which results in lower overall annual rainfall totals. During this same season, the islands of Southeast Asia receive a second sequence of monsoon rainfall, this time on the north-facing slopes. The Indonesian island of Sumatra, for example, receives the bulk of its rain on north-facing slopes during the December-to-February period as winds blow southward from mainland Southeast Asia across the South China Sea. Between June and August, in contrast, rain falls on south-facing slopes, when monsoon winds sweep northward from the Indian Ocean.

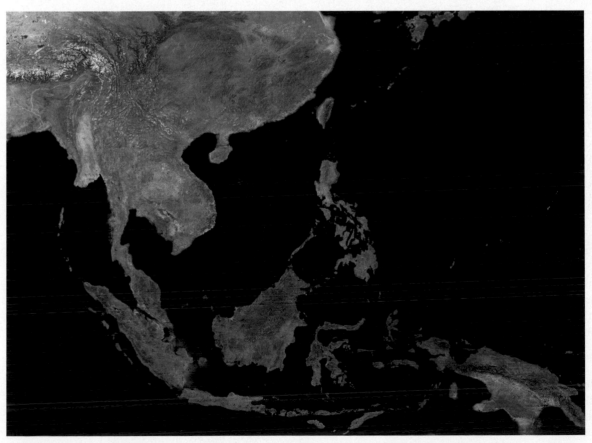

▲ **FIGURE 10.2 Southeast Asia from Space** This image shows the main islands and major peninsulas that make up Southeast Asia.

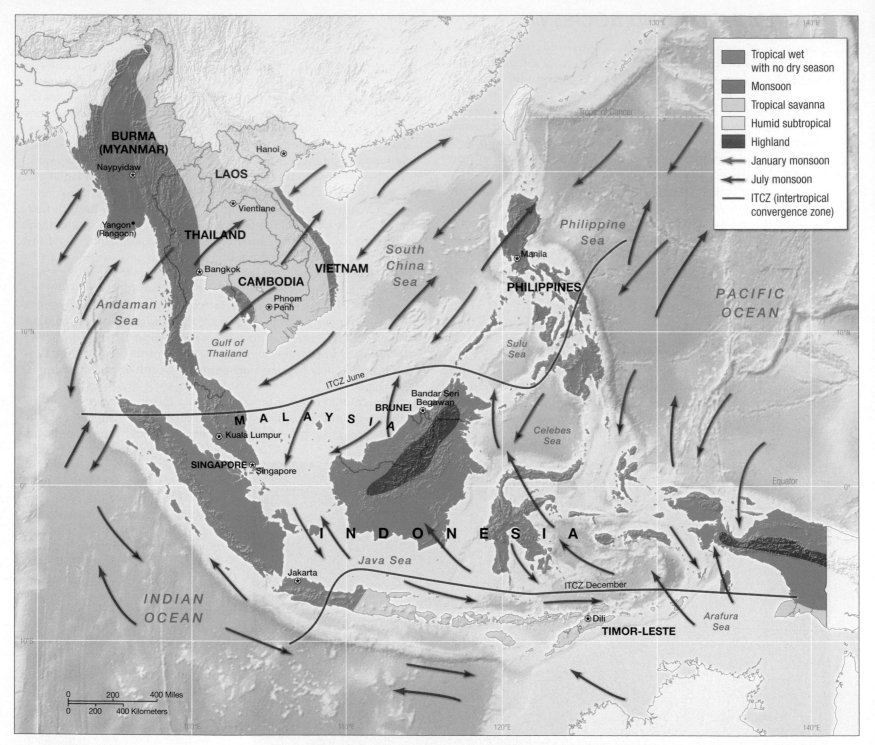

▲ **FIGURE 10.3 Climate Map of Southeast Asia** Southeast Asia has a hot, wet climate and seasonal monsoon winds that bring heavy precipitation. The mainland is drier than the islands, with more seasonal rainfall. Climates within the region can vary from year to year.

The equatorial location of the Southeast Asian islands and their relation to the ITCZ means that the islands experience daily tropical thunderstorms throughout the year. Warm ocean temperatures from August to October also combine with eddies in the trade winds to produce the large rotating storms known as **typhoons** in Pacific Asia, **hurricanes** in the Americas, and **cyclones** in the Indian Ocean. Typhoons most commonly develop east of the Philippines and move westward into the South China Sea, whereas cyclones begin in the Indian Ocean and sweep into Southeast Asia from the west (**Figure 10.4**). The combined effect of monsoons, the ITCZ, and typhoons/cyclones means that the islands of Southeast Asia are among the wettest regions in the world with lush forest vegetation and fairly consistent annual temperature.

Despite the challenges produced by the dynamic climatic forces in this region, humans have effectively adapted to these conditions for a very long time. About 5,000 years ago, the selective breeding of a grass with edible seeds produced rice, the basis of human diets in Southeast Asia. Domesticated in several parts of Asia, rice originally flourished where rainfall was evenly distributed through the year and totaled more than 120 centimeters (80 inches). Rice complements fish and vegetables in the diet and is served locally with other crops that were probably domesticated in Southeast Asia, including taro, sago, bananas, mango, and sugar. The rice plant is also used for fodder and thatch in buildings.

The region's unique position in relation to global climatic wind patterns has afforded local people and travelers the opportunity to sail throughout this region and beyond. The summer onshore winds have brought traders and migrants to Southeast Asia for thousands of years, and with them their religious beliefs—first Hinduism and Buddhism and later Islam. New philosophical systems, such as Confucianism from China (see Chapter 8, p. 304), also found their way into the region, influencing Vietnamese culture. The shift of winds in winter carried the traders back home along with spices and other products, which globalized Southeast Asia's contributions to food production and consumption. Imagine, for example, a person in Spain in the 16th century buying nutmeg in the local market, a product that came from the famed **"Spice Islands"** of Southeast Asia, which are the Molucca Islands of modern-day Indonesia. Even though that person in Spain knew little or nothing about the islands and the people who lived there, his or her tastes were changed as a result of these global connections. These connections were quite expansive; spices were traded in parts of Africa, the Middle East, and Europe and the transoceanic trade networks of the Indian Ocean.

Humans have also modified the region's physical landscape for rice production using the water provided by the monsoon rains. Agricultural adaptations include the construction of terraces, paddies, and irrigation systems (**Figure 10.5**, p. 388). Terraces cut into steep hillsides provide level surfaces that facilitate water control and reduce erosion. The construction of dikes (ridges) allows fields to be flooded, plowed, planted, and drained before harvest in a system called **paddy farming**.

Rice is one of the few major crops that can grow in standing water and is suited to the flooding that accompanies heavy monsoon rains. Irrigated rice, or *sawah*, requires labor throughout the season to prepare and maintain fields, transplant seedlings, weed, and harvest each stalk by hand. Facilitated by the gendered division of labor in the region, rice production has utilized men and women in different ways in the highly labor intensive work. In some Southeast Asian societies, women were historically responsible for the majority of the planting and care of rice cultivation as well as other domesticated crops, while men were responsible for clearing fields, felling trees, and burning plots in preparation of planting. This has changed over time, as mechanization of agricultural production and communal systems of farming (where entire villages participated in rice production together) have given way to large-scale industrialized agricultural production systems.

There is the possibility that global warming will increase the land that can come under rice cultivation in some areas and cause extended periods of drying and drought, reducing rice yields, in other places. Changes such as these have already taken place in the periods of extended drought that have occurred in parts of Northeast Thailand over the last two decades. Large-scale rice agribusiness, which now keeps rice under cultivation year-round in parts of Southeast Asia, may be affecting microclimatic patterns by producing increased moisture in the air around large fields. Increases in global temperature also affect fisheries in the region and the long-term viability of many island communities, which are subject to problems associated with increasing sea levels. Broader global dynamics, such as deforestation related to increases in agricultural production in sub-Saharan Africa and South Asia, also affect global climate patterns. These kinds of dynamics may have a long-term impact on rainfall and the monsoon patterns in the region.

◀ **FIGURE 10.4 Cyclone Strikes the Philippines** People in the Greater Manila area of the Philippines tried to escape flooding in low-lying areas caused by Typhoon Washi, which struck on December 16, 2011. The typhoon killed more than 1,000 people and left tens of thousands homeless.

▲ **FIGURE 10.5 Terrace Agriculture in Luzon, the Philippines** Terracing affords a number of advantages in hilly areas that experience heavy rainfall. The practice allows people to control the flow of water through irrigation systems between different parts of the field; it protects topsoil from eroding during periods of heavy rainfall, sustaining the nutrient levels of the field; and it allows farmers to divide the land into paddies that can be used to grow rice seedlings, transplant seedlings into rice beds, and grow other basic fruits and vegetables to accompany this staple crop.

APPLY YOUR KNOWLEDGE Consult Figure 10.3 on p. 386 and describe the pattern of rainfall across the region of Southeast Asia. Identify and explain the global processes that leave parts of Southeast Asia wet all year and other parts seasonally dry.

Geological Resources, Risks, and Water Management

Plate tectonics have shaped the physical topography of Southeast Asia and influenced the historical configuration of land and oceans, highlands and lowlands, and geological hazards and resources. The mainland of Southeast Asia and the island of Borneo occupy the Sunda Shelf of the Eurasian Plate. The Sunda Shelf is now flooded by the shallow South China Sea between Borneo and the mainland, but it was exposed during the ice ages, forming a **land bridge**—land connecting two places that were once separated by water—to parts of Asia. Humans probably reached the islands of Southeast Asia

from Eurasia about 60,000 years ago over this land bridge. These people were the ancestors of contemporary Indigenous groups in parts of Malaysia and Indonesia, as well as New Guinea, Australia, and the Pacific Islands. The Indonesian island arc from Sumatra to Timor-Leste lies along the edge of the Eurasian Plate. To the south and east, the Australian, Pacific, and Philippine Plates are moving toward, and colliding with, the Eurasian Plate and are being forced downward in deep-sea trenches along major subduction zones (see Chapter 1, p. 16). The Mariana deep-sea trench, between the Pacific and Philippine Plates in the Pacific Ocean southeast of the Mariana Islands, is the deepest in the world at 11,034 meters (36,201 feet). These collision zones are associated with mountain building and volcanic activity and have created the thousands of islands that form the Indonesian and Philippine **archipelagoes**—a series of islands that are co-located within a large body of water (**Figure 10.6**).

In general, mainland Southeast Asia is geologically older, but despite long-term weathering, it is still very mountainous. Peaks in the highlands of northern Burma reach beyond 5,000 meters (16,400 feet), and ridges link to the Himalayas. The relatively young islands of the region have high mountains, steep slopes, and

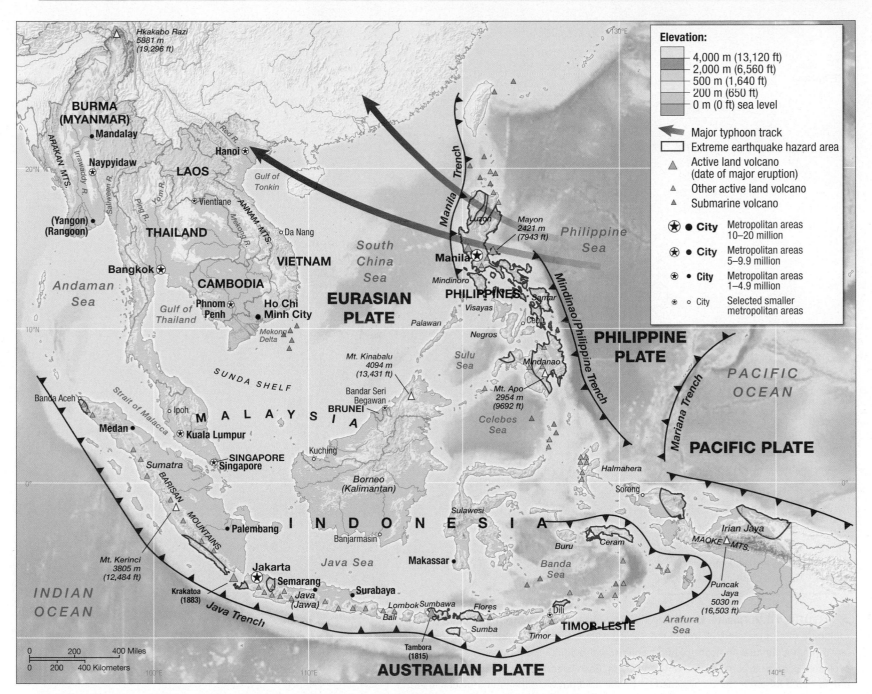

▲ FIGURE 10.6 Southeast Asia's Physiographic Regions Southeast Asia, one of the most geologically active regions of the world, experiences many earthquakes and volcanoes because it lies at the intersection of active tectonic plates. The region is very mountainous except for the major valleys of the Irrawaddy, Chao Phraya, Red (or Song Hong), and Mekong Rivers and some coastal plains.

generally narrow coastal plains. In Irian Jaya mountains can reach 5,300 meters (17,300 feet) and temperatures are so cold at these high elevations that permanent glaciers cap these mountains. More than a dozen other mountains and volcanoes in insular Southeast Asia reach 3,300 meters (10,800 feet).

Many countries of mainland Southeast Asia feature fragmented and elongated geographies that have historically created challenges to national integration, transportation, and economic development. Burma, Thailand, and Vietnam all include long narrow

segments that are less than 160 kilometers (100 miles) wide, although rivers cut through these countries, providing networks for transportation and large-scale agricultural developments. In the insular region, Indonesia has more than 13,600 islands (only half of them inhabited), and the Philippines is made up of more than 7,000 islands. As a result of these geographies, the economies and lifestyles of insular Southeast Asia have been oriented to the sea through ocean fishing and maritime trade for centuries (**Figure 10.7a**), and it has often been easier to use ocean or river

▲ **FIGURE 10.7 Coastal Landscapes of Southeast Asia** (a) This map illustrates the multitude of contradictory claims in the Spratly Islands in the South China Sea. International law regulates the use and control of ocean space, providing countries with a 200-mile exclusive economic zone around any land claim. Each island in the Spratly chain is important not so much for what can be provided for by the land but what the land provides in terms of access to the sea. (b) The many islands and long coastlines in much of Southeast Asia orient human activity toward the sea. Boats (called *junks*) in Vung Tau, Vietnam, are used for transport and fishing, for both subsistence and commercial sales.

transport than difficult overland routes. The complexity of coastlines and the many islands of some nations, together with the economic significance of ocean fishery and oil resources, have created conflicts over maritime territorial jurisdictions and boundaries. Countries within Southeast Asia and outside the region, including the People's Republic of China, lay claim to small island chains in the South China Sea and the ocean resources that surround those islands (**Figure 10.7b**).

Energy and Mineral Resources The geology and tectonics of Southeast Asia have created mineral and energy resources, such as tin and oil, but generally the region is not as resource-rich as

other regions, such as Africa or Latin America. Today, oil from Southeast Asia contributes about 5% of global production. Indonesia, Malaysia, and Brunei are the principal producers, exporting oil mainly to Japan from oil fields off the north coast of Borneo, Sumatra, and Java. With foreign assistance, Vietnam is expanding its oil production, and Burma is reinvigorating its energy sector through exploitation of gas reserves for domestic use and for sale to Thailand.

Tin mines, developed by European colonial administrations in the region during the 19th century, employed imported Chinese laborers because local people were reluctant to abandon subsistence rice production. This immigration produced an interesting

▲ **FIGURE 10.8 Palm Oil Collateral Damage** Conservationists believe that one of the biggest populations of wild orangutans on Borneo will reach extinction unless drastic measures are taken to stop the expansion of palm oil plantations.

social geography to Southeast Asia, as large Chinese communities were established throughout the region, particularly in Malaysia, Indonesia, Singapore, and Thailand. From colonial times, tin has been a major export from Malaysia, but exports have dropped in the last 20 years as deposits have become exhausted and competition has lowered prices. Thailand and Indonesia also have significant tin reserves and production. The Philippines has developed a wide range of mineral resources, including copper, nickel, silver, and gold, and Indonesia is developing new gold, silver, and nickel mines in Irian Jaya, in the face of opposition from environmentalists and Indigenous groups.

The newest energy sources to be developed in Southeast Asia are **biofuels**, especially palm oil, which can be converted into biodiesel. Indonesia and Malaysia accounted for over 80% of crude palm oil production in 2011, and Thailand is also starting to increase its production as the world's third-largest producer. In Thailand, this complements an already expansive production of sugarcane-based ethanol. Southeast Asian countries export a large portion of their palm oil production, sending much of it to India, China, and the European Union, making palm oil production not just a local issue but a global one. Concern is growing about the conversion of tropical forests to palm oil plantations across the region, particularly when expansion of this crop intrudes on space for **megafauna**—large-bodied mammals—such as orangutans (**Figure 10.8**). But as global demand for palm oil continues to increase, particularly from countries such as India and China, it is likely that palm oil will continue to be an important export crop for these countries into the future.

Risks and Benefits of Volcanic Activity The active volcanoes and earthquake zones of Southeast Asia pose great risks to the human populations of the region. The Indonesian and Philippine volcanoes are part of the Ring of Fire (see Chapter 8, p. 304), which surrounds the Pacific Ocean, linking this region with the seismically active regions of Japan, the western United States, and the Andes Mountains in South America. The tsunami of 2004 that struck off the coast of Indonesia and impacted Thailand and Burma, as well as Sri Lanka and India, represents one of the clearest examples of the risks related to tectonic activity in the region (**Figure 10.9**). The explosion of Mount Pinatubo in the Philippines in 1991 is another, as the area surrounding Manila was covered in ash.

The risks of volcanoes are balanced by the benefits they provide in terms of soil nutrients. As in other parts of the world, volcanic eruptions in Southeast Asia have deposited ash that contributes to fertile soils that can sustain high crop yields and associated population densities. In Java, the rich volcanic soils have contributed to productive agriculture and dense populations. Even without large rivers, the soil on the island of Java has sustained large populations historically.

River Systems Between the mountain ridges of Southeast Asia lie the major river valleys and deltas of the Irrawaddy, Chao Phraya, Mekong, and Red (or Song Hong) Rivers, all of which have been important to the growth of societies in this region. The Mekong River is the longest waterway in the region, flowing 4,000 kilometers (2,500 miles) from the highlands of Tibet to its delta in Vietnam and Cambodia. The Mekong forms the main transportation and settlement corridor for Laos and Cambodia, while also serving as a key trade network between these two countries and China, Burma, and Thailand (see "Emerging Regions: The Greater Mekong," on p. 394.). The river also provides water for irrigation and hydroelectric development and is a productive fishery. One of the most unusual physical features in the Mekong Basin is Tonle Sap in Cambodia.

▲ **FIGURE 10.9 Tsunami Damage in Banda Aceh, Indonesia** The city of Banda Aceh was the hardest hit by the tsunami that struck Indonesia in 2004. An estimated 387,000 people were left homeless in the city and surrounding areas.

This large lake acts as a safety-valve overflow basin for flooding. During the dry season, water flows out of the lake along the Sab River into the Mekong; during flooding on the Mekong, water flows back up into the lake, sometimes more than doubling the lake's area (**Figure 10.10**).

Soil fertility is very high in the river valleys and deltas that receive regular replenishment of river sediment and nutrients during annual floods. Because of these important sediment flows, major river systems have been developed over centuries to support large-scale urban life, and empires have flourished as a result in Burma, Thailand, Cambodia, Laos, and Vietnam. It is very possible that the domestication of rice might have first taken place in North Vietnam, in the Red River valley, while rivers in first Cambodia and later Thailand, Burma, and Laos fed the large-scale mainland empires of Angkor Wat, Cambodia; Sukhothai, Thailand; and Pagan, Burma. Regions farther from rivers in Southeast Asia experience soil limitations typical of tropical climates around the world; in these areas, heavy rain and warm temperatures wash nutrients through the soil, and organic material is broken down and recycled rapidly into forest vegetation.

Today, these complex river systems serve a vital role in the food production and transportation systems of the mainland Southeast Asian countries. Thailand, for example, still produces enough rice for domestic and export markets, while the fisheries of the larger and smaller river systems provide much-needed protein to millions of people. River systems in the region are also contested sites of local and regional politics, as people struggle to control precious water-related resources. Dam projects along river systems, such as the Mekong, have met with local resistance for extended periods of time.

Ecology, Land, and Environmental Management

The vegetation and ecosystems of Southeast Asia reflect its wet tropical climate. A natural land cover of dense forests historically dominated the region. Today, the two major types are evergreen tropical forests in the wetter areas and tropical deciduous, or monsoon, forests where rainfall is more seasonal or less intense and trees lose their leaves in the dry season (**Figure 10.11**). In drier regions, forests are less diverse, vegetation cover changes to savanna and grasslands, and rainfall is generally less or falls at higher elevations. **Mangrove forests**—groups of evergreen trees that form dense, tangled thickets in marshes and along muddy tidal shores—are found along regional coastlines, together with a rich offshore marine ecosystem that includes coral reefs and fisheries. Indonesia leads the world in mangrove area, with more than 4 million hectares (10 million acres). Mangrove forests, with their deep, strong root structures, serve a vital function, protecting areas from erosion while providing habitat for myriad animal species. Indonesia is also ranked second in the world in terms of its biodiversity: the country is home to at least 10% of the world's plant species, bird species, and mammal species.

Climate change and plate tectonics together have produced a fascinating division in the ecology of the Southeast Asia region. As discussed earlier in the chapter, the last ice age left the Sunda Shelf exposed between the mainland and Indonesia until about 16,000 years ago. Many megafauna, including tigers, elephants, and orangutans, migrated across this land bridge to the islands of Indonesia. To the south, animals such as kangaroos and opossums moved north across the land bridge formed by the Australian Plate to New Guinea and other islands. However, between Bali and Lombok is a deep ocean trench that remained ocean, even during ice ages, and prevented these two very different types of species communities from mixing. This created an ecological division called **Wallace's Line**, named after naturalist Alfred Wallace, who traveled extensively in the region in the 19th century.

In addition to the naturally occurring vegetation of this region, ecological systems have been introduced via European colonialism. Rubber production grew rapidly after 1876, when plants grown in Britain from seeds smuggled out of Brazil were introduced to Malaysia. As a result, rubber plantations grew from 800 hectares (2,000 acres) in 1898 to 850,000 hectares (2.1 million acres) in

◀ **FIGURE 10.10 Tonle Sap Lake, Cambodia** The Tonle Sap Lake is a very important part of the region's ecosystem and economy. In the dry season, the lake shrinks to under 3,000 square kilometers, but it grows to over 16,000 square kilometers in the rainy season. The lake provides fresh water for farming and an abundance of fish.

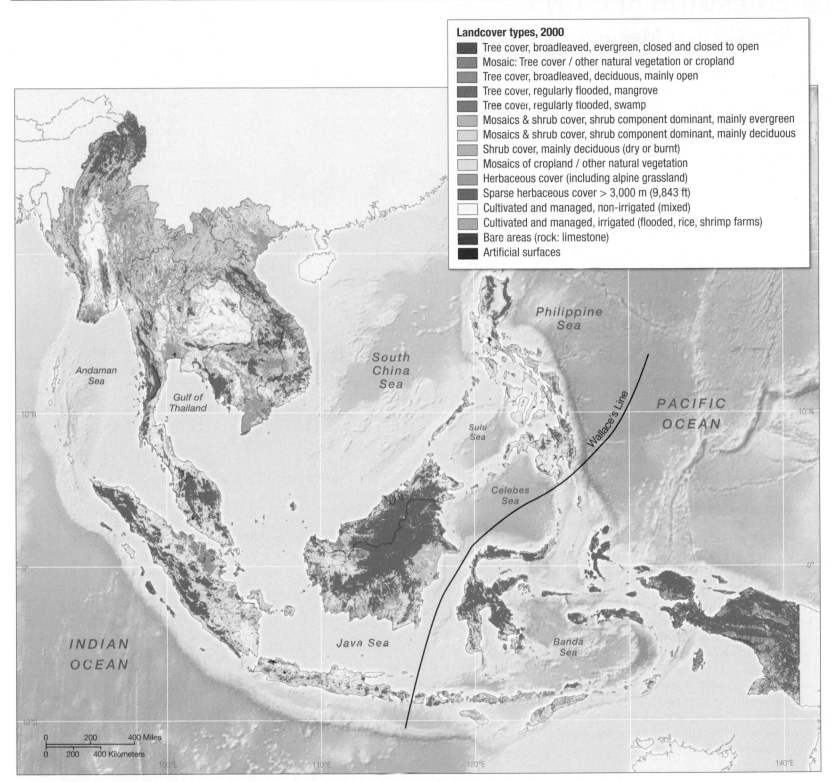

Landcover types, 2000
- Tree cover, broadleaved, evergreen, closed and closed to open
- Mosaic: Tree cover / other natural vegetation or cropland
- Tree cover, broadleaved, deciduous, mainly open
- Tree cover, regularly flooded, mangrove
- Tree cover, regularly flooded, swamp
- Mosaics & shrub cover, shrub component dominant, mainly evergreen
- Mosaics & shrub cover, shrub component dominant, mainly deciduous
- Shrub cover, mainly deciduous (dry or burnt)
- Mosaics of cropland / other natural vegetation
- Herbaceous cover (including alpine grassland)
- Sparse herbaceous cover > 3,000 m (9,843 ft)
- Cultivated and managed, non-irrigated (mixed)
- Cultivated and managed, irrigated (flooded, rice, shrimp farms)
- Bare areas (rock: limestone)
- Artificial surfaces

▲ **FIGURE 10.11 Southeast Asian Vegetation and Ecology** The land cover of most of Southeast Asia is forest—evergreen in the wetter areas and tropical deciduous in drier regions. This map shows the actual land cover in 2000 and the extensive conversion of forests to agriculture and other land uses. Wallace's Line indicates the location of a deep-ocean trench. The trench divided two sections of a great land bridge that rose above Southeast Asian oceans and seas 16,000 years ago.

1920, as the explosion of automobiles in North America and Europe increased demand for rubber tires. Even though there was a ban on production by local people to protect the profits and monopoly of European planters, rubber quickly became popular with local Malayan farmers because of the high price it brought and the limited labor its production required.

Deforestation, in part a result of colonial forest management practices and modern-day economic development practices, is

EMERGING REGIONS
The Greater Mekong

The Mekong River is one of the longest rivers in the world; its headwaters are in the Tibetan Plateau in China, and its delta is in southern Vietnam. In Southeast Asia, the Mekong meanders through Burma, Laos, Thailand, and Cambodia (**Figure 10.1.1**). A truly international river, it is used to move goods and peoples back and forth between China and Southeast Asia. This large river also marks the boundaries between a number of countries in the area. At its most famous point, one can view three countries at once from the shores of the Mekong: Burma, Thailand, and Laos. Known as the **Golden Triangle,** this meeting point has historically been a critical sector for the trade of opium; today it marks an important tourist site for people traveling through the region.

On any given morning in and around the Golden Triangle area, a visitor can see Chinese boats parked along the shores of the towns along the Mekong in Thailand, Burma, or Laos. One town in Thailand, Chiang Saen, just south of the Golden Triangle, is an important port for Chinese goods making their way into mainland Southeast Asia. The boats, which are very simply designed, serve as homes to sailors who move goods from China into Southeast Asia and then pick up goods to return to China. As the boats park at a shoreline, which is not heavily fortified, they are often greeted by a Thai customs officer, who examines the manifest and signs off on the goods to be brought into the country. It is a very relaxed international point of exchange, as merchants serve food within feet of the Chinese vessels parked along the shore. The fresh fruits and vegetables that come in from Yunnan Province in China are welcomed commodities in the local Thai markets. The peoples of mainland Southeast Asia have much in common historically with the peoples of Yunnan, as many of the ethnic groups found in Thailand and Laos today once lived in Yunnan Province.

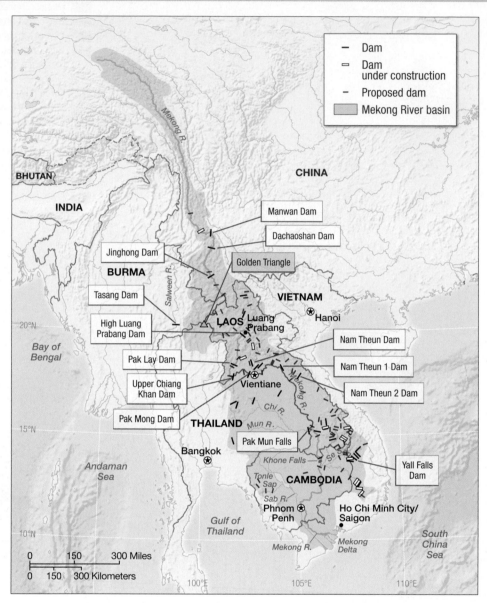

▲ **FIGURE 10.1.1 The Mekong River** The Mekong River links China in East Asia to mainland Southeast Asia, as it flows from a source on the Tibetan Plateau to the delta in Vietnam and Cambodia.

the most significant environmental problem in Southeast Asia today (**Figure 10.12a**, p. 396). Forests once dominated the region, providing habitat for a diverse ecology and food, medicine, fuel, fiber, and construction materials to local peoples. Traditional **swidden farming**—the practice of clearing small patches of forest at a time for agriculture—had minimal impact on the overall forest cover of the region. Widespread clearance began in the late 1800s with the expansion of rice production and export of tropical hardwoods under European colonial control. Most of the deforestation during this period was in lowland regions, such as the Irrawaddy Delta. After World War II, timber extraction expanded in the highlands, especially teak cutting in Thailand and Burma for export to the furniture industry. More recently, the growth of oil palm plantations, the pulp and paper industry, and cutting trees for plywood and veneers has

Even as boats move goods back and forth across the Mekong and through the two world regions of East Asia and Southeast Asia, there remains tension about the future of the river and its development. The **Mekong River Commission (MRC)**, formally established in 1995, coordinates the planning of flood control and dam projects and studies environmental issues within the basin. The 1995 accord provides mechanisms for resolving disputes within the river basin between Cambodia, Laos, Thailand, and Vietnam. In 1996, China and Burma joined as "dialogue partners" of the MRC. Several large dams have been constructed on the Mekong tributaries—among them the Manwan on the headwaters in China and the Nam Ngum in Laos, which generates several million dollars of hydroelectric sales to Thailand—and many others are proposed (**Figure 10.1.2**). These dams have not been erected without controversy, as local communities struggle to maintain control over local resources in the wake of post-dam-construction flooding and government appropriation of land resources for dam development. Often local interests are in conflict with global organizations that seek to profit from the dam ventures, as Vietnamese, Chinese, and Russian firms bid on building large hydroelectric projects along this very important waterway.

China is still not a formal partner of the MRC, but it controls the headwaters of the Mekong River and has its own dams along it. In 2010, the issue of China's damming projects came to a head as downstream droughts consumed northeast Thailand, Laos, Cambodia, and Vietnam. The 2010 disputes suggest that this emerging region will play a significant role for the 60 million people in Southeast Asia who are making their living from the Mekong River as well as the millions of Chinese people who may benefit from the hydroelectric power of the river in Yunnan Province, China. ■

▲ **FIGURE 10.1.2 Chinese Trading Boats in Chiang Saen, Thailand** Every day, boats travel up and down the Mekong River moving goods between China and Southeast Asia. The simple boat structure and docking area hides the reality that this place is an important site of global trade and connection.

placed even greater pressure on forests. The timber industry provides thousands of jobs and a significant amount of foreign capital in Southeast Asia, making it difficult for governments to eliminate.

Swidden clearing has also intensified in areas where population has increased, land is limited, and chain saws are available. The causes of extended swidden practices are tied as much to politics and the economy as they are to local practice. The extensive shifting of cultivation involved in swidden farming was sustainable when widely scattered villages controlled large areas of land and could use fields for a few years and then leave them for 10 to 30 years to recover. As access to land decreases and population and consumption increase, land is cleared more frequently, causing long-term declines in environmental productivity.

The impact of deforestation includes loss of species habitat, flooding and soil erosion, and smoke and pollution from forest burning. Local communities have also been marginalized from lands they once occupied, as logging concessions are given over to larger corporate or state entities. Deforestation has destroyed or is threatening the habitat of many species and Indigenous human communities in Southeast Asia. At the same time, the deep forests of mainland Southeast Asia are still relatively unexplored by biologists. New species are being discovered, including relatively large animals such as the Vu Quang ox and giant muntjac deer, identified as recently as the 1990s. But these isolated regions are coming under further attack, as poachers move into once-isolated forest lands to collect prized megafauna, such as tigers (**Figure 10.12b**).

Governments have responded by setting aside forest reserves and banning logging in many regions as well as insisting on local processing rather than export of raw logs. Local people and international environmentalists have responded to deforestation in Southeast Asia by forming social movements and alliances to protect forests. In Indonesia, protesters have called on the government to stop large-scale logging and mining projects that have rapidly depleted the country's forest resources (**Figure 10.12c**). The damage from forest clearing includes loss of biodiversity, erosion, and flooding, and loss of forest benefits to Indigenous populations. Deforestation and the fate of Indigenous populations on Borneo have garnered the attention

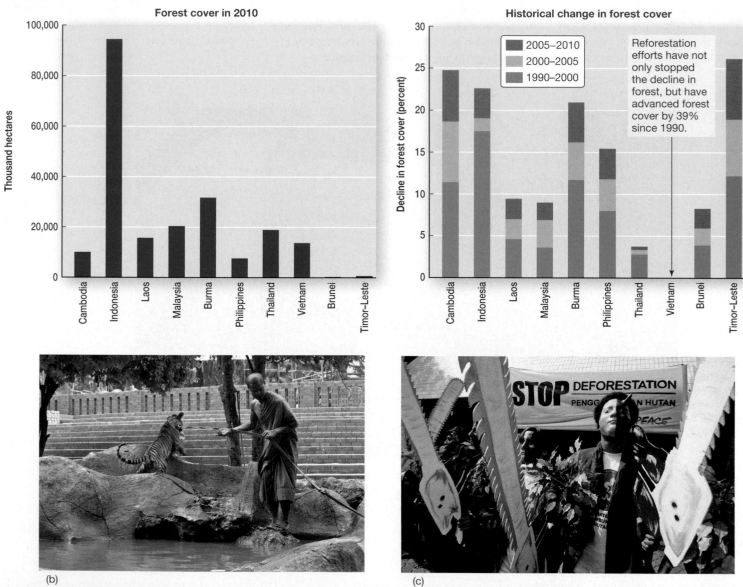

(a) **Deforestation in Southeast Asia**

▲ **FIGURE 10.12 Deforestation in Southeast Asia** (a) These graphs illustrate Forest cover in 2005 (in thousand hectares) and deforestation rates (for 1990–2000, 2000–2005 and 2005–2010 in percentages) for selected countries and periods of years in Southeast Asia. (b) A monk plays with a tiger cub at a monastery in Kanchanaburi Province, Thailand. Monks have been rescuing and caring for tigers who have been injured by poachers or dislocated from their homes because of deforestation. (c) Activists in Jakarta, Indonesia, wear masks and hold mock chainsaws as they call on the Indonesian government to stop deforestation in their country.

of international environmental and human rights groups. By the late 1990s, Borneo was providing more than half the world's tropical hardwoods, mainly through concessions given to multinational timber corporations, such as Weyerhaeuser, Georgia-Pacific, and Mitsubishi. Although bans on export of raw logs were put in place in 2002, the corresponding increase in wood processing on Borneo, including more than 50 plywood mills, keeps the pressure on the forests.

> **APPLY YOUR KNOWLEDGE** List and describe three impacts that deforestation has on local Indigenous human and animal populations in Southeast Asia. Conduct an Internet search to find out what steps are being taken at the regional and global level to address these impacts.

HISTORY, ECONOMY, AND TERRITORY

The historical geography of Southeast Asia is distinguished by its connections to other world regions, from East Asia (Chapter 8) to South Asia (Chapter 9) to Europe (Chapter 2) and the Americas (Chapters 6 and 7). Because Southeast Asia has long been connected through trade networks—both land-based and maritime—to other parts of the world, the economies of this region are related to broader global economic processes, including the ebbs and flows of trade in and out of India and China and the changing nature of global financial markets. Southeast Asian economies have entered into a more direct relationship with the financial markets of Japan, the United States, and Europe. Southeast Asia's diversity is a result of its relative location, and its practices have adapted to the broader global movements of peoples, ideas, and biological species.

Historical Legacies and Landscapes

Maritime trade brought merchants from the Middle East, China, and India to Southeast Asia about 2,000 years ago, together with new religions, crops, and technologies. Southeast Asian **entrepôt** (commercial trading) towns, such as Oceo near the modern-day Cambodia–Vietnam border, became points of connection between this region and India and China about 1,800 years ago. For most of Southeast Asia, early interaction was with India, whose Hindu and Buddhist religions and related forms of government were attractive to local chiefs who saw advantage in the divine privilege they could be granted as god-kings (*deva raja*). Over time, a number of urban-based kingdoms emerged in the region. Two of the most powerful were the mainland **Khmer** (or Cambodian) Empire and the island **Srivijaya** culture of Sumatra. The Khmer kingdom emerged in the 9th century near Tonle Sap Lake and built the magnificent 12th-century Hindu (and to a lesser extent Buddhist) temple compound of **Angkor Wat**, one of the largest religious structures ever built and a focus of tourism in contemporary Cambodia (see "Visualizing Geography: Angkor Wat," p. 398). Srivijaya ruled Sumatra and the southern Malay Peninsula from the 7th to the 12th centuries by controlling the region's long-distance maritime trade through the Strait of Malacca. This powerful state, which was based in Buddhist religious traditions, built numerous Buddhist monuments, such as Borobudor (**Figure 10.13**).

On the mainland, an increasingly powerful Thai kingdom, centered on the city of Ayutthaya on the Chao Phraya River, conquered the Khmer Empire in the 14th century and maintained control of parts of the Malaysian Peninsula for four centuries. The Thai competed with different empires—including the Burmese and to a lesser extent the kingdoms found in northern Thailand, Laos, and Cambodia—from the 14th to the 19th centuries for control of the people and resources of the mainland. These empires privileged **Theravada Buddhism**—a form of Buddhism that focuses on the importance of the monastic order and adherence to rituals written in Pali, a language that originated in South Asia. As Theravada Buddhism spread through the mainland, often blending with local **animism**—a belief system that focuses on the souls and spirits of nonhuman objects in the natural world—these empires consolidated their holdings around a number of very important city–states, such as Bangkok, Thailand, and Pagan, Burma.

By the 11th and 12th centuries, new religious systems, including Islam, also entered insular Southeast Asia through global trade from South Asia. By the 15th century, the political geography of insular

◀ **FIGURE 10.13 Ceremonial and Religious Space in Southeast Asia** Borobudor Temple, Indonesia, is one of the best examples of Buddhist architecture in the world. This site dates from the 8th and 9th centuries and represents the life of the Buddha as well as other important Buddhist ideals.

VISUALIZING GEOGRAPHY
Angkor Wat

▲ **FIGURE 10.2.1 Angkor Wat, Cambodia** By the 10th century, there were large kingdoms influenced by Indian culture and religion throughout mainland Southeast Asia. One of these different kingdoms is exemplified by the 12th-century temple of Angkor Wat and sacred city of Angkor Thom, which was constructed by the Khmer Empire just north of Tonle Sap Lake in Cambodia. The temple design features symbols from the Hindu cosmos, including mountains, artificial lakes, and sculptures.

Angkor Wat, a massive temple complex that served as a center point for the Khmer kingdom in Cambodia between the 9th and 15th centuries C.E., is one of the most significant historical sites in the world today (**Figure 10.2.1**). The greater Angkor Empire controlled not only Cambodia at its height, but also many parts of present-day Burma, Thailand, Laos, and southern Vietnam (**Figure 10.2.2**). One can find the influence of Cambodian architecture in many of these places, where Khmer-style **stupas**—mound-shaped monuments that hold sacred Buddhist relics—can still be found today. At the center of the Khmer Empire for many centuries was the larger Angkor Wat complex, which included hundreds of sites and buildings dedicated to the Hindu religious systems and the notion that Khmer kings were "god-kings" or *deva raja*, a concept borrowed from South Asian religious practice.

Today, Angkor Wat is a **United Nations Educational, Scientific, and Cultural Organization (UNESCO)** heritage site. Massive

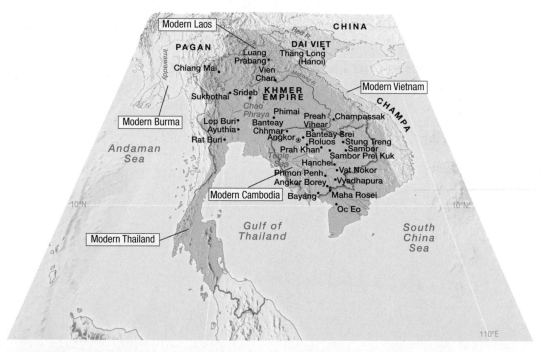

▲ **FIGURE 10.2.2 Map of Khmer Empire** At its height in the 12th century C.E., the Khmer Empire had either direct control or strong linkages to people in what is modern-day Burma, Thailand, Laos, and Vietnam, making it one of the largest empires in the region's history.

investment has been made in the complex's revitalization. Viewing this complex from space, one can see the city's expanse. With its moats, temples, and buildings dedicated to the Angkor kings, this city served as the center point of the kingdom for centuries. An extremely large reservoir and extensive canal system collected water to be used in agriculture during the dry seasons. The ability to control water clearly facilitated the kingdom's success; the Khmer leaders were able to extract loyalty and labor from its citizens in exchange for consistent harvests and access to needed resources for the largely agricultural economy (**Figure 10.2.3**). Like many complexes in other parts of Southeast Asia, the Angkor Wat complex was much more than just temples. It was a vibrant and robust complex urban society that, at its height, influenced peoples across the region. It may have been one of the largest cities in the world at its zenith, outstripping the much smaller cities that could be found in Europe or the Americas during the same period. ∎

▲ **FIGURE 10.2.3 Angkor Wat Complex from Space** Remotely sensed data produce amazing graphics of the historical geography of the Angkor Wat region. This image shows the expanse of this empire, the size of its reservoir, and the extensive canal system found in the region.

Southeast Asia was restructured around a series of Islamic-based **sultanates**—Muslim states ruled by supreme leaders or sultans—such as Malacca (in present Malaysia) and Brunei (on the island of Borneo). Malacca's strategic location controlling the trading route between India and China gave it great commercial power as a sea-based trading state, and after the rulers enthusiastically adopted Islam, Malacca disseminated Islamic beliefs and institutions throughout different parts of Southeast Asia.

European Colonialism in Southeast Asia Southeast Asia has had two overlapping but somewhat distinct colonial periods. The mercantile (or trade) period spanned from about 1500 to 1800. The industrial (or export-oriented) period lasted from about 1800 to 1945. A number of powers have sought to control the valuable markets in the region, including Portugal, Spain, the Netherlands, Britain, and France. The United States also played an important colonial role in the region in the 20th century (**Figure 10.14**).

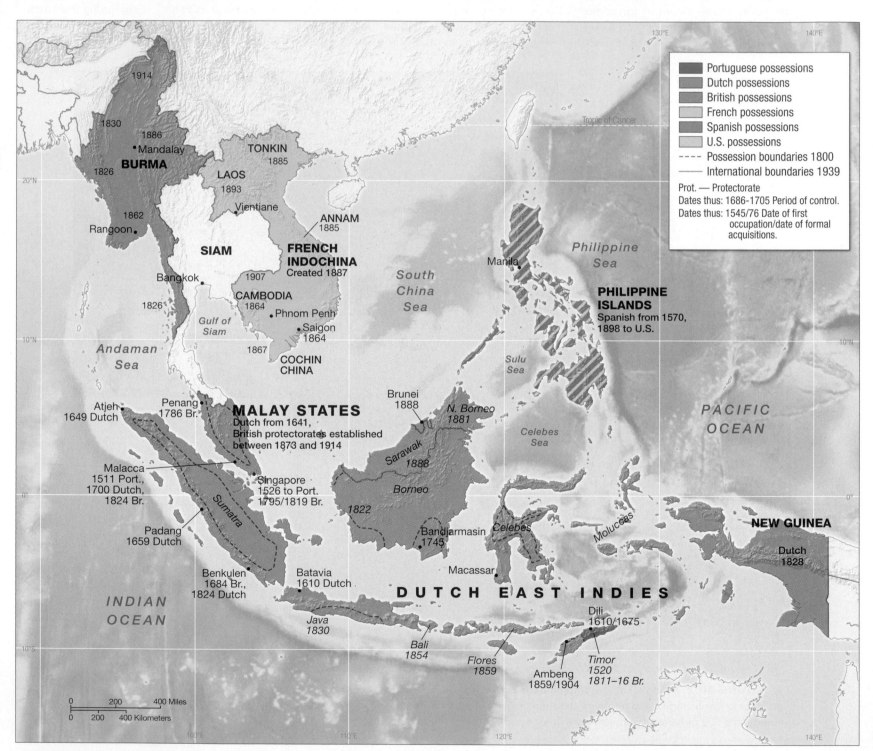

▲ **FIGURE 10.14 European Expansion into Southeast Asia** This map shows the areas controlled by the major colonial powers in Southeast Asia. The dates on the map represent the time at which a particular city or region within a country came under direct colonial control.

Portugal dominated early mercantile trade in the region. The Portuguese sailed around Africa, established a headquarters in Goa, India, and then moved to control the strategic port of Malacca in 1511 and Timor-Leste in 1515. To obtain commodities such as cloves, nutmeg, and pepper from across the Indonesian archipelago, they relied on Indigenous and other Asian producers and merchants. Meanwhile, the Spanish sailed across the Pacific, arriving in the Philippines and claiming several sizable islands and hundreds of smaller ones for themselves. The Spanish spread Catholicism and expanded their global empire by gaining access to trade in spices and other commodities in Southeast and East Asia. The city of Manila, founded in 1571, became the center of trade with Latin America. Galleons departed regularly for Acapulco, Mexico. The third colonial power to move into Southeast Asia was the Netherlands. The Dutch dominated trade from about 1600 to 1750; their initial focus was on the Molucca Islands of Indonesia—the famous Spice Islands. The formation of the Dutch East India Company in 1602, with its headquarters in Java, consolidated commercial interests to control trade in Indonesia (then called the East Indies) by restricting the production of valuable spices such as pepper and by destroying communities that ignored restrictions or participated in smuggling.

The British colonial effort in Southeast Asia was an extension of their activities in India. They focused on Burma, Malaya, and Borneo and on the control of strategic ports. Like the Dutch, the British were also led by a trading company, the British East India Company, formed in 1600 and committed to British domination of South Asia (see Chapter 9, p. 344). The British waited until the 19th century to make a move from India into Southeast Asia, beginning with control of the strategic ports of Penang (1786), Singapore (1819), and Malacca (1824) (**Figure 10.15**). They fought two wars with the Burmese and used a protectorate system to gain control of the Malay Peninsula. The British reoriented colonial economies to exports of tin, rubber, and tropical hardwoods and controlled trade routes between India and China.

French influence originated in the 18th century, and the French consolidated their holdings in the 19th century in response to rivalry with Britain over commercial links to China. The French governed Cambodia, Laos, and the districts of Tonkin, Annam, and Cochin China in Vietnam as the Union of French Indochina. They created a new writing system for Vietnamese, which is still used today, and engaged in a model of colonial rule that instituted French language training and French cultural practices for local elites.

The United States was a colonial power in Southeast Asia for less than 50 years, acquiring the Philippines in 1898 after victory in the Spanish–American War. Both Spain and the United States wanted to control the Philippines because of its location as a gateway to trade with Asia. Even after the United States granted independence to the Philippines in 1946, it continued to treat the islands as a strategic location, maintaining several military installations, such as Clark Air Force Base and Subic Bay Naval Base. Both bases were closed after the eruption of the volcano Mount Pinatubo in 1991.

Unlike other parts of Southeast Asia, Thailand (called Siam until 1939) was able to maintain its political independence throughout the colonial period. Thailand provided a buffer between British and French interests; Siam lost territory to the British in Malaya and Burma and to the French in Cambodia and Laos, but maintained an independent kingdom with Britain's tacit support. The ability of the Thai to adapt themselves to the changing geopolitical dynamics of the region also benefited their autonomy during this period. Although it remained independent, however, Siam was linked into colonial trading systems and was vulnerable to the policies of the European powers in the region.

During the 19th century, Southeast Asia was fully integrated into a European-centered global trading system. The colonial economies also altered land use practices in the region. The Dutch introduced the **Culture System** into Java in 1830, for example. This system required farmers to devote one-fifth of their land and their labor to export-crop production, especially coffee and sugar, with the profits going to the Dutch government. The British and French promoted rice production to feed laborers and growing populations. Rice farming expansion was especially dramatic in the Irrawaddy Delta of Burma and the Mekong Delta of Indochina. These colonial systems also created spatial inequalities. Key ports and trading cities such as Singapore, Batavia (present-day Jakarta), and Manila developed as regional cores and gateways to the world. Export agricultural regions were oriented to these cities, while remote rural peripheries remained in subsistence livelihoods with little investment in education or other services. Most of Southeast Asia remained under colonial control during the first part of the 20th century.

> **APPLY YOUR KNOWLEDGE** Compare and contrast the different colonial experiences of two nations in Southeast Asia. Using your understanding of historical geography, list the legacies the colonial experience has left in these two places.

▲ **FIGURE 10.15 Raffles Hotel in Singapore** The Raffles Hotel is named after Sir Thomas Stamford Raffles, who secured control of Singapore for the British in 1819. This building stands as both a symbol of British colonialism and European architectural influence in the region.

Postcolonial Southeast Asia European colonial power in Southeast Asia was diminished by the Japanese invasion of the region in World War II, which challenged the prevalent image of Western

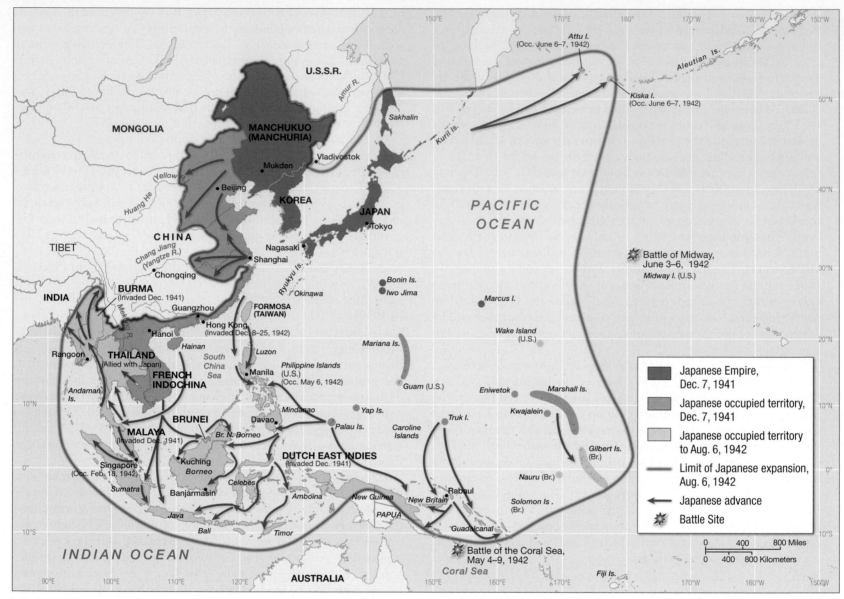

▲ **FIGURE 10.16 Japan's Occupation of Southeast Asia** The Japanese occupation of Southeast Asia during World War II had severe effects. Japan sought access to the natural resources of the region, claiming legitimacy from a shared set of cultural values with the regions that they captured under the slogan "Asia for the Asiatics." But the Japanese occupation cut off trade revenues, used forced labor, and diverted resources to Japan at the expense of local economies and food security. The destruction of bridges and roads during the conflict left Southeast Asia's infrastructure in ruins. In Vietnam, the Japanese requisitioned rice and forced farmers to grow jute fiber, resulting in a famine that killed more than 2 million people in 1944 and 1945. The Japanese were particularly harsh on the Chinese population of Southeast Asia.

racial superiority in Asia (**Figure 10.16**). The Japanese granted independence to the people of Burma in 1943 and promised independence to Indonesia on their retreat, fueling local desires for autonomy from their former European colonizers. Hastened by postwar global calls for decolonization and calls for the end of Japanese influence in the region, independence came to most of the region almost immediately following World War II (except for Indochina where extensive investments made the French reluctant to hand over power). The Philippines was granted independence from the United States directly after the war, in 1946. The British formally granted independence to Burma in 1948 and created the Federation of Malaysia in 1963.

Singapore left the federation to become an independent country in 1965. Brunei converted from a British protectorate to an independent nation in 1983. Indonesia was granted independence by the Dutch in 1949 only after a violent struggle following a declaration of independence in 1945. Western New Guinea, formerly Dutch New Guinea, became part of Indonesia as Irian Jaya in 1963.

In the postcolonial period, newly independent countries faced great challenges in forging a sense of national unity within a global framework of bounded nation–states. In Indonesia, the government promoted the concept of **pancasila**—unity in diversity through belief in one God, nationalism, humanitarianism, democracy, and social justice—as the national ideology. Pancasila and a singular Indonesian language were leveraged to try to unify the country across myriad islands and cultures. Fractures soon developed around religious differences and in response to the Indonesian government's repression of its critics and the political opposition. In 1965, a communist coup against Indonesian leader Sukarno was unsuccessful. At least 500,000 people associated with the communist movement were massacred in the aftermath. Similarly, the Philippines has struggled to develop a unified national identity, as religious and ethnic tensions remain, particularly on the island of Mindanao, where a majority Muslim population struggles against the centralized Catholic-dominated Philippine state. Equally important is the struggle in southern Thailand, where a minority Muslim population fights, sometimes violently, for religious freedom and political autonomy. In Burma, northern-based ethnic groups continue to resist the Burmese-dominated dictatorship in the central region of that country. Nationalist struggles intersected with the Cold War, which had its most global and long-lasting impacts in Vietnam, Laos, and Cambodia. Independence movements have also been economic struggles for the development of locally autonomous, planned, socialist economies (see the section "Territory and Politics," later in this chapter). All these struggles demonstrate how the global process of nation building, which has always depended on a rigidly defined set of borders, remains problematic in places where colonial borders and global geopolitical conflicts cut through ambiguous cultural spaces.

▲ **FIGURE 10.17 Coffee Plantation in Indonesia** Coffee was first planted in this area in the early 1700s by the Dutch, who began exporting coffee from Java to Europe in 1711. Today, coffee remains an important export crop for Indonesia.

Economy, Accumulation, and the Production of Inequality

The wars and instability in the countries of Southeast Asia after World War II limited economic development and trade with the world for more than 30 years. In some cases, such as in Laos, a noncapitalist economy intentionally interfered with global market forces by focusing attention on social equality rather than economic development. Other countries in Southeast Asia did reorient their capitalist economies, first to import substitution and later to export manufacturing in efforts to take advantage of the postwar global economy. In the mid- to late 20th century, government-led economic development policies in Malaysia, Singapore, and Thailand brought Southeast Asia into a new relationship with the global capitalist system.

Postcolonial Agricultural Development Although agriculture has decreased in economic significance across most of the region in the last 30 years, it is still a major employer and is essential to food security. Agriculture employs more than 40% of the economically active population in Indonesia, Thailand, and Vietnam and almost 70% in Cambodia and Laos. Rice production continues to dominate the land area of Southeast Asia, but the export of spices and plantation crops, such as rubber, tea, coffee, and sugar, are still important as well (**Figure 10.17**). Because of regional and international demand, the area planted with oil palm and pineapples has increased over the past 20 years, especially in Malaysia, the Philippines, and Thailand. Multinational corporations such as Del Monte and Dole are heavily involved in producing pineapples and other fruits in Thailand and the Philippines. Southeast Asian agriculture has also been transformed by the Green Revolution in the last 50 years.

The **Green Revolution** was kicked off by genetic modification research by U.S. food scientists in Mexico in the 1940s. This revolution involved the development of higher yielding seeds, especially wheat, rice, and corn, which when introduced in various countries in the developing world (in combination with irrigation, fertilizers, pesticides, and farm machinery) were able to increase crop production. Green Revolution technologies contributed to dramatic increases in rice and other cereal production in Indonesia, the Philippines, and Thailand. Indonesia and the Philippines almost tripled their production of rice per hectare between 1961 and 2005. The benefits of the Green Revolution in Southeast Asia and other world regions were unequally distributed. Some traditional communal land and rights were abandoned as agriculture shifted its orientation to exports and wage labor. The introduction of mechanical rice harvesting, which replaced the role of human labor in the picking and processing of rice, increased rural unemployment and landlessness in countries such as Malaysia. The rate of success was also higher among those with access to irrigation. When the new seed varieties were planted uniformly across large areas, crops became vulnerable to diseases and pests, making the use of toxic pesticides a logical next step. Rather than encourage pesticide use, the Indonesian government has sponsored the use of integrated pest management by small-scale farmers, which uses less chemically intense techniques.

Postcolonial Manufacturing and Economic Development From the 1950s through the 1970s, several Southeast Asian governments—Malaysia, Indonesia, Philippines, Thailand—developed their own local manufacturing industries in an attempt to diversify their economies away from agriculture and limit their reliance on imported manufactured goods. These countries focused on developing goods for their own markets instead of importing goods from outside their countries. In Malaysia, **tariffs**—taxes placed on goods from outside the region—on imports such as clothing and plastics were increased dramatically to protect locally produced goods from global competition. Local manufacturing was dominated by non-Malay (especially

▲ **FIGURE 10.18 Silk Textile Factory in Vietnam** Silk production has a long history in Vietnam. Today, silk production has been industrialized and depends on a relatively inexpensive female labor force, giving the product a comparative advantage in the world silk market.

Chinese) investment and produced low-value goods that quickly saturated domestic markets or faced, in the case of steel, a glutted global market. As a result, the development of local manufacturing industries did not benefit much of the population.

In the late 1970s and early 1980s, Malaysia, Singapore, and Thailand all shifted their focus to developing products that could be exported to other parts of the world in an effort to profit from their competitive advantages in the global economy. These advantages, which still distinguish much of this region today, include a relatively well-educated workforce that has been willing to work for lower wages. Countries that were at one time fairly isolated from the global economy—such as Vietnam—have now also taken advantage of similar competitive advantages to increase their share of the global manufacturing investments (**Figure 10.18**). The move toward export-oriented development—where countries rely heavily on producing goods for the global market—has benefited from very strong state involvement and high levels of **foreign direct investment (FDI)**. This is why Southeast Asia is also vulnerable to changes in the global economy, as economic downturns in the investing countries—for example, Japan, Germany, or the United States—limit investment in Southeast Asia.

To take advantage of FDI to the region, Malaysia implemented its **New Economic Policy** in 1971. The New Economic Policy intended to shift the economy from primary **commodities**, such as rubber, to higher value exports, such as computer chips. A goal of the policy was to distribute the benefits of economic development to the ethnic Malay population, called the **Bumiputra** ("those of the soil"), and this program explicitly discriminated against ethnic Chinese populations. From 1980 onward, foreign investment flowed into Malaysia, and the share of the country's manufacturing involved in producing

exports increased from 20% to 80% by 2003. Like Singapore, Penang in Malaysia has become a center for high-technology manufacturing for the computer industry. Government investment in training skilled labor and generous tax incentives have attracted multinationals such as Intel, Sony, Philips, Motorola, and Hitachi to Malaysia. The capital city, Kuala Lumpur, with a population of about 1.5 million, is the center of a manufacturing region called the Klang Valley conurbation. (An conurbation is an extended urban area that typically encompasses several towns or cities.) Manufacturing sectors that developed or relocated to Southeast Asia include automobile assembly, chemicals, and electronics. Japan has been a major source of foreign investment; many Japanese-owned firms have relocated seeking cheaper labor and land. For example, this "offshore manufacturing" included 250 Japanese firms in Malaysia by 1990.

The Little Tigers The rapid growth of a number of Southeast Asian economies, averaging 8% per year in the early 1980s, was seen as part of the larger East Asian economic miracle. The more successful Southeast Asian countries—Thailand, Malaysia, and Indonesia—joined the "Asian tigers" (see Chapter 8, p. 304) Hong Kong, South Korea, Taiwan from East Asia, and Singapore. These new "**little tigers**" were termed **newly industrializing economies (NIEs)**, rather than "developing" or "underdeveloped" countries. High rates of savings, balanced budgets, and low rates of inflation were indicators of a successful transition to industrial economies. Foreign investment flowed into Southeast Asia to take advantage of growing domestic markets for consumer goods, such as automobiles and soft drinks, and of valuable natural resources, such as oil and minerals. In some cases, new industries developed locally to meet national demand, as is the case with the Proton, Malaysia's "national car."

Perhaps the most successful of the "little tigers" has been Singapore, which is one of the 25 wealthiest countries in the world in terms of per capita gross domestic product (GDP) and has the highest standard of living in Asia (**Figure 10.19**). Singapore's superior

▲ **FIGURE 10.19 Singapore's Financial District** This view of the city of Singapore clearly shows its modern high-rise core.

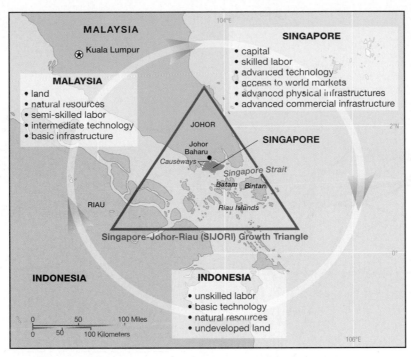

▲ **FIGURE 10.20 Malay Peninsula and Singapore** (a) The Malay Peninsula and Singapore are at the center of economic development in Southeast Asia, especially the SIJORI growth triangle. (b) This figure illustrates the economic complementaries of Singapore; Riau, Indonesia; and Johor, Malaysia.

infrastructure—especially the excellent port and international airport—has made it the import and shipping center for the region and the busiest port in the world. The international business executive flying into Singapore's modern airport sees towering high-rise offices, luxury hotels, and large-scale industrial facilities. The government of Singapore made the astute decision to move to activities that required a skilled workforce, such as precision engineering, aerospace, medical instruments, specialized chemicals, and most notably, the emerging computer industry. To support this policy, the government invested in education, especially engineering and computing, in high-technology industrial parks, and in state-owned pilot companies, such as Singapore Aerospace. The strategy was successful. Singapore is now a **world city** and a center for information technology and aerospace. Singapore has also taken advantage of its location in the region and is part of a new region called **SIJORI**, which includes Singapore (SI), Johor Baharu (JO) in Malaysia, and the Riau Islands (RI) of Indonesia (**Figure 10.20**).

Southeast Asia in the Global Economy The global economic crisis of 2008 had a significant impact on the economies of Southeast Asia (**Figure 10.21**, p. 406). Ten of the 11 countries in the region reported a negative change in GDP—the total value of all materials, foodstuffs, goods, and services produced in a country in a particular year—in 2009, and several countries have reported significant increases in inflation between 2007 and 2008. The economies of Southeast Asia remain intimately linked to global economy through the production and export of inexpensive goods. One overall indicator of this linkage is the relationship between the value of exports and the value of GDP. Singapore, with exports valued at 211% of GDP in 2010, and Malaysia, at 97%, are the most integrated into the global economy through exports, followed by Thailand (71%), the Philippines (35%), and Indonesia (25%). In terms of key global commodities, Southeast Asia produces more than half of the world's rubber, coconut, tin, palm oil,

and hardwoods. This integration means that many of the economies in this region are susceptible to global slowdowns or recessions.

It is likely that the continuing instability in the global economy will further impact the manufacturing and service sectors in the region, particularly in tourism, which is the largest foreign exchange earner in some of the countries in the region, including Thailand. Although export-led production helped propel Southeast Asia out of the crisis in the late 1990s, it is less likely that exports will now provide a similar boost because importing countries are enacting protectionist measures to defend local markets against less-expensive Southeast Asian commodities. The United Nations Economic and Social Commission for Asia and the Pacific has outlined the short-term and long-term impacts of this process for Southeast Asia, highlighting the fact that volatility in fuel and food markets coupled with declining demand (both locally and globally) places the region in a precarious position moving forward. Moreover, the Commission points out the longer-term impacts that global climate change will likely have on the economic and social outlook of the region, suggesting that these changes increase the vulnerabilities of people whose livelihoods depend on water resources for agriculture and fisheries development.

Social and Economic Inequality Although there have been economic gains of the last half of the 20th century in the region, there remain large variations in economic and social conditions between Southeast Asia and other world regions and within Southeast Asia itself. Southeast Asia includes countries that have some of the world's highest per capita GNIs (gross national incomes). For example, in 2010, the GNI per capita of Singapore was more than U.S. $40,000, while Brunei's GNI per capita was above U.S. $30,000. At the same time, Southeast Asia also boasts some of the lowest GNI per capita in the world. Cambodia's 2010 per capita GNI was only U.S. $750,

GDP, annual percent change

▲ **FIGURE 10.21 Gross Domestic Product** GDP annual percent change measures the change from one year to the next in total GDP. The crisis of 2008 is clearly represented on this graph; all countries in the region saw a fairly distinct decrease in relative GDP in 2009 as a result of the crisis.

and for Laos the per capita GDP for the same year was U.S. $1,050. The distribution of income tends to be most unequal in the strongest economies in Southeast Asia. For example, the wealthiest 20% control about half the wealth in Malaysia, the Philippines, Singapore, and Thailand. Wealth concentration in Southeast Asia has been associated with **crony capitalism**, in which leaders allow friends and family to control the economy, and **kleptocracy**, in which leaders divert national resources for their personal gain (**Figure 10.22**).

Definitions of poverty vary across the region, but even so there do seem to be general improvements in overall conditions. For example, Malaysia reduced the proportion of people living in poverty from 60% in 1970 to 8% in 2011. However, in Malaysia and Thailand, about a quarter of rural people live in poverty, compared to fewer than 10% of urban residents. In terms of economic and social indicators, women are more equal to men in Southeast Asia than in many other regions. Female literacy averages more than 85%, only 10% less than the male rate, compared to an average for women of less than 50% in sub-Saharan Africa and South Asia. Despite these relatively positive measures, many workers in Southeast Asia are exploited in terms of wages and labor conditions when compared to workers producing similar goods in North America and Europe. In the absence of unions, many work 12-hour days and seven days a week without benefits,

▲ **FIGURE 10.22 Kleptocracy in Action** Imelda Marcos, wife of Filipino leader Ferdinand Marcos who ruled the Philippines from 1966 to 1986, was famous for her extravagant shoe collection. Some see her tremendous wealth as a sign of the Marcos regime's corruption. Imelda Marcos and her husband appointed a number of their relatives to positions of power in the government.

making products such as clothing and electronics for global markets. International companies such as Nike have come under criticism for sweatshop labor practices in countries such as Vietnam and Cambodia, where wages are less than U.S. $2 a day, and workers earn less than 5% of what workers earn producing similar goods in the United States.

> **APPLY YOUR KNOWLEDGE** Provide three examples that illustrate how Southeast Asia is integrated into the global economy today. Briefly discuss who in the region has benefited the most from global integration and who has benefited the least. What are some of the explanations for why some places are successful while others are less so?

Territory and Politics

The intersections of local and regional politics and social action challenge some of this region's national governments. Legacies of this region's wider global interconnectivity tied to colonialism, territorial expansionism by some countries after independence, and Cold War conflicts have contributed to continuing instability in the region. The Cold War had the most global and long-lasting impacts in what was once French Indochina. Communist groups in Vietnam struggled against the French occupation of their country during colonialism, resisted Japanese occupation during World War II, battled French recolonization after the war, and struck out against the United States during the Cold War.

The Indochina Wars Led by **Ho Chi Minh**, the communists established a separate government in the northern Vietnamese city of Hanoi in September of 1945. From this base, they supported a guerrilla war against the French in southern Vietnam with assistance from the Soviets and Chinese. When the French withdrew from the region after a devastating loss at Dien Bien Phu in 1954, Laos and Cambodia became permanently independent. At that time, the Geneva Accords temporarily divided Vietnam into North Vietnam and South Vietnam. North Vietnam became an independent country, and thousands of refugees, especially Catholics, fled to South Vietnam, which established its capital in Saigon.

While this was taking place, communist guerilla rebels (called the **Vietcong**) took control of several regions of South Vietnam. Subscribing to the so-called **domino theory**, which held that the communist takeover of South Vietnam would lead to the global spread of communism throughout Southeast Asia, the United States sent military advisors to South Vietnam in 1962. This was followed by U.S. bombing of North Vietnam in 1964 and escalation to a full-scale land war—the **Vietnam War**—with more than half a million U.S. troops, in 1965 (**Figure 10.23**).

The Vietnam War was probably the most serious global manifestation of Cold War competition; it wrought terrible social and environmental effects in Southeast Asia as well as on U.S. domestic and international politics. More than a million Vietnamese people and 58,000 Americans died. U.S. forces sprayed 2 million hectares (5 million acres) of Vietnam with defoliants (chemicals that remove the foliage from plants, thus making it more difficult for ground

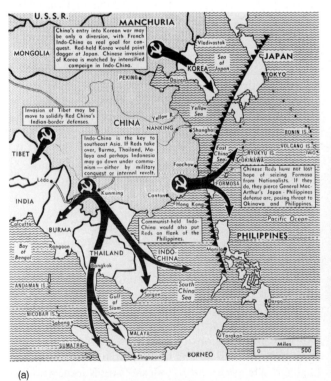
(a)

(b)

▲ **FIGURE 10.23 Domino Theory** (a) Fear of the expansion of communism in East and Southeast Asia promoted the use of map images to depict the communist threat. This map, which was produced in the 1950s, demonstrates the threat that a communist China and by association a communist Vietnam posed to U.S. interests in East Asia. Maps such as this were used later to justify the extensive, costly, and deadly Vietnam War. (b) U.S. troops wait to be evacuated at Khe Sanh, Vietnam, in 1968—a year during which domestic opposition to U.S. involvement in Southeast Asia grew dramatically. Images such as these helped fuel large-scale domestic opposition to the war.

forces to find cover). Defoliants such as **Agent Orange** poisoned ecosystems and caused irreparable damage to human health. Cambodia and Laos were also bombed and defoliated to disrupt communist supply lines and camps.

The war ended with the unification of the North and South Vietnam, while 2 million people left South Vietnam, fearing repression after unification. Many (referred to as the *Vietnamese boat people*) sailed away in small, fragile boats, forming a global Vietnamese diaspora (a term for the dispersion of people from their original homeland) in Southeast Asia, Europe, and the Americas. The new unified Vietnamese government confiscated farms and factories to create state- and worker-owned enterprises, resettled ethnic minorities into intensive agricultural zones, and moved 1 million people into new economic development regions. But U.S.-led economic sanctions from 1975 to 1993 limited the new nation's potential for exports and restricted some critical imports such as medicines. The pockmarked landscape created by the U.S. bombing campaign promoted the spread of disease such as malaria, because mosquitoes thrive in pools of standing water in old craters. On the other hand, the craters have also provided new economic opportunities for some who have converted them to fisheries (**Figure 10.24**).

During this same period, the **Khmer Rouge** overcame the U.S.-backed military government in Cambodia. In 1975, they instituted a cruel regime under the leadership of **Pol Pot**. The Khmer Rouge's interpretation of Mao Tse Tung's revolutionary approach in China (see Chapter 8, p. 304) led Pol Pot to suspend formal education, empty the cities, attack the rich and educated, and isolate his country from the world. The new government renamed the country Kampuchea and engaged in mass murder of those people it believed were not loyal to the new regime. A brutal death march out of the capital Phnom Penh in 1975–76 left thousands of people dead. Even more were killed through torture and in the farming collectives set up by the Khmer Rouge. The places where mass murders of Cambodians took place were later described as **killing fields**, a term that powerfully describes the farmlands that were turned into mass grave sites for the

millions of people murdered by the regime. A quarter of Cambodia's population—including most of the intellectuals and professionals—died between 1975 and 1979, when Vietnam invaded Kampuchea and installed a new government. Many Cambodians who escaped migrated to countries outside the region, including the United States. Large Cambodian populations live in U.S. cities such as Long Beach, California, and Boston, Massachusetts. Cambodians were not the only peoples affected by the war: thousands of Hmong (a minority ethnic group living in Laos, Thailand, Burma, and southern China) became refugees and fled Laos after fighting for the United States during the CIA-led covert war in that country. Today, Hmong populations can be found in places as diverse as Merced, California; Paris, France; and Guyana, South America.

APPLY YOUR KNOWLEDGE Examine the relationship between the Cold War and political conflict in Southeast Asia during the 1970s. List some of the consequences of the struggles that took place in Vietnam and Cambodia, in particular.

Postcolonial Conflict and Ethnic Tension In addition to the obvious legacies of the Cold War, some of the longest lasting conflicts in the region are those that involve ethnic minorities seeking political recognition or independence. Other conflicts have been tied to the fractured physical and cultural geographies that have made national unification more difficult in the region (**Figure 10.25**). Three key examples provided by Timor-Leste, Burma, and Thailand illustrate the relationship between territory, politics, and social action in the region. These examples are indicative of how Southeast Asia, as a regional space, is managing its own sociocultural and political-economic differences.

Timor-Leste is a mostly Christian, former Portuguese colony that occupies the eastern portion of the island of Timor, at the eastern end of the Indonesian archipelago. Initial hopes for independence in Timor-Leste were dashed when Indonesia occupied the island after Portugal gave up colonial control in 1976. Twenty years of resistance and more than 200,000 deaths finally resulted in a referendum for independence in 1999. After the vote, however, anti-independence militias went on a rampage, and thousands of refugees were forced to flee. In 2000, Timor-Leste was granted status as a newly independent country operating under initial United Nations administration and with a strong presence from Australia.

Ever since gaining its own independence in 1948 from Great Britain, the Burmese government has faced rebellion from several ethnic and religious minorities who resist the domination of the Burmese-speaking authority. Muslims on the northern Arakan coast have fled into neighboring Bangladesh and more recently into Thailand, Malaysia, and Indonesia. The Karen, who live along the border with Thailand and were favored by the British because of their Christian beliefs, have also been repressed by the Burmese government. With a population of more than 3 million, the Karen have been able to establish an insurgent state supported by smuggling diamonds and opium. The Shan, with a population of

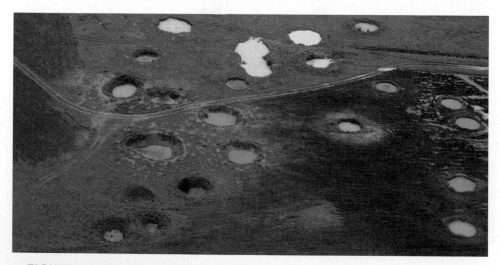

▲ **FIGURE 10.24 The Legacy of U.S. Bombing** The U.S. bombing campaign against the North Vietnamese as well as in Cambodia and Laos created a distinct landscape of bomb craters throughout the region. In many places, these craters remain to this day, and some are used (as pictured here in Vietnam along the Laos border) as fisheries. Craters have also led to expanses of standing water, which can be ideal breeding grounds for malarial mosquitoes and other pests.

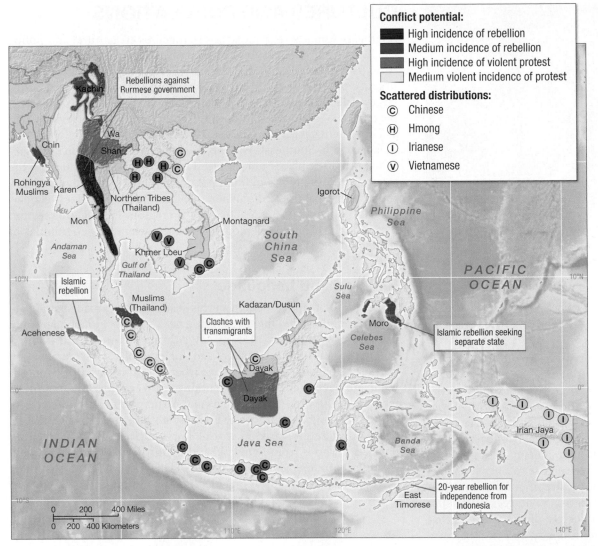

Conflict potential:

■ High incidence of rebellion
■ Medium incidence of rebellion
■ High incidence of violent protest
□ Medium violent incidence of protest

Scattered distributions:

© Chinese
Ⓗ Hmong
Ⓘ Irianese
Ⓥ Vietnamese

◀ **FIGURE 10.25 Map of Conflict Zones** This map shows some of the major zones of conflict in Southeast Asia, including separatist movements in the outer islands of Indonesia and the Philippines.

4.5 million in northern Burma, have similarly used drug profits to establish control of their territory and oppose the Burmese government.

The repressive authority of the Burmese military, which formally ruled Burma from 1962 until March 2011, when quasi-open general elections were held, has long been opposed by Burmese who want a more democratic government. Popular protests in 1988 resulted in martial law and the establishment of the State Law and Order Restoration Council (SLORC—now known as the State Peace and Development Council). Elections were held in 1990 in which an opposition party, the **National League for Democracy (NLD)**, won 60% of the vote, compared to 21% for the existing government party. SLORC cancelled the election results and remained in power. One of the leaders of the NLD, **Aung San Suu Kyi**, won the Nobel Peace Prize in 1991 for her work in bringing democratic reform to Burma. Resolute in her quest for democracy and in her opposition to the authoritarian approach of SLORC, Aung San Suu Kyi was placed under house arrest by the government in 1989. Her latest release was in 2011. Since her release, she has been on the campaign trail, running for parliament. She was successful in April 2012, winning a seat in an election in which the NLD won 43 of the 45 parliamentary seats available (**Figure 10.26**). While such a victory is hard won, the NLD's new seats still place the party in a minority; its 43 total seats is less than 10% of the entire parliament, suggesting that there

▲ **FIGURE 10.26 Ms. Aung San Suu Kyi, Campaigning for Political Office in Burma** Aung San Suu Kyi is a Nobel Peace Prize winner who has been under house arrest in Burma for most of the last twenty years. Recent changes in the politics of Burma have allowed her to run again for office. She was elected to Burma's parliament in 2012.

▲ **FIGURE 10.27 Flag of the Association of Southeast Asian Nations** This flag symbolizes the unification of the region through its supranational organization.

is still a long way to go before the country sees greater democratic participation.

Although Thailand is often described as the "land of smiles" in the literature of the global tourist economy, the country is not without political tensions within its own borders. Thailand has witnessed over 20 military coups since the absolute monarchy was overthrown in 1934. The most recent coup took place in 2006 and resulted in a military dictatorship that has only recently given rise to a newly elected government. Southern Thailand is also the site of conflict between the central government and local separatists, who seek autonomy from the Thai state. This conflict is reflective of the tensions that exist in the south between the Islamic minority and the majority Buddhist population. There have also been a number of recent high-profile trials in the country that charge Thai citizens with insulting the monarchy. For some of these citizens, this has led to extended jail terms; the penalty for insulting the monarchy in Thailand is one of the harshest in the world. These political tensions may be further exacerbated if and when the current king, who has reigned for over 60 years, passes his authority to his oldest son.

Regional Values and Cooperation Despite internal, regional, and international political conflicts, Southeast Asia provides a model for economic and, to a lesser extent, political cooperation in **ASEAN**— the Association of Southeast Asian Nations, which was formed in 1967 (**Figure 10.27**). The Asian Free Trade Association (AFTA) was created as part of ASEAN in 1993 to reduce national tariffs within the region. And a policy of constructive engagement with both military and socialist governments led to invitations to the rest of the Southeast Asian region to join ASEAN (Brunei, 1994; Vietnam, 1995; Burma and Laos, 1997; Cambodia, 1999). Despite some resistance from Malaysia, ASEAN has been open to discussions with other nations and groups, including China and Australia and more recently the United States and India. The development of ASEAN signals just how important and "real" the region of Southeast Asia has become over time. Although it remains a disparate and diverse region, Southeast Asia is not immune to the global trend toward regional organizations.

CULTURES AND POPULATIONS

Southeast Asia is a region marked as much by its cultural and political diversity as its similarity. For example, although insular Southeast Asia, particularly Indonesia, was for centuries characterized by the practice of Hinduism and Buddhism, today Indonesia is the world's largest Islamic country. At the same time, on the island of Bali, people continue to practice Hinduism. In other parts of the region, such as Laos, animism thrives (**Figure 10.28**). In studying religion as well as language in Southeast Asia, we must consider the region's interactions with other places and people. This section begins by tracing the modern-day religious and linguistic geographies of the region. Next, the section expands the discussion of culture to examine local gender, sexual, drug, and ethnic practices in the region. Finally, this section discusses the changing demographic and settlement geographies of the region.

Religion and Language

Although the maps presented in this section show broad spatial patterns of religious and linguistic differences, these regional maps only provide a partial story of everyday cultural religious and linguistic practice in Southeast Asia. Take, for example, an average day in northern Thailand, in a local village where a health-care worker might wake and begin her day with a Buddhist prayer. Next, she might attend a meeting with village healers and participate in an animistic ceremony blessing a local forest. Her day next turns to her work in a hospital, where she speaks in central Thai (*phasaa klang*) with the head physician, who is from Bangkok, and in northern Thai (*kam muang*) with her patients, who are from the local area. On an unusual day, she might also find herself in a meeting with international visitors to

▲ **FIGURE 10.28 Spirit House in Laos** Spirit houses are a common feature throughout Southeast Asia. They provide shelter to local animistic spirits, who are then less likely to interfere with the everyday life of people on that land. It is common to find spirit houses on the land of practicing Buddhists, as they represent the blending of animistic traditions with Buddhism.

discuss health-care programs for people living with HIV and AIDS in the area. In this meeting she would rely on translators and her own English-speaking skills.

The complex religious and linguistic geographies of this region are tied to the global processes that brought Buddhism, Hinduism, Islam, and Catholicism to the region. They are also tied to the very long-term migration of Malayo-Polynesian peoples (Malay, Indonesian, or Tagalog) and Austro-Asiatic speakers (Khmer, Mon, and Vietnamese) and to the more recent migration of Tai-speaking peoples (Thai, Lao, or Shan) as well as to the even more recent migration of Chinese (e.g., Hakka) and Indian (e.g., Tamil) speakers. English is also spoken widely throughout the region, particularly in urban areas and among the region's elite. For example, English is one of the four official languages of Singapore, along with Mandarin, Tamil, and Malay. Southeast Asia's regional geography of language and religion is a partial representation of the broader historical geographies of global interconnection that distinguish the cultural diversity of the region today.

Religion The contemporary religious geography of Southeast Asia reflects centuries of evolution under Indian, Chinese, Arab, and European influences (**Figure 10.29**). Generally, Buddhism tends to dominate the mainland region and Islam the islands

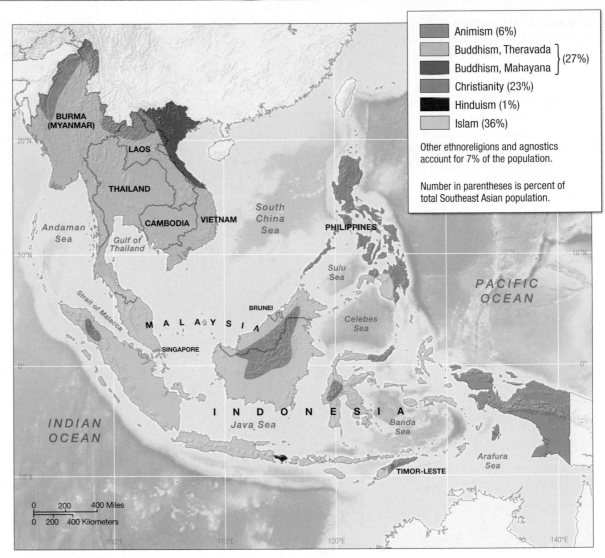

▲ **FIGURE 10.29 Map of Dominant Religions in Southeast Asia** Religious belief in Southeast Asia is dominated by versions of Buddhism on the mainland and Islam on the islands. Buddhism includes both Theravada and Mahayana, the former more conservative and found mostly in Burma, Thailand, and Cambodia, and the latter associated with Vietnam. Catholicism is important in the Philippines. Some ethnic minority groups in remote mountain and island areas maintain animist beliefs.

today. Islamic believers in Indonesia outnumber those in the Middle East, although Islamic practices in Southeast Asia are sometimes seen as more liberal. Islam is also widespread in Malaya and is growing in the Philippines. Intensification of Islamic belief has resulted in the seclusion and veiling of women; in political conflict over the enforcement of Islamic law, especially in diverse populations; and in terrorism linked to global tensions. In 2002, the bombing of a nightclub in Bali by Islamic terrorists killed 202 people. Further attacks in Indonesia's tourist regions have been linked to radical Islamicist groups, including attacks on hotels in Jakarta's tourist district in July 2009.

Hinduism is common in Bali, whereas Christianity is the religion of 85% of the Philippine population. In the Philippines, 400 years of domination by Spain and the United States resulted in several distinctive characteristics, including a majority Catholic religion, mainly Spanish names, highly concentrated land ownership, an agriculture oriented to exports of sugar, tobacco, and pineapples to the Americas, the widespread use of English, and a general orientation to the West, especially the United States. Christianity is also

found among the some of the hill tribes of Burma and in Vietnam, who were converted by French and British missionaries. Western missionaries did not make similar inroads into the Buddhist beliefs of lowland residents. Remote ethnic groups have maintained animistic beliefs, especially in the mountains of Burma, Laos, and Vietnam and in Borneo and Irian Jaya. Religion and cultural tradition intersect in the reverence for the monarchy adopted in countries such as Thailand and Cambodia. The Thai royal family is held in high, almost godlike esteem, with pictures of the king in the majority of homes. The royal family has selectively intervened during times of political crisis and plays a balancing and leadership role.

Language Southeast Asian cultural diversity is reflected in more than 500 distinct ethnic and language groups. Indonesia, for example, has more than 300 ethnic groups and languages, and the population of Laos speaks more than 90 different languages. No language has unified the region in the way that Arabic has in the Middle East or Spanish has in Latin America, although the various dialects of

▶ **FIGURE 10.30** Map of **Dominant Languages in Southeast Asia** Southeast Asia has hundreds of distinct languages, which fall into five major language families: Malayo-Polynesian, spoken in insular Southeast Asia and the Malay Peninsula; Tibeto-Burmese in Burma; Tai-Kadai in Thailand; Papuan in New Guinea; and the Austro-Asiatic languages of Vietnam and Cambodia.

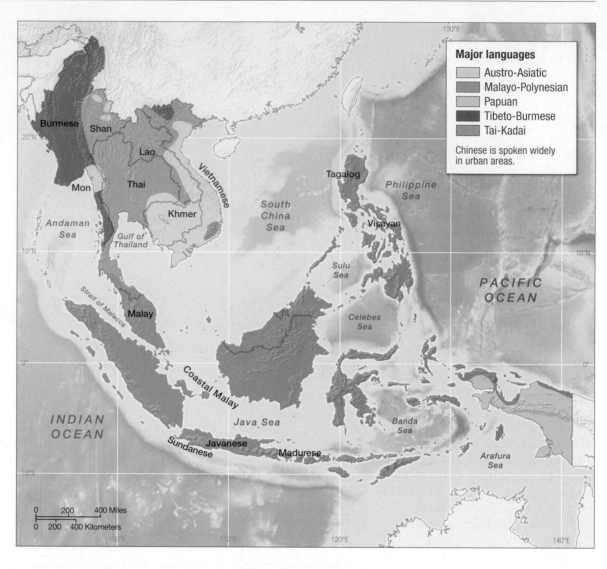

Malay share enough in common to form a broad *lingua franca* (trade language) in Malaysia, Indonesia, and Brunei.

Some national boundaries of Southeast Asia do enclose a dominant cultural and language group. Thailand (Thai), Burma (Burmese), Cambodia (Khmer), Vietnam (Vietnamese), Laos (Lao), and Malaysia (Malay) each have large majority populations that speak the same language and share cultural and religious traditions (**Figure 10.30**). The adoption of a common language in Indonesia is an attempt to integrate diverse cultural groups across this wide geographic space. However, there are significant minority populations in many of these countries, including the Karen and Shan in Burma and the Hmong in Laos, as well as large populations of Chinese and Indians. And within these groups there are distinct cultural and regional differences. For example, the Indian populations include Bengalis and Sikhs in Burma and Tamils in Malaya.

Cultural Practices, Social Differences, and Identity

Just as there is linguistic and religious diversity in the region, there is also cultural and social diversity. In Southeast Asia, for example, one can find both **matrilocality** (where married couples move into the wife's family home) and **patrilocality** (where married couples

move into the husband's family home). While in matrilocal families authority generally passes from father to son-in-law, not to daughters, the common practice of matrilocality does give women *some* power because they live with their own families rather than with in-laws. And, in general, women have more authority within Southeast Asian families and societies than in many other world regions. For example, in many parts of Indonesia, women often manage the family money. The value of daughters is reflected in a more equal preference for female and male children than is seen in either East or South Asia, where preferences are highly biased toward males. Women also have employment opportunities in services and manufacturing and are preferred by many high-technology companies because they accept lower wages and are perceived as more careful and docile. This has created some conflict in home communities, where men are "left behind" by female spouses who have more earning potential than their male counterparts.

Nevertheless, the broader cultures of Southeast Asia remain **patriarchal**: men have authority over the family and society in social and political systems, and socioeconomic conditions are generally better for men than for women. Women have access to work, but that work comes with a lower wage. Men tend to receive higher wages and more education. In poorer Theravada Buddhist communities on the mainland, for example, young boys have access to free education and housing through local monasteries, whereas girls do not (**Figure 10.31**).

▲ **FIGURE 10.31 Boys Sitting Exams at a Temple in Thailand** In Thailand, boys, particularly from poorer families, can become novice monks and obtain an education at a temple. This same privilege is rarely, if ever, extended to girls.

This provides boys with a double benefit, as they are granted the resources of temple life and are able to increase their merit as well as that of their parents by training as a Buddhist novice.

In recent years, **nuclear family** structures have become more common in Southeast Asia. Nuclear households contain just two generations, a pair of adults and their children. Households have also become more suburban, as agricultural land is slowly (or sometimes rapidly) being converted into single-family home communities, many of them gated. The family focus has shifted from parents and older people to children, and from farming or industrial labor to education and service-sector work. New commercial spaces such as malls, restaurants, and vacation resorts serve a growing middle class throughout the region. New globalized social identities are also emerging in the region that are centered not around production—agricultural or manufacturing—but around consumption of high-end consumer goods and luxury items, such as Starbucks coffee.

APPLY YOUR KNOWLEDGE Explain the difference between matrilocality and patrilocality. In what ways does the geographic organization of marriage impact gender relations in Southeast Asia?

Sexual Politics Southeast Asia, particularly Thailand, is well known for its commercial sex work industry. In Thailand, **polygyny**—the practice of having more than one wife at a time—was legal until the 1930s, when the new constitutional government formally outlawed the practice in an effort to appear more "Western" to the rest of the world. Although illegal, polygyny has persisted for some of Thailand's elite. This practice, referred to as *mia noi* (minor wife), has fostered an underground economy of commercial sex in Thailand. Thailand's role in the

Vietnam War as the rest and relaxation (R&R) capital of the region for U.S. soldiers also helped foster the image of Thailand as a global commercial sex capital, and by the 1990s, tours were being advertised in places such as Japan and Germany. The sex industry is internationally promoted with the image of passive and exotic Asian women.

Today, commercial sex work venues exist throughout Thailand, in both rural and urban areas. Women are trafficked into the country from neighboring countries, including Burma, Laos, Cambodia, and southern China. Some of the sex workers are very young women, sold into bondage by rural families or smuggled in to live in slave-like conditions. Studies suggest that for women from poor rural families, sex work may appear as a rational choice for making a living, providing opportunities for them to send money back to their villages and families. The explosion of commercial sex work in the region was also critical to the spread of HIV in the 1990s. A campaign to promote 100% condom usage in these illegal venues in the late 1990s has been credited with the sharp reductions in HIV rates in Thailand. This safe-sex campaign, however, has failed to address the underlying gender and sexual politics that promote this industry in the first place, ironically saving an industry that is illegal in Thailand from destruction. While AIDS rates have been reduced, human trafficking continues.

HIV/AIDS Politics Despite successful prevention efforts, according to U.N. estimates, between 420,000 and 660,000 people in Thailand were living with HIV in 2009—just under 1% of the total population—and an additional 21,000 to 37,000 died that same year (see **Figure 10.32**). This is by far the highest number of cases in Southeast Asia and is

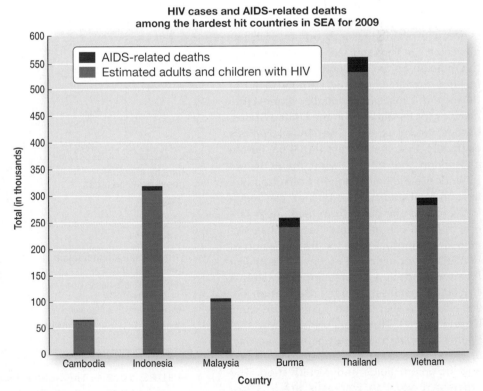

▲ **FIGURE 10.32 HIV Cases and AIDS-Related Deaths in 2009** This graph represents the total number of people living with HIV as well as AIDS-related deaths in the year 2009. Thailand clearly has the highest rate of HIV cases and AIDS-related deaths, but it is important to note that the other five countries reported their first HIV cases later than Thailand.

GEOGRAPHIES OF INDULGENCE, DESIRE, AND ADDICTION
Opium and Methamphetamine

The mist-shrouded highlands of Burma, Laos, and Thailand contain much of mainland Southeast Asia's remaining forest areas; they are also known for the fields of opium poppies that historically have provided as much as half the global supply of heroin. More recently, this region, known as the Golden Triangle (**Figure 10.3.1**), has also become an important global and regional center for the production of methamphetamines (called *speed* in English; *yaa baa* in Thai). The drug economy has helped local ethnic groups fight off Burmese oppression, funding rebellion against the Burmese state. Although tourists still visit the region, a recent Thai government-led "war against drugs" has led to an escalation of violence in this area, particularly in rural areas that are cross-border sites for drug trafficking.

The allure of the opium poppy is centuries old. The narcotic was a component of trade and conflict between China and Europe; the exchange of opium for Chinese tea, silk, and spices by the British created thousands of addicts in China (see Chapter 8, p. 304). In 1905, with heroin addiction on the rise, the U.S. Congress banned opium and eventually prohibited both opium and heroin sales. In Thailand, an active campaign to eradicate opium production was successful, but similar campaigns largely failed in Burma and Laos. Often global in scope, large markets for heroin exist in the United States, Europe, and Australia. The United Nations estimates that there are 11 million heroin and opium addicts worldwide, many unable to work productively and who turn to crime to fund their addiction. Addiction can result in overdoses and death from respiratory failure, and the use of shared needles has significantly contributed to the transmission of HIV/AIDS. Production of opium in Southeast Asia was in decline, decreasing from almost 110,000 hectares under production in 2000 to just over 20,000 hectares in 2006 (**Figure 10.3.2**). However, trends show that opium cultivation is once again on the rise, up 73% since 2006 to almost 40,000 hectares of opium fields under cultivation. The greatest production increases are in Laos, but there are increases in both Thailand and Burma as well.

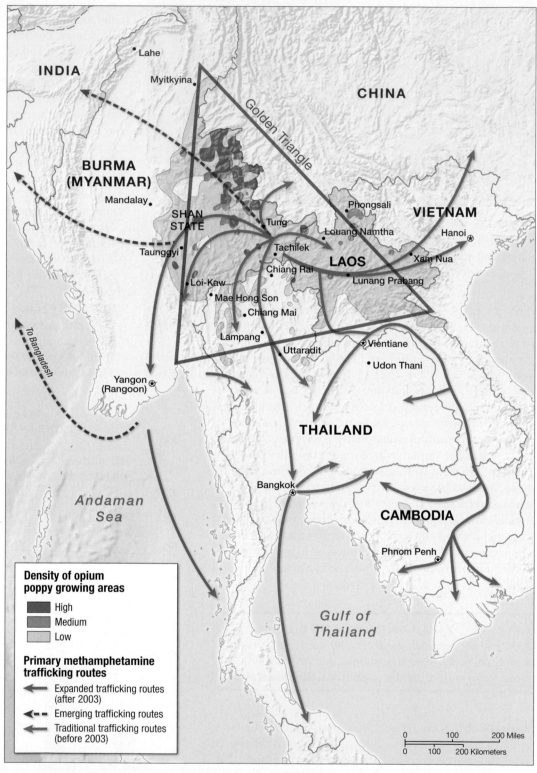

▲ **FIGURE 10.3.1 The Golden Triangle** This map highlights the density of poppy production in the region known as the Golden Triangle. The map of Thailand, which engaged in an active campaign to reduce poppy production, clearly shows the success of that program. Methamphetamine production takes place across Burma and parts of Laos, while Thailand serves as a key trans-shipment site for the drug.

In Southeast Asia, local drug warlords have turned to new markets and commodities in the wake of the competition from opium production in other global regions. Methamphetamines are highly sought in the urban and rural markets of the region, where they are used recreationally, as well as by workers in the manufacturing sector, who want to extend their workday. Seizure of methamphetamine pills has increased dramatically in the region, from just over 30 million pills seized in 2008 to almost 80 million in 2010 (**Figure 10.3.3**). While many people in Southeast Asia smoke methamphetamines, recent increases in injection methamphetamine use have been reported in Cambodia, Indonesia, Laos, Malaysia, and Thailand. Youth with disposable income, who participate in an active club scene in Thailand, often use methamphetamines in conjunction with alcohol and other "club drugs," such as ecstasy.

The resurgence of opium and heroin production in the region suggests that Southeast Asia's historical place as a key drug-producing and consumption region remains disturbingly strong. And the increase in the number of people injecting methamphetamines suggests that Southeast Asia must continue to examine how and why such drug-use production and consumption systems are sustained in the region. ■

◄ **FIGURE 10.3.2 Opium Harvest** Members of the Wa ethnic group in northern Burma harvest opium poppies. Ethnic minorities have long relied on growing poppies and harvesting opium on small plots in the highlands to bring in cash; historically, they used the drug medicinally. More recently, drug cartels control its production and distribution, using the opium "tar" to produce heroin.

◄ **FIGURE 10.3.3 Drugs Seized in the Golden Triangle** The methamphetamines, called *yaa baa* locally, pictured here were seized as part of a crackdown on drug trafficking in Thailand. Although heroin is still produced in the region, new drugs, such as *yaa baa*, are now taking its place.

the result of the prevalence of the sex industry and intravenous drug use (IVDU)—heroin and methamphetamines (see "Geographies of Indulgence, Desire, and Addiction: Opium and Methamphetamine," p. 414). UNAIDS reports that about 45% of the IVDU population in treatment for drug addiction in Thailand is HIV-positive. This high rate is partially the result of a strong antidrug government policy, which regulates and thereby limits access to clean needles, putting drug users at greater risk for HIV through the reuse and sharing of syringes.

Thailand is usually depicted as the epicenter of the HIV epidemic in the region, but other countries demonstrate alarming numbers as well. In Burma, Vietnam, and Cambodia, the HIV incidence rates—the rate of cases among the total population in a given year—in 2009 were 0.4%, 0.5%, and 0.6%, respectively. Although these rates seem relatively low, they show dramatic increases in the region's HIV-positive population since 2000. When these numbers are compared to the adequacy of the health care systems in the region, it is easy to see how the HIV epidemic could quickly outstrip the capacity of several other Southeast Asian countries, such as Burma and Cambodia, to deal with the epidemic. Unlike Thailand, which has both a long tradition of public health and strong connections to the international HIV/AIDS outreach and care networks, other countries in the region lack the basic infrastructure to handle the growing crisis. Thailand was also able to rely on its strong history of family planning, which dramatically lowered fertility rates in the 1960s and 1970s, as well as on a small existing cohort of activists and public health officials. Thanks to these groups, the country was able to expand its free condom campaign to the country's commercial sex work industry in the 1990s.

It is also quite possible that the underreporting of HIV cases is common, as repression and discrimination against HIV-positive people is widespread. This may be the case in Burma, whose actual incidence rate might be significantly higher than reported. Under-reporting of HIV cases is common globally and is partially a reflection of the strong stigma associated with being HIV-positive. For this reason, country-based and regional efforts have worked to develop social support and advocate networks for people living with HIV and AIDS. Thailand was one of the first countries in the region to fund non-governmental efforts to establish People Living with HIV and AIDS (PLWHA) support groups throughout the country. These efforts were particularly strong in the north, the region most dramatically affected by the epidemic, throughout the 1990s. Today, these support groups work in conjunction with other community-based groups, such as herbal medicine physicians and public health workers, to provide care and support for HIV-positive people and their families. In spite of these efforts, Thailand and other dramatically affected countries face a crisis of families in which one or more of the parents have died from HIV, leaving a legacy of "**AIDS orphans**" throughout the region. The rising number of orphans is placing a strain on both the social service sector and extended family networks, creating new layers of complexity for an already extensive social, political, and economic crisis of prevention and care.

Minority Politics Southeast Asia is home to a number of ethnic groups that historically traversed national boundaries. As a result, throughout Southeast Asia, there are minority groups who are not considered citizens of any particular nation. These include refugee groups displaced during the Vietnam War period, such as the Hmong, who now live in parts of Thailand, or the Karen, the ethnic group that has long fought the domination of Burma by the ethnic Burmese military government. Many of these groups in mainland Southeast Asia are euphemistically called **hill tribes**, or *chao khao* in Thai. The hill tribes dwell in the sparsely populated border regions of Thailand, Burma, Laos, and Vietnam and have historically participated in dry farming and slash-and-burn agriculture, supplementing these practices with hunting-and-gathering techniques (**Figure 10.33**).

In insular Southeast Asia, ethnic groups referred to as **sea gypsies** rely heavily on fishing and boat travel between islands to sustain their daily caloric intake. These populations predate many of the modern-day dominant ethnic groups in the region, such as the Thai and Burmese. Their ancestors probably arrived over 5,000 years ago. The Moken, one of these migratory groups, are finding themselves under increasing pressure to settle and participate in the modern cash economy. For many of these people, tourism provides a way to bring cash directly to their communities, although tourism economies are often controlled by people outside the communities. In recent years, advocates of sustainable tourism have pushed for greater autonomy for these groups. In the meantime, many of these people live in limbo: they are not citizens of the states of Southeast Asia and do not have the power to fully live their lives as they traditionally have.

APPLY YOUR KNOWLEDGE List three possible connections that might exist between the spread of HIV, drug use, and minority and ethnic politics in Southeast Asia.

▲ **FIGURE 10.33 Hmong Farmers in Laos** In villages throughout Southeast Asia, ethnic minorities practice traditional production techniques to cultivate, harvest, and process crops that came to the region through global trade. In this case, corn is milled using stone tools and human labor.

Demography and Urbanization

Population distributions in Southeast Asia remain uneven, as urbanization has taken hold at a faster rate in some parts of the region. Rural areas remain less populated, but no less affected by the changes taking place at the national, regional, and global levels. Global and regional economic processes constantly create new networks of migration in and out of the region, as people from Southeast Asia seek new lives in other regions of the world and people move within the region from places of high instability to those of relative stability. Life expectancies across the region also vary widely. Aging populations are being asked to take on new responsibilities as the changing realities of family life shift in relation to the distribution of epidemic diseases such as HIV. In parts of Southeast Asia, the HIV epidemic has left some towns and villages with few working-age adults, placing the burden of raising children in the hands of grandparents, who historically would have been cared for by their own children.

Population and Fertility The population of Southeast Asia was estimated to be about 600 million in 2011. Indonesia, whose population was 238 million in 2011, is the world's fourth-most populous country. The Philippines is home to about 95 million people, followed by Vietnam at 88 million, Thailand at 70 million, Burma at 54 million, and Malaysia at 28 million. Population growth and life expectancy in Southeast Asia surged with the eradication of malaria and improved medical care in the 1950s. After several decades of growth at more than 2% per year, overall population growth has slowed to 1.3% per year, primarily as a result of significant fertility declines in Indonesia, Thailand, Singapore, and Vietnam. In these countries, the total fertility rate (average number of children born to a woman of childbearing age) has fallen from more than six to about two. Indonesia's population policy is often promoted as a model for non-coercive family planning. Indonesia promotes two-child families through advertising, grassroots leadership training, and free distribution of birth-control pills and condoms. Thailand has also had an extensive family planning campaign that began in the 1970s, which has successfully reduced overall fertility in the country.

Fertility rates have remained higher, at 3.2, in the Philippines, partly as a result of opposition to birth control by the Catholic Church, and in Malaysia, where the government encourages the ethnic and Muslim Malay population to have at least five children per married couple. Fertility rates are also higher in the poorer countries of the region, including Laos, where the fertility rate is 3.9,

and Timor-Leste, where it is almost 5.7. Southeast Asian data support theories that fertility decline is associated with higher income, lower infant mortality, and higher status of women. In wealthy Singapore, where the fertility rate has fallen to 1.2—below the replacement level of 2.1—and population growth is negative, the government is promoting marriage and childbearing, especially among the highly educated. Cambodia, Laos, and Timor-Leste have lower GDP per capita, higher infant mortality, and lower levels of female literacy and schooling than do other countries in the region. Until recently, however, population growth in Cambodia, Laos, and Vietnam was reduced by high death rates from war and famine.

The map of population distribution in Southeast Asia shows people concentrated in the river valleys and deltas and on the island of Java (**Figure 10.34**). The highest population densities (people per hectare) are in Singapore, the Philippines, and Vietnam. Levels of urbanization range from 100% in the city–state of Singapore, to more than 60% in Malaysia and the Philippines, to less than 30% in Cambodia, Laos, Thailand, and Vietnam.

Southeast Asian Migration Three major types of migration exist within Southeast Asia. The first is the worldwide phenomenon of people flowing into the cities from rural areas. They are driven by landlessness and agricultural stagnation or pulled by the attractions

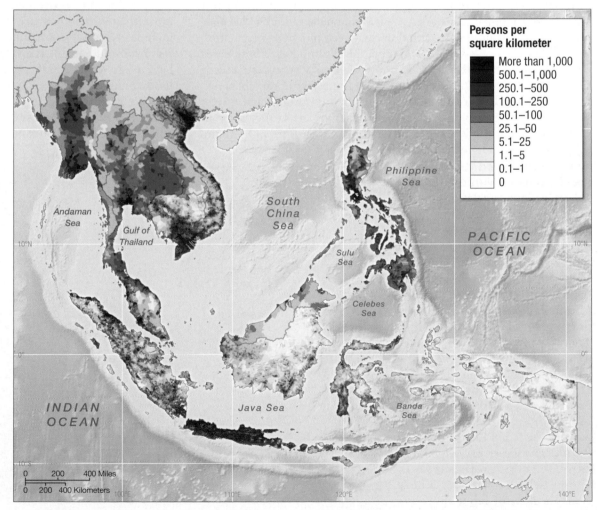

▲ **FIGURE 10.34 Population Density in Southeast Asia, 2012** The majority of people of Southeast Asia live in or near river valleys and deltas on the islands of Java, the Philippines, and Singapore.

▲ **FIGURE 10.35 Rohingya Refugees** In February 2012, 54 Rohingya refugees were rescued in the North Aceh Sea off the coast of Sumatra, Indonesia. These refugees, like thousands of others, were fleeing Burma's repressive government, even as elections in Burma provide some hope for change.

of urban areas, including job opportunities, education and health services, and access to consumer goods and popular culture. This phenomenon has commonly resulted in the development of megacities, such as Bangkok, Thailand, or Manila, the Philippines. These cities have significant populations and large slum areas where many recent rural migrants first settle (see further discussions later in this chapter).

The second type of migration stems from war and civil unrest within countries, such as the mass evacuations from cities in Cambodia under the Khmer Rouge and from war zones in Vietnam. War, ethnic and religious conflict, and poverty forced thousands of people to move within Southeast Asia between 1970 and 1995. Included in the total are more than 300,000 Laotians and 370,000 Cambodians who moved to Thailand, about 300,000 Vietnamese who moved to China and Hong Kong, 300,000 Muslims who moved from Burma to Bangladesh, 200,000 Muslims who moved from the Philippines to Malaysia, and 95,000 Burmese who moved to Thailand. The UN High Commission for Refugees reported that in 2010 there were 120,000 Burmese in refugee camps in Thailand; 29,000 Burmese refugees in Bangladesh; and 300,000 Vietnamese in China (**Figure 10.35**). There is also a flow of labor migrants among countries, with more than 500,000 Indonesians working in

the Malaysian construction industry and 400,000 Thais working in Singapore and Malaysia.

The third, and distinctively Southeast Asian, type of migration is the active resettlement of populations from urban to rural areas, especially in Indonesia. In 1950, the Indonesian government initiated a **transmigration** program, which redistributed populations from densely settled Java and the city of Jakarta to reduce civil unrest, increase food production in peripheral regions, and further goals of regional development, national integration, and the spread of the official Indonesian language. More than 4 million people moved to the Moluccas, Sulawesi, Sumatra, and Kalimantan, with 1.7 million of the migrants receiving official government sponsorship in the form of transport, land grants, and social services (**Figure 10.36**). The Philippines promoted similar resettlement programs, subsidizing migration from Luzon to frontier regions such as Mindanao. Problems with the transmigration program included the lack of infrastructure in the new settlements, conflict between the settlers and Indigenous groups, and the deforestation that occurred when relocated farmers accustomed to the fertile soils of Java struggled to make a living. Malaria, pests, and weed invasion also hindered the success of the program.

The enormous islands of Borneo, Sulawesi, and Irian Jaya, the Indonesian portion of New Guinea, have also become a focus for transmigration resettlement, mineral development, forest exploitation, and resistance by Indigenous groups. Borneo covers more than 750,000 square kilometers (290,000 square miles) and includes territory controlled by Indonesia (Kalimantan), Malaysia (Sarawak and Sabah), and Brunei. Borneo, Sulawesi, and Irian Jaya have more than 25,000 species of plants and 10% of the world's biodiversity. Elephants, tigers, rhinos, and orangutans reside in the mountainous and forested interiors, with human populations concentrated in the coastal plains and river valleys around Hulu Sungai in southern Kalimantan and Pontianak in western Kalimantan.

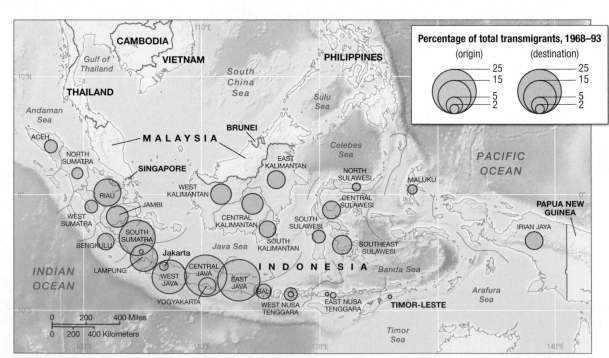

▲ **FIGURE 10.36 Transmigration Flows in Indonesia** The Indonesian government has relocated thousands of people from urban areas on Java to rural areas on the islands of Moluccas, Sumatra, Sulawesi, and Borneo.

International Migration International migration to and from the region has a long history. The most important flow into the region has been the centuries of movement of Chinese into Southeast Asia, beginning as early as the 14th century. Driven by civil wars, famine, and revolution in China, more than 20 million Chinese moved to Southeast Asia during the colonial period to work as contract plantation laborers harvesting rubber and to work in mines and railways. Imported labor cleared forests and built irrigation and drainage systems in these vast deltas, which became the so-called **rice bowls** of Southeast Asia. A persistent stream of migrants from India and China accelerated in the 19th century. Thousands of workers were brought in to convert the deltas to rice production, to mine tin and other minerals, to manage the services in major ports, and to develop small businesses to serve colonial and local demands for consumer goods.

To fill a shortage of European administrators, rights to collect taxes and harbor duties, to market opium, or to operate gambling were auctioned by colonial governments and purchased by the Chinese. Such activities were valuable revenue generators. Chinese migrants were seen as more entrepreneurial and fit to govern by some colonial administrators, and they became a core of the colonial economies. These so-called **overseas Chinese** became essential to the success of the colonial economy, and upon independence they became the entrepreneurs who ran banks, insurance companies, and shipping and agricultural businesses. As a result of this massive in-migration, ethnic Chinese make up a large percentage of the overall population in Singapore (77%), Malaysia (26%), and Thailand (10%). In other countries, such as Indonesia, ethnic Chinese make up a large percentage of the economic elite, even when they are a small percentage of total population. The Chinese generally live in urban areas, in separate neighborhoods often known as Chinatowns, with their own social clubs and schools.

Migration out of Southeast Asia includes large refugee flows and many thousands of labor migrants (**Figure 10.37a**). After 1974, 1.5 million people left Vietnam as refugees from war and communism. About half of the Vietnamese refugees went to the United States; France, Canada, and Australia accepted others. U.S. cities such as San Francisco have large Vietnamese populations living in distinctive neighborhoods. Some Vietnamese refugees remained in refugee camps in Hong Kong. After Vietnam invaded Laos in 1975, more than 300,000 Hmong fled to Thailand, fearing persecution because they had supported the Americans; they were subsequently resettled in the United States. Recently, with peace and reform, many Vietnamese are returning to Vietnam, and some bring capital earned abroad for investment.

The largest numbers of labor migrants from Southeast Asia work in the Middle East, especially in Saudi Arabia, Kuwait, and Oman, and in Hong Kong and Japan. Many Southeast Asians in the Middle East are Muslim women working in the service sector as nurses and maids. Philippine men have a tradition of joining the merchant marines of many countries and working as cooks, seamen, and mechanics. Thousands of Thais and Filipinos work in Hong Kong, where their wages are typically much lower than those of local laborers. Many Philippine women work as maids or nannies in North America, Europe, and Singapore.

The money sent back by these labor migrants (**remittances**) is very important to both national and local economies. For example, it is estimated that $23 billion was sent back to the Philippines in 2011, accounting for over 10% of the country's GDP, and $5 billion to Vietnam, accounting for over 5% of its GDP (**Figure 10.37b**).

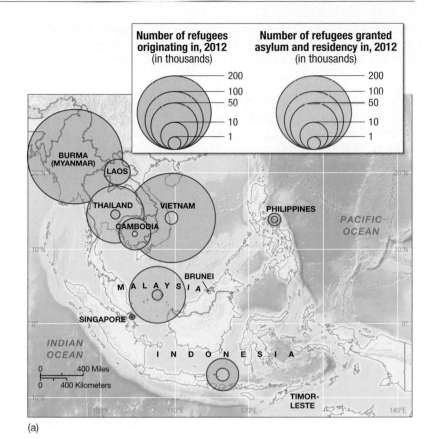

(a)

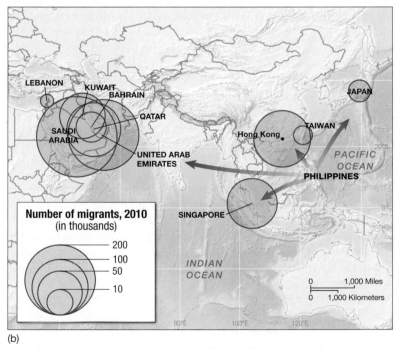

(b)

▲ **FIGURE 10.37 Migrations in Southeast Asia** (a) Southeast Asia has experienced enormous flows within and from the region, including migrants from Vietnam, Burma, Laos, and Cambodia to Thailand and beyond the region; from the Philippines to Malaysia; and from Timor-Leste to Indonesia. Between 1999 and 2007, the total refugee population was about 800,000. (b) This map shows where Philippine labor migrants were moving in 2010 and illustrates the significance of remittances from workers in Japan, Hong Kong, and the Middle East to the Philippine economy.

In contrast, Thailand's remittances totaled more than U.S. $2 billion (0.5% of the total GDP) in 2011, while Indonesia's remittance total was just over U.S. $7 billion (1.0% of the total GDP).

Urbanization About half of Southeast Asia's people live in cities, most of which have grown rapidly since 1950. The region is dominated by several enormous cities with large metropolitan-area populations, among them two of the world's top 20 largest cities: Manila (16.3 million) and Jakarta (18.9 million). Bangkok (7.2 million) and Ho Chi Minh City (6.4 million) also both rank in the top 35 worldwide. The unification of North and South Vietnam in 1976 left the new country with two major urban centers—Hanoi and Ho Chi Minh City (formerly called Saigon). These cities have grown so rapidly that they have serious problems of overurbanization—insufficient employment opportunities, an inadequate water supply, sewerage problems, and inadequate housing. With a wider metropolitan population of more than 9 million people, Bangkok is the current hub of mainland Southeast Asia and dominates Thailand with 32% of the country's urban population (more than 16 times the size of the next largest city), 90% of the trade and industrial jobs, and 50% of the GDP. There are more than 25,000 factories in the Bangkok metropolitan region, many of them labor-intensive textile producers.

Key secondary cities with more than 500,000 residents include Palambang, Medan, Bandung, Ujung Pandang, and Surabaya in Indonesia and Cebu and Davao in the Philippines. Many cities in Southeast Asia mix local design with that of colonial and modern planning, but they are also surrounded by unplanned growth, including desperately poor squatter settlements (**Figure 10.38**). Cities have extended into intensively farmed agriculture, especially rice paddies; town, industry, and agricultural villages have become intermixed. In Southeast Asia, boundaries between the city and the countryside are often blurred as they are in many world regions. There are numerous small industries, such as textiles, in rural areas and considerable agricultural production in the cities; new suburban developments are taking the place of rural farmland.

APPLY YOUR KNOWLEDGE Identify three places in the world today that are net recipients of Southeast Asian migrant populations. Why do Southeast Asians migrate overseas, and what are the benefits of those migrations to their home communities? What are some of the consequences of emigration to the countries in Southeast Asia?

▲ **FIGURE 10.38 Shantytowns in Manila, the Philippines** Squatter settlements are a common feature of large urban areas throughout the region. Often these settlements develop on the least desirable land. In this case, people live in tight quarters in close proximity to the railroad. These communities rarely have adequate services, such as water and sewage.

FUTURE GEOGRAPHIES

The region of Southeast Asia has emerged over time to become a relatively coherent economic entity dominated by the Association of Southeast Asian Nations (ASEAN). The longer history of the region highlights the global socioeconomic and political–economic processes that have brought the people of this region into contact with people from other parts of the world. Today, this region is even more intimately tied to the practices of distant others, as local environmental practices in the Indian Ocean region affect regional monsoon patterns and local agricultural practices such as deforestation change the patterns of water evaporation globally. The effects of the recent global recession on the region further illustrate the dependencies that Southeast Asia has on the economies of others.

Regional Cooperation and Conflict Regional cooperation is always a delicate balancing act between national and regional interests. Southeast Asia, despite the rise of ASEAN as a growing economic cooperative, is still fragmented across a diversity of cultural and political divisions. This is reflected in the politics of religion, the rise of Islamist practices and conservative Buddhist nationalism, for example, and in the tensions that are maintained between countries and subregions that have high levels of inequality. In the context of all this complexity, ASEAN is now negotiating free trade agreements with other countries and regional entities. What these negotiations will mean for the future regional geography of Southeast Asia is unclear, and ASEAN is not without its detractors. Some activists argue that reducing trade barriers within Southeast Asia and between ASEAN and larger economics, such as India or the United States, places local economies at a distinct competitive disadvantage. Other activists criticize ASEAN for doing little to intervene in countries that continue to oppress their populations. ASEAN's reluctance to involve itself in the local affairs of Burma and other errant member states is causing the organization to lose credibility both within and outside the region. These debates make it clear that although regional cooperation has produced some integrative regional effects, it has also created new regional and global dialogues that may challenge the stability of Southeast Asia as a region. Moreover, it is unclear if other partners, such as Papua New Guinea, can be brought into the broader ASEAN umbrella in the future.

Environmental Issues and Sustainability Future sustainable development in Southeast Asia must confront the impact of urban growth and land use change in much of the region. Pressing regional environmental issues include air pollution, spreading slums, inadequate infrastructure in the cities, and deforestation of the highlands of the mainland and many of the islands, such as Borneo. Managing the environment and land use more equitably and ecologically will contribute to maintaining and, it is hoped, improving social and economic conditions in this region (**Figure 10.39**). Additionally, questions over the management of international rivers, such as the Mekong as well as smaller river systems in the region, will test the ability of local governments and international organizations to manage the region's resources in relation to the longer-term environmental and cultural needs of all the people in the region.

▲ **FIGURE 10.39 Reforestation Project in the Philippines** The Philippines has the highest level of deforestation in Southeast Asia, with over one-third of its forest cover lost between the early 1990s and mid-2000s. Reforestation projects, such as this one, are an attempt to stabilize forest ecosystems in the country and mitigate the effects of long-term deforestation in the country.

Social and Health Issues While countries such as Thailand have managed the HIV crisis fairly well, other countries with weaker health-care infrastructures will continue to suffer from the spread of HIV and other infectious diseases. The spread of HIV/AIDS may limit the social and economic development potential of the region moving forward, as countries that are poised to see economic gains, such as Vietnam and Burma, are faced with this potential health crisis. For countries such as Cambodia that have long struggled with economic development and political conflict, the HIV epidemic may prove to be an even more serious concern. As a result, there will be increasing pressure on local governments and international organizations to provide both basic public health care and more expensive treatments for the growing HIV-positive population.

LEARNING OUTCOMES *REVISITED*

■ **Assess the relationship between global patterns of climatic and geological change over time, the modern-day map of physical landscapes in the region, and how people have adapted to this region's physical geographies.**

Southeast Asia is distinguished by its warm climates and high levels of rainfall, a product of its relationship with the intertropical convergence zone (ITCZ) and global wind patterns. The region is also subject to high levels of volcanic activity, particularly in the island areas. This activity has created uplift and high mountain peaks throughout the region. Unique physical geographies and ecologies result from the relationship that the region has with these global patterns. These patterns result in a varied landscape of dry, semiarid spaces, rain forests, and evergreen forests. Over time, people have adapted to the region's unique climate and physical landscape, producing large-scale agricultural systems and societies.

■ **Describe the historical and current relationship between local economic development and global economic issues and trends.**

Southeast Asia is well integrated into the global economy, and places in this region are impacted directly by global economic changes. Not every country in Southeast Asia has been able to or has chosen to participate in the global economy at the same level. Throughout the 1990s, many countries in the region grew rapidly, as worldwide markets aggressively purchased Southeast Asian products. Growth stalled with the global economic crisis of 2008, when many countries in the region (even those not fully integrated into the global economy) were subject to high levels of inflation.

■ **Assess the broader history of political tension in the region, noting the role global conflicts have played.**

Southeast Asia, like all world regions, has been subject to conflict for centuries. Since World War II, when Japan thrust the region into even wider global conflict, Southeast Asia has been the site of both international proxy conflict, such as the Vietnam War, and internal conflicts in countries including Cambodia and Burma. Ethnic tensions are present in the region, particularly in areas where minority populations seek autonomy, such as southern Thailand and in the Philippines.

■ **Trace the historical development of social life in the region, explaining how and why differences between places emerge over time.**

The history of this region is tied to both local developments and global flows of people and ideas. Migration into and through the region has brought important cultural and political systems from South Asia and East Asia, including forms of Buddhism and Hinduism. Global connections also introduced Islam to the region as well as Christianity, along with a diversity of economic and political systems. The differences within the region are a result of the relationship various places have to different global processes and local adaptations.

■ **Identify the major social issues in the region today and describe how they are the result of historical, geographical, and social patterns.**

Numerous social and environmental issues are key to understanding Southeast Asia. These issues are tied to a historical process of political and economic change in the region. For example, countries in Southeast Asia have long supplied the world with wood products, such as teak. This has led to rapid deforestation in much of the region. Moreover, the development of nation–states in the region beginning in the 1800s has precipitated conflict over stateless groups, such as hill tribe peoples, who find themselves without a country. Drug production and use is a problem in the region and related health issues, such as the HIV epidemic, also pose a serious concern.

■ **Explain the patterns of demographic change and migration in the region and their effects.**

Demographically, Southeast Asia is a region of dense urban populations and much less dense rural places. In the island region, particularly the Philippines, fertility rates have remained historically high, resulting in the large populations. Since the end of the Vietnam War, Vietnam has also witnessed extensive population growth, while countries such as Thailand and Indonesia, which have engaged in extensive family planning, have seen decreasing population growth rates. Migration in and out of the region is tied to a number of important factors, including the availability of jobs and the role that governments play in moving populations around the region.

KEY TERMS

Agent Orange (p. 408)

AIDS orphans (p. 416)

Angkor Wat (p. 397)

animism (p. 397)

archipelago (p. 388)

ASEAN (p. 410)

Aung San Suu Kyi (p. 409)

biofuel (p. 391)

Bumiputra (p. 404)

chao khao **(hill tribes)** (p. 416)

commodity (p. 404)

crony capitalism (p. 406)

Culture System (p. 401)

cyclone (p. 386)

domino theory (p. 407)

entrepôt (p. 397)

foreign direct investment (FDI) (p. 404)

Golden Triangle (p. 394)

Green Revolution (p. 403)

Ho Chi Minh (p. 407)

hurricane (p. 386)

Khmer (p. 397)

Khmer Rouge (p. 408)

killing fields (p. 408)

kleptocracy (p. 406)

land bridge (p. 388)

little tigers (p. 404)

mangrove forests (p. 392)

matrilocality (p. 412)

megafauna (p. 391)

Mekong River Commission (MRC) (p. 395)

monsoon (p. 385)

National League for Democracy (NLD) (p. 409)

New Economic Policy (p. 404)

newly industrializing economies (NIEs) (p. 404)

nuclear family (p. 413)

overseas Chinese (p. 419)

paddy farming (p. 387)

pancasila (p. 402)

patriarchal (p. 412)

patrilocality (p. 412)

Pol Pot (p. 408)

polygyny (p. 413)

remittances (p. 419)

rice bowls (p. 419)

sea gypsies (p. 416)

SIJORI (p. 405)

Spice Islands (p. 387)

Srivijaya (p. 397)

stupa (p. 398)

sultanates (p. 400)

swidden farming (p. 394)

tariff (p. 403)

Theravada Buddhism (p. 397)

transmigration (p. 418)

tsunami (p. 385)

typhoon (p. 386)

United Nations Educational, Scientific, and Cultural Organization (UNESCO) (p. 398)

Vietcong (p. 407)

Vietnam War (p. 407)

Wallace's Line (p. 392)

world city (p. 405)

THINKING GEOGRAPHICALLY

1. What is the main crop in Southeast Asia and the main systems by which it is produced? How and where did the Green Revolution affect this crop?

2. What roles did India and China play in Southeast Asia prior to the colonial period, and how did these roles influence culture and religion? What political and economic roles did Japan and the United States play in Southeast Asia during the 20th century?

3. What is the legacy of European and Japanese imperialism in Southeast Asia? How did European influence extend beyond the formal colonial period and into the Cold War in the region?

4. What is ASEAN? What role does it play in the region both politically and economically?

5. What are the most pressing social issues in Southeast Asia today? What efforts are being made by local governments to address these issues?

MasteringGeography™

Looking for additional review and test prep materials? Visit the Study Area in MasteringGeography™ to enhance your geographic literacy, spatial reasoning skills, and understanding of this chapter's content by accessing a variety of resources, including MapMaster interactive maps, videos, RSS feeds, flashcards, web links, self-study quizzes, and an eText version of *World Regions in Global Context.*

11 | Oceania

Sheep and cattle graze in the foothills of the mountains of the South Island, New Zealand. The island's tallest peak, Mt. Cook, with its stunning glaciers, looms the background.

IN NEW ZEALAND, THE MORE THAN 5.8 MILLION cows and 32 million sheep far outnumber the human population of 4.3 million and connect the environment and people to the wider world. Introduced by the British explorer James Cook in 1773, sheep altered the landscape through grazing and provided profitable exports of wool to the United Kingdom. The introduction of refrigerated transport in 1882 transformed global trade and allowed exports of frozen meat around the world, especially to Europe, where New Zealand lamb became popular with consumers.

Preferential trade links between New Zealand and the United Kingdom ended when Britain entered the European Economic Community in 1973. At about the same time, increased oil prices made long-distance transport more expensive. However, the New Zealand government subsidized sheep farming until 1984, when political shifts led to the end of domestic subsidies. Demand for wool also dropped in the 1980s as the international textile industry shifted to alternative materials. Sheep populations went into steep decline from their high of 70 million but have now stabilized thanks to contracts with big overseas

- Explain the factors that influence the climate of Oceania (including Antarctica and the Pacific Ocean) and the impacts of climate change and other environmental threats in the region.

- Describe the unique ecosystems and the impacts that the introduction of nonnative species have had in the region.

- Understand how European colonialism changed the region and describe the legacies of colonialism in terms of land use, demography, and trade.

- Explain how Australia and New Zealand are connected to the rest of the world through their exports and note some of the changes that have occurred in the region as a result of globalization, free trade, and a shift in demand from Europe to Asia.

- Consider how the region became an important site for geopolitical struggles, such as World War II, the Cold War, and nuclear policy, and describe the current international and regional political and economic alignments of major countries in the region.

- Explain the importance of tourism in the region and the economic, social, and environmental impacts of the tourist industry.

- Summarize the main shifts in Australian and New Zealand policies toward Indigenous peoples and on immigration issues.

- Compare and contrast the differences in culture and gender relations of different groups in Oceania.

supermarkets and manufacturers. These purchases are driven by demand from the high-end fashion industry for quality wool, a new interest in organic lamb and wool, and growing exports to China.

Meanwhile, prices and export opportunities for dairy and beef exports increased, and many sheep farms converted to dairy production. This conversion to dairy transformed landscapes and farm work in many regions of New Zealand. Trees and hedges were removed to facilitate irrigation, intensive cattle production polluted waterways, and the daily demands of milk production changed the rhythms of farm life.

The livestock industry in New Zealand is faced with new challenges as concerns about the health risks of meat-intensive diets have reduced demand for meat in some countries and as efforts to reduce greenhouse gases create pressure to reduce the emissions from animals and international transport. New Zealand provides a powerful illustration of how an economy built on agricultural exports has adjusted to changes in international trade and economics as well as shifts in domestic policy with widespread implications for national landscapes and lives. ■

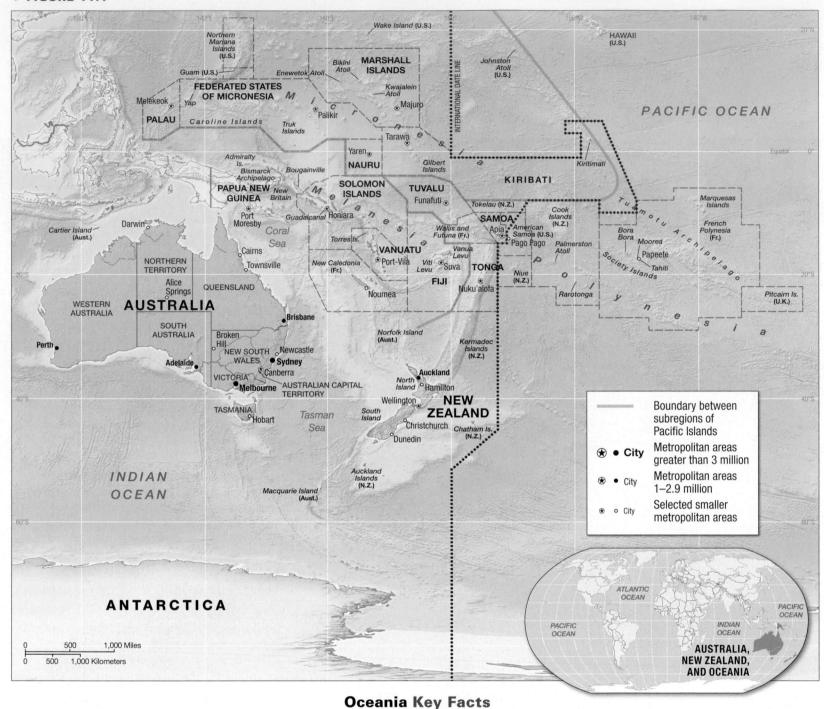

Oceania Key Facts

- Major Subregions: Australasia, Melanesia, Micronesia, Polynesia

- Major Physiogeographic Features: Great Barrier Reef, Great Dividing Range, Great Artesian Basin, North and South Island of New Zealand, Pacific Ocean, Island of New Guinea, Antarctica. Oceanic climate: mostly tropical and warm, rain from prevailing winds over oceans, very dry in Australian interior, cooler in New Zealand and Tasmania

- Major Religions: Christian, Indigenous (including animism), Buddhism and Islam (especially in immigrant communities)

- Major Language Families: English and French, Indigenous languages from Austronesian family.

- Total Area (total sq km): 8.8 million

- Population (2011): 37 million; Population under Age 15 (%): 24; Population over Age 65 (%): 11

- Population Density (per sq km) (2011): 4

- Urbanization (%) (2011): 66

- Average Life Expectancy at Birth (2011): Overall: 77; Women: 79; Men: 75

- Total Fertility Rate (2011): 2.5

- GNI PPP per Capita (current U.S. $) (2011): 27,470

- Population in Poverty (%, < $2/day): No data available for region as a whole, although poverty rates are high in some Pacific Islands, such as Micronesia (45%) and Papua New Guinea (57%)

- Internet Users (2011): 23,927,457; Population with Access to the Internet (%) (2011): 67.5; Growth of Internet Use (%) (2000–2011): 214

- Access to Improved Drinking Water Sources (%) (2011): Urban: 99; Rural; 64

- Energy Use (kg of oil equivalent per capita) (2009): 5001

- Ecological Footprint (hectares per capita consumed/hectares per capita available, global scale) (2011): 5.4/1.8

ENVIRONMENT AND SOCIETY

The Oceania region includes the continental landmass of Australia, the large islands of New Zealand and eastern New Guinea, and smaller nations and territories that are often referred to collectively as the Pacific Islands (**Figure 11.1**). One of the most isolated world regions in terms of its physical geography, Oceania occupies one-third of Earth's surface (**Figure 11.2**). The Pacific island groups are made up of more than 20,000 islands and include Fiji, the Solomon Islands, Vanuatu, and New Caledonia, which with Papua New Guinea are known as **Melanesia**, meaning black islands; Nauru, Kiribati, Palau, the Marshall Islands, the Federated States of Micronesia, Guam, and the Northern Mariana Islands, which are known as **Micronesia**, meaning small islands; and Samoa, Tonga, the Cook Islands, Wallis and Futuna, Easter Island, Tuvalu, Niue, Tokelau, and French Polynesia/Tahiti, which are known as **Polynesia**, meaning many islands (Polynesia is also sometimes defined as including Hawaii). In this chapter, we refer collectively to Melanesia, Micronesia, and Polynesia as the Pacific Islands. We also discuss Antarctica, the continent that lies to the south of the Pacific Ocean around the South Pole (see "Visualizing Geography: Antarctica," p. 428).

The countries in this region share an orientation to the ocean, relatively low populations, and world-famous, dramatic coastal and mountain landscapes. This region is also geologically rich in minerals, which brings it into the wider global economy of resource extraction. Its diverse ecosystems are highly susceptible to human-induced change; the region is also famous for the impact that invasive species—nonnative flora and fauna—have had on the Indigenous ecology. As such, it is an acknowledged hot spot for global climate change.

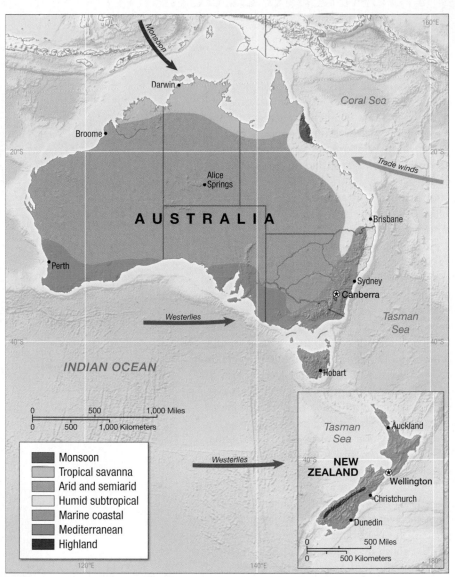

▲ **FIGURE 11.3 The Climate of Australia and New Zealand** Australia's interior is dominated by arid and semiarid climates. Coastal climates are much more diverse and include tropical savanna and some tropical wet regions as well as significant humid subtropical and Mediterranean areas. New Zealand is dominated by a marine coastal climate.

▼ **FIGURE 11.2 Oceania from Space** This image illustrates how the vast area of the Pacific Ocean dominates Oceania. Clouds cover the Antarctic continent at the bottom of the image.

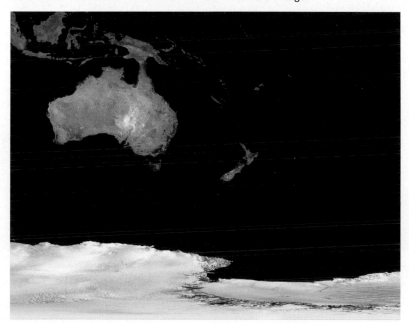

Climate, Adaptation, and Global Change

Most of Oceania lies within the tropics and is dominated by the warm seas and moisture-bearing winds of tropical latitudes. Australia and New Zealand reach farther south toward the South Pole and thus have climates that range from tropical in the north to the cooler temperate climates of the southern westerly wind belts, with annual average temperatures declining from north to south (**Figure 11.3**). The Australian climate is dominated by very dry and hot conditions that are associated with the interior of large continents in the tropics. It is one of the most arid areas on Earth, with two-thirds of the country receiving less than 50 centimeters (20 inches) of rainfall a year. This harsh climate limits human activity and has required complex adaptations from both people and ecosystems in the continent's interior. In interior Australian towns, such as Coober Pedy, which has been a center for the mining

VISUALIZING GEOGRAPHY
Antarctica

It is always a challenge to know where to place Antarctica within the groupings of world regions. However in a number of ways, it fits well within Oceania, because of the importance of the marine environment in Antarctica, its relationship with New Zealand and Australia, and its experience as a simultaneously globally interconnected and isolated regional space. Antarctica's relative inaccessibility is due in large part to its climate. Temperatures average –51°C (–60°F) during its six-month winter, when the sun does not rise above the horizon and the continent is in perpetual twilight. By September of each year, half the surrounding ocean is frozen, creating a vast mantle of Antarctic pack ice that has an area of 20 million square kilometers (32 million square miles). In the clear, bright light of unpolluted air, the snow and ice fields seem endless (**Figure 11.1.1**). Along parts of the coast, stark granite promontories provide fixed landmarks, but much of the coastline of Antarctica is constantly changing ice.

Although it is uninhabited by humans, there is a great deal of interest in Antarctica especially—in scientific research and in the exploitation of natural resources, such as deposits of iron ore, coal, gas, and oil. The **Antarctic Treaty** governs international relations on the continent. The treaty covers all the area south of 60°S and is one of the most successful international environmental and political agreements. Entered into force in 1961 and now signed by 49 countries, the treaty bans nuclear tests and the disposal of radioactive waste and ensures that the continent can be used only for peaceful purposes and is set aside for scientific research. In 1991, the treaty added a 50-year ban on mineral and oil exploration. Nevertheless, several countries—Australia, Argentina, Chile, France, New Zealand, Norway, and Great Britain—still claim specific slices of the Antarctic pie, hoping perhaps to be able to assert rights to offshore fisheries and onshore mineral exploitation (**Figure 11.1.2**).

Antarctica also figures in the wider global imagination. Each summer about 35,000 tourists visit, traveling mostly by ship to see the landscape and wildlife, which includes whales and millions of penguins. Unfortunately, human-caused environmental changes are having an impact. The ozone hole continues to

◀ **FIGURE 11.1.1**
The Antarctic Landscape A lone penguin wanders across a windswept and barren landscape of polar ice fields.

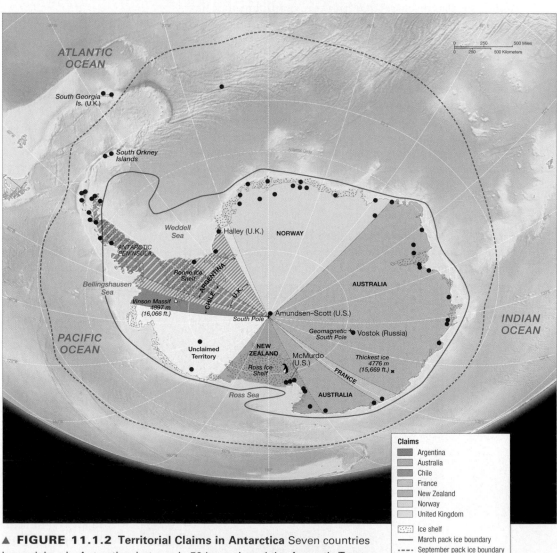

▲ **FIGURE 11.1.2 Territorial Claims in Antarctica** Seven countries have claims in Antarctica, but nearly 50 have signed the Antarctic Treaty. Sixteen countries maintain scientific research bases on the continent.

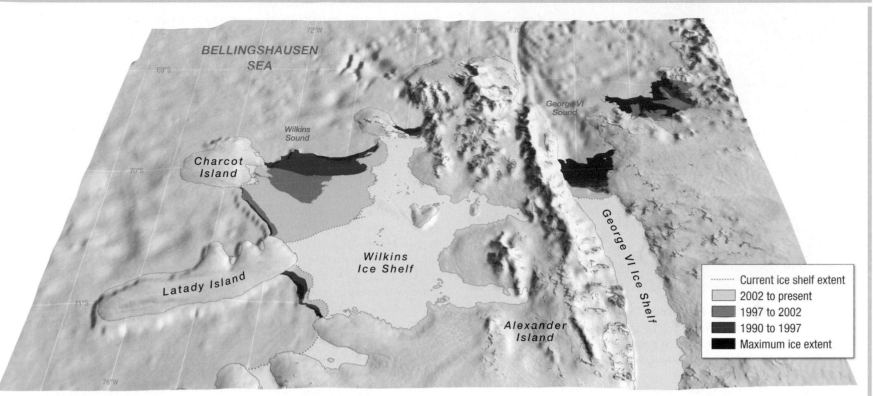

▲ **FIGURE 11.1.3 Ice Retreat around the Antarctic Peninsula** Using maps and satellites, scientists have mapped the retreat of ice caps and glaciers around Antarctica since 1947 and found that the edge of the ice has retreated by many miles in some parts of the continent.

appear over Antarctica despite international efforts to control ozone-depleting gases. Global climate change is also threatening Antarctica, especially the Antarctic Peninsula, where temperatures are warming, the ecology is changing, and ice cover is disappearing (**Figure 11.1.3**). If Antarctic ice continues to melt as a result of climate change, it may contribute to a worldwide rise in sea levels. About 5,000 scientists are based in Antarctica during the polar summer, working to advance our understanding of the challenges of ozone depletion and climate change, studying biodiversity, and exploring the geology beneath the ice (**Figure 11.1.4**). ■

▲ **FIGURE 11.1.4 Antarctic Science** There are several large scientific bases in Antarctica, each managed by a different country. The station at McMurdo Sound is operated by the U.S. National Science Foundation; it can house more than 1,200 residents and has a harbor (accessible during the Antarctic summer), aircraft runways, a chapel, and a greenhouse.

▲ **FIGURE 11.4 Bushfire in Australia** Drought, high temperatures, and combustible vegetation have set the stage for intense and devastating fires such as this fire that occurred in the Bunyip Sate Forest, 125 kilometers (78 miles) west of Melbourne, in February 2009.

of opals valued for jewelry, the temperatures are so intense that much of the town has been built underground. Droughts in Australia are associated with severe wildfires that can race across the tinder-dry bush vegetation, especially where oily eucalyptus fuels the fire. The Black Saturday bushfires in southern Australia in 2009 killed 173 people and burned more than a million acres (450,000 hectares). The fires occurred in the wake of a decade-long drought, during a heat wave in which temperatures reached 46°C (115°F), and were fueled by high winds, and perhaps were triggered by arson (**Figure 11.4**).

Most of the coastal perimeter of Australia has higher precipitation than its interior; Queensland in the northeast has heavy rainfall from the humid southeasterly trade winds that rise up over the highlands near the coast releasing their moisture. Northern Australia receives most of its rain from monsoon winds, drawn inland by the southward shift of the intertropical convergence zone (ITCZ) and heating of the landmass in the Southern Hemisphere summer from November to February (see Chapter 1, Figure 1.8, p. 10). The rainy season sometimes brings tropical cyclones and severe flooding to northern Australia. For example, 75% of Queensland was declared a flood disaster zone in 2011. Southern Australia receives rainfall from storms associated with the westerly wind zone, mostly during winter (June to August), when storm tracks shift northward. The southern coast around the cities of Adelaide and Perth has the mild temperatures and winter rainfall associated with the Mediterranean climate that is also found at this latitude in Chile and South Africa. The southernmost part of New South Wales, Victoria, and the island of Tasmania are wetter as a result of exposure to westerly rain-bearing winds. These areas also have considerable snowfall in the mountains.

New Zealand consists primarily of two major islands spanning 1,600 kilometers (976 miles) from north to south, which are called the North Island and the South Island, and a number of smaller islands. The two islands sit in the middle of the westerly wind belt,

and frequent storms release heavy rain on the western coasts as they rise over the high mountain ranges. The eastern coasts of New Zealand, however, are much drier because they lie in the rain shadow to the east of the mountains; this area can be sunny in summer when subtropical highs move southward and create clear and stable conditions. The North Island is generally much warmer than the South Island, but mountain climates are cooler and wetter throughout the country, often with heavy snow that favors ski tourism on the South Island from June to October.

The Pacific Islands are usually classified into the "high" islands and the "low" coral islands (which are called **atolls**). All the islands have warm temperatures associated with year-round high sun and the warmth of the tropical ocean. Islands with higher elevations experience the most substantial rainfall as moist winds rise over the land, cooling and releasing their moisture. Islands and seas throughout the region can also receive some rainfall from the towering cumulus clouds that occur for most of the year at the Equator, where intense heating creates rising air (convection) (see Figure 11.8, p. 438). The lower islands, however, are much drier because they do not benefit from orographic precipitation (see Chapter 1, p. 10) over mountain ranges and are small enough to escape the convective downpours. As a result, many low-lying islands experience near-desert conditions and shortages of freshwater. When there is an El Niño (see Chapter 7) and ocean currents shift direction and change sea-surface temperatures over the Pacific, changes in atmospheric circulation cause severe drought in Papua New Guinea, Australia, and Micronesia, resulting in crop failures, food shortages, and costly shipments of drinking water to smaller islands.

Antarctica has the coldest climate on Earth because polar latitudes receive little sun and its elevations average 2,286 meters (7,500 feet). The continent also includes a layer of ice that is more than 3 kilometers (2 miles) thick in some places. Temperatures can drop to –73°C (–100°F) during the dark winter. It is dry despite the water stored in the thick ice sheets that cover most of the area and can experience very high winds.

Oceania is especially vulnerable to global environmental changes. Scientists first noticed a dramatic drop in the amount of ozone over the South Pole in the 1980s, which is commonly referred to as the hole in the ozone layer. **Ozone depletion**—the loss of the protective layer of ozone gas—can result in higher levels of ultraviolet radiation and associated increases in skin cancer, cataracts, and damage to marine organisms; it results from the emission of human-generated pollutants into the atmosphere. Australia, with its southern latitude location, sunny days, and tradition of beach going and sunbathing is especially vulnerable to the effects of ozone depletion and has some of the highest levels of skin cancer in the world. However, the **Montreal Protocol**—a 1987 international treaty established to reduce chemical emissions that damage the ozone layer—has succeeded in preventing further ozone depletion.

In Oceania, the impacts of global warming resulting from an increase in carbon dioxide emissions worldwide include drier conditions in the already drought-prone interior of Australia, increased risk of forest fires, and the melting of the magnificent glaciers of New Zealand and Antarctica. A rise in global temperatures is also likely to produce a significant rise in sea levels, primarily because a warmer ocean takes up slightly more volume than a cooler one and also because global warming may melt glaciers and ice sheets. Sea-level rise is of urgent concern to many Pacific Islands, especially those on low coral atolls, where any increase in sea level may result in the disappearance of land, saltwater moving into drinking water supplies, and an increased vulnerability to storms. In response to the threat of global warming and sea-level rise, the Pacific Islands were early members of the **Alliance of Small Island States (AOSIS)** (see "Emerging Regions: Alliance of Small Island States (AOSIS)," p. 444).

Australia is a global leader in considering how to adapt to a warmer climate. Its plans include moving zones for key crops (including the vines that support the wine industry), relocating vulnerable species from warmer to cooler regions, building desalination plants to ensure urban drinking water supplies, and reviewing emergency plans for responding to wildfires and floods. However, Australia is also a major greenhouse gas emitter, using large amounts of coal and exporting it to other regions.

APPLY YOUR KNOWLEDGE Use the Internet to find a government Website of the smaller Pacific Islands, such as Tuvalu, Kiribati, or Samoa. How does the Website describe local vulnerability to climate change? What governmental policies are being proposed to reduce the risk of climate change?

Geological Resources, Risks, and Water Management

Geological history, especially tectonic activity, influences the geography of many of the major subregions of Oceania. Australia is a very old and stable landmass that sits on the Australian tectonic plate (see Chapter 1, Figure 1.14, p. 16). New Zealand, however, sits at the boundary where the Pacific Plate is subducting the Australian Plate and producing high levels of volcanic and earthquake activity.

The Australian landmass forms a continental shield of ancient stable rock with very little volcanic, earthquake, or other mountain-building activity. Australia, which has an area of 7.7 million square kilometers (3 million square miles), has three major physical divisions: the eastern highlands, the interior lowlands, and the western plateau. A series of interior, low-lying basins divide the western Australian Plateau from the uplands of eastern Australia (**Figure 11.5**). The eastern highlands of Australia are the remnants of an old folded mountain range with a steep escarpment on the eastern flanks. The highland crest is called the Great Dividing Range because it separates the rivers that flow to the east coast from those flowing inland or to the south.

Australia's interior lowlands were once flooded by a shallow ocean; however, only Lake Eyre now remains and is often dry, filled only occasionally by inland-draining rivers. A large part of the lowlands is referred to as the **Great Artesian Basin** because it holds the world's largest groundwater aquifer, a massive reservoir of underground water in porous rocks. The basin is described as artesian because overlying rocks have placed pressure on the underground water so that when a well is drilled, the water rises rapidly to the surface and discharges as if from a pressurized tap. The wells that tap the Great Artesian Basin are critical to the human settlements and livestock of arid east-central Australia, although the cost of drilling is increasing and the water is often warm and salty.

Although low rainfall and frequent drought are common in inland Australia, the exploitation of underground water from the Great Artesian Basin has allowed for the existence of scattered homesteads where ranchers (or stockmen) raise livestock on sheep and cattle stations (the name used for a farm or ranch). In these enterprises, cattle are left to fend for themselves, for the most part, and are only brought into the stations once or twice a year. Sheep are raised where rainfall is higher, to the east and west. The inland areas of Australia are often called the **Outback**, a term generally applied to the remote and drier inland areas of Australia. The Outback is often a long way from schools, shops, and hospitals. Adaptations to this isolation include distance education, with children taught through radio broadcasts (and more recently through the Internet), and a flying doctor service for emergency medical care.

The western plateau of ancient shield rocks with a few low mountains and large areas of flatter desert plains and plateaus occupies two-thirds of Australia. This region has numerous mineral deposits—the basis for Australia's mining industry—and old, weathered soils that are too nutrient poor or salty for agriculture. Western Australia and the interior lowlands

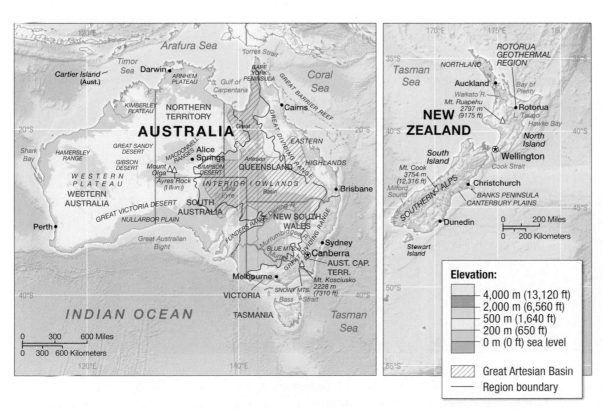

▲ **FIGURE 11.5 The Physical Landscape of Australia and New Zealand** In Australia, key features include the Great Dividing Range of the eastern highlands, central deserts, Great Artesian Basin, and Great Barrier Reef. In New Zealand, distinctive landforms include the Southern Alps and west coast fjords, such as Milford Sound.

▲ **FIGURE 11.6 Ayres Rock** The Ayres Rock tourist resort serves visitors who come to see the dramatic landscapes of central Australia and generates jobs for local Aboriginal peoples. Ayres Rock can be seen in the distance.

also contain impressive examples of desert landforms, including wind-shaped undulating ridges of sand dunes, stony plains with varnished rock fragments called desert pavement, and dry interior drainage basins called *playas*. Centuries of erosion by wind and water have left erosion-resistant domes of rocks standing above the surrounding landscape. The most famous of these isolated rock domes are visited by thousands of tourists and also serve as sacred sites for Australia's **Aborigines**—Indigenous peoples who have lived in Australia for more than 10,000 years. Famous rock domes include Ayres Rock, called *Uluru* by the Aborigines, and the Olgas, called *Kata Tjuta* (**Figure 11.6**).

In contrast to Australia, New Zealand is much younger geologically and more tectonically active because it is located where the Pacific Plate is moving under the Australian Plate. In 2011, a series of earthquakes caused severe damage in the city of Christchurch, killing 185 people and creating soaring insurance claims. The South Island has rugged mountains rising to more than 3,500 meters (11,500 feet) in the Southern Alps, dominated by Mount Cook at 3,754 meters (12,316 feet). The South Island is far enough south to have extensive permanent snowfields and more than 300 glaciers, some flowing almost to sea level. The west coast has stunning fjords, such as Milford Sound (**Figure 11.7a**), created when the sea flooded the deep valleys cut by glaciers. The east coast of the South Island has much gentler relief, with rolling foothills, long valleys with braided rivers and freshwater lakes, and plains formed from stream deposits, which are used for agriculture. The North Island has much more volcanic activity than the South Island. Volcanically heated water that emerges from hot springs, geysers, and steam vents is captured in geothermal facilities and used as a nonpolluting energy resource (**Figure 11.7b**) on the

(a)

(b)

▲ **FIGURE 11.7 New Zealand Landscapes** (a) Milford Sound is a beautiful deep fjord cut by former glaciers on the South Island of New Zealand. (b) The area around Rotorua, on the North Island of New Zealand, has numerous hot springs, geysers, and steam vents associated with volcanic activity. Geothermal energy sources, such as the Wairakei power plant shown in this photo contribute 10% of New Zealand's electricity production. The locale is also popular with tourists who are attracted by the landscape, spas, and thriving Maori culture.

North Island. The North Island also has extensive areas of rolling hills and valleys, where the warmer climate and rich volcanic and river-deposited soils nourish a productive agriculture.

The high Pacific Islands, which are mostly volcanic in origin, rise steeply from the sea and have very narrow coastal plains and deep narrow valleys (**Figure 11.8a**). Many high islands, such as those of Samoa, are created in linear chains as tectonic plates move over hot spots where molten rock reaches the surface. Other chains, such as the Marianas Islands and Vanuatu islands, form along the edge of tectonic plates. The heights of these islands promote heavy rainfall, and the cloud-capped mountains create spectacular landscapes, such as those of Tahiti, where the peaks rise to 2,100 meters (6,900 feet), and Bougainville (3,000 meters or 9,840 feet). The island of New Guinea is the second-largest island in the world (after Greenland; Australia is a continent) and—at more than 800,000 square kilometers (309,000 square miles)—is much larger than the country of New Zealand. The mountain spine of the island of New Guinea rises to more than 4,000 meters (13,000 feet) and features many extinct volcanoes and high isolated basins.

The low Pacific Islands are mostly atolls created from the buildup of skeletons of coral organisms that grow in shallow tropical waters. Atolls are usually circular, with a series of coral reefs or small islands ringing and sheltering an interior lagoon (**Figure 11.8b**). These lagoons may contain the remnants of earlier islands or a volcanic island that has sunk below the surface. Although in some cases, such as the islands of Nauru and Guam, tectonic activity may uplift coral reefs to create higher-elevation limestone plateaus, many of these islands are very low lying, with most of the land within a meter of sea level, making them vulnerable to storms, tidal waves, and rising seas. Pacific atolls include Kiribati, the Marshall Islands, and Tuvalu. The Kwajalein Atoll in the Marshall Islands is the largest in the world, with 90 small coral islands circling a 650-square-kilometer (251-square-mile) lagoon.

Valuable mineral deposits were identified in the 1840s in southern Australia, including silver, copper, and gold. In the 1850s, 40% of the world's gold was coming from Australia and immigrants were arriving to work in the mines. Famous mining centers include Broken Hill in New South Wales and Kalgoorlie in Western Australia and more recently Mount Isa (lead, zinc, copper) in Queensland and Pilbara (iron ore) in Western Australia. Other globally significant mineral deposits in the ancient rocks of Australia include nickel, opals, and uranium (see "Geographies of Indulgence, Desire, and Addiction: Uranium," p. 434). Coal is a major Australian export. Mined in Queensland, New South Wales, and Victoria, it provides 85% of the country's electricity and is a significant source of greenhouse gas emissions. New Zealand also has abundant mineral resources, including iron, coal, and gold.

Minerals are also significant to the economies of a distinctive set of islands across Oceania where mining has destroyed landscapes and created social tensions over the wealth that flows from mineral exports. Perhaps the most dramatic example is Nauru, an oval island only 21 square kilometers (8 square miles) in size that consists of an uplifted coral platform about 30 meters (100 feet) above sea level. Centuries of roosting seabirds had left most of Nauru covered with deep deposits of **guano** (bird droppings) producing the highest-quality phosphate rock in the world. Exploitation of the

(a)

(b)

▲ **FIGURE 11.8 Pacific Islands and Atolls** (a) The islands of French Polynesia, such as Moorea Island, shown here, have volcanic mountains that fall to narrow coasts. Clouds form over the mountains. (b) Low islands and reefs surround a lagoon in Tetiaroa Atoll in French Polynesia. These low islands are vulnerable to sea-level rise.

GEOGRAPHIES OF INDULGENCE, DESIRE, AND ADDICTION
Uranium

Much like oil in the Middle East, uranium links Oceania to the global hunger for cheap energy and to geopolitical conflicts. Uranium is a radioactive element. Controlled uranium reactions can generate electricity in nuclear power plants; uncontrolled reactions, which release enormous amounts of energy and radioactivity, can be used in bombs. Uranium became a valuable commodity in the wake of the U.S. bombings of Hiroshima and Nagasaki in World War II (see Chapter 8).

After World War II, the United States, Britain, and France joined the Cold War arms race and the effort to develop even more powerful weapons. They chose to test many of these uranium-based weapons on Pacific Islands, with devastating implications for local residents and environments. Radiation can have serious short- and long-term effects on people, including radiation poisoning, leukemia, and birth defects. The United States conducted nuclear tests between 1946 and 1958 in the Marshall Islands, relocating the residents of Bikini and Enewetak Atolls (**Figure 11.2.1**). Although the prevailing winds were supposed to carry the bomb's radioactive fallout away from inhabited islands, in 1954, radioactive ash dusted the island of Rongelap. The U.S. government evacuated its residents at short notice without warning them that they might not be able to return or sharing information about health risks.

Years later, in 1968, the residents of Rongelap and Bikini were told it was safe to return. However, people on Bikini subsequently had to be reevacuated when scientists discovered that dangerous levels of radioactivity persisted in food on the islands. Although the United States has monitored the health of the islanders and established a U.S. $90-million trust fund, many residents of the islands are angry and resentful about the experiments that disrupted their lives.

France similarly conducted more than 150 bomb tests on the tiny atolls of Moruroa and Fangataufa in French Polynesia beginning in 1966. The first bombs showered the surrounding regions with radioactivity, reaching as far as Samoa and Tonga, hundreds of miles to the west. Opposition from other Pacific Islands and New Zealand and Australia culminated in boycotts of French products, including wine and cheese, during the 1970s. Locals in Polynesia used the bomb tests as a reason to seek independence from France, and international activists have tried to halt testing (**Figure 11.2.2**). In 1985, the environmental group Greenpeace planned to protest tests by sailing its ship *Rainbow Warrior* to Moruroa, but French intelligence agents scuttled the ship while it was moored in the harbor of Auckland, New Zealand.

The resulting international scandal prompted New Zealand to take a strong stand against nuclear proliferation, banning all nuclear-powered and nuclear-armed vessels from its harbors (against the objections of the United States), breaking off diplomatic relations with France, and taking a leadership role in the antinuclear movement in the Pacific.

▲ **FIGURE 11.2.1 Bikini Atoll** The United States tested atomic bomb at Bikini Atoll in the Marshall Islands on July 25, 1946. Fallout from this and subsequent tests posed serious health risks to Pacific Islanders and resulted in the evacuation of several atolls.

phosphate for use as a fertilizer began in 1906, when it was exported to mainland Australia to make poor soils productive for crops and pasture. Phosphate dominated the economy of Nauru and was strip-mined, crushed, and sent by conveyor belts to ships that anchored outside the reef that surrounds the island. But the phosphate ran out in the 1990s, and the landscape of Nauru remains a desolate wasteland stripped of vegetation and soil, with cavernous pits dotting a rainless rocky plateau (**Figure 11.9**, p. 436). Drinking water comes from an aging desalination plant or is shipped in from Australia.

Mining also exacerbates problems on the island of Bougainville, which is controlled by Papua New Guinea but is geographically and culturally part of the Solomon Islands. The giant Panguna copper mine, owned by the multinational company Rio Tinto, was one of the world's largest open-pit mines and contributed as much as a quarter of Papua New Guinea's export earnings in the 1980s. The mine was developed in the forests without the participation of the resident Indigenous Nasioi and has polluted several rivers that provided fish and drinking water to other groups. Local people seeking a share of the copper revenues, concerned about the mine's environmental impacts, and demanding independence from Papua New Guinea, have joined a rebel movement. The conflict over mining on Indigenous lands in Bougainville mirrors the confllicts surrounding the Ok Tedi

New Zealand's actions contributed to the announcement by France in 1996, after riots in Tahiti, that it would end nuclear testing.

Australia has more than 30% of the world's uranium reserves. Mining uranium can release radioactivity into the landscape and creates risks for mineworkers. Some of the most important Australian resources of uranium are found on or near Aboriginal lands and in conservation areas (**Figure 11.2.3**). For example, the Ranger mine began operations in the Northern Territory in 1980 within the boundary of Kakadu National Park, a UNESCO World Heritage Site of great natural beauty and cultural value. (Recall that these sites are landscapes, buildings, or cities considered of international cultural or physical significance by the U.N.) The mine has produced more than 16 million tons (35 billion pounds) of radioactive waste and has created serious water-pollution problems in the area. ∎

▲ **FIGURE 11.2.2 Greenpeace Ship** A Greenpeace ship is moored off Tahiti during protests against nuclear testing. This ship replaced the *Rainbow Warrior*, scuttled by French agents in the harbor of Auckland, New Zealand, in 1985. New Zealand has declared its ports as nuclear-free zones and has joined Pacific Island nations in strongly protesting nuclear activities in Oceania.

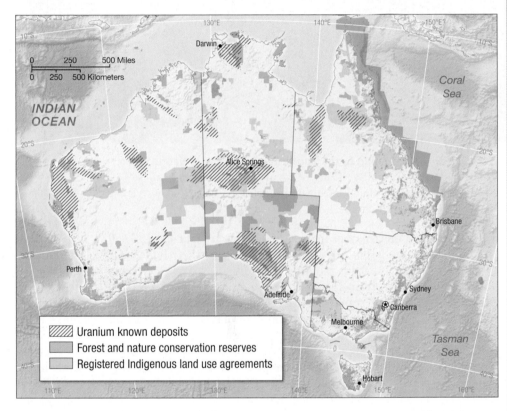

▲ **FIGURE 11.2.3 Uranium in Australia** Australia has internationally significant uranium resources that overlap with protected areas and Aboriginal land rights across the country. There are often conflicts over uranium development in these areas.

copper and gold mine on the mainland of Papua New Guinea, where the pollution of rivers by mine tailings has prompted international environmental concern.

APPLY YOUR KNOWLEDGE Conduct Internet research to find an example of how minerals in Oceania connect the region to the rest of the world while also causing conflict between local residents.

Ecology, Land, and Environmental Management

The relatively long isolation of Oceania from other regions and continents has contributed to the development of some of the world's most unique, diverse, and vulnerable ecosystems. Many of the species that evolved from the isolated populations have remarkable adaptations to the physical environment and are found only in that locality. These ecosystems were often significantly changed by the contact of the region with other regions through exploration, trade, and migration. As such, the region bears the ecological marks of both its isolation and global connectivity.

▲ **FIGURE 11.9 Mining in the Pacific** The landscape of Nauru has been devastated by the mining of the phosphate rock that had accumulated from the guano of roosting seabirds.

Australia is home to more than 20,000 different plant species, 650 species of birds, and 380 different species of reptiles partly because of its diverse climates that range from the tropics to cooler temperatures of southeastern Australia and Tasmania (**Figure 11.10a**). Australia has several remarkable species of **marsupials** and of **monotremes**, animals descended from mammals that died out on most other continents. A marsupial gives birth to a premature offspring that then develops and feeds from nipples in a pouch on the mother's body. Australian marsupials range in size from the large gray kangaroo (adults grow to 2 meters—6 feet) to koalas, wombats, and ferocious Tasmanian devils to small mice and voles (**Figure 11.10b** and **11.10c**). Monotremes are very unusual mammals in that they lay eggs rather than gestate their young within the body but then nurture the young with milk from the mother. Examples include the platypus, with a duck-like bill and beaver-like tail, and the spiny anteater.

Australian ecosystems are further defined by a number of dominant vegetation species, which include several types of forest and scrubland plants well adapted to dry climates and frequent fires. The most important type of tree is the eucalyptus, also known as the gum tree. The tree's leaves contain strong-smelling oils that are used to treat respiratory illness. Eucalyptus has been introduced into many other world regions for reforestation and pulp plantations, including California in the United States. Acacia, an important hardwood locally called *wattle*, is another dominant tree and shrub type, common especially in the drier woodlands. The densest forests are found along the wettest portion of the north eastern coast, with the largest remnants composed mainly of trees of Southeast Asian origin interspersed with woody vines (*lianas*). The forests of the southeast and southwest are dominated by drought-adapted eucalyptus, with more open woodlands in the transition to the dry interior. The northwestern and southern regions have shrub vegetation called *mallee* that includes low, multibranching eucalyptus and plains covered with

bushes similar to the sagebrush found in North America. The northern interior has extensive but sparse grasslands, and the driest zones have scattered grasses and shrubs characteristic of desert ecosystems (but without cacti).

Two-thirds of New Zealand was heavily forested prior to the arrival of humans about 1,000 years ago, with towering conifers (called kauri) in the north and beech in the cooler south. The remaining third of the land was covered with scrub, with grasses at drier, lower elevations and alpine grasslands (tundra) at high altitudes. Evolution in New Zealand produced no predators or carnivores, and several birds remained flightless, including the the kiwi bird and the now extinct moa, which was 3 meters (nearly 10 feet) tall.

On the islands of the Pacific, plant species that can be easily transported by ocean (for example, coconuts) or air (for example, fruit seeds eaten and excreted by birds) are more widely distributed. The variety declines as one moves eastward, away from the larger landmasses. Luxuriant rain forests are found on the wetter and higher islands; marshes and mangroves thrive along the coastal margins. The larger islands, such as New Guinea and Hawaii, also have extensive middle-elevation grasslands. The smaller and drier coral islands have much sparser vegetation, but coconut palms, which are a basis of human subsistence and grow along many beaches, are ubiquitous. There are few native mammals on the Pacific Islands (with the exception of New Guinea). The richest fauna include the birds that have been able to fly from island to island and marine organisms, especially those of reefs and lagoons, including turtles, shellfish, tuna, sharks, and octopus. New Guinea has more than 750 species of birds (**Figure 11.10d**).

Exotic and Feral Species Beginning with the introduction of the dingo from Southeast Asia by Australian Aborigines about 3,500 years ago, foreign species have devastated the native species of Oceania. The dingo, a canine similar to a coyote, is believed to have outcompeted and outhunted the marsupial predators, such as the now-extinct Tasmanian wolf. **Ecological imperialism**—the process of European organisms taking over the ecosystems of other regions of the world—led to the endangerment and extinction of numerous other native species in this region as introduced species came to dominate and local species lost out due to pressures from hunting, competition, and habitat destruction. Introduced species are also called **exotics** because they come from elsewhere.

The flightless birds of Oceania were the most vulnerable to exotic predators such as rats, cats, dogs, and snakes. Conservation efforts to protect birds today include the establishment of reserves and the careful monitoring and elimination of predators. On the Pacific Islands, there are great efforts to contain the spread of the introduced brown tree snake, the mongoose (a very aggressive small mammal), and a carnivorous snail, all of which prey on local species. Other threats to local species included European weeds, pests, and crops and escaped domesticated livestock. **Feral** animals of Australia—domesticated species that end up in the wild—include horses, cattle, sheep, goats, and pigs as well as camels (which were introduced to provide transportation across vast deserts). Escaped populations are as large as 23 million pigs, 2.6 million goats, 300,000 horses (called brumbies), and 300,000 camels and comprise the largest population of feral domestic animals in the world.

Another ecological disaster was the introduction of the European rabbit to Australia in 1859. Over the next 50 years, the rabbit

▲ **FIGURE 11.10 Animals in Oceania** (a) King penguins walk on the coast in the cool climate of Macquarie Island, Tasmania. (b) Wild kangaroos graze and play on a golf course in southern Australia. (c) A marsupial koala, which lives almost exclusively on eucalyptus leaves, is shown here with its young. A new baby is carried in its mother's pouch for about six months. (d) A colorful bird of paradise from New Guinea.

population exploded to plague proportions that devastated pasture-lands. The rabbit population was partially eradicated in the 1950s by the deliberate introduction of a disease called myxomatosis. Similarly, the introduction of the prickly pear cactus in the 1920s to create hedges led to the infestation of more than 20 million hectares (50 million acres) of Australian pasture before the cactus was eradicated by introducing a beetle that fed on the plants. The cane toad, introduced from Hawaii to control pests in the sugarcane district of Queensland, has destroyed native frogs, reptiles, and small marsupials; is highly toxic to predators; and has spread to northern Australia. Despite attempts to curb its expansion, it has gone largely unchecked

since its introduction; cane toads, it ironically turned out, did not actually eat the cane beetle, which it was originally imported to destroy.

APPLY YOUR KNOWLEDGE Use the Internet to research the current conservation and management status of one of Oceania's unique species (e.g., platypus, Tasmanian devil, etc.) or an invasive species in a specific country of Oceania (e.g., prickly pear cactus, cane toad, etc.). What policies have been put in place, and have they been successful?

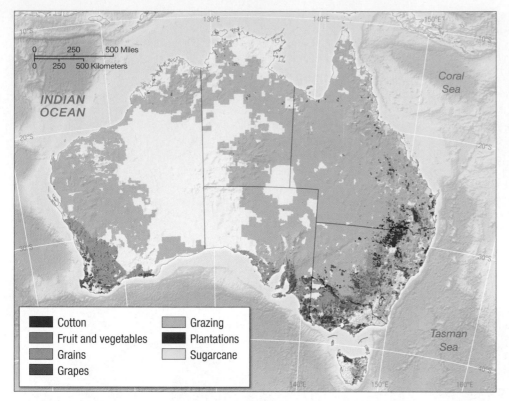

▲ **FIGURE 11.11 Australian Agriculture** While much of Australia is dry and used primarily for grazing, there are important agricultural areas across the country where grains, such as wheat, are grown. There are also areas along the tropical coast of Queensland that grow sugarcane. The Mediterranean climate of southern Australia supports cotton, rice, and a thriving wine industry.

this area and from Western Australia combine to support an export wheat industry that ranks with that of Canada and France (only the United States is ahead of this group). The irrigated farms of the Murrumbidgee Irrigation Area and the grazing lands of the Riverina are also very important to agriculture. Cotton from New South Wales contributes to Australia's dominance in world cotton exports. Southeastern Australia also includes several areas of intensive horticulture (fruits and vegetables) and vineyards that produce world-class wines (**Figure 11.12**).

Land use practices in New Zealand are similarly dominated by livestock development, particularly sheep, lamb, and dairy cattle farming as well as wheat and barley production (see chapter opener). In similar fashion to Australia, New Zealand has put some of its land into the development of vineyards and also invested in a reforestation program for productive woods. Across the Pacific Islands, the primary agricultural land uses are the production of local subsistence crops, as well as coconut, oil palm and sugar plantations.

APPLY YOUR KNOWLEDGE Why might countries in Oceania be worried about buy-local food movements in North America and Europe?

Agricultural Production and Land Use One of the most important agricultural regions in Oceania is the area of southeastern Australia that stretches from the Gold Coast in southeastern Queensland; south along the coast to the cities of Newcastle, Sydney, Wollongong, Melbourne, and Adelaide; and inland to Australia's capital city of Canberra and the agricultural regions of the Darling Downs, Murrumbidgee Irrigation Area, and the Barossa Valley (**Figure 11.11**). These rich agricultural lands host a livestock industry of milk, beef, and lamb production, with animals that graze on pastures improved by fertilizer (especially phosphate) and introduced grasses. The heart of the Australian wheat industry is located in a "fertile crescent" that stretches from the Darling Downs of southern Queensland to central South Australia. Wheat and other grains are often grown in rotation with sheep raising on larger farms. Wheat from

▲ **FIGURE 11.12 Wine in Australia** South Australian vineyards, such as the ones shown here in the Yarra Valley, Victoria, are an ideal spot to grow grapes, with the perfect soil, climate, and afternoon sea breeze.

Ocean Ecosystems The marine environment defines Oceania. Oceania provides many interesting examples of how communities manage renewable marine resources such as fisheries. Fisheries are often thought of as **common property resources**, which are managed collectively by a community that has rights to the resource, rather than owned by individuals. Strategies for managing marine resources in the Pacific include moratoriums—periods or places where fishing is not permitted by local communities—called *tabu* in the Pacific. Family or group access based on customary rights to harvest a resource is also recognized. Recently, some countries have brought fisheries under government control or have regulated harvesting more formally through permits and quotas.

The establishment of the international 200-nautical-mile (370-kilometer) **exclusive economic zone (EEZ)**, which was formalized in 1982 in the **United Nations Convention on the Law of the Sea (UNCLOS)**, was of tremendous significance to Oceania because it allowed countries with a small land area but many scattered islands, such as Tonga and the Cook Islands, to lay claim to immense areas of ocean. In addition to the EEZs, UNCLOS established the International Seabed Authority to manage seabed mining. The pattern of Pacific Islands is such that most of the region is now covered by their EEZs, with relatively little unclaimed ocean (**Figure 11.13**). These island nations can now demand licensing fees from the international fleet that seeks to catch tuna, for example, within their zones.

Outside of the EEZs, fisheries are open to all and are vulnerable to the so-called **tragedy of the commons**, in which an open-access common resource is overexploited by individuals who do not recognize how their own use of the resource can add to that of many others to degrade the environment—for example, overfishing a given

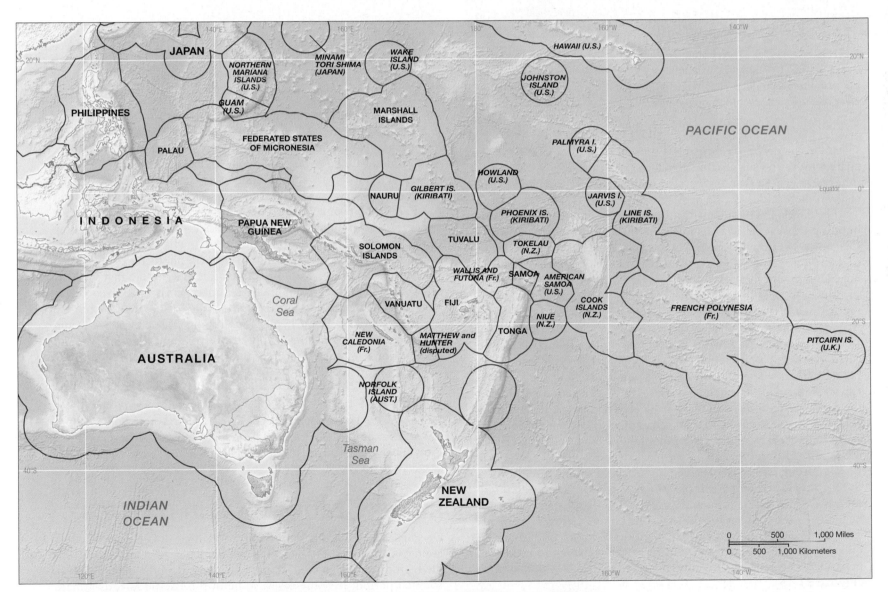

▲ **FIGURE 11.13 Marine Territorial Claims in the Pacific** The exclusive economic zones (EEZs) of Pacific nations and territories leave very little unclaimed ocean in the region. UNCLOS, the international treaty, which established these zones, was ratified by 119 member states of the United Nations in 1982 and now consists of more than 130 signatory members. One result of the treaty is that 90% of the world's fisheries are now claimed.

▲ **FIGURE 11.14 Overfishing** Greenpeace activists chase a fishing boat that they claimed was fishing illegally for yellowfin tuna using banned purse sein nets. They painted 'Pirate' on the side of the boat. Overfishing threatens tuna populations.

▲ **FIGURE 11.15 The Great Barrier Reef** The undersea landscape of the Great Barrier Reef has become a major destination for tourists who swim among the corals and grasses and view the colorful tropical fish.

species to the point of extinction (**Figure 11.14**). One of the greatest challenges in the sustainable management of fisheries is the lack of information about fish numbers, movement, and reproduction, especially in the Pacific.

The development of international law governing ocean space is also important to Pacific Islanders because they eat more fish per person than any other population, and fishing and coastal tourism are critical to the majority of smaller island economies. Marine resources include not only fish and shellfish but also valuable exports such as pearls and shell (mostly for shirt buttons), as well as products made from whales, the species that initially attracted many Europeans to the Pacific. At the same time, because of the warm water, the southern Pacific, where the islands are, is actually less biologically productive than the colder water that wells up near the continental shelves of Australia and New Zealand. The reefs near the low islands also limit the harvest of fish and other marine resources because coral organisms use up most of the nutrients, and the coral reefs snag nets. Island societies consume a very wide range of fish species as well as other marine organisms, such as sea cucumbers (which are also exported to China) and giant clams.

Within this region's massive marine ecosystems is the **Great Barrier Reef**, which is a UNESCO World Heritage Site. Fringing the northeast coast of Australia, the reef is more than 2,000 kilometers (1,250 miles) long and easily visible from space. The real beauty of the landscape is underwater. Diving below the surface, the visitor encounters intricate and colorful corals and waving sea grasses that are home to millions of brilliant fish and marine animals such as turtles, whales, and dugong (large seal-like mammals) (**Figure 11.15**).

The Great Barrier Reef consists of 3,400 coral reefs, incorporating more than 300 species of coral and hosting more than 1,500 species of fish and 4,000 different types of mollusks. The reefs were formed over millions of years from the skeletons of marine coral organisms in the warm tropical waters of the Coral Sea. The reef is Australia's second-most-popular foreign tourist destination (after Sydney). Tourists visit the reef and participate in activities that include fishing, diving, snorkeling, and reef walking. Today, the reef is under pressure from fishing with large nets that damage the reef, chemical and sediment pollution, climate change, and coastal development.

Concern about ocean pollution has recently crystallized around the **Pacific Garbage Patch**—an area where ocean currents circle and trap large amounts of plastic and other waste that has ended up in the Pacific (**Figure 11.16a**). The patch, which is in the Pacific, north of Hawaii, consists of millions of small plastic particles that do not degrade and can poison or choke birds and marine mammals. It symbolizes the more general human impact on the marine environment across Oceania that includes litter (such as the debris from the Japanese Tsunami—see Chapter 8), oil spills, plastic bags, waste and accidental losses from ships (**Figure 11.16b**), and the warming and acidification from higher levels of the greenhouse gas carbon dioxide.

APPLY YOUR KNOWLEDGE Conduct some Internet research to learn about efforts to reduce the pollution in the world's oceans. Who is undertaking these efforts? What do clean-up initiatives involve? How successful have they been to date?

HISTORY, ECONOMY, AND TERRITORY

Oceania's history is distinguished by a number of key phases, which include early human habitation, the extension of populations into the region, European contact and colonization, and a dynamic postcolonial period. In some cases, some of the Pacific Islands remain colonized—either formally or through association—by Australia, New Zealand, France, the United Kingdom, and the United States. The historical geography in the region has produced economies that are globally and regionally significant—such as Australia, which plays an important development role in other regions, such as Southeast

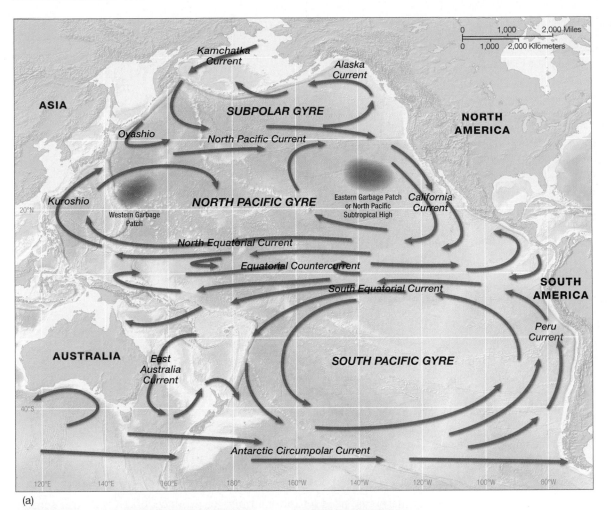

(a)

(b)

▲ FIGURE 11.16 The Pacific Garbage Patch (a) The currents that cross the Pacific Ocean in circular patterns, called gyres, trap plastic and other waste into accumulations of rubbish that damage marine ecosystems. So far, garbage has accumulated in both the western and eastern north Pacific. (b) Garbage that has washed up on the beach of the Pacific atoll, Kiribati, is shown here.

dispersal to the Pacific Islands, such as Fiji, Tonga, and Samoa, about 3,500 years ago.

The early inhabitants of Australia were the ancestors of the Aborigines, who still live in Australia today. Aborigines in Australia still preserve practices that reflect the early adaptations and modifications of the Australian environment. These practices include a complex spiritual relationship to the land, a nomadic lifestyle, and the use of fire in hunting. Roots, seeds, grubs, insects, and lizards were gathered for essential calories and proteins and people in some coastal areas also constructed traps for stonefish and eel. As Aborigine populations grew, they may have reduced local populations of major game species such as kangaroos, but the most significant environmental change they implemented was the transformation of vegetation through the use of fire. Fire was employed by Aborigines both to improve grazing for game and to drive animals to hunters. Ecologists believe that over thousands of years, some Australian vegetation became more resistant to these fires. The Aborigine worldview links people to each other, to ancestral beings, and to the land through rituals, art, and taboos. This worldview is associated with the **Dreamtime**, a concept that joins past and future, people and places, in a continuity that ensures respect for the natural world.

The **Maori**, the Indigenous people of New Zealand, arrived in New Zealand from eastern Polynesia sometime before 1300 C.E. and are believed to have caused much more widespread environmental transformations than Australian Aborigines, in part due to the smaller extent of the island environment. The Maori hunted the enormous moa bird to extinction, cleared as much as 40% of the original forests, and practiced agriculture based on shifting cultivation of sweet potatoes—a knowledge system they brought with them when they migrated to the country.

Asia—and underdeveloped—such as Papua New Guinea. Countries such as Papua New Guinea have remained underdeveloped because they have historically been positioned in a dependent relationship to larger global economic powers.

Historical Legacies and Landscapes

The early human history of Oceania is divided into two main phases: the migration of humans from Southeast Asia into Australia, New Guinea, and nearby islands about 40,000 years ago, and a second

European Exploration and Colonization of New Zealand and Australia Although Oceania had some early contact with other regions of the world, especially Southeast Asia, it was not until the mid-1700s that European explorers established the area for wider global trade and eventual colonization. Spanish and Portuguese

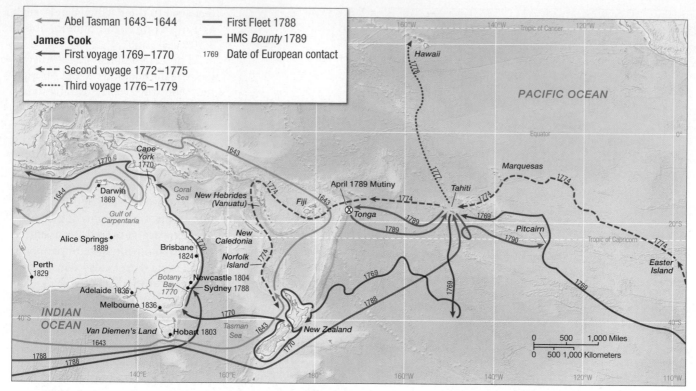

▲ **FIGURE 11.17 The European Exploration of Oceania** This map illustrates the sequence of European exploration and settlement of Oceania, including the voyages of Abel Tasman, Captain James Cook, and of the HMS *Bounty*. Cook made three voyages to Oceania, landing at Botany Bay and claiming Australia for Britain; he was ultimately killed in Hawaii.

sailors controlled the island of Guam, which was a stop for galleons traveling from Manila (Philippines) to Acapulco (Mexico), and may have encountered Australia when sailing around the area. The Dutch explored the west coast and south coasts of Australia and claimed it as Van Diemen's Land in 1642. The Dutch later named Tasmania after the Dutch explorer Abel Tasman (**Figure 11.17**). Tasman also approached New Zealand and made contact with the Maori in 1642, but the encounter was violent and the Dutch did not land. They did, however, begin calling the islands *Nieuw Zeeland* (after a region of the Netherlands). It was more than a hundred years later that the most enduring claim on the region was advanced by explorer Captain James Cook, who landed at Botany Bay, Australia, in 1770. Cook claimed the land for Britain and called the new territory New South Wales. Cook also explored New Zealand at this time and established harmonious relationships with the Maori.

Based on Cook's reports, the British government decided to populate New South Wales by using it as a penal colony. The British sent boatloads of convicts to the region to relieve pressure on their prisons and to reinforce territorial claims and provide cheap labor for economic development. More than 160,000 convicts were eventually transported to Australia by 1868, half to New South Wales and half to Van Diemen's Land (Tasmania). Most were not serious criminals, and many have been identified as petty thieves or Irish political activists. The convicts worked for the government or were assigned to private employers. Many eventually gained freedom through pardons and decided to stay in Australia. Pardons were often prompted by the

desperate need for more farmers to produce food. Other free settlers arrived when offered cash rewards for emigrating and assigned convicts as laborers.

The initial goal of the settlements was self-sufficiency, but many of them, including Sydney, were unable to produce an adequate range of foodstuffs because of poor soils, plant disease, and a variable climate. These problems persisted even after convicts were freed to increase the number of farmers, and many settlements depended on imports of food and other goods from Britain. The most successful agricultural areas were in coastal valleys, such as the Derwent Valley in Tasmania, which produced wheat. The main exports from the initial settlements were associated with the whaling trade, such as whale oil and seal pelts. Hobart, Tasmania, served as the main base of the Pacific Island whaling fleet.

A momentous shift took place with the import of the first livestock to Australia, especially the Merino sheep, which thrived in central New South Wales and Tasmania. The first wool shipment to Britain was in 1807, and high prices of wool globally encouraged expansion of the sheep industry in Australia. By 1860, exports totaled 16 million kilograms (35 million pounds), and there were 21 million sheep in Australia. The demand was driven by the success of the British textile industry (see Chapter 2) and tied the pastoral economy of Australia as well as New Zealand (see chapter opener) to manufacturing in Europe. The Australian Agriculture Company was established in 1824 and invested 1 million pounds (about U.S. $4.8 million) to promote wool and other agricultural exports. Although the British government was initially reluctant to permit frontier settlements away from the port communities, stockmen began to move inland, especially to the grasslands, such as the Bathurst Plain on the western slopes of the Great Dividing Range. As the European frontier expanded, European settlers came into conflict with the Aborigines, who defended their traditional lands and lost their lives in the process.

Commercial wheat production was centered in southeastern Australia, especially on the red-brown soils around Adelaide. The frost-free climate of the central coast of eastern Australia also allowed for sugar plantations around Brisbane beginning in the 1860s. These plantations depended on indentured laborers brought in from Vanuatu and the Solomon Islands. Cattle were also introduced into the warmer and drier regions of central and northern Australia after wells were drilled into the Great Artesian Basin in the 1880s. Cattle survived more easily than sheep on the sparse vegetation. The construction of railroads, which radiated from ports to terminuses at livestock yards, grain elevators, and mines, further facilitated development in the region. This transportation pattern increased the importance of the port cities but hindered later national integration in the 1890s, especially when it became evident that three different rail widths had been selected by different colonies.

Another major transformation of the Australian economy occurred with the discovery of gold in 1851 (see p. 433). The mid-1850s brought a great degree of self-government to Australia; at that time, two-thirds of the legislatures were elected by popular vote (and the rest appointed by the British) in the states of New South Wales, Victoria, South Australia, Queensland, and Tasmania. As in many other colonized regions, the boundaries between the Australian states were mostly drawn as straight lines, regardless of physical features or Indigenous land rights.

The history of European settlement and economic development in New Zealand began with the establishment of a sealing station on the South Island of New Zealand in 1792. However, British sovereignty and the first official settlers' colonies were not established until the 1840s in New Zealand. Initially it was treated by the British as part of New South Wales, Australia. Small whaling settlements established trading relationships with the Maori. This contact gave the Maori access to firearms, exposed them to disease, and ushered them into the capitalist economy. Missionaries had also established settlements in the early 1800s, contributing to the further transformation of Maori culture.

The British annexed New Zealand in 1840 through the **Treaty of Waitangi**, a pact with 40 Maori chiefs on the North Island **(Figure 11.18)**. This treaty purported to protect Maori rights and land ownership if the Maori accepted the British monarch as their sovereign, granted the British crown monopoly on land purchases, and became British subjects. At the last minute before the treaty was signed, land agents purchased large areas of land around the Cook Strait, often without identifying the true Maori owners. This land was held by the private New Zealand Association and included the sites of the cities of New Plymouth, Wellington, and Nelson.

Alarmed by European settlement, some Maori resisted the British and waged warfare for several years until suppressed in 1847. The introduction of sheep prompted further settler expansion in search of

▲ **FIGURE 11.18 Treaty of Waitangi** Waitangi Day commemorates the treaty signed by Britain's Queen Victoria and the Indigenous Maori peoples. Maori often protest on this annual holiday, claiming that many promises remain unfulfilled, including those addressing land and resource rights. Here, hundreds of demonstrators in a hikoi (march) head for the Waitangi Treaty Grounds in Waitangi, Bay of Islands, New Zealand.

pastures and a renewal of hostilities with the Maori during the Maori Wars of the 1860s.

After 1882, a technological innovation, the development of refrigerated shipping, allowed the economies of New Zealand and Australia to expand and shift from producing nonperishables such as wool, metals, and wheat to exporting the more profitable meat and dairy products. Trade was facilitated by the opening of the Suez Canal in 1867 and the Panama Canal in 1914, which reduced the time and cost of ocean transport to Europe. Australia and New Zealand became staple economies that depended on the export of a few natural resources.

European Exploration and Colonization of the Pacific Islands Initially the Pacific Islands were of little interest to European explorers, and they were drawn into the world economy more slowly than Australia and New Zealand. During the era of Spanish exploration in the 16th century, Guam, Palau, parts of the Federated States of Micronesia, and the Mariana Islands became Spanish colonies. More generally, explorers brought European diseases to local populations who had no resistance to them, so the most serious impact of early exploration was mortality.

But as Britain and France rose to power in Europe in the 18th century, a series of adventurers set out for the Pacific Islands. British explorer Samuel Wallis and Luis Antoine de Bougainville of France were made welcome by the people of Tahiti in the 1760s, and their reports of friendly people and abundance cultivated the myth of a tropical paradise.

EMERGING REGIONS
Alliance of Small Island States (AOSIS)

More than 40 countries and territories are members of the Alliance of Small Island States (AOSIS), which was established in 1990 to represent islands in international negotiations, especially those relating to climate change. Oceania is represented in AOSIS by 16 members, ranging from Papua New Guinea to tiny Niue and Nauru. In 1992, the United Nations recognized a similar grouping—**Small Island Developing States (SIDS)**—that includes 52 small islands that share similar sustainable development challenges. While SIDS only includes independent nation states, AOSIS includes islands that are still territories of other countries such as Guam (which is a territory of the United States).

These organizations provide an example of how shared geographic challenges can bring countries together around common interests and can constitute an emerging region with moral power in international negotiations. These islands face common development challenges that include small populations, limited resources, isolation, a dependence on tourism, high energy and transport costs, vulnerability to climate and economic shocks, and fragile environments.

They have lacked the power to influence international discussions as individual countries and joined together seeking a more powerful voice. AOSIS and SIDS cross lines of more traditional coalitions of nations within the United Nations, such as the African Group, the Least Developed Countries group, the European Union, or the members of the British Commonwealth.

One of the geographic challenges most important to island countries is climate change because of the risks they face from sea-level rise caused by warming oceans and melting ice (**Figure 11.3.1**). These small islands produce almost no greenhouse gas emissions but are tremendously vulnerable to climate change. Some islands, especially atolls, are so low that sea-level rise threatens their very existence. For example, groundwater resources on the islands of Kiribati are already becoming salty as sea-level rise allows seawater to penetrate the freshwater basins used for drinking water (**Figure 11.3.2**).

In addition, many tropical islands, especially in the Caribbean, lie in the path of hurricanes, which may intensify as a result of climate change. Because of the importance

of their marine environments in terms of food security and tourism, small islands are also concerned about **ocean acidification**. Acidification is caused when ocean waters take up excess environmental carbon dioxide and become more acidic as a result. The acidity damages marine organisms, especially corals and shellfish.

The shared interests in climate change and the development aspirations of small island states in the Pacific region can be seen in their contributions to the **RIO+20 United Nations Conference on Sustainable Development** that took place in June 2012. At RIO+20, AOSIS, and SIDS promoted sustainable development that addresses the unique and particular vulnerabilities of small island states and advocated real political action to reduce the risks of climate change through emission reductions and adaptation assistance. A growing number of environment and development organizations support the positions and claims of AOSIS and SIDS, such as the **Blue Economy**, which supports the sustainable management and equitable sharing of marine and ocean resources. ∎

▲ **FIGURE 11.3.1 Sea-Level Rise and the Pacific Islands** The Tarawa Climate Change Conference was held in 2010 on the Pacific island of Kiribati to express concerns over global warming, sea-level rise, and the lack of action on emission reductions by industrial countries.

▲ **FIGURE 11.3.2 Desalination on Tuvalu** Groundwater supplies on low-lying islands such as Tuvalu are becoming salty as sea level rises. One adaptation to this problem is the construction of desalination plants that produce clean water, but only at considerable cost.

The imposition of colonial rule met with resistance in the form of armed uprisings, alternative trading networks, political movements, and defiant behavior. The colonial powers restructured the economies and people of the Pacific Islands in ways that left enduring legacies. For example, the British brought large numbers of contract workers from India to work on plantations in Fiji, creating a divided society of ethnic Asians and Pacific Islanders that produces significant political tensions even today.

Political Independence, World War II, and Global Reorientation Independence was granted to Australia in 1901. The Commonwealth of Australia was established with a federal structure governing the six states of Western Australia, South Australia, Queensland, New South Wales, Victoria, and Tasmania. The Commonwealth also administered the federal territories of the Northern Territory and (after 1908) the Australian Capital Territory around Canberra. New Zealand declined to join this new nation, instead choosing dominion status as a self-governing colony of Britain in 1907. Both countries gained their own colonies in the Pacific in 1901, when Britain turned over Papua New Guinea to Australia and the Cook Islands to New Zealand. New Zealand was granted a mandate over Samoa in 1920 and jurisdiction over Tokelau in 1925.

By taking over as regional powers, both Australia and New Zealand also established themselves globally. They further exercised this new global authority by fighting on the side of the Allies in World War I in the Australian and New Zealand Army Corps (Anzac). At Gallipoli, Turkey, in 1915, more than 33,000 Anzac soldiers died. The war contributed to the development of a domestic steel and auto industry in Australia. The difficulties the war created in international trade also increased the prices for agricultural exports.

World War II further influenced Australia's geopolitical orientation. Although Australia fought in defense of Britain in Europe and North Africa, the rapid advance of the Japanese in Asia and the Pacific, including the capture of 15,000 Australians in Singapore and the bombing of the northern Australian city of Darwin, caused Australia to look to the United States more directly as an ally. Because the United States was also fighting the Japanese in the region, the alliance arose naturally. American and Australian troops fought together in New Guinea and the Pacific Islands, forging enduring bonds that were formalized in the ANZUS (Australia, New Zealand, United States) security treaty between Australia, New Zealand, and the United States, signed in San Francisco in 1951. During the Cold War, East and Southeast Asia became the focus of U.S. concern about communist expansion from China, and the United States became more involved in East Asia (see Chapter 8, p. 304) and then Indochina (see Chapter 10, p. 382). Both Australia and New Zealand sent troops to Korea and Vietnam.

World War II also marked a critical turning point in the history of the Pacific Islands. During the war, thousands of foreign soldiers fought and constructed military bases on the islands. In the process, many islanders lost their lives. The most significant casualties were sustained as the Japanese advanced from their colonies in Micronesia (such as Palau) to Guam, New Guinea, and the Solomon Islands and attacked the U.S. base at Pearl Harbor in Hawaii in 1941. The islands endured three years of intense and bitter warfare on land, air, and sea. Many famous battles, such as those of Guadalcanal and Saipan, occurred as the Allies, especially the United States, struggled to retake the islands from Japan. The damage from bombing was extensive.

▲ **FIGURE 11.19 Whaling Station** Whaling stations were established in New Zealand from the early 1800s. Thousands of whales were landed and processed at stations such as Perano's in the Cook Strait, shown here, which operated until the disappearance of whales caused the closure of this station in 1964.

At the beginning of the 19th century, the Pacific Islands came into further contact with missionaries, whalers, and traders. Hundreds of whaling ships called regularly at islands such as Tahiti, Fiji, and Samoa for supplies and maintenance (**Figure 11.19**). The London Missionary Society was very active in the Pacific, seeking conversions in Tahiti, Samoa, Tuamotu, Tuvalu, and the Cook Islands. The Methodists focused on Tonga and Fiji. The missionaries often worked through native chiefs, who were advised to alter local laws and traditions to conform to European principles, sometimes provoking local rebellions. Missionary activity altered traditional social ties, beliefs, and political structures.

Coconut, the staple product of the Pacific, became part of European trade from about 1840 in the form of copra, dried coconut meat, used to make coconut oil for soaps and food. Some Europeans sought to establish plantations on islands such as Fiji and Samoa. A scarcity of cotton during the U.S. Civil War prompted the establishment of cotton plantations on Fiji in the 1860s. The same scarcity increased the need for laborers for the cotton plantations in Queensland, Australia, and workers came from the islands. From the 1840s to 1904, people from the Pacific were kidnapped and enslaved in a process called **blackbirding** that brought thousands of laborers, collectively called *kanakas* (because many of them were of Kanak origin from the islands now known as Vanuatu), to Australian cotton and sugar plantations.

The islands were governed according to the different colonial styles of European nations modified to local conditions. Britain ruled through governors who incorporated native leadership into their administrations in a form of indirect rule. The Germans administered their Pacific island colonies through commercial companies, and the French practiced direct rule and assimilation into French culture and institutions.

▲ **FIGURE 11.20 U.S. Air Force Base on Guam** The United States established bases on islands taken from the Japanese during World War II and maintained a military presence through the Cold War that continues to the present day. This photo is of B-52 bombers at Andersen Air Force Base on the island of Guam in the northern Marianas.

After the war, the United States was determined to maintain military bases on islands such as Guam and American Samoa (**Figure 11.20**), especially in response to the Cold War. Despite this, self-government began with elected governments and small independence movements. Western Samoa became independent from New Zealand in 1962 (and dropped the "Western" from its name in 1997); Nauru from Australia in 1968; Fiji and Tonga from the United Kingdom in 1970; Papua New Guinea from Australia in 1975; and the Solomon Islands, Kiribati, Tuvalu from the United Kingdom, and Vanuatu from the United Kingdom and France by 1980 (**Figure 11.21**).

APPLY YOUR KNOWLEDGE Visit the library or a film archive to look for stories about World War II set in Oceania, such as the classic 1949 Broadway musical South Pacific, which was made into a film released in 1958. How do the stories portray local residents? Do you think these portrayals are accurate?

Economy, Accumulation, and the Production of Inequality

Within the context of Oceania's historical geography, there have been numerous attempts to regulate economic production and accumulation in relation to the global economy. Some countries, such as Australia and New Zealand, have benefited more from global economic integration at some times, and others have benefited less.

Countries that remain dependent on primary commodities, for example, have often found themselves in a position of relative disadvantage. Like countries in other world regions, however, the countries of Oceania have attempted to position themselves in relation to the wider global economy either through the development of protectionist measures or through resistance to global capitalist development.

Isolation and Integration in Australia and New Zealand Recall from Chapter 7 and Chapter 8 that import substitution is a trade and economic policy that supports the development of domestic industries with government protection. Beginning in the 1920s, regional policies to substitute expensive imports with domestic production resulted in heavy subsidies of manufacturing and high tariffs on imported goods, especially in Australia. The goal was to create new jobs, diversify the economy, and reduce the sensitivity to global demand for wool and minerals. As in other regions, these import substitution policies had mixed success. Although Australia and New Zealand had middle-class populations that generated a demand for manufactured goods, the overall market was small, and labor costs were high as a result of a strong tradition of labor union activism. The new industries were often inefficient because they were protected from competition in the world market. New Zealand had a particularly high level of government involvement in the economy, with state-run marketing boards controlling the export of commodities, such as wool, meat, dairy products, and fruit, and government ownership of banking, telecommunications, energy, rail, steel, and forest enterprises.

Although Australia protected manufacturing and subsidized agriculture, the government there set up few barriers to foreign investment. Many sectors, including minerals and land, had high levels of foreign ownership, especially by British firms. During the 1970s, investment patterns began to change, when Asian, especially Japanese, capital starting to flow into Australia. Another major change occurred when the United Kingdom decided to join the European Community in 1973 and was forced to end preferential trade relationships with British Commonwealth nations, including Australia and New Zealand (**Figure 11.22**, p. 448).

Both Australia and New Zealand decided to reduce government intervention in and regulation of the economy in the 1980s. Beginning in about 1983, the Australian government deregulated banking, reduced subsidies to industry and agriculture, and sold off public-sector energy industries. The move to policies of free trade and reduced government was even more dramatic in New Zealand. There, sweeping policy reversals eliminated agricultural subsidies, removed trade tariffs, reduced welfare spending, and privatized government-owned enterprises, including airlines, postal services, and forests. Although these policies succeeded in reducing the national debt, they also increased economic inequality and unemployment.

The most rapidly growing sector of major regional economies has been services. Employment in this sector has increased from 58% in 1980 to 75% in 2010 in Australia and from 58% to 71% in New Zealand during the same time period. Finance, tourism, and business services expanded the most and were associated with an increase

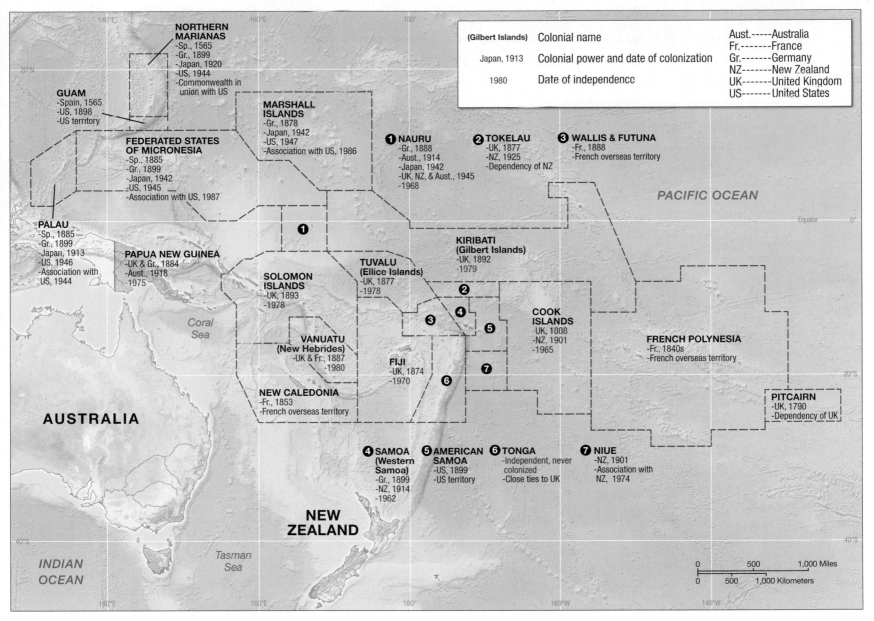

▲ **FIGURE 11.21 The Colonization and Independence of the Pacific** Many Pacific Islands are now independent from their former colonial rulers. This map shows the pattern and dates of colonial control, dates of independence of Pacific island countries, and current affiliations of territories. Note that only Tonga resisted colonization and that some countries, such as Nauru and Palau, were handed from one European power to another.

in female employment and considerable foreign investment. Oceania has become a major international tourist destination. Of the more than 11.7 million visitors in 2011, many came from Asia, especially Japan, China, and Singapore, as well as from the United Kingdom.

The future of agriculture in Australia and New Zealand in the wake of economic restructuring and global economic change is unclear, as competition from Latin America and Asia, the high cost of inputs, the loss of subsidies, and the changing structure of demand create a new geography of international agricultural trade (see chapter opener). That said, Australian wine has gained an excellent reputation in world markets, especially the wines of the Barossa Valley north of Adelaide and the Hunter Valley of New South Wales. Australian wine exports are the fourth largest in the world (behind France, Italy, and Spain). Wine is produced with the most advanced technology and innovative marketing and supports a thriving tourist industry.

Economic Development and Tourism in the Pacific Islands Many Pacific Islands are now integrated into the world system through their dependence on imported goods, transfer payments, improvements in transportation, and the emergence of international tourism. With their long experience of global trade, investment, and migration, the Pacific Islands have in many ways been less affected by more recent globalization of trade, capital flows, culture, and labor than have other world regions. They are sometimes referred to as **MIRAB (migration, remittances, aid, and bureaucracy) economics** because of their

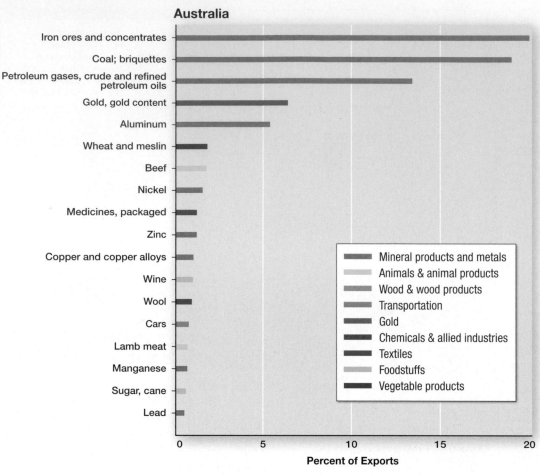

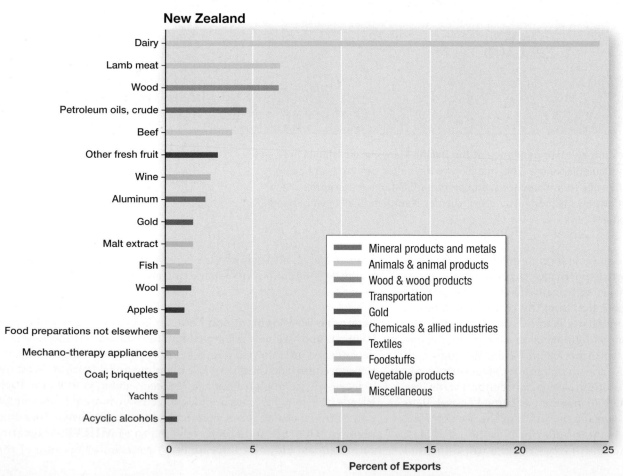

◄ **FIGURE 11.22 Australia and New Zealand Exports** These diagrams show the structure of Australian (a) and New Zealand (b) exports by value in 2009. (a) Australia's exports are dominated by coal, iron, gas, minerals, and wheat, and (b) New Zealand's by dairy products, lamb, wine, and timber.

dependence on labor migration, money sent back from overseas workers, foreign aid, and jobs in government.

Most of the nations in Polynesia spend twice as much on imports as they gain from exports, and they finance the deficit through borrowing, foreign aid, or remittances from citizens working elsewhere. However, compared to countries in Latin America and the Caribbean or Southeast Asia, total international debt is relatively low in Oceania. This low level of debt has been maintained despite the trade deficits because of the large flows of official aid, remittances from overseas workers, and interest on savings held outside the country.

Tourism is extremely important for the Pacific Islands. The widespread image of the islands as a vacation paradise originated in the accounts of the first European visitors, who described tropical abundance and peaceful locals. Island women were often inappropriately portrayed in these accounts as exotic and available partners. The appeal of the region as a tourist destination grew as international air and cruise routes included stops at island groups such as Hawaii and Fiji, often en route to Australia or Asia. But the big boom in Pacific tourism occurred from about 1980 onward when air travel became more accessible, and increased numbers of North Americans and Asians (especially Japanese) sought luxury tropical vacations.

The most popular Pacific island destinations, after the U.S. state of Hawaii, are Guam, Fiji, and Tahiti; the total number of tourists to the Pacific Islands (Hawaii excepted) reached more than 3 million a year in the late 1990s. The significance of international tourism to economies and employment in individual countries is tremendous; tourism is the major source of foreign exchange for the Cook Islands, Fiji, French Polynesia, Samoa, Tonga, Tuvalu, and Vanuatu (**Figure 11.23**). Challenges faced by these tourism-dependent economies include vulnerability to international trends in tourism, political unrest that dissuades tourists, ensuring that the benefits of tourism reach throughout the population, and minimizing negative effects on local cultures and environments. These effects can include overuse of resources, damage to ecosystems, disrespect for local culture, and revenue that is accrued mostly by companies outside the region.

▲ **FIGURE 11.23 Bora Bora** Luxury tourist resorts of Bora Bora in French Polynesia attract tourists with descriptions such as the "islands of dreams" and an "emerald in a setting of turquoise, encircled by a necklace of pearls." Visitors bring valuable foreign exchange to the economy, but resorts put pressure on the local resource base and encourage local people to modify their livelihoods and rituals to cater to foreign visitors.

> **APPLY YOUR KNOWLEDGE** What are the risks for countries whose economies depend on tourism? Pick a country in the region and research economic data on how its tourism industry fared during the recent global economic downturn.

Poverty and Inequality Although Oceania has generally higher incomes and better living conditions than many other world regions, there are significant differences between and within countries of the region. Australia and New Zealand are distinctive for their very high average incomes. The per capita annual gross national income (GNI) was almost U.S. $36,910 (in PPP) for Australia and U.S. $29,140 (in PPP) for New Zealand in 2010. Some Pacific Islands have an average GNI in PPP near or above U.S. $10,000 per person as a result of their associations and jobs with the United States (Guam) or France (French Polynesia, New Caledonia). Others, especially Papua New Guinea and the Solomon Islands, register less than U.S. $3,000 in GNI per capita per year. Although there is little published information on the distribution of incomes within most of the smaller countries, and although inequality is apparently less than in many other world regions, poverty persists throughout the region, including in Australia and New Zealand, where the Aborigine and Maori populations are particularly disadvantaged.

Australia and New Zealand were for many years reputed to have strong welfare systems and equitable societies, at least for non-Indigenous populations. However, income inequality has increased in the last two decades, as governments have reduced or privatized social services, especially in New Zealand. Larger populations of single parents, the elderly, refugees, and workers in low-paid service-sector jobs are also diminishing the overall ranking of Australia and New Zealand as places where everyone can make a good living.

Monetary measures such as GDP and PPP are of limited use where many people are living in economies based on exchanges and barter or on subsistence. The concept of **subsistence affluence** has been used to describe Pacific island societies. In these societies, monetary incomes may be low, but local resources such as coconut and fish provide a reasonable diet, and extended family and reciprocal support prevent serious deprivation. Adequate diets and relatively effective health and education systems contribute to comparatively high life expectancies and literacy and low infant mortality throughout the Pacific Islands. Literacy is above 90% for both men and women in much of the region. Notably, Papua New Guinea, Kiribati, and Vanuatu have higher infant mortality and lower literacy than other parts of Oceania.

Territory and Politics

Oceania has seen substantial political changes in recent years, including the shift in alignment from Europe to North America and Asia and the challenges of coping with its relative geographic isolation within a global economy. The stability of some independent democracies and dependencies has been threatened by internal tensions. Political and economic integration has been sought through regional cooperation agreements. And inequalities within Australia and New Zealand have highlighted the conditions of Indigenous groups, while at the same time, these countries have embraced multicultural identities. Oceania's political geography reflects the complex relationship between geopolitical forces and locally mediated conflicts and tensions.

Regional Cooperation Regional agreements include the South Pacific Commission, founded in 1947, which focuses on social and economic development and includes 21 island nations and territories, Australia, New Zealand, the United States, France, and the United Kingdom. The **Pacific Islands Forum**, established in 1971 as the South Pacific Forum, promotes discussion and cooperation on trade, fisheries, and tourism among all the independent and self-governing states of Oceania. It has supported maritime territorial rights and a nuclear-free Pacific as well as the goals of independence of French Polynesia and New Caledonia. In 2000, the Forum legitimized peacekeeping military operations led by New Zealand and Australia in the Solomon Islands and Papua New Guinea. There are also dozens of nongovernmental organizations and intergovernmental agencies that operate across the region, especially among the smaller Pacific Islands. For example, the University of the South Pacific fosters higher education across 12 countries through distance education and three main campuses in Fiji, Samoa, and Vanuatu.

Australia and New Zealand are members of larger economic and political alliances such as the Asia-Pacific Economic Cooperation group (APEC), which also includes Papua New Guinea and focuses on improving transportation links and liberalizing regional trade around the Pacific Rim (see "Emerging Region: The Pacific Rim" in Chapter 8, pp. 322–323). Both Australia and New Zealand have been able to take advantage of APEC to increase exports to Asia, especially to Japan. In attempts to foster regional markets, Australia and New Zealand created the free trade–focused **Closer Economic Relations (CER) agreement** in 1983. CER set out to remove all tariffs and restrictions on trade between the two countries. The resulting increased trade has been especially beneficial to New Zealand, which has the smaller domestic market of the two countries and has doubled its exports to Australia.

Independence and Secessionist Movements Oceania is relatively politically stable compared to many other regions. A number of smaller or resource-poor islands in the region have maintained close associations with, or are still under the control of, the United States or New Zealand. U.S. dependencies receive financial subsidies, called transfer payments, in return for military base sites and security control. France has also maintained its control over several islands, insisting that they are integral parts of the French state.

The most serious recent political conflicts in Oceania have involved encounters between ethnic groups in Fiji and demonstrations by independence or secessionist movements in New Caledonia and Papua New Guinea. The conflict in Fiji is a legacy of the British colonial policies that brought Asian Indians as indentured workers to local sugar plantations from 1879 to 1920. By the 1960s, Indo-Fijians almost outnumbered the ethnic Fijian population, dominating commerce and urban life and maintaining a separate existence with little intermarriage and continued cultural and religious segregation. The Indigenous Fijians took over the government at the time of its independence in 1970, but subsequent elections have produced contested wins for Indo-Fijian parties, and military takeovers occurred in 1987, 2000, and 2006. Because of recent failures to hold democratic elections, Fiji was suspended from the Commonwealth and the Pacific Island Forum.

In the nickel-rich islands of New Caledonia, the Indigenous Melanesian population, known as Kanaks, has been militantly pressing for independence for years but has been outvoted by those of French descent (called *demis*), who prefer to remain part of France. The Nouméa Accord of 1998 promises to grant independence by 2018. In the Solomon Islands, residents of Bougainville are trying to secede from Papua New Guinea, claiming ethnic affiliation with the other Solomon Islands that are independent and complaining that they do not receive a fair share of the profits from local mines. Within the Solomon Islands, there are conflicts between ethnic groups over land, such as those between longtime residents of Guadalcanal and immigrants from the neighboring island of Malaita. Although independence movements and ethnic rivalries endanger regional peace, one of the greatest threats to stability may be the lack of jobs for young people on Pacific Islands.

Multiculturalism and Indigenous Social Movements Among the most passionately debated issues in contemporary Australia and New Zealand are those relating to the rights of Indigenous peoples and the creation of a multicultural society and national identity. The countries share a history of British colonialism and dispossession of Indigenous lands and cultures but have distinctly different contemporary approaches to intercultural relationships.

In New Zealand, the 1840 Treaty of Waitangi frames Maori rights (see Figure 11.18, p. 443). Although the Maori interpreted the treaty as guaranteeing their land and rights, the century that followed saw large-scale dispossession of Maori land and disrespect for Maori culture. Maori landholdings were reduced from 27 million hectares (100,000 square miles) to only 1.3 million hectares (5,000 square miles), which amounts to just 3% of the total area of New Zealand. A series of protests, court cases, and reawakening to Maori tradition led to the establishment of the Waitangi tribunal in 1975, which eventually reinterpreted the Treaty of Waitangi as more favorable to the Maori and investigated a series of Maori land and fishery claims. The Maori were established as *tangata whenua* (the "people of the land"), and Maori was recognized as an official language of New Zealand. Some land claims were settled or compensated through money or grants of government land. Others are too large or threatening to private interests to be easily recognized.

A bicultural Maori and *Pakeha* (a Maori term for whites) society has been adopted rather than a multicultural policy that would encompass other immigrant groups, such as Pacific Islanders and Asians, or recognize the differences within Maori cultures. New Zealand's recognition of Maori rights and culture as part of a national identity has not solved some of the deeper problems of racism toward the Maori or of their poverty and alienation. Maori unemployment is twice that of white residents; average incomes, home ownership, and educational levels are less than half; and welfare dependence is much higher.

Australian Aborigines, in contrast, have had no recourse to a treaty to assert their rights. The European colonists saw the Indigenous peoples as primitive and their land as *terra nullius*, owned by no one, and therefore freely available to settlers. Only in the 1930s were reserves set

▲ **FIGURE 11.24 Aboriginal Issues in Australia** Millions of Australians have signed a formal apology to the Aborigines for discrimination and the damage done to the "stolen generation" of Aboriginal children, who were taken forcibly from their homes. This apology was flown over the Sydney Opera House.

aside for Aboriginal populations, mostly in very marginal environments with little autonomy or access to services (see Figure 11.2.3, p. 435). In many ways, the Aboriginal population had been made "invisible"; they were not counted in the census or allowed to vote until the 1960s. It was also stereotyped as a primitive and homogeneous nomadic culture, when in fact the Aboriginal population encompassed many different cultures. One of the most misguided programs set out to assimilate the Aboriginal population from 1928 to 1964 by forcibly removing their children from their families and communities and placing them in white foster homes and institutions. This **stolen generation** of as many as 100,000 Aboriginal children was given voice and officially acknowledged by the Australian government in a national inquiry in the 1990s.

Growing awareness and regret at the treatment of Aborigines has led to efforts at apology and reconciliation by many white Australians. In 2000, more than 300,000 people marched across the Sydney Harbor Bridge in a walk of reconciliation, and more than 1 million Australians signed "Sorry Books" that stated, "We stole your land, stole your children, stole your lives. Sorry," as a way of apologizing for the treatment of Aboriginal peoples (**Figure 11.24**). In February 2008, Australian Prime Minister Kevin Rudd formally apologized for these atrocities for the first time, especially for the removal of Aboriginal and Torres Strait Islander children from their families and for the indignity and degradation.

The Aboriginal Land Rights Act of 1976 gave Aborigines title to almost 20% of the Northern Territory and opened government land to claims through regional land councils. The states of South Australia and Western Australia have also handed over land to Aboriginal ownership or leases. In 1992, the Australian High Court effectively overruled the doctrine of *terra nullius*, encouraging Aboriginal claims for land and compensation. Aboriginal control now extends over about 15% of Australia, and claims have been made to at least another 20%. The more contentious claims involve land with valuable mineral resources, especially uranium (see "Geographies of Indulgence, Desire, and Addiction: Uranium," p. 434), or where development threatens sites that are considered sacred or spiritual by the Aborigines.

Despite the apologies and recognition of land claims, Indigenous Australians remain disadvantaged on almost all economic and social indicators. Their unemployment rate is at four times the national average and they have much lower average incomes, housing quality, and educational levels and higher levels of suicide, substance abuse, disease, and violence. Australian Aborigines still have much less power and recognition than the Maori of New Zealand; this is reflected in Australia's adoption of a multicultural rather than bicultural policy of national integration.

Multiculturalism emerged in the 1970s and in an official effort to embrace the distinctive cultures of many different ethnic and immigrant groups. The National Agenda for a Multicultural Australia (published in 1989) set out to promote tolerance and cultural rights and to reduce discrimination, while still maintaining English as the official language and avoiding special treatment for any one group. In contrast to New Zealand, where Maori language and culture is an essential component of the national bicultural identity, in Australia, Aborigines are just one of many ethnic groups in a multicultural society.

There has been considerable opposition to immigration, Aboriginal rights, and multiculturalism in the last decade, much of this exacerbated by recent global economic conditions. In the 1990s, a new political party, the One Nation party, emerged in Australia with a platform that opposes immigration, multiculturalism, and any special preferences for Aborigines. It is clear that land rights and reconciliation for Aboriginal peoples will remain contested as Australian governments strive to balance competing interests and needs.

APPLY YOUR KNOWLEDGE Summarize the different ways that Australia and New Zealand have treated their Indigenous populations. Can you see differences or similarities to the way Indigenous peoples have been treated in other countries such as the United States, Canada, and China?

CULTURE AND POPULATIONS

The culture and politics of this region are, as we have already seen, intimately interconnected to the wider global processes of migration, colonialism, conflict, postcolonial development, and environmental change. The linguistic and religious practices of the people in this region, for example, are partially related to the region's colonial experience. And the culture of migration in the region is tied to both the long-term experience of movement and recent environmental and economic changes that are creating push and pull factors within and beyond the region.

Religion and Language

Many of the estimated 260 interrelated languages spoken by Australian Aborigines that were unique to the continent have become extinct or have only a few surviving speakers. Many Aborigines no

families—Austranesian and Papuan. It is believed that almost 20% of all living languages are spoken on the island of New Guinea (**Figure 11.26**). Most languages on the island are not understood by even the people in the next valley—creating a fascinating linguistic geography. To compensate for Papua New Guinea's 820 distinct living languages, a widespread trade language (or *lingua franca*) called *Tok Pisin* is used that combines local and English words. Other island groups also have a large number of languages. For example, Vanuatu has 110 living languages, many spoken by at least 500 people, and the Solomon Islands have 70 different languages, most with at least 500 speakers. This diversity of languages has been explained by the fragmentation and isolation of New Guinea's physical geography and Vanuatu's many islands.

▲ **FIGURE 11.25 Maori Canoe** Maori war canoes (*Waka Taua*) are large canoes, often made from a single tree, traditionally paddled by up to 80 tattooed warriors, and decorated with sacred symbols particular to a local tribe. These canoes enter Auckland harbor at the beginning of the Rugby World Cup in 2011.

Cultural Practices, Social Differences, and Identity

The Indigenous cultures of Australia, New Zealand, and the Pacific Islands have been disrupted by contact with the rest of the world but also romanticized and commodified by outsiders. Recall from earlier in the chapter that in Australia, Aboriginal cultures, based on strong spiritual ties to land, were marginalized and homogenized when Aborigines were removed from their ancestral lands and resettled on reserves. Today, however, Aboriginal art, often

longer speak anything but English; of the Indigenous languages, only Mabuiag (the language of the Western Torres Strait islands) and the Australian Western Desert languages are spoken by more than a few hundred people. British settlement in Australia and New Zealand resulted in English being the most widely spoken, as well as the official, language.

The colonial heritage of Australia and New Zealand is also reflected in the dominance of the Christian religion in these countries—mostly Protestant in New Zealand and about half Protestant and half Catholic in Australia. Although, today, many people report themselves as having no strong religious beliefs at all. Catholicism was brought to the region by Irish immigrants as well as by southern Europeans, particularly those from Italy. Missionaries were successful in converting most of the Pacific Islanders to Christianity; a range of Protestant denominations and Catholicism are prevalent in French Polynesia.

Non-Christian, local beliefs are still important in the Pacific Islands, especially in Papua New Guinea, and some communities have fused local and Christian practices by combining harvest rituals with the celebration of Christmas, for example. Hinduism is important in Fiji among the Asian Indian population. As a result of Asian immigration, both Buddhism and Hinduism are growing in significance. In New Zealand, Maori communities abandoned many local beliefs when they converted to Christianity and lost their land in the 19th century, but there is now a revival of religion, ritual, and the teaching of the Maori language, which was made an official language of New Zealand in 1987, creating New Zealand's modern bilingual society (**Figure 11.25**). Almost all Maori speak English, and about 100,000 also speak or understand Maori, a language related to those of Polynesia and especially to Hawaiian and Samoan.

The Pacific Islands have an enormous number and variety of languages, which are divided into two general

▲ **FIGURE 11.26 Traditional Village in New Guinea** The yam house of a chief in Tukwaukwa village on Kiriwina in the Trobriand Islands of New Guinea is shown here. The islanders traditionally use yams as currency and accumulate them as a sign of wealth and power.

▲ **FIGURE 11.27 Aborigine and Maori Culture** Contemporary Aboriginal artists based some of their work on traditional aboriginal "X-ray" painting of animals on rocks.

and swim in peaceful lagoons fringed with coral reefs. It is nearly impossible to generalize about the cultures of the Pacific Islands, which are as varied as their many languages and have been long studied by anthropologists. Natives have been stereotyped by some explorers, anthropologists, and tourists as sexually promiscuous and living an easy life of tropical abundance. More sensitive studies have revealed complex cultural traditions with deep social meaning.

For example, in Samoa, these traditions include communal living and eating arrangements, extensive symbolic tattoos, and a traditional system of behavior—the Samoan Way—that includes obligations to community and church and respect for elders. Other distinctive social and material forms include the strict separation of the male and female in much of New Guinea and Melanesia; the tradition of ritual warfare and reciprocal gift exchange; the importance of local leaders, or "big men"; and close links with kin and extended families. Pigs are still considered

based on rock art designs in dotted forms, silhouettes, and so-called X-ray styles, has become very popular in contemporary markets (**Figure 11.27**), and native dances and songs, such as those that are part of the social gathering known as *corroboree*, are performed for tourists. These representations fail to fully appreciate the complexity of Aboriginal life in Australia; however, a variety of grassroots movements and educational efforts have emerged to promote Aboriginal culture and traditions.

Maori tradition—such as the welcome ceremonies that include the *hongi* (pressing noses together) and the *haka* war dance (**Figure 11.28**) now performed at international sporting events—is celebrated as integral to New Zealand's official bicultural identity. Maori architecture includes distinctive carved and decorated meetinghouses called *wharenui* (big house). Artistic expressions include intricate carvings such as those found on war canoes and decorative masks and tattoos (*moko*), which are also found on other Pacific Islands.

Society in Australia and New Zealand is still influenced by British culture, although attempts have been made to distinguish Australians and New Zealanders as less hierarchical and more informal. Echoes of the country's colonial relationship with Britain provoke considerable ambivalence, and there are strong attempts to establish distinct identities by embracing Indigenous traditions and new immigrant cultures such as those from Asia. However, the British legacy endures in the significance of sports such as cricket and rugby, where the national teams, such as the New Zealand All Blacks rugby team (see Figure 11.28), have gained international renown and enjoy enthusiastic local support. Surfing is often seen as the sport that characterizes Australia.

The islands of the Pacific form a distinctive image in the minds of most of the world—tropical paradises where local people fish, collect coconuts, and make crafts while tourists relax on beautiful beaches

▲ **FIGURE 11.28 The All Blacks** The New Zealand national rugby team—the All Blacks—perform a traditional Maori *haka* dance prior to an international match against Wales in Cardiff, UK.

▲ **FIGURE 11.29 Kava Ceremony** Traditionally dressed Fijian warriors carry a large bowl of kava during a welcoming ceremony in Nadi, Fiji in 2006. The Kava ceremony plays a part in welcome rituals in Fiji.

a measure of wealth on many islands, including New Guinea, and a traditional plant, kava, is consumed as a recreational and ritual relaxant on many islands (**Figure 11.29**). Polynesian cultures have a strong orientation to the ocean. Contemporary cultures on the Pacific Islands reflect the tensions between the maintenance and revival of traditional cultures, their selective construction for the tourist industry, and the widespread penetration of global culture, especially formal education, television, and processed foods.

Gender roles in Oceania are as rapidly changing and complex as in any other world region. In Australia, the image of the frontier rancher or miner is associated with heavy drinking, gambling, male camaraderie, and a tough, laconic attitude. However, in recent years, Australian cities, such as Sydney, have celebrated gay and transgendered identities through events such as annual Mardi Gras parades. Across the Pacific Islands, traditional gender roles vary from cultures controlled by women (such as the matriarchal traditions of the Marshall and Solomon Islands), to those where males and females lead mostly separate lives (such as in parts of New Guinea), to others controlled by men.

Although it shares the image of the frontier strong male with Australia, New Zealand has a long history of promoting women's equality and was the first country to extend the right to all adult women to vote (1893). Many women in Australia and New Zealand have shifted from roles as traditional housewives into a multitude of careers and to senior political positions (such as former Prime Minister Helen Clark of New Zealand and Prime Minister Julia Gillard of Australia), and their efforts are supported by a strong feminist movement.

Some of the more serious social problems in Oceania include alcohol abuse and high levels of domestic violence. Papua New Guinea has some of the highest levels of violence against women in the world, and violence and alcoholism are also a problem in some Aboriginal and Maori communities. With shelters and support groups, women in Papua New Guinea are building "safe houses" for women who may be subject to violence.

Oceania has an international reputation in film, literature, art, and music. The film industry in Australia and New Zealand is especially renowned. The most famous movies associated with New Zealand are probably the *Lord of the Rings* series directed by New Zealander Peter Jackson. These Oscar-winning spectacles feature the country's dramatic landscapes, and their popularity has attracted many tourists to the region (**Figure 11.30**). Australia is known for its actors who have become international stars, such as Cate Blanchett and the late Heath Ledger, or Nicole Kidman and Hugh Jackman, who together starred in *Australia*, a film that fictionalizes the history of the country.

> **APPLY YOUR KNOWLEDGE** See or read a recent film or novel about Australia or New Zealand and assess how it portrays culture, gender, and ethnic relations in the region.

Demography and Urbanization

Oceania is one of the least-populated world regions. Of its population of just over 37 million people, 22.7 million live in Australia, almost 7 million live in Papua New Guinea, and 4.4 million live in New Zealand. Most people live near water on the larger land areas in the region, especially along the coasts of southeast Australia and coastal Papua New Guinea. Overall population densities of the larger countries are very low: 3 people per square kilometer in Australia (7.8 people per square mile), 15 per square kilometer in Papua New Guinea (39 people per square mile), and 17 per square kilometer (44 people per square mile) in New Zealand in 2010. The smaller islands, in contrast, can have fairly high population densities, reaching

▲ **FIGURE 11.30 Lord of the Rings** Peter Jackson chose his native country, New Zealand, to film Tolkien's epic *Lord of the Rings*. The trilogy earned box office revenues of more than U.S. $3 billion, exposing millions of people to New Zealand landscapes that were often modified with special effects in the films. Tourists flocked to tours featuring the film. Tourist interest has been rekindled with Jackson's filming of *The Hobbit*.

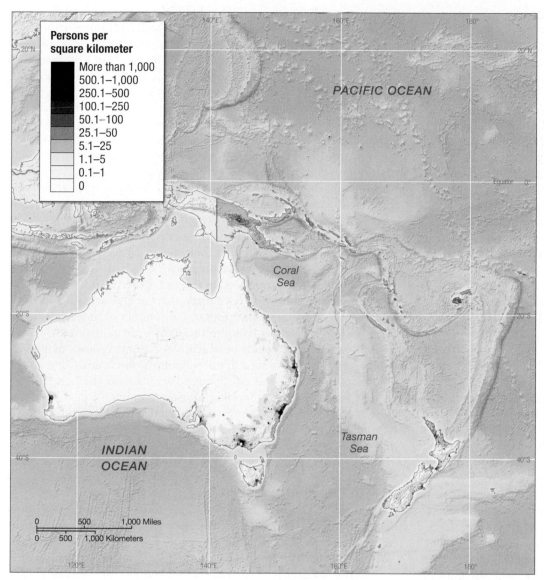

Persons per square kilometer

- More than 1,000
- 500.1–1,000
- 250.1–500
- 100.1–250
- 50.1–100
- 25.1–50
- 5.1–25
- 1.1–5
- 0.1–1
- 0

▲ **FIGURE 11.31 Population Density Map of Oceania, 2010** Although most of Oceania appears sparsely populated on this map, several small islands have high population density. Australia's population is concentrated in the cities and in the southeastern region of the country.

in Oceania when the Europeans arrived. The introduction of European diseases and the violence of some colonial encounters reduced native populations significantly and resulted in majority European populations in Australia and New Zealand within a century or so of initial settlement. The contemporary ethnic composition of Australia and New Zealand is strongly influenced by a larger history of immigration. Almost a quarter of Australia's current residents were born elsewhere. As noted earlier, the first European settlers were mostly English. Some Chinese arrived in Australia during the gold rush of the 19th century as well as migrants of Irish, Scottish, and German origins.

Australia sought to maintain a European "look" through the adoption of the **White Australia policy** after its independence in 1901. This policy essentially restricted immigration to people from northern Europe through a ranking that placed British and Scandinavians as the highest priority, followed by southern Europeans. By 1936, however, more people were coming from southern Europe, especially Italy and Greece. After World War II, the government made extra efforts to attract immigrants as labor for manufacturing, as a source of military volunteers, and in response to humanitarian pressure to resettle refugees from Eastern Europe.

In 1973, the racist restrictions on immigration were removed and replaced with skills criteria, and a new wave of immigration from Asia began. Large numbers of migrants from Vietnam, Hong Kong, and the Philippines arrived as a result. By the end of the 20th century, Australia's population was much more diverse, with about 73% of the population of British or Irish heritage, 20% from elsewhere in Europe, 5% from Asia, and only 2% Aborigine. The economic crisis of 2008 to 2009 slowed Australian immigration. As unemployment levels increased throughout the country, the Labor Party government limited migration.

New Zealand also adopted an immigration policy designed to attract white immigrants. Prior to World War II, this included a bias not just against nonwhites but also against Irish Catholics. After the war, labor shortages were met by temporary labor migration from the Pacific Islands. More recent changes in immigration policy have opened New Zealand to more permanent settlers from the Pacific as well as to Asian immigrants who can bring capital with them. According to the most recent census, New Zealand was still dominated by people of English and Scottish ancestry, but with a growing percentage of people identifying with Pacific Island (about 4.6%) or Asian (about 8%) heritage and a significant group (more than 15% who are full or part) of Maori residents.

The Pacific Islands have a much higher proportion of Indigenous people than Australia or New Zealand, averaging more than 80% Indigenous, 14% Asian, and 6% Europeans. The exceptions are Fiji,

more than 150 per square kilometer (388 people per square mile) in Guam, the Marshall Islands, Nauru, and Tonga (**Figure 11.31**).

The overall population of the region is growing quite slowly (at 1% per year), and it will take 65 years for the population to double at current growth rates. This slow growth is in large part due to the very low growth and fertility rates in Australia and New Zealand. Women in these two countries bear an average of less than two children each during their reproductive years. The low birthrate, combined with long life expectancy, is creating a growing proportion of older residents and raising concerns about the provision of services such as social security, health care, and pensions for the aging population. Fertility rates are higher in the Pacific Islands, ranging from 2.1 children per woman during her childbearing years in French Polynesia to 3.5 children per woman in Kiribati.

Immigration and Ethnicity There were at least half a million Australian Aborigines and perhaps a quarter of a million New Zealand Maori

which has a very large Asian Indian population associated with the importation of contract labor during colonial times (see p. 445); New Caledonia, where about 40% of the population are Indigenous, 40% French, and 20% descendants of Southeast Asian contract workers; and Guam, which has significant American and Filipino populations.

Oceanic Migration and Diasporas One of the largest diasporas within Oceania has been the emigration of Pacific Islanders to work and live in New Zealand and Australia. This emigration has been encouraged in the last 50 years by labor needs, refugee movements, and lenient rules allowing migration from current and former colonies. It has also been encouraged more recently by global climate change. More people from Niue, Tokelau, and the Cook Islands now live in Auckland, the capital of New Zealand, than remain on those islands. About one-eighth of all Samoans and migrants from Tonga and Fiji also live in Auckland. Australia has accepted thousands of refugees from Timor-Leste in Indonesia and from Fiji.

A second diaspora is from the Pacific Islands to North America, especially to Hawaii and California. Many Samoans and Tongans have moved to the United States, and unrest in Fiji sent a wave of several thousand Indo-Fijians to the United States and Canada. There is also a small emigration flow of Australians and New Zealanders to Europe and North America, including people of British heritage returning to retire in Britain and young people seeking educational and work opportunities in England, Canada, or the United States. Within Australia, there is an internal migration flow from the population concentration in southeastern Australia northward along the east coast and to western Australia as economic opportunities open up, air travel and the Internet expand, and life becomes more expensive in the southeast.

Urban Areas Oceania is highly urbanized in contrast to other world regions, with an urban population that averages about 66% of the total. Australia is about 89% urban, and New Zealand is 87% urban. Three-quarters of future population growth in these two countries is forecast to be in cities. This contrasts with other parts of the region. The average urban population for Pacific Island States is only 34% and Papua New Guinea is only 12.5% urban. The overall high levels of urbanization in Australia and New Zealand are due in part to the settlement patterns of European colonists and migrants and the harsh climatic conditions of much of Australia, which has limited large-scale rural development. In fact, despite popular media images of the Australian Outback, most Australians live close to the coast. One exception is the capital of Australia, Canberra, which is found inland between Melbourne and Sydney, Australia's two largest cities. This planned city was chosen as a site for the Australian capital because of its proximate relationship between the country's two major cities. Even though it is inland, Canberra is just a few hours drive from the coast.

Within the region, Sydney is the largest metropolitan space. It has a population of almost 4.4 million people; one in five Australians lives in and around Sydney. High levels of car ownership characterize the city, and the population sprawls into surrounding suburbs. The wealthier areas of the city are to the east, and poorer residents concentrate in the western suburbs. As immigrants from different regions settled in groups based on their origin, Sydney developed several ethnic neighborhoods, including Greek, Italian, and Vietnamese. Many Aborigines who migrated to the city settled in one suburb, and poverty and discrimination have made their neighborhood a focus of Indigenous action and social programs. A number of older downtown neighborhoods, such as Paddington, have been renovated through **gentrification**, the process in which older working-class neighborhoods are converted to serve higher-income households. Some of the older warehouse and manufacturing districts near the harbor have been redeveloped into shopping, museum, convention, and entertainment areas such as the Darling Harbor district. Sydney is also surrounded by parks and protected areas and by beautiful beaches that encourage swimming, surfing, and sailing (**Figure 11.32**).

Melbourne is Australia's second-largest city and was the manufacturing center of southeastern Australia (**Figure 11.33**). Located on a sheltered harbor on Port Phillip Bay, Melbourne has a local population of almost 4 million people. The city first developed as the transport hub for the 19th-century gold rush; it grew further when thousands of refugees and migrants were sponsored to come to Australia after World War II and sent to work in Melbourne's industrial sector, which manufactured textiles and clothing and processed metal. Contemporary industry includes chemicals, food processing, automobiles, and computers. The city has developed a reputation as a cultural center and liveable city.

▲ **FIGURE 11.32 Sydney Harbor** This panoramic view of Sydney shows the downtown business core and the Sydney Opera House, with its brilliant white sail-shaped roof. Although Sydney has some manufacturing, the city economy is overwhelmingly service oriented, focusing on trade, banking, and tourism.

▲ **FIGURE 11.33 Melbourne** Australia's second-largest city, Melbourne, has a downtown core surrounded by industrial activities and neighborhoods that attract many young people.

With 1.3 million residents, the city of Auckland remains the largest in New Zealand. Other cities, such as Wellington and Christchurch, have fewer than 400,000 residents. Cities in New Zealand, as well as the cities of Sydney, Melbourne, and Adelaide in Australia, are considered some of the world's most livable based on education, health and other services, personal risk, environmental quality, and access to recreation.

APPLY YOUR KNOWLEDGE Visit the website of a major city government in Australia or New Zealand. How does the city promote itself online? How does it discuss its problems and plans for the future?

FUTURE GEOGRAPHIES

Often viewed as a relatively isolated world region globally, Oceania, which stretches across a vast waterscape, is dynamically interconnected to the rest of the world in terms of its physical and human geographies. The future of this region depends on economic and environmental scenarios. In the future, Oceania may bear witness to continuing reorientation to Asia and serious impacts from global climatic change as well as tensions over economic and cultural policies within and between countries.

The Challenges of Climate Change and Ocean Sustainability For some countries in Oceania, geography will literally change as sea-level rise drowns their coasts and reshapes their maps. Others will be challenged by increasing risks of extreme events, such as drought, and the impact of warmer temperatures on their ecosystems. Australia and New Zealand, with coal exports and high livestock numbers, will need to consider how to reduce greenhouse gas emissions while adapting their agriculture and settlements to warmer and drier climates. The islands of the Pacific are likely to make climate adaptation a priority

and hope for assistance from the international community. The populations of some islands may need to migrate to other countries as seas rise. Another shared environmental challenge for the future is the protection of ocean and marine environments from pollution, overexploitation, and acidification to ensure the sustainability of economies that are dependent on the sea for food and tourist revenue.

Connections to the Global Economy This chapter has already pointed out how the economic orientation of the region has shifted from the colonial connections to Europe to closer links with Asia. Trade with Asia is growing, not only with China, but also with India and Southeast Asia. The agricultural economies of Australia and New Zealand are shifting to serve the Asian market while still remaining important to global commodity prices, especially those for grain and meat. Across the Pacific Islands, challenges include the diversification of economies, especially those that depend on foreign military bases and tourism, and finding ways to take advantage of the opportunities offered for commerce and services in new media and communications technologies. For example, Tuvalu was given the Internet domain ".tv" and has licensed it internationally to receive U.S. $50 million in royalties since 2000. Island economies will probably continue to rely on the remittances from migrants—unless migration policies elsewhere restrict opportunities.

Traditional Cultures in Contemporary Times The future of traditional cultures in Oceania, as is the case in many other world regions, is at risk from globalized culture, economic development, and political conflicts. While efforts at reconciliation and poverty alleviation in Australia and New Zealand have fostered growing respect for Aboriginal and Maori cultures, poverty, inequality, and racism continue. In the Pacific, local cultures are disappearing as young people are attracted by global popular culture or choosing to migrate because of unemployment and the attraction of urban life. Tourism and television also play a role in changing and commodifying local cultures (**Figure 11.34**). As people continue to move within and into the region, immigration policies and intercultural tensions will be important issues in national politics and identities.

▲ **FIGURE 11.34 Portrayal of Pacific Island Culture** Participants in the reality show *Survivor: South Pacific* wear their versions of native costume while competing near Upolu, Samoa, in 2011. This example demonstrates the way local cultures have been represented, caricatured and commodified by global popular culture.

LEARNING OUTCOMES REVISITED

■ **Explain the factors that influence the climate of Oceania (including Antarctica and the Pacific Ocean) and the impacts of climate change and other environmental threats in the region.**

Most of Oceania has a tropical climate where rains are associated with major wind belts such as the trades, monsoons, and westerlies. The interior of Australia is extremely dry because of its distance from the sea and the subtropical high-pressure zone. Climate change is already increasing the risks of heat waves and droughts in Oceania, especially in Australia, and sea-level rise, ocean acidification, and extreme weather events associated with climate change are a particular risk for islands in the Pacific. Adaptation may moderate theses impacts. In addition to climate change, the Pacific and Antarctica are threatened by pollution and the disturbance of fragile ecosystems, by the overharvesting of fish and whales, and by mineral exploitation. The Law of the Sea (UNCLOS) and the Antarctic Treaty provide some protection.

■ **Describe the unique ecosystems and the impacts that the introduction of nonnative species have had in the region.**

Oceania is home to many species that evolved on isolated islands that lacked predators. It is also home to species with unique characteristics or adaptations such as marsupials. Many of these are vulnerable to competition and predation from exotic nonnative plants and animals introduced during colonization, such as the rabbit in Australia or mongoose on the Pacific Islands.

■ **Understand how European colonialism changed the region and describe the legacies of colonialism in terms of land use, demography, and trade.**

In addition to ecological changes, Europeans introduced grains and livestock that altered land use, expropriated land from native peoples, and ruthlessly pursued minerals to export. Colonists settled in Australia and New Zealand, and people of European ancestry have dominated population, economy, culture, and politics ever since.

■ **Explain how Australia and New Zealand are connected to the rest of the world through their exports and note some of the changes that have occurred in the region as a result of globalization, free trade, and a shift in demand from Europe to Asia.**

The most important exports from the region include minerals, such as gold, copper, and coal; agricultural products, such as meat, dairy, sugar, and grain; and manufactured goods. The end of preferential trade links to Europe and the adoption of free-trade policies reoriented economies to global markets, with their vulnerabilities and opportunities, and to Asia, with its growing demand for food, minerals, and other goods.

■ **Consider how the region became an important site for geopolitical struggles, such as World War II, the Cold War, and nuclear policy, and describe the current international and regional political and economic alignments of major countries in the region.**

The region was mostly allied with Europe and North America in World War II, and Australia also sent troops to support the United States in Vietnam. The Japanese took over many of the Pacific Islands during World War II. The United States and its allies eventually reclaimed most of these and established lasting key military bases, territories, and cultural influences across the Pacific. France and other countries used the Pacific islands for testing nuclear weapons; these tests were met with opposition from antinuclear activists and the New Zealand government. The region still has strong links to Europe and the United States but is creating new alignments with Asia. Small islands in the region have joined with other islands as a lobbying group within the United Nations; their efforts are especially focused on issues relating to climate change.

■ **Explain the importance of tourism in the region and the economic, social, and environmental impacts of the tourist industry.**

Tourism is critical to the economies of most of the Pacific Islands and is also important to Australia and New Zealand. Visits are increasing, especially from Asia, but can be an unpredictable and unreliable revenue stream. Tourism can have some negative environmental and social impacts in the region if not managed carefully; for example, tourism can damage reefs. Visitors contribute money to the economy, but the resorts drain local resource bases and encourage local people to modify their livelihoods and rituals to appeal to foreign tourists' idealized images of the region.

■ **Summarize the main shifts in Australian and New Zealand policies toward Indigenous peoples and on immigration issues.**

European colonists who came to Australia did not recognize the land and cultural rights of Aborigines and claimed much of the best land. Starting in about the 1930s, the Australian government, dominated by whites of European origin, created reserves in marginal areas for Indigenous people and removed children from their families to assimilate them. In the late 20th century, Aborigines were given title to about 15% of the land in Australia, and in 2008, the government apologized for the earlier treatment they had received. The Maori of New Zealand were granted rights under the Treaty of Waitangi and are recognized in the country's bicultural policy. Aboriginal languages are in decline in Australia, but Maori is sustained as a part of New Zealand's bilingual policy. After a century of immigration, mostly from Europe, Australia adopted the White Australia policy in 1901 that prioritized northern European immigrants. This ended in 1973, opening the door for immigrants to arrive from Asia.

■ **Compare and contrast the differences in culture and gender relations of different groups in Oceania.**

Cultures of Australia and New Zealand include the Aborigines and Maoris, groups that have traditionally held strong spiritual beliefs and fostered connections to the land. Aboriginal languages are in decline in Australia, but Maori is sustained as a part of New Zealand's bilingual policy. The culture, religion, sport, and language of both Australia and New Zealand were heavily influenced by the British and by the image of the rugged Outback male. In recent years, more emphasis has been placed in Australia on a multicultural society that respects the religion, language, and traditions of newer immigrants as well as Aboriginal practices. The Pacific Islands are home to an enormous number of languages and cultures, with varied gender relations, especially in New Guinea. Many islands have cultural traditions tied to the sea. Traditional island dress, dances, and foods have often been appropriated to encourage tourism.

KEY TERMS

Aborigines (p. 432)

Antarctic Treaty (p. 428)

AOSIS (Alliance of Small Island States) (p. 431)

atoll (p. 430)

blackbirding (p. 445)

Blue Economy (p. 444)

Closer Economic Relations (CER) agreement (p. 450)

common property resources (p. 439)

Dreamtime (p. 441)

ecological imperialism (p. 436)

exclusive economic zone (EEZ) (p. 439)

exotics (p. 436)

feral (p. 436)

gentrification (p. 456)

Great Artesian Basin (p. 431)

Great Barrier Reef (p. 440)

guano (p. 433)

Maori (p. 441)

marsupial (p. 436)

Melanesia (p. 427)

Micronesia (p. 427)

MIRAB (migration, remittances, aid, and bureaucracy) economies (p. 447)

monotreme (p. 436)

Montreal Protocol (p. 430)

ocean acidification (p. 444)

Outback (p. 431)

ozone depletion (p. 430)

Pacific Garbage Patch (p. 440)

Pacific Islands Forum (p. 450)

Polynesia (p. 427)

RIO+20 United Nations Conference on Sustainable Development (p. 444)

Small Island Developing States (SIDS) (p. 444)

stolen generation (p. 451)

subsistence affluence (p. 449)

tragedy of the commons (p. 439)

Treaty of Waitangi (p. 443)

United Nations Convention on the Law of the Sea (UNCLOS) (p. 439)

White Australia policy (p. 455)

THINKING GEOGRAPHICALLY

1. How have the large expanse of ocean and the general isolation of the region defined the culture and the economy in Oceania?

2. Describe some of the physical, climate-related, and ecological differences between high volcanic islands and low coral atolls.

3. Compare and contrast the different ways in which European colonial powers changed land use across the Pacific.

4. How have changes in global trade in the past 75 years altered the economies of Oceania?

5. How did World War II and the Cold War affect the Pacific Islands politically, economically, and demographically?

6. Compare and contrast the ways Australia and New Zealand have dealt with Indigenous peoples, immigration, and multicultural movements.

MasteringGeography™

Looking for additional review and test prep materials? Visit the Study Area to enhance your geographic literacy, spatial reasoning skills, and understanding of this chapter's content by accessing a variety of resources, including **MapMaster** interactive maps, videos, RSS feeds, flashcards, web links, self-study quizzes, and an eText version of *World Regions in Global Context*.

APPENDIX Maps and Geographic Information Systems

Maps are representations of the world. They are usually two-dimensional, graphic representations that use lines and symbols to convey information or ideas about spatial relationships. Maps express particular interpretations of the world, and they affect how we understand the world and see ourselves in relation to others. As such, all maps are social products. In general, maps reflect the power of the people who draw them. Just including things on a map—literally "putting something on the map"—can be empowering. The design of maps—what they include, what they omit, and how their content is portrayed—inevitably reflects the experiences, priorities, interpretations, and intentions of their authors. The most widely understood and accepted maps—"normal" maps—reflect the view of the world that is dominant in universities and government agencies.

Maps that are designed to represent the *form* of Earth's surface and to show permanent (or at least long-standing) features such as buildings, highways, field boundaries, and political boundaries are called *topographic maps* (see, for example, **Figure A.1**). The usual device for representing the form of Earth's surface is the *contour*, a line that connects points of equal vertical distance above or below a zero data point, usually sea level.

Maps that are designed to represent the spatial dimensions of particular conditions, processes, or events are called *thematic maps*. These can be based on any of a number of devices that allow cartographers or map makers to portray spatial variations or spatial relationships. One device is the *isoline*, a line (similar to a contour) that connects places of equal data value (for example, air pollution, as in **Figure A.2**, p. A-2). Maps based on isolines are known as *isopleth maps*.

Another common device used in thematic maps is the *proportional symbol*. For example, circles, squares, spheres, cubes, or some other shape can be drawn in proportion to the frequency of occurrence of some phenomenon or event at a given location. Figure 9.19, p. 357, shows an example using proportional circles. Symbols such as arrows or lines can also be drawn proportionally to portray flows between particular places. Simple distributions can be effectively portrayed through *dot maps*, in which a single dot represents a specified number of occurrences of some phenomenon or event.

Yet another type of map is the *choropleth map*, in which tonal shadings are graduated to reflect area variations in numbers, frequencies, or densities (see, for example, Figure 1.22, p. 23).

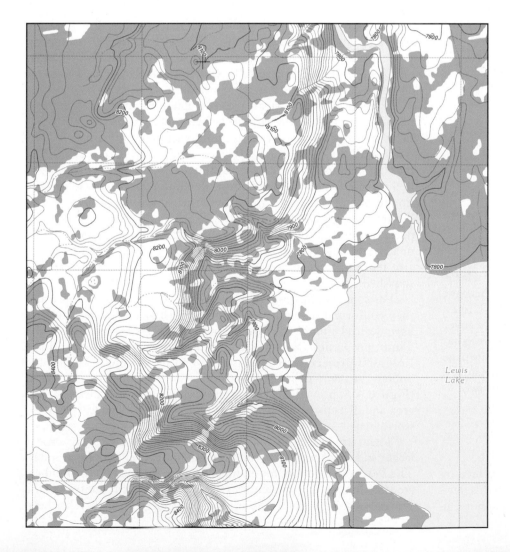

◀ **FIGURE A.1 Topographic Maps** Topographic maps represent the form of Earth's surface in both horizontal and vertical dimensions. This extract is from a Swiss map of Lugano at the scale of 1:25,000. The height of landforms is represented by contours (lines that connect points of equal vertical distance above sea level), which on this map are drawn every 20 meters (65 feet). Features such as roads, power lines, built-up areas, and so on are shown by stylized symbols. Note how the closely spaced contours of the hill slopes represent the shape and form of the land.

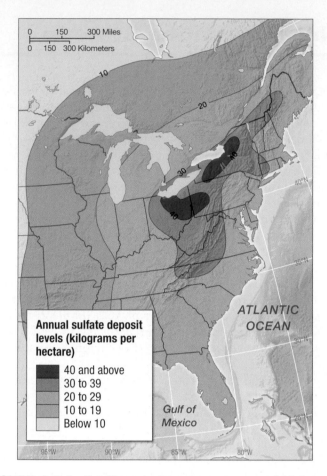

▲ **FIGURE A.2 Isoline Maps** Isoline maps portray spatial information by connecting points of equal data value. Contours on topographic maps (see Figure A.1) are a type of isoline. This map shows one type of air pollution in the eastern United States.

MAP SCALES

A *map scale* is simply the ratio between linear distance on a map and linear distance on Earth's surface. It is usually expressed in terms of corresponding lengths, as in "one centimeter equals one kilometer," or as a *representative fraction* (in this case, 1/100,000) or ratio (1:100,000). *Small-scale* maps are based on small representative fractions (for example, 1/1,000,000 or 1/10,000,000). They cover a large part of Earth's surface on the printed page. A map drawn on this page to the scale of 1:10,000,000 would cover about half of the United States; a map drawn to the scale of 1:16,000,000 would easily cover the whole of Europe. *Large-scale* maps are based on larger representative fractions (e.g., 1/25,000 or 1/10,000). A map drawn on this page to the scale 1:10,000 would cover a typical suburban subdivision; a map drawn to the scale of 1:1,000 would cover just a block or two.

MAP PROJECTIONS

A **map projection** is a systematic rendering on a flat surface of the geographic coordinates of the features found on Earth's surface. Because Earth's surface is curved and not a perfect sphere, it

is impossible to represent on a flat plane, sheet of paper, or monitor screen without some distortion. Cartographers have devised a number of techniques for projecting **latitude** and **longitude** (**Figure A.3**, p. A-3) onto a flat surface, and the resulting representations each have advantages and disadvantages. None can represent distance correctly in all directions, though many can represent compass bearings or area without distortion. The choice of map projection depends largely on the purpose of the map.

Projections that allow distance to be represented as accurately as possible are called **equidistant projections**. These can represent distance accurately in only one direction (usually north–south), although they usually provide accurate scale in the perpendicular direction (which in most cases is the Equator). Equidistant projections are often more aesthetically pleasing for representing Earth as a whole, or large portions of it. An example is the Polyconic projection (**Figure A.4**, p. A-4).

Projections on which compass directions are rendered accurately are known as **conformal projections**. On the Mercator projection (see Figure A.4), for example, a compass bearing between any two points is plotted as a straight line. As a result, the Mercator projection has been widely used in navigation for hundreds of years. The Mercator projection was also widely used as the standard classroom wall map of the world for many years, and its image of the world has entered deeply into general consciousness. As a result, many Europeans and North Americans have an exaggerated sense of the size of the northern continents and underestimate the size of Africa.

Some projections are designed such that compass directions are correct from only one central point. These are known as **azimuthal projections**. They can be equidistant, as in the Azimuthal Equidistant projection (see Figure A.4), which is sometimes used to show air-route distances from a specific location.

Projections that portray areas on Earth's surface in their true proportions are known as **equal-area projections**. Such projections are used when the cartographer wishes to compare and contrast distributions on Earth's surface—the relative area of different types of land use, for example. Examples of equal-area projections include the Eckert IV projection, Bartholomew's Nordic projection (used in Figure 1.25, p. 26), and the Mollweide projection (see Figure A.4). Equal-area projections such as the Mollweide are especially useful for thematic maps showing economic, demographic, or cultural data. Unfortunately, preserving accuracy in terms of area tends to result in world maps on which many locations appear squashed.

For some applications, aesthetic appearance is more important than conformality, equivalence, or equidistance, so cartographers have devised a number of other projections. Examples include the Times projection, which is used in many world atlases, and the Robinson projection, which is used by the National Geographic Society in many of its publications. The **Robinson projection** (**Figure A.5**, p. A-5) is a compromise projection that distorts both area and directional relationships but provides a general-purpose world map.

There are also political considerations. Countries may appear larger and thus more important on one projection than on another. The **Peters projection**, for example (**Figure A.6**, p. A-5), is a deliberate attempt to give prominence to the underdeveloped countries of the Equatorial regions and the Southern Hemisphere. As such, it has been

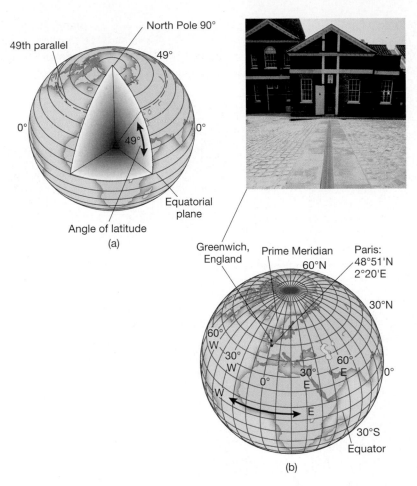

The Prime Meridian at the Royal Observatory in Greenwich, England. The observatory was founded by Charles II in 1675 with the task of setting standards for time, distance, latitude, and longitude—the key components of navigation.

▲ **FIGURE A.3 Latitude and Longitude** Lines of latitude and longitude provide a grid that covers Earth, allowing any point on Earth's surface to be accurately referenced. Latitude is measured in terms of angular distance (that is, degrees and minutes) north or south of the Equator, as shown in (a). Longitude is measured in the same way, but east and west from the Prime Meridian, a line around Earth's surface that passes through both poles (North and South) and the Royal Observatory in Greenwich, just to the east of Central London, in England. Locations are always stated with latitudinal measurements first. The location of Paris, France, for example, is 48°51′ N and 2°20′ E, as shown in (b).

officially adopted by the World Council of Churches and numerous agencies of the United Nations and other international institutions. Its unusual shapes give it a shock value that gets people's attention. For some, however, those shapes are ugly and many have rejected the Peters projection. This emphasizes both the aesthetic and political nature of map projections and decisions surrounding their use.

One kind of projection that is sometimes used in small-scale thematic maps is the *cartogram*. In this projection, space is transformed according to statistical factors, with the largest mapping units representing the greatest statistical values. **Figure A.7a**, p. A-6, shows a cartogram of the world in which countries are represented in proportion to the threatened amphibian species that are present there. This sort of projection is particularly effective in helping us visualize relative inequalities in the global distribution of biodiversity and endangered species. **Figure A.7b**, p. A-6, shows a cartogram of the world in which the cost of telephone calls made from the United States has been substituted for linear distance as the basis of the map. The deliberate distortion of the shapes of the continents in this sort of projection dramatically emphasizes spatial variations.

Finally, the advent of computer graphics has made it possible for cartographers to move beyond two-dimensional representations of Earth's surface. Computer software that renders three-dimensional statistical data onto the flat surface of a monitor screen or a piece of paper facilitates the **visualization** of many aspects of human geography in innovative and provocative ways (**Figure A.8**, p. A-7).

GEOGRAPHIC INFORMATION SYSTEMS

Geographic information systems (GIS)—organized collections of computer hardware, software, and geographic data that are designed to capture, store, update, manipulate, and display geographically referenced information—have rapidly grown to become a predominant method of geographic analysis, particularly in the military and commercial worlds. The software in GIS incorporates programs to store and access spatial data, to manipulate those data, and to draw maps.

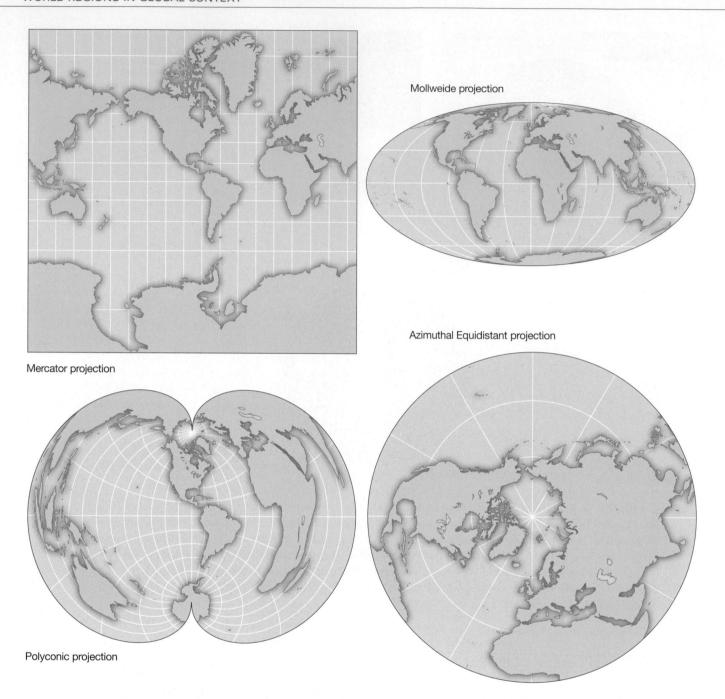

Mollweide projection

Azimuthal Equidistant projection

Mercator projection

Polyconic projection

▲ **FIGURE A.4 Comparison of Map Projections** Different map projections have different properties. The Polyconic projection is true to scale along each east–west parallel and along the central north–south meridian. It is free of distortion only along the central meridian. On the Mercator projection, compass directions between any two points are true, and the shapes of landmasses are true, but their relative size is distorted. On the Azimuthal Equidistant projection, distances measured from the center of the map are true, but direction, area, and shape are increasingly distorted with distance from the center point. On the Mollweide projection, relative sizes are true, but shapes are distorted.

As an industry, GIS is enormous, since GIS systems and analysis underpin many decisions in business and government. In 2010, GIS worldwide was a U.S. $4.4 billion industry, forecast to grow to U.S. $10.6 billion by 2015. Jobs in GIS are numerous and growing and include cartographic design, data analysis, programming, and the management of geographic analysis projects and databases. The number of jobs in geospatial industries is growing at approximately 35% annually, with much of that growth coming from the commercial subsection of the market.

The primary requirement for data to be used in GIS is that the locations for the variables are known. Location may be annotated by x, y, and z coordinates of longitude, latitude, and elevation or by such systems as zip codes or highway mile markers. Any variable that can be located spatially can be fed into a

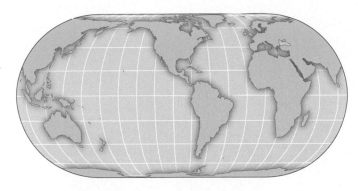

▲ **FIGURE A.5 The Robinson Projection** On the Robinson projection, distance, direction, area, and shape are all distorted in an attempt to balance the properties of the map. It is designed purely for appearance and is best used for thematic and reference maps at the world scale.

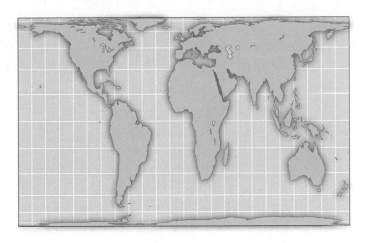

▲ **FIGURE A.6 The Peters Projection** This equal-area projection offers an alternative to traditional projections, which, Arno Peters argued, exaggerate the size and apparent importance of the higher latitudes—that is, the world's core regions—and so promote the "europeanization" of Earth. While it has been adopted by the World Council of Churches, the Lutheran Church of America, and various agencies of the United Nations and other international institutions, it has been criticized by cartographers in the United States on the grounds of aesthetics: one consequence of equal-area projections is that they distort the shape of landmasses.

GIS. Data capture—putting the information into the system—is the most time-consuming component of GIS work. Different sources of data, using different systems of measurement, scales, and systems of representation, must be integrated with one another; changes must be tracked and updated. Many GIS operations in the United States, Europe, Japan, and Australia have begun to contract out such work to firms in countries where labor is cheaper; India has emerged as a major data conversion center for GIS.

Applications of GIS

GIS is now common in most people's daily lives, through the use of online mapping and navigation technology in smartphones and automobiles. Beyond these daily applications, however, GIS can be used to answer complex questions, by merging data from different sources, on different topics, and at different scales. A geographic information system makes it possible to link, or integrate, information that is difficult to associate through any other means. For example, using data on levels of income, reported health problems, and the distribution of infrastructure and roads in a rural area, GIS analysis can reveal the best and most effective locations for new hospitals or clinics.

GIS can also be used to support complex decision making and prioritizing among diverse stakeholders. For example, in trying to determine locations for economic development near a sensitive environment or national park, diverse stakeholders (like business owners, local residents, and environmentalists) can be brought together to examine different possible outcomes and scenarios using computer-generated maps. As each group stresses different priorities, new maps can be generated for further discussion, allowing compromise and further discussion. In this way, GIS can be a tool for democratic decision making.

GIS technology can render visible many aspects of geography that were previously unseen. For example, it can produce incredibly detailed maps based on millions of pieces of information—maps that could never have been drawn by human hands. At the other extreme of spatial scale, GIS can put places under the microscope, creating

detailed new insights using huge databases and effortlessly browsable media (**Figure A.9**, p. A-7).

Many advances in GIS have come from military applications. GIS allows infantry commanders to calculate line of sight from tanks and defensive emplacements, allows cruise missiles to fly below enemy radar, and provides a comprehensive basis for military intelligence. Beyond the military, GIS technology allows an enormous range of problems to be addressed. It can be used, for example, to decide how to manage farmland; to monitor the spread of infectious diseases; to monitor tree cover in metropolitan areas; to assess changes in ecosystems; to analyze the impact of proposed changes in the boundaries of legislative districts; to identify the location of potential business customers; to identify the location of potential criminals; and to provide a basis for urban and regional planning.

Critiques of GIS

In just the past few years, GIS has resulted in the creation of more maps than were created in all previous human history. One result is that as maps have become more commonplace, more people and more businesses have become more spatially aware. Nevertheless, some critics have argued that GIS has a number of drawbacks. First, the quality and utility of maps are only as good as the data and decisions that lie behind them. Since all maps are only partial representations of the world, greater ease of map making provides new opportunities for propaganda, dissimulation, and deceit. Though GIS is available to many groups in society, moreover, the expertise and expense of GIS often mean that the power of map making has remained among elites, companies, and governments.

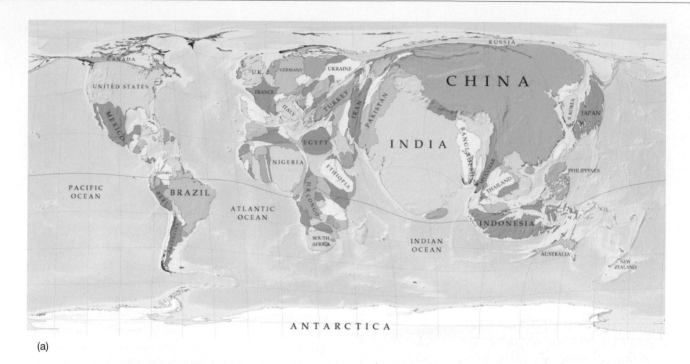

(a)

(b)

▲ **FIGURE A.7 Cartograms** In a cartogram, space is distorted to emphasize a particular attribute of places or regions. (a) This example shows the relative size of countries based on their population rather than their area; the cartographers have maintained the shape of each country as closely as possible to make the map easier to read. As you can see, population-based cartograms are very effective in demonstrating spatial inequality. (b) In this example, countries are sized based on the number of threatened amphibian species that are present there. In this case, the countries of the northern latitudes disappear while the countries of Central and South America stand out. These are centers of some of the world's most important biodiversity.

GIS is also an unquestionable part of the increasing surveillance of populations, made further possible by high-resolution satellite imagery, surveillance cameras, and other technology. Knowing and tracking where people are and where they are going is a powerful tool for corporations, governments, and other powerful institutions. A further fear is that GIS may be helping create a world in which people are not treated and judged by who they are and what they do, but more by where they live. People's credit ratings, ability to buy insurance, and ability to secure a mortgage, for example, are all routinely judged, in part, by GIS-based analyses that take into account the attributes and characteristics of their neighbors and neighborhoods. Like many new technologies, therefore, GIS has the power to support decision making and community empowerment, even while it can be a dangerous tool in the hands of the powerful.

◄ **FIGURE A.8 Visualization** A three-dimensional visualization of global deforestation using Google Earth, developed by David Tryse.

May River Estuary Predicted Land Cover in 2100 with 1.0 ft. of Sea-level Rise

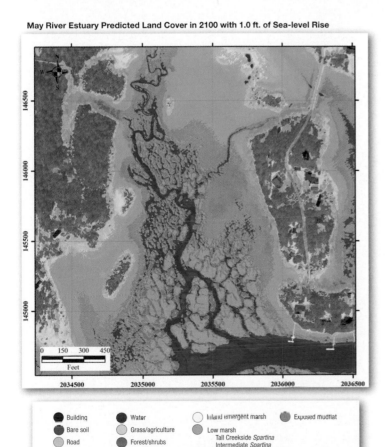

◄ **FIGURE A.9 GIS-Derived Map of Sea-Level Rise** This detailed map shows the predicted land cover of the May River Estuary, assuming a sea-level rise of 1 foot, using a composite digital terrain model obtained through remote sensing of multiple types of terrain.

GLOSSARY

Aborigines: Indigenous peoples of Australia.

acid rain: precipitation that has mixed with air pollution to produce rain that contains levels of acidity—often in the form of sulfuric acid—that are harmful to vegetation and aquatic life.

adivasi: an umbrella term for a heterogeneous set of ethnic and tribal people in India who represent a significant part of the manual labor force in the regions where they predominate.

African Union: based in Addis Ababa, Ethiopia, an organization that promotes solidarity among African states, the elimination of colonialism, and cooperative development efforts.

Agent Orange: a defoliant; a chemical agent that removes the foliage from plants, making it more difficult for military ground forces to find cover. U.S. forces sprayed 20 thousand square km (5 million acres) of Vietnam, as well as Laos and Cambodia, with Agent Orange during the U.S.-Vietnam War of the 1960s–1970s.

AIDS: acquired human immunodeficiency syndrome is the final stages of HIV disease and has resulted in millions of deaths worldwide.

AIDS orphan: a child who has lost one or more parents to an AIDS-related death.

ALBA (Alianza Bolivariana para los pueblos de nuestra América): Bolivarian Alliance for the Peoples of Our America is an intergovernmental organization that includes Bolivia, Cuba, Ecuador, and Nicaragua and opposes free trade and U.S. influence in the Latin America.

al-Qaeda: a global militant Islamist organization that operates as a network composed of a multinational, stateless army and a radical Muslim movement.

altiplano: high-elevation plateaus and basins that lie within even higher mountains, especially in Bolivia and Peru, at more than 3,000 meters (9,500 feet) in the Andes of Latin America.

altitudinal zonation: vertical classification of environment and land use according to elevation based mainly on variations in climate and vegetation.

Americanization: a process by which a generation of individuals born elsewhere felt less loyalty and fewer cultural ties to their countries of origin and developed a new ethos with ties to the United States.

ancestor worship: belief that the living can communicate with the dead and that dead spirits, to whom offerings are ritually made, have the ability to influence people's lives.

Angkor Wat: a 12th-century Hindu (and Buddhist) temple compound that is one of the largest religious structures ever built and a focus of tourism in contemporary Cambodia.

anime: a distinctive stylized form of Japanese animated film.

animism: a belief system that focuses on the souls and spirits of nonhuman objects in the natural world.

animist: a person who worships according to the premises of animism.

Antarctic Treaty: Since 1961, this agreement governs international relations on the Antarctic continent. Forty-nine nations have agreed to ban nuclear testing and waste disposal and mineral and oil exploration, setting aside the area south of 60°S latitude for peaceful purposes and scientific research.

Anthropocene (the "human era"): a term used to describe the current geological epoch in which human activity dominates the planet, beginning with the Industrial Revolution.

Anthropogenic: created or retained by human beings.

AOSIS: Alliance of Small Island States is an association of more than 40 low-lying, mostly island, countries that have formed an alliance to combat global warming, which threatens their existence through sea-level rise.

apartheid: South Africa's policy of racial separation that, prior to 1994, structured space and society to separate black, white, and colored populations.

apparatchik: a state bureaucrat in the Russian Communist Party during the Soviet era.

Arab League: a regional alliance in the Middle East founded in 1945 to strengthen ties amongst member states, set policies, and promote common interests. Member states include Egypt, Iraq, Lebanon, Saudi Arabia, Syria, Jordan, Yemen, Algeria, Bahrain, Kuwait, Morocco, Oman, Qatar, Sudan, Tunisia, the United Arab Emirates, and the Palestinian Authority.

Arab Spring: a term used to refer collectively to the numerous recent protests against dictatorships across the Middle East region in which people motivated by economic inequalities and lack of access to basic resources have taken to the streets to voice their concerns.

archipelago: a group of islands or expanse of water with many islands.

aridity: characteristic of a climate with insufficient moisture to support trees or woody plants.

Aryan: a group of livestock-keeping peoples from the northwest who populated the plains of the Ganga (Ganges) River in India from 1500–500 B.C.E.

Asian Brown Cloud: a blanket of air pollution 3 kilometers (nearly 2 miles) thick that hovers over most of the tropical Indian Ocean and South, Southeast, and East Asia, stretching from the Arabian Peninsula across India, Southeast Asia, and China almost to Korea.

Asian tigers: newly industrialized territories of Hong Kong, Taiwan, South Korea, and Singapore that have experienced rapid economic growth.

aspect: the direction in which a sloping piece of land faces.

assimilation: the process by which peoples of different cultural backgrounds who occupy a common territory achieve sufficient cultural solidarity to sustain a national existence.

Association of Southeast Asian Nations (ASEAN): an international organization of the nations of Southeast Asia established to promote economic growth and regional security.

atmospheric circulation: a global movement of air that transports heat and moisture and explains the climates of different regions.

atoll: a low-lying island landform consisting of a circle of coral reefs around a lagoon, often associated with the rim of a submerged volcano or mountain.

Aung San Suu Kyi (1945–): one of the leaders of the National League for Democracy in Burma and a recipient of the Nobel Peace Prize in 1991 for her work on bringing democratic reform to Burma.

Azimuthal Equidistant projection: A class of map projection where distances from the center of the map are true, but shapes and areas are distorted increasingly while moving outward. These are especially useful for images of Earth's poles.

Balfour Declaration: a 1917 British mandate that required the establishment of a Jewish national homeland.

balkanization: the division of a territory into smaller and often mutually hostile units.

baobab: a long-lived tree that is often considered symbolic of the savanna landscape in Africa with a thick trunk that stores water.

Berlin Conference: a meeting convened in 1884–1885 by German Chancellor Otto von Bismarck to divide Africa among European colonial powers.

biodiversity: the variety in the types and numbers of species in particular regions of the world.

biofuel: fuels produced from raw, often renewable, biological resources, such as ethanol made from corn or sugar.

biogeography: the study of the spatial distribution of vegetation, animals, and other organisms.

bioprospecting: search for plants, animals, and other organisms that may be of medicinal value or may have other commercial use.

birthrate: measure of the number of births in a population, usually expressed as births per 1,000 people per year or as a percentage.

blackbirding: the kidnapping and enslavement of people from the Pacific collectively called kanakas (because many of them were of Kanak origin from the islands now known as Vanuatu) who were captured to work in Australian cotton and sugar plantations from the 1840s to 1904.

blue economy: the use, conservation, and equitable sharing of marine and ocean resources through economic activities such as sustainable fisheries and tourism.

Boers: the Dutch word for farmers; the Boers were early Dutch European settlers in South Africa and the ancestors of people who now identify as Afrikaans.

Bolshevik: the majority faction of the Communist Party that led a successful revolution against the tsar in Russia in 1917.

bonded labor: labor that is pledged against an outstanding debt.

bracero: a guest worker from Mexico given a temporary permit to work as a farm laborer in the United States between 1942 and 1964.

brain drain: a situation the occurs when the most talented and well-educated young people emigrate out of their home nation in search of opportunities elsewhere.

Buddhism: a belief system that originated in South Asia and is traced to Prince Siddhartha, a religious leader who lived in northern India prior to the 6th century B.C.E. and came to be known as Buddha (the Enlightened One). Buddhism stresses nonviolence, moderation, and the cessation of suffering.

buffer zone: a group of smaller or less powerful countries situated between larger or more powerful countries which are geopolitical rivals.

Bumiputra: literally translated as "those of the soil"; a term used to define the ethnic Malay population in Malaysia, which excludes non-Malays (particularly people of Chinese ancestry living in Malaysia).

bureaucratic class: nonelected government officials.

bush fallow: the modification of shifting cultivation, where crops are rotated and land is left fallow (uncultivated) for short periods.

bushmeat: the flesh of wild animals hunted for human consumption. The sale of bushmeat has created controversy in regions where endangered species are eaten for food.

call center: an industry in which workers respond to sales and customer service phone calls from around the world and work long hours for modest pay.

cap-and-trade program: regulatory systems in which (1) the "cap" is a government-imposed limit on carbon emissions, (2) emission permits or quotas are given or sold to states (or firms), and (3) the "trade" occurs when unused permits are sold by those who have been able to reduce their emissions beyond their quota to those who are unable to meet their quota.

capitalism: a form of economic and social organization characterized by the profit motive and the control of the means of production, distribution, and exchange of goods by private ownership.

carbon footprint: a method of estimating how much an individual, a region, or a country is contributing to global climate change through its greenhouse gas emissions.

carbon market: a system in which countries and companies are required to reduce emissions and can earn so-called carbon credits toward meeting their reduction obligations by investing in reducing emissions in developing countries; this system allows developing countries that have ratified the Kyoto Protocol agreement to receive payments as incentives for investments in climate-friendly projects that reduce greenhouse gas emissions.

cartel: a collection of independent businesses formed to regulate production, prices, and marketing.

cartogram: a map projection that is transformed in order to promote legibility or to reveal patterns not readily apparent on a traditional base map.

caste: a system of kinship groupings that is reinforced by language, religion, and occupation.

chador: a loose, usually black, robe worn by Muslim women that covers the body, including the face, from head to toe.

chaebol: a South Korean term for the very large corporations in that country that, with government help, control numerous businesses and dominate the national economy.

Chao Khao (hill tribes): ethnic minorities living in mainland Southeast Asia, particularly, but not exclusively, in the northern more mountainous parts of the region.

charismatic megafauna: Large species with popular appeal (such as elephants or gorillas or pandas) that are often used by conservationists as symbols for activism in worldwide campaigns

child soldier: a child under 18 years of age who is forced or recruited to join armed groups.

circle of poison: the use of imported pesticides on export crops in developing countries to improve appearance; the countries then export back the contaminated crops to the regions where the pesticides were manufactured.

cirque: a deep, bowl-shaped basin on a mountainside, shaped by ice action.

civil society: a network of social groups and cultural traditions that operate independently of the state and its political institutions.

climate: typical conditions of the weather that is expected at a location, often measured by long-term averages of temperature and precipitation.

climate system: interactions of air, water, and the Sun's energy circulating around the globe; weather and climate are the product of the climate system. A climate system consists of five major components: the atmosphere, hydrosphere, cryosphere, land surface, and biosphere.

Closer Economic Relations (CER): a 1983 free trade agreement between Australia and New Zealand to remove tariffs and trade between the two countries.

Cold War: the period between World War II and 1990 in which the United States and the Soviet Union established their role as global powers; struggles between these two powers took place through a variety of proxy conflicts.

colonization: the establishment of a settlement in a place or region.

Columbian Exchange: the interchange of crops, animals, people, and diseases between the Old World of Europe and Africa and the New World of the Americas that began with the voyages of Christopher Columbus in 1492.

command economy: a national economy in which all aspects of production and distribution are centrally controlled by government agencies.

commodity: anything useful that can be bought or sold.

commodity chain: a network of labor and production processes that originates in the extraction or production of raw materials and whose end result is the delivery and consumption of a finished commodity.

Common Agricultural Policy (CAP): a system of European Union support of wholesale prices for agricultural produce created to bolster the agricultural sector of the EU.

common property resource: resources, such as fish or forests, that are managed collectively by a community that has rights to the resource, rather than their being owned by individuals.

Commonwealth of Independent States (CIS): a forum for discussing the management of economic and political problems, including defense issues, transport and communications, regional trade agreements, and environmental protection; members are the Russian Federation, Belarus, Ukraine, the Central Asian states, and some Transcaucasus states.

communism: a form of economic and social organization characterized by the common ownership of the means of production, distribution, and exchange.

Communist Council for Mutual Economic Assistance (COMECON): an organization established in 1949 to reorganize Eastern European economies in the Soviet mold; member countries were to pursue independent, centralized plans designed to produce economic self-sufficiency.

Confederate States of America: the government founded by the southern states of the United States in 1861 with the goal of seceding from the Union; the division from the northern states was rooted in slavery and economic issues.

conflict diamond: a diamond mined in a war zone and used to finance conflicts.

conformal projection: a map projection on which compass bearings are rendered accurately.

Confucianism: a spiritual and philosophical tradition native to China expressed in a philosophy that emphasizes the importance of ethics, good governance, education, family, and hard work.

Convention on International Trade in Endangered Species (CITES): a global treaty first signed in 1990 that restricts trade in rare or endangered animals and their parts between member countries.

convergent plate boundary: an areas where tectonic plates of Earth's crust meet and create subduction zones.

core region: a region that dominate trade, control the most advanced technologies, and have high levels of productivity within diversified economies.

counterurbanization: movement of population from cities to smaller towns and rural areas.

creative destruction: the withdrawal of investments from activities (and regions) that yield low rates of profit in order to reinvest in new activities (and new places).

crony capitalism: a system of capitalism in which leaders and the friends of leaders are privileged.

cultural hearth areas: centers or regions of sociocultural and political–economic production from which major, distinct cultures—ways of life, modes of agriculture, forms of craft and industry, language, and religions—historically emerged.

Cultural Revolution: a sustained attack on Chinese traditions and cultural practices in which millions of people were displaced, tens of thousands lost their lives, and much of urban China was plunged into a terrifying climate of suspicion and recrimination; launched in 1966 by Mao Zedong, who called it a "Great Proletarian Cultural Revolution."

culture: a shared set of attitudes and behavior characteristic of a particular social group that also includes shared symbols and everyday practice and activities.

Culture System: the Dutch colonial policy from 1830 to 1870 that required farmers in Java to devote one-fifth of their land and their labor to production of an export crop.

cyclone: a large rotating storm in the Indian Ocean, produced by warm ocean temperatures and eddies in the trade winds from August to October.

Dalai Lama: the spiritual leader of Tibet; the latest reincarnation of a series of spiritual leaders who, according to Tibetan Buddhism, have chosen to be reborn in order to enlighten others.

death rate: the measure of deaths in a population expressed as either 1,000 deaths per population or as a percentage.

deforestation: the clearing, thinning, or elimination of tree cover in historically forested areas, most typically referring to human-caused tree-cover loss.

deindustrialization: the decline in industrial employment in core regions as firms scale back activities in response to lower levels of profitability.

demilitarized zone (DMZ): a designated area between two or more countries where military activity is prohibited.

democracy: an egalitarian form of government in which all citizens determine public policy and the laws and the actions of their state and all citizens have a right to express their opinions.

demographic collapse: the rapid die-off of the Indigenous populations of the Americas that began occurring in about 1500 as a result of diseases introduced by the Europeans to which Indigenous peoples had little or no resistance.

demographic transition: the replacement of high birthrates and death rates by low birthrates and death rates.

Deng Xiaoping (1904–1997): the leader of China from 1978 to 1989; he embarked on a reorientation of the Chinese economy, dismantling central planning in favor of private entrepreneurship and market mechanisms and integrating China into the global economy.

dependency: an economic system in which resources flow from poorer to richer nations as a result of unfair trade, colonial control, and other unequal power structures.

dependency school (of development theory): a school of thought based on the premise that the economic progress of peripheral or less-developed countries is constrained by the economic and political power of affluent, core countries.

desalinization: a process that converts saltwater into drinking water.

desertification: the process by which arid and semiarid lands become degraded and less productive, leading to desertlike conditions.

development: the improvement in the economic well-being and standard of living of people.

development theory: an analysis of social change that assesses the economic progress of individual countries in an evolutionary way.

dharma: a core aspect of Hinduism and Buddhism; the principle of cosmic order, in which actions are aligned with a right way of being and doing in the universe (virtue, righteousness, social duty).

dialect: regional variations within a language.

diaspora: the spatial dispersion of a previously homogeneous group.

disinvestment: the sale of assets, such as factories and equipment.

distributary: river branch that flow away from the main stream.

divergent plate boundary: an area where tectonic plates of Earth's crust are pulling apart.

domestication: the adaptation of wild plants and animals through selective breeding for preferred characteristics into cultivated or tamed forms.

domino theory: the view that political unrest in one country can destabilize neighboring countries and start a chain of events.

dreamtime: the aboriginal worldview that links past and future and people and places in a continuity that ensures respect for the natural world.

drone: an unmanned, remote-controlled aerial military vehicle used to spy on and strike enemy targets.

dry farming: arable farming techniques that allow the cultivation of crops without irrigation in regions of limited moisture (50 centimeters or 20 inches per year).

ecological imperialism: a concept developed by historian Alfred Crosby to describe the way European organisms, including diseases, pests, and domestic animals, took over the ecosystems of other regions of the world, often with devastating impacts on local peoples, flora, and fauna.

ecosystem: a complex of living organisms, their physical environment, and all their relationships in a particular place.

ecosystem services: resources and processes that are supplied by natural ecosystems that benefit people, regions, and the planet as a whole, such as water and soil protection provided by wetlands or the removal of greenhouse gases from the atmosphere provided by forests.

ecotourism: tourism designed to be environmentally conscious and sensitive as well as provide economic opportunities for local people.

ECOWAS (Economic Community of West African States): an organization established in 1975 to promote trade and cooperation in West Africa; includes the countries of Benin, Burkina Faso, Cape Verde, Côte d'Ivoire,

Gambia, Ghana, Guinea, Guinea-Bissau, Liberia, Mali, Mauritania, Niger, Nigeria, Senegal, Sierra Leone, and Togo.

egalitarian society: a society based on the belief in equal social, political, and economic rights and privileges.

El Niño: the periodic warming of sea-surface temperatures in the tropical Pacific off the coast of Peru that results in worldwide changes in climate, including droughts and floods.

emerging region: an area where loosely connected locations are developing greater internal coherence.

emissions trading scheme (ETS): a system that allows energy-intensive facilities (e.g., power generation plants and iron and steel, glass, and cement factories) to buy and sell permits that allow them to emit carbon dioxide (CO_2) into the atmosphere.

enclave: a culturally distinct territory that is encompassed by a different cultural group or groups.

encomienda: a system by which groups of Indigenous people were "entrusted" to Spanish colonists who could demand tribute in the form of labor, crops, or goods and in turn were responsible for the Indigenous groups' conversion to the Catholic faith and for teaching them Spanish; designed to meet the needs of the colonies' early mining economy in Colonial Spanish America in the 1500s.

endogamous: a description of culture where families are expected to find marriage partners for their children among other members of the kinship group.

Enlightenment: 18-century movement in Europe and North America marked by a belief in the sovereignty of reason and empirical research in the sciences.

entrepôt: a seaport that is an intermediary center of trade and shipment.

equal-area (equivalent) projection: a map projection that portrays areas on Earth's surface in their true proportions.

equidistant projection: a map projection that allows distance to be represented as accurately as possible.

ethnic cleansing: the systematic and forced removal of members of an ethnic group in order to change the ethnic composition of a region.

ethnic group: a group of people whose members share cultural characteristics and commonalities, such as language or religion.

ethnicity: state of belonging to a social group that has a common national or cultural tradition; socially-created system of rules about who belongs to a particular cultural group.

ethnocentrism: the attitude that one's own race and culture is superior to that of others.

ethnonationalism: nationalism based on ethnic identity.

Europeanization: a process by which a non-European subject gradually adopts the cultural norms of European countries.

European Union (EU): a supranational economic and political organization made up of 27 European countries founded to recapture prosperity and power through economic and political integration; it is formerly known as the European Economic Community.

exclave: a portion of a country or a cultural group's territory that lies outside its contiguous land area.

exclusive economic zone (EEZ): an international 200-nautical-mile (370-kilometer) area that was formalized in 1982 in the United Nations Convention on the Law of the Sea (UNCLOS); of tremendous significance to Oceania because it allows countries with a small land area but many scattered islands, such as Tonga and the Cook Islands, to lay claim to immense areas of ocean.

exotic: *see* invasive species.

extractive reserve: areas of land, often forest, that are protected for extractive uses by local groups; one of the best known is in the Amazon Rainforest

and named for Chico Mendes, a rubber tapper who organized resistance to deforestation by ranchers and was murdered in 1988.

failed state: a nation in which the central government is so weak that it has little control over its territory.

fair trade movement: a movement concerned with ensuring that producers are paid a reasonable wage and that crops, including flowers and coffee, are produced with more sustainable methods.

fascism: a political philosophy characterized by a centralized, autocratic government that values nation and race over the individual.

favela: the Brazilian term for informal settlements lacking good housing and services that grow up around urban cores.

federal system: political authority divided between autonomous sets of governments, one national and the others at lower levels, such a state/province, county, city, and town; in a federal system, many decisions are made at the local level.

feng shui: the belief that the physical attributes of places can be analyzed and manipulated to improve the flow of cosmic energy, or *qi*, that binds all living things. Feng shui involves strategies of siting, landscaping, architectural design, and furniture placement to direct energy flows.

feral: a domesticated species that ends up in the wild.

Fertile Crescent: a region arching across the northern part of the Syrian Desert and extending from the Nile Valley to the Mesopotamian Basin in the depression between the Tigris and Euphrates Rivers.

feudal system: regional hierarchies composed, at the bottom, of local nobles and, at the top, of lords or monarchs owning immense stretches of land. In this system, landowners delegated smaller parcels of land in return for political allegiance and economic obligations in the form of money dues or labor.

First Arab-Israeli War: a conflict that took place in 1948 when British forces withdrew from Palestine. The war pitted Arab forces against the recently established Jewish state with the goal of eradicating Israel.

Five-Year Plan: a social and economic development initiative that was part of Communist Party Chairman Mao's "Great Leap Forward" in 1958 intended to accelerate economic growth in the People's Republic of China.

fjord: a steep-sided, narrow inlet of the sea formed when deeply glaciated valleys are flooded by the sea.

flexible production region: a region with a concentration of small- and medium-sized firms whose production and distribution practices take advantage of computerized control systems and local subcontractors in order to quickly exploit new market niches for new product lines.

foreign direct investment (FDI): an active financial stake in production in a country by a company or an individual located in another country.

formal region: a region with a high degree of homogeneity in terms of particular distinguishing features.

fossil fuel: deposits of hydrocarbon that have developed over millions of years from the remains of plants and animals, which have been converted to potential energy sources under extreme pressure below Earth's surface; oil and natural gas are two examples.

free trade association: an association whose member countries eliminate tariff and quota barriers against trade from other member states but continue to charge regular duties on materials and products coming from outside the association.

free trade zone: an area within which goods may be manufactured or traded without customs duties.

French and Indian War: a conflict in North America (1754–1763) among European colonial groups who were vying for control over Indigenous lands; also called the **Seven Years' War**.

functional region: an area characterized by a coherent functional organization of human occupancy.

fundamentalism: the belief in or strict adherence to a set of basic principles; often arises in reaction to perceived doctrinal compromises with modern social and political life.

G8: Group of Eight countries (Canada, France, Germany, Italy, Japan, Russia, the United Kingdom, and the United States) whose heads of state meet each year to discuss issues of mutual and global concern.

Gaza Strip: the part of the Occupied Territories in which Palestinians live, located on the Mediterranean coast.

gender: social differences between men and women rather than the anatomical differences related to sex.

Gender-Related Development Index (GDI): a composite indicator of gender equality assessing the standard of living in a country that aims to show the inequalities between men and women in the following areas: long and healthy life, knowledge, and a decent standard of living.

genetically modified organisms (GMOs): an organism that has had its DNA modified in a laboratory rather than through cross-pollination or other forms of evolution.

genocide: deliberate effort to destroy an ethnic, tribal, racial, religious, or national group.

gentrification: the renovation of older, centrally located, working-class neighborhoods by higher-income households.

geographic information system (GIS): An integrated computer tool which includes software programs as well as hardware devices for the handling, processing, and analyzing of geographic data.

geography: the study of the physical features of Earth and its atmosphere, the spatial organization and distribution of human activities, and the complex interrelationships between people and the natural and human-made environments in which they live.

geomorphology: the study of landforms.

geosyncline: a geological depression of sedimentary rocks.

Mohandas Gandhi (1869–1948): inspirational leader who made a case of India's independence from Great Britain through nonviolence.

glasnost: Mikhail Gorbachev's official policy change initiated in 1985 in the Soviet Union that stressed open government and increased access to information as well as a more honest discussion about the country's social issues and concerns.

globalization: the increasing interconnectedness of different parts of the world through common processes of economic, environmental, political, and cultural change.

global north: those parts of the globe experiencing the highest levels of economic development, which are typically—though by no means exclusively—found in northern latitudes.

global south: those parts of the globe experiencing the lowest levels of economic development, which are typically—though by no means exclusively—found in southern latitudes.

global warming: the increase in world temperatures and change in climate associated with increasing levels of carbon dioxide and other gases resulting from human activities, such as deforestation and fossil-fuel burning.

Golden Triangle: an area, which includes the highlands of Burma, Laos, and Thailand, that contains most of the region's forest areas and fields of opium poppies—producing as much as half of the world's supply of heroin—and more recently, a global and regional center for methamphetamines.

Gondwanaland: also known as Gondwana, this is a vast continental area believed to have formed 200 million years ago when the southern part of the ancient supercontinent Pangaea broke off and which resulted in the continents now mostly in the Southern Hemisphere.

government: an entity that has the power to make and enforce laws.

Grameen Bank: a nongovernmental grassroots organization formed to provide small loans—referred to as microfinance—to the rural poor in Bangladesh, which has grown into an international institution that lends worldwide to poor borrowers who have no credit.

Grand Canals: a series of canals built in the 7th century that linked the northern and southern regions of Inner China.

gravity system: a type of water mining where holes are drilled into the ground at the foothills of a mountain or hill to tap local groundwater; known as *qanat* in Iran, *flaj* on the Arabian Peninsula, and *foggara* in North Africa.

Great Artesian Basin: the world's largest reserve of underground water; it is under pressure so that water rises to the surface when wells are bored; located in central Australia.

Great Barrier Reef: a giant coral reef in the western Pacific Ocean, off the coast of Queensland, Australia, that is more than 2,000 kilometers (1,250 miles) long and consists of 3,400 coral reefs, incorporating more than 300 species of coral and hosting more than 1,500 species of fish and 4,000 different types of mollusks. The reefs were formed over millions of years from the skeletons of marine coral organisms in the warm tropical waters of the Coral Sea.

Great Depression: a severe decline in the world economy that lasted from 1929 until the mid 1930s.

Great Game: a conflict between Great Britain and Russia over the border region of Afghanistan and China in the 1800s and 1900s.

Great Leap Forward: Chinese policy scheme launched in 1958 to accelerate the pace of economic growth by merging land into huge communes and seeking to industrialize the countryside; largely viewed as a massive failure and partly responsible for starvation in the period that followed.

Green Belt Movement: a nongovernmental organization created to protect trees in and around Nairobi, Kenya, led by Nobel Prize–winning environmental and political activist Wangari Maathai.

greenhouse effect: the trapping of heat within the atmosphere by water vapor and gases, such as carbon dioxide, resulting in the warming of the atmosphere and surface.

greenhouse gases (GHGs): emissions such as methane and nitrous oxide produced by usage of fossil fuels that contribute to global climate change.

Green Revolution: a technological package of higher-yielding seeds, especially wheat, rice, and corn, that in combination with irrigation, fertilizers, pesticides, and farm machinery was able to increase crop production in the developing world after about 1950.

griot: a respected storyteller and singer in West Africa.

gross domestic product (GDP): an estimate of the total value of all materials, foodstuffs, goods, and services that are produced in a country in a particular year.

gross national happiness: a term coined by the king of Bhutan to encourage the measurement of the spiritual development and quality of life of the population as opposed to per capita economic activity.

gross national income (GNI): an estimate similar to GDP but including the net value of income from abroad—flows of profits or losses from overseas investments.

guano: bird-dropping deposits that can result in phosphate rock which is used in fertilizers.

guest worker: a foreigner who is permitted to work in another country on a temporary basis.

Gulf Cooperation Council (GCC): an organization that coordinates political, economic, and cultural issues of concern to its six member states—Saudi Arabia, Kuwait, Bahrain, Qatar, the United Arab Emirates, and Oman. The members of the GCC have joined to coordinate the management of their income from oil reserves and address problems of economic development as well as discuss social, trade, and security issues.

Gupta Empire: the empire that united northern India from 320–480 C.E. and produced the decimal system of notation, the golden age of Sanskrit and Hindu art, and contributions to science, medicine, and trade.

hacienda: a large agricultural estate in Latin America and Spain that grows crops mainly for domestic consumption (e.g. for mines, missions, and cities) rather than for export.

hajj: a pilgrimage to Mecca required of all Muslims.

Hamas: the Islamic Palestinian party founded in 1987 that largely governs the Gaza Strip and whom the Israeli government refuses to recognize or negotiate with.

Harappan Civilization: considered the first urban civilization in the Indus River Valley; employed sophisticated agricultural techniques to produce surpluses, primarily in cotton and grains.

harmattan: a hot, dry wind that blows out of inland Africa.

Hasidim: a mystical offshoot of Judaism, practiced by a very small percentage of Jews in the Middle East and North Africa.

hate crime: an act of violence committed because of prejudice against women; homosexuals; or ethnic, racial, and religious minorities.

heathland: open, uncultivated land with poor soil and scrub vegetation.

Hezbollah: a political and military organization based in Lebanon, in which some members are part of militias with a main goal of destroying the state of Israel.

Hinduism: a major religious and cultural tradition, developed from Vedic religion; not a single organized religion with one sacred text or doctrine. Hindusim has no unifying organizational structure, worship is not congregational, and there is no agreement as to the nature of the divinity. Hinduism exists in different forms in different communities and is practiced by roughly a billion people in South Asia.

Hindutva: a social and political movement that calls on India to unite as an explicitly Hindu nation; it has given birth to a range of political parties over the years, including the Bharatiya Janata Party (BJP).

HIV: human immunodeficiency virus usually transmitted through sexual contact, blood transfusions, and needle sharing and from mother to child, causing severe damage to the immune system and making it difficult for the body to fight infections.

Ho Chi Minh (1890–1969): the Vietnamese communist leader who led the Indochina Wars, starting in 1945.

Holocaust: Nazi Germany's systematic genocide of various ethnic, religious, and national minority groups during the 1930s and 1940s. As part of the Holocaust, the Germans and their collaborators killed nearly two out of every three European Jews as part of the "Final Solution," the Nazi policy to murder the Jews of Europe.

homelands: segregated areas set aside in South Africa for black residents under apartheid from 1959 where they were given limited self-government but no vote and limited rights in the general politics of South Africa.

Human Development Index (HDI): a United Nations metric based on measures of life expectancy, educational attainment, and personal income.

hurricane: a large rotating storm in the Americas, produced from August to October by warm ocean temperatures and eddies in the trade winds.

hydraulic civilization: large state societies hypothesized to have arisen from the needs of organizing massive irrigation systems; civilizations that survive, thrive, and expand based on their capacity for controlling water.

ideograph: a linguistic symbol representing a single idea or object rather than a sound.

imperialism: the extension of the power of a nation through direct or indirect control of the economic and political life of other territories.

import substitution: a process by which domestic producers provide goods or services that were formerly bought from foreign producers.

indentured servant: an individual bound by contract to the service of another for a specified term.

Indian National Congress Party: a party established in India to promote democracy and freedom in 1887 led by Mohandas Gandhi.

Industrial Revolution: the rapid development of mechanized manufacturing that gathered momentum in the early 19th century.

informal economy: economic activities that take place beyond the official record and not subject to formalized systems of regulation or remuneration (e.g., street selling and petty crime).

information technology (IT): the use of computer systems for storing, retrieving, and sending information.

intercropping: a mixing of different crop species that have varying degree of productivity and drought tolerance to produce a greater yield.

Intergovernmental Panel on Climate Change (IPCC): the Intergovernmental Panel on Climate Change, an international group of scientists that, since 1990, have prepared regular assessments on climate change.

intermontane: a set of basins, plateaus, and smaller ranges that lie between mountains.

internally displaced person: an individual who is uprooted within his or her own countries due to civil conflict or human rights violations.

internal migration: the movement of populations within a national territory.

internationalist: a person who believes in equal rights for all nations and wants to break down national barriers and end ethnic rivalries.

International Monetary Fund (IMF): an international organization that monitors the international financial system and provides loans to governments throughout the world.

intertropical convergence zone (ITCZ): a region where air flows together and rises vertically as a result of intense solar heating at the Equator, often with heavy rainfall, and shifting north and south with the seasons.

intifada: an uprising of Palestinians in the 1980s against the rule of Israel in the Occupied Territories.

invasive species: plants or animals from one place that "hitchhike" to new locations on human transport, through human-built canals, or are deliberately introduced and thrive at their destination, interacting with native species and habitats to create new ecological mixes, land covers, and novel ecosystems.

Iranian Green Revolution: the 2009 uprising in Iran based on allegations of election fraud against President Mahmoud Ahmadinejad.

Iron Curtain: the militarized frontier zone across which Soviet and East European authorities allowed the absolute minimum movement of people, goods, and information.

irredentism: the assertion by the government of a country that a minority living outside its borders belongs to it historically and culturally.

Islam: the religion of the Muslims that is based on submission to God's will according to the Qur'an.

Islamism: political movement or political identity that promotes Islamic law, pan-Islamic unity, and rejection of Western influences in the Muslim world.

Janjaweed: the militia of African Arabs supported by the authoritarian Islamic Sudanese government to fight the rebellion of native Africans in Darfur (beginning in 2003); this resulted in one of the most brutal campaigns of ethnic cleansing in African history.

Jasmine Revolution: the protest against dictatorial government policies in Tunisia in 2011 that was part of the Arab Spring movement around the Middle East and North Africa.

jati: kinship groups associated with Hinduism and therefore most common in India and Nepal.

jihad: a sacred struggle or striving to carry out God's will according to the tenets of Islam; term connotes both an inward spiritual struggle to attain perfect faith as well as an outward material struggle to promote justice and the Islamic social system.

jihadist: a member of the global movement that seeks war on behalf of Islam against those who oppose the religion.

karma: the idea (in Hinduism and Buddhism) of cosmic responsibilities for actions and deeds visited upon eternal souls throughout an endless cycle of reincarnated lifetimes.

keiretsu: Japanese business networks facilitated after World War II by the Japanese government in order to promote national recovery.

khan: the ruler or leader in the Central Asian Muslim kingdoms (khanates).

Khmer: Cambodian; the Khmer Empire emerged in the 9th century and later constructed the Angkor Wat.

Khmer Rouge: the U.S.-backed military government in Cambodia (1975–1979) led by Pol Pot; he isolated the country, suspended education, and committed the mass murders of millions.

killing fields: fields found throughout Cambodia where millions of people were buried during the reign of the Khmer Rouge.

Kim Il Sung (1912–1994): anti-Japanese Marxist–Leninist nationalist leader who came to power in 1949 in Korea and imposed an austere regime and regimented way of life.

Kim Jong Il (1942–2011): a son of Kim Il Sung who succeeded his father and was known as the "Dear Leader"; during his regime, food shortages increased between 1995–2005.

Kim Jong-un (1983–): a son of Kim Jong Il who succeeded him in 2012; the youngest head of state in the world, he governs one of the most highly militarized countries in the world.

kinship: the shared notion of relationship among members of a group often, but not necessarily, based on blood, marriage, or adoption.

kleptocracy: a form of government in which leaders divert national resources for their personal gain.

Kurds: non-Arabic people who are mostly Sunni Muslims and who have been struggling with both the Turkish and Iranian governments for autonomy; twenty million Kurds live in the mountainous region along the borders of Iraq, Iran, Syria, Turkey, and a small area in Armenia called Kurdistan by the Kurds.

Kyoto Protocol: an international treaty providing legally binding rules that seeks to limit and reduce greenhouse gas emissions in signatory countries in order to avert dangerous climate change; adopted in Kyoto, Japan, in 1997 and entered into force in 2005.

land bridge: a dry land connection between two continents or islands, exposed, for example, when sea level falls during an ice age.

land grabs: large-scale land purchases made by foreign investors in developing countries.

language: a method of human communication, either spoken or written, consisting of the use of words in a structured and conventional way. Language is a central aspect of cultural identity, reflecting the ways that different groups understand and interpret the world around them.

La Niña: the periodic abnormal cooling of sea-surface temperatures in the tropical Pacific off the coast of Peru that results in worldwide changes in climate, including droughts and floods that contrast with those produced by El Niño.

latifundia: huge estates in the Roman Empire typically worked on by slaves; or large rural landholdings established under Spanish colonialism.

latitude: the angular distance of a point on Earth's surface, measured north or south from the Equator, which is 0°.

Lenin , Vladimir Ilyich (1870–1924): a revolutionary who led the Bolshevik takeover of power in Russia in 1917, and was the architect and first head of the U.S.S.R.

liberation theology: a Catholic movement originating in Latin America focused on social justice and on helping the poor and oppressed.

lingua franca: a common language used to communicate among people of different backgrounds and languages, often for trading purposes.

Little Ice Age: the period of cooler climate that significantly reduced the growing season that occurred in about 1300 C.E.

little tigers: descriptive nickname for the economies of Thailand, Malaysia, and Indonesia, which have had relatively high rates of economic growth since the early 1980s.

local food: usually organically grown food that it is produced within a fairly limited distance from where it is consumed.

loess: a surface cover of fine-grained silt and clay deposited by wind action and usually resulting in deep layers of yellowish, loamy soils.

longitude: the angular distance of a point on Earth's surface, measured east or west from the prime meridian (the line that passes through both poles and through Greenwich, England, and given the value of 0°).

machismo: a Spanish word that constructs the ideal Latin American man as fathering many children, dominant within the family, proud, and fearless.

mafiya: organized crime groups in nations such as Chechnya, Azerbaijan, and Georgia.

malaria: a disease transmitted to humans by mosquitoes that causes fever, anemia, and often-fatal complications.

mandate: a delegation of political power over a region, province, or state.

Mandela, Nelson (1918–): anti-apartheid movement leader in South Africa; imprisoned for 27 years under apartheid, he became the first black president of South Africa in 1994 and leader of the African National Congress.

manga: Japanese print cartoons books (or komikku), which date at least to the last century.

mangrove forests: groups of evergreen trees that form dense, tangled thickets in marshes and along muddy tidal shores; found along regional coastlines.

Maori: Indigenous peoples of New Zealand.

Mao Zedong (1893–1976): a communist leader and founder of the People's Republic of China; responsible for the disastrous policies of the Great Leap Forward and the Cultural Revolution.

map projection: the systematic rendering on a flat surface of the geographic coordinates of the features found on Earth's surface.

maquiladora: industrial plant in Mexico, originally within the border zone with the United States and often owned or built with foreign capital, that assembles components for export as finished products free from customs duties.

marianismo: the veneration of the ideal woman in the image of the Virgin Mary; she is chaste, submissive, maternal, dependent on men, and closeted within the family.

market economy: an economy in which goods and services are produced and are distributed through free markets.

maroon communities: settlements in the Caribbean and Latin America in the 1700s and 1800s created by escaped and liberated slaves.

Marshall Plan: the strategy (named after U.S. Secretary of State George Marshall) of quickly rebuilding the West German economy after World War II with U.S. funds in order to prevent the spread of socialism or a recurrence of fascism.

marsupial: an Australian mammal such as the kangaroo, koala, and wombat that gives birth to premature infants that develop and drink milk from nipples in a pouch on the mother's body.

massif: mountainous block of Earth's crust bounded by faults or folds and displaced as a unit.

matrilocality: the cultural practice in which a married male–female couple lives with the family of the woman.

Mauryan Empire: the first empire to establish rule across South Asia; promoted a policy of "conquest by dharma."

Mecca: the city in present day Saudi Arabia where Muhammad was born in 570 C.E.; Muslim religious practice includes praying five times a day facing in the direction of Mecca.

megacity: one of the world's largest metropolitan areas.

megafauna: large-bodied mammals such as orangutans.

megalopolis: a continuous chain of metropolitan development, such as the urban corridor that extends from Boston to Washington, D.C., on the Eastern Seaboard of the United States (also used for any megacity of more than 10 million people).

Mekong River Commission (MRC): the intergovernmental organization that coordinates the management of the Mekong River Basin; its members include Cambodia, Laos, Thailand, and Vietnam, as well as "dialogue partners" Burma and the People's Republic of China.

Melanesia: the region of the western Pacific that includes the westerly and largest islands of Papua New Guinea, the Solomon Islands, Fiji, Vanuatu, and New Caledonia.

Mercator projection: A classic cylindrical map projection used for navigation; preserves the shapes and angles of continents and features perfectly, but exaggerates sizes grossly for continents and features increasingly distant from the equator. This leads to deceptively oversized northern and southern features, like Greenland, relative to equatorial continents like Africa.

merchant capitalism: a form of capitalism characterized by trade in commodities and a highly organized system of banking, credit, stock, and insurance services.

mestizo: a term used in Latin America to identify a person of mixed white (European) and American Indian ancestry.

microfinance: programs that provide credit and savings to the self-employed poor, including those in the informal sector who cannot borrow money from commercial banks.

Micronesia: the region of island states in the South Pacific that includes Guamn, Kiribiti, the Marshall Islands, the federate states of Micronesia, Nauru, Northern Mariana Islands, and Palau.

Millennium Development Goals (MDGs): eight goals to be met by the year 2015, agreed to by members of the United Nations, that include the eradication of poverty, universal primary education, gender equality, the reduction of child mortality, the improvement of maternal health, the combating of disease, environmental sustainability, and the creation of global partnerships.

minifundia: very small parcels of land farmed by tenant farmers or peasant farmers in Latin America.

MIRAB (migration, remittances, aid, and bureaucracy economies): economies such as those of many Pacific islands that depend on labor migration, money sent back from overseas workers, foreign aid, and jobs in government.

modernity: a forward-looking view of the world that emphasizes reason, scientific rationality, creativity, novelty, and progress.

Mollweide projection: A class of map projection, typically characterized by an oval shape, that tends to preserve sizes of continents but distorts their shapes.

monoculture: an agricultural practice in which one crop is grown intensively over a large area of land.

monotreme: an egg-laying mammal, such as the platypus, most often found in Australia and New Guinea.

Monroe Doctrine: the proclamation of U.S. President James Monroe in 1823 stating that European military interference in the Western Hemisphere, including the Caribbean and Latin America, would no longer be acceptable.

monsoon: a seasonal reversal of wind flows in parts of the lower to middle latitudes. During the cool season, a dry monsoon occurs as dry offshore winds prevail; in hot summer months, a wet monsoon occurs as onshore winds bring large amounts of rainfall.

Montreal Protocol: a 1987 international treaty established to reduce chemical emissions that damage the ozone layer.

moraine: the accumulation of rock and soil carried forward by a glacier and eventually deposited at the glacier's frontal edge or along its sides.

Mughals: 15th-century clan of Turks from Persia who conquered Afghanistan and most of India from 1526–1707; their rule promoted Islam in India, and they were famous for spectacular architecture such as the Taj Mahal.

mujahideen: a zealous group of fundamentalist Islamic tribal leaders in Afghanistan trained and armed by Pakistan.

multiculturalism: the process of immigrant incorporation in which each ethnic group has the right to enjoy and protect its officially recognized "native" culture.

Muslim: a member of the Islamic religion.

nation: a group of people often sharing common elements of culture, such as religion, language, a history, or political identity.

nationalism: the feeling of belonging to a nation as well as the belief that a nation has a natural right to determine its own affairs.

nationalist movement: organized groups of people sharing common elements of culture, such as language, religion, or history, who wish to determine their own political affairs.

nationalization: the process of converting key industries from private to governmental organization and control.

National League for Democracy (NDL): an opposition party to the military regime in Burma that won parliamentary seats in 2012, including one for leader and Nobel Peace Prize–winner Aung San Suu Kyi.

nation–state: the ideal form of a state, consisting of a homogeneous group of people living in the same territory.

naxalite: a member of an armed revolutionary group in the Indian subcontinent advocating communism.

Near Abroad: independent states that were formerly republics of the Soviet Union.

neocolonialism: economic and political strategies by which powerful states in core economies indirectly maintain or extend their influence over other areas or people.

neoliberalism: an economic doctrine based on a belief in a minimalist role for the state, which assumes the desirability of free markets as the ideal condition not only for economic organization but also for social and political life.

Neolithic Period: the era from 10,000 B.C.E. to 4,000 B.C.E. in which humans began to develop longer-term patterns of settlements; also referred to as the "New Stone Age."

New Economic Policy: an economic plan implemented in Malaysia in 1971 to take advantage of foreign direct investment (FDI) and move from primary commodities (like rubber) to high-value exports (computer chips).

newly industrializing economies (NIE): countries whose economies feature high rates of savings, balanced budgets, and low rates of inflation, which are indicators of a successful transition to industrial economy; a term used to describe "little tigers" in Southeast Asia.

Nollywood: the nickname for Nigeria's film industry.

nomadic pastoralist: a person who herds animals by moving from place to place and carefully and deliberately following rainfall and plant growth to maintain their flocks.

nongovernmental organization (NGO): a formally constituted organization that is not a part of the government and are not conventional for-profit business (e.g., environmental and humanitarian groups).

nontraditional agricultural exports (NTAEs): newer export crops, such as vegetables and flowers, that contrast with the traditional exports such as sugar and coffee, and often require fast, refrigerated transport to market.

North America Act of 1867: a law that created the Dominion of Canada, dissolving its colonial status and effectively establishing it as an autonomous state with its own constitution and parliament.

North American Free Trade Agreement (NAFTA): a 1994 agreement among the United States, Canada, and Mexico to reduce barriers to trade among the three countries, through, for example, reducing customs tariffs and quotas.

Northwest Passage: the ice-choked waterway spanning the Arctic Sea between the Atlantic and Pacific Oceans, north of the Canadian and Russian mainlands.

nuclear family: a household that contains just two generations, a set of parents and their children.

oasis: a spot in the desert made fertile by the availability of surface water.

Occupied Territories: a region under Israeli occupation that includes the West Bank, Gaza Strip, and the Golan Heights where many Palestinians live as refugees.

Occupy Wall Street: a social movement founded in Canada to express dissatisfaction with the growing wealth gap between a small number of superrich individuals and the rest of the population.

ocean acidification: caused when ocean waters take up excess environmental carbon dioxide and become more acidic as a result; the acidity damages marine organisms, especially corals and shellfish.

oil sands: deposits of sand, clay, other minerals, and water that are saturated with bitumen, which is oil or petroleum in a solid or extremely viscous state; also known as tar sands.

old-age dependency ratio: the number of people age 65 and older compared with the number of working-age people (age 15 to 64).

oligarch: a business leader who wields significant political and economic power.

one-child policy: a Chinese policy introduced in 1978 involving rewards for families that give birth to only one child, including work bonuses and priority in housing.

Opium Wars: conflicts between China and Great Britain in the middle of the 1800s that resulted in China's defeat and the signing of the Treaty of Nanking.

organic farming: farming or animal husbandry that occurs without commercial fertilizers, synthetic pesticides, or growth hormones.

Organization of Petroleum Exporting Countries (OPEC): a specialist economic organization with the central purpose of fixing crude oil prices among its member states. OPEC has 12 member states and is dominated by Middle Eastern Arab states.

orographic effect: the influence of hills and mountains in lifting airstreams, cooling the air, and inducing precipitation.

orographic precipitation: rain or snow that falls when moisture-laden air, which has blown over warm oceans, encounters a landmass, especially coastal mountains; results in the formation of a dry rain shadow region on the inland, or lee, side of the mountains, where sinking air that has lost its moisture becomes even drier.

Orthodox Judaism: a group that represents a small percentage of practicing Jewish people in the Middle East and North Africa who live according to strict adherence to the religious texts of the Old Testament.

Outback: remote, drier, and thinly populated interior of Australia.

overseas Chinese: migrants from China who settled in Southeast Asia as early as the 14th century but mainly during the period of European colonialism, when they arrived as contract plantation, mine, and rail workers and then moved into clerical and business roles.

ozone depletion: the loss of the protective layer of ozone gas that prevents harmful ultraviolet radiation from reaching Earth's surface and causing increases in skin cancer and other ecological damage.

Pacific Garbage Patch: trash accumulation in both the western and eastern Pacific Ocean caused by currents, called gyres; these currents trap plastic and other waste into accumulations of rubbish that damage marine ecosystems.

Pacific Rim: a loosely defined region of countries that border the Pacific Ocean.

paddy farming: a system of farming in which terraces are cut into steep hillsides to provide level surfaces for water control; the construction of dikes (ridges) allows fields to be flooded, plowed, planted, and drained before harvest.

pancasila: the Indonesian, postcolonial nation-building ideology whereby all Indonesians are connected through unity in diversity through belief in one god, nationalism, humanitarianism, democracy, and social justice.

Pangaea: an ancient supercontinent, comprising all the continental crust of Earth, of which Africa is the heart.

partition: the British division of India and Pakistan in 1947 along ethnic lines, particularly language and religion.

pass laws: policies of strict racial segregation between blacks and Afrikaners established by the Boers in South Africa during the 18th and 19th centuries and continuing under apartheid in the 20th century, which included the establishment of native reserves and the mandate that blacks needed permission to enter or live in white areas.

pastoralism: a system of farming and way of life depending on the environment and based on keeping herds of grazing animals (e.g., cattle, sheep, goats, horses, camels, yaks).

patriarchal: a description of society where men have authority over family and society in social and political systems and socioeconomic conditions are generally better for men than women.

patrilocality: the cultural practice by which a married male–female couple lives with the family of the man.

peripheral region: regions that are characterized by dependent and disadvantageous trading relationships, by inadequate or obsolescent technologies, and by undeveloped or narrowly specialized economies with low levels of productivity.

permafrost: a permanently frozen subsoil, which may extend for several meters below the surface layer and may defrost up to a depth of a meter (3 feet) or so during summer months.

Peters Projection: An equal-area map projection designed to show the relative size of continents to one another correctly, but marked by severe distortions in shape.

petrodollar: revenue generated by the sale of oil.

petroleum: a liquid compound that can be converted into fuel and developed into energy sources, lubricants, waxes, asphalt, and medicine.

physical geography: a branch of geography dealing with natural features and processes (*see also* physiography).

physiographic region: a broad region within which there is a coherence of geology, relief, landforms, soils, and vegetation.

physiography: the branch of geography dealing with natural features and processes; another word for physical geography.

pinyin: a system of writing Chinese language using the Roman alphabet.

place: a specific geographic setting with distinctive physical, social, and cultural attributes.

plantation: a large agricultural estate that is usually tropical or semitropical, monocultural (one crop), and commercial or export oriented; most plantations were established in the colonial period.

plantation economy: an economic system typical of colonial trade made up of extensive, European-owned, operated, and financed enterprises where single crops were produced by local or imported labor for a world market.

plate tectonics: the theory that Earth's crust is divided into large solid plates that move relative to each other and cause mountain building and volcanic and earthquake activity when they separate or meet.

pluralist democracy: a society in which members of a diverse group continue to participate in their traditional cultures and special interests.

polder: an area of low land reclaimed from a body of water by building dikes and draining the water.

Pol Pot (1925–1998): the leader of the brutal U.S.-backed military regime in Cambodia, starting in 1975. Under his leadership, formal education was suspended, cities were emptied, the rich and educated were attacked, the country was isolated from the rest of the world, and mass murder was committed against the Cambodian people.

Polyconic projection: A class of map projection where scale is preserved along each east-west parallel and along the central north-south meridian. Distortion increases moving outwards, east or west, from the central meridian.

polygyny: the practice of having more than one wife.

Polynesia: central and southern Pacific islands that include the independent countries of Samoa, Tonga, the Cook Islands, Niue, and Tuvalu; the U.S. territory of American Samoa; the French territories of Wallis, Fortuna, and French Polynesia (including the island of Tahiti, the Society Islands, the Tuamotu archipelago, and the Marquesas Islands); the New Zealand territory of Tokelau; and the British territory of the Pitcairn Islands.

primary sector: refers to economic activity that is concerned directly with natural resources of any kind.

pristine myth: the erroneous belief that the Americas were mostly wild and untouched by humans prior to European arrival. In fact, large areas had been cultivated and deforested by Indigenous populations.

privatization: the turnover or sale of state-owned industries and enterprises to private interests.

purchasing power parity (PPP) per capita: a measure of how much of a common "market basket" of goods and services a currency can purchase locally, including goods and services that are not traded internationally; PPP makes it possible to compare levels of economic prosperity between countries where the price of goods might be relatively much higher or lower.

Qital: the Arabic word for fighting or warfare; refers to a form of jihad in terms of conquest or conversion against nonbelievers.

quaternary sector: refers to economic activity that deals with the handling and processing of knowledge and information.

Qur'an: the Islamic sacred book; Muslims believe the contents are the direct spoken words of God to Muhammad.

race: a problematic and illusory classification of human beings based on skin color and other physical characteristics; biologically speaking, no such thing as race exists within the human species.

racialization: the practice of creating unequal castes where whiteness is considered the norm, despite the biological reality that no such thing as race exists within the human species.

rain shadow: a phenomenon that occurs when mountains cause most of the moisture contained in the air masses passing over them to condense and fall as rain before it can reach the parched interior deserts of the region.

Raj: the rule of the British in India.

REDD (Reducing Emissions from Deforestation and forest Degradation): programs that allow countries and companies with high greenhouse gas emissions to get credit for emission reductions by providing financial and other incentives for forest protection in the developing world.

region: a large territory that encompasses many places, all or most of which share similar attributes in comparison with the attributes of places elsewhere.

regional geography: study of the ways in which unique combinations of environmental and human factors produce territories with distinctive landscapes and cultural attributes.

regionalism: strong feelings of collective identity shared by religious or ethnic groups that are concentrated within a particular region.

regionalization: the process through which distinctive areas come into being.

religion: belief systems and practices that recognize the existence of powers higher than humankind.

remittances: money sent home to family or friends by people working temporarily or permanently in other countries.

remote sensing: a collection of information about parts of Earth's surface by means of aerial photography or satellite imagery designed to record data on visible, infrared, and microwave sensor systems.

reservation/reserve: an area of land managed by an Indigenous tribe under the United States Department of the Interior's Bureau of Indian Affairs, or in Canada, under the Minister of Indian Affairs.

Revolutionary War with Britain: the war between Great Britain and the U.S. colonies that led to the creation of a new, independent nation in the late 18th century.

rice bowl: commonly refers to regions in Asia where large-scale, wet-rice agriculture production takes place, providing a continuous source of this staple crop.

rift valley: a block of land that drops between two others, forming a steep-sided trough, often at faults on a divergent plate boundary.

Ring of Fire: a chain of seismic instability and volcanic activity that stretches from Southeast Asia through the Philippines, the Japanese archipelago, the Kamchatka Peninsula, and down the Pacific coast of the Americas to the southern Andes in Chile. It is caused by the tension built up by moving tectonic plates.

RIO+20 United Nations Conference on Sustainable Development: an international U.N. conference held in 2012 in Rio de Janeiro, Brazil, with the goal of promoting sustainable development, especially through strengthening international institutions and building a green economy.

Robinson Projection: A map projection that balances the preservation and distortion of distance, direction, area and shape, leading to an overall clear image of Earth's features, though a necessarily slightly distorted one.

Royal Geographical Society (RGS): a British geographic organization that awarded medals of honor to many explorers and well-known geographers, including David Livingstone, Henry Stanley, Richard Burton, John Speke, and Diana Liverman.

Rust Belt: a core of North American industrialization in the northeast and Midwest; also called the Snow Belt.

Sahel: an area at the southern border of the Sahara Desert in Africa with highly variable rainfall and a population dependent on pastoralism and a history of famine.

samizdat publications: dissident or banned literature produced through systems of clandestine printing and distribution in the Soviet Union and communist countries of Eastern Europe.

satellite state: a state that is economically dependent and politically and militarily subservient to another state.

savanna: grassland vegetation found in tropical areas with a pronounced dry season and periodic fires.

sea gypsies: nomadic fisherpeoples and ethnic minorities who commonly make their living and homes in the coastal waters of South and Southeast Asia.

secondary sector: economic activity involving the processing, transformation, fabrication, or assembly of raw materials or the reassembly, refinishing, or packaging of manufactured goods.

secular: nonreligious.

Semitic: referring to a language family including Arabic, Hebrew, and Aramaic.

serfdom: a social practice whereby members of the lowest class were attached to a lord and his land.

Seven Years' War: *see* French and Indian War.

sexuality: a set of practices and identities related to sexual acts and desires.

Shari'a: Islamic canonical law and the foundation of political institutions in Saudi Arabia, Iran, Oman, and Yemen.

Shi'a: a minority sect of Islam whose beliefs are based on different interpretations of Islam in 7th century C.E.; adherents are mostly located in Iran and Iraq. The Shi'a argue that political leadership is divine and leaders must be descendants of the Prophet.

shifting cultivation: an agricultural system that preserves soil fertility by moving crops from one plot to another.

Shinto: a Japanese Indigenous religious culture, which stresses a belief in the nature of sacred powers that can be recognized in existing things, and which include practices entailing ritual purification, the offering of food to sacred powers, sacred music and dance, solemn worship, and joyous celebration.

shogun: a local noble lord in dynastic Japan.

SIJORI: an economically integrated growth region that includes Singapore, Johor Baharu (Malaysia), and the Riau Islands (Indonesia).

Silk Road: an ancient east–west trade route between Europe and China.

Single European Act of 1985 (SEA): an act ratified by the European Community, which affirmed the alliance's ultimate goal of economic and political harmonization within a single supranational government.

site: physical attributes of a location that could include terrain, soil, vegetation, and water sources.

situation: the location of a place relative to other places and human activities.

slash-and-burn: an agricultural system often used in tropical forests that involves cutting trees and brush and burning them so that crops can benefit from cleared ground and nutrients in the ash.

slow food: a movement to resist fast food by preserving the cultural cuisine and the associated food and farming of an area.

Small Island Developing States (SIDS): Fifty-two small islands that share similar sustainable development challenges, especially climate change and sea-level rise.

Snow Belt: *see* Rust Belt.

social housing: rental housing that is owned and managed by a public institution or nonprofit organization.

social movement: organized political activism by groups or individuals.

Southern Crescent: a secondary, prosperous, emergent region that straddles the Alps, running from Frankfurt through Stuttgart, Zürich, and Munich, and finally to Milan, Turin, and northern Italy.

sovereign state: a political unit that exercises power over a territory and its people and is recognized by other states; a sovereign state's independent power is codified in international law.

soviets: a network of grassroots councils of workers that emerged in Russia at the turn of the 20th century.

spatial diffusion: a way that things spread through geographic space over time.

spatial justice: the fairness of the distribution of society's burdens and benefits, taking into account spatial variations in people's needs and in their contribution to the production of wealth and social well-being.

special economic zones (SEZs): carefully segregated export-processing areas in China that offer cheap labor and land, along with tax breaks, to transnational corporations.

Spice Islands: an archipelago in Southeast Asia (modern-day eastern Indonesia in particular) where items, such as nutmeg, pepper, and mustard, were domesticated and globally traded to other world regions, particularly beginning with the European colonial period, when the Portuguese and Dutch occupied the region.

Srivijaya: the island culture of Sumatra, an urban-based kingdom that emerged in Southeast Asia through maritime trade.

Stalin, Joseph (1869–1953): leader of the Soviet Union who developed a command economy, employed police terror for state compliance, and led the U.S.S.R. during World War II.

staples economy: a financial system based on natural resources that are unprocessed or minimally processed before they are exported to other areas where they are manufactured into end products.

state: an independent political unit with territorial boundaries that are internationally recognized by other states.

state capitalism: a market-based economy with private ownership and investment in which the state continues to own some firms, to seek and obtain technology, and to carefully control the value of its currency.

state socialism: an economy based on the principles of collective ownership and the administration of the means of production and distribution of goods, dominated and directed by state bureaucracies.

steppe: semiarid, treeless, grassland plains.

stolen generation: Aboriginal children who were forcibly removed from their homes in Australia from 1928 to 1964 and placed in white foster homes or institutions.

Structural Adjustment Programs (SAPs): economic reforms in the 1980s and 1990s that involved the removal of subsidies and trade barriers, the privatization of government-owned enterprises such as telephone and oil companies, reductions in the power of unions, and an overall focus on export expansion. These policies, while reducing inflation and debt, had negative effects on the poor.

stupa: mound-shaped monuments that hold sacred Buddhist relics.

subduction zone: an area where one tectonic plate of Earth's crust moves under another.

subsistence affluence: a decent standard of living achieved with little cash income through reliance on local foods and community resources.

subtropical high: a zone of descending air, which results in dry, stable desert conditions over the Sahara and Kalahari Deserts.

suburbanization: the growth of population along the fringes of large metropolitan areas.

Sudd: the vast wetland in South Sudan into which the Nile flows.

sultanate: a Muslim state ruled by supreme leaders or sultans.

Sun Belt: the region of the United States that experienced substantial growth (due to the growth of the computer and information technology economy) during the decline of industrialization in the Rust Belt during the 1960s and the 1970s.

Sundarbans: the distinctive ecology of untouched mangrove and tropical swamp forest stretching from West Bengal in India into coastal Bangladesh; it is home to crocodiles, Chital deer, and one of the largest single populations of Bengal tigers in the world.

Sunna: a set of practical guidelines for Islamic behavior—the body of traditions derived from the words and actions of the prophet Muhammad; not a written document.

Sunni: the sect of Islam practiced by the majority of Muslims in the Middle East and North Africa.

superfund sites: locations in the United States designated by the federal government to be extremely polluted and requiring extensive, supervised, and subsidized cleanup.

supranational organization: a collection of states with a common economic and/or political goal.

sustainability: the ability to meet current and future human needs, while simultaneously preserving precious environmental resources.

sustainable development: a vision of development that seeks a balance among economic growth, environmental impacts, and social equity.

swidden farming: form of agriculture in which land is cleared for cultivation by cutting and burning shrubs or trees, allowing multiple years of planting until forest regrowth occurs.

syncretic: refers to religious practices that have co-evolved and merged with one another over the centuries.

system: a set of elements linked together so that changes in one element often result in changes in another.

taiga: an ecological zone of boreal coniferous forest.

Taliban: a fundamentalist Muslim group that ruled much of Afghanistan between 1996 and 2001.

tariff: a tax placed on goods from outside the region to be paid on a particular class of imports or exports.

tavy: a Madagascan term for a slash-and-burn technique to clear forests.

technology system: a cluster of interrelated energy, transportation, and production technologies that dominates economic activity for several decades; since the beginning of the Industrial Revolution, we can identify four of them.

terms of trade: the relationship between the prices a country pays for imports compared to the price it receives for its exports; poor terms of trade are when import prices are much higher than export prices.

terracing: the creation of a distinctive landscape of stepped and reinforced flat agricultural fields cut into steep slopes in order to stabilize the land for cropping in the face of potentially catastrophic soil erosion, which typically allows sustained agricultural yields amid torrential rainfall in mountainous areas.

territorial production complex: regional groupings of production facilities in Soviet state socialism; complexes were based on local resources that were suited to clusters of interdependent industries.

tertiary sector: activity involving the sale and exchange of goods and services.

Theravada Buddhism: a conservative form of Buddhism practiced in Burma, Thailand, and Cambodia.

Three Gorges Dam: the largest electricity-generating facility in the world, based on an enormous hydroelectric river dam spanning the Chang Jiang (Yangtze River) in central China.

tipping point: places that have been identified as areas where climate could shift suddenly as a result of climate and environmental change.

topographic map: A type of map characterized by high levels of details, especially including relief (elevation) displayed in contour lines and both natural and man-made features, like rivers, roads, and buildings.

total fertility rate: the average number of children a woman will bear throughout her childbearing years (from approximately age 15 through 49).

trade winds: prevailing westerly winds in the tropics that blow toward the Equator from the northeast in the Northern Hemisphere or the southeast in the Southern Hemisphere.

trading empire: large-scale political economies that emerged as the industrial nations of Europe pursued overseas expansion in the early 19th century.

tragedy of the commons: when an open-access common resource is over exploited by individuals who do not recognize how their own use of the resource can add to that of many others to degrade the environment, for example, overfishing a given species to the point of extinction.

transform plate boundary: an area where the tectonic plates of Earth's crust are sliding past each other horizontally.

transhumance: the action of moving herds according to seasonal rhythms: in Europe to warmer, lowland areas in the winter and cooler, highland areas in the summer and in Africa from wetter to drier areas.

transmigration: a policy of resettling people from densely populated areas to less populated, often frontier regions.

Treaty of Nanking: the 1842 treaty that ended the first Opium War between China and Great Britain and ceded the island of Hong Kong to the British; the treaty allowed European and American traders access to Chinese markets through a series of treaty ports (ports that were opened to foreign trade as a result of pressure from the major powers).

Treaty of Tordesillas: an agreement to divide the world between Spain and Portugal along a north–south line 370 leagues (about 1800 kilometers, or 1100 miles) west of the Cape Verde Islands. Approved by the pope in 1494, Portugal received the area east of the line, including much of Brazil and parts of Africa, and Spain received the area to the west.

Treaty of Waitangi: the 1840 agreement in which 40 Maori chiefs gave the queen of England governance over their land and the right to purchase it in exchange for protection and citizenship; reinterpreted by the Waitangi tribunal in the 1990s, this treaty provides the basis for Maori land rights and New Zealand's bicultural society.

tribe: a group that shares a common set of ideas about collective loyalty and political action; in tribes, group affiliation is often based on shared kinship, language, and territory.

tsar: a ruler of the Russian Empire.

tsetse fly: a blood-sucking flying insect that lives in African woodland and scrub regions; associated with both human and livestock diseases such as sleeping sickness (tryptosomiasis) and nagana.

tsunami: a sometimes-catastrophic coastal wave created by offshore seismic (earthquake) activity.

tube well: water sources intended to provide drinking water free of bacterial contamination, but that are subject to high levels of arsenic contamination; installed in Bangladesh as a result of a campaign in the 1970s by the United Nations Children's Fund (UNICEF).

tundra: an arctic wilderness where the climate precludes any agriculture or forestry; permafrost and very short summers mean that the natural vegetation consists of mosses, lichens, and certain hardy grasses.

typhoon: a large rotating storm in Pacific Asia, which is produced by warm ocean temperatures and eddies in the trade winds from August to October.

UNCLOS (United Nations Convention on the Law of the Sea): formalized in 1982, this U.N. convention established a 200-mile exclusive economic zone (EEZ) for all nations, while also providing for the protection of international waters and the creation of rights of passage through key strategic waterways throughout the world.

Union for the Mediterranean: a multilateral partnership that encompasses 43 countries from Europe and the Mediterranean Basin established to form links across the Mediterranean with the European Union; formerly the Euro-Mediterranean Partnership, relaunched in 2008 as the Union for the Mediterranean.

Union of Soviet Socialist Republics (U.S.S.R.): a federal system created from the Russian empire in the aftermath of the 1917 Russian Revolution and formally dissolved in 1991; initiated by Lenin in 1922 to recognize the regional nationalities, which were to unite as a single Soviet people.

United Nations (U.N.): a supranational organization founded in 1945 aimed at facilitating cooperation in international law, security, economic development, human rights, and world peace; there are currently 193 member states in the U.N.

United Nations Educational, Scientific, and Cultural Organization (UNESCO): the arm of the U.N. whose mission is to contribute to the building of peace, eradication of poverty, sustainable development, and intercultural dialogue through education, the sciences, culture, communication, and information.

uplands: high, hilly land.

urban bias: the tendency to concentrate investment and attention in urban rather than rural areas.

urban primacy: a condition in which the population of the largest city in an urban system is disproportionately large in relation to the second- and third-largest cities in that system.

U.S. Civil War: the war between the North and the South in the United States from 1861–1865; precipitated the official end of slavery in the United States.

Vietcong: communist guerilla rebels who took control of portions of South Vietnam during the Indochina wars and fought the South Vietnamese government forces from 1954–1975 and U.S. forces during the Vietnam War with the support of the North Vietnamese army.

Vietnam War: a conflict between communist North Vietnam and U.S.-backed South Vietnam that began in 1964 after the United States bombed North Vietnam; the war resulted in over a million Vietnamese and 58,000 American deaths and ended with the unification of North and South Vietnam.

Wallace's Line: an imaginary line drawn in 1859 that serves as a division between species; the line is associated with the deep-ocean trench between the islands of Bali and Lombok in Indonesia that could not be crossed by animals and plants even during the low sea levels of the ice ages.

War on Terror: the U.S. response to the terrorist attacks of September 11, 2001; a global war against terrorism in which first Afghanistan and then Iraq and Pakistan were identified as the greatest threats to U.S. security.

watershed: the drainage area of a particular river or river system.

weather: the instantaneous or immediate state of the atmosphere (e.g., it is raining).

welfare state: a system in which the government undertakes to protect the health and well-being of its citizens with the aim of distributing income and resources to the poorer members of society.

westerlies: air in the midlatitudes flowing poleward from the tropics from west to east.

wet farming: agriculture that involves irrigation.

White Australia policy: a government policy in effect in Australia until 1975 that restricted immigration to people from northern Europe through a ranking; British and Scandinavian immigrant candidates were given the highest priority, followed by southern Europeans.

World Bank: a development bank and the largest source of development assistance in the world.

world city: a city in which a disproportionate share of the world's most important business—economic, political, and cultural—is transacted.

World Health Organization (WHO): the directing and coordinating authority for health within the United Nations system.

World Heritage Site: a place that has been formally identified as a protected site by the United Nations Educational, Scientific, and Cultural Organization (UNESCO).

world region: a large-scale geographic division based on continental and physiographic settings that contain major clusters of humankind with broadly similar cultural attributes.

world religion: belief systems that have adherents worldwide; Islam, Christianity and Judaism are world religions.

xenophobia: a hatred and/or fear of foreigners.

zero population growth (ZPG): a demographic state where the number of births match the number of deaths in a population in such a way that no natural population growth occurs.

Zionism: a movement whose chief objective has been the establishment of a legally recognized home in Palestine for the Jewish people.

zone of alienation: the area surrounding the Chernobyl reactor in Ukraine where radiation levels remain high as a result of the 1986 nuclear reactor accident; only a small number of residents and scientific teams reside in the zone of alienation.

PHOTO, ILLUSTRATION, AND TEXT CREDITS

CHAPTER 1

CO.1: AP Images/Imaginechina
1.2: NASA Goddard Space Flight Center Image by Reto Stöckli (land surface, shallow water, clouds). Enhancements by Robert Simmon (ocean color, compositing, 3D globes, animation). Data and technical support: MODIS Land Group; MODIS Science Data Support Team; MODIS Atmosphere Group; MODIS Ocean Group Additional data: USGS EROS Data Center (topography); USGS Terrestrial Remote Sensing Flagstaff Field Center (Antarctica); Defense Meteorological Satellite Program (city lights).
1.3: Greatstock Photographic Library/Alamy
1.4: Reuters/Thomas Mukoya
1.5a: Ted Foxx/Alamy
1.5b: Amana Images Inc./Alamy
1.6: Daniel Leppens/Shutterstock
1.7: AGUADO, EDWARD; BURT, JAMES E., UNDERSTANDING WEATHER AND CLIMATE, 2nd Ed., © 2001. Reprinted and Electronically reproduced by permission of Pearson Education, Inc. Upper Saddle River, New Jersey.
1.8: CHRISTOPHERSON, ROBERT W., GEOSYSTEMS: INTRODUCTION TO PHYSICAL GEOGRAPHY, 6th Ed., © 2006. Reprinted and Electronically reproduced by permission of Pearson Education, Inc. Upper Saddle River, New Jersey.
1.9b: Flickr/Tracy Packer Photography/Getty Images
1.10: Based on R.W. Christopherson, *Elemental Geosystems*, 7th ed., Upper Saddle River, NJ: Pearson Education, 2013, pp. 220–221.
1.12: Data from "GISS Surface Temperature Analysis, NASA, available at http://data.giss.nasa.gov/gistemp/graphs_v3/ and from The World Bank.
1.13: Climate Change 2007: Synthesis Report. Contribution of Working Groups I, II and III to the Fourth Assessment Report of the Intergovernmental Panel on Climate Change, Figure SPM.6. IPCC, Geneva, Switzerland.
1.15b: Alison Wright/Corbis
1.17: Noboru Hashimoto/AFP/Getty Images/Newscom
1.18: From "The Human Footprint Index," Center for International Earth Science Information Network (CIESIN), available at http://sedac.ciesin.columbia.edu/downloads/maps/wildareas-v2/wildareas-v2-human-footprint-geographic/World.pdf.
1.19: Millennium Ecosystem Assessment 2005, Living Beyond Our Means: Natural Assets and Human Well-Being (Statement from the Board). World Resources Institute, Washington, DC.
1.20: Mona Makela/Shutterstock
1.22: Based on B. Crow and A Thomas, *Third World Atlas*. Milton Keynes: Open University Press, 1982, pp. 37, 41.
1.23: Based on B. Crow and A. Thomas, *Third World Atlas*. Milton Keynes: Open University Press, 1982, pp. 37, 41.
1.24: Adapted from B. Crow and A. Thomas, *Third World Atlas*. Milton Keynes: Open University Press, 1982, p. 27.
1.25: Redrawn from Armesto, *The World: A History*, 2010.
1.26a: Nik Wheeler/Alamy
1.26b: CSI Productions/Alamy
1.26d: Fancy/Alamy
1.26c: Orestis Panagiotou/EPA/Newscom
1.27a: From the World Trade Organization.
1.27b: Michael Reynolds/EPA/Newscom
1.29: The World Bank: HID Per Capita: http://data.worldbank.org/indicator/NY.GNP.PCAP.PP.CD/countries?display=default
1.30: Nicolas Asfouri/AFP/Getty Images/Newscom
1.31: Antonio Bat/EPA/Newscom
1.32: Narciso Contreras/ZUMA Press/Newscom
1.35: LHB Photo/Alamy
1.36: Katarina Premfors / arabianEye / Getty Images
1.37: Based on Knox/Marston, *Human Geography: Places and Regions in Global Context* 6e, © 2013 Pearson Education.
1.38: Based on Knox/Marston, *Human Geography: Places and Regions in Global Context* 6e, © 2013 Pearson Education.
1.39: Adapted from E.F. Bergman, *Human Geography: Cultures, Connections, and Landscapes*, Upper Saddle River, NJ: Prentice Hall, 1994; Western Hemisphere after J.H. Greenberg, LANGUAGE IN THE AMERICAS. Stanford, CA: Stanford University Press, 1987; Eastern Hemisphere after D. Crystal, *The Cambridge Encyclopedia of Language*. Cambridge: Cambridge University Press, 1997.
1.40: Based on Knox/Marston, *Human Geography: Places and Regions in Global Context* 6e, © 2013 Pearson Education.
1.41: Moviestore Collection Ltd/Alamy
1.42: AP Images/Mark Lennihan
1.43: Peter Horree/Alamy
1.44: Xu Lei/Snider Images/The Image Works
1.45: From the Center for International Earth Science Information Network (CIESIN), Columbia University; and Centro Internacional de Agricultura Tropical (CIAT). 2005. Gridded Population of the World, Version 3 (GPWv3). Palisades, NY: Socioeconomic Data and Applications Center (SEDAC), Columbia University. Available at http://sedac.ciesin.columbia.edu/gpw.
1.46: From the Center for International Earth Science Information Network (CIESIN), Columbia University and Centro Internacional de Agricultura Tropical (CIAT), 2005, Griddled Population of the World,. Version 3 (GPWv3), Palisades, NY: Socioeconomic Data and Application Center (SEDAC) Columbia University. http://sedac.ciesin.columbia.edu/gpw/
1.48: From the U.S. Energy Information Administration (Oct 2008).
1.49: Data from Peace Direct, "Insight on Conflict," http://www.insightonconflict.org; Department of Peace and Conflict Research, "Uppsala Universitet," http://www.pcr.uu.se/research/ucdp/definitions; William Mitchell School of Law at UMN, World Without Genocide, http://www.worldwithoutgenocide.org.
1.1.1: Based on Global Warming Art by Robert A. Rhode.
1.1.2: Based on Weiss, Overpeck and Strauss, 2011.
1.1.3: Based on cleanair-coolplanet.org.

CHAPTER 2

CO.2: Yang Ling/Xinhua/Photoshot
2.2: NASA Earth Observing System
2.4: Adapted from R. Mellor and E.A. Smith, *Europe: A Geographical Survey of the Continent*. London: Macmillan, 1979.
2.5: Galyna Andrushko/Shutterstock
2.6: Matteo Volpone/Shutterstock
2.7: David Hughes/Shutterstock
2.8: photontrappist/Alamy
2.9: Redrawn from R. King et al., *The Mediterranean*. London: Arnold, 1997, pp. 59 and 64.
2.10: PStar/Shutterstock
2.11: The Print Collector/Alamy
2.13: Keystone Archives/HIP/The Image Works
2.14: Mike Abrahams/Alamy
2.15: Petrut Calinescu/Alamy
2.16: Adapted from The European Union.
2.1.1: Courtesy of Paul Knox
2.1.2: Iuri/Shutterstock
2.17: Courtesy of Paul Knox
2.18: Courtesy of Paul Knox
2.19: Marka/Alamy
2.20: Adapted from "High Speed Railroad Map, Europe 2008," by Bernese Media, July 2009, http://commons.wikimedia.org/wiki/File:High_Speed_Railroad_Map_Europe_2008.gif.
2.21: imagebroker/Alamy
2.22: Adapted from "Europe Main Map at the Beginning of the Year 1500," Euratlas- Nüssli, Rue du Mileu 30, 1400 Yverdon-les-Bains, Switzerland, available at http://www.euratlas.net/history/europe/1500/index.html.
2.23: Adapted from "Europe Main Map at the Beginning of the Year 1900," Euratlas - Nüssli, Rue du Mileu 30, 1400 Yverdon-les-Bains, Switzerland, available at http://www.euratlas.net/history/europe/1900/index.html.

2.24: Adapted from P. Knox, *Geography of Western Europe*. London: Croom Helm, 1984, p. 69. Courtesy of Paul Knox.

2.25: Dimitar Dilkoff/AFP/Getty Images/Newscom

2.26: Redrawn from R. Mellor and E.A. Smith, *Europe: A Geographic Survey of The Continent*. London: Macmillan, 1979, p. 22.

2.27: Pictorial Press Ltd/Alamy

2.28: From Center for International Earth Science Information Network [CIESIN], Columbia University, *Gridded Population of the World* [Gpw], Version 3. Palisades, NY: CIESIN, Columbia University, 2005. Available at http://sedac.ciesin.columbia.edu/gpw.

2.29: Data from UN Population Division. J. McFalls, Jr., "Population: A Lively Introduction," *Population Bulletin*, 46[2], 1991, p.1.

2.30: Michael Kemp/Alamy

2.31: Courtesy of Paul Knox.

2.32: QQ7/Shutterstock

2.33: Courtesy of Paul Knox

2.34: Courtesy of Paul Knox

2.36: Konstantinos Tsakalidis/Alamy

2.35: Courtesy of Paul Knox

2.2.1: Data from "Muslims Worldwide," Associated Press Interactive, available at http://hosted.ap.org/interactives/2011/islam/ and from "Muslims in Europe: Country Guide," BBC News, available at http://news.bbc.co.uk/2/hi/europe/4385768.stm.

CHAPTER 3

CO.3: epa european pressphoto agency b.v./Alamy

3.2: NASA Earth Observing System

3.1.3: AP Images/Association of Russian Polar Explorers

3.5: Steve Morgan/Alamy

3.6: Pavel Filatov/Alamy

3.7: BigJoker / Alamy

3.9: Dave and Sigrun Tollerton/Alamy

3.1: Maria Stenzel/National Geographic/Getty Images

3.11: Daniel Prudek/Shutterstock

3.12: ITAR-TASS Photo Agency/Alamy

3.13: Reuters/Ilya Naymushin

3.2.1: Tatiana Kolesnikova/PhotoShot

3.2.2: Rueters/Viktor Korotayev

3.14: Ivan Vdovin/Alamy

3.15: Redrawn from D.J.B. Shaw, *Russia in the Modern World*. Oxford: Blackwell, 1999, p.7.

3.16: Fine Art Images/Heritage Imagestate/Glow Images, Inc.

3.17: Redrawn from *Atlas of Twentieth Century World History*. New York: HarperCollins Cartographic, 1991, pp. 86-87.

3.18: Redrawn from P.L. Knox and J. Agnew, *The Geography of the World Economy*, 3rd ed. London: Arnold, 1998, p. 168.

3.20: Redrawn from J.H. Bater, *Russia and the Post-Soviet Scene*. London: Arnold, 1996, p. 314.

3.21: Dean C.K. Cox/WpN/PhotoShot

3.22: Adapted from "Facing Water Challenges in the Aral Sea, Uzbekistan: A WWDR3 Case Study," Waterwiki.net, http://waterwiki.net/index.php/Facing_Water_Challenges_in_the_Aral_Sea,_Uzbekistan:A_WWDR3_Case_Study and "The Disappearance of the Aral Sea," United Nations Environment Programme, http://www.unep.org/dewa/vitalwater/article115.html.

3.23: gopixgo/Shutterstock

3.24: Sergey Yakovlev travel/Alamy

3.25: Valery Markov/Fotolia

3.26: David Litschel/Alamy

3.27: Redrawn from G. Smith, *The Post-Soviet States: Mapping the Politics of Transition*. London: Arnold, 1999, p. 129.

3.28: AP Images/Sergey Ponomarev

3.29: Sergei Supinsky/EPA/Newscom

3.30: ITAR-TASS Photo Agency/Alamy

3.31: Redrawn from R. Miller-Gulland and N. Dejevsky, *Cultural Atlas of Russia* rev. ed. New York: Checkmark Books, 1998, pp. 26-27.

3.3.2: Stringer Russia/Reuters

3.32: Batareykin/Alamy

3.33: AP Images/Alexander Zemlianichenko

3.34: Reuters/Denis Sinyakov/Landov

3.35: From Center for International Earth Science Information Network [CIESIN], Columbia University; International Food Policy Research Institute [IFPRI]; and World Resources Institute [WRI], 2000. *Gridded Population of The World [Gpw]*, Version 3. Palisades, NY: CIESIN, Columbia University, 2005. Available at http://sedac.ciesin.columbia.edu/gpw/.

3.36: Updated from J.H. Bater, *Russia and the Post-Soviet Scene*. London: Arnold, 1996, p. 101.

3.37: Alan Solomon/KRT/Newscom

3.38: Adapted from "Russian International Migration," by LokiiT, November 2009, available at http://en.wikipedia.org/wiki/File:Russian_international_migration.PNG.

3.39: ZUMA Wire Service/Alamy

3.1.2: Based on map in *Newsweek*.

CHAPTER 4

CO.4: Lewis Houghton/Alamy

4.2: NASA Earth Observing System

4.4: Kevin Foy/Alamy

4.6: Redrawn from C.C. Held, *Middle East Patterns: Places, Peoples and Politics*, 3rd ed. Boulder: Westview Press, 2000, p. 23.

4.7: Frédéric Soreau/Glow Images, Inc.

4.8: Redrawn from P. English, *City and Village in Iran*. Madison: University of Wisconsin Press, 1966, p. 31.

4.9: Sola/parasola.net/Alamy

4.10: Walid Nohra/Shutterstock

4.11: Redrawn from C.C. Held, *Middle East Patterns: Places, Peoples and Politics*, 3e Boulder: Westview Press, 2000, p. 16.

4.12: Redrawn from C.C. Held, *Middle East Patterns: Places, Peoples and Politics*. 3rd ed. Boulder: Westview Press, 2000, p. 23.

4.1.2: Yann Arthus-Bertrand/Corbis

4.13: Pius Lee/Fotolia

4.14: Redrawn from *Hammond Times Concise Atlas of World History*. Maplewood, NJ: Hammond, 1994, pp. 100-1.

4.17: Ludovic/REA/Redux

4.16: Reuters/Raheb Homavandi

4.18: marka/Marka/SuperStock

4.19: Martin Harvey/Alamy

4.20: Based on University of Texas Press, available at http://www.lib.utexas.edu/maps/middle_east_and_asia/kurdish_lands_92.jpg.

4.21: From "Sudan: Aid agencies face extremely limited access in conflict-torn southern states," United Nations Office for the Coordination of Humanitarian Affairs (UNOCHA), available at http://www.unocha.org/top-stories/all-stories/sudan-aid-agencies-face-extremely-limited-access-contested-southern-areas. Map provided courtesy of the UN Office for the Coordination of Humanitarian Affairs

4.22: Data from [1] ProCong.org, 2008, http://israelipalestinian.procon.org/viewsource.asp?resourceID=000639. ProCon.org is a non-profit public charity with no government affiliation. It contains an amalgam of population and statistical data on deaths for both Israelis and Palestinians, including multiple sources (e.g., U.N., Israeli Ministry of Foreign Affairs and so on). [2] ProCon.org, http://israelipalestinia.procon.org/viewresource.asp?resourceID=00063. [3] BBC News, http://news.bbc.co.uk/1/shared/sp1/hi/middle_east/03v3_ip_timeline/html/19678.stm. [4] Palestinian Centre for Human Rights 2009, http://www.pchrgaza.org/files/PressR/English/2008/36-2009.html. [5] BBC News 2009, http://news.bbc.co.uk/1/hi/world/middle_east/7838618.stm. [6] Council on Foreign Relations 2009, http://www.cfr.org/publications/15268/. [7] The Electronic Intifada 2007, http://electronicintifada.net/bytopic/197.shtml. [8] UNWRA 2008, http://www.un.org/unrwa/publications/pdf/rr_countryandarea.pdf. [9] Congressional Research Service 2008, http://www.un.org/unrwa/publications/pdf/rr_countryandarea.pdf. All accessed June 29, 2009. [10] "Palestinian prisoners on mass hunger strike." AP, May 7, 2012, www.huffingtonpost.com/huffwires/20120507/ml-israel-palestinian-hunger-strike/

4.23: RUBENSTEIN, JAMES M., THE CULTURAL LANDSCAPE: AN INTRODUCTION TO HUMAN GEOGRAPHY, 9th ed., c.2008. Reprinted and Electronically reproduced by permission of Pearson Education, Inc. Upper Saddle River, New Jersey.

4.24: Adapted from *The Guardian*, October 14, 2000, p.5 and from the United Nations Relief and Works Agency for Palestine Refugees.

4.25: Based on BBC News, Web edition, available at http://news.bbc.co.uk/1/hi/world/middle_east/3111159.stm; accessed June 19, 2009.

4.26: Ryan Rodrick Beiler/Alamy

4.27: UPPA/Photoshot

4.28: Adapted from Iraq Body Count, available at http://www.iraqbodycount. org/analysis/numbers/2011/ and Iraq Coalition Casualty Count, available at http://icasualties.org/iraq/index.aspx.

4.29: Daily Mail/Rex/Alamy

4.30: EPA/epa european pressphoto agency b.v./Alamy

4.31: Redrawn from D. Hiro, *Holy Wars*. London: Routledge, 1989, frontispiece.

4.32: Photosindia/Alamy

4.33: ARCO/Svarc, P/Glow Images, Inc.

4.34: ZUMA Press, Inc./Alamy

4.36: Arco Images GmbH/Alamy

4.37: From Socioeconomic Data and Applications Center. Columbia University, Center for International Earth Science Information Network [CIESIN]. *Gridded Population of the World* [Gpwv3]. Palisades, NY: CIESIN, Columbia University. Available at http://sedac.ciesin.org/plue/gpw.

4.35: Clover/Superstock

4.38: Craig Ruttle/Alamy

4.39: Aurora Photos/Alamy

4.41: Redrawn and modified from A. Segal, *An Atlas of International Migration*. London: Hans Zell, 1993, pp. 95 and 103.

4.42: Hackenberg-Photo-Cologne/Alamy

4.43: Adapted from P. Rekacewicz, World Resources Institute.

4.1.1: From WRRG Economics, www.wtrg.com. Used with permission.

4.1.3: Adapted from "Energy Watch Group, Crude Oil: The Supply Outlook," 2007, p. 12. Available at http://www.energywatchgroup.org/fileadmin/global/pdf/EWG_Oilreport_10-2007.pdf.

CHAPTER 5

CO.5: Barry Lewis/In Pictures/Corbis

5.2: NASA Earth Observing System

5.4: Ariadne Van Zandbergen/Alamy

5.5: UNEP/GRID-Arendal, http://www.grida.no/graphicslib/detail/climate-change-vulnerability-in-africa_7239. Designed by Delphine Digout, Revised by Hugo Ahlenius, UNEP/GRID-Arendal. Used with permission.

5.6: ARYEETEY-ATTOH, SAMUEL A, THE GEOGRAPHY OF SUB-SAHARAN AFRICA, 1st Ed. c.1997. Reprinted and Electronically reproduced by permission of Pearson Education, Inc. Upper Saddle River, New Jersey.

5.7a: Blaine Harrington III/Corbis

5.7b: Panoramic Images/Getty Images

5.8b: BH Generic Stock Images/Alamy

5.8a: Photoromano/Fotolia

5.9: Based on "Forest Cover of Madagascar, 1950s to ~2000," in *Our Changing Planet: The U.S. Climate Change Science Program of Fiscal Year* 2007, figure 23, available at http://www.usgcrp.gov/usgcrp/Library/ocp2007/OCP07-Fig-23.htm.

5.1.1: AP Images/Tsvangirayi Mukwazhi

5.1.2: DeBeers

5.10: Reuters/Scanpix Denmark

5.11: Vadim Petrakov/Shutterstock

5.12: Based on Malaria Atlas Project, available at http://www.map.ox.ac.uk/browse-resources/endemicity/Pf_mean/world/; "Yellow Fever Vaccination Recommendations" available at http://www.itg.be/itg/GeneralSite/MedServ/Images/Gele%20koorts%20Afrika%2002.jpg; "Sleeping Sickness: The Socioeconomic Impacts & Possible Prevention Methods" available at http://sitemaker.umich.edu/section003group6/impact_on_food_production_and_security; "Classification of human African trypanosomiasis-endemic countries according to cases reported in 2009," available at http://www.plosntds.org/article/info%3Adoi%2F10.1371%2Fjournal.pntd.0001007; "Map of the estimated prevalence of eye worm history in Africa" available at http://www.who.int/apoc/raploa/en/index.html.

5.13: Rob Howard/Corbis

5.14: Based on I. L. L. Griffiths, *An Atlas of African Affairs*; D. L. Clawson and J. S. Fisher (eds.), *World Regional Geography: A Developmental Approach*; C. McEvedy, *The Penguin Atlas of African History*.

T5.2a: Blickwinkel/Hauke/Alamy

T5.2b: Espeel Pieter/Arterra Picture Library/Alamy

T5.2e: Andreas Rose/imagebroker/Alamy

T5.2c: arniepaul/Fotolia

T5.2d: Karen Roush/Fotolia

T5.2f: KsenyaLim/Shutterstock

5.15a: Sandro Vannini/Corbis

5.15: Nik Wheeler/Corbis

5.16: Based on A. Thomas and B. Crow (eds.), *Third World Atlas*; J. F. Ade, A. Crowder, M. Crowder, P. Richards, E. Dunstan, and A. Newman (eds.), *Historical Atlas of Africa*.

5.17a: Hulton-Deutsch Collection/Corbis

5.17b: London Stereoscopic Company/Hulton Archive/Getty Images

5.18: Adapted from A. Thomas and B. Crow (eds.), *Third World Atlas*. Buckingham, UK: Open University Press, 1994, p. 35.

5.19a: Michael Lewis/National Geographic/Getty Images

5.19b: Nomad/SuperStock

5.20: Data from: http://databank.worldbank.org

5.21: Jon Hrusa/EPA/Newscom

5.22: Redrawn from Wolfram Alpha LLC, available at http://www.wolframalpha.com/input/?i=africa+cell+phone+users.

5.23: The World Bank, http://siteresources.worldbank.org/INTAFRICA/Resources/Africas-Pulse-brochure_Vol5.pdf. Used with permission.

5.24: Based on "Sub-Saharan African Merchandise Trade with China," available at http://blogs.ft.com/beyond-brics/files/2010/12/africa-china-trade.jpg.

5.25: AP Images/Khalil Senosi

5.26a: Peter Titmuss/Alamy

5.26b: Reuters/Juda Ngwenya

5.28: lynn hilton/Alamy

5.27: imagebroker/Alamy

5.29: British Ministry of Defence/Handout/EPA/Newscom

5.31: AP Images

5.30: Reuters/Akintunde Akinleye

5.32: Based on S. Aryeetey-Attoh (ed.), *The Geography of Sub-Saharan Africa*. Upper Saddle River, NJ: Prentice hall, 1997, Fig. 4.6.

5.33b: kevin moloney/aurora photos/alamy

5.33a: Mandel Ngan/AFP/Getty Images/Newscom

5.34: Michael Regan/Getty Images

5.35: caroline penn/impact/hip/The Image Works

5.36: From Center for International Earth Science Information Network [CIESIN], Columbia University; International Food Policy Research Institute [IFPRI]; and World Resources Institute [WRI], *Gridded Population of the World* [GPW], Version 2, Palisades, NY: CIESIN, Columbia University, 2000. Available at http://sedac.ciesin.org/plue/gpw/index.html?main.html&2.

5.37: Data from "Ghana: Demographic and Health Survey," 2008, available at http://www.measuredhs.com/publications/publication-FR221-DHS-Final-Reports.cfm.

5.38: From the World Bank Group, "Intensifying Action Against HIV/AIDS in Africa: Responding to a Development Crisis," 1998, available at http://www.worldbank.org/html/extdr/offrep/afr/aidstrat.pdf; and from "Annual AIDS-related deaths by region, 1990-2009," Figure 2.3, p 22, available at www.unaids.org/documents/20101123_GlobalReport_Chap2_em.pdf.

5.39: Based on "Slum population in urban Africa," UNEP, available at http://www.grida.no/graphicslib/detail/slum-population-in-urban-africa_d7d6#.

5.40c: Gideon Mendel/ActionAid/Corbis

5.40b: Peter Macdiarmid/Getty Images News/Getty Images

5.40a: Kim Ludbrook/epa/Newscom

5.41: Liba Taylor/Corbis

5.42: From A Belward et al., "Renewable energies in Africa," JRC Scientific and Technical Reports, European Commission, available at http://publications.jrc.ec.europa.eu/repository/bitstream/111111111/23076/1/reqno_jrc67752_final%20report%20.pdf.

5.42c: Snapperuk/Alamy

5.2.1: Based on "Satellite Image Map," "Ethnic Group Distribution Map," "Distribution of Religion Map" and "Oil Infrastructure in Sudan Map," from South Sudan Info Maps, all available at http://southsudaninfo.net/maps/.

Table 5.2.1: From "Key Indicators for South Sudan," CIA Fact Book.

CHAPTER 6

CO.6: Accent Alaska.com/Alamy

6.2: NASA Earth Observing System

6.4: From "How Will Climate Change Impact the EPA Region 6 Area," US Environmental Protection Agency, available at http://www.epa.gov/region6/climatechange/maps.htm.

6.6: US Coast Guard Photo/Alamy

6.7: Adapted from E. Homberger, *The Historical Atlas of North America*. London: Penguin, 1995, p. 21.

6.8: Dave Reede/AgStock Images/Corbis

6.9: Jim Wark/Agstockusa/Age Fotostock
6.10: Omar Torres/AFP/Getty Images
6.2.2: AP Images/Darren Stone
6.11: MAY FOTO/Age Fotostock
6.12: Stock Connection Blue/Alamy
6.13: North Wind/North Wind Picture Archives
6.14: Caro/Alamy
6.15: Erin Paul Donovan/Alamy
6.16: Data from the U.S. Census Bureau, Economic Census 2010.
6.17: KNOX, PAUL L.; MCCARTHY, LINDA M., URBANIZATION: AN INTRODUCTION TO URBAN GEOGRAPHY, 3rd, © 2012. Printed and Electronically reproduced by permission of Pearson Education, Inc., Upper Saddle River, New Jersey.
6.18: Inge Johnsson/Alamy
6.3.3: Shi Sisi/ZUMA Press/Newscom
6.19: Norma Jean Gargasz/Alamy
6.20: Ethan Miller/Getty Images News/Getty Images
6.21: AP Images/The News-Gazette, Heather Coit
6.22: Adapted from "America's Wars: U.S. Casualties and Veterans," infoplease, data from the Department of Defense and Veterans Administration, available at http://www.infoplease.com/ipa/A0004615.html.
6.23: US Air Force Photo/Alamy
6.24: Michael Reynolds/EPA/Newscom
6.25: CTK/Alamy
6.26: Based on "2011-2012 MLB Player Map" available at http://www.everyoneelseisdoingit.com/maps/MLB_2012.html.
6.27: Igor Vidyashev/Alamy
6.28: Data from "Yearbook of Immigration Statistics," U.S. Department of Homeland Security, available at http://www.dhs.gov/yearbook-immigration-statistics.
6.29: Based on *The Historical Atlas of North America* and USGS National Atlas.
6.30: James Quine/Alamy
6.31: Gunter Marx/Alamy
6.32: gZUMA Press/Newscom
6.1.1: Based on "North America Energy Resources," Maps of the World, available at http://www.mapsofworld.com/north-america/energy-resources.html.
6.1.2: Data from FAOSTAT, 2010, available at http://faostat.fao.org/site/339/default.aspx.
6.1.3: Based on "National footprints as a proportion of the global Footprint," *Ecological Footprint* (EF), July 2010, available at http://action-town.eu/marketplace/indicators/ecological-footprint-ef/.
6.2.1: From UNODC, World Drug Report 2011 (United Nations Publication, Sales No. E.11.XI.10).
6.3.2: Based on "The Triangular Trade," Alistair Boddy-Evans, available at http://0.tqn.com/d/africanhistory/1/0/7/M/TriangleTrade001.jpg.

CHAPTER 7

CO.7: Robert Harding Picture Library/SuperStock
7.2: NASA Earth Observing System
7.3: Greg Probst/Corbis
7.6: Bernard Bisson/Sygma/Corbis
7.7: Niv Koren/Shutterstock
7.8: Diana Liverman
7.10: Tommy E Trenchard/Alamy
7.11a: RODRIGO ARANGUA/AFP/Getty Images/Newscom
7.11b: Zuma Press/Newscom
7.12: Joseph Sohm/Visions of America/Corbis
7.13a: Aivar Mikko/Alamy
7.13b: GlowImages/Alamy
7.13c: think4photop/Shutterstock
7.14c: Science Photo Library/Alamy
7.14b: Ted Spiegel/Corbis
7.1.2: Jeff Schmaltz/Nasa Images
7.16: Henry, P/Arco Images/Glow Images, Inc.
7.2.2: China Photos/Alamy
7.2.1: Ian Cumming/Glow Images, Inc.
7.18: Based on P. L. Knox and S. A. Marston, *Human Geography: Places and Regions In Global Context*, 3rd ed. Upper Saddle River, NJ: Prentice Hall, 2004.
7.20: ZUMA Wire Service/Alamy
7.21: Data from The World Bank.
7.3.1: Harish Tyagi/EPA/Newscom
7.22a: From the Food and Agriculture Organization of the United Nations, FAOSTAT, available at http://faostat.fao.org/#.
7.22b: The Fairtrade Foundation/Sue Atkinson
7.23a: Reuters/Rafael Perez
7.23b: Bettmann/Corbis
7.23c: Transcendental Graphics/Getty Images
7.23d: Bettmann/Corbis
7.24: Sergio Pitamitz/Terra/Corbis
7.25: AP Images/Dado Galdieri
7.26a: Based on the map "Mexican drug cartels' main areas of influence, 2010-11" in "Q&A: Mexico's drug-related violence," BBC NEWS, October 2012, available at http://www.bbc.co.uk/news/world-latin-america-10681249.
7.26b: Keith Dannemiller/Alamy
7.28: Steve Bly/Alamy
7.29: From Center for International Earth Science Information Network [CIESIN], Columbia University; International Food Policy Research Institute [IFPRI]; and World Resources Institute [WRI]. *Gridded Population of the World* [Gpw], Version 2, Palisades, NY: CIESIN, Columbia University, 2000. Available at http://sedac.ciesin.org/plue/gpw.
7.32: Paulo Whitaker/Reuters/Corbis
7.31: Jorge Uzon/AFP/Getty Images/Newscom
7.33: Sergio Pitamitz/Terra/Corbis
7.1.1: Based on G. Knapp and C. Caviedes, *South America*. Englewood Cliffs, NJ: Prentice Hall, 1995, p. 233; and from "Controversial Infrastructure Projects Proposed or Underway in the Amazon Basin," available at http://www.amazonwatch.org.
7.1.3: Data from "Projecto Prodes Monitoramento Da Floresta Amazonica Braseileira Satelite," from the Ministerio da Ciencia e Tecnologia, available at http://www.obt.inpe.br/prodes/index.php.
7.3.2: Data from "What does Brazil export?," The Observatory of Economic Complexity, available at http://atlas.media.mit.edu/explore/tree_map/export/bra/all/show/2009/.

CHAPTER 8

CO.8: MPAK/Alamy
8.2: NASA Earth Observing System
8.4: Nico Smit/Alamy Limited
8.5: Reuters/Shanghai
8.7: Mainchi Shimbun/Reuters/Landov
8.9: javarman/Fotolia
8.8: PRILL Mediendesign und Fotografie/Shutterstock
8.10: Bruno Morandi/Robert Harding
8.11: Keren Su/Getty Images
8.12: Alex Hofford/EPA/Newscom
8.13: Kwak-Sung-Ho/AFP/Getty Images/Newscom
8.1.2: David White/Alamy
8.1.1: AFP/Getty Images
8.2.1: AP Images/NASA
8.14: Izmael/Shutterstock
8.15: Based on I. Barnes and R. Hudson, *History Atlas of Asia*. New York: Macmillan, 1998, pp. 45 and 46–47.
8.16: Tito Wong/Shutterstock
8.17: akg-images/Newscom
8.3.2: AFP/Getty Images
8.18: Based on I. Barnes and R. Hudson, *History Atlas of Asia*. New York: Macmillan, 1998, p. 129.
8.19: Bill Pierce/Time Life Pictures/Getty Images
8.21: TPG Top Photo Group/Newscom
8.22: Rob Crandall/Newscom
8.23: Pedro Ugarte/AFP/Getty Images/Newscom
8.24: Chuck Nacke/Alamy
8.25: Chuong/Shutterstock
8.27: Colin Sinclair/Dorling Kindersley, Ltd.
8.26: Craig Hanson/Shutterstock
8.28: Peter Parks/AFP/Getty Images/Newscom
8.29: Alamy Creativity/Alamy

8.30: Reuters/Charles Platiau

8.31: An Qi/Alamy

8.32: From Center for International Earth Science Information Network [CIE-SIN], Columbia University; International Food Policy Research Institute [IFPRI]; and World Resources Institute [WRI], 2000. *Gridded Population of the World* [Gpw], Version 3, Palisades, NY: CIESIN, Columbia University, 2005. Available at http://www.sedac.ciesin.columbia.edu/gpw/.

8.34: Alain Le Garsmeur/Age Fotostock

8.33: Edmund Sumner/VIEW Pictures Ltd/Alamy

8.35: David Sacks/Stone/Getty Images

8.36: Based on G. Chaliand & J-P Rageau, *The Penguin Atlas of Diasporas*, 1995. Viking Penguin, Inc.

8.37: Lou Linwei/Alamy

8.38: Based on the chart "China's trade with Africa," in "The Chinese in Africa: Trying to Pull Together," from *The Economist*, April 2011, available at http://www.economist.com/node/18586448.

8.39: AP Images/Imaginechina

8.2.2: Based on map in "NASA Satellite Measures Pollution From East Asia to North America," available at http://www.nasa.gov/topics/earth/features/pollution_measure.html.

CHAPTER 9

CO.9: sZUMA Press/Newscom

9.2: NASA Earth Observing System

9.4: AP Images/Bikas Das

9.6: Robert Harding Picture Library Ltd/Alamy

9.7: Based on "Impact of sea level rise in Bangladesh," Vital Water Graphics, UNEP, available at http://www.unep.org/dewa/vitalwater/article116.html.

9.8: AP Images/Biswaranjan Rout

9.9: Adapted from B. L. C. Johnson, *South Asia*, 2nd ed. London: Heinemann, 1982, p. 9.

9.10: India Images/Dinodia Photos/Alamy Limited

9.11: Russell Kord/Alamy

9.12: Finnbarr Webster/Alamy

9.13: Based on "Flooding in Pakistan," Der Spiegel (2010), available at http://www.spiegel.de/panorama/a-711885.html.

9.14: johnnychaos/Fotolia

9.15: Manjeet & Yograj Jadeja/Alamy

9.16: AP Images/Mustafa Quraishi

9.17: ian howard/Fotolia

9.19: Robert Harding World Imagery/Getty Images

9.18: ErickN/Shutterstock.com

9.20: Redrawn from: J. Keay, *India: A History*. New York: Atlantic Monthly Press, 2000, p. 314.

9.21: Redrawn from I. Barnes and R. Hudson, *The History Atlas of Asia*, New York: MacMillan, 1998, pp. 118–119.

9.22: World History Archive/Alamy

9.23: Based on maps in "Partition of India," Wikipedia, available at http://en.wikipedia.org/wiki/Partition_of_india.

9.1.1: Martin Puddy/Getty Images

9.24: David Pearson/Alamy

9.25: Fredrik Renander/Alamy

9.27: blickwinkel/Alamy

9.26: Maygutyak/Fotolia

9.28: Randy Olson/National Geographic Image Collection/Alamy

9.30: Eye Ubiquitous/Glow Images, Inc.

9.29: Philippe Lissac/Godong/Photononstop/Glow Images, Inc.

9.31: Based on "India asks China to halt PoK projects," Hotgurgaon, October 2009, available at http://www.hotgurgaon.com/news/news.aspx?id=42234; and on "Chinese Troops in Pakistan-Occupied Kashmir: Indian General VK Singh," *The Economic Times*, October 2011, available at http://www.sott.net/article/236943-Chinese-Troops-in-Pakistan-occupied-Kashmir-Indian-General-VK-Singh.

9.32: Harish Tyagi/EPA/Newscom

9.34: Roberto Fumagalli/Alamy

9.35: Mariia Pazhyna/Fotolia

9.36: AP Images/Saurabh Das

9.37: Based on "Two Languages or One?" Hindu Urdu Flagship, University of Texas at Austin, available at http://hindiurduflagship.org/about/two-languages-or-one/.

9.38: Antoine Serra/In Visu/Corbis

9.39: AF archive/Alamy

9.40: BOISVIEUX Christophe/hemis.fr/Alamy

9.41: Joe Gough/Shutterstock

9.42: From the Center for International Earth Science Information Network [CIESIN], Columbia University; International Food Policy Research Institute [IFPRI] and World Resources Institute [WRI]. 2000. *Gridded Population of the World* [Gpw], Version 2, Palisades, NY: CIESIN, Columbia, University. Available at http://sedac.ciesin.org/plue/gpw.

9.43: Based on "National Family Health Survey, India."

9.44: AP Images/Aijaz Rahi

9.2.1: Based on Rocketjack, "World Happiness" in "Satisfaction with Life Index," September 2007, available at http://en.wikipedia.org/wiki/File:World_happiness.png.

9.2.2: Based on N. DeGuerre, "Happy Planet" in "Happy Planet Index," February 2007, available at http://en.wikipedia.org/wiki/File:Happy_Planet.PNG.

CHAPTER 10

CO.10: AP Images

10.2: NASA Earth Observing System

10.3: Redrawn from R. Ulack and G. Pauer, *Atlas of Southeast Asia*. New York: Macmillan, 1988, p. 6.

10.4: Alexander Widding/Alamy

10.5: Skip Nall/Corbis

10.6: Based on information from T.R. Leinbach and R. Ulack, *Southeast Asia: Diversity and Development*. Upper Saddle River, NJ: Prentice Hall, 2000, Map 2.1 and H.C. Brookfield and Y. Byron, *South-East Asia's Environmental Future: The Search for Sustainability*. New York: United Nations University Press, 1993, Figure 13.1.

10.7a: Redrawn from M. J. Valencia, J. M. Van Dyke, & N. A. Ludwig, *Sharing The Resources of the South China Sea*. Honolulu: University of Hawaii Press, 1999, Plate I.

10.7b: Chris Stowers/Dorling Kindersley

10.2.3: GeoEye/Photo Researchers, Inc.

10.9: WEDA/EPA/Newscom

10.8: JKlingebiel/Shutterstock

10.10: Simon Bowen/Alamy

10.11: Based on maps in R. Ulack and G. Pauer, *Atlas of Southeast Asia*. New York: Macmillan, 1988, p. 11; and T.R. Leinbach and R. Ulack, *Southeast Asia: Diversity and Development*. Upper Saddle River, NJ: Prentice Hall, 2000, Map 2.6b.

10.1.2: Ian Cruickshank/Alamy

10.12a: 1990-2000, 2000-2005 percent forest loss and 2005 forest cover: FAO State of the World's Forests. Rome, FAO, 2009; 1973-85 percent forest loss: Rain Forest Report Card, East Lansing, MI: Michigan State University, 2000.

10.12b: f4foto/Alamy

10.12c: AP Images/Dita Alangkara

10.13: NaughtyNut/Shutterstock

10.2.1: Simon Podgorsek/iStockphoto

10.14: Redrawn from B. Crow and Thomas [eds.], *Third World Atlas*, Philadelphia: Open University Press, 1984, p. 39.

10.15: teoyeekhai / Fotolia

10.16: Redrawn from M. Dockrill, *Atlas of 20th Century World History*, New York: Harper, 1991, pp. 74–75.

10.17: Hemis/Alamy

10.18: imagebroker/Alamy

10.19: mybeginner/Shutterstock

10.20: Based on Leinback, Thomas R.; Ulack, Richard, *Southeast Asia: Diversity and Development*, 1st Ed., © 2000; and on M. Sparke, T. Bunnell, and C. Grundy-Warr; "Triangulating the Borderless World: Geographies of Power in Indonesia-Malaysia-Singapore Growth Triangle," *Transactions of the Institute of British Geographers*, NS29, 398-485.

10.21: From the International Monetary Fund, World Economic Outlook Database, April 2009, http://www.imf.org/external/pubs/ft/weo/2009/01/weodata/index.aspx. Used with permission.

10.22: AFP Photo/FILES/Joel Nito/Newscom

10.23b: The Art Archive/Alamy

10.24: AP Images/David Longstreath

10.25: LEINBACH, THOMAS R.; ULACK, RICHARD, SOUTHEAST ASIA: DIVERSITY AND DEVELOPMENT, 1st ed., © 2000. Reprinted and Electronically reproduced by permission of Pearson Education, Inc. Upper Saddle River, New Jersey.

10.26: Photoshot

10.27: Reuters/Zainal Abd Halim

10.28: Luca Tettoni/Robert Harding Picture Library Ltd/Alamy

10.29: Redrawn from R. Ulack and G. Pauer, *Atlas of Southeast Asia*, New York: Macmillan, 1988, p. 29.

10.30: Redrawn from R. Ulack and G. Pauer, *Atlas of Southeast Asia*. New York: Macmillan, 1988, p. 27.

10.31: ACE Stock Limited/Alamy

10.3.2: Piers Cavendish/Imagestate Media Partners Limited - Impact Photos/ Alamy

10.3.3: Rungroj Yongrit/EPA/Newscom

10.32: From UNAIDS, 2008, *Report on the Global Aids Epidemic*. http://www .unaids.org/en/KnowledgeCentre/HIVData/GlobalReport/2008/2008_ Global_report.asp.

10.33: John Elk III/Alamy

10.34: From Socioeconomic Data and Applications Center. Columbia University, Center for International Earth Science Information Network [CIESIN]; International Center for Tropical Agriculture [CIAT], 2005. *Gridded Population of the World* [GPWv3], Palisades, NY: CIESIN, Columbia University. Available at: http://sedac.ciesin.org/plue/gpw.

10.35: MUKTAR/EPA/Newscom

10.36: Redrawn from T.R. Leinbach and R. Ulack, *Southeast Asia: Diversity and Development*, Upper Saddle River, NJ: Prentice Hall, 2000, Map 12.7.

10.37a: Based on UN High Commission on Refugees *Statistical Online Population Database*, 2007.

10.37b: Based on Philippines Overseas Employment Administration, 2000-2005.

10.38: Peter Treanor/Alamy

10.39: Nigel Hicks/Alamy

10.1.1: Redrawn from Mekong River Commission http://www.mr-cmekong.org/ img/.

10.2.2: Redrawn from Cambodian Guide, available at http://www.cambodian- guide.com/images/dyimg/090913115958Map6.jpg.

10.3.1: Adapted from "The Opium Kings," PBS Frontline. Available at http:// www.pbs.org/wgbh/pages/frontline/showsherion/maps/shan.html/.

CHAPTER 11

CO.11: Moyu0816/Fotolia

11.2: NASA Earth Observing System

11.1.1: Radius Images/Alamy

11.1.4: Dan Leeth/Alamy

11.3: Based on T. McKnight, *Oceania: The Geography of Australia, New Zealand and the Pacific Islands*, Englewood Cliffs, NJ: Prentice Hall, 1995, Figure 2.2e; and G.M. Robinson, R.J. Loughran, and P.J. Tranter, *Australia and New Zealand: Economy, Society and Environment*. New York: Oxford University Press, 2000, Figure 3.1.

11.4: AP Images

11.6: David Wall/Alamy

11.7b: Tim Graham/Alamy

11.7a: Marcel Hurni/Fotolia

11.8b: Douglas Peebles Photography/Alamy

11.8a: SuperStock/AGE Fotostock

11.2.1: AP Images

11.2.2: Jacques Langevin/Sygma/Corbis

11.9: Sinartus Sosrodjojo/Zuma Press/Newscom

11.10b: WorldFoto/Alamy

11.10c: Ben Twist/Fotolia

11.10a: BMJ/Shutterstock

11.10d: Szefei/Shutterstock

11.11: Based on "Land Use of Australia," Version 4, 2005/2006 (September 2010 release), Australian Department of Agriculture, Fisheries and Forestry, available at http://adl.brs.gov.au/anrdl/metadata_files/pa_luav4g9abl07811a00. xml; and on the "National Land and Water Resources Audit 2001," © Commonwealth of Australia, 2001.

11.12: Steve Lovegrove/Shutterstock

11.13: Redrawn from M. Rapaport (ed.), *The Pacific Islands: Environment and Society*. Honolulu: The Bess Press, 1999, 30.3.

11.14: Alex Hofford/Greenpeace

11.15: Jeff Hunter/Photographer's Choice RF/Getty Images

11.16b: Caroline Penn/Impact/HIP/The Image Works

11.17: Based on C. McEvedy, *The Penguin Historical Atlas of the Pacific*, New York: Penguin Books, 1998, pp. 49, 63, 65, 90.

11.18: AP Images/Malcolm Pullman

11.3.2: AP Images/Alastair Grant

11.3.1: Nicky Park/Epa/Landov

11.19: Christchurch City Libraries, File Reference: CCL PhotoCD 7, IMG0071

11.20: National Archives at College Park

11.21: Data from the Australian Broadcasting Company, available at http:// www.abc.net/au/ra/carvingout/maps/statistics.htm.au.

11.22: Based on A. Simoes, "What does Australia Export?" and "What does New Zealand Export?," The Observatory of Economic Complexity, available at http://atlas.media.mit.edu/explore/tree_map/export/aus/all/show/2010/ and http://atlas.media.mit.edu/explore/tree_map/export/nzl/all/show/2010/.

11.23: Douglas Peebles/Corbis

11.24: Avalon/Alamy

11.25: AP Images/New Zealand Herald, Steven McNicholl

11.26: ERIC LAFFORGUE / Alamy

11.27: EcoView/Fotolia

11.28: Martin Hayhow/AFP/Getty Images/Newscom

11.30: AF archive/Alamy

11.29: Reuters/David Gray

11.31: From Columbia University, Center for International Earth Science Information Network [CIESIN]; International Center for Tropical Agriculture [CIAT]. 2005 *Gridded Population of the World* [GPWv3], Palisades, NY: CIESIN, Columbia University. Available at http://sedac. ciesin.org/plue/gpw.

11.32: EastVillageImages/Fotolia

11.33: pp76/Fotolia

11.34: CBS/Monty Brinton/Landov

11.1.2: Based on G. Lean and D. Hinrichsen (eds.), *Atlas of the Environment*. Santa Barbara, CA: ABC-CLIO, 1994, pp. 182-83; and Terraquest, *Virtual Antarctica Expedition*.

11.1.3: Based on the "Coastal--Change and Glaciological Map of the Palmer Land Area, Antarctica: 1947----2009," (map I--2600--C), U.S. Geological Survey, available at http://pubs.usgs.gov/imap/i--2600--c/.

11.2.3: Based on maps from Australian Mines Atlas, http://www.australian- minesatlas.gov.au, © Commonwealth of Australia 2012 and Indigenous Land Use Agreements, National Native Title Tribunal, http://www.nntt.gov. au/Mediation-and-agreement-making-services/Documents/Quarterly%20 Maps/ILUAs_map.jpg, © Commonwealth of Australia 2008-2011. All rights reserved.

APPENDIX

A.1: Based on "Lewis Falls Quadrangle," U.S. Department of the Interior, U.S. Geological Survey, available at http://nationalmap.gov/ustopo/UST_slide- show/lewis_falls/2012_Lewis_Falls_WY_USTopo_image-off.html.

A.2: Based on J.M. Rubenstein, *The Cultural Landscape: An Introduction to Human Geography*, 1996, p. 584; and W.K. Stevens, "Study of Acid Rain Uncovers Threat to Far Wider Area," *New York Times*, January 16, 1990, p. 21.

A.5: Based on E.F. Bergman, *Human Geography: Cultures, Connections, and Landscapes*, © 1995, p. 12.

A.6: Based on E.F. Bergman, *Human Geography: Cultures, Connections and Landscapes*, © 1995, p. 13.

A.7a: Reprinted by permission of Mark Newman, University of Michigan.

A.7b: © Copyright SASI Group (University of Sheffield) and Mark Newman (University of Michigan).

A.8: © TerraMetrics (Google) 2012.

A.9: JENSEN, JOHN R.; JENSEN, RYAN R., INTRODUCTORY GEOGRAPHIC INFORMATION SYSTEMS, 1st Ed., © 2011. Reprinted and electronically reproduced by permission of Pearson Education, Inc. Upper Saddle River, New Jersey.

INDEX

A

Aborigines, Australian, 432, 441, 450–51, 453, 455
Acid rain, 58
Adaptation. *See also* specific countries and regions
    East Asia, 308
    Europe, 52–53
    in global context, 8–13
    Latin America and the Caribbean, 269
        Mayan, Incan, Aztec environmental adaptations, 274–75
    Middle East and North Africa (MENA), 132–33
    Oceania, 427–31
    Russian Federation, Central Asia, and Transcaucasus, 90–91
    South Asia, 347–49
    Southeast Asia, 387–88, 401
    Sub-Saharan Africa, 179, 187–88
    United States and Canada, 225–26
Adivasi, 372
Afforestation, 137
Afghanistan, 360, 368, 379
African Union, 201
Afrikaners, 192
Agent Orange, 408
Age of Discovery, 23, 59–60
Aging populations, European, 82–83
Agriculture. *See also* specific countries and regions
    Australia, 438
    bush fallow, 188
    China, 324
    East Asia, 308, 313, 322
        revolutions, 314
    Eastern Mediterranean Crescent, 146
    Europe
        Common Agricultural Policy (CAP), 68
        Golden Triangle, 69
    Fair Trade movement, 197
    Green Revolution, 278, 314, 403
    intercropping, 188, 349
    irrigation, 140–41
    Latin America, 278
    Latin America and the Caribbean, 284
    Middle East and North Africa, 135, 140–41, 146, 166
    New Zealand sheep farming, 424–25, 438
    paddy farming, 387
    pastoralism, 54–55
    polder landscape, 58
    Russian Federation, Central Asia, and the Transcaucasus, 90, 92, 96, 98
    South Asia, 355, 379–80
    Southeast Asia, 387
        Green Revolution, 403
        Opium crops, 414–15
    Sub-Saharan Africa, 187–88
    sheep farming, 424–25, 438
    swidden, 394–95
    terracing, 312, 349, 388
    transhumance, 188, 349
    United States and Canada, 232–33
Ahmadinejad, Mahmoud, 157
AIDS. *See* HIV/AIDS
AIDS orphans, 416
Akbar, 358
Akosombo Dam, 182
ALBA group, 292
Alexander II, 104
Alliance of Small Island States (AOSIS), 431, 444
Alpine Europe, 48–49, 55–56
Al-Qaeda, 249

B

Alsaciens, 73
Altiplano, 269
Altitudinal zonation, 267
Amazon deforestation, 276–77
Americanization, 239
American Revolution, 239
Amritsar, 370
Amundsen, Roald, 92
Ancient period, 22
Angkor Wat, 397, 398–99
Angola, 202, 203
Animals, animal parts, and exotic pets, 316
Anime, 35, 333
Animism, 329
Antarctica, 428–29
Antarctic Treaty, 428
Anthropocene, 8
Anthropogenic forest, 355
Apartheid, 201–2
Appalachian mountain chain, 228
Apparatchiks, 106
Arab Caliphate, 117
Arab League, The, 157–58
Arab Spring, 148
ArcelorMittal, 363
Archipelagoes, 388
Arctic, the, 92–93
Arctic National Wildlife Refuge (ANWR), 223
Argentina, 292
Arid climates, 10
Aridity, 89
Aridity, MENA, 131–33
Armstrong, Louis, 251
Arsenic contamination, 354
ASEAN (Association of Southeast Asian Nations), 32, 410
Ashoka, 358
Asian Brown Cloud, 315, 317
Asian Free Trade Association (AFTA), 410
Asian Tiger, 326
Aspect (exposure), 57
Assimilation, 250
Association of South East Asian Nations (ASEAN), 32
Atacama Desert, 272–73
Atlas Mountains, 132, 146
Atmospheric circulation, 8–10, 10f
Atolls, 430, 433
Aung San Suu Kyi, 409
Azimuthal projections, A-2
Aztec empire, 275f

Babylon, 142
Balfour Declaration, 151
Bali, 411
Balkanization, 74
Balkans, ethnic conflict in, 74
Ballet, 119
Baluchistan, 373
Bangkok, Thailand, 41, 420
Bangladesh, 350f, 353, 354
Baobab trees, 185–86
Basques, 73
Bengal Delta, 353
Benguela Current, 181
Berbers, 163
Bering, Vitus, 100
Berlin Conference, 193
Berlin Wall, 63, 248

Bharatiya Janata Party (BJP), 370
Biafra War, 205
Bible Belt, 251
Bikini Atoll, 434
bin Laden, Osama, 249, 250
Biodiversity, 19–20
Biofuels, 278, 391
Bioprospecting, 279
Bird of Paradise, 437
Birthrates, 42
Bitov, A., 93
Blackbirding, 445
Black Death, 57–58
Black Forest, 56
Black Sea, 95
Blue Economy, 444
Boers, 192, 201
Bollywood, 39, 373
Bolsa Familia (family allowance) program, 289
Bolsheviks, 104, 119, 120
Bonded labor, 366
Borobudor Temple, Indonesia, 397
Bosnia, 64
Botswana, 200, 216, 219
Braceros, 298
Brain drain, 215, 378
Brazil, 276–77, 278, 285, 289
BRICS countries, 286–87
Brighton Beach neighborhood, 123
British East India Company, 358–59
British Empire
    in Africa, 195
    in India, 358–60
Buddhism
    origin of, 35, 36f, 369–70
    South Asia, 369–70
    Southeast Asia, 410–11
    Theravada, 397
Bumiputra, 404
Bureaucratic class, 63
Burundi, 203, 217
Bush, George W., 156
Bush fallow, 188
Bushmeat, 190

C

Call centers, 363
Calligraphy, 331
Calvinist religion, 208
Canada. *See also* United States and Canada
    cultural nationalism, 252
    economy, 241–42
    environmental movement, 250
    immigration, 254
    multiculturalism, 250–51
    social welfare, 248
Canadian Shield, 227, 239
Cap-and-trade programs, 236
Cape of Good Hope, 190
Cape Town, 192, 201
Capitalism
    defined, 24–25
Carbon dioxide (CO_2), 12
Carbon footprint, 226, 231
Carbon market, 133
Caribbean, 244–45
CARICOM, 285
Cartels, 234, 293
Cartograms, A-6
Caste system, 371–73

Tsetse fly, 186–87
Tsunami, 17, 304–5, 385
Tuaregs, 205
Tube wells, 354
Tundra, 54, 96–97
Turkey
  forced migration, 78
  squatter settlements in Istanbul, 170
  Syrian refugees, 167
Turkish Empire, 98
Turkmenistan, 109
Typhoons, 386
Typhoon Washi, 387
Tyumen, Russia, natural gas pipeline, 125

U

Ukraine, 64
UNCLOS (United Nations Convention on the Law of
    the Sea), 439
U.N. Framework Convention on Climate Change
    (UNFCCC), 133
Union for the Mediterranean, 147
United Nations. *See also* World Health Organization
    (WHO)
  Convention to Combat Desertification, 179
  Development Programme (UNDP), 29
  Environment Program, 215
  Food and Agriculture Organization (FAO), 211
  High Commission on Refugees (UNHCR),
      156–57, 217
  Kyoto Protocol, 53, 133
  member countries, 34f
  nuclear tensions in Middle East, 157
  peacekeeping forces in sub-Saharan Africa,
      205, 208
United States and Canada, 222–61
  climate, adaptation, and global change, 225–26
    climate patterns, 225
    environmental modifications and impacts,
        225–26
  climate change, corporate and government
      response to, 259f
  cultural practices, social differences, and identity,
      251–52
    arts, music, and sport, 251–52
    Canadian cultural nationalism, 252
    sex, gender, and sexuality, 252
  demography and urbanization, 253–57
    immigration, 253–54
    immigration reform, protesting, 254
    internal migration, 254
    urbanization, industrialization, and new growth,
        254, 256
    urban to suburban migration, 257
    U.S. immigration, 1800–2000, 253–54
  ecology, land, and environmental management, use
      value, 229, 232–33, 236
    alternative food movements, 233, 236
    colonial land use, 232
    contemporary agricultural landscapes,
        232–33
    environmental challenges, 236
    Indigenous land use, 229, 232
  economy, accumulation, and production of
      inequality, 240–43, 246–47
    economies of United States and Canada,
        241–42
    Europe versus, 68
    global economic crisis, 246–47
    income disparity in U.S., 243t
    new economy, 242–43
    transforming economies, 242
    wealth and inequality, 243, 246
  future geographies, 257–59
  geological resources, risk, and water management,
      226–29

geologic and other hazards, 228
  mineral, energy, and water resources,
      228–29
  physiographic regions, 227–28
historical legacies and landscapes, 237–40
  colonization and independence, 238–39
  European settlement, 239–40
  Indigenous histories, 237–38
  slavery in United States, legacy of, 239
key facts, 224
main points, 223
maps
  Caribbean, 244f
  climate, 226f
  frontier trails, 256f
  global cannibis use, 234f
  major league baseball, globalization of, 252f
  migration of Neolithic hunters into
      Americas, 229f
  mineral wealth, 230f
  overview, 224f
  physiographic regions of United States, 227f
  from space, 225f
  temperature and precipitation due to climate
      change, 226f
  triangular trade routes, 244f
marijuana, 234–35
natural resources, 230–31
religion and language, 250–51
territories and politics, 247–50
  social movements, 250
  states and government, 247–48
  U.S. political and military influence, 248
  U.S. war on terror, 248–50
  U.S. interventions in Latin American revolutions,
      262–63
Uranium, 434–35
Urban bias, 214
Urbanization
  demography and, 41–43
  Europe, 63–64, 70, 81–82
  Latin America and the Caribbean, 298–99
  Middle East and North Africa, 142
  Oceania, 456–57
  Russian Federation, Central Asia, and the
      Transcaucasus, 121
  South Asia, 377–78
  Southeast Asia, 420
  United States and Canada, 254, 256
    urban to suburban migration, 257
Urban primacy, 298
Uruguay, 292
U.S. Agency for International Development, 145
Uyghurs, 332
Uzbekistan, 109

V

Vallee du Falgoux, 56
Vanuatu, 452
Veld, 192
Victoria Falls, 182, 183
Vietcong, 407
Vietnam War, 407–8
Virginia colony, 238
Volcanic hazards, 17, 271
  benefits of, 391

W

Wallace's Line, 392
Walloons, 73
War. See also Cold War; World War I;
    World War II
  American Revolution, 239
  Biafra, 205

Dirty, 292
European imperialism, 62–63
Iraq, 156–7
Indochina, 407–8
Korea, 327–28
Opium, 320, 339
Spanish-American, 244–45
World War II and independence in
    Pacific, 445–46
War on Terror, U.S., 248–49, 368
Water management, 17. See also specific countries
    and regions
  East Asia, 311–12
  Europe, 53–57
  Latin America and the Caribbean, 272
  map of world scarcity, 171f
  Middle East and North Africa (MENA), 131,
      134–36
    irrigation, 140–41
  Oceania, 431–32, 457
  Russian Federation, Central Asia,
      Transcaucasus
    Aral Sea, 109–10
    the Arctic, 92–93
    rivers and seas, 95
  South Asia
    arsenic contamination, 354
    energy and mineral resources, 354
    flooding hazard, 353–54
  Southeast Asia, 382–83, 391–92
  Sub-Saharan Africa, 182–83
  United States and Canada, 228–29
Watershed, 51
Weather, 8
Welfare state, 65
Westerlies, 10, 266
West Germany, 68
Wet farming, 140–41
White Australia policy, 455
Women
  East Asia, 331–32
  Europe, 77
  geographies and culture, 40
  Latin America and the Caribbean, 296
  Middle East and North Africa (MENA),
      162–66
  Oceania
    gender roles, sexuality, and identity, 454
  Russian Federation, Central Asia, Transcaucasus,
      120–21
  South Asia, 365–66, 377, 379
  Southeast Asia, 387, 413
  Sub-Saharan Africa, 210–11
    Millenium Development Goals
        (MDGs), 200
  United States and Canada, 252
World Bank, 145, 197, 285
World city, 81
World Health Organization (WHO), 186–87
World regions in global context, 5–47
  climate, adaptation, and global change, 8–13
    atmospheric circulation, 8–10
    climate change, 12–13, 12f, 13f
    climate regions, 11f
    climate zones, 10–12
  cultural practices, social differences, and identity,
      38–41
    culture and identity, 39–41
    genders, 40
    globalization, regionalization, and culture,
        38–39
    race, 40
    sexuality, 40–41
  culture, religion, and language, 34–38
    language, geographies of, 37–38
    religion, geographies of, 35–37
  demography and urbanization, 41–43

World – Physical

Great Basin	Land features
Caribbean Sea	Water bodies
Aleutian Trench	Underwater features

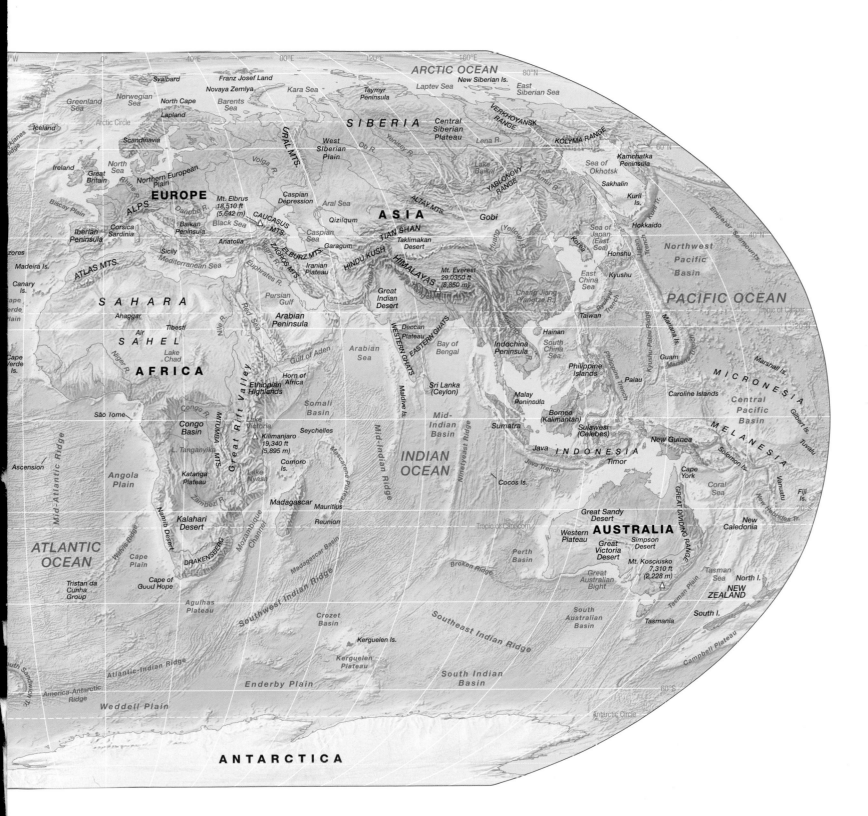